Metalorganic Vapor Phase Epitaxy (MOVPE)

Wiley Series in Materials for Electronic & Optoelectronic Applications

www.wiley.com/go/meoa

Series Editors

Professor Arthur Willoughby, *University of Southampton, Southampton, UK*

Dr Peter Capper, *Ex-Leonardo MW Ltd, Southampton, UK*

Professor Safa Kasap, *University of Saskatchewan, Saskatoon, Canada*

Published Titles

Bulk Crystal Growth of Electronic, Optical and Optoelectronic Materials, Edited by P. Capper
Properties of Group-IV. III–V and II–VI Semiconductors, S. Adachi

Charge Transport in Disordered Solids with Applications in Electronics, Edited by S. Baranovski

Optical Properties of Condensed Matter and Applications, Edited by J. Singh

Thin Film Solar Cells: Fabrication, Characterization, and Applications, Edited by J. Poortmans and V. Arkhipov

Dielectric Films for Advanced Microelectronics, Edited by M. R. Baklanov, M. Green, and K. Maex

Liquid Phase Epitaxy of Electronic, Optical and Optoelectronic Materials, Edited by P. Capper and M. Mauk

Molecular Electronics: From Principles to Practice, M. Petty

Luminescent Materials and Applications, A. Kitai

CVD Diamond for Electronic Devices and Sensors, Edited by R. S. Sussmann

Properties of Semiconductor Alloys: Group IV, III–V, and II–VI Semiconductors, S. Adachi

Mercury Cadmium Telluride, Edited by P. Capper and J. Garland

Zinc Oxide Materials for Electronic and Optoelectronic Device Applications, Edited by C. Litton, D. C. Reynolds, and T. C. Collins

Lead-Free Solders: Materials Reliability for Electronics, Edited by K. N. Subramanian

Silicon Photonics: Fundamentals and Devices, M. Jamal Deen and P. K. Basu

Nanostructured and Subwavelength Waveguides: Fundamentals and Applications, M. Skorobogatiy

Photovoltaic Materials: From Crystalline Silicon to Third-Generation Approaches, Edited by G. Conibeer and A. Willoughby

Glancing Angle Deposition of Thin Films: Engineering the Nanoscale, Matthew M. Hawkeye, Michael T. Taschuk, and Michael J. Brett

Physical Properties of High-Temperature Superconductors, R. Wesche

Spintronics for Next Generation Innovative Devices, Edited by Katsuaki Sato and Eiji Saitoh

Inorganic Glasses for Photonics: Fundamentals, Engineering and Applications, Animesh Jha

Amorphous Semiconductors: Structural, Optical and Electronic Properties, Kazuo Morigaki, Sandor Kugler and Koichi Shimakawa

Microwave Materials and Applications 2 Vol sel, Edited by Mailadil T. Sebastian, Rick Ubic, and Heli Jantunen

Molecular Beam Epitaxy: Materials and Applications for Electronics and Optoelectronics, Edited by Hajime Asahi and Yoshiji Korikoshi

Metalorganic Vapor Phase Epitaxy (MOVPE)

Growth, Materials Properties, and Applications

Edited by

STUART IRVINE

Centre for Solar Energy Research, College of Engineering, Swansea University, OpTIC Centre, St. Asaph, UK

PETER CAPPER

Ex-Leonardo MW Ltd, Southampton, UK

WILEY

This edition first published 2020

Registered Offices
John Wiley & Sons, Inc., 111 River Street, Hoboken, NJ 07030, USA
John Wiley & Sons Ltd, The Atrium, Southern Gate, Chichester, West Sussex, PO19 8SQ, UK

Editorial Office
The Atrium, Southern Gate, Chichester, West Sussex, PO19 8SQ, UK

For details of our global editorial offices, customer services, and more information about Wiley products visit us at www.wiley.com.

Wiley also publishes its books in a variety of electronic formats and by print-on-demand. Some content that appears in standard print versions of this book may not be available in other formats.

Library of Congress Cataloging-in-Publication Data
Names: Irvine, Stuart, editor. | Capper, Peter, editor.
Title: Metalorganic vapor phase epitaxy (MOVPE) : growth, materials properties, and applications / edited by Stuart Irvine and Peter Capper.
Description: First edition. | Hoboken, NJ : John Wiley & Sons, Inc., 2020. |
Series: Wiley series in materials for electronic & optoelectronic applications ; 7593 | Includes bibliographical references and index. |
Identifiers: LCCN 2019017934 (print) | LCCN 2019019971 (ebook) | ISBN 9781119313045 (Adobe PDF) | ISBN 9781119313038 (ePub) | ISBN 9781119313014 (hardback)
Subjects: LCSH: Metal organic chemical vapor deposition.
Classification: LCC TS695 (ebook) | LCC TS695 .M478 2020 (print) | DDC 621.3815/2–dc23
LC record available at https://lccn.loc.gov/2019017934

Cover Design: Dan Jubb
Cover Image: © molekuul_be/Shutterstock

Set in 10/12pt Times by SPi Global, Pondicherry, India

10 9 8 7 6 5 4 3 2 1

Dedication

S.J.C.I. – This book is dedicated to my wife Fiona and our family for their patience and support in completing this project at a difficult time for our family. I also commit this book to the memory of Primrose.

P.C. – This book is dedicated to my wife Marian and our sons Samuel and Thomas for all their love and support. I also wish to dedicate it to the memory of my brother Ken.

Contents

Contents

List of Contributors

I.M. Baker Leonardo MW Ltd, Southampton, UK

P. Capper Ex-Leonardo MW Ltd, Southampton, UK

A. Dadgar Otto-von-Guericke Universität Magdeburg, Germany

T. Detchprohm Georgia Institute of Technology, School of Electrical and Computer Engineering, Atlanta, USA

R.D. Dupuis Georgia Institute of Technology, School of Electrical and Computing Engineering, Atlanta, USA

C.M. Fetzer Boeing-Spectrolab, Sylmar, California, USA

H. Hardtdegen Ernst Ruska-Centre, Forschungszentrum Jülich GmbH, Germany

A. Hospodková Department of Semiconductors, Institute of Physics, Czech Academy of Sciences, Prague, Czech Republic

E. Hulicius Department of Semiconductors, Institute of Physics, Czech Academy of Sciences, Prague, Czech Republic

S.J.C. Irvine Centre for Solar Energy Research, College of Engineering, Swansea University, OpTIC Centre, St. Asaph, UK

H.J. Joyce Department of Engineering, Cambridge University, UK

N.H. Karam Karamco USA Inc., La Cañada, California, USA

G. Kartopu Centre for Solar Energy Research, College of Engineering, Swansea University, OpTIC Centre, St. Asaph, UK

B. Kunert imec, Leuven, Belgium

X. Li Electrical Engineering Program, Computer, Electrical, Mathematical Science and Engineering Division, King Abdullah University of Science and Technology (KAUST), Thuwal, Makkah, Saudi Arabia

X.-Q. Liu Boeing-Spectrolab, Sylmar, California, USA

W. Lundin Ioffe Institute, Russia

C.D. Maxey Leonardo MW Ltd, Southampton, UK

M. Mikulics Peter Grünberg Institute, Forschungszentrum Jülich GmbH, Germany

V. Muñoz-Sanjosé Dpto. Física Aplicada y Electromagnetismo, University of Valencia, Spain

D. Nelson IQEP, Cardiff, Wales

J.-H. Ryou Department of Mechanical Engineering, The University of Houston, Texas, USA

K.L. Schulte National Renewable Energy Laboratory, Golden, Colorado, USA

D.V. Shenai-Khatkhate Electronics & Imaging, DuPont de Nemours, Inc., Marlborough, Massachusetts, USA
(Formerly known as Dow Electronic Materials, Rohm and Haas Company, and Morton Metalorganics)

M.A. Steiner National Renewable Energy Laboratory, Golden, Colorado, USA

G.B. Stringfellow The University of Utah, Salt Lake City, USA

R. Talalaev STR Group Ltd, Russia

K. Volz Material Sciences Center and Faculty of Physics, Philipps-Universität Marburg, Germany

M. Weyers Ferdinand-Braun-Institute, Leibniz-Institut für Höchstfrequenztechnik, Berlin, Germany

M. Zíková Department of Semiconductors, Institute of Physics, Czech Academy of Sciences, Prague, Czech Republic

Foreword

It gives me immense pleasure to introduce this book on the technology of Metalorganic Vapor Phase Epitaxy (MOVPE), edited by Stuart Irvine and Peter Capper, and published by John Wiley. It is one of the most comprehensive books to have ever been published on the subject and will serve to educate, inform, train, and inspire a new generation of engineers and scientists of many disciplines who become involved in the semiconductor industry in the coming years and decades ahead.

The publication of such a comprehensive collection of chapters, covering the growth, materials properties, and applications enabled by MOVPE, and written by subject-matter experts, could not be more timely. It is no exaggeration to say that MOVPE has changed the world we all live in and the way we work, live, and play.

The entire global communications network, from the devices that provide the light signalling, switching, and detection in high-speed fiber optic systems and the optical infrastructure for dramatically increasing the information-handling capacity of todays' data centers, to the critical radio frequency (RF) components in handsets, base stations, and satellites that enable mobile communications in all its forms, would not be practically possible without MOVPE technology providing the manufacturing of the fundamental materials structures from which the key devices are made. Without MOVPE, we would have no high-speed broadband, no internet as we know it, no global mobile communications as we know it.

MOVPE has enabled a global lighting revolution to take place over the last two decades, in the form of light-emitting diodes (LEDs), which save enormous amounts of energy every year because they are up to 10 times more efficient at converting electrical energy into light energy than conventional incandescent light bulbs. Lighting accounts for around one-third of all energy use on our planet, so the replacement of incandescent bulbs with LED equivalents is cutting down carbon emissions enormously by reducing the need for power stations. LEDs are also being used to transform whole industries. From the way in which retailers use light to promote their products – be they clothing, cars, or perishable foods – to the advent of hydroponic and vertical farming to grow plants more efficiently and closer to the point of use, thereby reducing transport needs, are all contributing to carbon-emission reduction on our planet. The increasing production of electric vehicles is reliant upon materials made by MOVPE, as are the new wave of power-efficient devices used for a myriad of electrical switching applications such as mobile phone chargers, motors, inverters for solar farms, and a host of other energy-hungry applications.

All satellites launched today are powered in space by highly efficient solar cells made by MOVPE. Almost all large screens in stadiums, advertising hoardings, concert, and entertainment lighting use high-brightness LEDs made by MOVPE. High-power laser welding used in the automotive, aerospace, and other industrial sectors use materials made by MOVPE. 3D sensing now appearing on mobile handsets, gesture recognition in cars, and LiDaR (Light Detection and Ranging, the fundamental sensing technology for autonomous drive vehicles) all rely on MOVPE.

Looking forward, MOVPE will be used for many more applications and is now poised to enable the increasingly rapid adoption of compound semiconductors within the overall semiconductor industry. This is a $400 billion business spanning the entire globe and providing almost half of all global GDP growth directly and through its impact on information and communications technology (ICT). In other words, MOVPE is an absolutely core enabling technology for future global growth.

The MOVPE industry is growing rapidly but continues to require significant numbers of engineers, scientists, physicists, chemists, chemical engineers, operators, managers, and leaders. They all need to be well informed, well trained, knowledgeable, and inspired and motivated to help further develop the technology. By doing so, they will play a significant role, not only in bringing more efficient, more powerful, higher-speed, lower-cost products to market, thus enhancing the way we live our lives, how we work, and how we spend our leisure time; but also contributing to the reduction of greenhouse-gas emissions, to the benefit of our planet.

This book provides a comprehensive overview of MOVPE and should be used to train, inform, educate, and inspire this new generation of industrial and academic participants. I am proud to be associated with it, and I commend it wholeheartedly to the semiconductor community. No self-respecting technology bookshelf should be without it.

Drew Nelson
IQEP
Cardiff, Wales

Series Preface

Wiley Series in Materials for Electronic & Optoelectronic Applications

This book series is devoted to the rapidly developing class of materials used for electronic and optoelectronic applications. It is designed to provide much-needed information on the fundamental scientific principles of these materials, together with how these are employed in technological applications. The books are aimed at (postgraduate) students, researchers, and technologists, engaged in research, development, and the study of materials in electronics and photonics, and industrial scientists developing new materials, devices, and circuits for the electronic, optoelectronic and, communications industries.

The development of new electronic and optoelectronic materials depends not only on materials engineering at a practical level, but also on a clear understanding of the properties of materials, and the fundamental science behind these properties. It is the properties of a material that eventually determine its usefulness in an application. The series therefore also includes such titles as electrical conduction in solids, optical properties, thermal properties, and so on, all with applications and examples of materials in electronics and optoelectronics. The characterization of materials is also covered within the series in as much as it is impossible to develop new materials without the proper characterization of their structure and properties. Structure–property relationships have always been fundamentally and intrinsically important to materials science and engineering.

Materials science is well known for being one of the most interdisciplinary sciences. It is the interdisciplinary aspect of materials science that has led to many exciting discoveries, new materials and new applications. It is not unusual to find scientists with a chemical engineering background working on materials projects with applications in electronics. In selecting titles for the series, we have tried to maintain the interdisciplinary aspect of the field, and hence its excitement to researchers in this field.

Arthur Willoughby
Peter Capper
Safa Kasap

Preface

Whole industries currently rely on metalorganic vapor phase epitaxially (MOVPE) grown layers of a wide variety of materials. These industries range from lighting, based on light-emitting diodes made from III-V compounds (currently worth over $26Bn), to UV emitters, solid-state laser diodes, multijunction solar cells, high-frequency devices, and photovoltaic/thermovoltaic devices, again all mainly based on III-V compounds, to infrared imaging, based on mercury cadmium telluride. In addition, much MOVPE R&D activity is taking place in many III-V and II-VI compound systems as well as oxide materials systems throughout the world.

This book is an attempt to summarize the position in a number of these areas where MOVPE-grown layers are central to particular industries. The book is aimed at senior under- and post-graduates in physics, chemistry, materials science, electrical engineering, and optical engineering disciplines, as well as those employed in the various fields of thin-film crystal growth within the relevant industries. It is hoped that the former group will find the book readable both as an introductory text and as a useful guide to the literature. Workers in industry will hopefully find the book useful in bringing them up-to-date information in both their own and other areas of interest. To both groups of readers, we trust that the book will prove interesting and a spur to further progress in this key area of technology.

The first chapter is an introductory chapter covering historical background, basic reaction mechanisms, chemical precursors, growth substrates, reactors, main applications areas, and finally some health and safety considerations of MOVPE. The next chapter discusses the fundamental aspects of MOVPE in some detail. Areas covered include thermodynamics, kinetics, surfaces, and a wide range of specific materials systems.

Chapter 3 outlines the current situation with regard to the growth, properties, and applications of a range of Group III phosphides, arsenides, and antimonides. Topics covered include relevant precursors, doping issues, heterojunctions, and devices in a wide range of binary, ternary, and quaternary compounds. In situ optical characterization as a means to control growth is also discussed. The next chapter deals with the nitride semiconductors, a family of materials that has received much attention recently. After describing the properties of these materials, the main challenges in the growth of such materials are outlined. Doping is once again a key area in the growth and applications of these material as is the issue of suitable substrates, and these two topics are discussed at length. These materials are used for short-wavelength optoelectronic devices and for high-power, high-temperature electronic devices.

Chapter 5 deals with multijunction III-V solar cells. Various types of metamorphic solar cells are described, and details of the growth, properties, and applications of each are given. The next chapter describes the current position in the field of quantum dots (QDs) grown by MOVPE. The various types of QDs, e.g. embedded, are outlined in the various material systems. A comparison is given between MOVPE and molecular beam epitaxy (MBE) grown QDs. The importance of in situ measurements is stressed. The range of devices made in materials produced by MOVPE is outlined and some comparisons made to their MBE-produced counterparts. A brief list of future possibilities is also given.

Chapter 7 deals with the area of III-V nanowires and related nanostructures made by MOVPE. The crucial crystallographic properties are discussed in some detail before a description of the particle-assisted growth of nanowires is covered. Selective-area growth and the various relevant growth modes are outlined. Potential applications for these structures range from photovoltaics to topological quantum computing. Chapter 8 deals with the fascinating topic of integrating III-V compounds onto silicon substrates in monolithic structures. This combination would open up new device opportunities benefiting from optoelectronic properties of III-Vs and mature, cost-effective silicon-based integrated circuit process technology. Integration methods that have been recently researched are the focus – particularly, but not exclusively, those based on MOVPE. A range of devices has been made in such material, and some details are given in the chapter.

There is a change of focus in the next three chapters away from III-V compounds to several II–VI compound systems.

Chapter 9 deals with the pre-eminent infrared material, mercury cadmium telluride (MCT). Details of the growth process, including precursors, substrates, reactor design, and in situ monitoring are given, together with a discussion of doping and uniformity issues of heterostructures. Various device structures/types are discussed, including advanced technologies, such as small pixel sizes, higher operating temperature, two-color, and nonequilibrium devices. Recent work on avalanche photodiode arrays in astronomical applications ends the chapter. Chapter 10 describes the growth of CdTe and related materials for use in solar cells and radiation detectors. A variety of substrate types have been employed in this system, as has low-temperature and photo-/plasma-assisted growth. Another family of II–VI materials, those based on ZnO, is the subject of the next chapter. The potential uses of these materials range from photonic to piezoelectric applications. Details of the numerous precursor varieties and substrate types, together with their pretreatments, are given. As in the previous chapter, several variants of MOVPE growth techniques, such as low-temperature and photo-/plasma-assisted growth, have been applied to this material system. Growth of various nanostructures has also been demonstrated.

Chapter 12 goes into the details of MOVPE reactor design and its evolution over the past 25 years or so. Areas covered include MO sources (precursors), gas-handling systems, reactor designs, and exhaust systems. A range of reactor designs from horizontal flow, showerhead, and rotating disc to planetary reactors are discussed for III-V systems. In addition, the differences needed in reactor design for III-N systems are detailed.

Precursors are the subject of Chapter 13. The requirements, synthesis, and purification strategies for these starting chemicals is outlined. This is followed by details of Group III, V, II, and VI precursors, together with various dopant precursors. This chapter concludes with some comments on the important field of environmental, health, and safety aspects.

The final chapter of the book looks to the future of MOVPE as a technology. One example used is that of production reactor design for III-N growth. Another example used is that of growth of Ga_2O_3 and its alloys for a range of device types. Two-dimensional materials, such as graphene but also binary systems such as BN, are being researched by MOVPE. Other nanopatterned materials, poly- and amorphous materials, various oxide systems, high-temperature superconductors, and silicides are also outlined. The future of various device types, such as light-emitting diodes, lasers, optoelectronic integrated circuits, high-speed electronics, high-power electronics, solar cells, and detectors (both radiation and particle varieties) is also discussed.

Finally, we would like to sincerely thank all the contributors to the book, as well as Jenny Cossham, Emma Strickland, and Elsie Merlin of John Wiley & Sons Ltd, for their help in various forms and their patience throughout the course of the book preparation and production stages.

Stuart Irvine, Swansea University, UK
Peter Capper, Southampton, UK
April 2019

Safety and Environment Disclaimer

1

Introduction to Metalorganic Vapor Phase Epitaxy

S.J.C. Irvine[1] *and P. Capper*[2]
[1]*Centre for Solar Energy Research, College of Engineering, Swansea University, OpTIC Centre, St. Asaph, UK*
[2]*Ex-Leonardo MW Ltd, Southampton, UK*

1.1 Historical Background of MOVPE

The technique of metalorganic chemical vapor deposition (MOCVD) was first introduced in the late 1960s for the deposition of compound semiconductors from the vapor phase, a variant of chemical vapor deposition (CVD) but with the advantage over the then-existing methods for compound semiconductor epitaxy of only requiring a single temperature for reaction and film deposition. The pioneers of the techniques, Manasevit and Simpson [1], were seeking a method for depositing optoelectronic semiconductors such as GaAs onto different, nonlattice-matched substrates, including spinel and sapphire. The near-equilibrium techniques such as liquid phase epitaxy (LPE) and chloride vapor phase epitaxy (VPE) were not suitable for nucleation onto a chemically very different surface than the compound being deposited. The first paper [2] reported on the single-crystal growth of GaAs on various oxide substrates. The process was based on a combination of a volatile alkyl organometallic for the Group III element and a hydride gas for the Group V element. This basic approach has remained for all the III-V compounds, with a few exceptions where arsine or phosphine are replaced by liquid sources, tertiarybutylyarsine (TBAs) and tertiarybutylphosphine (TBP) [3]. For the Group II-VI compounds, the hydrides were less useful where low-temperature growth was considered to be important to control native defect concentrations. However, early work using H_2Se and H_2S for sources to grow ZnSe and ZnS was successful at higher temperatures [4]. For the tellurides, the hydride route was not practical as H_2Te is not stable under ambient conditions, so alkyl tellurium was used from the outset.

It was more than a decade after the first reports of MOCVD that the international conference series started with the conference in Ajaccio in 1981, when the "father" of MOCVD, Hal Manasevit

Metalorganic Vapor Phase Epitaxy (MOVPE): Growth, Materials Properties, and Applications,
First Edition. Edited by Stuart Irvine and Peter Capper.

gave an invited talk that covered a wide range of compound semiconductor materials that had been successfully grown by MOCVD, including III-V, II-VI, IV-VI, and II-IV-V_2 [5]. Perhaps the most significant developments over that first decade, represented at that conference, were the progress with improved purity of the organometallic precursors and low-defect-density, lattice-matched epitaxy, which led to the first demonstration of electron mobility >100 000 $cm^2/V\,s$ [6]. This made MOCVD competitive with other epitaxial techniques for GaAs, such as the emerging molecular beam epitaxy (MBE), and paved the way for high-performance optoelectronic devices, including quantum-well lasers and high-mobility transistors. The issues of purity of the early organometallics is addressed again in Section 14.1, which takes a historical perspective before looking forward to future prospects for MOCVD. The improved purity of the organometallic sources led to the demonstration of low-threshold lasers that opened the way for commercialization of MOCVD. Purification of precursors is covered in more detail in Chapter 13, where the challenges with traditional distillation methods were overcome using adduct purification.

The high-quality epitaxial nature of the films was emphasized by more commonly adopting the name of the growth method to be metalorganic vapor phase epitaxy (MOVPE) or organometallic VPE (OMVPE). There was some debate at the first international conference on what it should be called, which may seem strange now; but in those formative years, even the name was a point of discussion. The title of the conference was "IC-MOVPE I," which settled the issue. For the benefit of newcomers to MOVPE, all of these variants of the name can be found in the literature and, in most cases, can be used interchangeably. MOCVD can be considered broader, as it includes polycrystalline growth that is appropriate to the photovoltaic thin films covered in Chapter 10. The early niche applications of MOVPE were with GaAs photocathodes [7], GaAs heterojunction bipolar transistor (HBT) lasers [8], and GaInAsP lasers and detectors for 1.3-μm optical fibre-optic communications [9].

The ease with which GaAs could be grown using trimethylgallium (TMG) and arsine was not readily replicated by all the semiconductor materials of interest. Indium phosphide had proved to be particularly challenging due to a polymerization prereaction that occurred between trimethylindium (TMI) and phosphine [10]. This was overcome, initially, by going to low-pressure MOVPE and replacing TMI with triethylindium (TEI). Later improvements in the purity of TMI and in overcoming issues with it being a solid source at room temperature have led to TMI now being the precursor of choice. Atmospheric-pressure MOVPE of InP was achieved by Moss [11] using an elegant solution that took advantage of the adduct formation of In alkyls. A room-temperature stable adduct, TMIn.TEP (where TEP is triethylphosphine), was formed that prevented the polymerization reaction with PH_3 but the adduct decomposed over the hot substrate to yield TMIn and TEP. The TEP was stable and was exhausted from the reactor without further decomposition, while the TMIn and PH_3 or AsH_3 for the arsenic compounds reacted to form the compound semiconductor on the substrate. Adducts have also been used in the purification of the organometallic precursors, as will be described in Chapter 13, and for preventing prereactions with ZnO, as described in Chapter 11. The significance of growing InP was to be able to then grow ternaries and quaternaries for infrared lasers and detectors, lattice matched to InP. This enabled access to the growing market for 1.5-μm wavelength devices for long-range optic fibre-optic communications [12]. The purification of Group III precursors is discussed in Chapter 3 with regard to the device application of these materials. The decomposition chemistry and role of the Group V hydride is important to minimize the incorporation of carbon. In some devices, this is now used as an intentional P-type dopant, substituting on the Group V site.

The antimonides (InSb, GaSb, AlSb) are an important class of narrow-bandgap semiconductors for infrared detectors, long-wavelength lasers, and thermoelectric devices. Unlike GaAs- and InP-based semiconductors, where hydrides are normally used as the Group V source, for the antimonides it is necessary to use alkyl sources such as TMSb. The growth of the antimonides is described in Section 3.5.

The II-VI alloy mercury cadmium telluride (MCT) had also proved to be difficult to produce by MOVPE, and the growth in the interest in MCT for long-wavelength thermal imagers (operating around 10 μm) was stimulating research in different epitaxial techniques, including LPE, MBE, and MOVPE, all of which are used in production today. The challenge was from the very-high-equilibrium vapor pressure of mercury in MCT at growth temperatures ranging from 200 °C to 500 °C. Early success in the 1980s with an MOVPE approach using a liquid Hg source ensured a very fertile two decades of research that will be described in more detail in Chapter 9 [13]. This helped to demonstrate at an early stage that MOVPE can be a very versatile technique that has been proven again many times for different compound semiconductors over the intervening years.

Thus, the technique of MOVPE was born, but it was not until the late 1980s that MOVPE became a production technique of any significance. This success depended on painstaking work on improving the impurity of the organometallic precursors and development of MOVPE equipment to improve uniformity and deposit onto multiple substrates in the reaction chamber. The emergence of commercial equipment suppliers took over from the "home-built" reactors and provided much-needed standards to which production of epi-wafers could be benchmarked. By this time, the focus was on high-quality epitaxial layers on lattice-matched substrates, in contrast with the early work. One exception was the growth of CdTe epitaxial layers onto C-plane sapphire as a substrate for LPE growth of MCT for mid-wave infrared (MWIR) detectors. Ironically this was manufactured by Rockwell International, where Hal Manasevit carried out his original research. The topic of homoepitaxy versus heteroepitaxy has been a continual balance between achieving very high-quality epitaxial layers and materials functionality. To this day, the growth of high-quality epitaxy on silicon substrates has remained the ultimate challenge, through not only providing a ready supply of high-quality and cheap substrates but also the integration of optoelectronic and electronic devices. By the late 1980s and early 1990s, there was a commercial supply of GaAs on Si substrates to be used as a substitute for GaAs substrates. Improvements in the quality and size of GaAs substrates made this approach uncompetitive. Today, this is attracting a new generation of research activity; the topic of monolithic III-V integration on Si(001) substrates is covered in Chapter 8. New understanding of epitaxial processes and new applications are driving this resurgence of activity.

The characteristics of MOVPE that have taken it from a research curiosity to production have been in the simplicity of delivery of the reactive vapors and the versatility of selecting different alloy compositions, dopants, and layer thicknesses. These basic attributes have enabled the same basic technique to be used for narrow-bandgap semiconductors such as the infrared detector materials MCT and GaInSb and now for wide-bandgap semiconductors such as GaN and ZnO. Indeed, the success of GaInN in the 1990s for high-brightness blue light-emitting diodes (LEDs) has now led to this being the most commonly produced MOVPE material with the growth of the market for high-brightness white-light LEDs for displays and lighting [14]. A new set of challenges for MOVPE equipment has led to a generation of commercial reactors designed to cope with the higher temperatures of GaN and AlN epitaxy. For AlN, the temperature required for growth is over 1200 °C, and reactors capable of over 1400 °C are now being manufactured. Prereaction is also more of an issue than with GaAs. The nitrides now represent over half the total commercial output of MOVPE, and research on AlN is pushing emitters to shorter wavelengths, into the UV, where they could eventually replace mercury lamps as a source of UV radiation for sterilization [15]. The nitrides are discussed in Chapter 4 in some depth, covering GaInN, AlGaN, and substrates for nitride epitaxy.

The early strength of MOVPE was its ability to grow onto different substrates, but this was later abandoned in favor of the more conventional homoepitaxy; however, the nitrides rely on heteroepitaxy onto sapphire and SiC substrates, bringing MOVPE back to its roots with the early work of Manasevit. The goal of more recent work has been to achieve a reduced density of dislocations in the GaN layer using patterned substrates [16]. Heteroepitaxial growth onto silicon substrates has been widely studied and is of growing importance for optoelectronic device integration with silicon devices. The challenge is not only lattice mismatch but also the fundamental issue of growing a polar material onto a nonpolar substrate. If this is not properly understood and controlled, it will lead to antiphase domain boundaries. The fundamental considerations of MOVPE growth of III-V materials onto silicon substrates is discussed in Chapter 8. There is also a growing realization that some of the lattice-mismatch problems can be overcome by growing nanowires, which has been an intensive topic of research across all the material systems over the past 10 years and will be discussed in Chapter 7 [17]. The versatility of substrate materials presents MOVPE with the ultimate challenge of mating high-performance optoelectronic materials with silicon substrates to combine the best of optoelectronic and electronic performance.

1.2 Basic Reaction Mechanisms

The early progress of MOVPE was based on the availability of room-temperature volatile but thermally stable precursors for the Group II source and trihydrides for the Group V source. Both have the essential properties of being volatile in a suitable carrier gas stream (usually hydrogen) and being chemically stable at ambient temperature. The temperatures of the organometallic precursors are carefully controlled, normally below ambient temperature to avoid condensation in the lines, to provide a fixed saturated vapor pressure (SVP). This is combined with a measured volumetric flow of the carrier gas through the liquid organometallic source (contained in a stainless-steel bubbler) to yield a known molar flow as shown in Eq. (1.1):

$$F_m = \frac{p_{svp} \times v_b}{p_{STP} \times 22.4} \tag{1.1}$$

where F_m is the molar flow, p_{svp} is the saturated vapor pressure in the bubbler, v_b is the flow of carrier gas through the bubbler in l/min, and p_{STP} is the pressure at STP.

The organometallic precursors and hydrides are normally mixed outside the reaction chamber where additional dilution of the carrier gas is carried out and then introduced into the reaction chamber through a suitable injector arrangement and directed onto a hot substrate. This is shown schematically in Figure 1.1. The alternative design of reactors will be discussed later in this chapter. The reaction of the precursors to yield the III-V compound on the substrate can occur either in the hot vapor above the surface or on the hot surface. The stoichiometric reaction for GaAs growth is given as follows:

$$(CH_3)_3Ga + AsH_3 \rightarrow GaAs + 3CH_4 \tag{1.2}$$

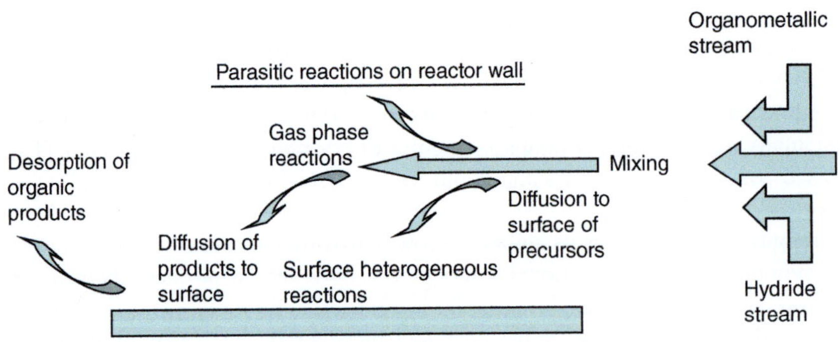

Figure 1.1 *Schematic of MOVPE gas transport and reaction process.*

This reaction has been the most widely studied of all the MOVPE reactions and was one of the original processes reported by Manasevit and Simpson [1]. One reason it works so well is that the hydrogen required to satisfy the ·CH_3 radical bond is supplied from the arsine hydride and at normal growth temperatures (around 700 °C) avoids unsatisfied carbon bonds that could lead to carbon incorporation. In reality, this very simple picture covers a complexity of reaction steps that have to take place and has been discussed in great detail by Chernov [18]. However, some of the important reaction steps will be described here as an introduction to the kinetics of GaAs MOVPE and will be an introduction to the more detailed coverage of kinetics in Chapter 2.

The schematic shown in Figure 1.2 gives some of the important reaction steps that have been identified for the reaction of GaAs. This gives some insight into the complexity of the reaction kinetics and alternative reaction paths that is a characteristic of all the MOVPE compound semiconductor reaction paths. Combined gas flow and reaction kinetic modeling in horizontal and vertical reaction chamber geometries by Jensen et al. used a total of 17 different gas-phase and 26 surface reactions to correctly predict the growth rate [19]. This work went on to model carbon incorporation from the MOVPE reaction, where it was known empirically that a high V/III ratio of precursors was needed to ensure low carbon incorporation from the reaction process. This showed that the increase in AsH_3 reduces the concentration of the carbene (CH_2) reaction product. This helps to illustrate that detailed modeling can give useful insights, but empirical determination of the best reaction conditions for high-quality epitaxial layers has played a significant role in the progress made with MOVPE. An understanding of the reaction kinetics does, however, enable some of the features of MOVPE to be understood, particularly when these relatively simple precursors are replaced by more complex precursors.

To look at the possible reaction paths in more detail, it can be seen from Figure 1.2 that the reaction process is started by a gas-phase homolysis of TMGa ($Ga(CH_3)_3$) to yield dimethylgallium ($Ga(CH_3)_2$) and methyl radicals (·CH_3). There are two important roles that the methyl radicals can take, and this is generally important in all MOCVD processes for deposition of III-V semiconductors:

Methyl radicals can react with the ambient hydrogen carrier gas to yield stable methane and a hydrogen radical.

Methyl radicals can react with arsine (AsH_3) to yield stable methane and AsH_2.

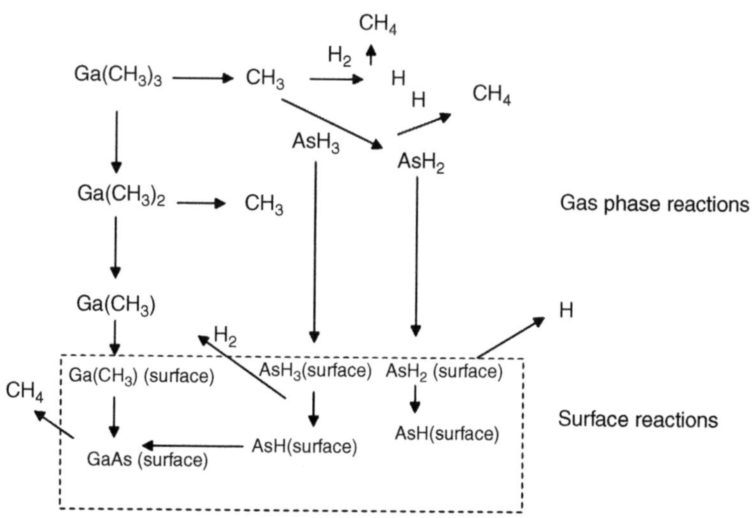

Figure 1.2 *Schematic of some of the key reaction steps in MOVPE of GaAs.*

Both of these steps can initiate the decomposition of arsine through either a methyl radical or a hydrogen radical removing a hydrogen atom from the arsine; these reaction steps can be seen in Figure 1.2. This helps to illustrate that the hydrogen carrier gas is not included in the stoichiometric reaction (reaction (1.2)) but can be important in the reaction process. Another consequence of the reaction kinetics lies in the interdependent nature of the reaction process on both precursors. This is fundamental to the success of MOVPE because it is unlikely that organometallic and hydride precursors have the same thermal decomposition characteristics, but they can be used together to react at the same temperature through radical intermediates. The latter has been illustrated in many studies of different precursors in simple reaction-tube experiments. For example, in the case of GaAs, the pyrolysis of TMGa will occur at 500 °C; but for arsine decomposition, a temperature of around 700 °C is required. A detailed study of TMGa and AsH_3 decomposition in H_2 by DenBaars et al. [20] showed that the reaction rate of AsH_3 not only was faster in the presence of TMGa but also was fast in the presence of a GaAs surface, which illustrates the importance of the surface-reaction mechanisms.

The stepwise removal of ligands from TMGa and arsine can occur in the vapor or on the surface. Figure 1.2 shows both alternatives; the dominant path will depend on both the surface temperature and vapor temperature above the substrate. A general rule is that more of the reaction process will occur on the surface at lower temperatures when the vapor reaction rate is slower and the surface adsorption is more efficient. It is also important to note that there are two different adsorption sites on the surface: the Ga sites that will take the anion species and the As sites that will take the cation species.

The kinetics of II-VI MOVPE are similar to those of III-Vs, but the use of hydrides for the Se and S sources, used in the early days, has been replaced with organometallic sources for the Group VI elements. This leads to the hydrogen-radical path becoming more important to satisfy the organic bonds and avoid carbon incorporation. For example, a proposed reaction path for the growth of CdTe is as follows:

$$(CH_3)_2 Cd + H_2 \rightarrow (CH_3)Cd + CH_4 + H \tag{1.3}$$

$$(C_2H_5)_2 Te + H \rightarrow C_2H_6 + (C_2H_5)Te \tag{1.4}$$

In these reaction steps, the less-stable dimethylcadmium (DMCd) undergoes a thermal bond homolysis where the methyl radical reacts with a hydrogen molecule (not shown). This produces a methyl cadmium radical (that is unstable and will rapidly lose another methyl ligand), a methane molecule, and a hydrogen radical. The hydrogen radical is then free to react with the diethyltelluride (DETe) precursor, yielding ethane and an unstable ethyl tellurium radical. This ensures that the Cd and Te atomic species are delivered to the surface at the same temperature, despite the Te precursor being more stable [21]. The MOVPE of II-VI semiconductors, described in Chapters 9, 10, and 11, tends to operate more in the kinetic regime due to the desire for low epitaxial growth temperatures. This makes the deposition rate more temperature sensitive, and hence temperature uniformity across the substrate is more critical than for higher-temperature growth. The low-temperature growth of these compounds has attracted interest in various forms of energy assisted-MOVPE such as photoassisted, photocatalytic, and photosensitized processes. Rather than replacing the thermal kinetic processes, these energy-assisted processes tend to be inserted in key parts of the reaction chain where there is a kinetically limiting step. Using short-wavelength UV radiation (typically less than 260 nm), it is possible to remove the ligands through photolysis in the vapor. This has been the approach used for HgTe, CdTe, and ZnSe to achieve epitaxial growth at temperatures below 400 °C. A more detailed description is given of the mechanisms for HgTe and CdTe photoassisted epitaxy in Chapter 10. Due to the limitation on space and the more recent decline in interest in ZnSe and ZnS, these materials are not covered in any detail in this book. However, the work carried out using photoassisted epitaxy provides an excellent background for the application of these energy-assisted mechanisms in MOVPE: for example, the work of Ando et al. [22] on photolysis of diethylzinc (DEZn) and dimethylselenide (DMSe) and the work by Fujita et al. on surface photocatalysis of ZnSe epitaxial films [23]. The kinetics of narrow-bandgap II-VI semiconductors will be covered in Chapters 9 and 10, with wider-bandgap II-VIs, including the oxides ZnO and CdO, covered in Chapter 11.

The details of the reaction kinetics change with substrate temperature, but as the temperature is increased, a point is reached where the rate of epitaxial growth is no longer determined by the overall reaction rate but is rather determined by the supply of precursors to the substrate. This is reflected by a depletion of the precursor concentration immediately above the surface and a gradient in precursor concentration toward the undepleted free stream. The limitation on the rate of epitaxial growth then becomes the rate of diffusion through the depleted boundary layer to the substrate. This is called *transport-limited growth* and is characterized by high growth rates and only a weak dependence of growth rate on substrate temperature. This is shown schematically in Figure 1.3. The plot is of log(growth rate) versus $1/T$ because of the expected Arrhenius relationship in the rate constants. This really only applies to the low-temperature (kinetic regime) where the growth rate can be expressed as:

$$\text{rate} = Ae^{-E_a/RT} \tag{1.5}$$

where A is a constant, E_a is an activation energy, R is the gas constant, and T is the temperature in Kelvin. The value of E_a can be determined experimentally from an Arrhenius plot of growth rate versus $1/T$ and can be attributed to the limiting kinetic step in the precursor-decomposition process.

In the transport-limited regime, there will still be a small dependence of growth rate on temperature due to the increase in diffusion rate with temperature; this is illustrated in Figure 1.3. Most MOCVD growth processes will take place in the transport-limited regime where it is easier to control growth rate, and modern reactor designs are optimized to work efficiently in this regime. However, a number of growth processes occur at lower temperatures in order to control properties such as native defect concentration of the epitaxial films. This is generally the case with II-VI semiconductors but can also apply to the formation of thermodynamically unstable III-V alloys such as the antimonides, which are discussed in Chapters 2 and 3.

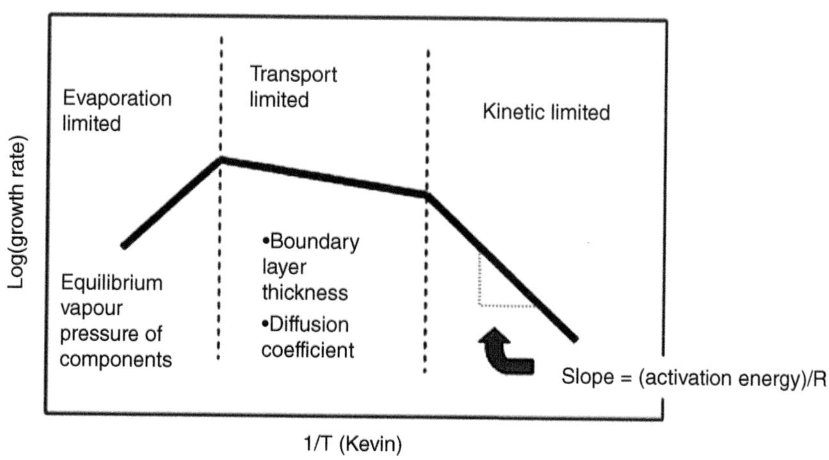

Figure 1.3 *Schematic of three different temperature regimes for MOVPE.*

In the high-temperature regime, the growth rate decreases with temperature as the equilibrium vapor pressure of the constituent elements in the film increases and gives desorption rates similar to the deposition rate, leading to significant loss of material through evaporation to the gas stream. Other mechanisms can also cause a decrease in growth rate at higher temperatures, such as free-radical etching and precursor desorption.

1.3 Precursors

Precursors provide the key to the success of MOVPE in providing a very-high-purity source of the components of the compound semiconductor in a convenient form that can be delivered at (or close to) room temperature in precise molar concentrations. The choice of precursors is not confined to simple alkyls and hydrides but can extend to almost any volatile organometallic as a carrier for the elemental components of a film. For III-V compounds, it is usual for hydrides to be used (NH_3, PH_3, AsH_3 and SbH_3) for the Group V source. However, extensive research has been carried out in finding organometallic alternatives, and TBP and TBAs have both been successfully used, as mentioned earlier in this chapter. Similarly, a number of liquid amine and azide sources have been used for GaN, but ammonia gas remains the firm favorite. In the case of II-VI semiconductors, it is usual to use an alkyl for both the Group II and Group VI elements. Hydrides have been used as Se and S sources, but prereaction makes it difficult to control the growth process and in particular can make it difficult to incorporate dopants. For II-VI oxides, liquid or gaseous sources such as tertiarybutanol (tBuOH) and nitrous oxide (N_2O) can be used. This enables a wide range of growth temperatures to be used, depending on the required film properties. The tBuOH would be used for low-temperature planar growth and the N_2O for higher-temperature quantum-wire growth.

The important properties of liquid precursors, and their selection, can be generalized and provides a basis for optimizing the MOCVD process. These properties can be summarized as follows:

1. SVP should ideally be in the range of 1–10 mbar in the temperature range of 0–20 °C. Outside of this range, this can be managed by selection of the bubbler temperature and, for less-volatile precursors, heated lines.
2. Stable for long periods at room temperature – can be stored when not in use.
3. Will react efficiently at the desired growth temperature – the combination of precursors may need to be considered here.
4. The reaction produces stable leaving groups – low carbon incorporation.
5. Avoids unwanted side reactions such as polymerization.

According to the Clausius–Clapeyron equation, the SVP of a liquid is given by an exponential relationship:

$$\mathrm{SVP} = e^{-\Delta G/RT} \tag{1.6}$$

where ΔG is the change in Gibbs free energy on evaporation, R is the gas constant, and T is the temperature of the liquid in the bubbler in Kelvin. This can be expressed as the heat of evaporation ΔH and the entropy for evaporation ΔS, where $\Delta G = \Delta H - T\Delta S$, which gives the familiar form of the SVP equation:

$$\mathrm{SVP} = Ae^{-\Delta H/RT} \tag{1.7}$$

This is of the form

$$\log_e(\mathrm{SVP}) = A - B/T \tag{1.8}$$

where A and B are constants given by $A=\Delta S/R$ and $B=\Delta H/R$. Manufacturers of the precursors will generally give the SVP data in the form of the constants A and B in Eq. (1.8). It can also be given in the form

$$\log_{10}(\text{SVP}) = A' - B'/T \tag{1.9}$$

To convert the constants in Eq. (1.9) to Eq. (1.8), simply multiply by ln10. For example, for TMGa, the values for A and B are 9.07 and 1703, respectively, giving an SVP at 20 °C of 182 torr; reducing the temperature of the TMGa source to 0 °C reduces the SVP to 68 torr. Using higher alkanes reduces the SVP and reduces the decomposition temperature. So, replacing the methyl ligands with ethyl ligands gives triethylgallium (TEGa) and reduces the SVP (20 °C) to 5 torr. More details on precursor choices are given in Chapter 13, and on the growth of III-arsenides, phosphides, and antimonides in Chapter 3. The choice of precursors, their synthesis, and their purification are covered in more detail in Chapter 13.

1.4 Types of Reactor Cell

Reactor cells have evolved enormously since the first demonstration of MOVPE using either a vertical quartz tube in Manasevit's experiments or a horizontal quartz tube (the Bass cell [24]). The common characteristic of these cells was the "cold" wall, with the only heated part of the cell being a graphite susceptor, normally RF heated, to hold the substrate. The design of reactor cells has attracted intense research effort to achieve uniform films with low contamination and low defect concentrations arising from particulates. The early research reactors quickly evolved to designs that were easier to scale up to production [25–27].

The horizontal and vertical reactor designs are shown schematically in Figure 1.4. The substrate is placed onto a graphite susceptor that is heated by either RF coupling via a coil surrounding the reactor, a resistance heater underneath the susceptor, or lamps placed underneath the susceptor. The reactor wall can be water cooled or gas cooled to minimize reaction and deposition onto these surfaces. Either of these reactor cells could be operated at atmospheric or reduced pressure. For reduced-pressure operation, the reactor-cell pressure would be typically a tenth of an atmosphere,

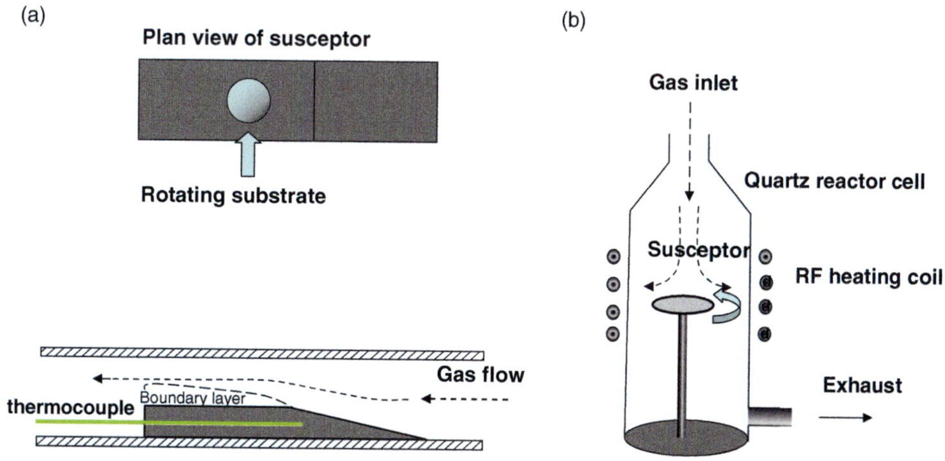

Figure 1.4 Schematic of (a) a horizontal Bass-type reactor cell and (b) a vertical Manasevit type reactor cell.

but a wide range of different pressures have been successfully used to optimize flow and minimize prereaction. Reduced pressure increases the gas velocity and helps to overcome the effects of free convection from a hot substrate that creates unwanted vortex rolls. The forced-convection parameter that is often quoted is the Reynolds number and is proportional to gas velocity. A high Reynolds number will ensure streamline flow; at a low Reynold's number, the buoyancy effects of the hot substrate will take over, and the gas flow will be dominated by free convection (characterized by the Grashof number) and become disorganized with recirculation cells. In transport-limited growth, the exact nature of the gas flow determines the uniformity of deposition and can also affect the defect concentration in the films due to particulates and reaction products being swept back across the growing surface. It is not normal to achieve the very high flow velocities and Reynold's numbers associated with turbulent flow in an MOVPE chamber; however, the disorganized flow due to free convection is often, wrongly, referred to as *turbulent flow*.

Another reason for using high flow velocities is to overcome the effects of depletion of the precursor concentration at the downstream end of the deposition region. For transport-limited growth, the growth rate is limited by the rate of diffusion from the free stream to the substrate. This region is called the *boundary layer*, and it increases in thickness going downstream from the leading edge of the susceptor, as shown in Figure 1.5. Some horizontal reactors are designed with a tilt in the susceptor so that the free cross-sectional area decreases and hence the flow velocity increases going downstream. This helps to flatten the boundary layer and ensure better uniformity. In later designs of horizontal reactors, the substrate is rotated to improve uniformity, and this concept of smoothing the thickness uniformity by substrate rotation has been the basis for achieving high-uniformity growth in advanced horizontal reaction cell designs such as the AIXTRON planetary reactor [28]. With rotation, it is no longer necessary to use high flow velocity, which is wasteful of gases and precursors, but the reactor can be designed to produce linear depletion that results in a flat thickness profile after rotation.

As research reactors have given way to production reactors, maintaining a uniform boundary layer with high flow velocity has not been practical for multiwafer reaction cells and for efficient gas usage. These problems have been resolved with different solutions in the vertical and horizontal reactor configurations with the Emcore (now Veeco) TurboDisc reactor that creates a thin boundary layer by high-speed rotation of the substrate, the Thomas Swan (now AIXTRON) close-coupled showerhead reactor, which has multiple injectors in a vertical configuration close to the substrates; and the AIXTRON planetary reactor, briefly described earlier. Each of these reactor designs will be described in greater detail in Chapter 12.

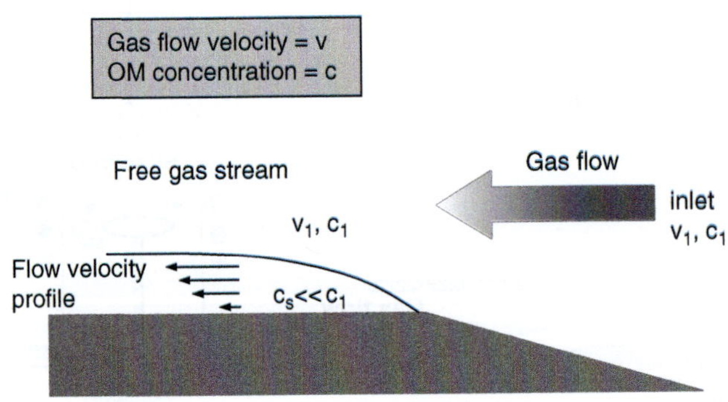

Figure 1.5 *Schematic of the boundary layer in a horizontal MOVPE reactor.*

1.5 Introduction to Applications of MOVPE

MOVPE has had a huge shift over the past 10 years from research into production as the success of GaInN LEDs has been revolutionizing the lighting market; in 2016 it was worth \$26Bn annually and set to grow to \$94Bn by 2022. This success is based not just on the efficiency of white-light LEDs but on the robustness of GaInN that enables long lifetime, now around 50 000 hours. At the heart of a white-light LED is a blue LED, surrounded by high-efficiency phosphors to emit a balanced white-light spectrum. The most efficient GaInN LEDs are based on low indium content alloys, at the blue end of the spectrum, around 400–450 nm. The entire alloy range will cover the whole of the visible spectrum but at lower efficiency toward the red. However, blue emission is suitable for driving high-efficiency phosphors, such as rare earth doped $Y_3Al_5O_{12}$ [29]. The wavelength of the blue LED and the emission band of the phosphor can be tuned to give a different balance of white light and optimum efficiency, lumen/W.

GaN has a hexagonal wurtzite lattice with lattice parameters of $a = 0.3186$ nm and $b = 0.5186$ nm. This is not readily lattice matched to suitable substrate materials, and the preferred substrates for LEDs have been sapphire and SiC. Heteroepitaxy of GaN has required a two-stage nucleation process, with nucleation at low temperature to achieve surface coverage followed by recrystallization and epitaxial film growth at over 1000 °C. Minimizing the dislocations propagating from the heterointerface has been a focus of attention for MOVPE growth to enable high-efficiency LEDs and laser diodes [30].

GaN-based power and high-frequency devices have developed in parallel with the light emitters but have benefitted from the materials improvements achieved for LED and lasers in GaN. GaN high-frequency and high-power devices require high electron mobility and are suitable for high-electron-mobility transistors (HEMT) that operate at very high frequencies, i.e. tens of GHz. GaN FETs are being developed for power switches and could displace silicon power switches for high current densities. GaN MOVPE on Si is attractive for these applications [31]. More will be discussed about substrates for GaN MOVPE in Chapter 4.

1.5.1 AlN for UV Emitters

The growth of Al-containing alloys is desirable to achieve wider bandgaps for applications such as UV LEDs and solar-blind detectors. The UV LEDs can operate down to 200 nm and have the potential to replace more costly mercury lamps for water purification and medical sterilization. The MOVPE growth of AlGaN alloys is more challenging than for GaInN; and in the case of Al-Ga alloys, the stability of AlN is much greater than that of GaN, which tends to form AlN clusters. The solution is growth temperatures above 1200 °C, and a new generation of MOVPE reactors are being built that can achieve temperatures over 1400 °C. Deep-UV LEDs have been demonstrated using MOVPE with emission wavelengths less than 250 nm [32]. High-temperature growth brings its own challenges with increased risk of prereaction, thus reducing growth rate and possibly creating particulates. This is addressed in Chapter 4, where the particular requirements for high-temperature growth are explained. This is also discussed in Section 12.4.1 in the context of reactor chamber design, which becomes very important for operation under more extreme conditions. The need for AlGaN devices is pushing the capability of MOVPE technology.

1.5.2 GaAs/AlGaAs VCSELS

Solid-state laser diodes are based on lateral propagation of light, with a cleaved crystal providing the reflectors, and are referred to as *edge emitters*. They have limitations in terms of processing, testing, and beam profile. The vertical cavity surface emitting laser (VCSEL) has the advantage of

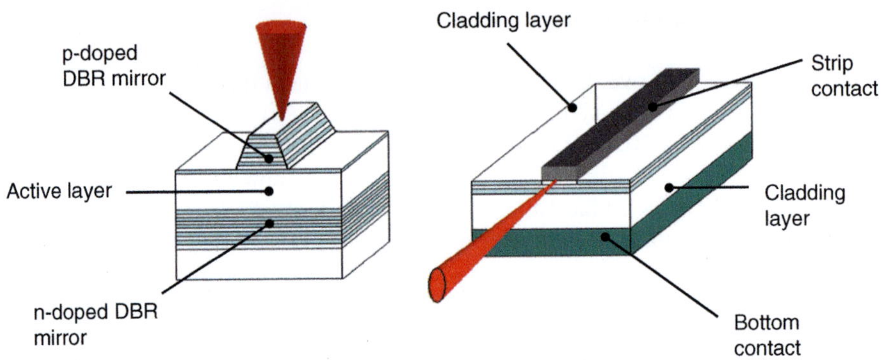

Figure 1.6 *Schematic of a VCSEL structure with the DBR mirrors in comparison with an edge emission laser that uses crystal facets for the mirrors (after K.J. Weeks, Bangor University, PhD Thesis, 2002).*

producing an array of lasers on each substrate that can be tested on the substrate. The beam profile is circular, and the laser diodes are more easily coupled into fibers for fiber-optic communications. However, the epitaxial structure, such as GaAs/AlGaAs or GaInAsP/InP, is more complex, with a stringent requirement for thickness uniformity. The mirrors in the vertical cavity are distributed Bragg reflectors (DBRs) that rely on a periodic structure of lattice-matched layers with a refractive index difference. This provides wavelength-selective reflection between the mirrors encapsulating a quantum well (QW) diode as the emitter, shown schematically in Figure 1.6. The ever-increasing range of application for VCSELs are reviewed by Harris et al. [33], and more details on these materials and laser applications are given in Chapter 3. The MOVPE of these materials is covered in Chapter 3 and includes a description of in situ monitoring that enables effective control over composition and thickness in the periodic structure in the Bragg reflectors. The challenge for the MOVPE engineer is not only achieving the correct center wavelength in the optical stack but also good electrical conduction. Success with solving these problems has led to high-volume manufacture of these devices.

1.5.3 Multijunction Solar Cells

The fundamental efficiency limit of a single-junction solar cell, known as the Shockley–Queisser limit, is around 30%, but this can be exceeded by stacking the semiconductor junctions, shown schematically in Figure 1.7. Each junction is designed to capture a different waveband to absorb a broader range of the solar spectrum more efficiently than would have been achieved with a single junction. The junctions are connected, in the two-terminal device, by tunnel junctions that are designed to minimize series resistance. These very efficient solar cells have found an application in space for powering satellites and for ground-based concentrator solar cells where the solar radiation is concentrated using Fresnel lenses in excess of $\times 500$. A solar conversion efficiency of over 44% has been achieved with a four-junction device, as reported by Dimroth et al. [34]. In this example's top cell, GaInP captures wavelengths shorter than 600 nm, the next cell (GaAs) covers the 600–900 nm range, followed by GaInAsP capturing 900–1100 nm, and the bottom cell GaInAs covering the 1100–1700 nm range. Low dislocation density is required for high-efficiency cells, so lattice matching of the different layers in the stack is very important. The three-junction GaInP/GaAs/Ge (where the Ge bottom cell is also the substrate) is a strained lattice-matched structure, but the four-junction cell requires lift-off techniques and wafer bonding of the GaAs cell to the GaInAsP cell. Chapter 5 considers the issues of lattice matching in some detail and describes

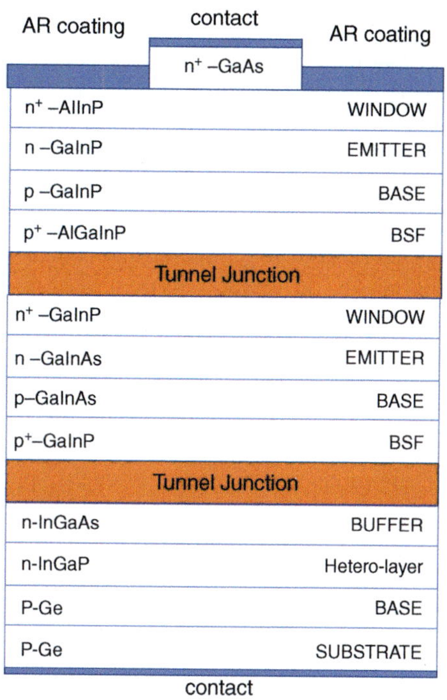

Figure 1.7 *Schematic of a triple junction III-V solar cell.*

methods for strain relief using metamorphic structures where strained layers are intentionally introduced. These devices have achieved impressive performance through precise control of alloy composition, doping, and layer thickness, where MOVPE has become the chosen technique for production. The challenges and opportunities with multijunction III-V solar cells are described in Chapter 5, where different approaches for using MOVPE to achieve efficient multijunction solar cell devices are presented.

1.5.4 GaAs and InP Transistors for High-Frequency Devices

Although the original motivation for MOVPE of III-V semiconductors was for optoelectronic applications, the application for high-frequency devices has established a large market that has grown rapidly in recent years with operational frequencies up to several hundred GHz. These devices include HEMTs, metal-oxide semiconductor field effect transistors (MOSFETS), and HBTs [35]. The attraction for high-frequency devices made from III-V semiconductors is that they have much higher electron mobility than silicon, but they must be of very high purity and have a low defect concentration to realize this potential. Excellent interface control (morphology and low defect concentration) is also needed for the devices. Although many of these devices are now in mass production for applications such as mobile phones, satellite communications, and solid-state radar, there are new opportunities with MOVPE of III-V nanowires. As devices shrink to increase packing density, the power dissipation rises unless higher-mobility materials can be used. GaAs and InAs nanowires provide a solution for narrow gate widths (the width of a nanowire) and high-quality single crystals. There are various ways of forming these nanowires by MOVPE, as will be described in Chapter 7; and now doped junctions can be formed within the length of the nanowire, opening the possibility for new designs of nanostructured devices [36].

1.5.5 Infrared Detectors

MOVPE of infrared detectors has included MWIR devices made from InSb, as described in Chapter 3, and short-wavelength, mid-wavelength and long-wavelength detectors made from the II-VI alloy, MCT, which is covered in Chapter 9. Both have had their own challenges, where MOVPE has proven to be a resilient process for producing high-quality epitaxial layers and devices. MCT is an alloy with bandgap ranging from zero up to 1.5 eV covering far-infrared to near-infrared detection, and will be discussed in some detail in Chapter 9. The applications of MCT range from numerous military uses through search and rescue, surveillance, earth observation, astronomy, and even sports (i.e. the Hotspot system used in international cricket). The antimonides are restricted in their bandgap, with alloying from the MWIR to near infrared (NIR). However, type-II superlattices of GaSb/InSb can extend to the long-wavelength infrared (LWIR) [37]. Both these material systems encounter the problem of lattice mismatch with usable substrates. MCT has a 14% lattice mismatch with GaAs and a 20% lattice mismatch with silicon. Remarkably, research has enabled heteroepitaxial growth of MCT onto these highly mismatched substrates. This is significant, as the major drive for applications in the infrared has been for focal plane arrays (FPAs) where it is necessary to collect charge from each detector into a silicon multiplexer. Despite lattice mismatch and misfit dislocations, high-performance FPAs of MCT detectors have been reported on silicon substrates using MOVPE [38].

1.5.6 Photovoltaic and Thermophotovoltaic Devices

The applications of MOVPE in photovoltaic (PV) and thermophotovoltaic (TPV) extend beyond the very successful multijunction solar cells discussed earlier. The drive for improved efficiency and lower-cost energy generation has stimulated different avenues of research with different materials systems. For example, the use of nanowires has been an active topic of research across all PV materials systems, as covered in Chapter 7, with the attraction of reducing the amount of semiconductor material while increasing the light trapping [39]. Nanowires also have the benefit of removing some of the lattice parameter matching that can place a restriction on the substrates used for planar epitaxial growth. The remarkable success of nanowires has been demonstrated with nanowire InP and GaAs photovoltaic devices, with solar conversion efficiencies of 13.8% and 15.3%, respectively [40, 41]. In the work by Wallentin et al. [40], they point out that although the InP nanowires (NWs) cover only 12% of the surface, they produce 83% of the photocurrent density of planar InP solar cells. This is due to the light trapping of the NW and shows the potential for more efficient materials utilization and more flexible device design. In Chapter 12, a novel core-shell nanowire structure is described for CdTe thin-film solar cells. More advanced PV structures based on NWs could take advantage of the materials control that is enabled by MOCVD and is an active area of research that is fully described in Chapter 12.

Narrow-bandgap InSb and GaSb can be used as photovoltaic devices to capture infrared radiation from hot sources (approximately 900 °C) with a bandgap around 0.5 eV, as an alternative route to energy conversion. The range of bandgaps available in this system stretches from 0.75 eV for GaSb down to 0.18 eV for InSb. The MOVPE growth of these alloys has to overcome the thermodynamic limitation of spinodal decomposition, where the alloy will try to separate into distinct phases at normal growth temperatures or as it cools from the deposition temperature. Further challenges in the MOVPE of these materials are covered in Chapter 3, where the low melting point of InSb (527 °C) necessitates low growth temperatures. There are also similarities to the chemistry of CdTe MOVPE, where an organometallic Sb precursor is used. In the absence of a hydride source, there is no readily available hydrogen radical supply to remove oxides from the substrate surface. A more detailed account of the thermodynamic limitations on the growth of unstable alloy compositions and how this can be overcome with MOVPE kinetics is given in Chapter 2. The device structures can be lattice matched by tuning the quaternary InGaAsSb composition and bandgap on GaSb substrates to fabricate high-quality p-n diodes. Other material combinations and strained superlattice structures have also been used to make devices in the mid-infrared waveband range. For example, Pitts et al. [42] have grown InAsSb/InPSb heterojunctions for detection in the 3–5 μm range. This work illustrates the need to carefully control the V/II ratio to obtain device-quality epitaxial layers. A further variant on the antimonide alloys is the introduction of dilute concentrations of nitrogen to obtain greater flexibility with bandgap bowing, but the solubility of nitrogen into these alloys remains a challenge to obtain sufficient nitrogen concentration [43].

The MOVPE of CdTe for photovoltaic devices has been confined more to research in understanding fundamental issues of composition and doping. Production of CdTe solar cells relies on vapor transport of Cd and Te_2 to a glass-based substrate, producing polycrystalline devices. This has been replicated with MOCVD of CdTe photovoltaics that can offset the higher cost of manufacture with greater materials versatility for more specialist applications. One example is the MOCVD of CdTe/CdZnS onto ultrathin glass for ultra-lightweight solar cells for space applications [44]. Chapter 10 covers the application of MOCVD for thin-film PV as well as the fundamental aspects of CdTe epitaxy.

1.6 Health and Safety Considerations in MOVPE

The development of MOVPE from a laboratory curiosity to full-scale production has gone hand in hand with development of safe practices in handling the precursors, safe design of MOVPE reactors, exhaust gas abatement, and waste products. There are two key chapters where these aspects are embedded: Chapter 12, on reactor design, and Chapter 13, on MOVPE precursors.

The organometallic sources are toxic: the body can readily absorb hazardous heavy metals in an organic form. The organometallic precursors are also highly flammable and, in some cases will spontaneously combust in contact with air. If there is an organometallic fire, the application of water will only fuel the fire, so dry powder extinguishers are needed. From the point of view of the MOVPE grower, the hazards are minimized by the organometallics being supplied in air-tight stainless-steel bubblers, with a strict procedure of pumping and purging used to change bubblers. Similar precautions must be taken in the manufacture of the precursors, but because the batch sizes are much larger, the potential scale of the hazard is also larger; in the early days, a fire at a synthesis facility for TMG or TMI was not unusual. One advantage of organometallic precursors is that the vapor pressures are low so an accidental escape leading to a fire can normally be readily contained.

The hydride sources, arsine and phosphine, present a larger potential toxic hazard, as they are contained in gas cylinders and an accidental escape can more readily exceed the safe threshold limits. Safe handling requires containment of the cylinders in a suitably extracted cabinet and an appropriate monitoring system wired to alarms and an external monitoring panel.

The range of risks and mitigation of these risks is reviewed by Shenai-Khatkhate et al. [45]. It is important to understand the safe exposure limits for each of the precursors used; these are supplied by the precursor manufacturers. Various measures can be used to determine toxicity, such as the permissible exposure limit (PEL) and the threshold limit values (TLVs). For example, the TLV is the level to which a worker can be exposed on a day-to-day basis over a long period of time without any adverse effect. For typical organometallics used in MOVPE, the TLV is in the range of 0.1 to 1 mg/m^3, but for Cd compounds it is 0.01 mg/m^3. For phosphine and arsine, the TLV is 0.42 mg/m^3 and 50 ppb, respectively. Environmental monitoring systems for these precursors have to be sensitive and accurate to a level significantly below these limits. The combination of procedures in place for safe handling of the precursors and operation of the MOVPE equipment with suitable fail-safe mechanisms has given the MOVPE industry an excellent safety record.

1.7 Conclusions

This chapter has set the scene for the subsequent chapters in this book. It has given some brief historical comments on MOVPE and its development over its ~50-year history. In addition, it has outlined the key features associated with MOVPE:

- Suitable substrate materials, both lattice matched and non-lattice matched, on which to grow the required films.
- Chemical precursors to provide the elements needed for each film composition required, which are either organometallic compounds or gases, the latter particularly for dopant elements.
- An understanding of the chemistry involved in the reactions applicable to each growth system.
- Design and construction of the equipment needed for the various growth systems, including the gas supply trains, reactor cells, abatement systems, etc. Some comments are also included on the necessary health and safety issues related to some of the toxic materials used in MOVPE, although it should be stressed here that workers must check on and apply local, i.e. country/state, regulations for their particular kit/growth system combination.
- Applications of the various materials systems produced using MOVPE, including cross-references to other chapters within this book.

The range of applications have turned MOVPE-based materials from early R&D samples into multibillion-dollar (e.g. III-nitride based LEDs) and multimillion-dollar (e.g. GaAs/InP-based laser diodes and high-frequency devices [for mobile phone and satellite communications uses], solar cells, infrared detectors, and photovoltaic and thermophotovoltaic devices) businesses over the past 50 years. The remaining chapters in this book go into the details of these exciting materials systems and the applications made from each of them, concluding with Chapter 14, which both reviews progress over the past five decades and gives some intriguing insights into future possible developments in MOVPE.

References

1. Manasevit, H.M. and Simpson, W.I. (1969). The use of metal-organics in the preparation of semiconductor materials: I. Epitaxial gallium-V compounds. *Journal of the Electrochemical Society* 116: 1725.
2. Manasevit, H.M. (1968). Single-crystal gallium arsenide on insulating substrates. *Applied Physics Letters* 12: 156.
3. Horita, M., Suzuki, M., and Matsushima, Y. (1992). MOVPE growth of InGaAsP using TBA and TBP with extremely low V/III ratio. *Journal of Crystal Growth* 124: 123–128.
4. Wright, P.J. and Cockayne, B. (1982). The organometallic chemical vapour deposition of ZnS and ZnSe at atmospheric pressure. *Journal of Crystal Growth* 59: 148–154.
5. Manasevit, H.M. (1981). Recollections and reflections of MO-CVD. *Journal of Crystal Growth* 55: 1.
6. Dapkus, P.D., Manasevit, H.M., and Hess K.L. (1981). High purity GaAs prepared from trimethylgallium and arsine. *Journal of Crystal Growth* 55: 10.
7. Andre, J.P., Guittard, P., Hallais, J. et al. (1981). GaAs photocathodes for low light level imaging. *Journal of Crystal Growth* 55: 235–245.
8. Frijlink, P.M., and Maluenda, J. (1982). MOVPE growth of $Ga_{1-x}Al_xAs$–GaAs quantum well heterostructures. *Japanese Journal of Applied Physics* 21: L574.
9. Razeghi, M., Hersee, S., Hirtz, P. et al. (1983). Very low threshold GaInAsP/InP double-heterostructure lasers grown by LP MOCVD. *Electronics Letters* 19: 336–337.
10. Duchemin, J.P., Hirtz J.P., Razeghi, M. et al. (1981). GaInAs and GaInAsP materials grown by low pressure MOCVD for microwave and optoelectronic applications. *Journal of Crystal Growth* 55: 64–73.
11. Moss, R.H. (1984). Adducts in the growth of III-V compounds. *Journal of Crystal Growth* 68: 78–87.
12. Razeghi, M., Hirtz, P., Larivain, J.P. et al. (1981). 1.5 μm room-temperature pulsed operation of GaInAsP/InP double heterostructure grown by LP MOCVD. *Electronics Letters* 17 (18): 643–644.

13. Mullin J.B., Irvine, S.J.C., and Ashen, D.J. (1981). Organometallic growth of II–VI compounds. *Journal of Crystal Growth* 55: 92.
14. Pimputkar, S., Speck, J.S., DenBaars, S.D. et al. (2009). Prospects for LED lighting. *Nature Photonics* 3, 180–182. Nakamura, S., Pearton, S., and Fasol, G. (2013). The Blue Laser Diode: The Complete Story. Berlin: Springer.
15. Shur, M.S. and Gaska, R. (2010). Deep-ultraviolet light-emitting diodes. *IEEE Transactions on Electron. Devices* 57 (1): 12–25.
16. Kim, Y.H., Ruh, H., Noh, Y.K. et al. (2010). Microstructural properties and dislocation evolution on a GaN grown on patterned sapphire substrate: A transmission electron microscopy study. *Journal of Applied Physics* 107:063501.
17. Caroff, P., Messing, M.E., Borg, B.M. et al. (2009). InSb heterostructure nanowires: MOVPE growth under extreme lattice mismatch. *Nanotechnology* 20: 1–7
18. Chernov, A.A. (1994). Kinetic processes in vapor phase growth. In: Handbook of Crystal Growth (ed. D.T.J. Hurle). Amsterdam: Elsevier.
19. Jensen, K.F., Fotiadis, D.I., and Mountziaris, T.J. (1991). Detailed models of the MOVPE process. *Journal of Crystal Growth* 107:1–11.
20. DenBaars, S.P., Maa, B.Y., Dapkus, P.D. et al. (1986). Homogeneous and heterogeneous thermal decomposition rates of trimethylgallium and arsine and their relevance to the growth of GaAs by MOCVD. *Journal of Crystal Growth* 77: 188–193.
21. Irvine, S.J.C., Mullin, J.B., Giess, J. et al. (1988). Recent developments in the pyrolytic and photolytic deposition of (Cd,Hg)Te and related II-VI materials. *Journal of Crystal Growth* 93: 732–743.
22. Ando, H., Inuzuka, H., Konagai, M. et al. (1985). Photoenhanced metalorganic chemical vapour deposition of ZnSe films using diethylzinc and dimethylselenide. *Journal of Applied Physics* 58: 802–805.
23. Fijita, S., Takeuchi, Y., and Fujita, S. (1988). Gas-phase and surface reactions in xenon lamp-assisted organometallic vapor-phase epitaxy of ZnSe. *Japanese Journal of Applied Physics* 27: L2019–L2021.
24. Bass, S.J. (1975). Device quality epitaxial gallium arsenide grown by the metal alkyl-hydride technique. *Journal of Crystal Growth* 31: 172–178.
25. Bergunde, T., Dauelsberg, T.M., Kadinski, L. et al. (1997). Heat transfer and mass transport in a multiwafer MOVPE reactor: modelling and experimental studies. *Journal of Crystal Growth* 170: 66.
26. Zhang, X., Moerman, I., Sys, C. et al. (1997). Highly uniform AlGaAs/GaAs and InGaAs(P)InP structures grown in a multiwafer vertical rotating susceptor MOVPE reactor. *Journal of Crystal Growth* 170: 83.
27. Thompson, A.G., Stall, R.A., Kroll, W. et al. (1997). Large scale manufacturing of compound semiconductors by MOVPE. *Journal of Crystal Growth* 170: 92.
28. Frijlink, P.M., Nicolas, J.L., and Suchet, P. (1991). Layer uniformity in a multiwafer MOVPE reactor for III-V compounds. *Journal of Crystal Growth* 107: 166–174.
29. George, N.C., Denault, K.A., and Seshadri, R. (2013). Phosphors for solid-state white lighting. *Annual Review of Materials Research* 43: 481–501.
30. Nakamura, S. (1991). In situ monitoring of GaN growth using interference effects. *Japanese Journal of Applied Physics* 30: 1620.
31. Ikeda, N., Niyama, Y., Kambayashi, H. et al. (2010). GaN power transistors on Si substrates for switching applications. *Proceedings of the IEEE* 98: 1151–1161.
32. Hirayama, H., Yatabe, T., Noguchi, N. et al. (2010). Development of 230–270 nm AlGaN-based deep-UV LEDs. *Electronics Communications Japan* 93: 24–33.
33. Harris, J.S., O'Sullivan, T., Sarmiento, T. et al. (2011). Emerging applications for vertical cavity surface emitting lasers. *Semiconductor Science and Technology* 26: 014010.
34. Dimroth, F., Grave, M., Beutel, P. et al. (2014). Wafer bonded four-junction GaInP/GaAs//GaInAsP/GaInAs concentrator solar cells with 44.7% efficiency. *Progress in Photovoltaics: Research Applications* 22: 277–282.
35. Ajayan, J. and Nirmal, D. (2015). A review of InP/InAlAs/InGaAs based transistors for high frequency applications. *Superlattices and Microstructures* 86: 1–19.
36. Zhang, C. and Li, X. (2016). III-V nanowire transistors for low-power logic applications: A review and outlook. *IEEE Transactions on Electron Devices* 63 (1): 223–234.
37. Liu, C., Li, Y., and Zeng, Y. (2010). Progress in antimonide based III-V compound semiconductors and devices. *Engineering* 2: 617–624.
38. Hall, D.J., Buckle, L., Gordon, N.T. et al. (2004). High-performance long-wavelength HgCdTe infrared detectors grown on silicon substrates. *Applied Physics Letters* 85: 2113; doi:10.1063/1.1791740.
39. Garnett, E.C., Brongersma, M.L., Cui, Y. et al. (2011). Nanowire solar cells. *Annual Review of Materials Research* 41: 269–95.
40. Wallentin, J., Anttu, N., Asoli, D. et al. (2013). InP nanowire array solar cells achieving 13.8% efficiency by exceeding the ray optics limit. *Science* 339: 1057–1060.

41. Aberg, I., Vescovi, G., Asoli, D. et al. (2016). A GaAs nanowire array solar cell with 15.3% efficiency at 1 Sun. *IEEE Journal of Photovoltaics* 6 (1): 185.
42. Pitts, O.J., Lackner, D., Cherng, Y.T. et al. (2008). Growth of InAsSb/InPSb heterojunctions for mid-IR detector applications. *Journal of Crystal Growth* 310: 4858–4861.
43. Huang, J.Y.T., Mawst, L.J., Jha, S. et al. (2008). MOVPE growth of Ga(As)SbN on GaSb substrates. *Journal of Crystal Growth* 310: 4839–4842.
44. Irvine, S.J.C., Lamb, D.A., Clayton, A.J. et al. (2014). Cadmium telluride solar cells on ultrathin glass for space applications. *Journal of Electronic Materials* 43: 2818–2823.
45. Shenai-Khatkhate, D., Goyetta, R.J., DiCarlo Jr., R. L. et al. (2004). Environment, health and safety issues for sources used in MOVPE growth of compound semiconductors. *Journal of Crystal Growth* 272: 816–821.

2

Fundamental Aspects of MOVPE

G.B. Stringfellow
The University of Utah, Salt Lake City, USA

2.1 Introduction

The fundamental aspects of MOVPE growth have been considered in a large number of papers during the nearly 50 years since the process was first used. There have been a number of thorough reviews published during the last 20 years, including my own book, *Organometallic Vapor Phase Epitaxy: Theory and Practice* [1], as well as a number of other, more specialized reviews [2–10].

The MOVPE technique has advanced tremendously in recent years and is today the major commercial manufacturing method [11] for several important devices, including injection lasers, light-emitting diodes (LEDs) [12, 13] and multijunction solar cells [14, 15]. This chapter will consider advances in our understanding of the fundamental aspects of MOVPE, with emphasis on topics of interest for application to the systems AlGaInN, used for high-performance blue, green, and white LEDs; and AlGaInP, used for high-efficiency red, orange, and amber LEDs and also for very high-efficiency solar cells. Fundamental aspects of the MOVPE growth of several other novel metastable materials used for laser, LED, and solar-cell devices will also be included where they give basic insights into the MOVPE growth process and the characteristics of the resulting materials.

The basic aspects of MOVPE growth and the properties of the resultant III/V alloys and other materials have been well studied, as indicated; however, the complexity of the underlying fundamental issues has prevented us from reaching a complete and detailed understanding. The first topic to be addressed in this chapter will be the thermodynamic aspects of MOVPE, the bedrock of our fundamental understanding, treated in Section 2.2. Emphasis will be on reviewing recent advances in our understanding of the thermodynamics of both the growth process and the mixing of the atoms in these alloys, particularly the resulting microstructures. The kinetic limitations to reaching thermodynamic equilibrium will be reviewed briefly in Section 2.3. Our understanding of the complexities involved in attempting to understand the thermodynamics and kinetics of the processes occurring at the growing surface has advanced significantly in the last two decades and

Metalorganic Vapor Phase Epitaxy (MOVPE): Growth, Materials Properties, and Applications,
First Edition. Edited by Stuart Irvine and Peter Capper.

so will be treated in some detail in Section 2.4. Finally, in Section 2.5, recent studies of several materials of interest for device applications will be considered in light of the basic ideas developed in Sections 2.2–2.4.

2.2 Thermodynamics

The basic thermodynamic concepts underlying the MOVPE process have been established for more than 20 years [16]. This section consists of a brief review of the fundamental concepts, with a more detailed treatment of relatively new developments. The emphasis will be on the III/V materials, since they are by far the most widely studied. However, the basic principles developed here will apply to virtually all materials grown by MOVPE. After the first three sections, thermodynamics, kinetics, and surface processes, several systems that are of prime importance for large-scale, commercial applications, including AlGaInP, the III/V nitrides, and several novel metastable systems containing elements such as B, N, and Bi added to GaAs and other compounds will receive an up-to-date review, with an emphasis on the thermodynamics underlying the MOVPE growth and microstructures of these fascinating materials.

Thermodynamic principles underlie the MOVPE process, as well as all other growth processes, since epitaxial growth is basically a highly controlled phase transition. Thus, thermodynamics determines the maximum growth rate possible for a particular set of growth conditions, although the actual growth rate is lower and typically controlled by mass transport, as discussed in Section 2.3. Very importantly, thermodynamics can often be used to predict the composition of solid alloys, including dopant incorporation, from the growth parameters. As we will see, it also drives the formation of nonrandom structures in ternary and quaternary alloy semiconductors.

During initial research on the use of semiconductor alloys for particular devices, it was commonly assumed that any semiconductor alloy could be produced, with no limitations as to growth technique. It was also assumed that the distribution of atoms within the alloy would be completely random; i.e. that the alloys were ideal. In this case, any property would be a function of only the solid composition. Neither of these assumptions is true.

The tacit assumption that any alloy could be produced with no limitations as to the growth technique has proven to be distinctly false for many alloys. In this section, we will explore the limitations imposed by thermodynamics. However, as will become evident, kinetic constraints may sometimes be used to allow the growth of metastable alloys. As might be expected from an understanding of materials in general, the microstructure, in particular deviations from a random arrangement of the atoms, has first-order effects on many, if not all, properties of vital importance for semiconductors, such as the bandgap energy, carrier mobility, and electron-hole recombination rates. The thermodynamics of the growth process and the resultant alloys gives a useful basis for understanding the MOVPE process as well as the microstructures of the resulting alloys. This will be the emphasis of this section and, indeed, the entire chapter.

2.2.1 Thermodynamics of MOVPE Growth

The first step in approaching the thermodynamics of MOVPE growth is to briefly review the basic concepts. The system consists of at least two phases, the vapor and the solid. A liquid and an additional solid phase may also appear under some conditions. For two phases in equilibrium, the value of the Gibbs free energy per mole, G, is at a minimum. G is defined as:

$$G = H - TS \tag{2.1}$$

where H is the enthalpy, S is the entropy, and T is the temperature. An alternate description of equilibrium is that the movement of an infinitesimal amount of material from the vapor to the solid phase will result in no change in the total Gibbs free energy per mole. The partial derivative of G with respect to the number of moles of component i, n_i, in a particular phase is termed the *chemical potential*, μ_i. The equilibrium condition for a two-phase system is then simply that the chemical potential of element i is the same in both phases. The value of μ_i for a particular element in both the vapor and solid phases of the MOVPE system can be expressed in terms of the activity, a_i, which is a product of the mole fraction, x_i, and a nonideality factor, the activity coefficient, γ_i:

$$\mu_i = \mu_i\left(\text{standard state}\right) + RT\ln a_i = \mu_i^o + RT\ln x_i\gamma_i \tag{2.2}$$

Of course, a condition of equilibrium throughout the entire MOVPE system would guarantee a growth rate of identically zero, since there would be no driving force for transfer of nutrients from the vapor to the solid phase. Typically, during the *equilibrium* liquid phase epitaxy (LPE) process, thermodynamic equilibrium is, indeed, established before growth begins. The liquid is held at a constant temperature for an extended period in contact with the solid substrate. Subsequently, the system is slowly cooled to introduce a small supersaturation in the liquid that drives the transfer of material from the liquid into the solid phase, i.e. epitaxial growth. Thus, it was natural to assume thermodynamic equilibrium throughout the system to determine the composition of a solid semiconductor alloy grown from a particular liquid composition at a given temperature using the temperature-composition phase diagram. The growth rate is determined by diffusion through the liquid phase near the solid/liquid interface [17].

The situation is not so obvious for MOVPE where the input vapor phase is intentionally highly supersaturated. Kinetic factors, particularly mass transport, prevent the entire vapor phase from reaching equilibrium with the growing solid. Thus, MOVPE is often referred to as a *nonequilibrium* process. In terms of the simple boundary-layer model, to be discussed briefly here and in Section 2.3, the growth process can be thought of as a highly supersaturated input vapor phase separated from the growing solid/vapor interface by a stagnant boundary layer. The large supersaturation of the input vapor is dissipated across this boundary layer. The chemical potential gradient drives the diffusion of the nutrients across the boundary layer and into the solid. The chemical potentials of the nutrients are thus reduced dramatically across the boundary layer, and the MOVPE process can be understood thermodynamically in terms of the concept of local equilibrium at the solid/vapor interface [16]. This typical situation occurs when the interface reaction kinetics are much more rapid than the diffusion kinetics and is termed *diffusion-limited growth*. Powerful thermodynamic factors control much of the growth process and particularly the materials properties, as discussed here and in Section 2.4. This analysis allows us to understand the growth rate, the composition of the solid, and much about the microstructure of the resultant alloys.

The activity coefficient, γ, referred to previously, is a function of the enthalpy of mixing in the alloy, ΔH^{M}. For ternary alloys, the regular solution model (RSM) [18] has frequently been used to interpret ternary III/V phase diagrams [16, 19]. This simple model describes solutions that are nonideal, i.e. the enthalpy of mixing is nonzero, but the arrangement of atoms on the respective sublattices is assumed to be random [16]. Using these assumptions, the entropy of mixing, ΔS^{M}, is simply the ideal configurational entropy of mixing:

$$\Delta S^{\mathrm{M}} = -R\left[x\ln x + \left(1-x\right)\ln\left(1-x\right)\right] \tag{2.3}$$

where R is the gas constant.

In the RSM, the enthalpy of mixing is calculated assuming that each atom interacts only with its Z nearest neighbors. By summing the individual bond energies, the enthalpy of mixing can be expressed by the symmetric relationship

$$\Delta H^{\mathrm{M}} = x\left(1-x\right)\Omega \tag{2.4}$$

Ω is the interaction parameter, typically with units of calories/mol. This RSM interaction parameter has no physical basis for semiconductor alloys, so it is treated as an adjustable parameter to fit experimental data to the calculated phase diagram in order to interpolate and extrapolate existing data.

For the near-equilibrium LPE growth technique, the liquid/solid phase diagram is essential for the prediction of the growth parameters giving rise to the particular alloy desired. Early calculations of the III/V ternary phase diagrams used the RSM to treat the enthalpy of mixing. This allowed the phase diagrams for several systems, including AlGaAs, GaInAs, InAsSb, and GaInP to be calculated using the solid-phase interaction parameter, Ω, as an adjustable parameter to match the calculated phase diagram to available experimental data [16, 19]. These will be termed the "experimental" values of Ω in what follows.

The RSM has several disadvantages. For semiconductor alloys, the interaction parameter is imbued with no physical basis. Thus, RSM calculations cannot be predictive, a distinct disadvantage when new systems are being explored. This led to the search for physical models that allow the calculation of independent values of Ω without the need for experimental phase-diagram data.

To provide a simple, physically predictive model that could be used to anticipate the behavior of new systems, the delta-lattice-parameter (DLP) model was developed in the early 1970s [20–22]. This model is based on the Phillips–Van Vechten dielectric theory of electronegativity, a model developed to allow the simple calculation of bonding energies in semiconductors [23]. The DLP model yields the interaction parameter

$$\Omega = 4.375K\Delta a^2 / a_{\text{avg}}^{4.5} \tag{2.5}$$

where Δa is the lattice parameter difference between the two constituent compounds and a_{avg} is the average lattice parameter. The constant K is the same for all III/V alloys and was determined to have a value of 1.15×10^7 cal/mol/Å$^{2.5}$ from the values of interaction parameter previously determined empirically from RSM phase diagrams calculated for systems with existing experimental data, the "experimental" values. At this point we had a predictive model for the enthalpy of mixing of the III/V alloys that could be used with the entropy of mixing, from Eq. (2.3), to predict the phase diagrams for new systems. Note that the value of Ω in Eq. (2.5) increases as the square of the difference in lattice constants. It is worth noting that the value of Ω also increases rapidly as the average lattice constant decreases.

More recently, Ho and Stringfellow [24, 25] found that the DLP value of the enthalpy of mixing of GaInN alloys is too large. Matsuoka [26] re-evaluated the value of K in the DLP model, as applied to the wurtzite AlGaInN ternary systems, and found that a value of 0.7×10^7 cal/mol/ Å$^{2.5}$ was more suitable.

The form of Eq. (2.5) suggests that the major factor affecting the enthalpy of mixing is the elastic energy [22]. This was confirmed by later calculations using the valence-force-field (VFF) model, where the enthalpy of mixing is explicitly assumed to be due to the microscopic stretching and bending of the bonds in the zincblende or wurtzite lattice, composed of tetrahedral cells [16, 24, 25]. The work of Ho and Stringfellow [24, 25] demonstrates that a modified version of the VFF model, where relaxation is allowed to shells beyond the nearest-neighbor atoms, can be used to accurately describe the GaInN phase diagram. As mentioned previously, it also shows that the original DLP model gives interaction parameters that are too large for the nitride alloys.

Since the development of the simple DLP model, large-scale, first-principles calculations have been used to calculate the free energy of mixing of III/V alloy systems [27–30]. However, these calculations are complicated by the inhomogeneous nature of the alloys, which necessitates the use of large cells containing many atoms. In addition, the entropy must be considered for calculations relevant to predicting the nature of the alloys produced during epitaxial growth. The common zero-K, first-principles energy calculations that ignore the entropy are of little or no value in this regard.

Comparison of values of Ω for the various models described previously shows results fairly consistent with the DLP model, as seen in Table 2.1. A large value of Ω leads to solid-phase immiscibility, to be discussed in more detail later. Another way of comparing data regarding systems with miscibility gaps is to use the critical temperature, T_c, the temperature below which the miscibility gap exists, which is equal to $\Omega/2R$ [16]. As discussed previously, for the nitride alloys, the values of Ω are too large by a factor of less than 2. Examining the values for GaInN, the results obtained using the modified VFF model [24, 25] and the modified DLP model are in approximate agreement with the results of first-principles calculations of Teles [28]. Recent first-principles calculations using the Stillinger–Weber potentials, modified to be applicable to the nitride alloys, are included in Table 2.1 [30]. The values for GaInN and AlInN are in agreement with the DLP model modified for the nitride alloys, as discussed previously. However, the results of Escalanti and Hart [29] give values of Ω much lower than obtained from the modified DLP model. For the GaAsN system, the comparative results are similar: The first-principles calculations of Neugebauer et al. [27] and Escalanti and Hart [29] yield values of the interaction parameter that are approximately 2× lower than for the original DLP calculation.

Recently, the DLP model has been used to calculate the enthalpy of mixing of novel alloys, where Bi is added to GaAs, GaP, and InAs [31]. The experimental results for the GaAsBi alloys are discussed in Section 2.5.3.3. The DLP results were compared with discrete points calculated using the density-functional theory (DFT). Quantitative agreement to within 5% was found for each of the alloys GaAsBi, GaPBi, and InAsBi [31]. The results for GaAsBi, InAsBi, and GaPBi are included in Table 2.1.

Table 2.1 also demonstrates that the values of the interaction parameters calculated using the DLP model for the conventional ternary alloys composed of Al, Ga, and In Group III atoms combined with P, As, and Sb Group V atoms are in general agreement with "experimental" values

Table 2.1 *Interaction parameters for selected III/V alloys.*

Alloy	Ω kcal/mol (DLP)	T_c (K) (DLP)	Ω (calc)	T_c (K, calc)	Ω (exp.)
InAsSb	2.29[a]	578			2.23–2.90[a]
GaAsSb	3.36[a]	848			4.00–4.50[a]
GaInAs	2.9[a]	467			1.65–3.00[a]
GaInP	3.65[a]	922			3.25–3.50[a]
GaInN	9.60 [24]	2424	5.98 [24]	1495 [24]	6.0[b]
			6.56 [26]	1295 [28]	
			6.61 [30]	1718 [30]	
			7.81 [32]		
AlInN	18.1	4530	11.37 [26]	1485 [28]	
			10.53 [30]	2717 [30]	
GaPN	33.1	8280			30.0[b]
GaAsN	44.5	11 100	20 [29]	4611 [27]	
GaAsBi	7.18	1800	8.5 [31]		
	8.5 [31]				
BGaAs	21.1	5280	10 [29]		
BGaP	30.1	7540			
BGaN	70.6	17 700	45.2 [32]	9000 [28]	
			23.4 [29]		
GaPBi	15.2 [31}		15.2 [31]		
InAsBi	5.6 [31]		5.6 [31]		

[a]Reproduced from Stringfellow 1999 [16].

[b]Based on agreement with experimental binodal. After Ho and Stringfellow 1996 [24].

determined empirically from experimental phase-diagram data. Guan et al. [31] compared the results of DFT calculations for a broad range of III/V and Group IV alloys (not including the Group III nitrides) and found excellent agreement.

A dramatic system is GaPN, where the very large enthalpy of mixing limits N incorporation into GaP during near-equilibrium LPE and melt growth to doping concentrations. The DLP model correctly predicts the observed solubility, as seen in Figure 2.1 [33].

From these comparisons between the results of the DLP model with experimental results and the results of large-scale, first-principles calculations, it can be concluded that the DLP model can act as a simple, surprisingly accurate and reliable guide to understanding the thermodynamics of mixing for a broad range of III/V and other tetrahedral alloys. This allows an understanding of the consequences of thermodynamics for epitaxial growth, including the solid composition and microstructure, as discussed in the following sections.

Of course, the simple thermodynamics of mixing in bulk (i.e. unstrained) alloys fails to account for other effects. One that has been studied in some detail in recent years is the effect of strain due to lattice-parameter mismatch with the substrate on the extent of the miscibility gap. This is discussed in detail in Section 2.2.3.

2.2.2 Solid Composition

Calculations using the DLP model have been used to predict the phase diagrams for many III/V alloys, which assists with the exploration of novel alloy systems for growth by equilibrium techniques such as LPE or VPE. Even more important is the use of this model for the calculation of the

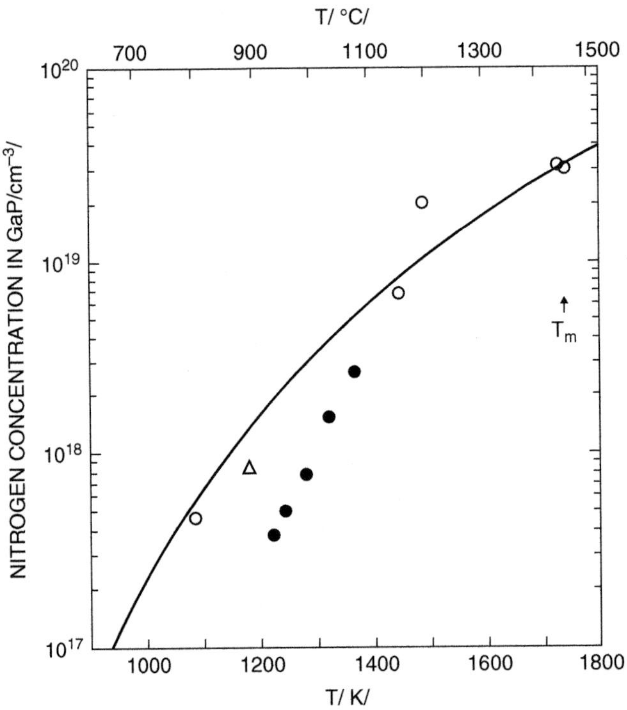

Figure 2.1 *Nitrogen concentration in GaP grown from the liquid as a function of temperature. The experimental data are from references [33] (o), [34] (Δ), and [35] (•). The line represents the equilibrium solubility limit calculated by Stringfellow [20, 21]. Source: After Karpinski et al. 1985 [33]; reprinted from the Journal of Crystal Growth with permission from Elsevier Science.*

solid composition from the growth parameters even for the so-called *nonequilibrium* MOVPE technique [16]. Such calculations are based on the premise, discussed previously, that even though the input vapor phase is highly supersaturated, local equilibrium can be assumed at the solid/vapor interface for nearly all growth conditions [37], as will be demonstrated later.

First, consider the thermodynamic analysis of the MOVPE growth of alloys containing both Al and In, such as AlGaInP and AlGaInN, perhaps the most important III/V alloys, due to their use in all commercial LEDs and high-efficiency solar cells. Thermodynamic factors dictate that they cannot be grown by equilibrium techniques, such as LPE or vapor phase epitaxy (VPE), where the Group III atoms are transported as halides [37]. This limitation is due to the very high formation energies of AlP and AlN as compared with those of and InP and InN, respectively. This dictates that controlled growth of these alloys is possible only using nonequilibrium techniques such as either MOVPE or molecular beam epitaxy (MBE) [38]).

In this section, we will explore the use of the DLP and VFF models for the calculation of the solid composition from the growth parameters for the nonequilibrium MOVPE technique [16]. Such calculations are based on the premise that local equilibrium can be assumed at the solid/vapor interface for mass-transport limited growth. This means the thermodynamic constraint is simply that the mass action expressions are nearly satisfied at the solid/vapor interface. As an example, for the growth of GaAs, assuming complete decomposition of the precursors:

$$a_{\mathrm{GaAs}} / p^{\mathrm{i}}_{\mathrm{Ga}} \left(p^{\mathrm{i}}_{\mathrm{As4}}\right)^{1/4} = K_{\mathrm{GaAs}} \tag{2.6}$$

where K_{GaAs} is the equilibrium constant for the reaction of elemental Ga plus As_4 to form GaAs at a particular temperature, and the superscript i indicates partial pressures at the solid/vapor interface. Since MOVPE is a highly nonequilibrium technique,

$$p^*_{\mathrm{Ga}} \left(p^*_{\mathrm{As4}}\right)^{1/4} >> p^{\mathrm{i}}_{\mathrm{Ga}} \left(p^{\mathrm{i}}_{\mathrm{As4}}\right)^{1/4} \tag{2.7}$$

where the star (*) superscript indicates input partial pressures. Typically, MOVPE growth is performed with a V/III ratio far in excess of unity in order to avoid the formation of relatively nonvolatile Group III droplets on the growing surface. In that case, the combination of Eqs. (2.6) and (2.7) means that the Ga is nearly depleted at the interface, while the As_4 partial pressure remains essentially unchanged, since the same number of Ga and As atoms are removed from the vapor to form the solid epitaxial GaAs layer.

The growth rate can now be written

$$J = D_{\mathrm{Ga}} \left(p^*_{\mathrm{Ga}} - p^{i}_{\mathrm{Ga}}\right) / RT\delta_{\circ} \tag{2.8}$$

where D_{Ga} is the diffusion coefficient of the fragment of TMGa diffusing through the boundary layer, with thickness δ_0. This expression represents the essential features of the growth rate even when the hydrodynamics and mass transport are governed by more physically realistic models than the overly simplistic boundary layer model: the GaAs growth rate is proportional to the input partial pressure of the Ga precursor, since the interface partial pressure is much smaller [16].

The same argument related to growth rate can be made for the growth of InAs or AlAs. This results in a particularly simple expression for the composition of the solid AlGaInAs alloy: The ratios of the Group three constituents in the solid are identical to the ratios of those Group three elements in the input vapor. In other words, the Group III distribution coefficient, defined for a ternary alloy as

$$k = \left[x / (1-x)\right] / \left[p_{\mathrm{a}} / p_{\mathrm{b}}\right] \tag{2.9}$$

is unity. This is also true for the commercially important quaternary AlGaInP and AlGaInN alloys described previously. There are, however, two constraints on this simple analysis: i) pyrolysis of the Group III precursors is complete; and ii) the vapor pressures of the three Group III elements are relatively low: i.e. the Group III elements cannot vaporize significantly before incorporation into the solid. A well-studied example is the growth of GaInN. Matsuoka et al. [26] grew high-quality GaInN alloys by MOVPE. The In distribution coefficient was found to be nearly unity for growth at 500 °C, as seen in Figure 2.2, and decreased to 0.1 at 800 °C due to the increased volatility of In when the temperature was increased from 500 to 800 °C. Even in this case, the distribution coefficient can be described thermodynamically using the DLP model [39], but the value of k_{In} is less than unity at higher temperatures [40, 41]. This situation also applies to Ga at the very high temperatures of 1000 to 1100 °C used to grow the nitrides. Ga leaves the surface, resulting in a Ga distribution coefficient of <1 [41].

For the AlGaAs, InGaAs, and AlGaSb systems, the distribution coefficients under typical MOVPE growth conditions are also nearly unity, as seen from Figure 2.2. A more recent example of this behavior comes from the work of Sommer et al. [44]. Novel BGaP and BGaAs alloys were grown using the precursors TBAs, TBP, TEB, TEGa, and TMIn. For BGaP, the B distribution coefficient was found to be unity, indicative of 100% incorporation of B, as seen in Figure 2.2. The distribution coefficient was found to be slightly higher at 575 °C than at 525 °C, which was interpreted to be due to the TEB precursor stability, i.e. incomplete pyrolysis at the lower temperatures. Kunert et al. [45] observed similar behavior for B concentrations of up to 7%. Several of their data points are also included in Figure 2.2. The solubility of B in BGaP is limited, as described later.

For III/V alloys with mixing on the anion sublattice, the analysis of the distribution coefficient is complicated partially due to the use of high input V/III ratios, required by the low volatility of the cation elements and the typically high volatility of the anion elements. The other factor complicating the analysis is the incomplete pyrolysis of the anion precursors. For this reason,

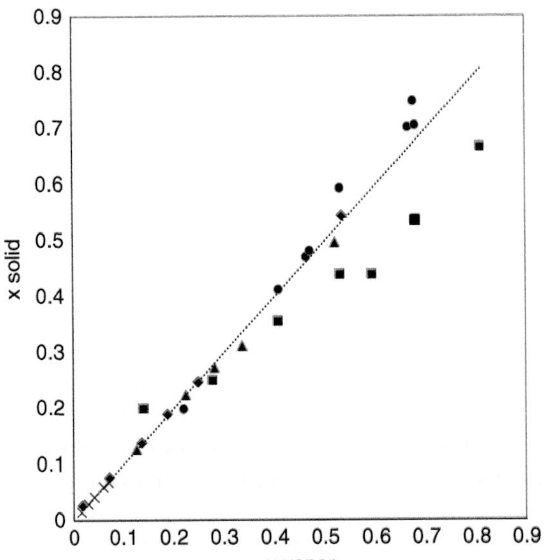

Figure 2.2 *Solid versus vapor concentration for the III/V alloys: (●) $Al_xGa_{1-x}As$ (data from Mori and Watanabe 1981 [42]); (■) $Al_xGa_{1-x}Sb$ (data from Cooper et al. 1982 [48]); (◆) $In_xGa_{1-x}As$ (data from Ludowise et al. 1981 [43]); (▲) $In_xGa_{1-x}N$ (data from Matsuoka 1998 [26]; (X) $B_xGa_{1-x}P$ (data from Kunert et al. 2008 [45]).*

we will first discuss the cases involving systems where all precursors decompose at the growth temperatures studied. An early example is the MOVPE growth of InAsSb. The experimental data of Fukui and Horikoshi [46] are plotted in Figure 2.3. The thermodynamic calculation of the Sb distribution coefficient uses the DLP value of the interaction parameter and assumes that both Group V precursors decompose to the tetramers [47]. For these calculations, equilibrium is assumed at the interface for both InAs and InSb. Two additional conservation constraints are imposed, one on composition and the other on stoichiometry [16]. The agreement between the calculations and the experimental results is an impressive validation of the use of thermodynamics for the analysis of the solid compositions of III/V alloys with mixing on the Group V sublattice grown by MOVPE.

A more complex system is GaAsSb. The thermodynamic analysis described previously, again with the DLP value of the interaction parameter, yields the Sb mole fraction in the solid versus the ratio of Sb to As + Sb in the input vapor phase for several values of the input V/III ratio. The calculated results are compared with experimental results from Cooper et al. [48] and Stringfellow and Cherng [49] in Figure 2.4. The GaAsSb system is predicted to have a miscibility gap, as indicated by the broken lines. As is seen, solid compositions throughout the miscibility gap can be obtained by using near-unity V/III ratios. In this case, both the Group III and the Group V elements are completely incorporated into the solid due to the very high supersaturation of the input vapor. Surface diffusion limits the amount of rearrangement of the random Group V elements into two separate phases, yielding a homogeneous solid. This is an example of kinetic hindrance preventing the establishment of thermodynamic equilibrium during growth. The occurrence of phase separation as well as the use of kinetics to suppress phase separation in the solid will be discussed in some detail later and in Sections 2.4 and 2.5 for several metastable alloys.

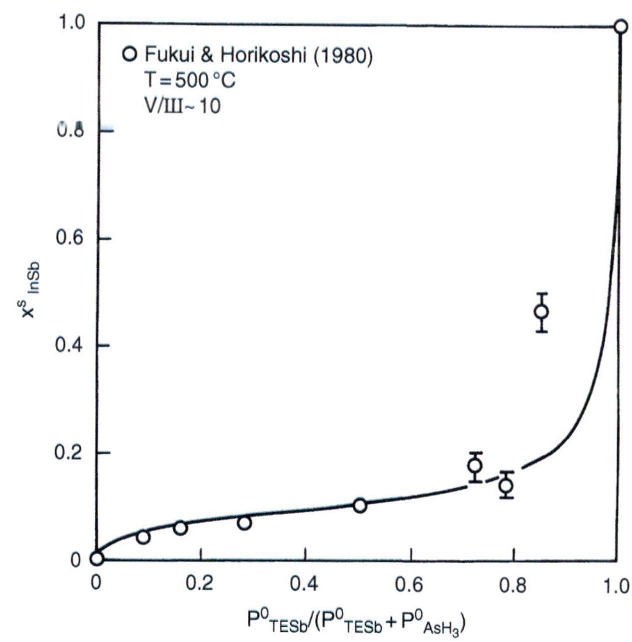

Figure 2.3 *Solid versus vapor composition for the alloy $InAs_{1-x}Sb_x$. The data points are from Fukui and Horikoshi 1980 [46]. The solid line was calculated with no adjustable parameters. Source: After Stringfellow 1983 [47]; reprinted from the Journal of Crystal Growth with permission from Elsevier Science.*

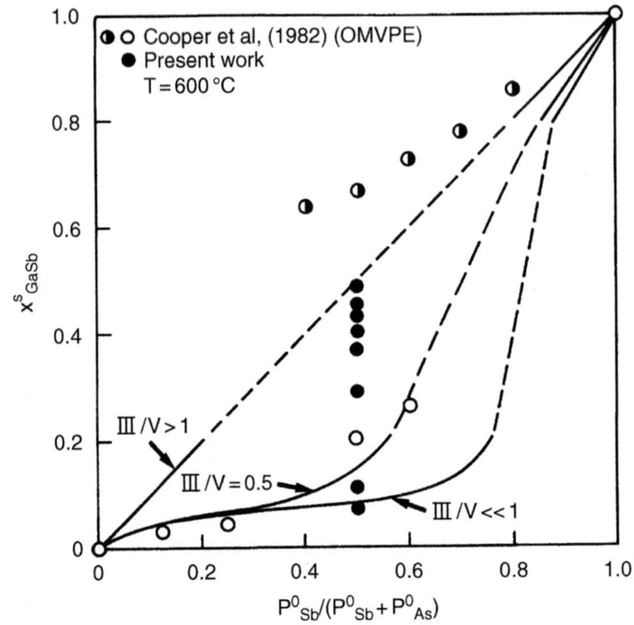

Figure 2.4 *Solid versus vapor composition for the alloy $GaAs_{1-x}Sb_x$. The data points are from the work of Cooper et al. 1982 [48] for V/III=2.0 (○) and V/III=0.5 (◑) and from Stringfellow and Cherng 1983 [49] (●). The curves were calculated for various V/III ratios. The broken sections of each curve represent the calculated regions of solid immiscibility. Source: After Stringfellow and Cherng 1983 [49]; reprinted from the Journal of Crystal Growth with permission from Elsevier Science.*

Biefeld [50] has summarized the literature for the antimonides grown by MOVPE. He found that the thermodynamic model described previously (equilibrium at the solid/vapor interface plus the DLP model) describes the experimental distribution coefficients for the InAsSb and GaAsSb systems, considered earlier, as well as the InGaSb, AlGaSb, and AlAsSb systems. In other cases, such as GaPSb and InPSb, the incomplete pyrolysis of one or more of the Group V precursors complicates the analysis of the distribution coefficient. The most prominent examples of this behavior are those where the relatively stable phosphine is used as the P precursor. This topic is discussed in more detail in Section 2.3.3.

The simple DLP calculations, described previously, ignore several important factors, most notably the macroscopic strain produced in coherent systems due to lattice mismatch between the epitaxial layer and the substrate. Not surprisingly, considering the elastic-energy basis of the enthalpy of mixing in the DLP and VFF models, the strain energy related to epilayer/substrate lattice parameter mismatch perturbs the equilibrium solid composition significantly. Since this effect always moves the equilibrium solid composition in a direction to minimize the lattice mismatch, it is termed compositional *lattice latching* or preferably *lattice pulling*. It was first reported for GaInP alloys grown by LPE [51], but has since been observed in many alloy systems grown by all of the major growth techniques [52]. Particularly striking recent examples for MOVPE growth include results for GaAsSb, discussed in Section 2.3.3, and GaInN, discussed in Section 2.5.2.

Kunert et al. [45] studied the MOVPE growth of GaPN on GaP and Si substrates. They report that the N incorporation is 1.6 times more efficient when depositing on Si substrates. They interpret their results to indicate that this is due to the effect of strain on the adsorption/desorption process, resulting in more efficient NH_3 pyrolysis on the more nearly lattice-matched Si substrates. This is another potential impact of strain on the MOVPE process.

GaAsP alloys have a small positive enthalpy of mixing that causes no noticeable effects on materials properties or device performance. However, a major problem with GaAsP epitaxially grown on GaAs substrates is that the lattice parameter mismatch between the relaxed epilayer and substrate results in the production of large numbers (approximately 10^6 cm^{-2}) of dislocations threading up through the epilayer for thick, incoherent layers. When the lattice parameter of the epitaxial layer does not match that of the substrate, the difference must be accommodated. As discussed in Section 2.4.2, for very thin layers, the epilayer remains coherent with the substrate, and the mismatch is accommodated elastically, mostly by deformation of the thin epitaxial layer. However, for thick layers, the elastic energy, which is proportional to the epilayer thickness, becomes so large that it becomes energetically favorable for "mismatch" dislocations to be generated at the substrate/epilayer interface. Some of these dislocations propagate up through the epilayer, where they act as nonradiative recombination centers and have been shown to reduce the GaAsP LED efficiencies to values of approximately 0.1% [53]. Similar affects have been observed in the entire range of conventional III/V alloys (Al, Ga, and In combined with P, As, and Sb) [53]. This has fueled the search for semiconductor materials where bandgap and lattice parameter can be chosen independently, including quaternary alloys where the two compositional degrees of freedom allow for the independent choice of these two materials parameters. A thorough recent review of the importance of thermodynamics, including the strain energy, on the growth and properties of III/V alloys for optical devices can be found in [10].

2.2.3 Phase Separation

An important feature of the DLP calculations, supported by the results summarized in Table 2.1, is that the value of Ω is always positive and increases dramatically as the difference in lattice constant between the constituent binary compounds increases. From solution thermodynamics [16, 19, 54] it is well known that for alloys with large values of enthalpy of mixing, the phase diagram includes a region of solid immiscibility, i.e. a miscibility gap for temperatures less than the critical temperature, T_c. Table 2.1 gives the values of T_c calculated for several systems of interest here.

The simple thermodynamics of mixing in bulk (i.e. unstrained) alloys fails to account for other effects. As discussed previously, the strain energy is a significant factor in the enthalpy of an epitaxial alloy layer [51, 52] and, hence, the solid composition, e.g. lattice pulling. Thus, it will have a major effect on phase separation.

The coherency strain energy for spinodal decomposition must be included in the total energy for realistic calculations of the solid-phase miscibility gap and the critical temperature. In general, when like atoms having different sizes congregate to form small clusters in a coherent system, i.e. one with no dislocations formed to relieve the strain, the strain energy of the system increases significantly. This acts to suppress compositional fluctuations. In fact, coherent phase separation is completely suppressed in many systems [16].

The other type of strain energy that must be considered is from the lattice mismatch with the substrate for coherent epi-layers. The effect of the mismatch strain on the extent of the miscibility gap has been studied in some detail, both theoretically and experimentally, in recent years due to the increasing importance of *metastable* alloys for important devices.

However, the simple calculations often overestimate the strain effects because they ignore the non-uniform strain in the layer due to relaxation at surfaces, layer thickness non-uniformity, island formation during growth, and other factors [16, 52]. Thus, they typically overestimate the suppression of spinodal decomposition. This is discussed in detail in Section 2.4.3.2 and in Section 2.5 for specific systems. An interesting case is when a substrate is chosen that is lattice matched to a particular metastable alloy. The strain energy is zero for the lattice-matched alloy, which acts to stabilize the uniform material. This is closely related to the lattice-pulling phenomenon discussed previously. This effect will be discussed further for specific alloys in Section 2.5.

A striking demonstration of this comes from the near-field scanning optical microscopy results of Kim et al. [55], who found that they could observe luminescence features associated with two types of bandgap fluctuations due to compositional clustering in InGaN/GaN quantum wells (QWs). They found larger potential fluctuations in the regions near V-pits, associated with threading dislocations. Strain relaxation near the walls of the V-pits allows a decrease in strain energy and more pronounced compositional fluctuations, as expected from the earlier discussion.

The presence of a miscibility gap has great practical significance. The growth of alloys within the miscibility gap is not possible by equilibrium techniques, such as LPE. A well-known example is the GaAsSb system, where a wide range of solid compositions cannot be grown at normal temperatures, due to the miscibility gap [16]. Another striking example is the early attempts to add N to GaP during LPE growth, described previously [20], where the amount of N that can be incorporated is very small due to the miscibility gap.

Using MOVPE or MBE, kinetic constraints may prevent the atomic rearrangement needed for establishment of a thermal equilibrium distribution of the atoms in the solid. The use of high growth rates and low growth temperatures both limit the ability of the adatoms arriving from the vapor to rearrange at or near the surface to form the lower free energy, inhomogeneous or 2-phase solid. This technique has been widely used to grow alloys within the region of solid immiscibility.

An important question at this point is how a metastable alloy, grown by MOVPE or MBE within the miscibility gap, reacts to the thermodynamic driving force for phase separation. Experimental observations indicate that for these metastable alloys, the thermodynamic driving force leads to clustering of like atoms, the initial stage of spinodal decomposition and phase separation. These compositional nonuniformities are important features of the microstructure. Perhaps not surprisingly, they produce significant changes in the materials properties that, in turn, have dramatic effects on the performance of devices such as visible LEDs and lasers. The detailed description of these metastable materials will be delayed until after Sections 2.3 and 2.4, since surface and kinetic processes are important elements in determining the microstructures formed during growth.

The atomic-level rearrangement of atoms on a particular sublattice to form the inhomogeneous solid, either due to solid phase immiscibility or ordering, to be discussed later, must occur on or near the surface during growth. Clearly, surface diffusion coefficients are orders of magnitude larger than in the bulk. However, even in the top few layers of the solid, diffusion coefficients are expected to be much higher than in the bulk, because the stoichiometry is not perfect in these layers due to the weaker bonding at the surface. In addition, the subsurface atoms can communicate with the surface atoms. For example, during Ge deposition on Si, Ge diffuses down to the fourth layer [36].

Perhaps the most dramatic examples of the importance of the microstructures are provided by considering the ternary GaInP and GaInN alloy systems. These alloys are used for commercial LEDs, high-efficiency solar cells, and a number of other devices. As seen in Table 2.1, the GaInN critical temperature, calculated using the VFF model, is 1222 °C, well above the typical MOVPE growth temperatures of 500–700 °C. The compositions of interest for green and blue LEDs (0.15 to 0.2) are, thus, within the region of solid immiscibility, as seen in Figure 2.5. Some evidence of phase separation was demonstrated in very early experiments where polycrystalline thin films of GaInN were produced over the entire composition range. Annealing for extended periods resulted in new X-ray peaks, suggesting that the samples decomposed during annealing to form new phases [63].

AlGaInN alloys have no lattice-matched substrate. Thus, they are typically grown on sapphire substrates. The differences in crystal structure and the atomic spacings lead to highly imperfect epitaxial layers. Dislocation densities are typically in the range of 10^9 to $10^{10}\,cm^{-2}$. As discussed previously, in a typical III/V semiconductor, this would have a catastrophic effect on the LED efficiency.

Following the commercial success of high-efficiency blue, green, and white LEDs made in this material by MOVPE, more than a thousand studies have probed the microstructure. The addition

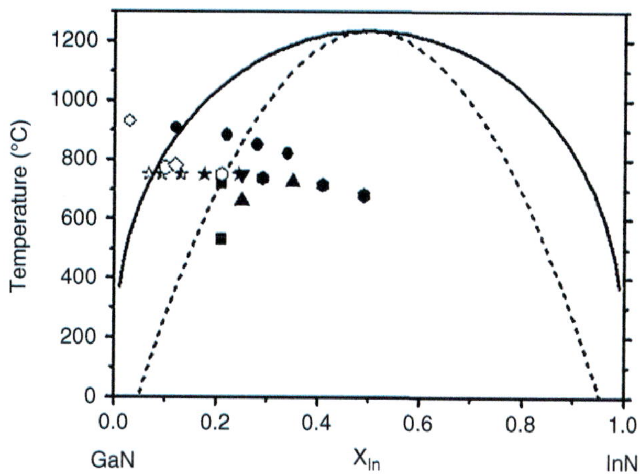

Figure 2.5 *Calculated bimodal (solid) and spinodal (dashed) curves (Ho and Stringfellow 1996 [24]) compared with experimental data (Piner et al. 1998 [56], Ponce et al. 2003 [57], Rao et al. 2004 [58], Doppalapudi et al. 1998 [59], Kar et al. 2008 [60], Faleev et al. 2009 [61], and Potin et al. 2004 [62]) for GaInN/ Open data points = single phase; half-filled data points = metastable: filled data points = 2 phase mixtures. Source: Reproduced from Stringfellow 2010 [52]; reprinted from the Journal of Crystal Growth with permission from Elsevier Science.*

of In to GaN has been repeatedly demonstrated to lead to multiple order of magnitude improvements in the LED efficiencies. This work has led to the widespread belief that compositional modulation due to the miscibility gap produces In-rich, low-energy-gap regions that "trap" minority carriers before they can recombine at dislocations. This yields a two to three orders of magnitude increase in the quantum efficiency, as discussed in Section 2.5.2. In fact, for epitaxial layers having extremely high dislocation densities, in excess of 10^9 dislocations per square centimeter, blue LED efficiencies have exceeded 80% [62], an amazing result.

2.2.4 Ordering

Another deviation from a random distribution of atoms in semiconductor alloys is ordering. The observation of atomic-scale ordering in alloys such as AlGaAs, InGaP, and GaAsSb in the 1980s was unexpected. An *ordered* structure refers to an alloy consisting of alternating monolayers (MLs) of atoms on one sublattice having different compositions along a particular crystallographic direction [65]. The ordering phenomenon was observed many years earlier in metal alloy systems; hence the nomenclature used refers to metallic systems, e.g. Cu-Pt (with ordering of {111} planes) and Cu-Au (with ordering of {100} planes). In metallic systems, where diffusion in the solid is more rapid, the ordering process is driven by the thermodynamics of the bulk alloys. For systems with *negative* values of the ΔH^M, ordering is predicted to occur on the basis of simple thermodynamic calculations [16, 18]. As mentioned previously, the enthalpy of mixing in III/V alloys is always *positive*. Thus, one expects these materials to exhibit compositional clustering and not ordering when the lattice-parameter difference is large. Only for ideal alloys, with $\Omega=0$, does one expect a random arrangement of the atoms on each sublattice.

2.2.4.1 Surface Thermodynamics

Eventually, this apparent paradox was explained in terms of surface (as opposed to bulk) thermodynamics [66]. As discussed in Section 2.4.2, low-index surfaces in semiconductor materials reconstruct to lower the energy of the surface atoms, where dangling bonds exist for a nonreconstructed surface [67]. For growth by MOVPE where the V/III ratio is typically much greater than unity, the surface is terminated by Group V atoms. The formation of dimer bonds between the Group V atoms to eliminate dangling bonds produces a (2×4) type of reconstruction. The [110] rows of orthogonally oriented Group V dimers lead to alternating [110] rows of compressive and tensile stress in the third buried layer. This is because the Group V atoms move together when the dimer bond is formed. The resulting alternating compressive and tensile stress leads to an energetic driving force for formation of the CuPt ordered structure. For alloys with mixing on the Group III sublattice, such as GaInP, this produces the [110] rows of alternating large and small atoms that form the CuPt-B variants. These calculations also predict that the surface structure of alloys with mixing on the Group V sublattice, such as GaAsP, will produce the CuPt-B variants, in agreement with experimental observations [68]. This surface thermodynamic origin of CuPt ordering is supported by experimental observations for a wide range of III/V alloys with mixing on both the Group III and Group V sublattice.

Ordering is an example of a self-assembly process occurring during growth, driven by surface thermodynamics. Samples grown by different epitaxial techniques are now known to produce different degrees of order, in part due to differences in surface bonding. For example, ordering is virtually never observed in layers grown by LPE [65, 69]. The liquid phase apparently passivates the surface, thus removing the need for surface reconstruction, which diminishes the driving force for the formation of ordered structures.

For layers grown by MOVPE, drastic differences in order parameter are observed using different growth parameters, as discussed in Section 2.4.3.1. Since ordering alters the bandgap energy, this means the ubiquitous plot of bandgap energy versus lattice constant (or solid composition) is basically flawed, since it assumes that the bandgap energy is a unique function of solid composition, as discussed previously. However, as with most materials, the properties are partially determined by the microstructure: e.g. ordering or large, nonrandom compositional fluctuations due to incipient phase separation.

Ordering has extremely important practical consequences. The order-induced property change that elicits the most interest by the device community is the reduction of the bandgap energy. More than 10 years before ordering was discovered, careful researchers in different laboratories measured bandgap energies of GaInP layers grown by various techniques, but all lattice matched to GaAs substrates, i.e. all with the same composition, to vary by more than 100 meV [70]. The variation of bandgap energy was a decade later found to be due to CuPt ordering [71]: In GaInP, ordering causes a large shrinkage of the bandgap. The first attempts to quantitatively determine the effect of order on the bandgap energy were theoretical [66]. The reduction in bandgap energy has the following dependence on the degree of order, S [72],

$$E_g = E_g(S=0) - \Delta E_g S^2 \tag{2.10}$$

A value of $S=0$ indicates disorder, and a value of 1 indicates a perfect superlattice of alternating (111) monolayers of pure GaP and InP. Ernst et al. [73] examined the bandgap energy versus order parameter experimentally for GaInP, yielding a value of ΔE_g of 471 meV. The largest experimentally measured change in bandgap energy due to CuPt ordering is 160 meV [74]. Transmission electron diffraction superspot intensity measurements yield values of S as large as 0.7 [65, 67, 75]. Recent spin echo nuclear magnetic resonance (NMR) measurements of ^{71}Ga [75] yield the best independent value to date of 0.6 for a sample grown to maximize the order parameter. Thus, as would be expected, due to kinetic limitations, the material is never fully ordered. The large bandgap shrinkage due to order in GaInP is extremely important for devices, including visible LEDs and injection laser diodes [78]. To produce the shortest-wavelength (most visible) devices, ordering must be avoided.

Ordering has been observed in a number of III/V alloys [65]. In fact, CuPt ordering has been observed in virtually all III/V alloys studied when the layers are grown by MOVPE on (001)-oriented substrates [79]. Bandgap narrowing due to CuPt ordering has been experimentally observed in GaInAs alloys lattice matched to (001) InP substrates [76]. A maximum bandgap reduction of 65 meV was observed. Experimental determination of the effect of CuPt order on the bandgap of InAsSb alloys reveals a bandgap shrinkage of approximately 45 meV [77]. In InAsSb alloys, the shrinkage of bandgap energy is potentially beneficial, since it moves the wavelength further into the infrared where an atmospheric window exists between 8 and 12 μm [76]. Thus, ordered InAsSb has the potential to be a useful material for IR detectors.

2.2.4.2 *Control of Ordering Using Growth Parameters*

As discussed, ordering is driven by surface thermodynamics; however, kinetic factors can limit the degree of ordering observed. Low growth temperatures and high growth rates produce more random mixtures simply because the atoms are less able to rearrange themselves near the surface as growth proceeds. Of course, taking low growth temperatures and high growth rates too far produces highly defected material that is not suitable for minority carrier devices. An understanding of the thermodynamics and kinetics of growth and the formation of microstructures has led to the ability to control the degree of order in epitaxial layers using parameters such as the growth temperature, V/III ratio, growth rate, and substrate misorientation [65, 79–81].

The effects of temperature during MOVPE growth on the degree of CuPt order have been studied extensively for GaInP. For example, a systematic study using the reactants trimethylgallium, trimethylindium, and PH_3 [65, 82] shows a clear maximum in the degree of order at temperatures near 640 °C, with a continuous decrease in order with increasing temperature until at 720 °C the material is essentially disordered. Similarly, a decrease in growth temperature to 520 °C produces material with only a small degree of order. Nearly identical results were obtained for the growth of GaInP using tertiarybutylphosphine (TBP) rather than phosphine [83].

Determination of the concentration of $[\bar{1}10]$ P dimers on the surface using surface photoabsorption spectroscopy (SPA) revealed that the major cause for the decrease of ordering at high temperatures was loss of the (2×4)-like surface reconstruction, a surface thermodynamic effect due to a reduction in the thermodynamic driving force for ordering. However, the loss of order at low temperatures was determined to be due to kinetic effects, i.e. the inability of the cation adatoms to rearrange themselves on the surface at low temperatures [65, 82].

The flow rate of the Group V precursor is also found to have a significant effect on the ordering process. The degree of order is found to decrease for low input partial pressures of the P precursor. The data from the studies of changing temperature (620–720 °C) and TBP partial pressure (<200 Pa) are combined for the plot of the degree of order versus the SPA signal in Figure 2.6. The one-to-one relationship between the degree of order and the concentration of $[\bar{1}10]$ P dimers on the surface is observed for changes in both temperature and TBP partial pressure. Thus, these are purely surface thermodynamic effects. The effects of Sb and Bi surfactants are also thermodynamic, as discussed in Section 2.4.4.

The dependence of order parameter and hence bandgap energy on growth parameters has been exploited to produce a novel type of heterostructure in GaInP by simply changing either the growth temperature or the input V/III ratio. This technique has been used for the production of abrupt heterostructures with no change in solid composition and, hence, no mismatch dislocations. The PL from a heterostructure using a change in temperature from 740 C to 620 °C is an amazingly large 160 meV [82, 89]. The utilization of changes in the phosphine flow rate to control ordering has also been used to produce InGaP QWs having the same composition throughout [90]. Such structures have also been used to produce back surface field (BSF) multijunction solar cells [91]. Amber-green (566 and 600 nm) emission from AlInP was found to be increased by a factor of three

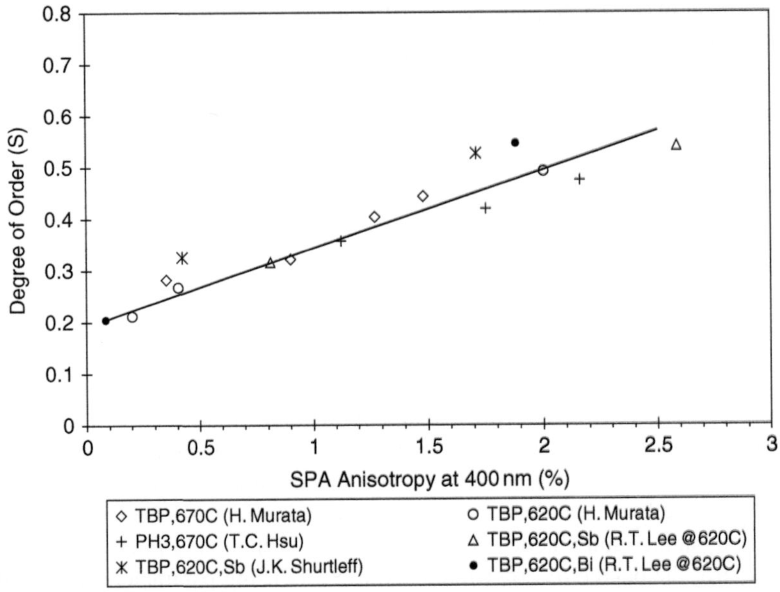

Figure 2.6 *Degree of order, determined from the 20-K PL peak energy, versus the SPA anisotropy intensity at 405 nm due to the [110]-oriented P dimers. Data from Murata et al. 1996 [83], Hsu et al. 1998 [84], and Lee et al. 2000 [85] are for samples grown using variations in the temperature and group P partial pressure. Data from Shurtleff et al. 1999 [86], Stringfellow et al. 2000 [87], and Jun et al. 2000 [88] show the effects of adding a small amount of surfactant Sb and Bi, respectively. Source: After Stringfellow et al. 2000 [96]; reprinted from the Journal of Crystal Growth with permission from Elsevier Science.*

when electron confinement was controlled using an order–disorder structure produced using control of the order parameter [92].

The results of growth-rate studies reinforce the observation that kinetic factors can be significant in the ordering process. Considering only thermodynamic factors, the growth rate should not affect the ordering in GaInP for MOVPE growth at high V/III ratios. Changing the partial pressures of the Ga and In precursors will change the growth rate, but will not affect the P partial pressure at the interface, as discussed in Section 2.3.1. Thus, the P coverage of the surface and the surface reconstruction should be independent of growth rate. For MOVPE growth of GaInP at a temperature of 670 °C, increasing the growth rate from 0.25 to 2 μm/h was found to have no detectable effect on the degree of order [93]. This clearly demonstrates the lack of a kinetic factor in the ordering process under these conditions. However, at higher growth rates, from 4 to 12 μm/h at a temperature of 680 °C [94] a marked decrease in the degree of order was observed with increasing growth rate. The ordering is virtually eliminated at a growth rate of 12 μm/h (about 10 monolayers/s). The reduction in order parameter with increasing growth rate seen for rates above 4 μm/h gives a rough measure of the time constant of the ordering process. The data indicate that the ordering time constant, attributed to surface diffusion, is approximately 0.25 s at this temperature.

Other factors can also affect the degree of order for GaInP layers grown by MOVPE. For example, variation of the misorientation of the substrate reveals three factors [65]: (i) Small misorientation angles in the $[\bar{1}10]$ direction increase the degree of order. However, for misorientation angles exceeding 6 degrees, an increasing misorientation angle leads to a reduction in the order parameter [95]. The latter effect has been shown to be due to a reduction in the concentration of P dimers at the surface. (ii) $[\bar{1}10]$ steps, themselves, have a disordering effect and, conversely, [110] steps are found to increase ordering [65]. These effects are presumably kinetic in nature. (iii) Furthermore, studies of the effect of the addition of the n-type dopant Te during MOVPE growth of GaInP clearly show that the Te disorders the material by increasing the step velocity [96, 97]. This effect appears to be similar to that of increasing the growth rate; increasing the step velocity leaves less time for rearrangement of the Ga and In atoms at the step edge.

The choice of Ga precursor has also been found to affect the degree of order in GaInP [98]. Perhaps the most useful and interesting method of controlling the degree of CuPt order in GaInP grown by MOVPE is by the use of surfactants. This topic will be discussed in detail in Section 2.4.4.

2.3 Kinetics

Thermodynamic equilibrium determines the state of a closed system given very long times. However, in the real MOVPE system, the times allowed for equilibrium to be established are typically less than a second. This means kinetic constraints are superposed upon the thermodynamic driving force. The two must be superposed to understand the real MOVPE processes and resulting microstructures. This is sometimes approached using the concept of kinetically hindered equilibrium. There are three aspects of kinetics that must be considered in any treatment of the MOVPE process: mass transport, homogeneous reaction kinetics to treat gas-phase pyrolysis reactions, and heterogeneous kinetics for reactions occurring on the surface. This section will give a brief overview of these topics. The basic concepts have been treated in detail in previous reviews [99, 100]. Heterogeneous reactions will be treated in Section 2.4.

2.3.1 Mass Transport

The MOVPE process is virtually always carried out in a cold-wall system, since we do not want the precursors to decompose on the reactor walls. The precursors are typically diluted in H_2 or N_2, so no condensation of the reactants occurs upstream of the substrate on the room-temperature walls. The precursors are most often delivered to the growing layer by diffusion through a *boundary layer*, discussed briefly in Section 2.2.2. The hydrodynamic aspects of the gas flow are, in real systems, highly complex and dependent on the details of the reactor geometry [99]. Many reactor configurations are used in current research and commercial reactors; thus, no attempt will be made here to analyze the details of the actual mass-transport processes occurring. Instead, an effort will be made to acquaint the reader with the general features of the processes occurring in the reactor, concentrating on the delivery of the reactants to the growing surface.

The gases flow through the reactor by forced convection, where the fluid flow is due to an imposed pressure gradient. Beginning with conventional tube reactors, the low flow velocities give rise to laminar flow: The velocity vectors are parallel to the walls. They have a value of zero at the reactor walls and vary as a smooth function of position within the reactor. Two factors complicate this simple picture: obstacles or abrupt changes in diameter and buoyancy-driven or *natural* convection. Even considering these complications, the fluid flow can be calculated by solving the equations for conservation of mass, conservation of momentum (the Navier–Stokes equation), conservation of energy, and the equation of state, with appropriate boundary conditions [101].

However, to give an intuitive feel for the fluid-flow and mass-transport processes occurring in a MOVPE reactor, the simple *boundary layer* model is useful. Based on early observations, a stagnant *boundary* layer is postulated in the vapor phase adjacent to the growing epitaxial layer. The gases are assumed to be well mixed in the region outside of the boundary layer having reactant input partial pressures determined by the relative flow rates of the precursors and the carrier gas. This simplifies the equation controlling the mass-transport process to the one-dimensional Fick's first law. For the growth of III/V semiconductors using a V/III ratio in the input gas phase of much greater than unity and nearly complete depletion of the Group III element at the solid/vapor interface, this gives the growth rate from Eq. (2.8), which is proportional to the input partial pressure of the Group III precursor and is independent of the input Group V partial pressure. This simple model describes many of the basic features of the MOVPE growth process [102]. For II/VI semiconductors, the more volatile Group II precursor is often present in excess, making p^* in Eq. (2.8) the input Group VI precursor.

This boundary layer model is certainly not justified in terms of the actual hydrodynamics of real reactors. However, if the details of the model are not taken literally, it has proven to be a useful zeroth-order approximation for the analysis and interpretation of experimental observations.

One approach to improving the model is to simply improve the hydrodynamic treatment of the gas flow in a cold-wall tube reactor [103, 104]. The opposite approach is to perform a full hydrodynamic analysis of gas flow in a simple reactor geometry. An example of this approach, applied to a horizontal rectangular reactor, is described by Moffat and Jensen [105].

Diffusion-limited growth, described previously, gives a growth rate that is a weak function of temperature. The growth rate is also independent of substrate orientation. Factors that act to decrease the boundary-layer width, such as increasing the gas velocity or susceptor rotation, result in increased growth rates. To achieve high uniformity of growth rate and solid composition over many substrates, commercial production reactors frequently utilize a vertical flow to the surface with the substrates rotated to give a thin, uniform boundary-layer thickness [106]. To minimize nonuniformities due to depletion of the reactants, *showerhead* reactors have been developed where the reactants are injected from many ports situated very near the substrate surfaces [107].

2.3.2 Precursor Pyrolysis

At lower temperatures, the growth rate decreases and is thermally activated [102]. In this region, the growth rate is controlled by precursor pyrolysis. For example, the low-temperature growth of GaAs is controlled by the rate of pyrolysis of the Ga precursor. A change from TMGa to the more labile TEGa results in an increase in the GaAs growth rate [102].

Eyring's theory of absolute reaction rates [108] states that the rate of any reaction is proportional to the rate of formation of an activated complex. The formation of the excited state requires the input of energy, E^*, resulting in a reaction rate constant, k, for a homogeneous, unimolecular reaction. For example,

$$\mathrm{Ga}(\mathrm{CH_3})_3 = \mathrm{Ga}(\mathrm{CH_3})_2 + \mathrm{CH_3} \tag{2.11}$$

of

$$k = Ae^{-E^*/kT} \tag{2.12}$$

Simple homolysis reactions are the most commonly observed pyrolysis reactions for precursors used in MOVPE growth. The activation energy for homolysis will be determined by the strength of the metal–radical bond. For the Group III elements, the trimethyl precursors decompose at significantly higher temperatures than the triethyl compounds, as mentioned previously, and the pyrolysis temperature decreases as the atomic number of the Group III element increases; thus, TMGa decomposes at higher temperatures than TMIn. Kuech [109] gives ranges of the temperatures required for pyrolysis of various precursors commonly used for the MOVPE growth of several compound semiconductors.

Chain reactions have also been observed for MOVPE precursors. For example, in reaction (2.11), the methyl radical can subsequently interact with the H_2 carrier gas to form CH_4 and H, with the H then attacking the TMGa [100]. This is the main factor accounting for the ambient dependence of pyrolysis temperatures typically observed [109].

Additional complex reactions may be observed to control the pyrolysis of relevant precursors. For example, alkyl exchange reactions are observed where complexes formed in the vapor phase allow Group III alkyls to exchange radicals [100]. Thus, two Group III precursors, such as TMIn and TEGa, can undergo a bimolecular reaction where the two precursors exchange radicals to form EDMIn and MDEGa.

In addition, monomolecular pyrolysis reactions may occur, particularly in precursors having larger radicals, such as tertiarybutylarsine (TBA) and tertiarybutylphosphine (TBP). So-called β-*elimination* reactions, producing alkenes and intermediate molecules with M–H bonds, are sometimes observed [100].

2.3.3 Control of Solid Composition

As discussed in Section 2.2.2, the solid composition of alloys is often controlled by thermodynamics. However, for alloys with mixing on the Group V sublattice, the solid composition is frequently controlled by the precursor pyrolysis rates. The results for the growth of InAsP using the hydrides of both As and P are shown in Figure 2.7 for growth over a range of temperatures. We see that the P distribution coefficient is approximately 0.05 for growth at 600 °C. At a growth temperature of 850 °C, the distribution coefficient approaches unity. The phosphides are more thermodynamically stable than the arsines, so k_P should be greater than unity. The value of k appears to be controlled by incomplete PH_3 pyrolysis. Similar behavior is observed for GaAsP [113]. The results for GaAsP have been found to change dramatically when PH_3 is replaced by TBP, which pyrolyzes at much lower temperatures. Chen et al. [114] demonstrated that this caused significant increases in k_P, as expected due to the more rapid pyrolysis of TBP. However, even in this case, the calculated distribution coefficient was found to be significantly below the

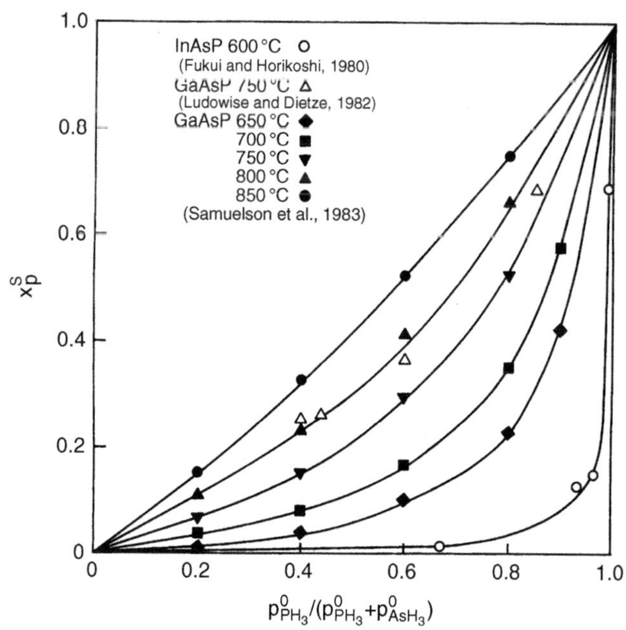

Figure 2.7 *Solid versus vapor composition for the III/V alloys with As and P mixing on the Group V sublattice. The experimental data are from Fukui and Horikoshi 1980 [110], Ludowise and Dietze 1982 [111], and Samuelson et al. 1982 [112]. Source: After Stringfellow 1983 [47]; reprinted from the Journal of Crystal Growth with permission from Elsevier Science.*

values predicted from the thermodynamic calculations. This was attributed to complexities related to the P species formed at the interface [100, 114].

Clearly, the reactions involved in MOVPE growth are complex. One factor making the situation even more difficult to analyze is that the reactions occur in an inhomogeneous, nonisothermal system of flowing gases. Thus, reactor geometry and pressure will have first-order effects on the growth process. Even today, our understanding is incomplete. Gas-phase reactions include the pyrolysis reactions yielding the components of the epitaxial layer, as well as complex reactions involving adduct formation in the vapor, due to the Lewis-acid and Lewis-base natures of many of the respective Group III and Group V precursor molecules. As a further complication, the gas-phase pyrolysis reactions are seldom complete, so heterogeneous pyrolysis reactions occurring on the growing surface often play a key role in the pyrolysis and growth reactions [99, 100, 109]. Thus, it is expected that the chemical and physical state of the surface will play an important role. This is a topic that is somewhat neglected. Nevertheless, it is clear that surface reconstruction, as controlled by the temperature and gas-phase composition as well as the presence of surfactant, will play an essential role in the overall kinetics of the growth process. This will be discussed in detail in Section 2.4.

First-principles calculations are frequently used to help sort out these complex problems. This topic is treated in some detail in the literature [99] and so will not be treated further here. Such calculations are often used as an aid in reactor design and operation and are expected to become even more useful as we unravel the complexities of the homogeneous and heterogeneous chemical reactions occurring during deposition.

The growth of the nitride alloys has become a prominent MOVPE growth topic due to the advantages of these materials for many applications, including green, blue, and white LEDs, blue and green injection lasers, high-power electronic devices, and solar-blind detectors. Great progress has been achieved empirically to yield high-quality materials for device applications. Unfortunately, the basic kinetics of the MOVPE growth of these materials is extremely complex; thus, an understanding of the growth process has lagged practical advances. The growth of AlGaInN alloys is found to occur by a daunting series of gas-phase reactions in addition to the final heterogeneous reactions occurring at the solid/vapor interface [109]. A recent review article [115] gives considerable insight into the GaN growth process based on a combination of detailed experimental results, reactor modeling, and quantum-chemical calculations. Creighton et al. [115] conclude that parasitic reactions occur in the boundary layer leading to the formation of nanoparticles containing 10^4 to 10^7 atoms. These 2–50 μm particles are driven away from the substrate by thermophoresis, resulting in problems with reactor efficiency as well as control of growth rate and solid composition. The reactions occurring in the boundary layer are observed to involve homolysis as well as concerted reactions involving the interaction of Group III and nitride precursors to form adducts, which subsequently decompose to form oligomers. The key to minimization of these parasitic reactions is to decrease the reactor pressure and the time the atoms spend in the boundary layer. One way to accomplish the latter is to increase the rate at which the substrate is rotated, which decreases the boundary-layer thickness. Creighton et al. [115] were able to develop a model that accurately describes the growth rate versus spin rate. The results are reproduced in Figure 2.8.

In a thorough review article on the MOVPE growth of the Sb alloys, Biefeld [50] notes that thermodynamics controls the solid composition in many of these alloys, including InAsSb, GaAsSb, InGaSb, AlGaSb, and AlAsSb, as discussed in Section 2.2.2. As expected from the earlier discussion of composition control for systems involving precursors that do not pyrolyze homogeneously, complications related to the kinetics of pyrolysis have important effects on the solid composition for other alloys, such as InPSb and GaPSb. Kandekar [116] studied the effects of precursor chemistry on alloy composition in GaAsSb. They determined that solid composition control is complex, involving the basic thermodynamics of mixing, the effects of strain (lattice

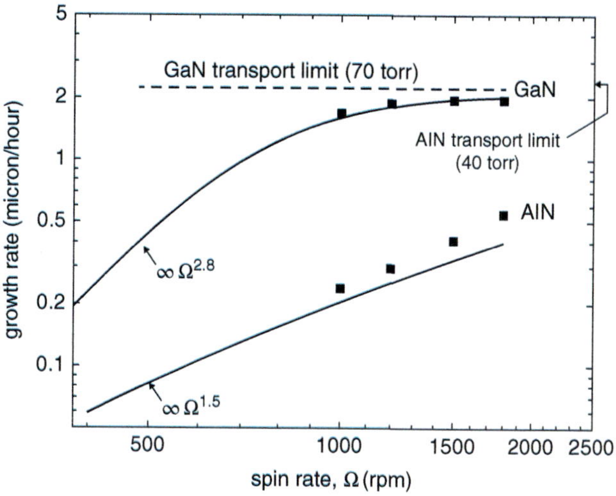

Figure 2.8 Experimental (solid points) and theoretical (solid curves) growth rates as a function of spin rate at matched flow conditions for AlN (40 Torr) and GaN (70 Torr). The power-law dependence for the theoretical results is shown near 500 rpm. Source: After Creighton et al. 2007 [115]; reprinted from the Journal of Crystal Growth with permission from Elsevier Science.

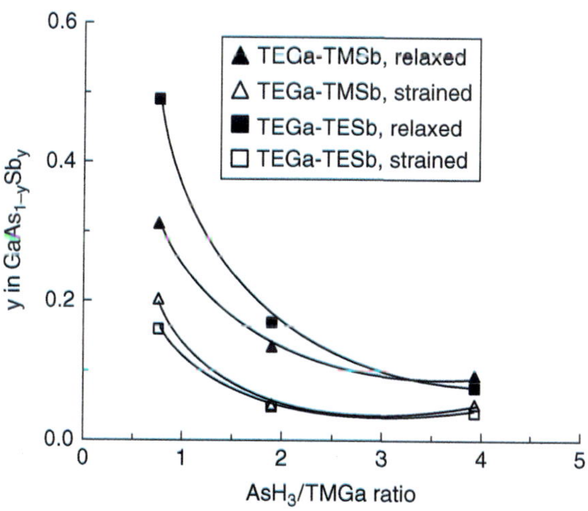

Figure 2.9 Variation in the Sb-mole fraction with inlet AsH_3/TMGa ratio for the TMGa-TMSb and TMGa-TESb growth of GaAsSb on GaAs substrates: open symbols represent relaxed and closed symbols represent strained GaAsSb layers. All samples were grown at 530 °C and at a constant Sb:TMGa-precursor ratio of 1.55. Source: After Khandekar et al. 2007 [116]; reprinted from the Journal of Crystal Growth with permission from Elsevier Science.

pulling), surface composition (Sb coverage), and precursor pyrolysis. The experimentally observed Sb-mole fraction is plotted versus the AsH_3/TMGa ratio in Figure 2.9. The data is for the MOVPE growth of GaAsSb using several precursors and As/Ga ratios in the input gas flow. The results show pronounced lattice pulling, discussed in Section 2.2.2. The layers elastically strained to coherently lattice match the substrate all have significantly lower Sb concentrations, demonstrating the large effect of lattice pulling.

Huynh et al. [117] studied the MOVPE growth of highly lattice mismatched InGaSb alloys on GaAs substrates using TMGa, TMIn, and TMSb at temperatures between 500 and 625 °C. Thermodynamically, we would expect the In distribution coefficient to be unity, as discussed in Section 2.2.2. However, at 500 °C, TMIn is completely decomposed while TMGa is not. Thus, the In content demonstrates an In distribution coefficient well above unity. It decreases as the growth temperature increases due to the increased TMGa pyrolysis. Of course, this means that in the temperature range 500 to 600 °C, the growth rate increases with increasing temperature.

A striking example of the effects of complex pyrolysis reactions is the effect of the addition of a small concentration of the p-dopant DEZn on the growth of InP and GaInP nanowires [118]. The DEZn caused several effects, including a change from a mixed zincblende/wurtzite structure to pure zincblende (ZB). Heurlin et al. [118] also discovered a marked increase in the nanowire growth rate as DEZn was added to the system. The In for the nanowire growth comes mainly from the pyrolysis of TMIn on the planar surface surrounding the nanowire. The increase in nanowire growth rate was interpreted to be due to an increase in the In diffusion length on the planar substrate. This was attributed strictly to the effect of Zn on the surface. The work of Otnes et al. [119], from the same group, for GaInP nanowires grown at 440 °C showed that the introduction of DEZn leads to smaller-diameter, predominately ZB nanowires. In addition, the nanowires became significantly more Ga rich. This was attributed to the increased TMGa pyrolysis rate due to the ethyl radicals released during DEZn pyrolysis. At this growth temperature, the TMIn pyrolysis is nearly complete, but the TMGa pyrolysis is normally much slower. The authors postulate an effect similar to that discussed previously for the increased pyrolysis of TMGa due to the release of CH_3 radicals that react with the H_2 ambient to produce a chain reaction that increases the TMGa pyrolysis rate. The ethyl radicals are postulated to enhance TMGa pyrolysis via chain reactions, with a consequent increase in the amount of Ga incorporated into the growing nanowire.

2.4 Surface Processes

As discussed in the last section, many pyrolysis reactions occur as the precursors transit the boundary layer. However, some of the precursor molecules as well as their homogeneous pyrolysis products reach the surface. The subsequent processes occurring on the surface are major factors determining the outcome of the MOVPE growth process. In the MOVPE environment, it is difficult to observe the processes occurring on the surface directly. However, for many decades [120, 121] an idea of these surface processes has been deduced from macroscopic observations, simple calculations, and informed imagination. During MOVPE, the majority of the surface consists of atomically flat regions separated by steps, which may be introduced due to misorientation of the substrate from the typical low-index planes or by 2D nucleation and lateral growth. The surface during growth cannot be absolutely smooth due to entropic considerations, with concentrations of adatoms and advacancies that increase exponentially with temperature. The adatoms themselves can condense into two-dimensional clusters, due to both thermodynamic and kinetic factors. These general observations have led to two common model schemes developed to explain the general features of growth: namely *step flow* or *Frank–van der Merve growth*, and *Stranski–Krastanov* or *three-dimensional island nucleation and growth* [122, 123]. Today, we are able to image the atomic-scale features using scanning probe microscopy techniques. The results are surprisingly similar to the features expected from the early models. Now that some of the physical processes can be viewed directly, our fundamental understanding of MOVPE is advancing rapidly.

The important surface processes include adsorption and desorption of surface species, heterogeneous pyrolysis of the species reaching the surface, two-dimensional surface reactions, and the processes leading directly to formation of the solid: adatom diffusion, nucleation of islands, and attachment of adatoms at step edges. In addition, subsurface diffusion within the top one to four layers of the solid may allow atomic rearrangement of the solid during growth.

2.4.1 Surface Reconstruction

A natural starting point for a discussion of surface processes is to consider the configuration of atoms on the surface and how the adatoms bond to each other and to the underlying layer. An obvious first step is to consider the surface as if the bulk solid were simply truncated with the atomic positions frozen in the same positions as in the bulk solid. However, it was realized decades ago that this would result in a very high-energy surface, due to the high density of dangling bonds. This would result in very strong binding of adatoms to the surface, leading to very little surface diffusion and, consequently, statistically rough surfaces. This is contrary to the common observation of smooth surfaces and atomically abrupt, heterostructure interfaces. This can all be rationalized if we consider that the surface atoms are allowed to rearrange themselves to form a low-energy "reconstructed" surface. In fact, in UHV environments, the surface reconstructions can be experimentally characterized using electron diffraction and scanning tunneling microscopy (STM) techniques. Even in an atmospheric-pressure MOVPE reactor, STM observation of the surface can now be done at temperatures as high as 650 °C [124].

Elementary thermodynamic considerations suggest that the surface atoms will rearrange to lower the surface energy by the elimination of dangling bonds. The *electron-counting rule* is widely used to identify potentially low-energy reconstructions [125]. This yields reconstructions where all of the electrons, including those from the surface atoms, are in bonding configurations. Formation of these reconstructions will, of course, lead to changes in the positions of the surface atoms, which will add energy due to the stretching and bending of their covalent bonds. However, these energies are much smaller than those of the dangling bonds. For example, the most commonly observed structure for (001) GaAs terminated by As atoms under UHV conditions is the (2×4) reconstruction, a terminology that refers to the periodicity of the (001) surface in the two <110> directions orthogonal to (001) [125, 126]. The other, even more anion-rich reconstruction is the (4×4) structure [125, 126]. During MOVPE growth, direct measurement techniques are typically unavailable, due to the atmosphere that is necessarily present. However, as mentioned previously, STM techniques can be used in some cases, and optical techniques have proven useful for identifying the surface reconstructions by comparison with results from the direct techniques. For MOVPE, (2×4) and (4×4) are the two most commonly observed reconstructions for (001) GaAs and many other compounds and alloys [127].

The equilibrium reconstructed semiconductor surface formed during growth is determined by the extensive thermodynamic variables of the system. Thus, the occurrence of the stable reconstructions can be represented by a surface phase diagram where the surface phase (or reconstruction) is plotted versus the Group V pressure at the interface and the temperature [16]. The phase rule applies to the surface phases in exactly the same way as for the bulk. For example, when the system consists of a single surface reconstruction (or surface phase) plus the vapor, the system has two degrees of freedom. Thus, fixing two independent variables, typically temperature and the Group V partial pressure, completely defines the thermodynamic state of the surface. Changes in either of the two independent variables may result in a change in the surface reconstruction.

Clearly, the surface reconstruction will affect the key surface processes, since it determines the bonding of adatoms to the surface. Adsorption and desorption rates will be determined by the bonds formed with the surface atoms. As expected, reconstruction has a direct effect on heterogeneous pyrolysis. For the MOVPE growth of (Al)GaAs on (111)B substrates, formation of the (2×2) reconstruction suppresses TMGa pyrolysis, resulting in low growth rates [128]. Certainly, the barriers to surface diffusion will also depend on the surface reconstruction and will typically be anisotropic; i.e. they will depend on the direction of motion, since the formation of dimers alters the surface symmetry. Attachment at step edges would be expected to depend on the reconstruction occurring there. Thus, surface thermodynamic factors will control not only the details of the surface reactions, but also the growth mode: i.e. step flow or 2D island nucleation and growth. Field et al. [129] found that surface reconstruction strongly affects dopant incorporation in MBE-grown GaAsBi alloys. For (1×3) and (2×3) surfaces, they found Si incorporates more easily onto Group III sites, giving n-type conductivity, due to the difficulty in breaking the surface As dimers. However, for the (2×1) reconstructed surface, Si incorporation into Group V sites is easier, due to the lack of As dimers, leading to p-type conductivity.

2.4.2 Atomic-Level Surface Processes

Next, we consider the atomic-level processes occurring on the reconstructed surface. As precursor molecules and their decomposition products reach the surface, they must first adsorb to begin the growth process. Once on the surface, the precursor molecule or partially decomposed precursor can decompose heterogeneously. Making a bond to the surface weakens the remaining metal–ligand bonds, leading to rapid pyrolysis at normal growth temperatures. Subsequently, the adatoms diffuse to sites where they can be incorporated into the solid, typically by attachment at a "half-crystal" site, most often at a step edge. Alternatively, a group of adatoms may aggregate to form a two-dimensional nucleus. Both processes are driven by the supersaturation of adatoms on the surface [65].

The adsorption and desorption processes are often treated using the Langmuir model [65, 130]. It is widely believed that adsorption of adatoms consists of two steps. First, the atom or molecule is physisorbed: i.e. attached to the surface only by weak van der Waals forces. The physisorbed entity can then either desorb or find a site where it can form a chemical bond to the surface, where it is chemisorbed. For compound semiconductors, two distinct sets of surface sites are identified for the cation and anion atoms based on where the maximum number of bonds can be formed. Thus, the adsorption processes for the cation and anion atoms forming the solid are not competitive. This is described in terms of the Langmuir–Hinshelwood adsorption isotherm [67]. For the low surface coverages expected for MOVPE growth, the growth rate is simply proportional to the product of the partial pressures of the components in the vapor adjacent to the interface. Such behavior has been observed repeatedly for most normal growth conditions [4, 67, 131, 132]. In the typical case of an excess of one component, the component with the lower partial pressure controls the growth rate, as discussed in Section 2.2.2. For the growth of III/V semiconductors, the Group V input partial pressure is much higher than for the Group III element; thus, the growth rate is proportional to the input Group III partial pressure. The opposite occurs for the MOVPE growth of II/VI semiconductors, simply because the Group II elements are more volatile than the Group VI elements, and consequently, the materials are grown in a Group II-rich environment.

For very high surface coverages, which may occur at very low temperatures or when the surface sites are blocked by surfactant atoms, the Langmuir isotherm saturates, and the growth rate will no longer be a linear function of the input partial pressure of the minority precursor. This is seldom observed. However, Forghani et al. [133] observed that the GaAs growth declines significantly for high concentrations of the surfactant Bi. Bi is slow to desorb at low temperatures, does not significantly participate in growth, and has a low solubility in the GaAs. By occupying all of the surface sites, it stops the adsorption of the Ga precursor.

Direct observation of adatom surface diffusion and the behavior of step motion during MOVPE growth is difficult. Thus, we must infer the adatom diffusion lengths and the behavior of steps from information that we can observe *ex situ*. A simple approach is to observe the geometry of island development for growth on patterned substrates as a function of growth parameters, such as temperature and input reactant partial pressures, as well as the presence of surfactants. Asai [134] pioneered this approach for MOVPE growth. He created steps on the (001) GaAs surface by using patterned substrates and observed the growth rates in several directions as the growth parameters were varied. He found that circular islands became ovals after growth, demonstrating that the lateral growth rate depends on step orientation, with [110] steps propagating at higher lateral velocities than [$\bar{1}$10] steps. Asai explained the results using a simple model of adatom attachment at step edges in terms of the bonds formed during adatom attachment, assuming no reconstruction of the surface or of the step edges. This simple model has also been used to describe the basic features of a number of other growth phenomena, including ordering and dopant incorporation. However, in reality, the lateral growth rate must be dependent on both the sticking coefficient at the step edge and the diffusion coefficient. Both would be expected to be anisotropic on a reconstructed surface. Reichert and Cohen [135] studied the effects of substrate misorientation and growth temperature on the lateral growth rates for nominally (001) GaAs. They postulated that anisotropic lateral growth is caused by anisotropies in surface diffusion due to the presence of steps and surface reconstruction. Wixom et al. [136] studied the effects of the surfactants Sb and Bi on the lateral growth rates in the two orthogonal <110>(or {110}) directions for patterned, singular (001) GaAs grown by MOVPE. Both surfactants were found to increase the [110] lateral growth rate by nearly 300%, with no change in the [$\bar{1}$10] direction. They used kinetic simulations to conclude that the effect is due to an increase in the frequency of [110] diffusion events: i.e. an increase in the surface diffusion coefficient in the [110] direction, due to a decrease in the Ehrlich–Schwoebel [137, 138] barrier at the step edge, which results in a higher sticking coefficient for atoms approaching from the lower terrace than from the upper terrace. Step dynamics similar to those reported for macroscopic GaAs islands, described previously, are also seen for microscopic islands formed naturally by 3D nucleation during the MOVPE growth of GaInP on singular (001) GaAs substrates [139].

The interpretation of data related to surface diffusion and attachment at steps is complex due to the fact that both processes are typically involved in determining step flow. One method for deducing surface diffusion lengths is to observe the growth of nanowires as the spacing between the wires is altered. For example, Jensen et al. [140] studied InAs nanowires grown by chemical beam epitaxy (CBE) using TMIn and TBAs precursors at a temperature of 440 °C on singular (111)B substrates. For the growth of these nanowires, the As contributing to the wire growth is only that part impinging directly onto the Au particles at the top of the nanowires. For CBE growth, the TMIn reaches the planar surface and pyrolyzes heterogeneously. The In is assumed to be collected from the surface at the base of the nanowire from a circular area having a radius equal to the In diffusion length. From measurements of the growth rate of the nanowires versus the separation between wires, it is possible to make an estimate of the In diffusion length. They deduce a diffusion length of 0.65 μm. They also determined the diffusion length of In along the side facets of the wires to exceed 10 μm. However, this interpretation implicitly assumes an isotropic diffusion length. As discussed previously, this is typically untrue for reconstructed surfaces.

The microstructure of the surface during MOVPE depends on, among other things, the step densities, the sticking coefficients at step edges, the Ehrlich–Schwoebel barrier height, and the growth parameters such as the incoming flux of atoms. For vicinal substrates, where the substrate misorientation produces an array of steps of a particular geometry, the growth generally occurs by step-flow growth. Each adatom has the time and mobility to diffuse to a step where it is incorporated into the solid. On the other hand, when the step spacing is too large or the adatom mobility too small, the region between steps becomes highly supersaturated. When this supersaturation exceeds a critical level, the adatoms cluster to form 2D islands between steps. This results in layer-by-layer or Frank–van der Merve growth. This type of growth is nearly universal for singular substrates.

The bonding of adatoms at step edges is complicated due to both reconstruction that may occur at the step edge, again to reduce the number of dangling bonds, and the occurrence of step bunching. The growth parameters and the presence of surfactants have major influences on step bunching [141–144], as discussed later. It has been postulated that the reconstructions allowed at bilayer steps may lower their energy relative to monolayer steps that may contain dangling bonds [141]. The elimination of step bunching by surfactant Te, demonstrated in both GaAs [145] GaInP [139], has been attributed to reconstruction at the step edges.

In addition to the thermodynamic effects described previously, the origin of step bunching may alternatively be attributed to kinetic effects. The simplest example occurs when the sticking coefficients for adatoms are high and the surface mobilities low. This gives rise to *statistical roughening* where the roughness increases as the layer thickness increases. Another example is related to an asymmetry in sticking coefficients at step edges. It is likely that the sticking coefficient for an adatom approaching a step edge will be different for approach from the top or the bottom of the step. As discussed previously, an adatom arriving at a down step has been postulated to face an energy barrier because the bonding cannot be maintained as the adatom passes over the step. In this case, simple kinetics dictates that shorter terraces will become longer and longer terraces shorter: step ordering occurs. A reversal of the relative sticking coefficients at "up" and "down" steps will produce step bunching. This results in kinetic roughening and the formation of 3D "wedding cake" structures [146].

For heteroepitaxial growth when the lattice match between the epilayer and substrate is not perfect, elastic strain is built up as the layer becomes thicker. A thin layer will grow by the Frank–van der Merve, layer-by-layer mechanism and deform, like a drum head, to match the atom positions in the substrate. The elastic energy will be proportional to the layer thickness, while coherency is maintained. As the layer exceeds a certain critical thickness, the energy of the system can be reduced if the thin epitaxial layer separates into islands, since the islands can relax laterally. This is termed Stranski–Krastanov (S–K) growth [121–124]. The thermodynamics of S–K growth has been analyzed for a wide array of III-V ternary systems, resulting in *thickness–composition phase diagrams* [147]. Growth occurs by the S–K mechanism when the lattice-parameter mismatch is large and/or for thick layers. As a result of the balance between thermodynamic and kinetic effects, the islands assume a nearly uniform, kinetically determined, size and spacing that are determined kinetically. This may result in the spontaneous formation of quantum dots.

2.4.3 Effects of Surface Processes on Materials Properties

As discussed previously, surface thermodynamic and kinetic factors determine the microscopic growth process. This, in turn, will have major effects on the properties of the materials being grown, including solid composition, dopant incorporation, and microstructure. These topics will occupy the remainder of this section.

2.4.3.1 Ordering

A clear indication of the importance of surface thermodynamics for semiconductor alloys, including the elemental semiconductors, III/V semiconductors, and II/VI semiconductors, is the effect of the surface on the microscopic arrangement of the atoms in the solid, particularly the spontaneous formation of ordered structures, as discussed in Section 2.2.4.1. VFF calculations indicate that the (2×4) reconstruction energetically favors formation of the B-variants of the CuPt structure [148], which are not stable in the bulk.

For GaInP grown by MOVPE, the presence of $[\bar{1}10]$ P dimers is found to correspond directly to the degree of CuPt order observed experimentally [96]. As the concentration of P dimers is changed using either a change in growth temperature or input partial pressure of the P precursor, a corresponding change in the degree of CuPt order is observed, as seen in Figure 2.6 [96].

For GaInP grown on (001) GaAs, misorientation of the (001) substrate to produce [110] steps is found to enhance the ordering process, while $[\bar{1}10]$ steps are found to retard ordering, as discussed in Section 2.2.4.2. This suggests that surface steps also play a role in the ordering process. Although the explanation for this phenomenon is not certain, bonding at the step edge may be significant. Both experiment and theory suggest that the surface reconstruction may be different for the surface far from step edges and near them [149]. For (001) GaAs covered by Sb, near a step edge the GaSb (4×3) reconstruction forms, due to the ability of the surface atoms to relax laterally. Away from the step edge the compressive strain energy results in a (2×4) reconstruction.

A clear link between ordering and the surface reconstruction has been established. However, kinetic effects also play a role. For example, the addition of surfactants that increase step velocity has been observed to result in a decrease in the degree of order [65]. Ordering has also been shown to disappear at high growth rates, as discussed in Section 2.2.4.2. Both of these kinetic effects are due the atoms near the surface having insufficient time to rearrange from random to ordered.

2.4.3.2 *Phase Separation*

As discussed in Section 2.2.3, phase separation is predicted to occur in semiconductor alloys with a large difference in the size of atoms mixed on a particular sublattice. In practice, the degree to which phase separation actually occurs during growth is typically limited by kinetic factors, particularly by the ability of the atoms to rearrange themselves from the random arrangement of the arriving adatoms. In cases where the kinetic limitations are severe, such as low growth temperatures and high growth rates, the occurrence of phase separation may be completely suppressed, resulting in a random arrangement of the atoms in the solid. More often, the phase separation is limited to compositional clustering of like atoms. As discussed in Section 2.2.3, the rearrangement will occur in the layers at and just below the solid/vapor interface. Thus, it is clear that a major kinetic factor in determining the degree of compositional clustering for a fixed growth rate will be the mobility of the atoms on or near the growing surface. This will be a function of the surface reconstruction, as mentioned previously, as well as the temperature.

The remainder of this section will use as an example the effects of surface processes on phase separation during MOVPE growth of the extremely carefully studied and important GaInN alloys grown on GaN, which are, in turn, grown on sapphire substrates. A very thin GaInN layer, uniform in thickness and coherently strained to match the GaN lattice constant, as in the QW structures used in LEDs and lasers, will be under a significant compressive strain field. Even though the GaInN system has a miscibility gap, a highly simplified analysis indicates that this uniform, thin, strained layer may be stable over a range of In compositions up to values of 0.4 to 0.5, depending on the growth temperature [150]. Thus, phase separation would be suppressed, yielding compositionally uniform layers for the compositions of interest for blue and green LEDs. However, the actual situation is somewhat more complex. For example, thin strained layers can reduce their elastic energy by formation of thickness modulations. This has the effect of decreasing the dominance of the strain stabilization effect. For this reason, the widely cited Karpov [150] phase diagram likely overestimates the effect of strain on phase separation.

As discussed in Section 2.4.2, for layers over a certain thickness, S–K islands are formed [52, 67]. Near the top surface of the islands, the lattice constant will relax toward the equilibrium value. This will allow for the enhanced incorporation of In due to the lattice pulling effect [52]. Thus, the S–K growth mode will yield a two-dimensional array of In-rich areas in the thin QW layer. The island size and density will be determined by surface diffusion coefficients and the growth rate. The formation of these islands will depend on the nominal layer thickness, surface reconstruction, growth mode, and the growth parameters.

Ponce et al. [57] made a systematic, detailed study of the evolution of the cathodoluminescence (CL) spectrum and microstructure, measured by transmission electron microscopy (TEM), for approximately 100-nm thick GaInN films grown by MOVPE at temperatures between 700 and 900 °C. The films were coherent. For layers with In concentrations of less than 0.08, the composition was found to be uniform throughout. For In contents of 0.1 to 0.2, secondary phases were observed to nucleate near dislocations threading through the epilayer from the underlying GaN. As a result, a new, low-energy CL peak was observed, due to the In-rich second phase. This may be due to elastic relaxation near the dislocations that allows the incorporation of a higher In concentrations due to lattice pulling or, as discussed in Section 2.2.3, this may be due to less strain stabilization near the V-pits associated with threading dislocations [55]. The work of Ponce et al. [57] for layers with x_{In}>0.2, showed spontaneous phase separation using both TEM and CL techniques. All of the experimental observations of Ponce et al. [57] are in agreement with the calculated phase diagram, as seen in Figure 2.5.

Kanitani et al. [151] studied 3-nm thick GaInN QW layers with In contents of 0.08, 0.15, and 0.25. The atomic probe tomography (APT) In composition profiles reproduced in Figure 2.10 are nearly periodic. For the layer with a nominal composition of $Ga_{0.75}In_{0.25}N$, the In concentration varies periodically from approximately 0.2 to 0.3. The period is on the order of 5 nm and is found to depend exponentially on the growth temperature, as expected for a diffusion-limited formation mechanism. The formation of In-rich clusters for thin QW layers has been repeated in many hundreds of studies using various techniques, as summarized in [52]. In a recent review article, Specht and Kisielowski [152] critically examined the controversy regarding electron microscopy evidence of compositional clustering in InGaN alloys. They convincingly demonstrated that compositional fluctuations exceed those expected for random alloys for compositions within the miscibility gap from In concentrations of approximately 0.1 to 0.9, with the largest fluctuations for compositions near 0.5 (see Figure 8 in [152]). They concluded that the collected electron microscopy shows "undisputed" evidence of phase separation in GaInN alloys.

2.4.4 Surfactants

Clearly, control of surface reconstruction, step structure, attachment at step edges, and surface diffusion coefficients have first-order effects on the MOVPE growth process. A powerful method for controlling these surface processes is the addition of surfactants during the growth

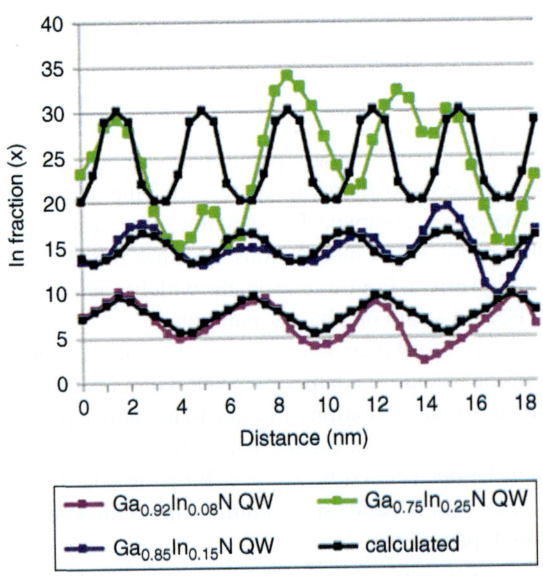

Figure 2.10 Local In composition profiles obtained using atomic probe tomography for 3-nm thick QW layers grown by MOVPE. Source: After Kanitani et al. 2016 [151]; reprinted with permission from the Japan Society of Applied Physics.

process. A *surfactant* is an element added during growth that segregates to the surface and affects the growth process. Surfactants are elements that are relatively nonvolatile and have a limited solubility in the growing solid, typically due to the large size difference between them and the atoms constituting the growing crystal. The use of surfactants has emerged in recent decades as a powerful tool for controlling epitaxial growth and achieving more desirable film qualities. They typically smooth the surface and prevent island formation [153]. The original use of surfactants during the MBE growth of semiconductors was for control of the structure and morphology of highly strained layers in both elemental and III/V semiconductors [154]. The addition of dopants during MBE growth was shown to affect both adatom attachment at step edges [155] and surface reconstruction [156]. It probably also affects the specific surface energy [157] and the diffusion of adatoms [158]. Table 2.2 gives an overview of the effects of surfactants on the properties of III/V materials grown by MOVPE. The remainder of Section 2.4.4 will summarize several of these effects.

Table 2.2 *Summary of effects of surfactants on MOVPE growth and materials properties.*

Effect	Surfactant	Semiconductor	Applications/(Notes)	Refs
Change in reconstruction	Sb	GaP, InP	(Theory)	160
	Sb, Bi	GaInP, GaAsSb	Change in ordered structure	161, 162
	Bi	GaInAs		163
Smoother surfaces	Te, Sb, Bi	GaAs		145, 168
		GaInP		88, 166, 173
		InAsP		167
		GaInAs		169
	In	GaN		205–206
		GaInN		207, 210, 211
		GaInAsN		203
		AlGaInN		215–218
Increased step velocity	Sb, Bi, Te	GaAs, GaInP		88, 136, 172–174
Smaller, denser S–K islands	Sb	GaInN		212
Reduction in ordering	Sb, Bi, As, Te	GaInP	Increase in bandgap energy	86, 88, 96, 186
				86
	Sb, Bi		O/D heterostructures	166, 195–197
	Sb	GaAs	Improved tandem solar cells	92, 189, 190
	Sb			
	Te		Improved LEDs	183
Increased PL intensity	Sb	GaInP		203
	Sb	AlGaInN		210, 211
Altered solid composition	Sb	GaInN		188, 202, 207, 210, 212
	Sb	GaInAs		199
Increased Zn doping	Sb, Bi	GaP, GaInP	Record highest level	166, 198
		GaInAs	Record highest level	199
		GaAs		201
Decreased C contamination	Sb, Bi	GaP, GaInP	>10× reduction	198
Decreased O contamination	Sb	GaInP		198
Increased H incorporation	Sb	GaP, GaInP		198

We would expect that at high surface coverages, surfactants would change the surface reconstruction, with attendant changes in the energetics and dynamics of MOVPE growth. Changes in reconstruction due to surfactants were first observed in both elemental and III/V semiconductors grown by MBE [159]. First-principles total-energy calculations indicate that the surfactant Sb changes the surface reconstructions of InP and GaP [160].

At very high levels, Sb produces a change in the surface reconstruction of GaInP from the commonly observed (2×4)-like to a (2×3)-like reconstruction. This results in formation of a new, triple-period ordered structure. [161]. Similar results were reported for MOVPE-grown GaAsSb alloys. The addition of Bi was shown to increase the typically observed CuAu and chalcopyrite ordering [162], due to a change from the (4×4) to the (2×3) reconstructed surface. These ordered structures have a three-fold periodicity in the [110] direction. Laukkanen et al. [163] found for (001) GaInAs that Bi stabilizes the (2×1) and sometimes the (2×4) reconstruction. Further examples of the close correlation between surface reconstruction and the ordered structure produced during epitaxial growth are given in a review article by Norman [164].

Surfactants would also be expected to accumulate at step edges and, consequently, change the energetics of steps as well as the dynamics of adatom attachment at step edges [155]. The calculations of Batyrev et al. [165] indicate that Sb will preferentially attach at surface steps, changing the step structure and disrupting the ordering processes, as discussed later.

Early work on the use of surfactants during MOVPE growth showed that the addition of Te during GaAs growth leads to smoother surfaces by the elimination of step bunching [145]. This was attributed to enhanced adatom attachment at steps. Monte Carlo simulations support this interpretation [136]. Smoothing was also reported for Te in GaInP grown by MOVPE [166]. Similar effects are reported for Te-doped InPAs alloys [167].

The use of surfactants also significantly alters the MOVPE growth of GaAs. Sb and Bi both result in a significant (3×) increase in the [110] step velocity [136]. Kinetic simulations indicate that this is due to an increase in the frequency of [110] diffusion events due to a decreased hop barrier. The addition of Te (which appears to function as a surfactant as well as an n-type dopant) was found to result in a significant smoothing of GaAs layers. The rms roughness was reported to decrease by an order of magnitude [145]. Sb surfactant was found to suppress three-dimensional growth for InGaAs QW lasers [169].

The use of surfactants is perhaps the most useful and interesting method of controlling the degree of CuPt order in GaInP grown by MOVPE. Several studies of GaInP grown by MOVPE have demonstrated a connection between ordering and n-type [170–174] or p-type [175–181] dopant concentration. The results show that a drastic decrease in ordering is caused by introducing a high concentration of dopants during growth. The effect for some impurities has been attributed to diffusion in the bulk [177, 180–183]. However, recent data for Te added during MOVPE growth of GaInP indicate that the disordering is due to a surfactant effect: an increase in the [$\bar{1}$10] step velocity [172–174]. The effect would be similar to that of an increased growth rate, as described in Section 2.2.4.2. The surfactant Te has also been seen to reduce ordering in GaAsP alloys [174].

The most important surfactant for the MOVPE growth of III/V materials is Sb, which has been shown to strongly affect the ordering in GaInP. Since it is isoelectronic with P, no change in the Fermi level position at the growth temperature is expected. In addition, because it is much larger than P, the solubility is small [184], as discussed in Section 2.2.3 [22]. Thus, it is likely to accumulate at the surface during growth. The addition of tiny amounts of TESb in the input vapor produces a marked increase in the low-temperature PL peak energy [85, 86, 96]. The results indicate that the GaInP layers are highly ordered without Sb and essentially disordered when grown using the same parameters, but with addition of a small amount of Sb. Figure 2.6 includes the SPA anisotropy intensity at 400 nm for several TESb flow rates during growth. The results show a clear reduction in the intensity of the peak near 400 nm in the SPA anisotropy spectra, as well as the degree of CuPt order, when Sb is added to the system during the growth of $Ga_{0.52}In_{0.48}P$ layers. The addition of Sb causes a decrease in the concentration of $[\bar{1}10]$ P dimers on the surface [85, 96], resulting in less ordering and higher bandgap energies. This is thought to be the first experimental evidence that an isoelectronic dopant, in this case Sb, can act as a surfactant during MOVPE growth to produce major changes in critical properties.

Since the $[\bar{1}10]$ P dimers produce the surface thermodynamic driving force for formation of the CuPt structure during growth, it is not surprising that the removal of $[\bar{1}10]$ P dimers by the addition of Sb eliminates ordering. This has been supported by the *ab initio* total-energy calculations of Chakrabarti and Kunc [185], which demonstrate that adsorbed Sb and Bi both result in less strain in subsurface GaInAs and GaInP layers, leading to a reduction in the amount of CuPt ordering. In fact, the degree of order versus 400-nm SPA anisotropy signal for the $Ga_{0.52}In_{0.48}P$ layers grown using Sb fall very near the data discussed previously for the effects of temperature and flow rate of the P precursor, as clearly demonstrated in Figure 2.6. An increase in bandgap energy of 135 meV is reported due to the addition of TESb during MOVPE growth [86]. We observe that the surfactant Bi produces effects similar to those of Sb. Jun et al. [186] also found that Bi yields an order of magnitude increase in step velocities and gives smoother GaInP surfaces. It has also been found to reduce the number of $[\bar{1}10]$ P dimers on the surface. Bi has been used to produce a 110-meV increase in the bandgap energy of partially ordered GaInP [186].

The addition of another Group V element, As, has been shown to produce similar reductions in ordering with attendant reductions in the GaInP bandgap energy [96]. Even though As is readily incorporated into GaInP, it has been shown to produce an independent decrease in order parameter by replacing P dimers on the surface.

Nitrogen could also be considered a surfactant in GaInP, since the solubility is even lower than in GaP (see Section 2.2.3). The addition of high concentrations of DMHy during the MOVPE growth of GaInP has been found to result in a decrease in order, due to the monatomic form of N on the surface [188]. It has also been observed to reduce In incorporation [191].

As discussed previously, Batyrev et al. [165] suggested that the amount of Sb on the surface necessary to reduce ordering is insufficient to cover the entire surface. They suggest, supported by first-principles calculations, that Sb adsorbed at step edges induces changes in the step structures that disrupt the ordering process. As discussed previously, step motion can be dramatically affected by surfactants. This leads to kinetic effects on ordering [96] similar to those caused by changes in growth rate and substrate misorientation, discussed in Section 2.2.4.2. Clearly, both thermodynamic and kinetic effects may lead to a decrease in the degree of order.

The ability to control ordering and, thus, the bandgap energy of GaInP layers by the addition of Sb allows the MOVPE growth of unique heterostructures. Since the addition of a surfactant such as Sb or Bi is found to completely eliminate ordering [96], as discussed previously, by simply adding the surfactant during MOVPE growth, the bandgap of $Ga_{0.5}In_{0.5}P$, lattice matched to the GaAs substrate, can be reduced markedly. This provides the opportunity to produce heterostructures and QWs without any change of solid composition. Figure 2.11 shows the PL spectrum for an O/D heterostructure grown in this way using surfactant Sb [86]. The remarkable 135 meV difference in the bandgap energy between the two layers is more than 5 kT at room temperature, which is sufficient for many heterostructure devices.

These results suggest a new and powerful concept: determination of the surface reconstruction during growth, using a surfactant, as a method of controlling the properties of the resultant epitaxial layer. This has been clearly demonstrated for the control of the bandgap energy in GaInP. However, the implications may be much broader. In fact, the use of surfactants has been shown to affect the growth mode and step velocity, and hence the morphology, as well as solid composition and incorporation of both intentional and background dopants, as described later.

Control of the bandgap energy is vital for red, orange, and yellow LEDs fabricated in GaInP. As an example, LEDs produced in disordered InGaP lattice matched to GaAs emit red light, whereas

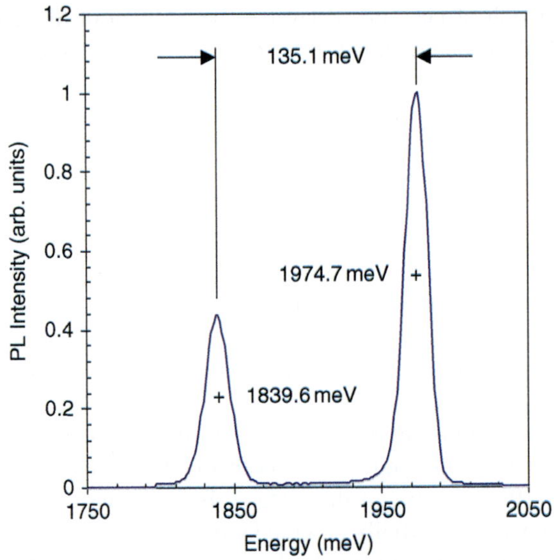

Figure 2.11 *20 K PL for disorder on order heterostructure produced by using a small concentration of surfactant Sb to produce the top disordered layer. The PL from the bottom (without Sb, ordered) layer occurs at 1.840 eV and that from the top, surface (with Sb, disordered) layer occurs at 1.975 eV. Source: After Shurtleff et al. 1999 [86]; reprinted from Applied Physics Letters with permission.*

those produced in ordered material produce infrared radiation. Thus, ordering must be eliminated in the materials used for the fabrication of these LEDs. Additionally, Lee et al. [189] used DOD structures in GaInP LEDs to increase the light intensity by a factor of 7× as compared to homojunction LEDs. Christian et al. [190] produced amber-green LEDs in AlInP grown mismatched to GaAs substrates. They used a change in growth temperature from 650 to 725 °C to create a heterostructure for electron confinement, which resulted in a threefold increase in light output. The combination of alloy composition, strain, and ordering may be useful for decreasing the wavelength in GaInP LEDs [191].

Ge/GaAs/GaInP multijunction solar cells have reached efficiencies exceeding 40% [192, 193]. For these cells, it is important, from current-matching considerations, to have the bandgap of the GaInP subcell as large as possible [194]. Thus, ordering must be eliminated. The best current solar cells use Sb to disorder the GaInP [166, 195–197].

For GaP grown by MOVPE, both Sb and Bi have been found to have major effects on dopant incorporation. Several groups have found that both surfactants increased the incorporation of Zn intentionally added for p-type doping during the MOVPE growth of both GaInP and GaP [166, 198]. Sb also increased the incorporation of Zn in GaInAs grown by MOVPE. At a growth temperature of 500 °C, the addition of surfactant Sb caused the Zn doping level to increase from $2\times10^{19}\,cm^{-3}$ to $6.5\times10^{19}\,cm^{-3}$, the highest level ever reported [199].

The increase of Zn doping caused by Sb in GaP and GaInP is accompanied by an increase in the level of H incorporated [198]. This has been interpreted, based on first-principles calculations, as being due to a dual surfactant effect involving both Sb and H on the surface [200]. However, the increase in Zn doping reported for GaInAs [199] is not accompanied by an increase in H incorporation. The possibility of these effects occurring due to surfactants at step edges has also been considered [198]. The addition of either Sb or Bi during the MOVPE growth of GaP and GaInP also produces a significant reduction in the incorporation of unintentional carbon and oxygen [198].

Sb also results in increases in both Zn and background In incorporation in GaAs grown by MOVPE [201]. However, Sb, Bi, and Tl were found to decrease N incorporation [202]. This was interpreted, using the simple Langmuir model, in terms of competition for surface sites.

Chen et al. [203] studied the effects of Sb on the growth of GaInAsN QW structures emitting at 1265 nm. Sb was found to improve PL intensity and reduce the peak half-width. As for other materials, discussed previously, it also smooths the surface, suppressing the formation of islands. Sb was also found to suppress localization effects due to fluctuations in alloy composition, as described in the last section.

As discussed in more detail in Section 2.5.3.3, the addition of Bi to GaAs to lower the bandgap energy is a topic of current interest. Since Bi has a limited solubility in GaAs, due to its large size, Bi tends to accumulate at the surface and act as a surfactant. Forghani et al. [204] found that for high Bi concentrations in the vapor during MOVPE growth, Bi acts to reduce the growth rate. This was attributed to the blocking of surface sites by $Bi(CH_3)_x$, which prevents Ga adsorption. At low TEBi vapor concentrations, no reduction in growth rate was observed.

The use of surfactants Sb and Bi during MOVPE growth of GaN gives results similar to those for other systems, summarized previously: i.e. their use resulted in smoother surfaces [205–207]. First-principles DFT calculations indicate that N diffusion on the surface limits lateral growth. The addition of surface Sb is found to produce intermediate SbN species that have much lower activation barriers, leading to more rapid N transport on the surface [206]. As expected from the discussion of thermodynamics in Section 2.2.3, both elements are rejected from the solid because they are so much larger than the GaN host [208]. A dramatic illustration of the importance of Sb for determination of the surface kinetic processes is the marked effect on epitaxial lateral overgrowth (ELO), where GaN is grown in holes on a patterned surface in order to reduce the dislocation density, which is otherwise extremely high in GaN grown on sapphire substrates. The presence of Sb was found to enhance the lateral overgrowth rate. It also strikingly altered the dominant facets formed [209].

For GaInN layers grown by MOVPE, the addition of surfactant Sb leads to smoothing of the layer [210, 211], as for many other materials. Merrell et al. [212] found that above a certain critical Sb concentration, the S–K islands formed on the surface become smaller and denser. This was interpreted as possibly being due to a change in surface reconstruction at high Sb levels. This is superficially similar to the results for GaInP and GaAsSb described previously, which were also postulated to be due to an Sb-induced change in surface reconstruction during growth. Improved morphology plus the suppression of V-pits has also been reported for thick GaInN layers [207]. They also interpreted these results in terms of the change in surface reconstruction at high Sb concentrations leading to a change in the growth mode.

The improvements in surface morphology due to Sb have been correlated with improvements in optical and structural properties. Sadasivam et al. reported that the addition of Sb resulted in a 3.3-fold increase in green PL intensity [210]. Baranowski et al. [211] also noted increased PL intensities as well as longer PL lifetimes for InGaN/GaN MQW structures used in LEDs.

Merrell et al. [212] also reported an abrupt increase in In incorporation above a specific Sb concentration. First-principles DFT calculations predict an increase in In incorporation at high Sb concentrations, ascribed to tensile sites in the cation layer produced by adsorbed Sb that are energetically favorable for incorporation of the larger In atoms [213]. However, Koch et al. [207] and Sadasivam et al. [210] reported reductions in In concentration at high Sb concentrations. Baranowski et al. [211] also reported a reduction in In incorporation for high Sb concentrations. This was attributed to enhanced In desorption due to the weak Sb–In bonds.

The surfactant Sb plays a major role in LED performance in InGaP, due to ordering, as described previously. Unlike the InGaP alloys, ordering does not appear to be a major factor determining the properties, in particular the bandgap energy, of InGaN, although it has been reported that it may adversely affect LED performance [214].

Indium, itself, has also been investigated as a surfactant during the MOVPE growth of GaN. Both Shu et al. [215] and Keller et al. [216] report altered morphologies with small amounts (<0.2%) of In incorporation. Keller et al. [216] reported reduced defect densities in AlN films, again with small amounts of In incorporation. As discussed in Section 2.2.3, In tends to be rejected from both GaN and AlN, due to its relatively large size. First-principles analysis of GaN grown on the c-plane suggests that In can diffuse with a low barrier in In-covered (0001) GaN grown by MOVPE, leading to improved morphologies and a reduction in the number of extended defects [217]. For AlGaN/GaN high-electron-mobility transistor structures, Feng et al. [218] report a change in growth mode to two-dimensional layer-by-layer growth due to surfactant In, which accounts for a decrease in surface roughness that leads to a decrease in electron scattering.

2.5 Specific Systems

The basic concepts discussed in Sections 2.2–2.4 govern the MOVPE growth of all materials. In this section, we will concentrate on relatively new systems, in an attempt to augment the discussions of more common systems found in references [1–15]. We will attempt to use these concepts to understand the MOVPE growth characteristics as well as the properties of the solid, including both the solid composition and the microstructure.

2.5.1 AlGaInP

As discussed in Section 2.2.2, non-nitride materials for high-efficiency LEDs must be lattice matched to the substrate. Otherwise, the dislocations generated kill the recombination efficiency. One of the early successful alloy systems that can be lattice matched to GaAs substrates is AlGaInP, where LED emission energies from the infrared to green are possible. The growth of alloys, heterostructures, and QWs lattice matched to the GaAs substrate have all been extensively developed. However, as discussed in Section 2.2.2, these alloys containing Al and In together can be effectively grown only by either MOVPE or MBE, and not by LPE or VPE. The best red AlGaInP double-heterostructure LEDs, grown by MOVPE, the sole commercial process, have external quantum efficiencies exceeding 50% [13, 78].

This alloy system is also very important for fabrication of the highest-performance multijunction solar cells, as discussed in Section 2.4.4. The GaInP system has a value of critical temperature, listed in Table 2.1, near typical MOVPE growth temperatures. Thus, phase separation would not be expected to have a major effect on the properties of the alloys produced for LEDs. However, evidence of compositional fluctuations has been reported [219]. An important feature of the microstructure of GaInP alloys is the low dislocation density, due to the lattice match to the GaAs (or Ge) substrates. Dislocations are very deleterious to the performance of all minority-carrier devices, including solar cells.

In disordered material, the GaInP LEDs emit red light. However, as discussed in detail in Section 2.2.4.2, the light emitted can have a much longer wavelength due to the spontaneous formation of the CuPt ordered structure. Thus, ordering must be avoided in both solar cells and LEDs. In some cases, the growth parameters and substrate misorientation are chosen to minimize ordering, as discussed in Section 2.4.3.1. Often, a more attractive alternative is to use the surfactant Sb, as discussed in Section 2.4.4. This is the approach adopted for commercial high-efficiency, multijunction solar cells [166, 195, 196]. Similarly, the use of ordering to produce the order/disorder heterostructures described in Sections 2.2.4.2 and 2.4.4 has been used to improve, by a factor of three, the light output of amber-green LEDs (566 to 600 nm emission wavelengths) in AlInP alloys grown on GaAs substrates [190].

2.5.2 Group III Nitrides

The use of surfactants Sb and Bi during MOVPE growth of GaN gives results similar to those for other systems, summarized previously: i.e. their use produces smoother surfaces [205–207]. As expected from the discussion of thermodynamics in Section 2.2.3, both elements are rejected from the solid because they are so much larger than the GaN host [220].

The MOVPE growth of InGaN on GaN (grown on sapphire substrates) shows pronounced lattice pulling [221]. In Figure 2.12, the In content of layers grown coherently on GaN is compared with that for thick layers where the strain has been relaxed. The In content for the relaxed layers is found to exceed that for the strained layers by factors of between two and three. At a growth temperature of 740 °C, for example, the In content of the strained layer is 10% and that for the relaxed layer is approximately 30%. The compositional pulling effect in GaInN alloys has recently been explained theoretically by Inatomi et al. [222].

As discussed in Section 2.2.3, in recent years AlGaInN alloys grown by MOVPE have become the dominant materials used for very-high-efficiency blue and green LEDs. The blue LEDs, when coated by a photon downconverting phosphor, are used to produce the white LEDs available commercially. For GaInN alloys, Table 2.1 shows that the value of the solid interaction parameter is fairly large. This leads to a miscibility gap predicted to extend to alloys with $x > 6\%$ InN at typical MOVPE growth temperatures. The solid phase diagram calculated using the value of enthalpy of mixing from the modified VFF model is shown in Figure 2.5. The calculated phase diagram was originally published in 1996 [24], before experimental data were available. As shown in this figure, a collection of representative experimental data published subsequently clearly shows the occurrence of spinodal decomposition and phase separation. Section 2.4.3.2 summarizes the analysis of the compositional fluctuations from the point of view of surface processes. For (0001)-oriented substrates, S–K growth appears to allow incipient phase separation even in thin strained GaInN layers. The detailed growth conditions determine the growth mode and the extent of kinetically limited compositional clustering, possibly leading to a wide range of experimental results for the magnitude and scale of the compositional modulations.

Compositional clustering due to the AlGaInN miscibility gap has been verified in many hundreds of studies using techniques such as transmission electron microscopy (TEM), Raman scattering,

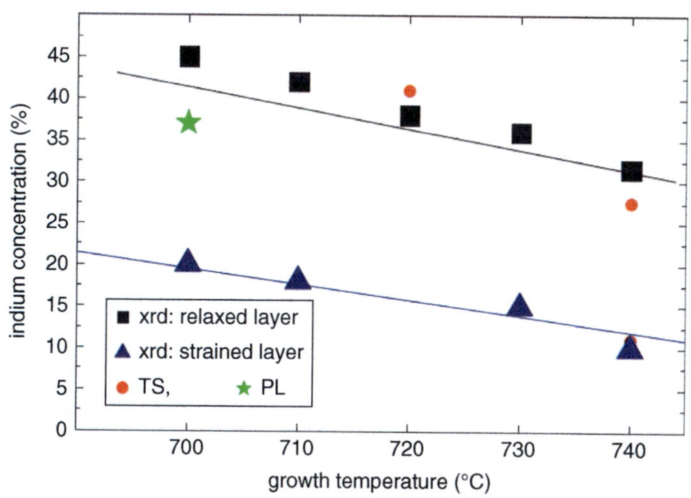

Figure 2.12 *Illustration of lattice pulling in GaInN grown on GaN by MOVPE. Source: After Leyer et al. 2008 [221]; reprinted from the Journal of Crystal Growth with permission from Elsevier Science.*

and detailed studies of the time and temperature dependence of absorption and photoluminescence spectra (see [52] and [152] for recent reviews). Perhaps most impressive are the results of atomic probe tomography (APT) and TEM studies of 3-nm QW layers with In concentrations of 0.08, 0.15, and 0.25, as reproduced in Figure 2.10. As discussed in Section 2.2.3, the APT results clearly show large In fluctuations, from $\Delta x_{In}=0.2$ to 0.3 in the layer with a nominal In content of 0.25, with a period of 3–5 nm [151]. The wavelength of the fluctuations is classically due to diffusion. The temperature dependence of the periodicity yields an activation energy of 2.2 eV, presumably for diffusion in the solid at or near the surface. It is worth repeating here that strain and kinetic factors, particularly the growth temperature, play a large role in the amount of clustering observed experimentally. This may account for a few reports of the absence of compositional clustering in excess of that seen in random alloys [222].

It is worth repeating here that strain and kinetic factors, particularly the surface reconstruction and step structure as well as the growth temperature, play a large role in the amount of clustering observed experimentally. This may account for a few reports where experimental data is interpreted to indicate the absence of compositional clustering in excess of that seen in random alloys [223]. A demonstration of the importance of surface processes in the occurrence of phase separation is the observation of Humphries et al. [224] that for GaInN grown by MOVPE, no phase separation is observed for material grown on (0001) substrates, but clear evidence of compositional clustering to form In-rich regions a few nanometers in diameter is seen for GaInN grown on non-polar $(11\bar{2}0)$-oriented substrates. Of course, the surface reconstruction and microscopic surface morphology will depend on the growing surface orientation. The authors suggest that this may be due to the lower growth temperatures used for the nonpolar substrates, although this would increase the thermodynamic driving force for phase separation. Surface kinetic effects including those due to the different surface reconstructions, surface morphologies, and the lower growth temperature may be important.

Unexpectedly, the presence of compositional inhomogeneities due to solid immiscibility has been widely observed to result in a significant improvement in the performance of blue and green LEDs [52]. The AlGaInN alloys grown on highly mismatched sapphire substrates have a very high dislocation density, in the range of 10^9–10^{10} cm^{-2}. As discussed in Section 2.2.2, such a high dislocation density would absolutely kill the radiative recombination efficiency in any other III/V alloy studied [52]. The cause of this apparently aberrant behavior appears to be that the compositional fluctuations give rise to low-bandgap, In-rich regions that trap the minority carriers where they recombine radiatively before they can diffuse to the harmful dislocations where they would recombine nonradiatively [52, 223, 227]. At a dislocation density of 10^9 cm^{-2}, the average spacing between dislocations is roughly 300 nm, whereas the spacing between In-rich clusters is only of the order of 5 nm, as seen in Figure 2.10. Thus, the probability of nonradiative recombination at the dislocations is very small. Figure 2.13 shows a plot of LED efficiency versus the In content of the GaInN alloy [52]. For GaN and alloys with low In concentrations, the external efficiency is very small. The efficiency increases by orders of magnitude as In is added to the system. For the higher In concentrations required for blue LEDs, efficiencies as high as 80% have been reported [224]. This is an amazing number for LEDs with such extraordinarily high dislocation densities.

An additional feature of Figure 2.13 is the decrease in blue LED efficiency seen for alloys with the higher In contents used for green, yellow, and red LEDs. Recent studies of MQW LEDs grown on (0001) substrates have reported improvements leading to the best red LEDs to date [229, 230], with quantum efficiencies of 2.9% [230]. However, the efficiency is still much lower than for blue LEDs.

The widely observed decrease in LED efficiency as the wavelength increases into the yellow and red regions of the spectrum is generally attributed to the increase in mismatch strain energy and the decrease in crystal quality, without specific identification of the nonradiative defects. A thorough review of these high-In alloys has recently been published by Damilano and Gil [228].

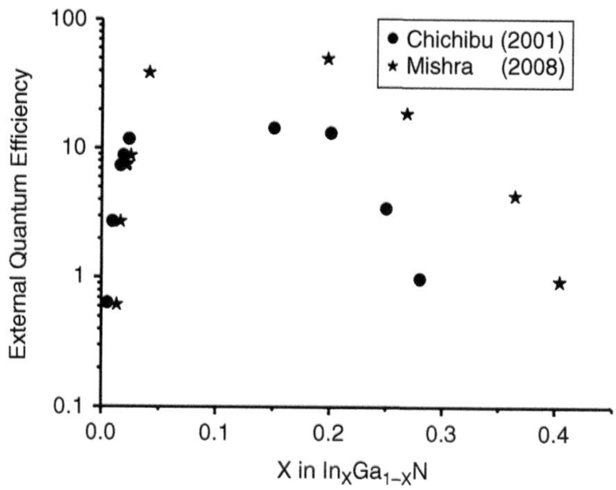

Figure 2.13 *External quantum efficiency of InGaN MQW LEDs versus x_{In}. The data points are from Mishra 2008 [226] and Chichibu et al. 2001 [227]. Source: After Stringfellow 2010 [52]; reprinted from the Journal of Crystal Growth with permission from Elsevier Science.*

It is generally concluded that reducing the amount of phase separation, in this case the formation of two separate phases, in high-In-content alloys produces higher-quality materials, more suitable for LEDs. This can be achieved by kinetic hindrance, frustrating the thermodynamic driving force for clustering and phase separation. Studies of the effects of kinetic parameters have shown that lower growth temperatures and higher growth rates reduce the compositional clustering and phase separation in high-x $Ga_{1-x}In_xN$ alloys. This is expected, since both factors will allow less redistribution of the In and Ga atoms at the growth front. For the growth of alloys with $x=0.65$ by MBE, Pantha et al. [231] found, mainly based on X-ray diffraction studies, that growth rates exceeding 0.5 μm/h were necessary to produce homogeneous and single-phase alloys. They also found higher electron mobilities for samples grown using the higher growth rates. Ju et al. [232] reported that MBE growth at low temperatures of 450–550 °C produced layers with smaller PL half-widths and higher PL intensities than those grown at 600–630 °C. They interpreted their results in terms of lower carrier localization energies due to In clustering for the layers grown at the lower temperatures. Papadomanolaki et al. [233] found that for thick layers with high In concentrations grown by MBE, low growth temperatures of 400 °C yielded single-phase materials with very high electron mobilities. Higher growth temperatures yielded alloys exhibiting phase separation. These combined results generally agree with those for other, even more metastable alloys, as discussed later. They suggest the use of low growth temperatures for the MOVPE growth of GaInN with high InN mole fractions. However, access to the very low growth temperature regime is limited in MOVPE due to the incomplete pyrolysis of the common N precursors, particularly NH_3. Recent work by Sterzer et al. [234] using novel more labile N precursors reported MOVPE growth of GaInN at temperatures as low as 490 °C.

The opposite approach has also been reported to be successful. An interesting parallel to the observations for the low In content GaInN alloys, described previously, is that controlling the growth process to *increase* the amount of In clustering yields higher efficiencies for yellow-green LEDs grown by MOVPE [235]. The main radiative recombination pathway for green InGaN/GaN MQW structures has been found to be due to excitons localized at In-rich regions due to compositional fluctuations [236]. This prevents nonradiative recombination at dislocations. They also found that fluctuations of QW thickness give smaller bandgap fluctuations, which are ineffective in trapping excitons at room temperature.

New insights into the effects of spinodal decomposition on the recombination efficiency have been obtained from recent studies of nanowires. Murotani et al. [237] studied GaInN nanowires grown by MBE over a wide range of InN concentrations. They reported that as the InN content increased, carrier localization was observed, and that the localization energy increased due to the reduction in bandgap in the high-In clusters formed due to spinodal decomposition. Very significantly, this resulted in high internal quantum efficiencies across the entire range of InN concentrations. Presumably, the increase in compositional fluctuations leading to exciton localization in the nanowires is due to the relaxation of the mismatch strain in nanowires. They concluded from PL results that the minority carriers migrate to the regions of highest In content, where they recombine, entirely consistent with the results discussed previously for thin GaInN layers with lower InN contents. This is a very interesting approach to the fabrication of LEDs covering the entire visible spectrum. It would not only ameliorate the "green gap" problem, but, perhaps, lead to efficient amber and, ultimately, red LEDs made from nanowires in the AlGaInN alloy system.

Further clarification of phase separation was provided by Cherns et al. [238], who reported that $In_{0.5}Ga_{0.5}N$ nanowires grown by MBE spontaneously produce a core/shell structure. In the core, compositional variations due to spinodal decomposition were observed using TEM techniques. They observed In-rich ($x=0.7$) and In-poor ($x=0.3$) platelets formed by phase separation. Again, the use of nanorods is significant here, because they are elastically relaxed.

The use of the microstructure produced spontaneously during epitaxial growth offers a promising approach for the production of white LEDs without the need for a downconverting phosphor [239]. Compositional clustering and phase separation in GaInN QWs can be used to produce In-rich regions producing yellow emission. Combined with the blue emission from the matrix, this produces light that appears white. This is promising because the phosphor inherently wastes energy in the downconversion process. The phosphor may also be the most rapidly degrading component of current white LEDs. Approaches to controlling the spinodal decomposition to enhance the formation of the In-rich areas include surface roughening and the use of underlying layers to reduce strain [239].

Another III/V semiconductor alloy predicted to have a miscibility gap is InAlAsSb, potentially useful for high-performance multijunction solar cells. Careful studies using quantitative scanning-transmission electron microscopy clearly show the presence of nonrandom compositional fluctuations [240].

2.5.3 Novel Alloys

A number of studies have explored the addition of nontraditional isoelectronic elements such as Tl [241], Bi [205, 242], N [20, 21, 234], and B [243] to the more familiar III/V compounds in order to produce improved materials for optoelectronic devices. All of these elements have limited solubilities due to the size difference between them and the common Group III and Group V elements that they replace in the lattice. Thus, the successful MOVPE growth of these alloys must typically rely on kinetic factors, particularly low growth temperatures and high growth rates, to allow the production of single-phase, metastable alloys. Two of the more interesting systems currently being explored are GaAs: N and GaAs: Bi.

The drive to reach an efficiency exceeding 50% for multijunction solar cells has concentrated on the optimization of the bandgaps of the several p/n junctions [192–194]. Theoretical studies indicate that one promising approach is to replace GaAs by a lower-bandgap material with the same lattice constant [14]. The addition of a few percent N to GaAs results in a reduced bandgap energy: 1% N reduces the bandgap by nearly 200 meV [244]. A larger element, such as In, could also be added to compensate for the reduction in lattice constant. $GaAs_{0.8}N_{0.2}$ alloys are also potentially useful for semiconductor devices, such as lasers, lattice matched to Si substrates [245], thus enabling Si-based optoelectronic devices and photonic integrated circuits.

2.5.3.1 GaAsN

As seen in Table 2.1, the critical temperature for GaAsN alloys is nearly 10,000 °C, indicating a very low solubility of N in GaAs [20, 21]. Zhang and Wei [246] have pointed out that the strain energy of a coherent GaN-rich second phase on a GaAs substrate during epitaxial growth will dramatically increase the thermodynamic limit of N incorporation in GaAs to approximately 4% at 650 °C. Neugebauer and Van de Walle [27], using first-principles total-energy calculations, determined a lower limit for the critical temperature in this system of 4611 K, a factor of two lower than predicted from the DLP model, as seen in Table 2.1. Experimentally, low-temperature MBE [247] has been used to produce epitaxial layers with up to 15% N. However, even low N concentrations (<1%) are reported by Wang et al. [248] to lead to increased electron scattering and decreased PL intensity for samples grown by CBE at 440 °C. Wang et al. attributed this to a spatially fluctuating N content, perhaps due to the miscibility gap. Low-temperature MOVPE has been used to produce layers with 5% N [234] using the novel N precursor di-tertiary-butyl-arsino-amine (DTBAA). Temperatures of >490 °C resulted in a rapid reduction in N incorporation. The same group [249] has suggested that compressive strain from the Si substrate enhances N incorporation due to the lattice-pulling effect discussed in Section 2.2. This strain is also calculated to stabilize the single-phase alloy for N concentrations of 15%.

A number of experimental observations indicate that compositional nonuniformities and phase segregation occur under some conditions. Almosni et al. [250] used low-temperature MBE to grow GaAsPN lattice matched to Si substrates. After rapid thermal annealing, the temperature-dependent PL results showed that nitrogen aggregates to create cluster states. Levander et al. [251] were able to grow amorphous layers with compositions spanning the complete GaAsN composition range by MBE at 200 °C. Annealing produced complete phase segregation into GaAs:N and GaN:As regions, a clear indication of the large miscibility gap.

On the other hand, Beaudry et al. [252] grew GaAsN by MOVPE at temperatures ranging from 500–650 °C. They observed a uniform N distribution for compositions as high as 3.75%. Attempts to produce layers with higher N concentrations resulted in layers exhibiting phase separation. Ludwig et al. [245] used OMVPE at temperatures from 525–700 °C on Si substrates to produce QW layers with up to 17% N having uniform compositions with no signs of phase separation. Thomas et al. [253] grew GaAsN layers by CBE at 500 °C. Layers with 4% N showed no evidence of phase separation. At a nitrogen concentration of 10%, the layers exhibited phase separation.

The extremely high critical temperature for these alloys, seen in Table 2.1, suggests that compositionally uniform, high-quality GaAsN layers, suitable for minority carrier devices, may be difficult to grow. However, as discussed in Section 2.2.3, growth on a substrate having a smaller lattice constant will stabilize the lattice-matched alloy against spinodal decomposition, even though the critical temperature is quite high. Wegele et al. [254] performed DFT calculations that show that alloys lattice matched to Si substrates are stable. This is consistent with the experimental observations of GaAsN layers grown by MOVPE reported by Ludewig et al. [245]. An additional factor may be kinetics, which can apparently be used to produce metastable single-phase alloys having a uniform composition.

2.5.3.2 GaPN

The GaPN alloys have received significant interest in recent years, because the addition of N to GaP changes the band structure from indirect to direct [255]. Small concentrations of N lead to a reduction in the bandgap energy and a decrease in lattice constant. This makes this material a candidate for visible LEDs [256]. Because of the large miscibility gap, exhibited in Figure 2.1, nonequilibrium conditions, in particular low growth temperatures, have been used during gas-source MBE growth to produce N concentrations as high as 3.1% [256]. The ability to grow these materials on Si substrates could provide an economic incentive for such LEDs as well as for lasers for on-chip optical links [245]. Nanowires of these alloys have also been explored [257].

An interesting observation by Kunnert et al. [45] is that the N incorporation in GaNP grown by MOVPE is 1.6 times more efficient when depositing on Si substrates than on GaP. This might be attributed to lattice pulling, but they attribute it to strain-induced desorption.

The effect of lattice pulling is clearly demonstrated for the MOVPE growth of InSbN alloys, used for long-wavelength IR detectors [269].

2.5.3.3 GaAsBi

Like N, Bi addition also reduces the GaAs bandgap, in this case by 88 meV/%Bi [258]. GaAsBi alloys are being explored for high-efficiency solar cell and other device applications. Bi was earlier added to InAs and InAsSb alloys in an effort to produce IR detectors functioning in the atmospheric window at 12 μm [242]. For GaAsBi, the DLP model gives a value of the critical temperature of 1800 K, as seen in Table 2.1. Bennarndt et al. [258] have used the DLP model to calculate the binodal line in the phase diagram and determined a Bi solubility of roughly 10% at 500 °C. Similar results were obtained using ab initio calculations including strain effects, but only when the fictitious GaBi phase is used as the reference state [259], which is implicitly the same assumption as is used for the DLP calculations. A systematic agreement between DFT and DLP calculations has been found for a number of Bi-containing alloys [31]. As expected from the DLP model, and confirmed by DFT calculations [259], coherent growth on a GaAs substrate strained to increase the lattice constant is found to produce a significant increase in the Bi solubility.

Using low-temperature (270 °C) MBE, GaAsBi alloys with Bi concentrations as large as 20% were reported by Bennarndt et al. [258], with no signs of spinodal decomposition. This appears to be an example of the suppression of decomposition using kinetic effects. Temperatures of greater than 330 °C resulted in rejection of Bi from the solid. At 420 °C, no Bi was incorporated. No Bi droplets were observed, so the Bi was assumed to simply evaporate from the surface. Alloys with 32% Bi were grown and were found to show clear evidence of two phases. MOVPE GaAsBi growth studies have been made of the effects of growth temperature [260, 261], the precursors [133], and the ratio of precursors in the vapor [260, 261]. In general, the Bi distribution coefficient is found to be constant at low Bi precursor flow rates. Ludwig et al. [260] reported a Bi distribution coefficient of nearly unity. The Bi concentration in the solid saturates at approximately 3–8%, depending on growth conditions. Ludwig et al. [260] reported degradation in the properties at high Bi concentrations, possibly due to Bi clustering.

Natterman et al. [262] codoped GaAsBi with a Group V element smaller than As to compensate for the local strain induced by the large Bi atoms, which is expected to increase the solubility [31]. Thus, in the alloys Ga(NAsBi) and Ga(PAsBi) grown by MOVPE at 400 °C, the Bi content at which the Bi distribution coefficient begins to decrease, i.e. the Bi content in the solid saturates, was found to increase with the introduction of either P or N into the solid. This is consistent with the effect of strain discussed in Section 2.2.3: i.e. strain in the lattice due to Bi incorporation decreases the solubility limit.

The more conventional technique of increasing the substrate lattice parameter is also expected to increase the Bi solubility in GaAs, as discussed in Section 2.2.3. Guan and coworkers [31] calculated the Bi solubility including the strain energy in DLP phase diagram calculations. They studied the pseudomorphic growth of InGaPB alloys on several different substrates and concluded, as expected from earlier discussions, that coherency strain stabilizes the alloy against spinodal decomposition under normal growth conditions. As discussed previously, this conclusion is similar to that reached by Jacobsen et al. [259] using DFT calculations for GaAsBi grown on tensile strained GaAs substrates.

Kuech et al. [263] suggested a specific surface reaction mechanism for the suppression of compositional clustering in GaAsBi alloys with compositions deep within the miscibility gap during low-temperature MOVPE growth of these alloys using TMBi as the Bi precursor. At the low temperatures used, the weak Bi–methyl bonds will break. This will provide a supply of methyl radicals that can adsorb, providing a high surface coverage at the low temperatures employed. The adsorbed methyl radicals will decrease the surface diffusion coefficients of the Group III elements, thus slowing the rate of atomic rearrangement to produce the compositional fluctuations. This will result in a "frozen-in," uniform solid composition even though thermodynamic factors drive compositional clustering.

Very recent results by Luna et al. [264] clearly show the formation of uniformly spaced Bi-rich regions approximately 5–6 nm on a side in quantum wells (QWs). The Bi content was seen to vary from 3.5–6% in layers with an average Bi concentration of 4.9% grown by MBE at temperatures of from 220 to 370 °C. TEM cross sections are similar to those observed for GaInN QWs, as discussed previously.

2.5.3.4 BGaAs, BGaP, and BGaN

Another alloy examined in order to produce lower-bandgap materials lattice matched to GaAs for solar cell applications is BGaInAs. Being smaller than In, B can be added during MOVPE growth to compensate for the increase in lattice constant as In is added to GaAs. Since it increases the bandgap by only 4–8% per %B, the decrease in bandgap due to In allows the growth of materials with bandgaps smaller than GaAs [243]. Materials with up to 4% B were produced using diborane as the B precursor [43, 243]. The B distribution coefficient in BaGaAs is only 0.065. This was attributed to the formation of a B-rich surface layer that eventually blocks Ga incorporation, preventing further BGaAs growth [45].

BGaP alloys can be used to produce materials for visible LEDs and lasers lattice matched to Si substrates for "silicon-phonics" applications, including optical interconnects between chips. Kunert et al. [45], were able to grow epitaxial layers of BGaP with B concentrations of up to 7% using triethylboron (TEB) as the boron precursor. At a growth temperature of 575 °C, the B distribution coefficient was found to be unity, exactly as expected for nonvolatile elements that pyrolyze completely, as described in Section 2.2.2. B produces a rapid reduction in the bandgap energy. The bowing parameter has been determined to be 6.2 eV [265], allowing an (indirect) bandgap reduction of 0.4 eV for a B concentration of 6%. High crystalline quality layers of $B_{0.03}Ga_{0.97}P$, lattice matched to Si substrates, were used as strain-compensating layers to produce GaNAsP/BGaP MQW structures on (001) Si substrates. As seen in Table 2.1, due to the large size difference between B and Ga, BGaP alloys are prone to phase separation. Thus, low growth temperatures are required to avoid compositional clustering [260].

Very-high-bandgap BGaN alloys are interesting for UV and high-temperature device applications. They have been grown by low-pressure MOVPE [266] using the precursors TMGa, TEB, and NH_3 at a temperature of 1050 °C to produce alloys with B contents as large as 3.6%. At a very low B concentration of 0.7%, uniform alloys were reported [267]. However, the same group reported that at 1.7% B, coherent B-rich clusters were formed, containing 20% B. Phase separation was earlier reported by Wei et al. in layers grown at 1000 °C by MOVPE [268]. This is consistent with the very high calculated interaction parameters listed in Table 2.1.

2.6 Summary

In this chapter, the fundamental concepts upon which our understanding of the MOVPE process and the microstructures of the resulting semiconducting materials have been reviewed. The emphasis is placed upon recent developments, including our increased understanding of the surface processes occurring during growth and the control of these processes, including the use of surfactants. The basic framework established was then used to understand and explain recent results obtained in important alloy systems. Emphasis was placed on the most commercially important systems, AlGaInP and AlGaInN used for LEDs and solar cells. However, recent results on new systems, where the large size differences between host and alloying elements lead to large regions of solid immiscibility, such as GaAsN, GaAsBi, GaPB, and GaNB, have also been discussed. These novel alloy systems potentially allow the growth of materials that expand the bandgap energy–lattice constant space allowed for common alloys. However, the growth conditions that allow penetration of the miscibility gap, as well as the thermodynamics of the alloys themselves, are frequently found to lead to significant deviations from the properties expected for high-quality, homogeneous alloys. Sometimes these effects are deleterious, but in other cases, such as GaInN for blue LEDs, the effects are decidedly positive.

References

1. Stringfellow, G.B. (1999). *Organometallic Vapor Phase Epitaxy: Theory and Practice*, 2e. San Diego: Academic Press.
2. Stringfellow, G.B. (2007). Fundamentals of vapor phase epitaxial growth processes. *AIP Conference Proceedings* 913: 48.
3. Kuech, T.F. (2007). Surfactants in semiconductor epitaxy. *AIP Conference Proceedings* 913: 288.
4. Kisker, D.W. and Kuech, T.F. (1994). In: *Handbook of Crystal Growth, Volume 3* (ed. D.T.J. Hurle), Ch. 3. Amsterdam: Elsevier.
5. Stringfellow, G.B. (1994). In: *Handbook of Crystal Growth, Volume 3* (ed. D.T.J. Hurle), Ch. 12. Amsterdam: Elsevier.
6. Jensen, K.F. (1994). In: *Handbook of Crystal Growth, Volume 3* (ed. D.T.J. Hurle), Ch. 13. Amsterdam: Elsevier.
7. Stringfellow, G.B. (2001). Fundamental aspects of organometallic vapor phase epitaxy. *Material Science and Engineering B* 87: 97.
8. Stringfellow, G.B. (2004). *Crystal Growth from Fundamentals to Technology* (eds G. Muller, J. Metois, and P. Rudolph), 1. Amsterdam: Elsevier.
9. Kuech, T.F. (2016). III-V compound semiconductors: Growth and structures. *Progress in Crystal Growth and Characterisation of Materials* 62: 352.
10. Kuech, T.F., Mawst, L.J., and Brown, A.S. (2013). Mixed semiconductor alloys for optical devices. *Annual Review of Chemical and Biomolecular Engineering* 4: 187.
11. Nelson, D. (2016). Invited talk presented at the 18th International Conference on MOVPE, San Diego, July 2016.
12. Tsao, J.Y., Han, J., Haitz, R.H. et al. (2015). The blue LED Nobel Prize: Historical context, current scientific understanding, human benefit. *Annalen der Physik* 527: A53–A61;Weisbuch, C.W. (2018). Historical perspective on the physics of artificial lighting. *Compte Rendue Physique* 19: 89.
13. Krames, M. et al. (1999). High-power truncated-inverted-pyramid $(Al_xGa_{1-x})_{0.5}In_{0.5}P/GaP$ $(Al_xGa_{1-x})_{0.5}In_{0.5}P/GaP$ light-emitting diodes exhibiting >50% external quantum efficiency. *Applied Physics Letters* 75: 2365.
14. Law, D.C. et al. (2010). Future technology pathways of terrestrial III–V multijunction solar cells for concentrator photovoltaic systems. *Solar Energy Materials and Solar Cells* 94: 1314.
15. Geisz, J.F. et al. (2008). 40.8% efficient inverted triple-junction solar cell with two independently metamorphic junctions. *Applied Physics Letters* 93: 123505.
16. Stringfellow, G.B. (1999). *Organometallic Vapor Phase Epitaxy: Theory and Practice*, 2e. San Diego: Academic Press; Ch. 2.
17. Small, M.B., Giess, E.A., and Ghez, R. (1994). In: *Handbook of Crystal Growth* 3a (ed. D.T.J. Hurle), Ch. 6. Amsterdam: Elsevier;Small, M.B. and Ghez, R. (1979). Growth and dissolution kinetics of III-V heterostructures formed by LPE. *Journal of Applied Physics* 50: 5322.
18. Swalin, R.A. (1962). *Thermodynamics of Solids*. New York: John Wiley and Sons;Gokcen, N.A. (1986). *Statistical Thermodynamics of Alloys*. New York: Plenum Press.
19. Stringfellow, G.B. (1975). *International Review of Science, Inorganic Chemistry* Series 2, Vol. 10. London: Butterworths; 111.
20. Stringfellow, G.B. (1972). Calculation of the solubility and solid-gas distribution coefficient of N in GaP. *Journal of the Electrochemical Society* 119: 1780.
21. Stringfellow, G.B. (1972). Calculation of ternary phase diagrams of III–V systems. *Journal of the Physics and Chemistry of Solids* 33: 665.
22. Stringfellow, G.B. (1974). Calculation of ternary and quaternary III–V phase diagrams. *Journal of Crystal Growth* 27: 21.
23. Phillips, J.C. (1973). *Bonds and Bands in Semiconductors*. New York: Academic Press.
24. Ho, I.H. and Stringfellow, G.B. (1996). Solid phase immiscibility in GaInN. *Applied Physics Letters* 69: 2701.

25. Ho, I.H. and Stringfellow, G.B. (1997). Solubility of nitrogen in binary III–V systems. *Journal of Crystal Growth* 178: 1.
26. Matsuoka, T. (1998). Phase separation in wurtzite $In_{1-x-y}Ga_xAl_yN$. *MRS Internet Journal of Nitride Semiconductor Research* 3: 54.
27. Neugebauer, J. and Van de Walle, C.G. (1995). Electronic structure and phase stability of GaAs1 – xNx alloys. *Physical Review B* 51: 10568.
28. Teles, L.K. et al. (2002). Phase separation and gap bowing in zinc-blende InGaN, InAlN, BGaN, and BAlN alloy layers. *Physica E* 13: 1086.
29. Escalanti, L. and Hart, G.L.W. (2004). Boron alloying in GaN. *Applied Physics Letters* 84: 705.
30. Mohamad, R., Bere, A., Chen, J. et al. (2017). Investigation of strain effects on phase diagrams in the ternary nitride alloys (InAlN, AlGaN, InGaN). *Physica Status Solidi A* 214: 1600752;Mohamad, R. et al. (2018). A theoretical investigation of the miscibility and structural properties of InAlGaN alloys *Physica Status Solidi B* 255: 1700394.
31. Guan, Y., Luo, G., Morgan, D. et al. (2017). Paper presented at OMVPE Workshop, Santa Fe, New Mexico, August 2017; Guan, Y. (2017). Thermodynamic and kinetic factors controlling the Bi incorporation in III-V thin films grown by MOVPE, Ph.D. thesis. University of Wisconsin, Madison, USA.
32. Takayama, T., Yuri, M., Itoh, K. et al. (2001). Analysis of phase-separation region in wurtzite group III nitride quaternary material system using modified valence force field model. *Journal of Crystal Growth* 222: 29.
33. Karpinski, J., Jun, J., Grzegory, I. et al. (1985). Crystal growth of GaP doped with nitrogen under high nitrogen pressure. *Journal of Crystal Growth* 72: 711.
34. Thierry-Mieg, V., Marbeuf, A., Chevallier, J. et al. (1983). Determination of the nitrogen doping of liquid phase epitaxy of GaP and $Ga_xIn_{1-x}P$ alloys by optical absorption and photoluminescence. *Journal of Applied Physics* 54: 5358.
35. Hayes, T.J., Mottram, A., and Peaker, A.R. (1979). Nitrogen doping profiles in gallium phosphide grown by liquid phase epitaxy. *Journal of Crystal Growth* 46: 59.
36. Uberuaga, B.P., Leskovar, M., Smith, A.P. et al. (2000). Diffusion of Ge below the Si(100) surface: Theory and experiment. *Physical Review Letters* 84: 2441.
37. Stringfellow, G.B. (1984). A critical appraisal of growth mechanisms in MOVPE. *Journal of Crystal Growth* 68: 111.
38. Stringfellow, G.B. (1978). VPE growth of III/V semiconductors. *Annual Review of Materials Science* 8: 73.
39. Koukitu, A., Takahashi, N., Taki, T. et al. (1997). Thermodynamic analysis of the MOVPE growth of $In_xGa_{1-x}N$. *Journal of Crystal Growth* 170: 306.
40. Rossow, U., Jönen, H., Brendel, M. et al. (2011). Growth of the active zone in nitride based long wavelength laser structures. *Journal of Crystal Growth* 315: 250.
41. Mauder, C. et al. (2011). Effect of indium incorporation on optical and structural properties of m-plane InGaN/GaN MQW on $LiAlO_2$ substrates. *Journal of Crystal Growth* 315: 246.
42. Mori, Y. and Watanabe, N. (1981). AlGaAs frown by metalorganic chemical vapor depostion for visible lasers. *Journal of Applied Physics* 52: 2792.
43. Ludowise, M.J., Cooper, C.B., and Saxena, R.R. (1981). The growth and characterization of uniform Ga1-xInxAs (x ≤0.25) by organometallic VPE. *Journal of Electronic Materials* 10: 1051.
44. Sommer, N. et al. (2013). Growth of (BGa)As, (BGa)P, (BGa)(AsP) and (BGaIn)P by MOVPE. *Journal of Crystal Growth* 370: 191.
45. Kunert, B., Zinnkann, S., Volz, K. et al. (2008). Monolithic integration of Ga(NAsP)/(BGa)P multi-quantum well structures on (001) silicon substrate by MOVPE. *Journal of Crystal Growth* 310: 4776.
46. Fukui, T. and Horikoshi, Y. (1980). Organometallic VPE growth of $InAs_{1-x}Sb_x$ on InAs. *Japanese Journal of Applied Physics* 19: L53.
47. Stringfellow, G.B. (1983). Thermodynamic aspects of organometallic vapor phase epitaxy. *Journal of Crystal Growth* 62: 225.
48. Cooper, C.B., Saxena, R.R., and Ludowise, M.J. (1982). The organometallic VPE growth of GaSb and GaAsl – xSbx using trimethylantimony. *Journal of Electronic Materials* 11: 1001.
49. Stringfellow, G.B. and Cherng, M.J. (1983). OMVPE growth of $GaAs_{1-x}Sb_x$: solid composition. *Journal of Crystal Growth* 64: 413.
50. Biefeld, R. (2002). The metal-organic chemical vapor deposition and properties of III–V antimony-based semiconductor materials. *Materials Science and Engineering R* 36: 105.
51. Stringfellow, G.B. (1972). The importance of lattice mismatch in the growth of $Ga_xIn_{1-x}P$ epitaxial crystals. *Journal of Applied Physics* 43: 3455.
52. Stringfellow, G.B. (2010). Microstructures produced during the epitaxial growth of InGaN alloys. *Journal of Crystal Growth* 312: 735.
53. Lester, S.D., Ponce, F.A., Craford, M.G. et al. (1995). High dislocation densities in high efficiency GaN-based light-emitting diodes. *Applied Physics Letters* 66: 1249.
54. Gibbs, J.W. (1876). On the equilibrium of heterogeneous substances. *Transactions of the Connecticut Academy of Arts and Sciences*, III, 108;Gibbs, J.W. (1878). On the equilibrium of heterogeneous substances (contd.). *Transactions of the Connecticut Academy of Arts and Sciences*, III, 343.

55. Kim, M.K., Choi, S., Lee, J.H. et al. (2017). Investigating carrier localization and transfer in InGaN/GaN quantum wells with V-pits using near-field scanning optical microscopy and correlation analysis. *Nature Scientific Reports* 7: 42221.
56. Piner, E.L., El-Mastry, N.A., Liu, S.X. et al. (1998). Phase separation in InGaN grown by metalorganic chemical vapor deposition. *Materials Research Society Symposia Proceedings* 482: 125.
57. Ponce, F.A. et al. (2003). Microstructure and electronic properties of InGaN alloys. *Physica Status Solidi B* 240: 273.
58. Rao, M., Kim, D., and Mahajan, S. (2004). Compositional dependence of phase separation in InGaN layers. *Applied Physics Letters* 85: 1961.
59. Doppalapudi, D., Basu, S.N., Ludwig, K.F. et al. (1998). Phase separation and ordering in InGaN alloys grown by molecular beam epitaxy. *Journal of Applied Physics* 84: 1389.
60. Kar, A., Alexson, D., Dutta, M. et al. (2008). Evidence of compositional inhomogeneity in $In_xGa_{1-x}N$ alloys using ultraviolet and visible Raman spectroscopy. *Journal of Applied Physics* 104: 073502.
61. Faleev, N., Jampana, B., Jani, O. et al. (2009). Correlation of crystalline defects with photoluminescence of InGaN layers. *Applied Physics Letters* 95: 051915.
62. Potin, V., Hahn, E., Rosenauer, A. et al. (2004). Comparison of the In distribution in InGaN/GaN quantum well structures grown by molecular beam epitaxy and metalorganic vapor phase epitaxy. *Journal of Crystal Growth* 262: 145.
63. Osamura, K., Naka, S., and Murakami, Y. (1975). Preparation and optical properties of $Ga_{1-x}In_xN$ thin films. *Journal of Applied Physics* 46: 3432.
64. Narukawa, Y., Ichikawa, M., Sanga, D. et al. (2010). White light emitting diodes with super-high luminous efficacy. *Journal of Physics D, (Applied Physics)* 43: 354002.
65. Stringfellow, G.B. (1998). Ordering in III/V semiconductor alloys. In: *Thin Films: Heteroepitaxial Systems*, (ed. M. Santos and W.K. Liu), 64–116. World Scientific Publishing;Stringfellow, G.B. (1993). In: *Common Themes and Mechanisms of Epitaxial Growth* (ed. P. Fuoss, J. Tsao, D.K. Kisker et al.), 35–46. Pittsburgh: Materials Research Society.
66. Wei, S.H. and Zunger, A. (1994). Optical properties of zinc-blende semiconductor alloys: Effects of epitaxial strain and atomic ordering. *Physical Review B* 49: 14337.
67. Stringfellow, G.B. (1999). *Organometallic Vapor Phase Epitaxy: Theory and Practice*, 2e. San Diego: Academic Press; Ch. 3: Niu, X. Stringfellow, G.B. and Liu, F. (2011). Appl. Phys. Lett. 99: 213102.
68. Chen, G.S., Jaw D.H., and Stringfellow, G.B. (1991). Atomic ordering in GaAsP. *Journal of Applied Physics* 69: 4263.
69. Ueda, O., Hoshino, M., Takechi, M. et al. (1990). Comparative study of atomic ordering in InGaP crystals grown by metalorganic vapor phase epitaxy, chloride-vapor phase epitaxy, and liquid phase epitaxy. *Journal of Applied Physics* 68: 4268.
70. Stringfellow, G.B., Lindquist, P.F., and Burmeister, R.A. (1972). Liquid phase epitaxial growth of GaxIn1 − xP. *Journal of Electronic Materials* 1: 437.
71. Gomyo, A., Iobgayashi, K., Kawata, S. et al. (1987). Studies of $Ga_xIn_{1-x}P$ layers grown by metalorganic vapor phase epitaxy; Effects of V/III ratio and growth temperature. *Journal of Crystal Growth* 77: 367.
72. Laks, D.B., Wei, S.H., and Zunger, Z. (1992). Evolution of alloy properties with long-range order. *Physical Review Letters* 69: 3766.
73. Ernst, P., Geng, C., Scholz, F. et al. (1997). Band-gap reduction and valence-band splitting of ordered $GaInP_2$. *Applied Physics Letters* 67: 2347.
74. Su, L.C., Ho, I.H., and Stringfellow, G.B. (1994). Analysis of the scanning electron microscope mirror image based on the dielectric surface microstructure. *Journal of Applied Physics* 75: 5135.
75. Mao, D., Taylor, P.C., Kurtz, S.R. et al. (1996). Average local order parameter in partially ordered GaInP2. *Physical Review Letters* 76: 4769.
76. Arent, D.J., Bode, M., Bode, K.A. et al. (1993). Band-gap narrowing in ordered $Ga_{0.47}In_{0.53}As$. *Applied Physics Letters* 62: 1806.
77. Kurtz, S.R., Dawson, L.R., Biefeld, R.M. et al. (1992). Ordering-induced band-gap reduction in InAs1 − xSbx (x≊0.4) alloys and superlattices. *Physical Review B* 46: 1909.
78. Stringfellow, G.B. (1997). In: *High Brightness Light Emitting Diodes* (eds. G.B. Stringfellow, and M.G. Craford). Boston: Academic Press.
79. Mascarenhas, A. (2002). *Spontaneous Ordering in Semiconductor Alloys*. New York: Kluwer Academic/Plenum Publishers.
80. Wang, T.Y., Kimball, A.W., Chen, G.S. et al. (1991). GaInP/AlGaInP strained quantum wells grown using atmospheric pressure organometallic vapor phase epitaxy. *Journal of Crystal Growth* 109: 279.
81. Stringfellow, G.B. (2002). In: *Spontaneous Ordering in Semiconductor Alloys* (ed. A. Mascarenhas), Ch. 3. New York: Kluwer Academic/Plenum Publishers.
82. Su, L.C., Ho, I.H., Kobayashi, N. et al. (1994). Order/disorder heterostructure in $Ga_{0.5}In_{0.5}P$ with $\Delta E_g = 160\,meV$. *Journal of Crystal Growth* 145: 140.
83. Murata, H., Ho, I.H., Su, L.C. et al. (1996). Surface photoabsorption study of the effects of growth temperature and V/III ratio on ordering in GaInP. *Journal of Applied Physics* 79: 6895.
84. Hsu, T.C., Hsu, Yu, and Stringfellow, G.B. (1998). Effect of P precursor on surface structure and ordering in GaInP. *Journal of Crystal Growth* 193: 1.

85. Lee, R.T., Shurtleff, J.K., Fetzer, C.M. et al. (2000). Surfactant controlled growth of GaInP by organometallic vapor phase epitaxy. *Applied Physics Letters* 87: 3730.
86. Shurtleff, J.K., Lee, R.T., Fetzer, C.M. et al. (1999). Band-gap control of GaInP using Sb as a surfactant. *Applied Physics Letters* 75: 1914.
87. Stringfellow, G.B., Lee, R.T., Fetzer, C.M. et al. (2000). Surfactant effects of dopants on ordering in GaInP. *Journal of Electronic Materials* 29: 134.
88. Jun, S.W., Fetzer, C.M., Lee, R.T. et al. (2000). Bi surfactant effects on ordering in GaInP grown by organometallic vapor-phase epitaxy. *Applied Physics Letters* 76: 2716.
89. Su, L.C., Ho, I.H., and Stringfellow, G.B. (1994). Kinetically controlled order/disorder structure in GaInP. *Applied Physics Letters* 65: 749.
90. Hsu, Y., Stringfellow, G.B., Inglefield, C.E. et al. (1998). Quantum wells due to ordering in GaInP. *Applied Physics Letters* 73: 3905.
91. Garcia, I., Rey-Stolle, I., Algora, C. et al. (2008). Influence of GaInP ordering on the electronic quality of concentrator solar cells. *Journal of Crystal Growth* 310: 5209.
92. Christian, T.M., Beaton, D.A., Makherjee, K. et al. (2013). Amber-green light-emitting diodes using order-disorder $Al_xIn_{1-x}P$ heterostructures. *Journal of Applied Physics* 114: 074505.
93. Chun, Y.S., Lee, S.H., Ho, I.H. et al. (1997). Effect of growth rate on step structure and ordering in GaInP. *Journal of Applied Physics* 81: 646.
94. Cao, D.S., Reihlen, E.H., Chen, G.S. et al. (1991). Effect of growth rate on properties of GaInP grown by OMVPE. *Journal of Crystal Growth* 109: 279.
95. Murata, H., Lee, S.H., Ho, I.H. et al. (1996). Correlation between surface structure and ordering in GaInP. *Journal of Vacuum Science and Technology B* 14: 3013.
96. Stringfellow, G.B., Shurtleff, J.K., Lee, R.T. et al. (2000). Surface processes in OMVPE – the frontiers. *Journal of Crystal Growth* 221: 1.
97. Lee, S.H., Fetzer, C.M., and Stringfellow, G.B. (1999). Te doping of GaInP: Ordering and step structure. *Journal of Applied Physics* 85: 3590.
98. Hageman, P.R., Bauhuis, G.J., and Olsthoorn, S.M. (1998). Influence of Ga precursor choice on ordering degree of MOVPE grown $Ga_{0.5}In_{0.5}P$. *Journal of Crystal Growth* 194: 272.
99. Jensen, K. F. (1994). *Handbook of Crystal Growth* 3b, (ed. D.T.J. Hurle), Ch. 13. Amsterdam: Elsevier.
100. Stringfellow, G.B. (1999). *Organometallic Vapor Phase Epitaxy: Theory and Practice*, 2e. San Diego: Academic Press; Ch. 5.
101. Stringfellow, G.B. (1999). *Organometallic Vapor Phase Epitaxy: Theory and Practice*, 2e. San Diego: Academic Press; Ch. 6.
102. Stringfellow, G.B. (1999). *Organometallic Vapor Phase Epitaxy: Theory and Practice*, 2e. San Diego: Academic Press; Ch. 7.
103. Giling, L.J. (1987). In: *Advanced Crystal Growth*, (eds P.M. Dryburgh, B. Cockayne, and K.G. Barraclough), 309–336. New York: Prentice Hall.
104. van de Ven, J., Rutten, G.M.J., Raaijmakers, M.J. et al. (1986). Gas phase depletion and flow dynamics in horizontal MOCVD reactors. *Journal of Crystal Growth* 76: 352.
105. Moffat, H.K. and Jensen, K.F. (1986). Complex flow phenomena in MOCVD reactors: I. Horizontal reactors. *Journal of Crystal Growth* 77: 108.
106. Frijlink, P.J., Nicolas, J.L. et al. (1991). The radial flow planetary reactor: low pressure versus atmospheric pressure MOVPE. *Journal of Crystal Growth* 115: 203.
107. Hui, L. (2011). Mass transport analysis of a showerhead MOCVD reactor. *Journal of Semiconductors* 32: 033006.
108. Glasstone, S., Leidler, K.J., and Eyring, H. (1941). *The Theory of Rate Processes*. New York: McGraw-Hill.
109. Kuech, T. (2016). III-V compound semiconductors: Growth and structures. *Progress in Crystal Growth and Characterisation of Materials* 62: 352.
110. Fukui, T. and Horikoshi, Y. (1980). Organometallic VPE Growth of $InAs_{1-x}Sb_x$ on InAs. *Japanese Journal of Applied Physics* 19: L53.
111. Ludowise, M.J. and Dietze, W.T. (1982). The effect of trimethylaluminum concentration on the incorporation of P In AlxGa1 – xPyAs1 – y grown by organometallic vapor phase epitaxy. *Journal of Electronic Materials* 11: 59.
112. Samuelson, L., Omling, P., Titze, H. et al. (1982). Organometallic epitaxial growth of GaAsP. *Journal of Physics C (Solid State Physics)* 12: 323.
113. Smeets, E.T.J.M. (1987). Solid composition of $GaAs_{1-x}P_x$ grown by organometallic vapour phase epitaxy. *Journal of Crystal Growth* 82: 385.
114. Chen, C.H., Cao, D.S., and Stringfellow, G.B. (1988). Use of tertiarybutylphosphine for the growth of InP and GaAs1-xPx. *Journal of Electronic Materials* 17: 67.
115. Creighton, J.R., Wang, G.T., and Coltrin, M.E. (2007). Fundamental chemistry and modeling of group-III nitride MOVPE. *Journal of Crystal Growth* 298: 2.

116. Khandekar, A.A., Yeh, J.Y., Mawst, L.J. et al. (2007). Effects of Ga- and Sb-precursor chemistry on the alloy composition in pseudomorphically strained $GaAs_{1-y}Sb_y$ films grown via metalorganic vapor phase epitaxy. *Journal of Crystal Growth* 303: 456.
117. Huynh, S.H., Ha, M.T.H., Do, H.B. et al. (2016). Impact of interfacial misfit dislocation growth mode on highly lattice-mismatched InxGa1-xSb epilayer grown on GaAs substrate by metalorganic chemical vapor deposition. *Applied Physics Letters* 109: 102107.
118. Heurlin, M., Anttu, N., Camus, C. et al. (2015). In situ characterization of nanowire dimensions and growth dynamics by optical reflectance. *Nano Letters* 15: 3597.
119. Otnes, G., Heurlin, M., Zeng, X. et al. (2017). InxGa1–xP nanowire growth dynamics strongly affected by doping using diethylzinc. *Nano Letters* 17: 702.
120. Kossel, W. (1927). Zur theorie des kristallwachstum. *Nachrichten Gesellschaft der Wissenschaften in Gottingen*, 135.
121. Stranski, I.N. (1928). Zur theorie des kristallwachstums. *Zeitschrift fur Physikalische Chemie* 136: 259.
122. Venables, J.A. (1998). Atomic Interactions and surface processes in heteroepitaxy. In: *Thin Films: Heteroepitaxial Systems*, (eds. M. Santos and W.K. Liu), 1–63. World Scientific Publishing.
123. Tsao, J.Y. (1993). In: *Materials Fundamentals of Molecular Beam Epitaxy*. Boston: Academic Press, Ch. 6.
124. Pristovsek, M.B., Rähmer, M., Breusig, R. et al. (2007). In situ scanning tunnelling microscopy during metal-organic vapour phase epitaxy. *Journal of Crystal Growth* 298: 8.
125. Srivastava, G.P. (2006). The electron counting rule and passivation of compound semiconductor surfaces. *Applied Surface Science* 252: 7600.
126. Ohtake, A. (2008). Surface reconstructions on GaAs(001). *Surface Scientific Reports* 63: 295.
127. Kamiya, I., Mantese, L., Aspnes, D.E. et al. (1996). Optical characterization of surfaces during epitaxial growth using RDS and GIXS. *Journal of Crystal Growth* 163: 67.
128. Sudijono, J., Johnson, M.D., Snyder, C.W. et al. (1992). Surface evolution during molecular-beam epitaxy deposition of GaAs. *Physical Review Letters* 19: 2811.
129. Field, R.L., Occena, J., Jen. T. et al. (2016). Influence of surface reconstruction on dopant incorporation and transport properties of GaAs(Bi) alloys. *Applied Physics Letters* 109: 252105.
130. Langmuir, I. (1918). The adsorption of gases on plane surfaces of glass, mica and platinum. *Journal of the American Chemical Society* 40: 1361.
131. Reep, D.H. and Ghandhi, S.K. (1983). Deposition of GaAs epitaxial layers by organometallic CVD: Temperature and orientation dependence. *Journal of the Electrochemical Society* 130: 675.
132. Reep, D.H. and Ghandhi, S.K. (1984). Electrical properties of organometallic chemical vapor deposited GaAs epitaxial layers: temperature dependence. *Journal of the Electrochemical Society* 131 2697.
133. Forghani, K., Guan, Y., Wood, A. et al. (2015). The effect of the Bi precursors, $(CH_3)_3Bi$ and $(C_2H_5)_3Bi$, on the metal-organic vapor phase epitaxy of $GaAs_{1-y}Bi_y$ films. *Chemical Vapor Deposition* 21: 166.
134. Asai, H. (1987). Anisotropic lateral growth in GaAs MOCVD layers on (001) substrates. *Journal of Crystal Growth* 80: 425.
135. Reichert, W. and Cohen, R.M. (2000). Effects of substrate orientation and surface reconstruction on patterned substrate OMVPE of GaAs. *Journal of Electronic Materials* 29: 118;Reichert, W. and Cohen, R.M. (2000). Lateral OMVPE growth of GaAs on patterned substrates. *Journal of Crystal Growth* 220: 364.
136. Wixom, R.R., Rieth, L.W., and Stringfellow, G.B. (2004). Sb and Bi surfactant effects on homo-epitaxy of GaAs on (001) patterned substrates. *Journal of Crystal Growth* 265: 367.
137. Ehrlich, G. and Hudda, F.G. (1966). Atomic view of surface self-diffusion: Tungsten on tungsten. *Journal of Chemical Physics* 44: 1039.
138. Schwoebel, R.L. (1969). Step motion on crystal surfaces. II. *Journal of Applied Physics* 40: 614.
139. Lee, S.H. and Stringfellow, G.B. (1998). Step structure and ordering in GaInP. *Journal of Applied Physics* 83: 3620.
140. Jensen, L.E., Bjork, M.T., Jeppesen, S. et al. (2004). Role of surface diffusion in chemical beam epitaxy of InAs nanowires. *Nano Letters* 4: 1961.
141. Tersoff, J., Phang, Y.H., Zhang, Z. et al. (1995). Step-bunching instability of vicinal surfaces under stress. *Physical Review Letters* 75: 2730.
142. Kasu, M. and Kobayashi, N. (1994). Scanning tunneling microscopy study of GaAs step structures on vicinal substrate grown by metalorganic chemical vapor deposition. *Japanese Journal of Applied Physics* 33: 712.
143. Stringfellow, G.B. and Su, L.C. (1996). Step structure during OMVPE growth of ordered GaInP. *Journal of Crystal Growth* 163: 128.
144. Ishizaki, J., Gota, S., Kishida, M. et al. (1994). Mechanism of multiatomic step formation during metalorganic chemical vapor deposition growth of GaAs on (001) vicinal surface studied by atomic force microscopy. *Japanese Journal of Applied Physics* 33: 721.
145. Lee, S.H. and Stringfellow, G.B. (1998). Influence of tellurium doping on step bunching of GaAs(001) vicinal surfaces grown by organometallic vapor phase epitaxy. *Applied Physics Letters* 73: 1703.

146. Yin, X., Shi, J., Niu, X. et al. (2015). Wedding cake growth mechanism in one-dimensional and two-dimensional nanostructure evolution. *Nano Letters* 15: 7766.
147. Nakajima, K. (1999). Equilibrium phase diagrams for Stranski-Krastanov structure mode of III–V ternary quantum dots. *Japanese Journal of Applied Physics* 38: 1875.
148. Zhang, S.B., Froyen, S., and Zunger, A. (1995). Surface dimerization induced CuPtB versus CuPtA ordering of GaInP alloys. *Applied Physics Letters* 67: 3141.
149. Bickel, J.E., Modine, N.A., Pearson, C. et al. (2008). Elastically induced coexistence of surface reconstructions. *Physical Review B* 77: 125308.
150. Karpov, A. (1998). Suppression of phase separation in InGaN due to elastic strain. *MRS Internet Journal of Nitride Semiconductor Research* 3: 5.
151. Kanitani, Y., Tanaka, S., Tomiya, S. et al. (2016). Atom probe tomography of compositional fluctuation in GaInN layers. *Japanese Journal of Applied Physics* 55: 05FM04.
152. Specht, P. and Kisielowski, C. (2017). On the chemical homogeneity of InGaN alloys – Electron microscopy at the edge of technical limits. *Materials Science in Semiconductor Processing* 65: 24.
153. Markov, I. (1996). Kinetics of nucleation in surfactant-mediated epitaxy. *Physical Review B* 53: 4148.
154. Tournie, E. and Ploog, K.H. (1993). Surfactant-mediated molecular beam epitaxy of strained layer semiconductor heterostructures. *Thin Solid Films* 231: 43;Reinking, D., Kammier, M., Horn-von Hoegen, M. et al. (1997). Enhanced Sb segregation in surfactant-mediated-heteroepitaxy: High-mobility, low-doped Ge on Si. *Applied Physics Letters* 71: 924;Cunningham, J.E,. Goossen, K.W., Jan, W. et al. (1994). *Journal of Vacuum Science and Technology B* 13: 646;Neves, B.R.A., Andrade, M.S., Rodriques, W.N. et al. (1998). Coherent-to-incoherent transition in surfactant mediated growth of InAs quantum dots. *Applied Physics Letters* 72: 1712.
155. Kandel, D. and Kaxiras, E. (1995). Surfactant mediated crystal growth of semiconductors. *Physical Review Letters* 75: 2742;Oh, C.W. Kim, E. and Lee, Y.H. (1996). Kinetic role of a surfactant in island formation. *Physical Review Letters* 76: 776.
156. Oigawa, H., Wassermeier, M., Behend, J. et al. (1997). Reconstruction of the GaAs(001) surface induced by submonolayer Be deposition. *Surface Science* 376: 185.
157. Jenkins, S.J. and Srivastava, G.P. (1998). Thermodynamic evidence for surfactant behaviour of Sb in the growth of Ge on Si(001). *Surface Science* 398: L308.
158. Voigtlander, B., Zinner, A., Weber, T. et al. (1995). Modification of growth kinetics in surfactant-mediated epitaxy. *Physical Review B* 51: 7583.
159. Norgami, J., Baski, A.A., and Quate, C.F. (1995). Structure of the Sb-terminated Si(100) surface. *Applied Physics Letters* 58: 475;LeGoues, F.K., Kesan, V.P., Iyer, S.S. et al. (1990). Surface-stress-induced order in SiGe alloy films. *Physical Review Letters* 64: 2038;Miki, K., Owen, J.H.G., Bowler, D.R. et al. (1999). Atomically perfect bismuth lines on Si(001). *Physical Review B* 59: 14868.
160. Wixom, R.R., Modine, N.A., and Stringfellow, G.B. (2003). Theory of surfactant (Sb) induced reconstructions on InP(001). *Physical Review B* 67: 115309.
161. Fetzer, C.M., Lee, R.T., Shurtleff, J.K. et al. (2000). The use of a surfactant (Sb) to induce triple period ordering in GaInP. *Applied Physics Letters* 76: 1440.
162. Jiang, W.Y., Liu, J.Q., So, M.G. et al. (2004). Effect of Bi surfactant on atomic ordering of GaAsSb. *Applied Physics Letters* 85: 5589.
163. Laukkanen, P. et al. (2007). Structural properties of Bi-stabilized reconstructions of GaInAs(100) surface. *Applied Physics Letters* 90: 082101.
164. Norman, A.G. (2002). In: *Spontaneous Ordering in Semiconductor Alloys* (ed. A. Mascarenhas), Ch. 2. New York: Kluwer Academic/Plenum Publishers.
165. Batyrev, I.G., McMahon, W.E., Zhang, S.B. et al. (2005). Step structures on III-V phosphide (001) surfaces: How do steps and Sb affect CuPt ordering of GaInP2? *Physical Review Letters* 94: 096101.
166. Olson, J.M., McMahon, W.E., and Kurtz, S. (2006). Effect of Sb on the properties of GaInP top cells. *Conference Record, 2006 IEEE World Conference on Photovoltaic Energy Conversion*, 787.
167. Cederberg, J.G. and Lee, S.R. (2007). Surfactant effects associated with Te-doped InPAs alloys. *Applied Physics Letters* 91: 201915.
168. Niu, X., Stringfellow, G.B., and Liu, F. (2011). Phase separation in strained epitaxial InGaN islands. *Applied Physics Letters* 99: 213102.
169. Sato, T., Mitsuhara, M., Watanabe. T. et al. (2005). Surfactant-mediated growth of InGaAs multiple-quantum-well lasers emitting at 2.1 μm by metalorganic vapor phase epitaxy. *Applied Physics Letters* 87 211903.
170. Gomyo, A., Hotta, H., Hino, I. et al. (1989). Silicon and selenium doping effects on band-gap energy and sublattice ordering in $Ga_{0.5}In_{0.5}P$ grown by metalorganic vapor phase epitaxy. *Japanese Journal of Applied Physics* 28: L1330.
171. Goral, J.P., Kurtz, S.R., Olson, J.M. et al. (1990). Transmission electron microscopy study of the effect of selenium doping on the ordering of GaInP2. *Journal of Electronic Materials* 19: 95.

172. Lee, S.H., Fetzer, C.M., Stringfellow, G.B. et al. (1999). Te doping of GaInP: Ordering and step structure. *Journal of Applied Physics* 85: 3590;Lee, S.H., Hsu, T.C., and Stringfellow, G.B. (1998). Step structure and ordering in Te-doped GaInP. *Journal of Applied Physics* 84: 2618.
173. Lee, S.H., Fetzer, C.M., and Stringfellow, G.B. (1998). Effect of Te doping on step structure and ordering in GaInP. *Journal of Crystal Growth* 195: 13.
174. France, R.M., Feifel, M, Belz, A. et al. (2019) *Journal of Crystal Growth* 506: 61.
175. Kurtz, S.R., Olson, J.M., Friedman, D.J. et al. (1994). Ordering and disordering of doped $Ga_{0.5}In_{0.5}P$. *Journal of Electronic Materials* 23: 431.
176. Gomyo, A., Suzuki, T., Kobayashi, K. et al. (1987). Evidence for the existence of an ordered state in Ga0.5In0.5P grown by metalorganic vapor phase epitaxy and its relation to band-gap energy. *Applied Physics Letters* 50: 673.
177. Kurtz, S.R., Olson, J.M., Friedman, D.J. et al. (1994). Ordering and disordering of doped Ga0.5In0.5P. *Journal of Electronic Materials* 23: 431.
178. Morita, E., Ikeda, M., Kumagai, O. et al. (1988). Transmission electron microscopic study of the ordered structure in GaInP/GaAs epitaxially grown by metalorganic chemical vapor deposition. *Applied Physics Letters* 53: 2164.
179. Dabkowski, F.P., Gavrilovic, P., Meehan, K. et al. (1998). Disordering of the ordered structure in metalorganic chemical vapor deposition grown $Ga_{0.5}In_{0.5}P$ on (001) GaAs substrates by zinc diffusion. *Applied Physics Letters* 52: 2142.
180. Suzuki, T., Gomyo, A., Hino, I. et al. (1988). p-Type doping effects on band-gap energy for $Ga_{0.5}In_{0.5}P$ grown by metalorganic vapor phase epitaxy. *Japanese Journal of Applied Physics* 27: L1549.
181. Lee, M.K., Horng, R.H., and Haung, L.C. (1991). Ordering effects on the electrical characteristics of Ga0.5In0.5P grown by metalorganic chemical vapor deposition. *Applied Physics Letters* 59: 3261.
182. Lee, S.H., Fetzer, C.M., Stringfellow, G.B. et al. (1999). Step structure and ordering in Zn-doped GaInP. *Journal of Applied Physics* 86: 1982.
183. France, R.M., Feifel, M., Belz, J. et al. (2019). Single- and dual-variant atomic ordering in GaAsP compositionally graded buffers on GaP and Si substrates. *Journal of Crystal Growth* 506: 61.
184. Jou, M.J., Jaw, D.H., Fang, Z.M. et al. (1990). Organometallic vapor phase epitaxial growth of a new quaternary semiconductor alloy $Ga_{1-x}In_xP_{1-y}Sb_y$. *Journal of Crystal Growth* 106: 208.
185. Chakrabarti, A. and Kunc, K. (2003). Atomic ordering in InxGa1−xAs alloy thin films: Action of surfactants. *Physical Review B* 68: 045304;Chakrabarti, A. and Kunc, K. (2004). Effect of strain on atomic ordering and action of surfactants in ternary alloy thin films. *Physical Review B* 70: 085313.
186. Jun, S.W., Lee, R.T., Fetzer, C.M. et al. (2000). Bi surfactant control of ordering and surface structure in GaInP grown by organometallic vapor phase epitaxy. *Journal of Applied Physics* 88: 4429.
187. Stringfellow, G.B. (2004). Development and current status of organometallic vapor phase epitaxy. *Journal of Crystal Growth* 264: 620.
188. Chapman, D.C., Reith, L.W., Stringfellow, G.B. et al. (2004). The use of nitrogen to disorder GaInP. *Journal of Applied Physics* 95: 6145.
189. Lee, M.K., Horng, R.H., and Haung, L.C. (1992). Disorder/order/disorder $Ga_{0.5}In_{0.5}P$ visible light-emitting diodes. *Journal of Applied Physics* 72: 5420.
190. Christian, T.M., Beaton, D.A., Makherjee, K. et al. (2013). Amber-green light-emitting diodes using order-disorder $Al_xIn_{1-x}P$ heterostructures. *Journal of Applied Physics* 114: 074505.
191. Bhusal, L., Fluegel, B., Steiner, M.A. et al. (2009). Ordering induced direct-indirect transformation in unstrained $Ga_xIn_{1-x}P$ for $0.76 \leq x \leq 0.78$. *Journal of Applied Physics* 106: 114909.
192. Geisz, J.F. et al. (2008). 40.8% efficient inverted triple-junction solar cell with two independently metamorphic junctions. *Applied Physics Letters* 93: 123505.
193. King, R.R., Law, D.C., Edmondson, K.M. et al. (2007). 40% efficient metamorphic GaInP/GaInAs/Ge multijunction solar cells. *Applied Physics Letters* 90: 183516.
194. Kurtz, S.R., Faine, P., and Olson, J.M. (1990). Modeling of two-junction, series-connected tandem solar cells using top-cell thickness as an adjustable parameter. *Journal of Applied Physics* 68: 1890.
195. Barrigon, E., Barrutia, L., Ochoa, M. et al. (2016). Effect of Sb on the quantum efficiency of GaInP solar cells. *Progress in Photovoltaics: Research and Applications* 24: 1116.
196. Fetzer, C.M., Ermer, J.H., King, R.R. et al. (2002). *Isoelectronic surfactant induced sublattice disordering in optoelectronic devices. US patent* 7126052.
197. Tong, S.C., Wang, J.S., Wu, C.B. et al. (2016). The reduction of antiphase boundary defects by the surfactant antimony and its application to the III–V multi-junction solar cells. *Solar Energy Materials and Solar Cells* 144: 418.
198. Chapman, D.C., Howard, A.D., and Stringfellow, G.B. (2006). Zn enhancement during surfactant-mediated growth of GaInP and GaP. *Journal of Crystal Growth* 287: 647;Howard, A.D., Chapman, D.C., and Stringfellow, G.B. (2006). Effects of surfactants Sb and Bi on the incorporation of zinc and carbon in III/V materials grown by organometallic vapor-phase epitaxy. *Journal of Applied Physics* 100: 044904;Howard, A.D. and Stringfellow, G.B. (2007). Effects of low surfactant Sb coverage on Zn and C incorporation in GaP. *Journal of Applied Physics* 102: 074920.
199. Sato, T., Mitsuhara, M., Iga, R. et al. (2011). Influence of Sb surfactant on carrier concentration in heavily Zn-doped InGaAs grown by metalorganic vapor phase epitaxy. *Journal of Crystal Growth* 315: 64.

200. Zhu, J., Liu, F., and Stringfellow, G.B. (2010). Enhanced cation-substituted p-type doping in GaP from dual surfactant effects. *Journal of Crystal Growth* 312: 174.
201. Shurtleff, J.K., Jun, S.W., and Stringfellow, G.B. (2001). Surfactant effects on doping of GaAs grown by organometallic vapor phase epitaxy. *Applied Physics Letters* 78: 3038.
202. Dimroth, F., Howard, A.D., Shurtleff, J.K. et al. (2002). Influence of Sb, Bi, Tl, and B on the incorporation of N in GaAs. *Journal of Applied Physics* 91: 3687.
203. Chen, W.C., Su, Y.K., Chuang, R.W. et al. (2007). Optical investigations on the surfactant effects of Sb on InGaAsN multiple quantum wells grown by MOVPE. *Journal of Crystal Growth* 298: 145.
204. Forghani, K., Guan, Y., Wood, A. et al. (2015). The effect of the Bi precursors, $(CH_3)_3Bi$ and $(C_2H_5)_3Bi$, on the metal-organic vapor phase epitaxy of $GaAs_{1-}yBiy$ films. *Chemical Vapor Deposition* 21: 166.
205. Zhang, L., Tang, H.F., Scieke, J. et al. (2002). Influence of Bi impurity as a surfactant during the growth of GaN by metalorganic vapor phase epitaxy. *Journal of Crystal Growth* 242: 302.
206. Gokhale, A.A., Kuech, T.F., and Mavrikakis, M. (2007). A theoretical comparative study of the surfactant effect of Sb and Bi on GaN growth. *Journal of Crystal Growth* 303: 493.
207. Koch, H., Pietzonka, I., Galler, B. et al. (2015). Effect of antimony on growth mode and properties of thick InGaN layers. *Journal of Crystal Growth* 414: 42.
208. Bernardini, F., Profeta, G., and Continenza, A. (2003). Bi incorporation in GaN and AlxGa1 − xN alloys. *Physical Review B* 68: 195205.
209. Zhang, L., Tang, H.F., Schieke, J. et al. (2002). The addition of Sb as a surfactant to GaN growth by metal organic vapor phase epitaxy. *Journal of Applied Physics* 92: 2304.
210. Sadasivam, K.G., Shim, J.I., and Lee, J.K. (2011). Antimony surfactant effect on green emission InGaN/GaN multi quantum wells grown by MOCVD. *Journal of Nanoscience and Nanotechnology* 11: 1787.
211. Baranowski, M. et al. (2012). Time-resolved photoluminescence studies of the optical quality of InGaN/GaN multi-quantum well grown by MOCVD—antimony surfactant effect. *Semiconductor Science and Technology* 27: 105027.
212. Merrell, J.L., Liu, F., and Stringfellow, G.B. (2013). Effect of surfactant Sb on In incorporation and thin film morphology of InGaN layers grown by organometallic vapor phase epitaxy. *Journal of Crystal Growth* 375: 90.
213. Zhang, Y. and Zhu, J. (2016). Surfactant antimony enhanced indium incorporation on InGaN $(000\bar{1}$ surface: A DFT study. *Journal of Crystal Growth* 438. 43.
214. Moustakis, T.D. et al. (2008). Growth of III-nitride quantum dots and their applications to blue-green LEDs. *Physica Status Solidi A* 205: 2560.
215. Shu, C.K., Ou, J., Lin, H.C. et al. (1998). Isoelectronic In-doping effect in GaN films grown by metalorganic chemical vapor deposition. *Applied Physics Letters* 73: 641.
216. Keller, S., Heikman, S., Ben-Yaacov, I. et al. (2001). Indium-surfactant-assisted growth of high mobility AlN/GaN multilayer structures by metalorganic chemical vapor deposition. *Applied Physics Letters* 79: 3449
217. Northrup, J.E. and Van de Walle, C.G. (2004). Indium versus hydrogen-terminated GaN(0001) surfaces: Surfactant effect of indium in a chemical vapor deposition environment. *Applied Physics Letters* 84: 4322.
218. Feng, Z.H., Cai, S.J., Chen, K.J. et al. (2006). Isoelectronic indium-surfactant-doped $Al_{0.3}Ga_{0.7}N$/GaN high electron mobility transistors. *Applied Physics Letters* 88: 122113.
219. Follstaedt, D.M., Schneider, R.P., and Jones, E.D. (1995). Microstructures of (In,Ga)P alloys grown on GaAs by metalorganic vapor-phase epitaxy. *Journal of Applied Physics* 77: 3077.
220. Novikov, S.V. et al. (2013). Molecular beam epitaxy of highly mismatched N-rich $GaN_{1-x}Sb_x$ and $InN_{1-x}As_x$ alloys. *Journal of Vacuum Science and Technology B* 31: 03C102.
221. Leyer, M., Stellmach, J., Meissner, C. et al. (2008). The critical thickness of InGaN on (0001)GaN. *Journal of Crystal Growth* 310: 4913.
222. Inatomi Y. et al. (2017). Theoretical study of the composition pulling effect in InGaN metalorganic vapor-phase epitaxy growth. *Japanese Journal of Applied Physics* 56: 078003.
223. Humphries, C.J. (2007). Does In form In-rich clusters in InGaN quantum wells? *Philosophical Magazine* 87: 1971;O'Niel, J.P., Ross, I.M., Cullis, A.G. et al. (2003). Electron-beam-induced segregation in InGaN/GaN multiple-quantum wells. *Applied Physics Letters* 83: 1965;Shivaraman, R., Kawaguchi, Y., Tanaka, S. et al. (2013). Comparative analysis of $202\bar{1}$ and $20\bar{2}\bar{1}$ semipolar GaN light emitting diodes using atom probe tomography. *Applied Physics Letters* 102: 251104.
224. Humphries, C.J. et al. (2017). The atomic structure of polar and non-polar InGaN quantum wells and the green gap problem. *Ultramicroscopy* 176: 93.
225. Narukawa, Y., Ichikawa, M., Sanga, D. et al. (2010). White light emitting diodes with super-high luminous efficacy. *Journal of Physics D (Applied Physics)* 43: 354002.
226. Kaneta, A., Hashimoto, T., Nishimura, K. et al. (2010). Visualization of the local carrier dynamics in an InGaN quantum well using dual-probe scanning near-field optical microscopy. *Applied Physics Express* 3: 102102; Mishra, U. (2008). Group III nitride optoelectronics. Invited talk, Electronic Materials Conference, Santa Barbara, CA.
227. Chichibu, S.F. et al. (2001). Optical and structural studies in InGaN quantum well structure laser diodes. *Journal of Vacuum Science and Technology B* 19: 2177;Chichibu, S.F., Kojima, K., Uedono. A. et al. (2017). Defect-resistant

radiative performance of m-plane immiscible AlInN epitaxial nanostructures for deep-ultraviolet and visible polarized light emitters. *Advanced Materials* 29: 1603644.
228. Damilano, B. and Gil, B. (2015). Yellow-red emission from (Ga,In)N heterostructures. *Journal of Physics D (Applied Physics)* 48: 403001.
229. Hwang, J., Hashimoto, R., Saito, S. et al. (2014). Development of InGaN-based red LED grown on (0001) polar surface. *Applied Physics Express* 7: 071003.
230. Shojiki, K., Tanikawa, T., Choi, J. et al. (2015). Red to blue wavelength emission of N-polar 0001 InGaN light-emitting diodes grown by metalorganic vapor phase epitaxy. *Applied Physics Express* 8: 061005.
231. Pantha, B.N., Li, J., Lin, J.Y. et al. (2010). Evolution of phase separation in In-rich InGaN alloys. *Applied Physics Letters* 96: 232105.
232. Ju, J., Loitsch, B., Stettner, T. et al. (2014). Trade-off between morphology, extended defects, and compositional fluctuation induced carrier localization in high In-content InGaN films. *Journal of Applied Physics* 116: 053501.
233. Papadomanolaki, E., Bazioti, C., Kazazis, S.A. et al. (2016). Molecular beam epitaxy of thick InGaN(0001) films: Effects of substrate temperature on structural and electronic properties. *Journal of Crystal Growth* 437: 20.
234. Sterzer, E. et al. (2016). Efficient nitrogen incorporation in GaAs using novel metal organic As–N precursor di-tertiary-butyl-arsano-amine (DTBAA). *Journal of Crystal Growth* 439: 19;Sterzer, E. et al. (2018). 1 eV Ga(NAsSb) grown by MOVPE using DTBAA. *AIP Advances* 8: 055329.
235. Jiang, Y. et al. (2015). Realization of high-luminous-efficiency InGaN light-emitting diodes in the "green gap" range. *Nature Scientific Reports* 5: 10883 DOI:10.2038/srep10883.
236. Lin, T., Kuo, H.C., Jiang, X.D. et al. (2017). Recombination pathways in green InGaN/GaN multiple quantum wells. *Nanoscale Research Letters* 12: 137.
237. Murotani, H., Yamada, Y., Tabata, T. et al. (2013). Effects of exciton localization on internal quantum efficiency of InGaN nanowires. *Journal of Applied Physics* 114: 153506.
238. Cherns, D., Webster, R.F., Novikov, S.V. et al. (2014). Compositional variations in $In_{0.5}Ga_{0.5}N$ nanorods grown by molecular beam epitaxy. *Nanotechnology* 25: 215705.
239. Park, I., Kwon, M., Baek, S. et al. (2005). Enhancement of phase separation in the InGaN layer for self-assembled In-rich quantum dots. *Applied Physics Letters* 87: 061906;Wang, X.H., Jia, H.Q., Guo, L.W. et al. (2007). White light-emitting diodes based on a single InGaN emission layer. *Applied Physics Letters* 91: 161912;Lu, C., Huang, C., Chen, Y. et al. (2008). Dependence of spectral behavior in an InGaN/GaN quantum-well light-emitting diode on the prestrained barrier thickness. *Journal of Applied Physics* 104: 043108.
240. Hirst, L.C. et al. (2017). Imaging atomic-scale clustering in III-V semiconductor alloys. *ACS Nano* 11: 2734.
241. Huang, K.T., Cohen, R.M., and Stringfellow, G.B. (1995). InTlSb growth by OMVPE. *Journal of Crystal Growth* 156 320.
242. Huang, K.T., Chiu, C.T., Cohen, R.M. et al. (1994). InAsSbBi alloys grown by organometallic vapor-phase epitaxy. *Journal of Applied Physics* 75: 2857.
243. Geisz, J.F., Friedman, D.J., Olson, J.M. et al. (2000). BGaInAs alloys lattice matched to GaAs. *Applied Physics Letters* 76: 1443.
244. Tisch, U., Finkman, E., and Salzman, J. (2002). The anomalous bandgap bowing in GaAsN. *Applied Physics Letters* 81: 463.
245. Ludewig, P., Reinhard, S., Jandieri, K. et al. (2016). MOVPE growth studies of Ga(NAsP)/(BGa)(AsP) multi quantum well heterostructures (MQWH) for the monolithic integration of laser structures on (001) Si-substrates. *Journal of Crystal Growth* 438: 63;Ludewig, P., Diederich, M., Janderi, K. et al. (2017). Growth of high N containing GaNAs/GaP/BGaAsP multi quantum well structures on Si (001) substrates. *Journal of Crystal Growth* 467: 61.
246. Zhang, S.B. and Wei, S.H. (2001). Nitrogen solubility and induced defect complexes in epitaxial GaAs:N. *Physical Review Letters* 86: 1789.
247. Bi, W.G. and Tu, C.W. (1997). Bowing parameter of the band-gap energy of GaN_xAs_{1-x}. *Applied Physics Letters* 70: 1608.
248. Wang, L., Elleuch, O., Shirahata, Y. et al. (2016). Inhomogeneous nitrogen incorporation effects on the transport properties of GaAsN grown by CBE. *Journal of Crystal Growth* 437: 6.
249. Stolz, W. (2016). Invited talk presented at ICMOVPE 18: San Diego, July.
250. Almosni, S. et al. (2016). Correlations between electrical and optical properties in lattice-matched GaAsPN/GaP solar cells. *Solar Energy Materials and Solar Cells* 147: 53.
251. Levander, A.X. et al. (2011). Thermal stability of amorphous $GaN_{1-x}As_x$ alloys. *Applied Physics Letters* 98: 161902.
252. Beaudry, J.N., Masut, R.A., and Desjardins, P. (2008). $GaAs_{1-x}N_x$ on GaAs(001): Nitrogen incorporation kinetics from trimethylgallium, tertiarybutylarsine, and 1,1-dimethylhydrazine organometallic vapor-phase epitaxy. *Journal of Crystal Growth* 310: 1040.
253. Thomas, S., White, S., Chalker, P.R. et al. (2002). Nitrogen incorporation in GaInNAs and GaAsN near the solubility limit. *Journal of Materials Science: Materials in Electronics* 13: 525.

254. Wegele, T., Beyer, A., Ludewig, P. et al. (2016). Interface morphology and composition of Ga(NAsP) quantum well structures for monolithically integrated lasers on silicon substrates. *Journal of Physics, D* 49: 075108.
255. Bellaiche, L., Wei, S.H., and Zunger, A. (1997). Band gaps of GaPN and GaAsN alloys. *Applied Physics Letters* 70: 3558.
256. Buyanova, I.A., Chen, W.M., Goldys, E.M. et al. (2001). Structural properties of a GaN_xP_{1-x} alloy: Raman studies. *Applied Physics Letters* 78: 3959.
257. Dobrovolsky, A., Sukrittanon, S., Kuang, Y.J. et al. (2014). Raman spectroscopy of GaP/GaNP core/shell nanowires. *Applied Physics Letters* 105: 193102.
258. Bennarndt, W., Boehm, G., and Amann, M.C. (2016). Domains of molecular beam epitaxial growth of Ga(In)AsBi on GaAs and InP substrates. *Journal of Crystal Growth* 436: 56.
259. Jacobsen, H., Puchala, B., Kuech, T.F. et al. (2012). Ab initio study of the strain dependent thermodynamics of Bi doping in GaAs. *Physical Review B* 86: 085207.
260. Ludewig, P., Bushell, Z.L., Nattermann, L. et al. (2014). Growth of Ga(AsBi) on GaAs by continuous flow MOVPE. *Journal of Crystal Growth* 396: 95.
261. Moussa, I., Fitouri, H., Rebey, A. et al. (2012). Atmospheric-pressure metalorganic vapour phase epitaxy optimization of GaAsBi alloy. *Thin Solid Films* 516: 8372.
262. Nattermann, L., Ludewig, P., Sterzerm E. et al. (2017). Exploiting strain to enhance the Bi incorporation in GaAs-based III/V semiconductors using MOVPE. *Journal of Crystal Growth* 470: 15.
263. Luna, E., Wu, M., Hanke, M. et al. (2016). Spontaneous formation of three-dimensionally ordered Bi-rich nanostructures within $GaAs_{1-x}Bi_x$/GaAs quantum wells. *Nanotechnology* 27: 325603.
264. Kuech, T.A., Guan, Y., Babcock, S. et al. (2017). Talk presented at the 18th Workshop on OMVPE.
265. Hossain, N., Hosea, T.J.C., Sweeney, S.J. et al. (2011). Band structure properties of novel $B_xGa_{1-x}P$ alloys for silicon integration. *Journal of Applied Physics* 110: 063101.
266. Ougaazzaden, A., Gautier, S., Sartel, C. et al. (2007). BGaN materials on GaN/sapphire substrate by MOVPE using N2 carrier gas. *Journal of Crystal Growth* 298: 316.
267. Gautier, S. et al. (2011). Deep structural analysis of novel BGaN material layers grown by MOVPE. *Journal of Crystal Growth* 315: 288.
268. Wei, C.H., Xie, Z.Y., Edgar, J.H. et al. (2000). MOCVD growth of GaBN on 6H-SiC (0001) substrates. *Journal of Electronic Materials* 29: 452.
269. Jin, Y.J., Tang, X.H., Ke, C. et al. (2018). Bandgap engineering of InSb by N incorporation by metal-organic chemical vapor deposition. *Journal of Alloys and Compounds* 756: 134.

3

Column III: Phosphides, Arsenides, and Antimonides

H. Hardtdegen[1] and M. Mikulics[2]
[1]Ernst Ruska-Centre, Forschungszentrum Jülich GmbH, Germany
[2]Peter Grünberg Institute, Forschungszentrum Jülich GmbH, Germany

3.1 Introduction

The epitaxy of zincblende crystallizing III-V alloys and their properties is greatly affected by the lattice constants of the deposited layer with respect to the substrate [1]. In general, for heteroepitaxial growth of a layer on a substrate, a close fit of the substrate in-plane lattice to the epitaxially deposited layer is desirable. A mismatch in the lattice will induce strain, dislocations, or both. Strain will be alleviated eventually, when the thickness of the layer exceeds a critical value [2]. The defects/dislocations formed have a detrimental effect on the properties of the material [3]. Since the lattice parameter of a ternary III-V alloy is directly related to the mole fraction of the respective binary compounds [4], the misfit of an epitaxial layer to a substrate can be adjusted by choosing the right composition of a ternary alloy. However, the alloy composition also determines the bandgap, its nature (direct/indirect) carrier mobility, and therefore the optical and electrical properties of the material. A further degree of freedom, i.e. bandgap engineering, can be achieved by depositing quaternary alloys. The trade-off, however, is a more difficult growth process and characterization of the composition. A very useful map of phosphide, arsenide, and antimonide can be employed to "plan" epitaxial growth and is presented in Figure 3.1 [5]. This map delivers information on the ability to deposit ternary alloys successfully and epitaxially on each other and on a chosen substrate with respect to in-plane lattice constants. These are presented for the compounds at room temperature and may differ at epitaxy temperatures due to different thermal

Metalorganic Vapor Phase Epitaxy (MOVPE): Growth, Materials Properties, and Applications,
First Edition. Edited by Stuart Irvine and Peter Capper.

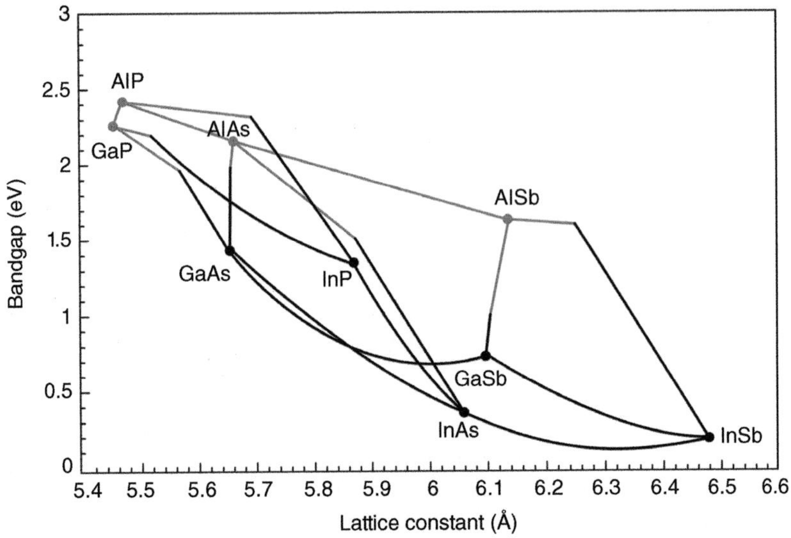

Figure 3.1 *Bandgap energy and lattice constant for different column III phosphide, arsenide, and antimonide alloys. The end point denotes the respective binary compounds. Gray lines denote an alloy with an indirect bandgap, black lines with a direct bandgap. (Source: Reproduced from Ryou et al. 2009 [5], with permission of the Royal Society of Chemistry.)*

expansion coefficients. The map also delivers information on the nature of the layer's expected bandgap – direct or indirect – and its size.

Before starting an epitaxial growth run, a structural design concept should be on hand. The substrate has multiple functions: on the one hand, it provides the in-plane lattice parameter; on the other hand, its physical properties such as reflectivity, absorbance, and thermal and electrical conductivity may also play an active role in the application or device. In addition to tuning the composition and thickness of a single layer, the optical and electrical properties of the material can be tailored by depositing multiple layers, varying the layer thickness down to the nanoscale (utilizing bandgap alignment and quantum confinement effects), and doping in them. The lateral and vertical control of layer thickness, composition, and doping are a prerequisite for successful and reproducible material deposition. The choice of the precursors as well as the growth conditions should be adapted to this end and also to the application envisaged. How important is the avoidance of deep- and shallow-level impurity incorporation? What effect may interface abruptness have? Is diffusion of dopants or constituents of the alloy considered a problem for the application/device characteristics? Which growth rate would be most appropriate? These are only a few of the questions that need to be answered and that play a major role when choosing the most suitable conditions for epitaxy. They will be addressed with respect to the applications selected. In general, a compromise needs to be made and priorities need to be set. Therefore, the definition of important characteristics for the application envisaged is a guideline to the choice of the most appropriate growth conditions.

First, the available precursors and their employment for the respective material systems will be discussed. Then, the specifics of their growth and their properties will be presented. Finally, the specifics of growth conditions and precursor choice for the application or device envisaged will be provided. The MOVPE of III-V compounds and their applications presented will be organized according to the substrate employed. Materials are dealt with that are (nearly) lattice-matched to

GaAs and InP. The antimonides are a special class of materials, since they are frequently deposited on substrates with a large lattice mismatch such as GaAs and InP in addition to GaSb, which is not obtainable as a semi-insulating substrate and is also expensive and less readily available.

3.2 Precursors for Column III Phosphides, Arsenides, and Antimonides

Suitable volatility and stability of the source molecules are important prerequisites for the choice of a precursor. To this end, tri-alkyl column III precursors R_3M (M=Al,Ga,In) are most widely used. They exhibit a high enough volatility and can be transported with a carrier gas, either hydrogen or nitrogen [6], without the need to heat the lines to the reactor. Primarily trimethyl column III precursors are employed. The trimethylaluminum (TMAl) compound is a dimer: this must be considered when adjusting the column III organic partial pressures for a specific composition. If alloys are to be deposited with a low carbon content and/or at a low growth temperature, triethylgallium (TEGa), triethylaluminum (TEAl), or tritertiarybutylaluminum (TtBAl) may be chosen as the Ga or Al precursor, respectively. The stability of the M–C bond decreases as the number of carbon-containing groups attached to the carbon bonded with the metal increases [7]. Unintentional carbon-impurity incorporation is then reduced by a more complete decomposition of the molecule. The use of alanes such as trimethylamine alane (TMAAl) and dimethylethylamine alane (DMEAAl) together with TEGa has proven to be suitable for the deposition of alloys containing low carbon and oxygen impurity. The deposition must then be carried out at low temperature. The arsenides and phosphides are preferably then grown using nitrogen as the carrier gas [8] to achieve homogeneous layers with good composition control.

With respect to the column V phosphorus and arsenic compounds, the sources are provided as hydrides. It was found that the column V source must deliver column VH_x species to the growth process or surface reaction that help remove the alkyl radical by producing alkanes [9]. Due to the high vapor pressure of the hydrides together with their high toxicity, alternative organic As and P precursors were developed. They exhibit a much lower vapor pressure and therefore increase the safety of the deposition process. Tertiarybutylarsine (TBAs) and tertiarybutylphosphine (TBP) have become the most common organic As and P sources, respectively. The alkyl hydrides have a lower stability than the hydrides and allow deposition at low growth temperatures and/or more efficient growth. Since the antimony hydride stibine is only stable at room temperature, organic antimony precursors are employed for growth. For those antimonides, which are deposited preferentially below 500 °C, triethylantimony (TESb) or trisdimethylaminoantimony (TDMASb) is employed. For higher-temperature growth, such as for GaSb, trimethylantimony (TMSb) is most often used. Overall, there is an endeavor to inject source compounds of matching stability. This is helpful in obtaining homogeneous films. Depending on the growth conditions, different choices of precursor combinations make sense.

Donor doping in III–V semiconductors is carried out based either on column IV elements, namely Ge, Si, or Sn replacing the column III element in the lattice; or on a column VI element, i.e. S or Te, substituted for the column V element in the lattice. The dopant Se is avoided due to well-known memory effects. Most commonly, Si and disilane, silane, and ditertiarybutylsilane (DtBSi) are used as the element and the precursor, respectively. However, in the $(Al_xGa_{1-x})_{0.52}In_{0.48}P$ material system, diethytelluride (DETe) is preferentially employed [10]. p-Type doping is carried out either by carbon replacing a column V element or by column II elements substituting for a column III element in the lattice. CBr_4 is the most commonly used carbon source, and diethylzinc (DEZn), dimethylzinc (DMZn), and biscyclopentadienylmagnesium (Cp_2Mg) are common column II sources.

3.3 GaAs-Based Materials

3.3.1 (AlGa)As/GaAs Properties and Deposition

By varying the ratio of aluminum to gallium in the alloy (Figure 3.1 [5]), a broad bandgap range from 1.424 eV to 2.168 eV is achieved, whereas the lattice parameter barely varies from 0.56533 nm to 0.56611 nm for GaAs and AlAs, respectively. A transition of the direct bandgap to the indirect bandgap is observed as the Al concentration exceeds values of about 45% [5, 11]. Due to its reactivity toward oxygen, the Al-containing compounds oxidize more easily, the higher the Al content. Therefore, they must be protected from the environment by a thin layer of GaAs. On the other hand, this property can be exploited for device processing [12]. Since the AlGaAs/GaAs material system is widely investigated and well understood with respect to the relationship between composition and lattice parameter, it is especially suited to demonstrate exemplarily how III-V material growth can be studied and optimized.

There are basically three growth regimes [13–15] in which the vapor phase epitaxy process of III-V compounds is carried out. They are well-demonstrated in the plot presented in Figure 3.2 [13]. In this plot, at lower temperatures, a linear increase of the growth rate versus the reciprocal

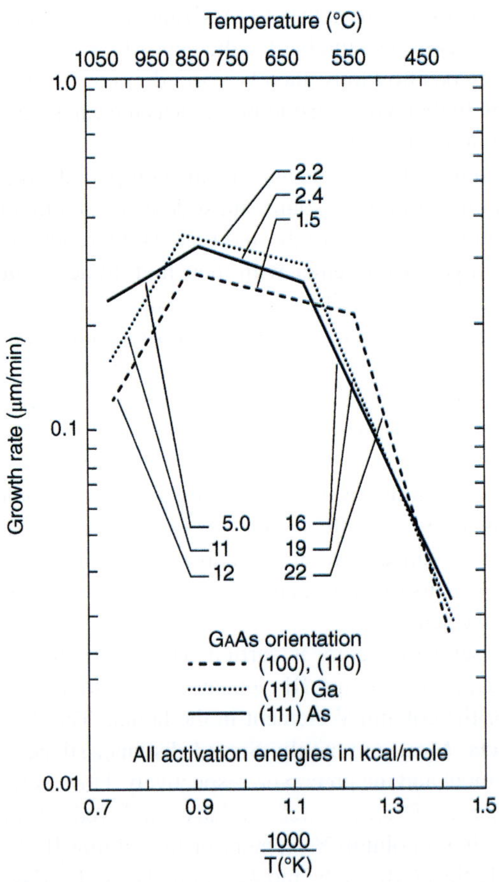

Figure 3.2 Temperature dependence of GaAs growth rate for four different substrate orientations. Three growth regimes are observed: at lower temperatures, a linear increase is observed, at intermediate temperatures, growth is only weakly temperature dependent and at higher temperatures, a linear decrease is observed. Source: Reproduced from Reep and Ghandi 1983 [13], with permission from the Electrochemical Society.

temperature is observed. In this regime, the slope gives information on the activation energy of the slowest (rate-determining) reaction. At somewhat higher temperature, the growth rate is nearly independent of the reciprocal temperature. Here, the diffusion of the reactive species through a boundary layer to the substrate surface determines the growth rate. At even higher temperatures, the growth rate decreases. In this regime, growth is thermodynamically controlled, and we approach an equilibrium between growth and decomposition. Furthermore, at higher temperatures, the temperature of the gas phase increases to such an extent that prereactions may take place, leading to parasitic deposition and depletion of reagents close to the substrate. For applications, growth is preferentially carried out in the diffusion-controlled growth regime, so that small changes in the temperature do not affect the growth rate.

The composition and the growth rate are determined by the partial pressures of the column III metalorganic compounds. The column V source is usually injected in excess (typically larger ratios than 10). In addition to its role in the formation of the (AlGa)As, it is needed to compensate for As loss at growth temperature from the substrate surface [16] due to its high volatility. In addition, it delivers column VH_x species [17] that react with the alkyl radicals to form alkanes and ensures pure materials.

When depositing alloys, the stability and reactivity of the precursors should be similar, i.e. ideally a similar growth behavior of the respective binary compounds using the precursors chosen should be observed. These are prerequisites for uniform and reproducible large-area deposition. An example of ideal [18] deposition behavior is presented in Figure 3.3. Growth of the binary compounds is similar throughout the temperature range under investigation. Furthermore, a wide temperature range is found in which growth is diffusion controlled and uniform layer deposition is to be expected for the precursors trimethylgallium (TMGa), trimethylaluminum (TMAl), and arsine. The uniformity of the ternary composition is exceptional for this set of precursors, due to their matching stability [18]. If precursors are less stable, such as is the case for triethylgallium (TEGa) or triisobutylgallium (TiBGa), the temperature range for diffusion-controlled growth is very small and shifted toward a lower temperature, as is presented in Figure 3.4 [19]. Such precursors are useful for epitaxy at low temperatures.

In addition to the stability of the precursors in MOVPE, the carrier gas also plays an important role in growth uniformity and material purity [20–22]. Even though hydrogen is often employed as

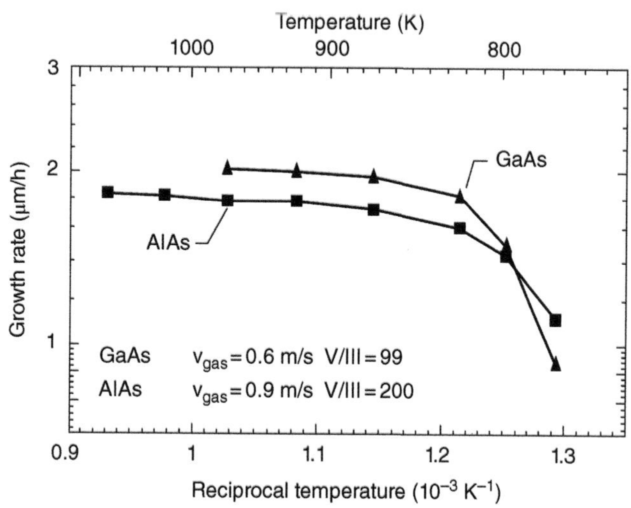

Figure 3.3 *Dependence of growth on temperature for the binary compounds AlAs and GaAs using the methyl column III precursors and arsine. A similar growth behavior leads to uniform (AlGa)As deposition. Source: Reproduced from Hollfelder et al. 1994 [18], with permission from Springer Nature.*

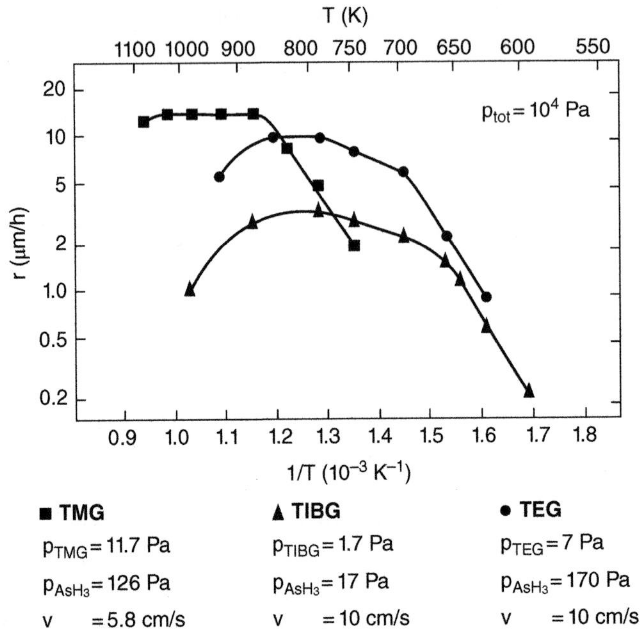

Figure 3.4 *Dependence of GaAs growth on temperature for different Ga precursors. The less stable the precursor, the narrower the temperature range and the lower the temperature in which growth is independent on temperature. Only the precursor TMGa exhibits a wide diffusion-controlled growth regime suitable for uniform deposition. Source: Reproduced from Plass et al. 1988 [19], with permission from Elsevier.*

the carrier gas for III-V semiconductor layers because of its easy purification, nitrogen has its advantages, too. Nitrogen is non-explosive (not reactive) and does not react with the GaAs surface, inducing As desorption. It is less thermally conductive, so parasitic deposition and a depletion of reagents is less pronounced. Its hydrodynamic properties also help enhance homogeneity during deposition. Therefore, more easily decomposable precursors can also be employed to achieve uniformly deposited layers [8].

The most prominent unintentional (intrinsic) impurities in (AlGa)As are oxygen and carbon. Both impurities increase with the concentration of Al in the alloy [23]. Oxygen forms a deep-level trap in (AlGa)As and is a nonradiative recombination center detrimentally affecting carrier concentration and mobility as well as luminescence efficiency [24]. Carbon is a shallow acceptor and is an intrinsic residual impurity in the layers. Besides possible leaks in the equipment, trialkyl aluminum compounds are the main suppliers of oxygen impurities in the layers [24]. They are reactive toward moisture and oxygen and form dialkylaluminum alkoxides. In the special case of trimethylaluminum, dimer gas molecules evaporate from the liquid phase. They exhibit a lower vapor pressure than the molecule dimethylaluminum methoxide, which is a monomer in the gas phase. Therefore, the alkoxide (and the impurity oxygen) is extracted first from the bubbler and is preferentially injected into the reactor. The higher their partial pressure in the reactor, the higher the intrinsic impurity content in the layers. The impurity content in (AlGa)As is strongly affected by the growth parameters. Once oxygen impurities are present, their incorporation can be reduced by using high V/III ratios (above 25) to increase the concentration of the reducing agent AsH_x on the surface. Furthermore, higher temperatures help reverse the exothermic reaction forming aluminum oxide back to the elements again. In addition, the suboxides are more easily desorbed at higher temperatures (above 700 °C). The intrinsic carbon concentration originates from incomplete decomposition of the metalorganic compound. Therefore, the oxygen impurity concentration is found to decrease with increasing growth temperatures [25]. The carbon concentration is naturally also reduced by increasing the V/III ratio, i.e. the concentration of AsH_x present on the surface. It is responsible for the formation of volatile alkanes from the surface and the removal of intrinsic carbon. Additionally, it can be reduced by using a slightly misoriented substrate surface [26]. The misorientation produces steps on the substrate surface at which the atomic hydrogen supplying AsH_x molecules adhere. Since arsine contains GeH_4 as an impurity and Ge forms a shallow-level donor in (AlGa)As layers, an increase of V/III ratio affects the background carrier concentration and type. At low V/III ratios, the layers are at first p-type (intrinsic carbon). The acceptor concentration gradually decreases as the V/III ratio increases. At an intermediate V/III ratio, the layers are compensated. A further increase in V/III ratio leads to n-type layers. This behavior is visualized in Figure 3.5 for GaAs layers [27] and also holds for (AlGa)As layers. Since GeH_4 decomposes more, the higher the temperature, the n-type impurity incorporation also increases. The effect of growth temperature on the incorporation of carbon impurities is different for GaAs and (AlGa)As. The carbon concentration decreases with temperature for GaAs, whereas it increases for (AlGa)As. The latter is due to the formation of stable AlC at higher temperatures (above 750 °C). Specular layers of high crystallinity can be obtained in a large temperature window between 600 and 700 °C und 650 and 800 °C for GaAs and (AlGa)As, respectively. Growth is carried out in a large range of reactor pressures between 2 and 1000 hPa. The effects of growth parameters on impurity incorporation and defect formation need to be considered, when the layers are to be employed for applications.

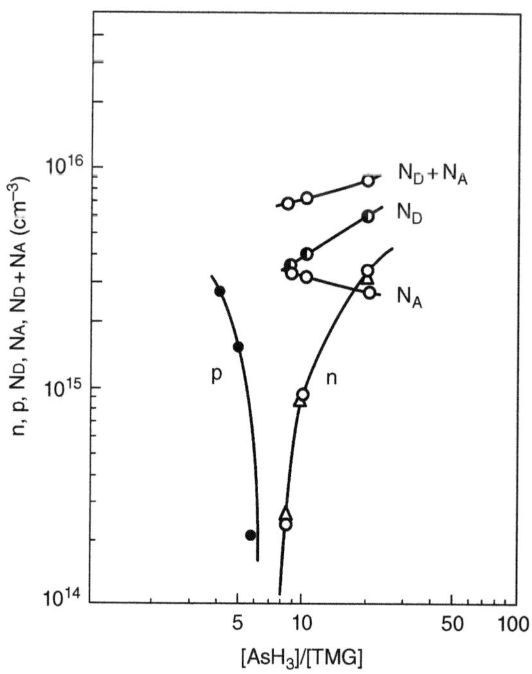

Figure 3.5 *Effect of AsH_3 and TMGa input molar flow ratio on the electron and hole concentration in GaAs. Source: Reproduced from Nakanisi 1984 [27], with permission from Elsevier.*

3.3.1.1 Doping in (AlGa)As Structures

The GaAs and (AlGa)As layers exhibit a background carrier concentration (depending on the growth conditions) in the range of below 10^{14} cm^{-3} and in the 10^{15} cm^{-3}, respectively. Si is the most frequently used donor for doping in (AlGa)As structures. Its major advantage is that it does not exhibit any memory effects. Mostly, silane or disilane are employed as the precursors. These two sources behave differently with respect to their incorporation. For silane, doping is kinetically limited, and the carrier concentration increases with the growth temperature [28]. In contrast, doping with disilane is virtually independent of growth temperature [29, 30]. A concentration of up to $\sim 8\times10^{18}$ cm^{-3} and levels up to 10^{19} cm^{-3} are achieved for GaAs and $Ga_{0.7}Al_{0.3}As$, respectively, before saturation is observed. No effect of V/III ratio on doping at a constant column III partial pressure was found. This is because Si incorporates on a column III lattice site and is not in competition with column V elements on the surface. The metalorganic precursor ditertiarybutylsilane (DtBSi) is a suitable alternative to the non-carbon-containing Si hydrides. It is a liquid at room temperature and has a lower vapor pressure than the Si hydrides. Nevertheless, the vapor pressure is still sufficient [31] for epitaxy. All three precursors exhibit a linear dependence of dopant incorporation on partial pressure at otherwise constant growth conditions. Therefore, the control of donor level is well managed. In general, disilane and ditertiarybutylsilane (DtBSi) are more suitable for growth at lower temperatures than silane. Donor-related traps (DX centers) [32–34] are found in $Al_xGa_{1-x}As$ layers for $x\geq0.2$. Their concentration increases with the Al content [35]. Therefore, efficient doping is usually carried out in layers for which the Al content does not exceed 30% of the column III sublattice sites.

Acceptor doping is performed preferentially with carbon. The advantages are that carbon does not tend to diffuse in the lattice, no memory effects are observed, and high incorporation levels up to $\sim1\times10^{21}$ cm^{-3} can be achieved. Two different methods are reported. On the one hand, carbon may be incorporated from the column III source molecule by avoiding any evolution of atomic hydrogen from the As hydride. Instead of injecting hydrogen containing arsine or tertiarybutylarsine (TBAs), trimethylarsenic (TMAs) is employed [36] or replaces part of the molar flow of arsine [37] (or TBAs). Alternatively, a carbon-containing precursor such as CBr_4 [38] could be added to the growth process. The first approach allows for very high incorporation rates, whereas the second gives better control over the whole doping range from $\sim1\times10^{16}$ cm^{-3} to $\sim1\times10^{19}$ cm^{-3}. Highly p-doped layers are passivated by hydrogen [39, 40]. A post-growth anneal in an atmosphere without hydrogen is then advantageous. Column II metalorganic sources such as diethylzinc (DEZn), dimethylzinc (DMZn), and biscylopentadienylmagnesium (Cp_2Mg) can also be employed. Their solubility in the lattice is, however, lower than for carbon, and memory effects and diffusion are more pronounced. Acceptor concentrations of up to $\sim1\times10^{20}$ cm^{-3} [41] and $\sim1\times10^{19}$ cm^{-3} [42, 43] were achieved with zinc and magnesium, respectively.

3.3.1.2 Deposition of (AlGa)As/GaAs Heterostructures

The optimal deposition conditions for heterostructure growth depend on the application envisaged. Generally, the GaAs substrate is first annealed in the carrier gas in which the column V precursor is injected. This helps remove any impurities (such as oxygen or carbon), which form volatile gaseous products with hydrogen radicals present in the reactor. Then, a roughly 300-nm thick GaAs buffer layer is deposited to improve the surface of the GaAs substrate. Typical growth rates are 1–2 μm/h. During growth, the column V precursor flow is injected continually. At all heterostructure interfaces, however, the growth is interrupted by stopping the column III precursor injection. By doing so, the surface can smoothen. This interruption, however, should not be too long, since impurities in the gas phase could accumulate on the surface. The growth-interruption times are related to the flow velocity in the reactor and the time it takes to exchange the total gas volume in the reactor. Growth is resumed by switching the column III precursors on again. The interface between GaAs and (AlGa)As is generally smoother than the interface from (AlGa)As to GaAs, owing to the slower diffusion of Al on the surface for the latter. This feature should be considered when planning a device structure, which is sensitive to the interface.

3.3.2 GaInP, (AlGa)InP/GaAs Properties and Deposition

The $Al_xGa_yIn_{1-x-y}P$ material system is widely employed for numerous photonic and electronic applications. In contrast to the (AlGa)As/GaAs material system, the lattice constant is strongly affected by the column III elemental composition. Lattice matching to GaAs can be achieved with a composition close to 50% for In on the column III sublattice (i.e. $x+y\sim0.5$ in $Al_xGa_yIn_{1-x-y}P$). A bandgap energy anomaly was observed for this material system, which was attributed to ordering on the column III sublattice [44]. The bandgap of the alloy deposited lattice matched to GaAs ranges from 2.3 eV for AlInP to 1.91 eV for GaInP for fully disordered material, i.e. a statistical distribution of the cations in the column III sublattice as presented in Figure 3.6 [45]. The bandgap may be reduced up to 160 meV [46] if the column III atoms are ordered [44] (see Section 2.4.3.1). In the extreme, an InP/(AlGa)P "superlattice" is formed on the {111} planes of the cubic zincblende structure, inducing a $CuPt_B$ structure [47]. In addition to the effect on bandgap, ordering is also associated with antiphase boundaries [48], which detrimentally influence the minority carrier lifetimes in the material. Besides the exact tuning of the composition, the control of the $Al_xGa_yIn_{1-x-y}P$ structure is mandatory for any application envisaged.

Therefore, the effect of experimental conditions on ordering has to be well understood. The influence of temperature and V/III ratio for the MOVPE of GaInP was investigated by Gomyo and co-workers [49–51]. They found that at lower temperatures, around ≤600 °C, no superstructure is present. Above this temperature, the superstructures become more pronounced until they reach a maximum content of ordering at 680–700 °C. At temperatures beyond 700 °C, the superstructure becomes less pronounced again. At 770 °C, a statistical distribution of the column III cations is observed. In the lower temperature range, the surface kinetics – and in the high temperature range, the entropy – induces the statistical distribution of the column III cations on the sublattice. Since ordering leads to band-edge shrinkage, a U-shaped behavior of the bandgap in dependence of the growth temperature is observed [52, 53] as presented in Figure 3.7. In addition, the V/III ratio also influences the degree of ordering, but its effect is dependent on the temperature: for example, at

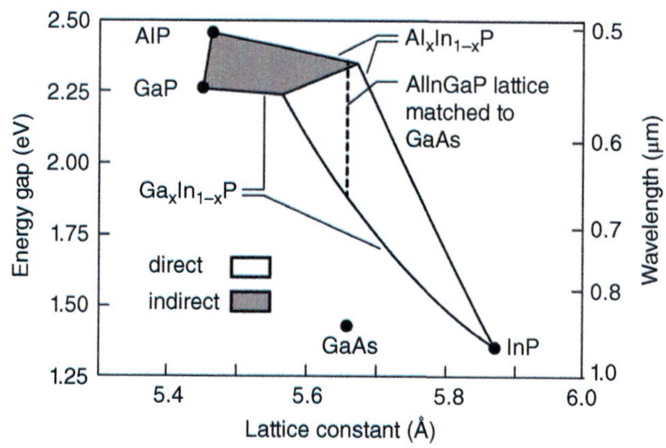

Figure 3.6 Energy gap as a function of lattice constant for the $Al_xGa_yIn_{1-x-y}P$ material system. A direct bandgap between and 1.9 eV to 2.26 eV is obtained for lattice-matched alloys, i.e. from the red to green spectral range. Source: Reproduced from Vanderwater et al. 1997 [45], with permission from IEEE.

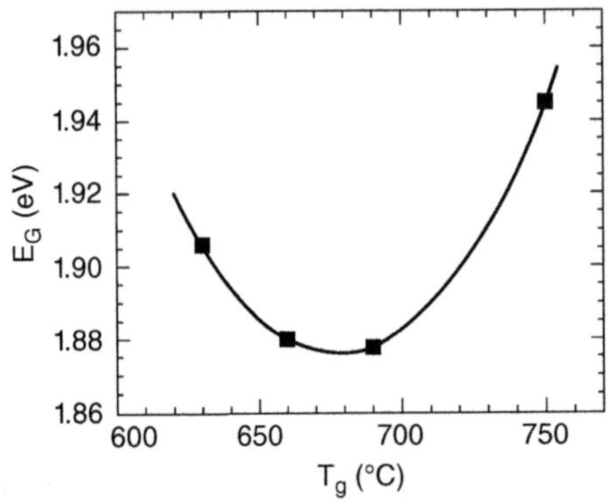

Figure 3.7 *Dependence of the bandgap energy of GaInP on growth temperature. The highest degree of ordering and the smallest bandgap occur at 680 °C. Source: Reproduced from Ernst et al. 1996 [53], with permission from AIP.*

700 °C, the degree of ordering increases with the V/III ratio [54]. Furthermore, it has been reported that ordering can be promoted by using nitrogen as the carrier gas [55]. Many effects reported for higher growth temperatures are related to the effective column V/column III element ratio (after decomposition of the precursors) and not to the column V/column III precursor partial pressure ratio. Ordered structures induced by the growth parameters have been associated with the formation of the 2×4 reconstruction during growth and the formation of P-dimers [54]. Ordering can also be impeded by using misoriented substrates and introducing $[\bar{1}10]$ steps on the substrate surface [46, 54, 56, 57]. Here, the steps on the surface prevent the formation of P–dimers, which enhance the ordering of the column III cations. Extrinsic impurities have also reported to be influential to column III element ordering on the sublattice. Surfactants [58] such as Sb [59], Te [60, 61], and especially dopants, which occupy column III sublattice sites [62, 63], are known to enhance disorder in the structure.

Ordering in the $(Al_xGa_{1-x})_{0.5}In_{0.5}P$ material system is also dependent on the Al content x. For $x \leq 0.5$, the effect of ordering on the bandgap gradually diminishes as the concentration increases. This holds for growth temperatures at which ordering in GaInP is observed (670 °C). The ordering and therefore the bandgap, however, are more weakly influenced than for GaInP. This may be due to the fact that the Al–P formation enthalpy is higher than for the other column III cations with phosphorus. A reduction of the mobility of the column III atoms on the surface can be expected, resulting in a reduced degree of ordering.

Unfortunately, ordered structures are most often only partially ordered. They contain ordered domains of variable size (depending on the growth conditions) in a disordered matrix. They exhibit "moving emission" and varying radiative lifetimes when the excitation intensity is varied [53]. Overall, this leads to undesired optical properties. Therefore, for most high-volume industrial applications such as LEDs and solar cells, fully disordered material is preferential. Fortunately, ordering can be impeded by choosing the appropriate growth conditions and orientation of the substrate [49, 50, 55, 64]. These conditions are, for example, a high deposition temperature of around 750 °C or higher, a V/III ratio of 200 (due to the high stability of phosphine compared to arsine), a high growth rate of 2–3 μm/h, and the use of hydrogen as the carrier gas. On the one hand, the high growth temperature helps reduce the incorporation of oxygen in the alloy similar as for (AlGa)As. High growth temperatures are, however, detrimental to In-containing compounds, since the gas phase tends to be increasingly depleted of the In species at higher temperatures. In addition, the (001) GaAs substrate is misoriented around 6° toward $\{111\}_A$, so that sufficient $[\bar{1}10]$ steps on the substrate surface are available. Due to the need for relatively high growth temperatures, growth is usually performed with the methyl column III sources and phosphine as the column V hydride. Growth is usually carried out in the diffusion-controlled growth regime. In this regime, the composition of the column III elements is almost linearly proportional to the input gas-phase composition [65].

3.3.2.1 *Doping in in the $(Al_xGa_{1-x})_{0.5}In_{0.5}P$ Material System*

Formerly, Se was used as the dopant, but memory effects were detrimental to applications [63, 66]. Si and Te [10] are both suitable dopants for GaInP and the quaternary $(Al_xGa_{1-x})_{0.5}In_{0.5}P$ alloys. Either silane, disilane, or diethyltelluride is employed as the precursor.

In contrast, p-doping is much more problematic. For p-doping of $(Al_xGa_{1-x})_{0.5}In_{0.5}P$ alloys, DMZn and Cp_2Mg were investigated as the precursors [67]. It was found that an appreciable acceptor concentration can only be achieved for Zn if the Al concentration in the alloy is small [67]: i.e. the range for doping is between ~ $2 \times 10^{17}\,cm^{-3}$ and ~ $2 \times 10^{18}\,cm^{-3}$ for $x = 1$ and $x = 0$, respectively. Similar to (AlGa)As, the Al-containing alloys are sensitive to oxygen. However, if high doping levels are to be achieved, Mg is the better choice. As is demonstrated by Figure 3.8, the maximum Mg concentration and acceptor concentration were found to be over ~ $2 \times 10^{19}\,cm^{-3}$ and over ~ $1 \times 10^{19}\,cm^{-3}$ for GaInP, and ~ 1×10^{19} and ~ $1 \times 10^{18}\,cm^{-3}$ for AlInP, respectively. Oxygen is a deep-level donor in $(Al_xGa_{1-x})_{0.52}In_{0.48}P$ [68]. Its incorporation leads to a compensation of the acceptors in p-doped material. This is probably the reason for the lower acceptor concentration in AlInP. The Mg precursor is much less reactive toward oxygen than dimethylzinc, will not be a supplier of unintentional oxygen, and is therefore extremely well suited for doping in this material system. Noticeable saturation of the Mg as well as acceptor concentration occurs at the same Mg–column III partial pressure ratio. Above this ratio, diffusion is observed in the epitaxial layers. This indicates that the solubility limit has then been exceeded [67].

Overall, doping affects the ordering in the alloys especially if the dopant occupies the column III lattice site. Then disorder is promoted [63].

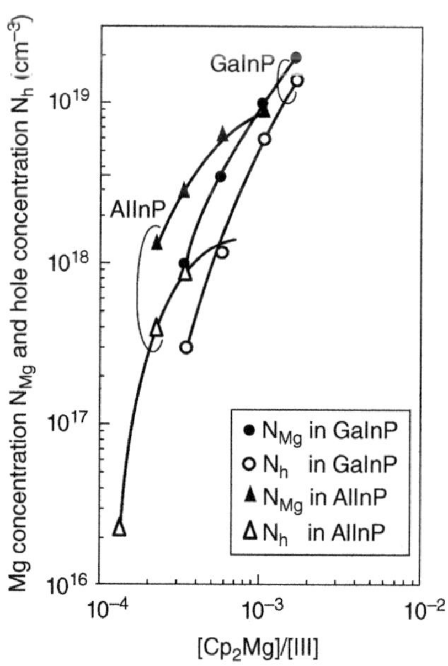

Figure 3.8 *Mg and acceptor concentration in GaP and AlInP as a function of Cp_2Mg / column III precursor partial pressure ratio. Saturation is observed at the same ratio in both alloys. Source: Reproduced from Ohba et al. 1988 [67], with permission from Elsevier.*

3.3.2.2 $(Al_xGa_{1-x})_{0.5}In_{0.5}P$/GaAs Heterojunctions

For applications that rely heavily on interface abruptness, optimization of the switching procedure at the heterointerface is extremely important. The exchange of the column V element from one layer to the next within a heterostructure or even from the substrate to the epitaxial layer is a special challenge. Due to the high volatility of the column V element, the substrate/growing surface has to be stabilized. A carry-over of the column V element may occur if the switching sequence of the precursors is not optimized. An interruption of the column V precursor may be detrimental to the material and surface properties. Also, a carry-over of In, especially at higher growth temperatures, can occur: for example, from $(Al_xGa_{1-x})_{0.5}In_{0.5}P$ to GaAs. In both cases, a carry-over, however, leads to the unintentional formation of alloys exhibiting different optoelectronic properties (i.e. bandgap, band discontinuity [69], etc.) and should be avoided. Therefore, the switching sequence of the column V precursors including a growth interruption by closing the lines to the column III precursors is usually part of the growth strategy [70]. Which procedure is best is a parameter of the MOVPE system. The growth temperature also plays an important role, since it determines the desorption of the volatile elements from the substrate surface. In fact, for applications that rely heavily on interface abruptness, a lower substrate temperature and a not-too-high V/III ratio are advantageous.

3.4 InP-Based Materials

Compounds deposited on InP substrates are predominantly applied in photonic devices suitable for data communication systems functioning in the telecommunication wavelength range 1310–1550 nm. To this end, quaternary $Ga_xIn_{1-x}As_yP_{1-y}$ alloys can be deposited, lattice matched to InP. InP substrates exhibit a higher thermal conductivity ($0.7\,W\,cm^{-1}\,K^{-1}$) than GaAs substrates ($0.46\,W\,cm^{-1}\,K^{-1}$) and are easily rendered semi-insulating by doping with Fe. Therefore, additionally, they form the basis for ultra-fast electronic device applications. InP exhibits a direct bandgap of 1.34 eV, which is the largest direct bandgap of all commercially available compound semiconductors substrates. Because of its large lattice constant of 5.87 Å, InP is chosen for the deposition of low-bandgap alloys with their relatively large lattice parameters and in which high electron mobilities are achieved at room temperature.

One of the challenges for electronic device applications is the preparation of highly resistive substrates. Specifically, an accumulation of the n-type impurity Si on the substrate surface is observed. It has been demonstrated that the InP surface has an affinity toward Si [71–73]. Si bonded to oxygen is present in molecular sieves used for water purification. It does not contribute to the conductivity of the water and may contaminate the surface of the substrate unnoticed during a cleaning and rinsing procedure. Therefore, great care has to be taken in substrate preparation for electronic device applications. The contamination becomes visible when Hall-effect measurements are carried out at room and liquid-nitrogen temperatures. The layers are conductive. Hall measurements then disclose a low electron mobility at room temperature and a freeze-out of carriers toward lower temperatures. This results in an unintentional parallel conductive channel at the substrate/epilayer interface, which for example cannot be controlled by a gate in transistor structures.

3.4.1 InP Properties and Deposition

Similar to the growth of $(AlGa)_{0.5}In_{0.5}P$ alloys, the MOVPE of InP is carried out at rather high V/III partial pressure ratios of > 200 due to the rather limited decomposition efficiency of PH_3. Quite early on, the MOVPE of InP was first investigated using triethylindium (TEIn) as the column III precursor [74, 75] and carried out at atmospheric pressure. Reproducible growth, however, was difficult to achieve. This is due to the fact that gas-phase interactions between both precursors were observed. Two successful methods can be carried out to alleviate the problem. The reactor pressure can be lowered to decrease the residence time of the precursors in the gas phase [76]. The other method is to choose precursors that are more stable than TEIn and do not prereact as easily with PH_3, such as trimethylindium (TMIn) [77, 78]. Naturally, both methods can be combined [79]. In addition, a carrier gas can be employed, which has a lower heat conductivity and helps reduce prereactions upstream of the susceptor [76, 80]. The growth-temperature range in which specular surfaces are obtained and optimal optical and electrical properties of the epilayers are achieved lies between 550 °C and 650 °C. In this range, growth is diffusion controlled and nearly independent of temperature. The background doping concentration is difficult to determine since it is so low ($<10^{14}\,cm^{-3}$) that even for a thickness of 4 μm, the layers are fully depleted. Lightly n-doped samples with a carrier concentration ~ $1\times10^{14}\,cm^{-3}$ exhibit a mobility of ~ $5500\,cm^2V^{-1}\,s^{-1}$. Deposition of InP is possible at somewhat lower growth temperatures and a lower column V/column III precursor partial pressure ratio of below 30 using the more easily pyrolyzed precursor tertiarybutylphosphine (TBP). InP deposited with TBP yields comparable electrical and optical properties [81] to the precursor PH_3.

3.4.2 AlInAs/GaInAs/AlGaInAs Properties and Deposition

The ternaries $Al_{0.48}In_{0.52}As$ and $Ga_{0.47}In_{0.53}As$ are employed in InP-based heterostructures as high- and low-bandgap layers, respectively. The bandgap of $Ga_{0.47}In_{0.53}As$ was determined to be 0.74 eV, making it suitable for fiber optic systems working in the telecommunication wavelength range. In addition, the electron mobilities and peak velocities are definitely higher than for GaAs [82, 83] so it is especially suitable for high-frequency electronic applications [84].

For both ternaries as well as for the quaternary, a strong dependence of lattice parameter on composition is observed. An exact control of composition is, therefore, necessary to achieve lattice matching, since undesirable strain build-up and its successive relaxation are detrimental to device applications. Growth is easily controlled by injecting the respective column III precursor pressure ratios and considering the slightly lower diffusion coefficient of TMIn (see Figure 3.9 [85]) and of TMAl compared to its Ga homolog. Just as for AlGaAs, the dimeric nature of TMAl also needs to be considered.

Deposition is carried out at elevated temperatures compared to InP for both the ternaries and the quaternary and is diffusion controlled [86]. Typically, the temperature range is between 600 °C and 710 °C. For AlInAs (as well as for the lattice-matched quaternary AlGaInAs), the unintentional impurity incorporation decreases with increasing temperature [87] in the optimal temperature range, and the concentration of deep-level defects also decreases with temperatures in that range [88]. The morphology improves markedly, if the V/III ratio is not too small (above 20) [89]. At atmospheric pressure, the optimal growth temperature is slightly lower, around 650 °C [90], than at reduced reactor pressure (680–710 °C) [87]. The layers exhibit a background carrier concentration and electron mobility of ~ $4\times10^{15}\,cm^{-3}$ and ~ $1900\,cm^2V^{-1}\,s^{-1}$ [90], respectively, at room temperature. The carrier concentration in lattice-matched GaInAs layers is lower, and the electron mobility is (naturally) higher than for AlInAs. Here, a carrier concentration of $5\times10^{14}\,cm^{-3}$ and a mobility $11{,}500\,cm^2V^{-1}\,s^{-1}$ were determined at room temperature. The growth conditions for the quaternary AlGaInAs are extrapolated from the conditions and precursor partial pressures for both ternaries [91]. By varying the Ga to Al ratio in the alloy, the bandgap can be varied from 0.74 eV to 1.45 eV [92], which relates to a wavelength between 0.86 μm and 1.68 μm. As for all ternary compound semiconductors containing Al, AlGaInAs is sensitive toward oxygen, which is a deep-level impurity. The concentration decreases as the growth temperature increases.

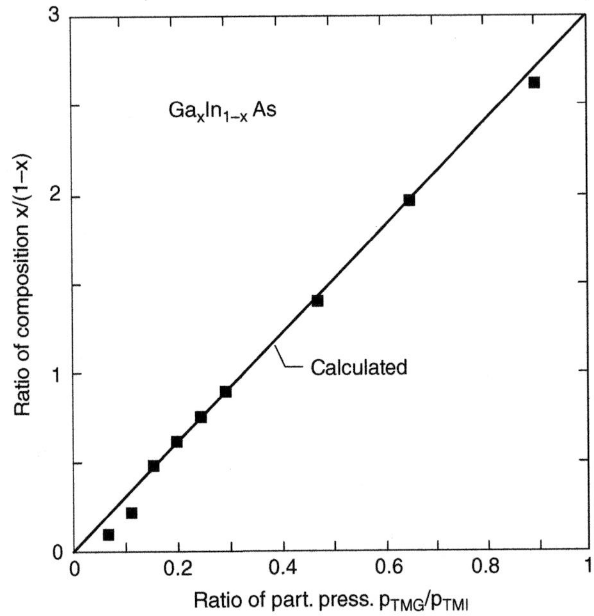

Figure 3.9 *Composition of $Ga_xIn_{1-x}As$ in the solid phase as a function of column III precursor partial pressure ratio. Linearity is observed. Source: Reproduced from Meyer et al. 1992 [85], with permission from Springer Nature.*

3.4.3 AlInAs/GaInAs/InP Heterostructures

Usually, an InP buffer layer is grown on the InP substrate before the active layers are deposited. The active layers usually contain arsenic as the column V element. The formation of abrupt interfaces between InP and AlGaInAs and vice versa is a similar challenge as for the AlGaInP/GaAs material system. Since P is much more volatile than As, the InP surface needs stabilization at temperatures around and above 300 °C (compared to 450 °C for GaAs), especially during low-pressure MOVPE. As described earlier (Section 3.3.2.2), the switching sequence of the precursors must be optimized. An interruption of the column V precursor flow is not feasible and is detrimental to the material and surface properties. A carry-over of As from AlGaInAs to InP is more probable than from phosphorus from InP to AlGaInAs. The carry-over, however, leads to the unintentional formation of alloys [85, 93, 94] exhibiting different optoelectronic properties (i.e. bandgap, band discontinuity, etc.) and should be avoided.

3.4.4 $In_xGa_{1-x}As_yP_{1-y}$ Properties and Deposition

For $In_xGa_{1-x}As_yP_{1-y}$, the bandgap and lattice parameter can be controlled separately [95] by adjusting the composition for the column III and the column V elements. A special challenge is that the size of the radii of the column III and column V elements in the alloy differ strongly. The incorporation of more gallium as well as more phosphorus in the lattice will reduce the lattice parameter and increase the bandgap. More than one characterization technique is then necessary to determine the composition of the alloy. In addition to the performance of X-ray diffraction and photoluminescence spectroscopy to determine the composition of the alloy, in situ optical methods are also useful [96]. The composition with respect to the ratio of the column III elements in the alloy is affected linearly by the ratio of the respective column III precursor partial pressures. The control is quite straightforward, since the decomposition efficiency of the In and Ga precursor conventionally used – TMIn and TMGa – is similar.

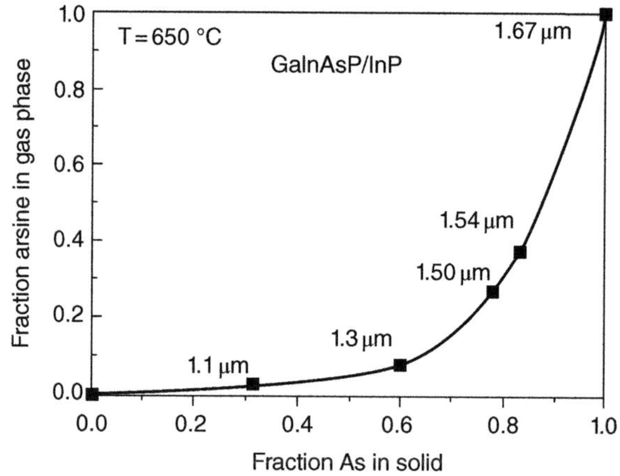

Figure 3.10 *Incorporation of As into GaInAsP as a function of the AsH_3 fraction of the column V hydrides in the gas phase. For emission wavelengths below 1.3 μm, control of composition becomes very challenging. Source: Reproduced from Nelson et al. 1988 [100], with permission from Elsevier.*

This has already been studied for GaInAs growth lattice matched to InP [86]. However, the stability of both hydrides, arsine and phosphine, differs strongly [97–99]. The decomposition efficiency of phosphine is much lower than for arsine at the temperature employed for obtaining layers of excellent surface morphology. Therefore, the composition of the gas phase, i.e. the hydride partial pressures, need to be controlled exactly. The probability of incorporating arsenic is much higher, as for phosphorus. A PH_3/AsH_3 partial pressure ratio of 50 is needed, for example, to achieve a composition of $In_{0.79}Ga_{0.21}As_{0.47}P_{0.53}$ in the quaternary [86]. In general, as presented in Figure 3.10, for lattice-matched InGaAsP epilayers capable of emitting in the range between ~1.00 μm and ~1.67 μm wavelength, it becomes increasingly difficult to control the wavelength below 1.3 μm [100]. The growth temperature must be very well controlled. At constant precursor partial pressures, the wavelength increases with increasing temperature due to the more complete decomposition of phosphine [101, 102]. The effect is a wavelength shift in the region of 1.8 nm/K [102]. The optimal growth temperature range is located between 620 °C and 660 °C. The photoluminescence wavelength uniformity of GaInAsP on large areas is also challenging due to the difference in decomposition efficiency. However, it was demonstrated that the uniformity is greatly improved if nitrogen is used as the carrier gas instead of hydrogen [103, 104], as presented in Figure 3.11. If tertiarybutylarsine (TBAs) and tertiarybutylphosphine (TBP) are employed as the column V precursors, the efficiency of the As and P precursors becomes similar, making compositional control and uniformity much easier [105, 106].

3.4.4.1 Doping in the $In_xGa_{1-x}As_yP_{1-y}$ and in AlGaInAs Material Systems

For the $In_xGa_{1-x}As_yP_{1-y}$ material system, Si and S are the most conventionally used n-type dopants, and disilane/silane and H_2S serve as the precursors. At constant column III and column V precursor partial pressures, the donor concentration varies with the square root of the input partial pressure and varies linearly with the input partial pressure for disilane on the one hand and silane and H_2S [107] on the other hand, respectively. When using disilane and hydrogen sulfide, the incorporation of the dopant is nearly temperature independent. No noticeable diffusion of the dopants is observed in the growth process. The incorporation range of Si and S is possible up to ~ $1 \times 10^{19}\,cm^{-3}$ [108]. The Al-containing alloys are preferentially doped with Si using disilane/silane [109]. A similar behavior of the Si incorporation is observed with respect to donor incorporation and diffusion as for the $In_xGa_{1-x}As_yP_{1-y}$ material system.

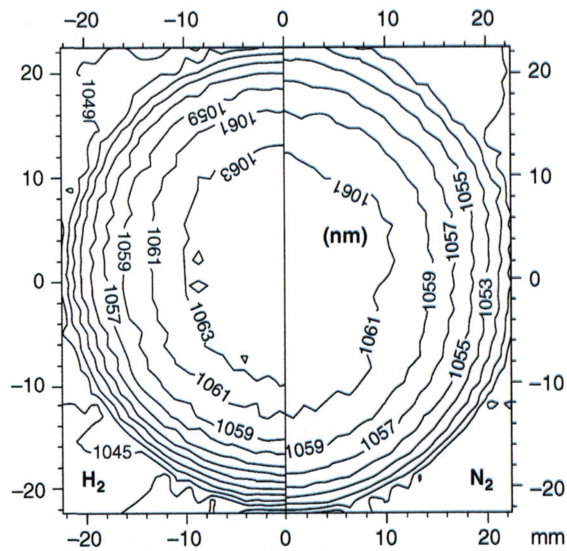

Figure 3.11 GaInAsP room-temperature photoluminescence wavelength distribution for deposition in hydrogen (left) and nitrogen (right) as the carrier gas. Wavelength contour lines are drawn with a spacing of 2 nm. Source: Reproduced from Roehle et al. 1997 [103], with permission from Elsevier.

Zn is the most usual p-type dopant for InP and is injected as dimethyl or diethyl zinc into the growth process. The acceptor concentration relates linearly to the input partial pressure until a concentration of $\sim 2 \times 10^{18}\ cm^{-3}$ [110] is achieved. Above this level, the acceptor concentration saturates. The same acceptor concentration and behavior are also observed if Mg doping is performed with Cp_2Mg [42]. As soon as the concentration reaches the saturation level, zinc will diffuse through the epilayer during growth, and no steep doping profiles are achievable. For heterojunction bipolar transistor (HBT) structures, high p-doping levels in AlInAs need to be mastered. The major difficulty is that a rather high concentration of residual deep-level donors is observed in AlInAs [111, 112]. The donor was identified to be oxygen. For very high acceptor concentrations, carbon is most suited: it is known not to diffuse at high incorporation concentrations and also known to be incorporated with high concentrations. CBr_4 is the preferential dopant source [113]. Acceptor concentrations of up to $\sim 3 \times 10^{19}\ cm^{-3}$ are reached [114] at low V/III ratios of 20 and low growth temperatures (530 °C). The disadvantage of the precursor is, however, that it reacts with AlInAs and specifically with In in the compound. The composition changes unintentionally, and the growth rate decreases. Therefore, the column III partial-pressure ratios have to be adjusted as soon as the dopant is added to the growth process. Highly p-doped layers are passivated similarly by hydrogen [113] as highly p-doped GaAs [40] layers. A post-growth anneal in a hydrogen-free atmosphere is therefore useful to activate all acceptors. There has been considerable interest in InP epilayers with high resistivity for a wide range of microwave and optoelectronic applications. It is possible to render the InP layers semi-insulating by doping them with iron using the precursor biscyclopentadienyl iron (Cp_2Fe) [115, 116]. A resistivity of $>2 \times 10^8\ \Omega$ cm is reached.

3.5 Column III Antimonides Properties and Deposition

Due to their high electron mobilities at room temperature and their narrow bandgaps, the antimonides have been mainly employed industrially as highly sensitive magnetic field sensors in the high field range (above 0.1 T, InSb) and for devices suitable for long-wavelength range applications.

Table 3.1 Melting point temperatures and bandgaps (according to D.W. Palmer, www.semiconductors.co.uk) for binary Al, Ga, and In phosphides, arsenides, and antimonides. The melting point temperatures decrease with increasing anion and cation size owing to the increase in relative covalent bond share and decreasing bond energy, respectively.

AlP	1827 °C	2.45 eV	GaP	1477 °C	2.27 eV	InP	1062 °C	1.34 eV
AlAs	1740 °C	2.15 eV	GaAs	1238 °C	1.42 eV	InAs	942 °C	0.36 eV
AlSb	1070 °C	1.62 eV	GaSb	712 °C	0.75 eV	InSb	527 °C	0.17 eV

Column III alloys of antimonides behave quite differently in comparison to their phosphide and arsenide counterparts, since they exhibit a higher relative covalent bond share and weaker bond energies. Therefore, a lower melting-point temperature (by a few hundred degrees) and smaller bandgaps are observed. A comparative overview of these properties is presented in Table 3.1. There is no stable antimony hydride (stibine) above room temperature. Due to its high vapor pressure and its purity, TMSb is the most commonly precursor used. However, it does not decompose noticeably below 500 °C. The same holds for the methyl Al and Ga precursors. As a consequence, at the low growth temperatures suitable for antimonides, deposition is performed in a regime in which growth is not mass-transport limited but rather in the more complex kinetically controlled growth regime. In contrast to the phosphides and arsenides, a precise control of the V/III partial pressure precursor ratio is mandatory due to the low equilibrium vapor pressure of Sb at growth temperature. Similar to the column III metals, any antimony reaching the substrate surface will then be incorporated [117]. Excess Sb will not desorb, affecting the morphology of the layers and the phases formed.

Due to the low bandgap of the binary antimonide compounds, it is difficult to obtain native semi-insulating substrates. Such substrates are needed especially for the characterization of intentional as well as unintentional doping in the binary layers. Even though InSb and GaSb [118] bulk crystals can be doped and are then highly resistive at low temperatures, room-temperature electrical properties of the unstrained layers cannot be determined. Semi-insulating GaAs substrates have been used instead. The difference in lattice constant between the antimonide layer and the substrate leads to crystal defects due to strain relaxation. The effect of crystal defects and of strain on the electrical properties of the compounds determined, however, cannot be neglected. All these characteristics have to be considered when planning growth experiments.

3.5.1 Deposition of InSb, GaSb, and AlSb

Homoepitaxial growth of the binary antimonides or of multinary compounds on antimonide substrates is a special challenge. The oxides formed on the substrate surface cannot be removed in the reactor upon baking in the column V/carrier gas mixture, because the column V precursor will not deliver the atomic hydrogen necessary for oxide removal. Furthermore, even if hydrogen is employed as the carrier gas, the H_2 molecules will not dissociate at the temperatures at which the crystalline substrates are stable or readily deoxidized. Therefore, the surface cannot be purified by heating and forming volatile oxygen (water) and hydrocarbon compounds. Special care must be taken to chemically etch the surface before growth, so that epitaxial growth will succeed. Similar to phosphide and arsenide wafer surfaces, the antimonide surface also needs to be protected from the loss of antimony at deposition temperature. However, in contrast to the phosphides and arsenides, excess antimony will not desorb and antimony crystals will grow, roughening the substrate surface. On the other hand, if the partial pressure of the antimonide precursor is too low, a column III rich surface evolves during pregrowth bakeout, and column III element droplets or crystals form.

The most important binary antimonide for industrial application is InSb. It is used in highly sensitive magnetic field sensors for automotive applications, is deposited by MOVPE, and is usually n-type. Growth of the layers is difficult because of InSb's low melting-point temperature (527 °C). Therefore, deposition must be carried out below 500 °C – a temperature at which the conventionally available Sb precursor TMSb is nearly undecomposed. It is very challenging to achieve a factual In:Sb ratio of unity on the growing surface and needs a careful fine-tuning of the precursor partial pressures to obtain smooth films. Even though alternative Sb precursors may have better matching pyrolysis characteristics to TMIn [119, 120], and growth control with respect to layer uniformity and morphology is more easily obtained, the highest electron mobilities – important for the application – are still achieved for the methyl In and Sb precursors (see overview in [121]).

GaSb is more easily deposited than InSb. Usually, TMGa and TMSb are chosen as the precursors. Since TMGa and TMSb decompose at nearly the same rate, partial-pressure ratios close to unity are injected, and a stoichiometric Ga:Sb ratio of unity is obtained at the growing surface. The optimal growth temperature is below 550 °C [121]. At this temperature, deposition is kinetically controlled, as can be seen in Figure 3.12. Therefore, if the temperature is altered during growth, the growth rate is strongly affected. This needs to be considered for growth control. A calibration of the growth rate is then mandatory. Since the last decomposition step takes place on the substrate surface, and no active hydrogen is present from the column V source to form alkanes, carbon is incorporated unintentionally in the layers. The thin films are p-type.

The sensitivity of Al toward oxygen and the high bond stability of Al to carbon makes the deposition of highly pure AlSb thin films difficult. In contrast to the phosphides and arsenides, an excess of the column V precursor is not recommended. The Sb-precursor cannot deliver atomic hydrogen to the process and remove the impurities. The injection of *in situ* electrochemically produced stibine does not lead to epitaxial AlSb thin films, since elemental antimony crystals form [121]. The employment of alternative, less stable Al precursors – for example, with triisobutyl and tritertiarybutyl [122] groups attached to Al or the use of alane adducts with amine [123] – however, was successful. The triisobutyl and tritertiarybutyl groups are easily homolytically separated from Al, and the adduct bond also dissociates easily and an Al–C bond is not maintained.

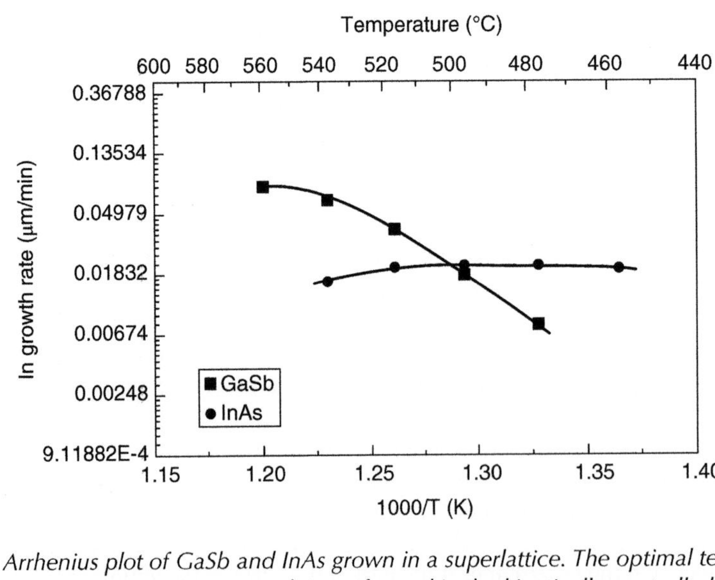

Figure 3.12 Arrhenius plot of GaSb and InAs grown in a superlattice. The optimal temperature for GaSb growth is below 550 °C. Here, growth is performed in the kinetically controlled growth regime. Source: Adapted from Aardvark et al. 1997 [121], with permission from Elsevier.

Preferentially, a less stable Sb precursor is then also employed, such as TESb [124]. The layers are p-type. The sensitivity of Al compounds toward oxygen and moisture originating from the precursors, the carrier gas, or insufficient leak-tightness of the equipment still remains. The incorporation of oxygen as an extrinsic deep level impurity then depends on the control of equipment and precursor and gas purity. Therefore, high Al-content layers are often highly resistive [124].

3.5.2 Deposition of Ternary Column III Alloys (AlGa)Sb and (GaIn)Sb

Successful growth of (AlGa)Sb is very difficult but can be carried out if the less stable Al precursor tritertiarybutylaluminum (TtBAl) is employed together with triethylgallium (TEGa) [122] or if a combination of triethyl Al, Ga, and Sb is employed [125]. Then, low carbon concentrations in the region of $\sim 10^{18}\,cm^{-3}$ are possible. Similar to the growth of (AlGa)As, the lattice parameter and therefore the strain barely changes in the alloy with the composition. In addition, the oxygen concentration in the thin films correlates with the Al content and is determined by the equipment and gas purity.

The deposition of the ternary (GaIn)Sb is more challenging [121]. Due to the disparity of the precursor pyrolysis of the Ga and In precursors at the low temperatures to be used for growth, the optimal growth temperature is dependent on the intended alloy composition. At given precursor input partial pressures, higher temperatures will increase the GaSb growth rate and therefore reduce the In content in the layer. For each compound, the antimony precursor partial pressure injected then needs to be adjusted accordingly to obtain smooth films of only one phase. As the InSb mole fraction in the layers increases, the layers change from p-type to n-type. The properties – especially with respect to morphology – of the ternary are sufficient up to a mole fraction of 20–25%.

3.5.3 Deposition of Ternary Column V Alloys In(AsSb), GaAsSb

In(AsSb) is of high interest for its long-wavelength applications as well as for being the III-V compound semiconductor with the smallest bandgap at room temperature [124, 126]. The bandgap is strongly affected by the alloy composition. It is therefore necessary to know this (nonlinear) dependence well – as presented in Figure 3.13 – and to carefully control the composition. A thermodynamic model was proposed [127–129] and was used to describe the growth of In(AsSb). The model predicts the composition of the ternary and predicts that the thermodynamically more stable binary compound will determine and control the composition of the ternary alloy. Since InAs is thermodynamically more stable than InSb in the optimal growth temperature range (around 500 °C), arsenic is preferentially incorporated into the solid if the V/III ratio >1. If the V/III ratio is ≤ 1, however, all the column V materials (As and Sb) are incorporated in the solid. Since growth is carried out at a V/III ratio close to or slightly greater than one, arsenic will be preferentially incorporated in the alloy. The growth rate is only dependent on the column III precursor TMIn [130].

Manesevit and Simpson were the first to report on the deposition of Ga(AsSb) in the late 1960s [74]. However, it took more than 10 years before growth of the compound was described more completely [131, 132]. Arsine as well as trimethylarsenic (TMAs) were employed together with trimethylantimony (TMSb) and TMG as the precursors. Typical growth temperatures were around 600 °C. It was found that different compositions are attainable throughout the miscibility gap. The gap is between 25 and 70% antimony content in the layers [133]. Both column V ternary alloys In(AsSb) and Ga(AsSb) can be deposited lattice matched to InP, which is a special advantage, since then semi-insulating substrates are available. Misfit dislocations do not affect the properties of the thin films. This special property of these two alloys allows the design and preparation of devices not attainable for other antimonides. Ga(AsSb) behaves similarly to In(AsSb) with respect to column V incorporation and V/III ratio.

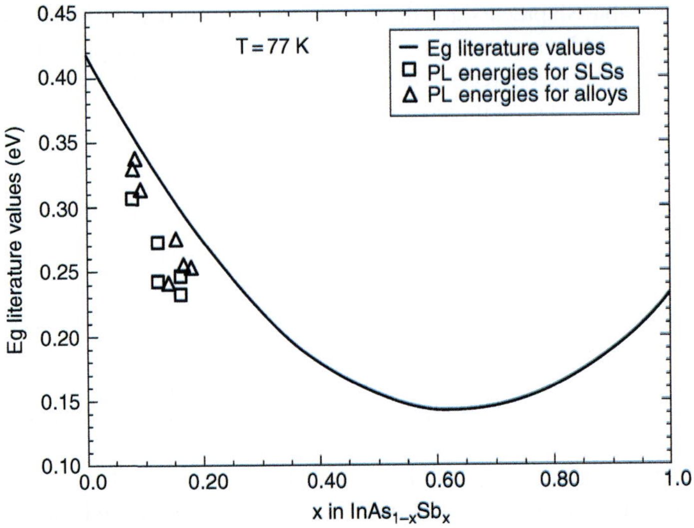

Figure 3.13 Variation of the bandgap energy as a function of antimony composition for $InAs_{1-x}Sb_x$. Source: Reproduced from Biefeld 2002 [124], with permission from Elsevier.

3.5.4 Deposition of Quaternary Alloys

Only two quaternary antimonide alloys are technologically important and have been studied intensively: InGaAsSb [134–136] and AlGaAsSb [137–140]. They have been investigated for their suitability in thermovoltaic applications. Basically, the growth conditions can be deduced for the parameters used for the ternaries.

3.5.5 Epitaxy of Electronic Device Structures

The simplest electronic device structures – metal-semiconductor field-effect transistor (MESFET) and junction field-effect transistor (JFET) structures – usually were not produced by epitaxial methods. As soon as heterojunctions are involved in the structures, and bandgap engineering is to be used, epitaxy comes into play. MOVPE and material-related aspects for two different heterojunction device types will be discussed in the following: on the one hand, the high-electron-mobility transistor; and on the other hand, the hetero-bipolar transistor. Selected examples for different material systems will be given and discussed with respect to material deposition.

3.5.5.1 High-Electron-Mobility Transistors (HEMT)

The high-electron-mobility transistor (HEMT) – it's all in the name – relies on achieving high electron mobilities. Due to the band alignment of a high- and low-bandgap semiconductor, a triangular quantum well (QW) forms in which carriers can move two-dimensionally with large diffusion lengths without scattering – a so-called *two-dimensional electron gas* (2DEG). Figure 3.14 depicts an example of the band profile of this device concept for the AlGaAs/GaAs material system and the influence of n- (Si) doping. It is very important that the interface is smooth and abrupt. The layer sequence consists of the low-bandgap undoped GaAs buffer and channel layer, and an undoped AlGaAs barrier layer in which a doped layer is inserted, which is separated from the channel layer by an undoped spacer layer and is covered with a thin doped GaAs cap layer for ohmic contact formation. Coulomb scattering of the carriers by impurities is conceptually avoided as long as the undoped layers are as pure as possible [141, 142]. Scattering in the potential well is prevented by providing a smooth and abrupt interface.

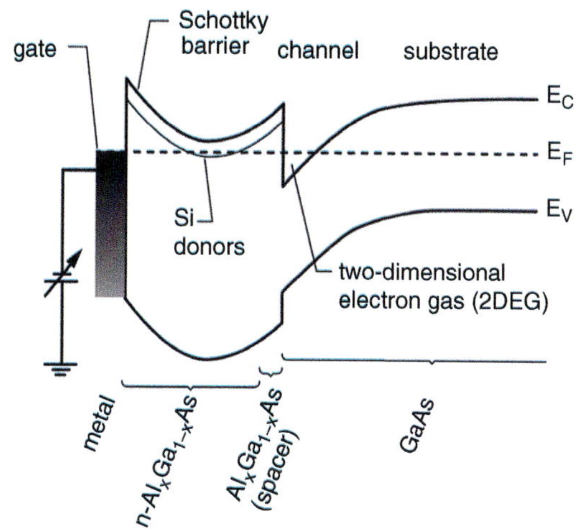

Figure 3.14 *Schematics for conduction- and valence-band profiles of a 2DEG AlGaAs/GaAs heterostructure.*

In general, for deposition of a HEMT structure, growth optimization with respect to the interface is carried out first. The growth parameters for interface abruptness are optimized toward achieving as high as possible mobility at liquid He temperature in a two-dimensional electron gas structure. In addition, an exact orientation of the (001) substrate surface is helpful, since any misorientation creates steps on the surface that contribute to carrier scattering [143]. Then, in a second step, the device structure parameters, such as layer thicknesses, composition, and doping, need to be optimized so that carrier confinement is achieved in the channel. By optimizing the device structure, remote-charged impurity scattering from the intentional dopant atoms is also minimized. The distribution of the carriers and their concentration are tuned by choosing the correct layer thickness, doping height of the carrier supply layer, and spacer layer thickness. Finally, the Schottky-barrier layer and its thickness are optimized.

3.5.5.1.1 GaAs-Based HEMT Structures

For the AlGaAs/GaAs material system, the GaAs/AlGaAs heterointerface is chosen for the formation of the 2DEG, since it is more abrupt and smoother than the AlGaAs/GaAs interface. The growth parameters need to be optimized [144]: ~30% Al content on the column III lattice site and sheet carrier concentrations in the range of $\sim 3\times10^{11}\,cm^{-2}$ are chosen. A spacer layer of ~40 nm is optimal between the hetero-interface and the carrier supply layer. With such structures, mobilities of up to $\sim 2\times10^{6}\,cm^2/V\,s$ at liquid He temperatures have been reached with MOVPE [143, 145]. The growth temperature should not be too high – in the range of 650–700 °C [143, 146] to reduce diffusion at the heterostructure interface and to reduce carbon incorporation, but high enough to reduce the incorporation of oxygen in the layers. Furthermore, at higher temperatures, a larger number of n-type impurities is also observed due to impurities such as Si and Ge in the precursors used. It may be helpful to deposit an oxygen-gettering layer as a buffer directly on the substrate surface before growing a thicker GaAs buffer layer, which also later serves as the 2DEG channel to the transistor [147]. The interface abruptness can also be characterized using photoluminescence spectroscopy [143, 148]. To this end, GaAs QWs of varying thickness are inserted into the $Al_{0.3}Ga_{0.7}As$ barriers. The full width at half-maximum (FWHM) of the emission peaks at liquid He temperature are studied, and the growth parameters are optimized to minimize the FWHM of the excitonic transition. For a transistor device structure, a high channel conductivity needs to be achieved, and parallel conduction in the dopant supply layer is to be avoided. Self-consistent

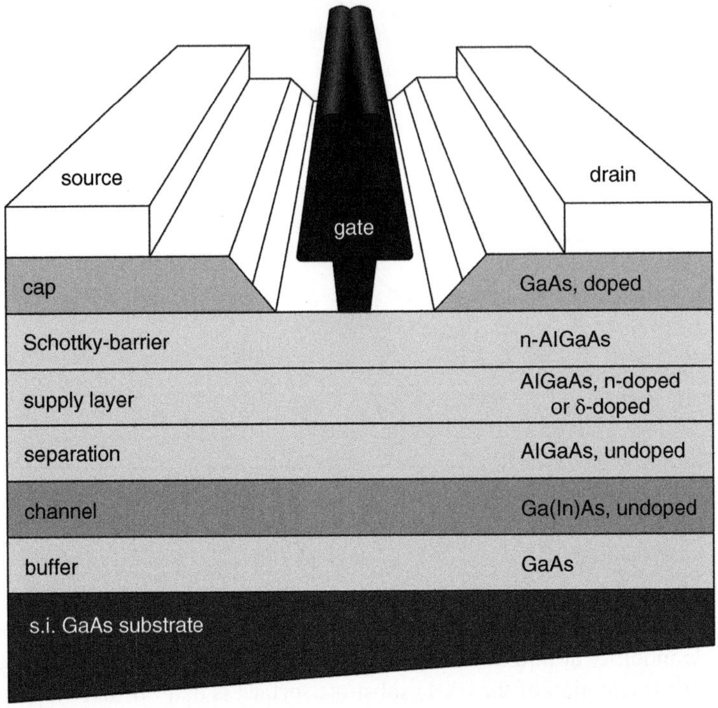

Figure 3.15 *Schematic of the HEMT transistor structure designed for obtaining high channel conductivity.*

calculations of the conduction-band profile are useful to find suitable layer structures, which are then assessed using temperature-dependent Hall-effect measurements with respect to channel conductivity. In comparison to the high-mobility structure, an increase in carrier concentration in the dopant supply layer and a reduction of the spacer layer thickness to close to 10 nm is necessary so that all the carriers are transferred from the dopant spike to the channel. Figure 3.15 depicts the general layer schematics of such a structure. If the optimization is successful, the carrier concentration is in the range of $\sim 1\times10^{12}\,cm^{-2}$ with a mobility of $\sim 7\times10^{3}\,cm^{2}/V\,s$ found in the channel, and no carriers are left in the dopant supply layer. The performance of the HEMT can be improved by inserting a 10-nm thick $Ga_xIn_{1-x}As$ layer with $0.15<x<0.27$ strained, i.e. pseudomorphic channel layer. Such HEMT structures are called pHEMT structures. The inserted thin layer exhibits a lower bandgap than GaAs and therefore higher electron mobilities. Additionally, the conduction-band offset is also larger than for structures with a GaAs channel, so that higher dopant concentrations can be chosen for the supply layer. This approach improves the conductivity of the HEMT structure. pHEMTs are now the standard for GaAs-based commercially available HEMTs.

3.5.5.1.2 InP-Based HEMTs

Due to the high conduction-band offset in $Al_{0.48}In_{0.52}As/Ga_{0.47}In_{0.53}As$ [149] heterostructures allowing high sheet carrier concentrations, and due to the high electron mobility in the high In content channel, $Al_{0.48}In_{0.52}As/Ga_{0.47}In_{0.53}As$ HEMT structures are extremely suitable for high-frequency, low-noise, and high-power applications [84]. A similar optimization of the layer thickness and doping height of the carrier supply layer and the spacer layer thickness needs to be carried out as for the GaAs-based HEMTs. In contrast to the GaAs-based system, the structure is inverted. The sequence of growth is as follows [84]: first an AlInAs buffer layer on the clean (see Section 3.4) InP substrate is deposited, followed by a GaInAs channel layer. Then, a 5-nm thick AlInAs spacer is grown, succeeded by an n-doped AlInAs supply layer and an undoped AlInAs Schottky barrier layer. Finally, an n-doped GaInAs layer is deposited to achieve low Ohmic contact resistance. For the initiation of AlInAs growth on the InP substrate, great care must be taken that the PH_3 is removed from the reactor, before AlInAs growth starts. Otherwise, the morphology of the sample surface deteriorates [150]. A further improvement of the channel conductivity can be achieved by increasing the In content up to ~80%, effectively producing a pHEMT structure. One disadvantage of material systems employing Al-containing compounds is that they are affected by aging, the formation of DX centers [34], and the kink effect. Therefore, there have been endeavors to replace AlInAs as the barrier layer by InP. Since the conduction-band offset to GaInAs is then smaller, the distribution of the dopants needs to be reconsidered. To this end, the channel layer was modulation-doped from both sides. Since the column V precursor changes at the low-bandgap/high-bandgap interface, growth optimization at the interface needs to be carried out carefully first with respect to achieving high electron mobility [151, 152]. In a second step, the structure is fine-tuned toward high channel conductivity [153]. The performance of the "Al-free" HEMTs is then comparable to that for AlInAs/GaInAs heterostructures [154].

3.5.5.2 *Heterojunction Bipolar Transistor (HBT)*

In the heterojunction bipolar transistor (HBT) – as the name implies – a heterojunction and both types of carriers are involved in the conduction process. Major ingredients of the NPN type of HBT transistor structure are the n-type emitter, the p-type base, and an n-type collector [155]. It is desirable that the valence-band offset is larger than the conduction-band offset in the heterojunction. Since the emitter layer exhibits a wider bandgap than that of the base, the p-type doping level in the base can be increased, and the injection of holes from the base into the emitter is limited without sacrificing electron-injection efficiency. A wider bandgap of the collector with respect to the base constitutes a further improvement of the device concept [156]. The device performance is then improved since, for example, the breakdown voltage is increased. Finally, a grading in the composition of the base layer [157] assists the transport of minority carriers by drift and, in doing so, creates a quasi-electric field across the base layer. The effect is that the base transit time, which is otherwise limited by diffusion, decreases [158]. The device concept calls for the optimization of growth parameters with respect to obtaining high base-level doping, high minority carrier lifetimes, a low level of unintentional shallow-level dopants, and the control of grading and thicknesses in the structure, without degrading the crystal structure.

Before choosing a material system for this device, the band-offset distribution needs to be considered beside the size of the bandgap: a small conduction-band offset combined with a large valence-band discontinuity are desirable. An overview of GaAs and InP lattice-matched alloys and their band-offsets distribution is presented in Figure 3.16. The most suitable material combinations in this respect are GaInP/GaAs and the InP/GaInAs heterojunctions.

3.5.5.2.1 GaAs-Based HBTs

Even though the HBT was originally proposed by Kroemer for the GaAs/$Ga_{0.5}In_{0.5}P$ material system [157], GaAs/AlGaAs was exploited first. Single heterojunction transistor structures were developed using Zn [159] and Mg [160] as the base dopants for GaAs. Doping concentrations of up to $\sim 1 \times 10^{19}\,cm^{-3}$ were achieved. Preferentially, the GaAs base layers were deposited at about 600 °C, which is about a 100 °C lower temperature than the adjacent AlGaAs emitter layer. Also, the reactor was preheated in arsine before starting (the lower-temperature) growth of the collector and base layers. By doing so, the unintentional oxygen uptake is reduced, which simultaneously increases the minority carrier lifetime. Endeavors to replace Zn or Mg with carbon using CCl_4 as

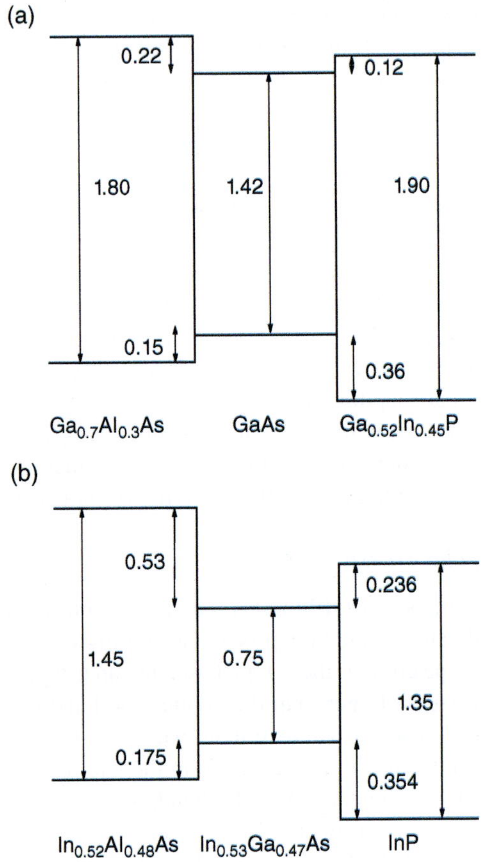

Figure 3.16 *Band-offset distribution for (a) GaAs and (b) InP lattice-matched alloys under consideration for HBT applications (energy values are in eV). Source: Reproduced from Alexandre et al. 1995 [158], with permission from Elsevier.*

the precursor were successful [161]. The long-term stability of the devices has proven to be much more stable than those produced with Zn or Mg as the base dopants. However, there were reports that unintentional incorporation of carbon in the successively deposited AlGaAs was found [162]. Furthermore, AlGaAs needs to be compositionally graded with an increase of the Al content toward the surface of the emitter (and a decrease of the Al content toward the contact layer), so that more efficient electron injection from the emitter to the base is provided. Both challenges can be overcome when using a GaAs/$Ga_{0.5}In_{0.5}P$ heterojunction. Due to its ideal band-offsets distribution with GaAs, no compositional grading in the emitter is necessary, and a higher acceptor concentration in the base is possible. Additionally, $Ga_{0.52}In_{0.48}P$ is not reactive to oxygen impurities in the gas phase, and unintentional carbon uptake is not easily achieved [162]. In addition to the advantages already mentioned, GaAs/$Ga_{0.5}In_{0.5}P$ HBT structures are much less complicated with respect to layer design and the epitaxy process and exhibit advantages with respect to processing and etching selectivity [163]. For the ternary, growth conditions are chosen that should produce disordered material (see Section 3.3.2, here <600 °C). Growth temperatures between 510 °C [164] and 580 °C [165] were reported to be optimal. After deposition, a post-growth anneal is necessary to achieve a high activation level of the p-type dopants layers [40, 165]. If the highly p-doped layer is inserted into structures that also contain n-doped layers (such as in HBTs), the annealing temperature may need to be as high as 700 °C to drive out the remaining hydrogen. The n-doped layer apparently hinders the dehydrogenation process [166].

3.5.5.2.2 InP-Based HBTs

Similar considerations with respect to the layer design as for the GaAs-based HBT structures were carried out for the InP material system. Here, only one heterostructure system comes into play: the GaInAs/InP system. Single HBT structures consist of a GaInAs collector and base and an InP emitter. However, these HBT structures were mainly useful for low-voltage and low-power dissipation applications, due to the small energy bandgap and the low thermal conductivity of GaInAs, respectively. An insertion of an InP layer instead of GaInAs as the collector improved the devices due to the higher thermal conductivity of InP and the increased breakdown voltage. However, the electron-blocking effect resulting from the conduction-band discontinuity of the base–collector (BC) heterojunction was still detrimental to device performance. To solve this problem, the design of the device structure was further optimized, including a double heterostructure at the collector and the base and the base and emitter transition [167], and also inserting thin compositionally step-graded InGaAsP layers at both interfaces [168]: the step-graded layers are effective in attenuating carrier blocking caused by the conduction band spike in the collector. This concept is presented in Figure 3.17 [169]. From a growth point of view, the control of doping is essential. Also, achieving high minority carrier lifetimes is extremely important. High acceptor doping of the base with carbon instead of the highly diffusive Zn or Mg is a challenge, since carbon as a group IV element presents a pronounced amphoteric behavior. This is due to the lower indium–carbon binding energy as compared to that of gallium–carbon. However, by adding CCl_4 or CBr_4 as the precursors and using suitable growth conditions, nevertheless sufficient p-type conductivity is achieved in GaInAs [167]. Maximum doping levels of $\sim 5 \times 10^{19}\,cm^{-3}$ were reached [169, 170]. Not all the layers in the device structure are especially sensitive to oxygen incorporation, making this material system robust. Growth is carried out at around 550 °C for the InP layers and 600 °C [171] for the InGaAs(P) layers.

3.5.6 Epitaxy of Optoelectronic Device Structures

GaAs- or InP-based semiconductors are employed for a broad range of optoelectronic devices and a large number of different applications such as optical fiber communication, infrared and visible light-emitting and laser diodes (LEDs and LDs), detectors, and high-efficiency solar cells [172]. Here, we will center on the epitaxy of LED and LD structures. From a materials point of view, for both devices, long minority carrier lifetimes are mandatory to achieve efficiency, since emission is based on minority carrier injection and recombination of excess carriers (doping). Therefore,

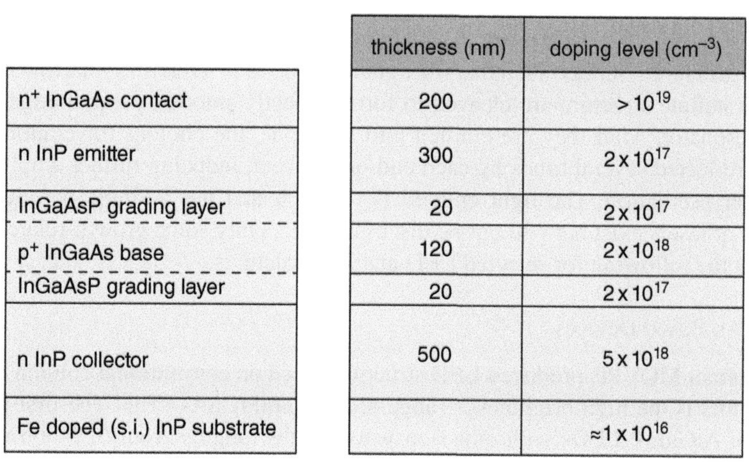

Figure 3.17 *Example of a schematic of an InP-based double heterostructure HBT with step grading. Source: Reproduced from Nottenburg et al. 1986 [169], with permission from IEEE.*

growth must be optimized to obtain structures with an as low as possible concentration of deep-level traps and defects that limit the minority carrier lifetime.

3.5.6.1 Light-Emitting Diodes (LEDs) and Laser Diodes

In 1962, Nick Holonyak – known to be the father of visible LEDs – discovered and invented the first practical visible spectrum LED. He was working on p-n junction *lasers* based on III-V semiconductors [173]. Most endeavors were directed toward infrared emitters, which were to be used for a large range of products in consumer electronics. One example is the remote control. Devices for telecommunication applications were developed later. However, first, red LEDs became available on the market in the late 1960s and were used in indicator lamps. Since then, the development of light emitters has been enormous.

With respect to LEDs, homo-junction p-n structures based, for example, on GaP were produced first, followed by single p-n heterostructures. Modern epitaxial techniques such as MOVPE came into play later, as more and more highly efficient LEDs were in demand [11]. In addition to indicator lights, street and automotive lighting, general lighting, and large-scale display applications were the focus of applications.

The LED efficiency and especially the internal quantum efficiency are determined by epitaxy. The epitaxial layer structure, the layer design parameters, the characteristics of the material system chosen, as well as the material quality determine the LED emission characteristics. Practical issues for growth include the ability to control doping, composition, thickness, and defect concentrations reproducibly in the layers and to realize homogeneity. All high-performance, high-brightness LEDs are based on double heterojunctions in which the active layer is sandwiched between two layers with a higher bandgap, which are p- and n-doped, respectively. They serve as cladding layers, confining the carriers in the active layer. The recombination of electrons and holes in the active layer occurs spontaneously. An overview of the conduction- and valence-band profile for homojunction, single heterojunction (SH) and double heterojunction (DH) LED structure design types is presented in Figure 3.18 [11]. Further developments have taken place: for example, with respect to using QW structures as the active region to tailor the wavelength [174], to include strained layers [175, 176], and to incorporate Bragg mirrors so that the emitted light is reflected and not absorbed by the substrate [177].

Laser (acronym for light amplification of stimulated emission of radiation) diode structures are very similar to LED structures (from an epitaxial grower's point of view). As for LEDs, the generation of light (radiation) is also induced by operating a p-n junction in the forward-biased direction and by the spontaneous recombination of electron–hole pairs. However, an optical cavity is necessary. To this end, an optical waveguide is needed. In the simplest approach, it is made on the crystalline structure's surface such that the light is confined to a relatively narrow line. The two ends of the crystalline structure are cleaved to form perfectly smooth, parallel edges, forming a Fabry–Pérot resonator. After they are emitted into the mode, the photons travel along the waveguide and are reflected several times by each end of the facet, inducing further amplification and generating laser oscillation. The light emitted is coherent and has a single wavelength. Many different types of lasers exist but will not be discussed here. Only some growth-related issues will be presented in the following for selected LED and LD structures.

3.5.6.1.1 GaAs-Based Devices

The most important MOVPE-produced LED structure based on conventional column III – column V semiconductors is the high-brightness orange-red AlGaInP/GaAs LED. The first paper on the performance of AlGaInP LEDs with emission wavelengths ranging from 621–566 nm was published in 1990 [178], soon after the growth conditions of $(Al_xGa_{1-x})_{0.5}In_{0.5}P$ QWs in high Al content

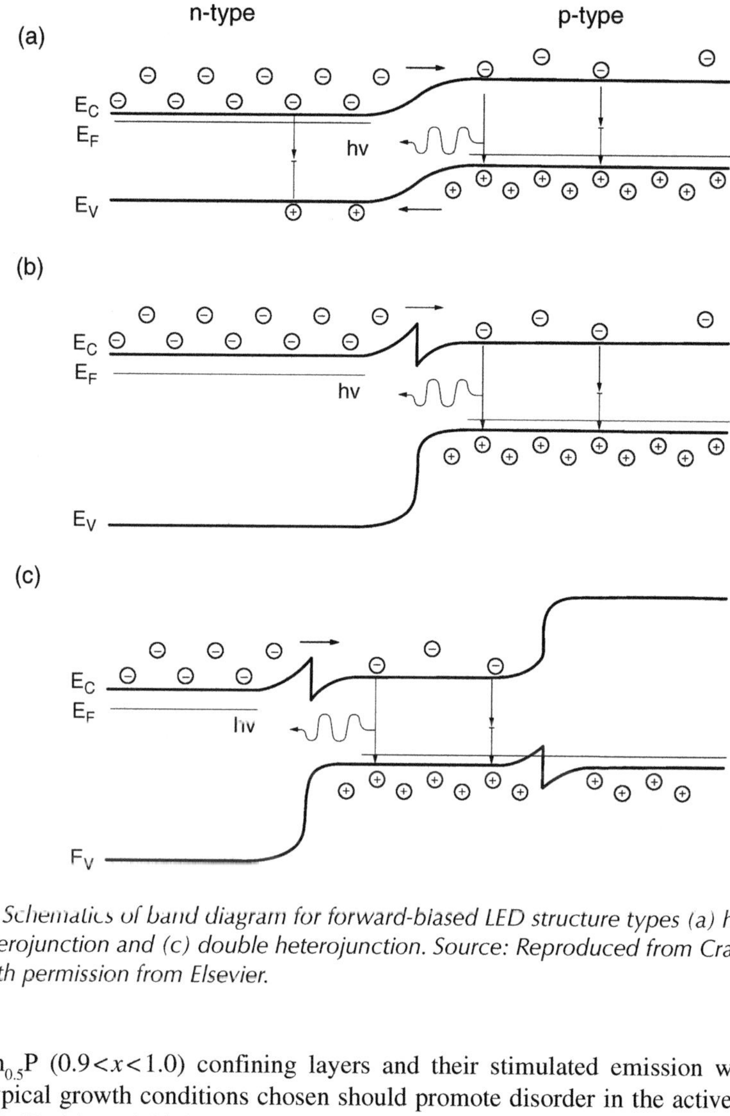

Figure 3.18 Schematics of band diagram for forward-biased LED structure types (a) homojunction (b) single heterojunction and (c) double heterojunction. Source: Reproduced from Craford et al. 1999 [11], with permission from Elsevier.

$(Al_xGa_{1-x})_{0.5}In_{0.5}P$ ($0.9<x<1.0$) confining layers and their stimulated emission were presented [179]. The typical growth conditions chosen should promote disorder in the active material (see Section 3.3.2). To this end, high growth temperatures are employed, between 750 °C and 780 °C. Such temperatures are also advantageous to reduce oxygen incorporation in the layers, which forms deep-level traps and would reduce minority carrier lifetimes detrimental to emission efficiency. The structure design for the LED presented in Figure 3.19 consists of an undoped $(Al_xGa_{1-x})_{0.5}In_{0.5}P$ active layer with $0.2<x>0.43$ embedded in n- and p-doped confining AlInP cladding layers. Further improvements were accomplished by depositing 15 μm of highly p-doped GaP on top of the structure as a wide-bandgap transparent window layer [180]. Finally, the light output can be increased by a factor of two if the absorbing n-doped GaAs substrate is selectively removed and replaced – by means of wafer bonding – with a transparent n-doped GaP substrate [181].

Simpler methods such as liquid phase epitaxy (LPE), molecular beam epitaxy (MBE), and vapor phase epitaxy (VPE) were first used predominantly for various GaAs-based LDs and LEDs. In principal, it was demonstrated quite early that the performance of LD structures deposited by MOVPE exhibit comparable device performance to those deposited by the other techniques. There are no fundamental obstacles, as presented in a review by Dupuis in 1981 [182]. Therefore,

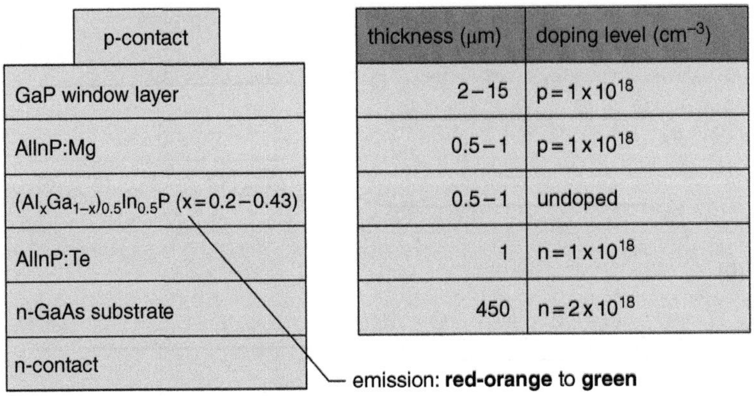

Figure 3.19 *Schematics of an $(Al_xGa_{1-x})_{0.5}In_{0.5}P$ LED with Al content x ranging from 0.2 (622 nm) to 0.43 (554 nm) and with a thick transparent GaP window. Source: Reproduced from Fletcher et al. 1991 [180], with permission from Springer Nature.*

later, MOVPE as a high-volume process came into play for more complex and sophisticated structures because of its versatility to deposit few-nanometer thin layers as well as μm thick layers within one growth run, and the ability to deposit compositionally graded as well as strained layers [183].

From a growth point of view, laser diode structures containing single or multiple QWs are not only challenging with respect to composition and thickness control but also if the growth conditions optimal for the barriers and wells differ. This is the case for InGaP/InGaAlP visible LD structures. Ishikawa reported continuous wave (cw) room-temperature operation of LDs based on this material system [184]. Even though InGaP is preferentially deposited around 600 °C to prevent ordering in the structures, and InGaAlP is deposited rather at temperatures at or above 750 °C to avoid oxygen deep-level incorporation, a compromise needs to be made, and 700 °C was chosen for epitaxy [10, 184, 185]. Similarly, the deposition of highly strained $In_{0.39}Ga_{0.61}As$/GaAs QW laser structures needs to be optimized. To avoid strain relaxation, growth should be carried out at lower temperatures and high column V/column III precursor ratios. However, at too low a temperature, the decomposition of the precursors may not be complete, leading to a higher unintentional carbon concentration and inferior crystal quality. An increase in reactor pressure from 50–150 mbar affecting the migration of especially the indium atoms on the growing surface allowed the deposition of high-quality quantum-well laser structures at a growth temperature of only 475 °C [186]. For the optimization of complex structures, an interplay between structure design and growth parameters is mandatory.

3.5.6.1.2 InP-Based Devices

Alloys such as InGaAsP and AlGaInAs lattice-matched to InP have been employed for laser diodes emitting in the wavelength range between ~ 1.0 μm and ~ 1.7 μm. There are numerous different device concepts that would go beyond the scope of this book. Nevertheless, an ingredient of InP-based laser diode structures contained in many modern LDs will be discussed with respect to growth optimization: the active region consisting of multiple quantum wells (MQWs). The emission wavelength is determined not only by the bandgap of the emitter but also by its thickness in the well. The effect was presented in a study: using an InGaAsP alloy composition with a nominal room-temperature bandgap of 0.8 eV ($\lambda = 1.55$ μm) as the well, stacks of different QW thicknesses were formed by 250 Å InP barriers. Photoluminescence studies disclosed that the recorded wavelength at 2 K decreases from 1449 nm to 905 nm when the well thickness decreases from 2200 Å (i.e. bulk material) to 4 Å [187]. In addition to exploiting this effect, a suitable growth optimization strategy with respect to growth temperature, column V/column III precursor ratios, substrate misorientation, and precursor switching sequence (especially if the column V precursor changes during MQW growth) should be carried out. The superlattice structures are then evaluated by X-ray diffraction, Raman spectroscopy, and photoluminescence spectroscopy [187, 188].

The growth rate especially of strained QWs may differ from bulk layer growth [183] and must also be controlled, before the MQWs are inserted as the active medium in the laser structure. An example of such a broad area laser structure is presented in Figure 3.20 [183]. The structure contains a fivefold MQW consisting of InGaAs/InGaAsP layers. The characterization of MQW structures can also be employed to study the uniformity across the wafer and will disclose whether the reproducibility from QW to QW changes during growth. In addition, in situ optical methods such as, for example, reflective anisotropy spectroscopy (RAS) are a wonderful supplement and can disclose problems directly during growth before structures are processed [96].

As was described in Section 3.4.4, the control of InGaAsP composition is a challenge owing to the difference in decomposition efficiency of the hydrides arsine and phosphine. By using the column V precursors tertiarybutylarsine and tertiarybutylphosphine with similar decomposition efficiency, Ae et al. [106] improved column V compositional control in the quaternary and were able to demonstrate state-of-the-art low-threshold $\lambda = 1.3\,\mu m$ MQW laser diodes.

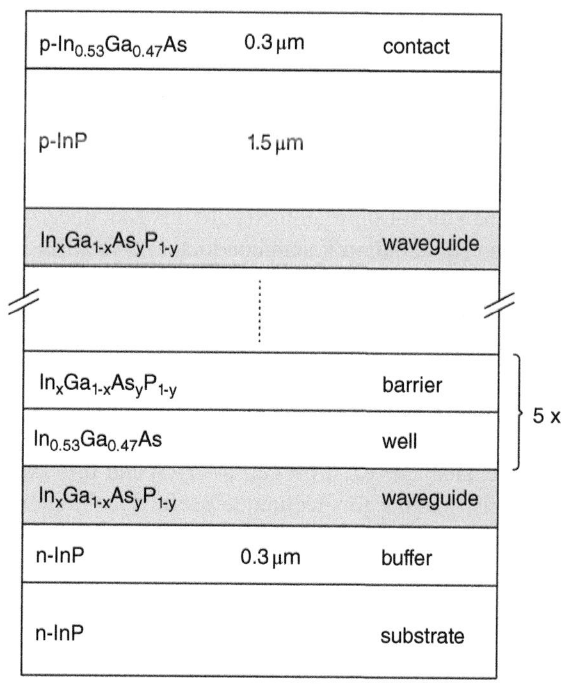

Figure 3.20 *Example of a multiple quantum-well broad-area laser diode structure suitable for data communication. Source: Reproduced from Speier 1992 [183], with permission from Elsevier.*

3.6 In Situ Optical Characterization/Growth Control

Complex material and layer systems such as those usually deposited by MOVPE call for an exact control of growth parameters that determine the layer thickness (i.e. growth rate) composition, doping level, and interface stoichiometry. Any unintentional changes would affect the device properties and become visible only after time-consuming processing. Therefore, it is advantageous to observe growth with a nonintrusive method. To this end, optical reflectance-based in situ techniques such as spectroscopic reflectance [189], (emissivity corrected) pyrometry [21, 190], and reflectance anisotropy spectroscopy [16, 21, 191, 192] have become indispensable. An optical window is incorporated into the MOVPE equipment, which is situated in such a way that light hits the surface in nearly normal incidence. It has to be ensured that deposition on the window is avoided.

For spectroscopic reflectance, white light is used as the light source. Changes from run to run are detectable, if the fingerprints of two growth runs are compared. Fabry–Perot oscillations are observed as the layers grow. They are used to calculate the growth rate. By recording the reflectance at different wavelengths, optical constants can be extracted. The data measured can be evaluated using the virtual interface model [193]. In addition to the growth rate, the composition can then also be determined. A roughening of the layer during growth affects the intensity of reflectance. This helps identify the growth step or interface, which must still be optimized.

The temperature of the substrate is one of the most important parameters in MOVPE since it affects the crystallinity of the samples, the composition, the doping, and the growth rate. Its control is challenging since the conventionally used thermocouples are often inaccurate. They make poor thermal contact with the sample. Due to susceptor and/or wafer rotation, the thermocouple is placed far from the growing layer. Furthermore, changes in the surface temperature on the wafer may go unnoticed, as parasitic deposition in the reactor increases or if the carrier gas or the gas used for gas foil rotation changes [21, 194]. Pyrometers are also sometimes employed; however, the emissivity changes during growth and then affects the intensity emitted from the growing surface. A correct temperature determination is then difficult. By combining reflectance measurements (that determine emissivity) with pyrometry – so-called *emissivity corrected pyrometry* – correct temperatures can be determined and recorded.

Reflectance anisotropy spectroscopy (RAS) exploits the anisotropic optical response of surfaces, which originates from their anisotropic geometry. To this end, linearly polarized light is used as the light source. The column III – column V semiconductors crystallize in the zincblende structure and are isotropic in the bulk but exhibit in general anisotropic surfaces. Such surfaces are highly sensitive to the growth conditions. The RAS signal is influenced by changes in the layer thickness (even monolayer oscillations are observable [192, 195]), the stoichiometry [96, 196], ordering [197], morphology parameters [198], and especially the doping concentrations (in the range from $\sim 10^{17}\,cm^{-3}$ – $\sim 10^{19}\,cm^{-3}$) [96, 196]. Even though the interpretation of the complete spectra is difficult, since the RAS signal measured is composed of a complex dielectric function of both the bulk and the surface, large databases exist for GaAs- [195] and InP- [96] based materials, and research has advanced strongly, making this technique useful for the online control of MOVPE device processes [192].

3.7 Conclusions

During the last four decades, MOVPE has matured into *the* high-volume industrial deposition technique for complex and sophisticated III-V heterostructures. The effect of its technically relevant parameters such as temperature, reactor pressure, precursor partial pressures, as well as carrier gas and precursors on the properties of GaAs-based, InP-based, and column III-antimonide (hetero) structures was reviewed here, together with the strategical choice of appropriate growth parameters for electronic and optoelectronic device applications. This knowledge and the development of *in situ* optical techniques ensured growth control and reproducibility and formed the basis for MOVPE's triumphal course.

Even though it seems as if "everything that can be invented has been invented" (attributed to Charles H. Duell, Commissioner of the US patent office in 1899), there is no doubt that conventional III-Vs will further impact the fields of wireless communication and optoelectronic as well as extraterrestrial applications during this century. Intensive efforts in the advancement of novel and/or alternative materials and device solutions with a focus on low energy consumption and "green" IT have already been presented for a large range of device applications. However, III-V's unique physical properties, lifetime, reliability as well as functionality at room temperature, cost-efficiency, and reproducibility still play a crucial role in the choice of materials and epitaxial techniques. In this competition, the MOVPE of III-V materials is still devoid of alternatives in the fields mentioned.

References

1. van der Merwe, J.H. (1978). The role of lattice misfit in epitaxy. *Critical Review of Solid State Material Science* 7: 209–31.
2. van der Merwe, J.H. (1991). Misfit dislocation generation in epitaxial layers. *Critical Review of Solid State Material Science* 17: 187–209.
3. Queisser, H.J. (1998). Defects in semiconductors: Some fatal, some vital. *Science* 281: 945–50.
4. Vegard, L. (1921). The constitution of mixed crystals and the space occupied by atoms. *Zeitschrift für Physik* 5. 17–26.
5. Ryou, J., Kanjolia, R., and Dupuis, R.D. (2009). *CVD of III-V Compound Semiconductors*. *Chemical Vapor Deposition*, Royal Society of Chemistry, Cambridge, 272–319.
6. Hardtdegen, H. and Giannoulés, P. (1998). MOVPE gets green signal. *III-Vs Review* 11: 34–9.
7. Jones, A.C. (1993). Metalorganic precursors for vapour phase epitaxy. *Journal of Crystal Growth* 129: 728–73.
8. Hardtdegen, H., Ungermanns, C., Hollfelder, M. et al. (1994). A new approach towards low-pressure metalorganic vapor phase epitaxy of (AlGa)As using triethylgallium and dimethylethylaminealane. *Journal of Crystal Growth* 145: 478–84.
9. Mountziaris, T.J. (1991). Gas-phase and surface reaction mechanisms in MOCVD of GaAs with trimethyl-gallium and arsine. *Journal of the Electrochemical Society* 138: 2426.
10. Bour, D.P. and Shealy, J.R. (1987). High-power (1.4W) AlGaInP graded-index separate confinement heterostructure visible ($\lambda \sim 658$ nm) laser. *Applied Physics Letters* 51: 1658–60.
11. Craford, M.G., Stockman, S.A., Peanasky, M.J. et al. (1999). Visible light-emitting diodes. *Semiconductors and Semimetals* 64: 1–47.
12. Choquette, K.D., Geib, K.M., Ashby, C.I.H. et al. (1997). Advances in selective wet oxidation of AlGaAs alloys. *IEEE Journal of Selected Topics in Quantum Electronics* 3: 916–26.
13. Reep, D.H. and Ghandi, S.K. (1983). Deposition of GaAs epitaxial layers by organometallic CVD. *Journal of the Electrochemical Society* 130: 675.
14. Shaw, D.W. (1975). Kinetic aspects in the vapour phase epitaxy of III–V compounds. *Journal of Crystal Growth* 31: 130–41.
15. Stringfellow, G.B. (1984). A critical appraisal of growth mechanisms in MOVPE. *Journal of Crystal Growth* 68: 111–22.
16. Hardtdegen, H., Pristovsek, M., Menhal, H. et al. (1998). In situ characterization of GaAs growth in nitrogen atmosphere during MOVPE: a comparison to hydrogen atmosphere. *Journal of Crystal Growth* 195: 211–6.
17. Jones, A.C. (1997). Developments in metalorganic precursors for semiconductor growth from the vapour phase. *Chemical Society Review* 26: 101.
18. Hollfelder, M., Hardtdegen, H., Meyer, R. et al. (1994). (AlGa)As grown by low pressure metalorganic vapor phase epitaxy using a N_2 carrier. *Journal of Electronic Materials* 23: 1061–5.
19. Plass, C., Heinecke, H., Kayser, O. et al. (1988). A comparative study of $Ga(CH_3)_3$, $Ga(C_2H_5)_3$ and $Ga(C_4H_9)_3$ in the low pressure MOCVD of GaAs. *Journal of Crystal Growth* 88: 455–64.
20. Hardtdegen, H., Hollfelder, M., Meyer, R. et al. (1992). MOVPE growth of GaAs using a N_2 carrier. *Journal of Crystal Growth* 124: 420–6.

21. Haberland, K., Kaluza, A., Zorn, M. et al. (2002). Real-time calibration of wafer temperature, growth rate and composition by optical in-situ techniques during $Al_xGa_{1-x}As$ growth in MOVPE. *Journal of Crystal Growth* 240: 87–97.
22. Dauelsberg, M., Hardtdegen, H., Kadinski, L. et al. (2001). Modeling and experimental verification of deposition behavior during AlGaAs growth: a comparison for the carrier gases N_2 and H_2. *Journal of Crystal Growth* 223: 21–8.
23. Kuech, T.F., Wolford, D.J., Veuhoff, E. et al. (1987). Properties of high-purity $Al_xGa_{1-x}As$ grown by the metalorganic vapor-phase-epitaxy technique using methyl precursors. *Journal of Applied Physics* 62: 632–43.
24. Kakinuma, H., Mohri, M., and Akiyama, M. (1997). Characterization of oxygen and carbon in undoped AlGaAs grown by organometallic vapor-phase epitaxy. *Japanese Journal of Applied Physics* 36: 23–8.
25. André, J.P., Boulou, M., and Micrea-Roussel, A. (1981). Luminescence of $Al_xGa_{1-x}As$ grown by MOVPE. *Journal of Crystal Growth* 55: 192–7.
26. Kondo, M. and Tanahashi, T. (1994). Dependence of carbon incorporation on crystallographic orientation during metalorganic vapor phase epitaxy of GaAs and AlGaAs. *Journal of Crystal Growth* 145: 390–6.
27. Nakanisi, T. (1984). The growth and characterization of high quality MOVPE GaAs and GaAlAs. *Journal of Crystal Growth* 68: 282–94.
28. Azoulay, R., Dugrand, L., Ankri, D. et al. (1984). MOCVD n-type doping of GaAs and GaAlAs using silicon and selenium and fabrication of double heterostructure bipolar transistor. *Journal of Crystal Growth* 68: 453–60.
29. Kuech, T.F., Veuhoff, E., and Meyerson, B.S. (1984). Silicon doping of GaAs and $Al_xGa_{1-x}As$ using disilane in metalorganic chemical vapor deposition. *Journal of Crystal Growth* 68: 48–53.
30. Moffat, H.K., Kuech, T.F., Jensen, F. et al. (1988). Gas phase and surface reactions in Si doping of GaAs by silanes. *Journal of Crystal Growth* 93: 594–601.
31. Leu, S., Protzmann, H., Höhnsdorf, F. et al. (1998). Si-doping of MOVPE grown InP and GaAs by using the liquid Si source ditertiarybutyl silane. *Journal of Crystal Growth* 195: 91–7.
32. Nelson, R.J. (1977). Long-lifetime photoconductivity effect in n-type GaAlAs. *Applied Physics Letters* 31: 351–3.
33. Mizuta, M., Tachikawa, M., Kukimoto, H. et al. (1985). Direct evidence for the DX center being a substitutional donor in AlGaAs alloy system. *Japanese Journal of Applied Physics* 24: L143–6.
34. Mooney, P.M. (1990). Deep donor levels (DX centers) in III-V semiconductors. *Journal of Applied Physics* 67: R1–26.
35. Tachikawa, M., Mizuta, M., Kukimoto, H. et al. (1985). A simple calculation of the DX center concentration based on an l-donor model. *Japanese Journal of Applied Physics* 24: L821–3.
36. Hardtdegen, H., Ungermanns, C., Wirtz, K. et al. (1994). Heavy carbon doping in low-pressure metalorganic vapor phase epitaxy of GaAs using trimethylarsenic — a comparison between the carrier gases N_2 and H_2. *Journal of Crystal Growth* 145: 440–6.
37. Kuech, T.F., Tischler, M.A., Wang, P.J. et al. (1988). Controlled carbon doping of GaAs by metalorganic vapor phase epitaxy. *Applied Physics Letters* 53: 1317–9.
38. Richter, E., Kurpas, P., Gutsche, D. et al. (1995). Carbon doped GaAs grown in low-pressure metalorganic vapor phase epitaxy using carbon tetrabromide. *Journal of Electronic Materials* 24: 1719–22.
39. Tavendale, A.J., Pearton, S.J., Williams, A.A. et al. (1990). Degradation of hydrogen-passivated p-type layers in GaAs by minority carrier injection and reverse bias annealing. *Materials Research Society Proceedings* 184: 87.
40. Liu, Q., Brennemann, A., Hardtdegen, H. et al. (1996). Characterization of hydrogen passivation and carbon self-compensation of highly C-doped GaAs by means of X-ray diffraction. *Journal of Applied Physics* 79: 710.
41. Glew, R.W. (1984). Zinc doping of MOCVD GaAs. *Journal of Crystal Growth* 68: 44–7.
42. Veuhoff, E., Baumeister, H., Rieger, J. et al. (1991). Comparison of Zn and Mg incorporation in MOVPE InP/GaInAsP laser structures. *Journal of Electronic Materials* 20: 1037–41.
43. Rask, M., Landgren, G., Andersson, S.G. et al. (1988). Abrupt p-type doping transitions using bis-(cyclopentadienyl)-magnesium in metal-organic vapor phase epitaxy of GaAs. *Journal of Electronic Materials* 17: 311–4.
44. Suzuki, T., Gomyo, A., Iijima, S. et al. (1988). Band-gap energy anomaly and sublattice ordering in GaInP and AlGaInP grown by metalorganic vapor phase epitaxy. *Japanese Journal of Applied Physics* 27: 2098–106.
45. Vanderwater, D.A., Tan, I.-H., Hofler, G.E. et al. (1997). High-brightness AlGaInP light emitting diodes. *Proceedings of the IEEE* 85: 1752–64.
46. Hsu, T.C., Stringfellow, G.B., Kim, J.H. et al. (1998). Surface photoabsorption transients and ordering in GaInP. *Journal of Applied Physics* 83: 3350–5.
47. Ueda, O., Takikawa, M., Takechi, M. et al. (1988). Transmission electron microscopic observation of InGaP crystals grown on (001)GaAs substrates by metalorganic chemical vapor deposition. *Journal of Crystal Growth* 93: 418–25.
48. Seong, T.-Y., Kim, J.H., Chun, Y.S. et al. (1997). Effects of V/III ratio on ordering and antiphase boundaries in GaInP layers. *Applied Physics Letters* 70: 3137–9.
49. Gomyo, A., Kobayashi, K., Kawata, S. et al. (1986). Studies of $Ga_xIn_{1-x}P$ layers grown by metalorganic vapor phase epitaxy; Effects of V/III ratio and growth temperature. *Journal of Crystal Growth*, 77: 367–73.
50. Gomyo, A., Suzuki, T., Kobayashi, K. et al. (1987). Evidence for the existence of an ordered state in $Ga_{0.5}In_{0.5}P$ grown by metalorganic vapor phase epitaxy and its relation to band-gap energy. *Applied Physics Letters* 50: 673–5.

51. Suzuki, T., Gomyo, A., and Iijima, S. (1988). Strong ordering in GaInP alloy semiconductors; Formation mechanism for the ordered phase. *Journal of Crystal Growth* 93: 396–405.
52. Hsu, T.C., Hsu, Y., and Stringfellow, G.B. (1998). Effect of P precursor on surface structure and ordering in GaInP. *Journal of Crystal Growth* 193: 1–8.
53. Ernst, P., Geng, C., Hahn, G. et al. (1996). Influence of domain size on optical properties of ordered $GaInP_2$. *Journal of Applied Physics* 79: 2633–9.
54. Murata, H., Hsu, T.C., Ho, I.H. et al. (1996). Surface photoabsorption study of the effect of V/III ratio on ordering in GaInP. *Applied Physics Letters* 68: 1796–8.
55. Schmidt, R., Ahe, M v. d., Dieker C. et al. (1998). Contributions to understanding the optical properties of partially ordered $(Al_{0.3}Ga_{0.7})_{0.52}In_{0.48}P$. *Journal of Crystal Growth* 195: 124–31.
56. Chun, Y.S., Lee, S.H., Ho, I.H. et al. (1997). Effect of growth rate on step structure and ordering in GaInP. *Journal of Applied Physics* 81: 646–9.
57. Murata, H. (1996). Correlation between surface structure and ordering in GaInP. *Journal of Vacuum Science and Technology B* 14: 3013.
58. Stringfellow, G.B., Lee, R.T., Fetzer, C.M. et al. (2000). Surfactant effects of dopants on ordering in GaInP. *Journal of Electronic Materials* 29: 134–9.
59. Shurtleff, J.K., Lee, R.T., Fetzer, C.M. et al. (1999). Band-gap control of GaInP using Sb as a surfactant. *Applied Physics Letters* 75: 1914–6.
60. Lee, S.H., Fetzer, C., and Stringfellow, G.B. (1998). Effect of Te doping on step structure and ordering in GaInP. *Journal of Crystal Growth* 195: 13–20.
61. Lee, S.H., Fetzer, C.Y., Stringfellow, G.B. et al. (1999). Te doping of GaInP: Ordering and step structure. *Journal of Applied Physics* 85: 3590–6.
62. Lee, S.H., Fetzer, C.M., Stringfellow, G.B. et al. (1999). Step structure and ordering in Zn-doped GaInP. *Journal of Applied Physics* 86: 1982–7.
63. Gomyo, A., Hotta, H., Hino, I. et al. (1989). Silicon and selenium doping effects on band-gap energy and sublattice ordering in $Ga_{0.5}In_{0.5}P$ grown by metalorganic vapor phase epitaxy. *Japanese Journal of Applied Physics* 28: L1330–3.
64. Ernst, P., Geng, C., Scholz, F. et al. (1996). Ordering in $GaInP_2$ studied by optical spectroscopy. *Physica Status Solidi* 193: 213–29.
65. Chen, C., Stockman, S.A., Peanasky, M.J. et al. (1997). In: Semiconductors and Semimetals (eds. G.B. Stringfellow and M.G. Craford), Ch. 4. New York: AcademicPress.
66. Lewis, C.R., Ludowise, M.J., and Dietze, W.T. (1984). H_2Se "memory effects" upon doping profiles in GaAs grown by metalorganic chemical vapor deposition (MO-CVD). *Journal of Electronic Materials* 13: 447–61.
67. Ohba, Y., Nishikawa, Y., Nozaki, C. et al. (1988). A study of p-type doping for AlGaInP grown by low-pressure MOCVD. *Journal of Crystal Growth* 93: 613–7.
68. McCalmont, J.S., Casey, H.C., Wang, T.Y. et al. (1992). The effect of oxygen incorporation in semi-insulating $(Al_xGa_{1-x})_yIn_{1-y}P$. *Journal of Applied Physics* 71: 1046–8.
69. Guimarães, F.E.G., Elsner, B., Westphalen, R. et al. (1992). LP-MOVPE growth and optical characterization of GaInP/GaAs heterostructures: interfaces, quantum wells and quantum wires. *Journal of Crystal Growth* 124: 199–206.
70. Nakano, T., Shioda, T., Abe, E. et al. (2008). Abrupt InGaP/GaAs heterointerface grown by optimized gas-switching sequence in metal organic vapor phase epitaxy. *Applied Physics Letters* 92: 112106.
71. Hollfelder, M., Hardtdegen, H., Breuer, U. et al. (1995). Unintentional impurity incorporation at the interface between InP substrate and buffer layer grown by LP-MOVPE. Proceedings of the Seventh International Conference on Indium Phosphide Related Materials. IEEE; p. 132–5.
72. Thomas, N., Jacob, G., and Hardtdegen, H. (1998). Improvement of the surface quality of InP wafers using TOF-SIMS as characterisation. Proceedings of the International Conference on Indium Phosphide Related Materials. IEEE; p. 96–9.
73. Jacob, G., Regreny, P., Thomas, N. et al. (1997). Improvement of the surface quality of polished InP wafers. Proceedings of the International Conference on Indium Phosphide Related Materials. IEEE; p. 420–3.
74. Manasevit, H.M. and Simpson, W.I. (1969). The use of metal-organics in the preparation of semiconductor materials. *Journal of the Electrochemical Society* 116: 1725.
75. Fukui, T. and Horikoshi, Y. (1980). Properties of InP films grown by organometallic VPE method. *Japanese Journal of Applied Physics* 19: L395–7.
76. Razeghi, M. and Duchemin, J.P. (1983). Growth and characterization of InP using metalorganic chemical vapor deposition at reduced pressure. *Journal of Crystal Growth* 64: 76–82.
77. Bass, S.J., Pickering, C., and Young, M.L. (1983). Metal organic vapour phase epitaxy of indium phosphide. *Journal of Crystal Growth* 64: 68–75.
78. Hsu, C.C., Cohen, R.M., and Stringfellow, G.B. (1983). OMVPE growth of InP using TMIn. *Journal of Crystal Growth* 63: 8–12.

79. Zhu, L.D., Chan, K.T., and Ballantyne, J.M. (1985). MOCVD growth and characterization of high quality InP. *Journal of Crystal Growth* 73: 83–95.
80. Hardtdegen, H. and Giannoules, P. (1995). An outstanding innovation in LP-MOVPE: use of nitrogen as the carrier gas. *III-Vs Review* 8: 34–9.
81. Saxena, R.R., Fouquet, J.E., Sardi, V.M. et al. (1988). High-quality InP layers grown by organometallic vapor phase epitaxy using tertiarybutylphosphine and phosphine. *Applied Physics Letters* 53: 304–6.
82. Yoon, K.S., Stringfellow, G.B., and Huber, R.J. (1987). Monte Carlo calculation of velocity-field characteristics in GaInAs/InP and GaInAs/AlInAs single-well heterostructures. *Journal of Applied Physics* 62: 1931–6.
83. Zhu, L.D., Sulewski, P.E., Chan, K.T. et al. (1985). Two-dimensional electron gas in $In_{0.53}Ga_{0.47}As$/InP heterojunctions grown by atmospheric pressure metalorganic chemical-vapor deposition. *Journal of Applied Physics*, 58: 3145–9.
84. Mishra, U.K., Brown, A.S., Jelloian, L.M. et al. (1988). High-performance submicrometer AlInAs-GaInAs HEMTs. *IEEE Electron Device Letters* 9: 41–3.
85. Meyer, R., Hardtdegen, H., Carius, R. et al. (1992). Optical and structural properties of MOVPE grown $Ga_xIn_{1-x}As$/InP strained multiple quantum well structures. *Journal of Electronic Materials* 21: 293–8.
86. Duchemin, J.P., Hirtz, J.P., Razeghi, M. et al. (1981). GaInAs and GaInAsP materials grown by low pressure MOCVD for microwave and optoelectronic applications. *Journal of Crystal Growth* 55: 64–73.
87. Davies, J.I., Hodson, P.D., Marshall, A.C. et al. (1988). Improvements in the structural quality of $Al_{0.48}In_{0.52}As$ grown by low pressure metal-organic vapour-phase epitaxy. *Semiconductor Science and Technology* 3: 223–6.
88. Luo, J.K., Thomas, H., Clark, S.A. et al. (1993). The effect of growth temperature on the electrical properties of AlInAs/InP grown by molecular beam epitaxy and metal-organic chemical-vapor deposition. *Journal of Applied Physics* 74: 6726–33.
89. Kuo, C.P., Cohen, R.M., Fry, K.L. et al. (1985). Characterization of $Ga_xIn_{1-x}As$ grown with TMIn. *Journal of Electronic Materials* 14: 231–44.
90. Aina, L. and Mattingly, M. (1987). High-quality InAlAs grown by organometallic vapor phase epitaxy. *Applied Physics Letters* 51: 1637–9.
91. Davies, J.I., Marshall, A.C., Scott, M.D. et al. (1988). Ga-Al-In-As ternary and quaternary alloys lattice matched to InP for electronic, optoelectronic and optical device applications, by LP-MOVPE. *Journal of Crystal Growth* 93: 782–91.
92. Wakefield, B., Halliwel, M.A.G., Kerr, T. et al. (1984). Direct energy gap of $Al_{1-x}In_xAs$ lattice matched to InP. *Applied Physics Letters* 44: 341–3.
93. Grützmacher, D., Hergeth, J., Reinhardt, F. et al. (1990). Mode of growth in LP-MOVPE deposition of GalnAs/lnP quantum wells. *Journal of Electronic Materials* 19: 471–9.
94. Meyer, R., Hollfelder, M., Hardtdegen, H. et al. (1992). Characterization of interface structure in GaInAs/InP superlattices by means of X-ray diffraction. *Journal of Crystal Growth* 124: 583–8.
95. Mircea, A., Mellet, R., Rose, B. et al. (1986). The growth and characterization of device quality InP/$Ga_{1-x}In_xAs_yP_{1-y}$ double heterostructures by atmospheric-pressure MOVPE using trimethylindium. *Journal of Electronic Materials* 15: 205–13.
96. Wolfram, P., Steimetz, E., Ebert, W. et al. (2003). Growth of InGaAsP/InP-laser structures monitored by using RAS techniques. *Journal of Crystal Growth* 248: 240–3.
97. Smeets, E.T.J.M. (1987). Solid composition of $GaAs_{1-x}P_x$ grown by organometallic vapour phase epitaxy. *Journal of Crystal Growth* 82: 385–95.
98. Biefeld, R.M. (1986). The use of metalorganic chemical vapor deposition to prepare device quality Ga(AsP) strained-layer superlattices. *Journal of Electronic Materials* 15: 193–9.
99. Jordan, A.S. (1995). Thermodynamic modeling of As and P incorporation in $Ga_xIn_{1-x}P_yAs_{1-y}$ epitaxial layers grown by organometallic vapor phase epitaxy. *Journal of Electronic Materials* 24: 1649–54.
100. Nelson, A.W., Spurdens, P.C., Cole, S. et al. (1988). The role of MOVPE in the manufacture of high performance InP based optoelectronic devices. *Journal of Crystal Growth* 93: 792–802.
101. Meyer, R., Grützmacher, D., Jürgensen, H. et al. (1988). Controlled uniform growth of GaInAsP/InP structures for laser application on 2 inch wafers by LP-MOVPE at 20 mbar. *Journal of Crystal Growth* 93: 285–91.
102. Reier, F.W., Harde, P., and Kaiser, H. (1991). LP-MOVPE growth and characterization of wide-gap InGaAsP/InP layers ($\lambda_g < 1.2\,\mu m$) for optical waveguide applications. *Journal of Crystal Growth* 107: 867–70.
103. Roehle, H., Schroeter-Janssen, H., and Kaiser, R. (1997). Large- and selective-area LP-MOVPE growth of InGaAsP-based bulk and QW layers under nitrogen atmosphere. *Journal of Crystal Growth* 170: 109–12.
104. Hardtdegen, H., Giannoulés, P., and Succi, M. (1998). MOVPE gets green signal. *III-Vs Review* 11: 34–9.
105. Horita, M., Suzuki, M., and Matsushima, Y. (1992). MOVPE growth of lnGaAsP using TBA and TBP with extremely low V/lll ratio. *Proceedings of the Sixth International Conference on Metal Vapor Phase Epitaxy*. IEEE; p. 191–2.
106. Ae, S., Terakado, T., Nakamura, T. et al. (1994). Low threshold $\lambda = 1.3\,\mu m$ multi-quantum well laser diodes grown by metalorganic vapor phase epitaxy using tertiarybutylarsine and tertiarybutylphosphine precursors. *Journal of Crystal Growth* 145: 852–7.

107. Cheng, W., Zrenner, A., Ye, Q.-Y. et al. (1989). Magnetotransport studies of δ-doping layers in MOCVD-grown InP. *Semiconductor Science and Technology* 4: 16–9.
108. Grützmacher, D. and Glade, M. (1992). MOVPE in Ga-In-As-P systems for opto-electronic applications. *Microelectronic Engineering* 18: 89–117.
109. Aina, L., Mattingly, M., Serio, M. et al. (1991). MOVPE of AlInAs HEMT structures. *Journal of Crystal Growth* 107: 932–41.
110. Chichibu, S., Kushibe, M., Eguchi, K. et al. (1990). High concentration Zn doping in InP grown by low-pressure metalorganic chemical vapor deposition. *Journal of Applied Physics* 68: 859–61.
111. Naritsuka, S., Noda, T., Wagai, A. et al. (1993). Electrical properties and deep levels of InAlAs layers grown by metalorganic chemical vapor deposition. *Journal of Crystal Growth* 131: 186–92.
112. Bhat, R., Koza, M.A., Kash, K. et al. (1991). Growth of high quality AlInAs by low pressure organometallic chemical vapor deposition for high speed and optoelectronic device applications. *Journal of Crystal Growth* 108: 441–8.
113. Ito, H. and Yokoyama, H. (1997). Carbon doping in InAlAs grown by metalorganic chemical vapor deposition. *Journal of Crystal Growth* 173: 315–20.
114. Ougazzaden, A., Holavanahalli, J., Geva, M. et al. (2000). Carbon doping of InAlAs in LP-MOVPE using CBr_4. *Journal of Crystal Growth* 221: 66–9.
115. Long, J.A., Riggs, V.G., and Johnston, W.D. (1984). Growth of Fe-doped semi-insulating InP by MOCVD. *Journal of Crystal Growth* 69: 10–4.
116. Franke, D., Harde, P., Wolfram, P. et al. (1990). Doping and diffusion behaviour of Fe in MOVPE grown InP layers. *Journal of Crystal Growth* 100: 309–12.
117. Haywood, S.K., Henriques, A.B., Mason, N.J. et al. (1988). Growth of GaSb by MOVPE. *Semiconductor Science and Technology* 3: 315–20.
118. Pino, R., Ko, Y., and Dutta, P.S. (2004). High-resistivity GaSb bulk crystals grown by the vertical Bridgman method. *Journal of Electronic Materials* 33: 1012–5.
119. Graham, R.M., Jones, A.C., Mason, N.J. et al. (1993). Improved materials for MOVPE growth of GaSb and InSb. *Semiconductor Science and Technology* 8: 1797–802.
120. Baucom, K.C. and Biefeld, R.M. (1994). Growth of InSb using tris(dimethylamino)antimony and trimethylindium. *Applied Physics Letters* 64: 3021–3.
121. Aardvark, A., Mason, N.J., and Walker, P.J. (1997). The growth of antimonides by MOVPE. *Progress in Crystal Growth and Characterisation of Materials* 35: 207–41.
122. Wang, C.A. (1997). Organometallic vapor phase epitaxial growth of AlSb-based alloys. *Journal of Crystal Growth* 170: 725–31.
123. Biefeld, R.M., Allerman, A.A., and Pelczynski M.W. (1996). Growth of n- and p-type Al(As)Sb by metalorganic chemical vapor deposition. *Applied Physics Letters* 68: 932–4.
124. Biefeld, R.M. (2002). The metal-organic chemical vapor deposition and properties of III–V antimony-based semiconductor materials. *Materials Science and Engineering R Reports* 36: 105–42.
125. Behet, M., Schneider, P., Moulin, D. et al. (1996). Low pressure metalorganic vapor phase epitaxy and characterization of (Al,Ga)Sb/GaSb heterostructures. *Journal of Crystal Growth* 167: 415–20.
126. Biefeld, R.M. and Kurtz, S.R. (2001). In: *Infrared Detectors and Emitters: Materials and Devices* (eds. P. Capper and C.T. Elliot). Boston: Springer US; p. 205–32.
127. Stringfellow, G.B. (1983). Thermodynamic aspects of organometallic vapor phase epitaxy. *Journal of Crystal Growth* 62: 225–9.
128. Stringfellow, G.B. (1984). A critical appraisal of growth mechanisms in MOVPE. *Journal of Crystal Growth* 68: 111–22.
129. Stringfellow, G.B. (1984). Thermodynamic aspects of OMVPE. *Journal of Crystal Growth* 70: 133–9.
130. Biefeld, R.M. (1986). The preparation of $InAs_{1-x}Sb_x$ alloys and strained-layer superlattices by MOCVD. *Journal of Crystal Growth* 77: 392–9.
131. Stringfellow, G.B. and Cherng, M.J. (1983). OMVPE growth of $GaAs_{1-x}Sb_x$: solid composition. *Journal of Crystal Growth* 64: 413–5.
132. Cooper, C.B., Saxena, R.R., and Ludowise, M.J. (1982). The organometallic VPE growth of GaSb and $GaAs_{1-x}Sb_x$ using trimethylantimony. *Journal of Electronic Materials* 11: 1001–10.
133. Stringfellow, G.B. (1999). *Organometallic Vapor-Phase Epitaxy*. Amsterdam: Elsevier.
134. Bougnot, G. (1988). Growth of GaInAsSb alloys by MOCVD and characterization of GaInAsSb/GaSb p-n photodiodes. *Journal of the Electrochemical Society* 135: 1783.
135. Wang, C. (1998). OMVPE growth of GaInAsSb in the 2–2.4 μm range. *Journal of Crystal Growth* 191: 631–40.
136. Wang, C.A., Choi, H.K., Oakley, D.C. et al. (1998). Recent progress in GaInAsSb thermophotovoltaics grown by organometallic vapor-phase epitaxy. *Journal of Crystal Growth* 195: 346–55.
137. Cao, D.S., Fang, Z.M., and Stringfellow, G.B. (1991). Organometallic vapor-phase epitaxial growth of $Al_xGa_{1-x}Sb$ and $Al_xGa_{1-x}As_ySb_{1-y}$. *Journal of Crystal Growth* 113: 441–8.

138. Cooper, C.B., Saxena, R.R., and Ludowise, M.J. (1980). OMVPE growth of AlGaSb and AlGaAsSb. *Electronics Letters* 16: 892.
139. Giesen, C., Szymakowski, A., Rushworth, S. et al. (2000). MOVPE of AlGaAsSb using TTBAl as an alternative aluminum precursor. *Journal of Crystal Growth*, 221: 450–5.
140. Koljonen, T., Sopanen, M., Lipsanen, H. et al. (1996). Metalorganic vapor phase epitaxial growth of AlGaSb and AlGaAsSb using all-organometallic sources. *Journal of Crystal Growth* 169: 417–23.
141. Dingle, R., Störmer, H.L., Gossard, A.C. et al. (1978). Electron mobilities in modulation-doped semiconductor heterojunction superlattices. *Applied Physics Letters* 33: 665–7.
142. Esaki, L. and Tsu, R. (1970). Superlattice and negative differential conductivity in semiconductors. *IBM Journal of Research and Development* 14: 61–5.
143. Pelucchi, E., Moret, N., Dwir, B. et al. (2006). Sub-meV photoluminescence linewidth and $>10^6 cm^2/Vs$ electron mobility in AlGaAs/GaAs quantum wells grown by metalorganic vapor phase epitaxy on slightly misoriented substrates. *Journal of Applied Physics* 99: 093515.
144. Kobayashi, N., Fukui, T., and Tsubaki, K. (1984). Selectively-doped GaAs/n-AlGaAs heterostructures grown by MOCVD. *Japanese Journal of Applied Physics* 23: 1176–81.
145. Chui, H.C., Hammons, B.E., Harff, N.E. et al. (1996). $2\times10^6 cm^2/V s$ electron mobility by metalorganic chemical vapor deposition with tertiarybutylarsine. *Applied Physics Letters* 68: 208–10.
146. Chui, H.C., Hammons, B.E., Simmons, J.A. et al. (1995). High quality single and double two-dimensional electron gases grown by metalorganic vapor phase epitaxy. *Applied Physics Letters* 67: 1911–3.
147. André, J.P., Brière, A., Rocchi, M. et al. (1984). Growth of heterostructures for HEMT devices. *Journal of Crystal Growth* 68: 445–9.
148. Hardtdegen, H., Hollfelder, M., Ungermanns, C. et al. (1995). In: *Institute of Physics Conference Series* No 141 (eds. H. Goronkin and U.K. Mishra), Ch. 2. Bristol and Philadelphia: Institute of Physics; p. 81–6.
149. People, R., Wecht, K.W., Alavi, K. et al. (1983). Measurement of the conduction-band discontinuity of molecular beam epitaxial grown $In_{0.52}Al_{0.48}As/In_{0.53}Ga_{0.47}As$, N-n heterojunction by C–V profiling. *Applied Physics Letters* 43: 118–20.
150. Aina, L., Mattingly, M., Serio, M. et al. (1991). MOVPE of AlInAs HEMT structures. *Journal of Crystal Growth* 107: 932–41.
151. Hardtdegen, H., Meyer, R., Løken-Larsen, H. et al. (1992). Extremely high electron mobilities in modulation doped $Ga_{1-x}In_xAs$/InP heterostructures grown by LP-MOVPE. *Journal of Crystal Growth* 116: 521–3.
152. Hardtdegen, H., Meyer, R., Hollfelder, M. et al. (1993). Optimization of modulation-doped $Ga_{1-x}In_xAs$/InP heterostructures towards extremely high mobilities. *Journal of Applied Physics* 73: 4489–93.
153. Meyer, R., Hardtdegen, H., Leuther, A. et al. (1993). Optimization of strained $Ga_xln_{1-x}As$/InP heterostructures towards high channel conductivity for HEMT application. Proceedings of the 5th International Conference on Indium Phosphide Related Materials. IEEE; p. 485–8.
154. Schimpf, K., Hollfelder, M., Marso, M. et al. (1993). 0.2 μm T-gate InP/InGaAs/InP pHEMT with an InGaP diffusion barrier layer grown by LP-MOCVD using an N_2-carrier. *54th Annual Device Research Conference Digest.* IEEE; p. 126–7.
155. Kroemer, H. (1957). Theory of a wide-gap emitter for transistors. *Proceedings of the IRE* 45: 1535–7.
156. Beneking, H. and Su, L.M. (1982). Double heterojunction NpN GaAlAs/GaAs bipolar transistor. *Electronics Letters* 18: 25.
157. Kroemer, H. (1983). Heterostructure bipolar transistors: What should we build? *Journal of Vacuum Science and Technology B* 1: 126.
158. Alexandre, F., Benchimol, J., Launay, P. et al. (1995). Modern epitaxial techniques for HBT structures. *Solid State Electronics* 38: 1667–74.
159. Fricke, K., Hartnagel, H.L., Lee, W.-Y. et al. (1992). AlGaAs/GaAs HBT for high-temperature applications. *IEEE Transactions on Electron Devices* 39: 1977–81.
160. Bottka, N., Gaskill, D.K., Wright, P.D. et al. (1991). Qualification of OMVPE AlGaAs/GaAs HBT structures using nondestructive photoreflectance spectroscopy. *Journal of Crystal Growth* 107: 893–7.
161. Cunningham, B.T., Stillman, G.E., and Jackson, G.S. (1990). Carbon-doped base GaAs/AlGaAs heterojunction bipolar transistor grown by metalorganic chemical vapor deposition using carbon tetrachloride as a dopant source. *Applied Physics Letters* 56: 361–3.
162. Abernathy, C.R., Ren, F., Wisk, P.W. et al. (1992). Improved performance of carbon-doped GaAs base heterojunction bipolar transistors through the use of InGaP. *Applied Physics Letters* 61: 1092–4.
163. Yang, Q., Scott, D., Chung, T. et al. (2000). Optimization of emitter cap growth conditions for InGaP/GaAs HBTs with high current gain by LP-MOCVD. *Journal of Electronic Materials* 29: 75–9.
164. Razeghi, M., Omnest, F., Defourt, M. et al. (1990). High performance GaAs/GalnP heterostructure bipolar transistors grown by low-pressure metal-organic chemical vapour deposition. *Semiconductor Science and Technology* 278: 278–80.

165. Brunner, F., Richter, E., Bergunde T. et al. (2000). Effect of high-temperature annealing on GaInP/GaAs HBT structures grown by LP-MOVPE. *Journal of Electronic Materials* 29: 205–9.
166. Kurishima, K., Yamahata, S., Nakajima, H. et al. (1998). Performance and stability of MOVPE-grown carbon-doped InP/InGaAs HBTs dehydrogenated by an anneal after emitter mesa formation. *Japanese Journal of Applied Physics* 37: 1353–8.
167. Matsuoka, Y. and Sano, E. (1995). Double-heterostructure bipolar transistors for high-speed ICs and OEICs. *Solid State Electronics* 38: 1703–9.
168. Chor, E.F. and Peng, C.J. (1996). Composite step-graded collector of InP/InGaAs/InP DHBT for minimised carrier blocking. *Electronics Letters* 32: 1409.
169. Nottenburg, R.N., Bischoff, J.C., Abeles, J.H. et al. (1986). Base doping effects in InGaAs/InP double heterostructure bipolar transistors. *International Electron Devices Meeting*, 278–81.
170. Velling, P. (2000). A comparative study of GaAs- and InP-based HBT growth by means of LP-MOVPE using conventional and non gaseous sources. *Progress in Crystal Growth and Characterisation of Materials* 41: 85–131.
171. Kurishima, K., Nakajima, H., Kobayashi, T. et al. (1994). Fabrication and characterization of high-performance InP/InGaAs double-heterojunction bipolar transistors. *IEEE Transactions on Electron Devices* 41: 1319–26.
172. Mokkapati, S. and Jagadish, C. (2009). III-V compound SC for optoelectronic devices. *Materials Today* 12: 22–32.
173. Holonyak, N. and Bevacqua, S.F. (1962). Coherent (visible) light emission from $Ga(As_{1-x}P_x)$ junctions. *Applied Physics Letters* 1: 82–3.
174. Sheu, J., Chi, G., Su, Y. et al. (2000). Luminescence of an InGaN/GaN multiple quantum well light-emitting diode. *Solid State Electronics* 44: 1055–8.
175. Morkoc, H., Sverdlov, B., and Gao, G.-B. (1993). Strained layer heterostructures, and their applications to MODFETs, HBTs, and lasers. *Proceedings of the IEEE* 81: 493–556.
176. Bugge, F., Zeimer, U., Staske, R. et al. (2007). MOVPE growth optimization for laser diodes with highly strained InGaAs MQWs. *Journal of Crystal Growth* 298: 652–7.
177. Kato, T., Susawa, H., Hirotani, M. et al. (1991). GaAs/GaAlAs surface emitting IR LED with Bragg reflector grown by MOCVD. *Journal of Crystal Growth* 107: 832–5.
178. Kuo, C.P., Fletcher, R.M., Osentowski, T.D. et al. (1990). High performance AlGaInP visible light-emitting diodes. *Applied Physics Letters* 57: 2937–9.
179. Kuo, C.P., Fletcher, R.M., Osentowski, T.D. et al. (1988). Stimulated emission in $In_{0.5}(Al_xGa_{1-x})_{0.5}P$ quantum well heterostructures. *Journal of Crystal Growth* 93: 389–95.
180. Fletcher, R.M., Kuo, C.P., Osentowski, T.D. et al. (1991). The growth and properties of high performance AlGaInP emitters using a lattice mismatched GaP window layer. *Journal of Electronic Materials* 20: 1125–30.
181. Kish, F.A., Steranka, F.M., DeFevere, D.C. et al. (1994). Very high-efficiency semiconductor wafer-bonded transparent-substrate $(Al_xGa_{1-x})_{0.5}In_{0.5}P$/GaP light-emitting diodes. *Applied Physics Letters* 64: 2839–41.
182. Dupuis, R.D. (1981). AlGaAs-GaAs lasers grown by metalorganic chemical vapor deposition — A review. *Journal of Crystal Growth* 55: 213–22.
183. Speier, P. (1992). MOVPE for optoelectronics. *Microelectronic Engineering* 18: 1–31.
184. Ishikawa, M., Ohba, Y., Sugawara, H. et al. (1986). Room-temperature cw operation of InGaP/InGaAlP visible light laser diodes on GaAs substrates grown by metalorganic chemical vapor deposition. *Applied Physics Letters* 48: 207–8.
185. Kaspari, C., Zorn, M., Weyers, M. et al. (2008). Growth parameter optimization of the GaInP/AlGaInP active zone of 635 nm red laser diodes. *Journal of Crystal Growth* 310: 5175–7.
186. Su, Y.K., Chen, W.C., Wan, C.T. et al. (2008). Optimization of the highly strained InGaAs/GaAs quantum well lasers grown by MOVPE. *Journal of Crystal Growth* 310: 3615–20.
187. Thijs, P.J.A., van Dongen, T., Kuindersma, P.I. et al. (1988). High quality InGaAsP-InP for multiple quantum well laser diodes grown by low-pressure OMVPE. *Journal of Crystal Growth* 93: 863–9.
188. Geurts, J., Finders, J., Woitok, J. et al. (1994). Characterization of the interface abruptness of $In_{0.53}Ga_{0.47}As$/InP multi quantum wells by Raman spectroscopy, X-ray diffractometry and photoluminescence. *Journal of Crystal Growth* 145: 813–8.
189. Nakamura, S. (1991). Analysis of real-time monitoring using interference effects. *Japanese Journal of Applied Physics* 30: 1348–53.
190. Breiland, W.G. (2003). Reflectance-correcting pyrometry in thin film deposition applications. Sandia National Laboratories Report SAND2003-1868.
191. Aspnes, D.E. (1995). Observation and analysis of epitaxial growth with reflectance-difference spectroscopy. *Materials Science and Engineering B* 30: 109–19.
192. Zettler, J.-T. (1997). Characterization of epitaxial semiconductor growth by reflectance anisotropy spectroscopy and ellipsometry. *Progress in Crystal Growth and Characterisation of Materials* 35: 27–98.
193. Breiland, W.G. and Killeen, K.P. (1995). A virtual interface method for extracting growth rates and high temperature optical constants from thin semiconductor films using in situ normal incidence reflectance. *Journal of Applied Physics* 78: 6726–36.

194. Hardtdegen, H., Kaluza, N., Steins, R. et al. (2005). Use of wafer temperature determination for the study of unintentional parameter influences for the MOVPE of III-nitrides. *Physica Status Solidi* 242: 2581–6.
195. Haberland, K., Bhattacharya, A., Zorn, M. et al. (2000). MOVPE growth of (Al,Ga)InP-based laser structures monitored by real-time reflectance anisotropy spectroscopy. *Journal of Electronic Materials* 29: 468–72.
196. Zorn, M., Haberland, K., Knigge, A. et al. (2002). MOVPE process development for 650 nm VCSELS using optical in-situ techniques. *Journal of Crystal Growth* 235: 25–34.
197. Zettler, J.-T., Haberland, K., Zorn, M. et al. (1998). Real-time monitoring of MOVPE device growth by reflectance anisotropy spectroscopy and related optical techniques. *Journal of Crystal Growth* 195: 151–62.
198. Steimetz, E., Zettler, J.-T., Schienle, F. et al. (1996). In situ monitoring of InAs-on-GaAs quantum dot formation in MOVPE by reflectance-anisotropy-spectroscopy and ellipsometry. *Applied Surface Science* 107: 203–11.

4

Nitride Semiconductors

A. Dadgar[1] *and M. Weyers*[2]
[1]*Otto-von-Guericke Universität Magdeburg, Germany*
[2]*Ferdinand-Braun-Institute, Leibniz-Institut für Höchstfrequenztechnik, Berlin, Germany*

4.1 Introduction

GaN was grown in the very early days of MOVPE by Hal Manasevit [1]. However, compared to the classical III-Vs, the III-nitrides (AlN, GaN, InN, and their alloys; BN is usually not considered in this context, but alloying B into GaN and AlN is a research topic [2, 3] and may also find its way into applications) initially played only a minor role. The book by Stringfellow only briefly mentions the nitrides [4], even in the second edition from 1999 [5]. Even Razeghi [6] later completely concentrated on the classical GaAs- and InP-based III-Vs. It took a while until the basic problems with high defect density due to heteroepitaxy (dominantly on sapphire) [7, 8] and the electrical activation of p-doping [9–11] were solved to a level that allowed for practical application in devices. These breakthrough achievements, together with the application of the double heterostructure concept with thin InGaN quantum wells (QWs) [12, 13], lay the ground for the development of blue light-emitting diodes (LEDs) and also laser diodes. These blue LEDs, which are the heart of white emitters for general lighting, are now the device for which the largest area of epitaxial wafers is produced by MOVPE and for which the biggest MOVPE (or MOCVD) reactors are on the market. The AIXTRON G5+ has a capacity of 8″×6″ (https://www.aixtron.com/en/products/compound-semiconductors-mocvd/aix-g5-c), and the Veeco EPIK 700 (http://www.veeco.com/products/turbodisc-epik-700-gan-mocvd-system-for-led-production) has a 12″×6″ capacity, and both can be highly automated. This economic success, which was made possible by their research, was a relevant factor in the decision to award the 2014 Noble Prize in Physics to the three pioneers Isamu Akasaki, Hiroshi Amano, and Shuji Nakamura [14], who achieved the breakthroughs mentioned here.

Blue LEDs consisting primarily of GaN with some AlGaN layers with moderate Al content and thin InGaN QWs with moderate In content posed significant challenges to be mastered on the way

Metalorganic Vapor Phase Epitaxy (MOVPE): Growth, Materials Properties, and Applications,
First Edition. Edited by Stuart Irvine and Peter Capper.

to technical and economic success. When adding more In, additional issues like miscibility gaps for InGaN [15] and AlInN [16], In desorption, and H etching arise on the way to lower bandgap and longer emission wavelengths. On the other side of the spectrum, low Al adatom mobility requiring high growth temperature and the high lattice mismatch between AlN and GaN leading to a pronounced tendency for cracking are issues to be tackled. Thus, this chapter first discusses the important properties of the III-nitrides and challenges for MOVPE growth. These challenges also have some impact on the MOVPE systems and precursors used. Finally, device applications are briefly addressed.

4.2 Properties of III-Nitrides

The properties of III-nitrides are the topic of books like [17–19]. Hence, here only some of the basic properties, especially those relevant for epitaxial growth, are summarized. The III-nitrides are direct semiconductors for all compositions. The bandgap ranges from 6.2 eV (200 nm in the UVC) through 3.4 eV (365 nm in the UVA) to around 0.7 eV. Hence, in principle, the whole spectral range from the IR through the visible down to the UVC can be covered by III-nitrides. In addition to the blue LEDs around 450 nm, which are produced in large quantities as the basis for white LEDs, green LEDs up to 570 nm (see the websites and catalogues of OSRAM, Nichia, and others) and UV LEDs [20] down to the UVC (265 nm (www.nikkisouvled.com) for disinfection [21] and also less than 235 nm [22] for absorption spectroscopy) are on the market. The possible adjustment of the bandgap also can be exploited for tailoring the sensitivity of photodetectors and, for example, to make them solar-blind [23].

The high bandgap of GaN and AlGaN comes with a high breakdown field strength (GaN: 3.3 MV/cm [24] versus Si: 0.3 MV/cm [25]), which is exploited in field-effect transistors to achieve high power density and high switching voltage.

The energetically most favorable crystal structure of the III-nitrides is the hexagonal wurtzite configuration, as depicted in Figure 4.1. This structure lacks inversion symmetry and features alternating sheets of metal atoms and N atoms in the c-direction (which is the usual growth direction).

This leads to intrinsic spontaneous polarization and also a strong difference between the metal-polar +c direction (which is the usual growth direction for most devices) and the N-polar –c direction. These differences result in different processes at the growing surface. N-polar growth in MOVPE usually is rougher and leads to higher impurity uptake than metal-polar growth [26]. The additional piezoelectric polarization occurring due to strain when growing, for example, a pseudomorphic AlGaN layer on GaN is important for the properties of heterostructures. The polarization fields occurring in the polar c-direction, on the one hand, are of benefit for certain applications since they

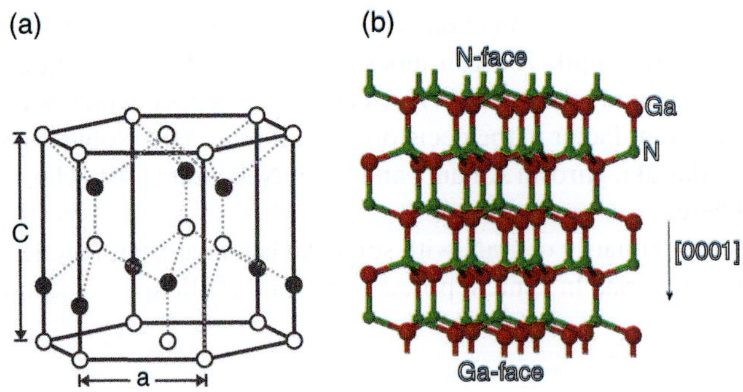

Figure 4.1 (a) Hexaogonal unit cell with lattice parameters c and a, white balls represent metal atoms, black balls N atoms. Top surface is metal face. (b) Ball-and-stick model of GaN with +c direction from top to bottom and opposing N and Ga faces.

Table 4.1 *Compilation of basic properties of binary nitrides relevant for epitaxial growth. The properties are given at room temperature but it has to be noted that, for example, the thermal expansion coefficients change considerably between room temperature and the typical nitride growth temperatures*

Material	AlN	GaN	InN
Energy gap (T = 300 K) [eV]	6.213 [30]	3.447 [30]	0.675 [31]
Lattice parameter c [Å]	4.982 [32]	5.185 [32]	5.703 [32]
Lattice parameter a [Å]	3.112 [32]	3.189 [32]	3.545 [32]
Lattice mismatch $\Delta a/a$ to GaN [%]	−2.5	-	+11
Thermal expansion coefficient α [10^{-6} K^{-1}]	4.2 [33]	5.59 [33]	3.8 [34]

allow for the growth of AlGaN/GaN heterojunction field effect transistors (HFETs) without using a dopant like Si for metal-polar growth, because their electrons accumulate in GaN under an AlGaN cap layer. N-polar growth allows for normally off transistors but is not pursued by many groups [27]. On the other hand, the polarization fields lead to separation of electrons and holes in QWs and thus reduced overlap for radiative recombination in LEDs. At heterojunctions, they can give rise to additional barriers for carrier injection or lead to carrier accumulation. This effect can be mitigated by using non- or semipolar orientations [28, 29].

Table 4.1 summarizes some basic properties of the binary nitrides AlN, GaN, and InN. These properties will be discussed in more detail in the following sections.

4.3 Challenges in the Growth of III-Nitrides

4.3.1 Lattice and Thermal Mismatch

As seen in Table 4.1, there is a considerable mismatch in the lattice constant *a*, which is the relevant in-plane lattice constant for the usual growth on the c-plane. This mismatch limits the thickness up to which, for example, AlGaN can be grown pseudomorphically on GaN or AlN. In addition to lattice mismatch within the III-Ns, a large lattice mismatch exists for the growth on heterosubstrates (see Table 4.2).

The considerable mismatch of the thermal expansion coefficients leads to strain and wafer bow when cooling from the growth temperature to room temperature. Cracking of wafers thus is a challenge, especially for thick heterostructures, as will be discussed in the section on substrates. This requires control of the bow during growth.

The bandgap of semiconductors is related to the bond strength. Hence, it is obvious that the Al–N bond strength is significantly higher than that of Ga–N. The In–N bond is even weaker. Diffusion on the growing surface to find the right place for incorporation into the lattice is essential for step-flow growth and thus smooth surfaces. AlN with its higher bond strength requires higher growth temperatures for sufficient mobility on the surface than GaN. However, the vapor pressure of Ga (over Ga) is roughly two orders of magnitude higher than that of Al over Al and approaches 1 Pa above 1000 °C. Hence, preferential Ga desorption limits the temperature at which AlGaN can be grown [41]. It is difficult to compare growth temperatures since often only the temperature of the susceptor is given, and pyrometric measurements are difficult for substrates and layer stacks that are transparent at the usual wavelengths [42]. However, using the same reactor, the growth temperature measured on the backside of the susceptor has to be lowered from the 1400 °C preferred for AlN [43] to below 1220 °C to obtain significant Ga incorporation and to around 1150 °C to make the Ga content controllable by the gas-phase composition and not by desorption (or etching by hydrogen) that is controlled by temperature.

Table 4.2 *Lattice parameters a and c thermal expansion coefficients of GaN and commonly used substrates for growth. Thermal expansion coefficients are also given as an average (**bold**) of the measured values for the temperature range from 300 to 1350 K (typical GaN growth temperature). *value for GaN calculated on base of reference [36].*

	a [Å]	c [Å]	Bond length [Å]	Mismatch of a(GaN) to a(substrate) [%]	In-plane thermal expansion coefficient [10^{-6} K^{-1}]	Substrate thermal mismatch to GaN [%]
GaN	3.189	5.185	1.95	–	3.9–6.0 (300–1000 K) [35] **Ø 5.5** *	–
Sapphire	4.758	12.991	N.A.	14.8 (30° rotation)	7.3 [36] **Ø 7.0** [37]	24.6 **27.3**
6H-SiC	3.081	15.092	1.89	−3.2	4.3 [38] **Ø 4.5** [39]	−21.8 **−18.2**
Si (111)	5.431	–	2.35	17.0	2.6–4.4 **Ø 3.97** [40]	−52.7 **−27.8**

Similar considerations apply for the growth of InGaN, since In has a significantly higher vapor pressure than Ga. For In, the effect of etching by hydrogen due to formation of volatile In-hydride is much more pronounced than for the less volatile Ga. This requires not only considerably lower growth temperatures for In-containing compounds but also the avoidance of hydrogen in the gas phase. InGaN and InAlN are grown under N_2 as carrier gas, since otherwise only insufficient In incorporation can be obtained [44].

The most extreme case with respect to differences in bond strength is AlInN. The main difficulty of growing AlInN is the choice of growth parameters. High-quality AlN is best grown at high temperatures well above 1000 °C, low pressure, and low V/III ratio; the latter two are important to reduce TMAl-NH_3 precursor prereactions that lead to low growth rates and can lead to particle formation in the gas phase. This is discussed in more detail in Section 4.3.3. High temperatures help to avoid nitrogen vacancies [45] and high carbon impurity concentrations stemming from the organic compounds of the metalorganics used.

InN is best grown with opposing growth parameters such as low temperature (<550 °C) [46] due to its high vapor pressure, high V/III ratios, and moderate to high reactor pressures. The latter two are required to avoid In droplet formation and because ammonia has a low decomposition rate at low temperatures, requiring high levels of ammonia. Also, nitrogen carrier gas is a must for InN because hydrogen inhibits growth.

These parameters seem incompatible, but for growing the alloy, the situation is not as bad as suggested by comparing these requirements. It turns out that AlInN with low In content, like lattice-matched growth on GaN, can be grown under AlN-like growth conditions but under nitrogen at lower temperature. Across a wide range [47–49] of In content, from 1% to greater than 25% [50–53], the only parameter that must be varied is temperature, with a change in absolute In content of ~1.1% for every 10 K temperature change [54], similar to the behavior for InGaN layers [5]. For a-planar layers the variation in In content is higher and amounts to ~2% per 10 K in the concentration range between ~ 10 to ~ 30% (700–800 °C growth temperature) [55].

Figure 4.2 gives an example of the impact of growth conditions on the In content for 50 pseudomorphically grown AlInN samples.

In detail, the growth temperature is not the only parameter that impacts the In content. Pressure, V-III ratio, and III-III ratio show a temperature-dependent influence on composition and/or growth rate. Hums et al. observed that for lower growth temperatures of 700 °C, a variation of the III-III ratio enables the growth of layers with different In content; while at 820 °C, the In content is nearly constant and only the growth rate changes [54].

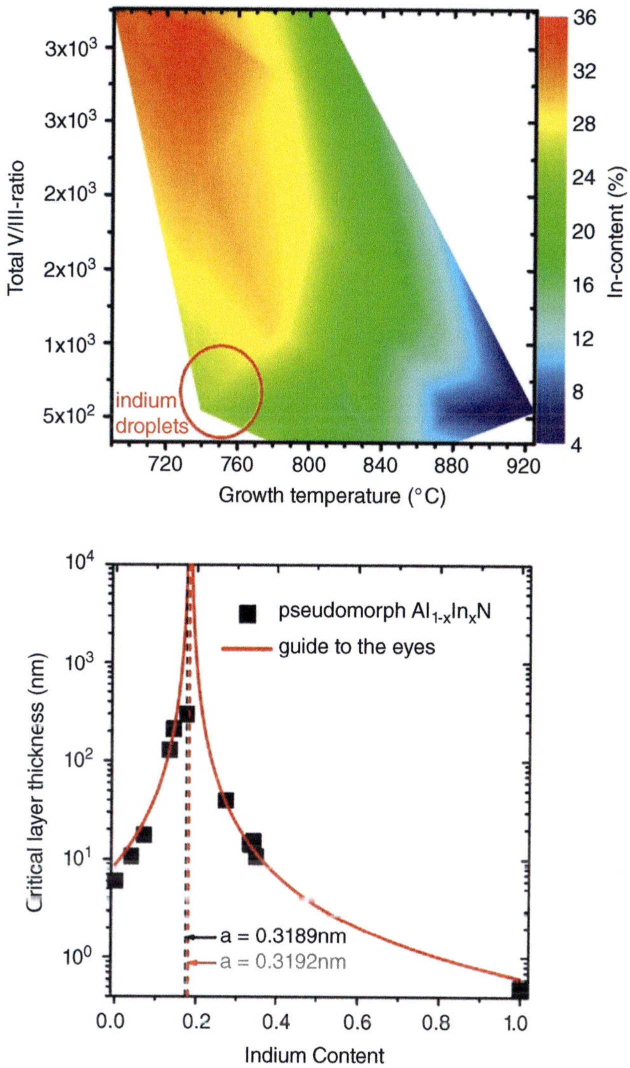

Figure 4.2 *Change in In content of AlInN by a variation of growth temperature and V/III-ratio (top) and experimentally determined critical thickness for AlInN on GaN grown on Si(111) (bottom) Top: For low temperatures and ammonia partial pressure In droplets form on the surface (Data from 50 pseudomorphically grown samples). Bottom: Pseudomorphic growth was determined by grazing incidence-ray diffraction of the (10$\bar{1}$0) reflection. The data collected deviates from the expected theoretical values of the critical thickness, therefore only a guide to the eye is drawn. Source: Hums 2012 [54]; reprinted with kind permission of Christoph Hums.*

4.3.2 Ternary Alloys: Miscibility and Compositional Homogeneity

4.3.2.1 AlGaN

AlN and GaN are miscible over the whole range of $Al_xGa_{1-x}N$ compositions from $x=0$ to 1. However, this does not imply that AlGaN layers have a homogeneous composition. The higher mobility of Ga on the surface can result in Ga enrichment at surface steps of sufficient height. Here, macrosteps of 10 nm height can be sufficient [56]. Figure 4.3 shows that the extent and shape

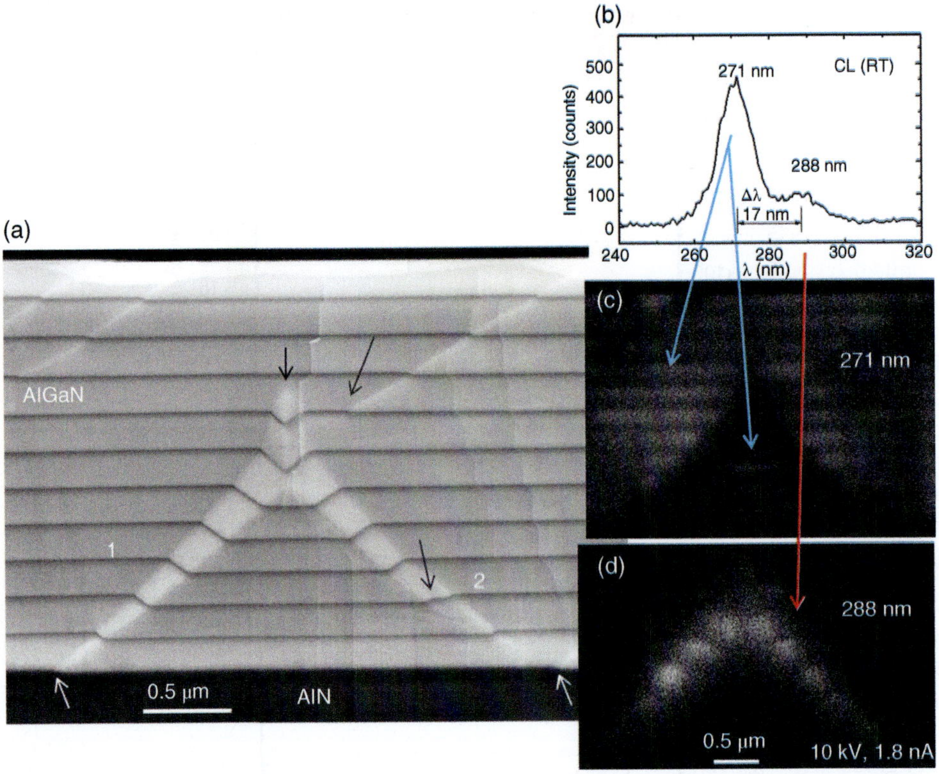

Figure 4.3 (a) SEM micrograph (BSE mode), spectrum (b) and monochromatic CL images (c, d) of $(1\bar{1}00)$ cross section of a nominally $Al_{0.5}Ga_{0.5}N$ layer with AlN marker layers (dark lines) grown at 0.75 µm/h on ELO-AlN. Source: Adapted from Knauer et al. 2014 [57]; reprinted with kind permission of Wiley-VCH.

of such areas with higher Ga content than the matrix depends on the growth temperature, growth rate, and V/III ratio, since these parameters impact surface diffusion [57].

4.3.2.2 In-Containing Ternaries

In-containing alloys are most difficult to grow by MOVPE because of the very different size and bonding of In to N in comparison to the Al–N and Ga–N bonds. This leads to a large miscibility gap and the relatively low thermal stability of these alloys. The miscibility gap has already been discussed in Chapter 2 of this book.

4.3.2.2.1 AlInN

The most common composition of AlInN grown is nearly lattice-matched to GaN [58]. This is achieved at an In content of approximately 18%, which is low enough to enable thin layers with homogeneous In content. Lattice-matched growth is preferred for Bragg mirror layers, e.g. for vertical cavity surface emitting lasers (VCSELs) [59–63] as well as for high-power FETs, where it offers low resistivities and also the potential benefit of reduced aging due to low lattice mismatch in the active region.

Spinodal decomposition of AlInN is the main factor limiting the growth of high-In-content layers on GaN [64, 65]. However, depending on the alloy composition, an ideal set of growth parameters can be found that lowers the tendency for spinodal decomposition. These ideal growth parameters differ strongly for layers with higher In content (» 30%) versus those with lower In content. Spinodal decomposition has also been described as strain dependent [66, 67, 16]. Nevertheless, in experiments, up to an In content of 49% – right in the center of the typically expected spinodal gap – has been achieved without spinodal decomposition for a 47-nm thick InAlN layer [52].

Pseudomorphic growth (Figure 4.2) is most important when it comes to devices. Gadanecz et al. have shown that only pseudomorphically grown AlInN has high thermal stability, even close to the growth temperature of the layer [68], which is beneficial for high-temperature operation [69–71]. The optical properties have been determined, for example, by Sakalauskas et al. [72], and QW structures have been demonstrated [73].

It has been observed that growing thicker layers, even when fully pseudomorphic or nominally lattice matched, leads to an In gradient with increasing In content. Thus, when growing thick layers, for example, either In precursor flow should be decreased or temperature increased with increasing thickness to avoid In accumulation. Also, heating after AlInN growth is critical for the composition in the interface region. In desorption is commonly observed if the capping layer has been grown at high temperature, which can be avoided when the initial part is grown at a temperature close to that during AlInN growth. These difficulties in controlling the growth of AlInN have up to now prevented its introduction into actual device technology.

4.3.2.2.2 InGaN

The differences between GaN and InN in bond length, bond strength, and thermal stability are smaller than for AlInN. However, there is no substrate or binary buffer for pseudomorphic growth, and InGaN growth in general starts on a GaN buffer. Hence, strain relaxation is an important issue: e.g. for green LEDs with highly In-containing QWs. In addition to the already-mentioned miscibility gap, which can result in phase separation, the higher mobility of the less strongly bonded In on the surface can lead to In enrichment at step edges, as already discussed for AlGaN. Hence, meandering inhomogeneous composition is common in thicker InGaN layers [74]. Fluctuations in In content do therefore differ in dependence on the used substrate [75] and growth conditions, but are regarded as beneficial for light emission from LEDs due to carrier localization.

In most devices, InGaN is only grown to a thickness around a few nanometers; but in stacked QWs, stress adds up and limits the number of QW stacks that can be grown free of relaxation. This accounts in particular for LEDs in the green wavelength region where high In content and/or thicker QWs are required. InGaN usually being applied in light-emitting layers, piezo- and pyroelectric fields omnipresent in *c*-axis oriented GaN have to be taken into account. These fields reduce quantum efficiency due to the quantum-confined Stark effect (QCSE). Growing on nonpolar planes has the difficulty of often-poor material quality due to stacking faults as well as relaxation of InGaN QWs by forming a stacking fault to relax the layer [76]. When trying to grow polarization-matched structures by adjusting the composition and strain of a quaternary barrier, it has been shown that apart from difficulties like relaxation of a multiquantum well (MQW), the reduction in QCSE leads to a significant blue shift, hampering green-light emission [77].

4.3.3 Gas-Phase Prereactions

NH_3 has a free electron pair in the second shell that likes to form bonds with empty orbitals on Group-III atoms in compounds like trimethylaluminum (TMAl), the usual Al precursor in MOVPE. This leads to the formation of adducts in the gas phase to a much larger extent than for the other group-V hydrides, AsH_3 and PH_3.

The simple GaN growth reaction is usually sketched by writing $(CH_3)_3Ga + NH_3 \rightarrow GaN + 3CH_4$. However, the reality is far from this simple view. In particular, in III-N epitaxy metalorganics tend to form adducts with ammonia that sometimes (for Ga adducts) decompose again at growth temperature or, in the case of TMAl and ammonia, are stable at the growth temperature (if $< 1100\,°C$) or even form nanocrystallites that can degrade layer quality. The first stage of adduct formation between trimethylgallium (TMGa) and NH_3 occurs at relatively low temperatures and can be written as $(CH_3)_3Ga + NH_3 \rightarrow (CH_3)_3Ga{:}NH_3$ [78]. At 90 °C, this adduct forms a six-membered ring, cyclo(trimmidohexamethyltrigallium) $3[(CH_3)_3Ga{:}NH_3] \rightarrow [(CH_3)_2Ga{:}NH_2]_3 + 3CH_4$, stable up to around 500 °C. As a consequence of adduct formation, very little TMGa is present at the growth surface at temperatures below 500 °C, but products of adducts and, at high temperatures, mostly Ga reach the surface [79]. However, TMAl–NH_3 adducts are much more stable [80, 81] and lead to a reduction in precursor efficiency and formation of AlN nanoparticles in the gas phase at typical MOVPE growth temperatures below 1100 °C. The pathways of prereactions of TMAl and ammonia are sketched in Figure 4.4.

TMAl–NH_3 adducts can also reduce the efficiency of TMGa in AlGaN epitaxy by the formation of mixed but stable TMAl–TMGa–NH_3 adducts. Such prereactions can significantly lower the efficiency of both MO-precursors in III-N MOVPE [81]. With TMGa or trimethylindium (TMIn) and ammonia, these problems are less severe; and at typical growth temperatures, adducts formed by the mixed gases in the colder gas region decompose again because of their limited thermal stability. Thus, the TMGa–NH_3 adduct formation pathway is not relevant, as long as the growth temperature is high, in contrast to TMAl-NH_3 adduct formation. However, if AlN particles form in the gas phase, they also consume Ga and thus reduce the TMGa source efficiency [82].

The formation of adducts between the p-type dopant $MgCp_2$ and NH_3 leads to two different adducts (NH_3-$MgCp_2$ and $(NH_3)_2$-$MgCp_2$) with low vapor pressure that can explain the delayed incorporation of Mg and memory effects due to, for example, condensation at colder parts of the reactor [83]. The simple approach to avoid this is to mix those compounds as late as possible and to use heated reactor walls to avoid the deposition of the adduct or to use reactor concepts that are nearly free of walls that can interact with the gas phase involved in layer growth. Because of deposits, Mg doping has an influence on layer structures grown directly after; for example, a Mg-doped p-cap layer of an LED will impact seeding in subsequent epitaxy experiments unless the reactor is conditioned by, for example, heating under H_2, etching using Cl_2, or additionally coating the Mg-"contaminated" interior with undoped GaN. In industrial production, this can be, in principal,

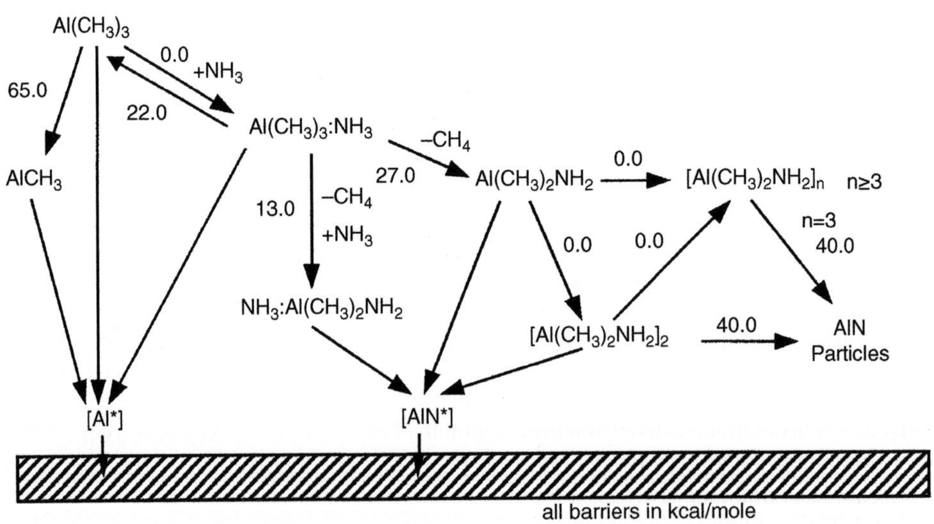

Figure 4.4 *Growth reactions of TMAl and ammonia. Source: Mihopoulos et al. 1998 [80]; reprinted with kind permission of Elsevier.*

compensated for by using an adapted seeding layer growth scheme or multiple chambers dedicated to, for example, the (thick) n-side, the active region, and the p-doped top part of the devices.

4.3.4 Doping of III-Nitrides

For most devices, doping is essential to enable or ensure device operation. For LEDs and Schottky diodes, controlled n- and/or p-type doping is required, while for many high-frequency and high-power-electronic devices, semi-insulating properties of the buffer layers are prerequisite. In general, the III-Ns suffer from high dislocation density and impurity concentrations of, for example, C, O, or Si in the $10^{15}\,cm^{-3}$ to $10^{17}\,cm^{-3}$ range, depending on growth parameters and purity of the precursors used. In addition, intrinsic defects can lead to acceptor and donor states in the material. Therefore, typical n-type concentrations are rarely below the $10^{16}\,cm^{-3}$ range for most heteroepitaxial GaN layers. The layer conductivity can, however, also switch to semi-insulating behavior, depending on the defect density of the layer, in particular the C concentration. Undoped GaN often shows low electron mobilities, which is due to a high concentration of compensating defects that, to some extent, originate from unintentional C doping, which is also responsible for yellow photoluminescence emission [84, 85].

The most difficult task in III-N doping is p-type doping, due to the high activation energy of the Mg acceptor level [86]. This increases with increasing bandgap energy and thus makes it very difficult to achieve high hole concentrations, especially in AlGaN alloys. Therefore, a contacting layer (usually a p-GaN layer) is grown atop AlGaN light-emitter structures. For InGaN alloys, in principle, p-doping should be simpler than in GaN due to an expected lower Mg activation energy. However, the high intrinsic electron concentration present in In-containing alloys makes p-doping difficult.

4.3.4.1 n-Type Doping

For n-type doping, Si has long been the only dopant used in III-Ns. It is available as pure or diluted silane (or disilane) at low cost and high purity, and an early investigation by Nakamura compared Si and Ge and concluded that Si is better suited for n-type doping [87]. This is true with regard to growth cost and precursor efficiency but not when it comes to high doping concentrations. At high Si concentrations above $1–2\times10^{19}\,cm^{-3}$, the surface of GaN layers usually roughens in MOVPE [87, 88, 98]. Indeed, depositing only Si in the presence of ammonia can be used to deposit an antisurfactant layer that can be applied to force 3D growth for defect reduction [89, 90]. In the case of a deposition in the monolayer range, the composition is not pure SiN but a SiGaN layer [91]. The antisurfactant behavior also seems to play a role for the resulting surface morphology when applying Si concentrations in GaN well above $10^{19}\,cm^{-3}$. However, Si-Ga alloy formation and GaN backetching have also been observed, so not only SiN formation plays a role (Figure 4.5) [92].

In molecular beam epitaxy (MBE), however, Si concentrations well above those in MOVPE are achievable in GaN. Higher Si concentrations in MOVPE are possible, though, for AlGaN at low Al-concentrations [93]. While the incorporation of Si into the growing surface apparently is facilitated by the addition of Al, the increasing bandgap results in a slight increase of the donor-activation energy from around 15 meV for GaN to around 50 meV for $Al_{0.8}Ga_{0.2}N$. This still allows for efficient n-doping and moderate sheet resistance for AlGaN with sufficiently low density of threading dislocations. For Al content above $x=0.8$, formation of DX centers starts, which leads to a strong increase in the effective donor-activation energy and thus a strong increase in sheet resistivity [94].

With maximum Si doping concentrations in GaN in the low $10^{19}\,cm^{-3}$, contacting and lateral current transport are still far from ideal. Also, for some device types, one aims for an n-type doping concentration that enables ohmic contacts without a high-temperature alloying process, to prevent diffusion processes in the grown structure. In order to understand the origin of tensile stress by Si doping, Ge doping was investigated in MOVPE again in 2011 [95]. These experiments proved that

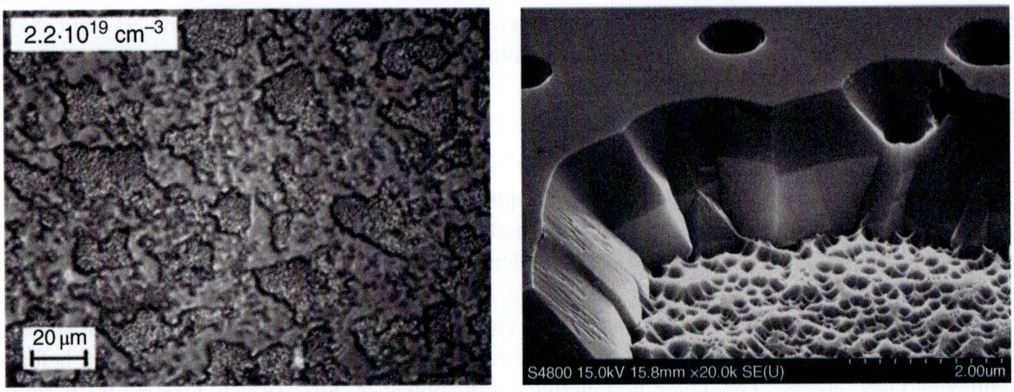

Figure 4.5 *SEM images of a GaN layer after growing a top GaN layer doped with 2.2 × 10*^{19}*cm*$^{-3}$ *on an undoped GaN buffer layer. In plan-view large islands are observed (left) while a higher magnification view reveals that the top layer (right darker region) and the buffer layer (lighter region) are uncovered after applying high Si doses. Source: Images from Fritze 2012 [92] with kind permission of Stephanie Fritze.*

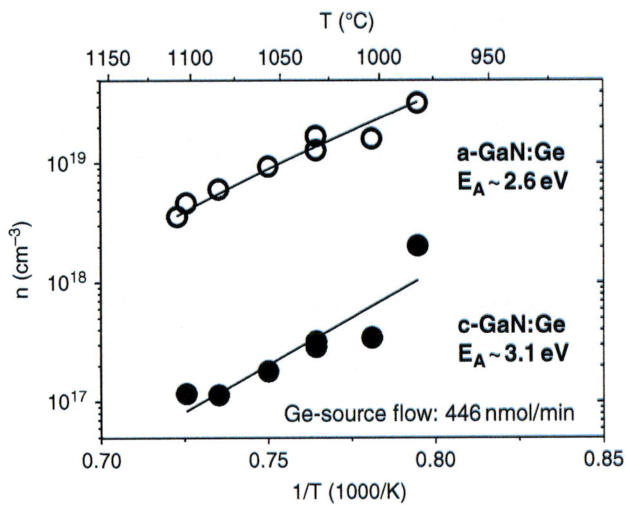

Figure 4.6 *Temperature and orientation dependence of germanium incorporation in GaN at constant Ge-source flow (TBGe or GeH*$_4$*). The activation energy for desorption is within error identical and around 2.6*$^{-3}$*.1 eV. Source: M. Wieneke and A. Dadgar, Otto-von-Guericke University, Magdeburg, unpublished results.*

dislocation climb is not related to the position of the Fermi level, and high n-type doping concentrations also were achieved. With doping concentrations in excess of 10^{20} cm^{-3} in GaN layers, Ge enables low-resistivity contacting and current-spreading layers, homoepitaxial Bragg mirrors [96], and also homoepitaxial MOVPE tunnel diodes [97]. Germanium nitride decomposes above 850 °C, while silicon nitride is stable up to 1900 °C. Most likely, the formation of a stable GeN compound is not possible at typical GaN growth temperatures, which could explain the absence of significant surface roughening. At concentrations in the 10^{20} cm^{-3} region, pit formation (most likely correlated to screw-type dislocation decoration) is visible, though [98]. Also, step bunching has been observed [99]. In our experience, this can be reduced by growing at slightly higher temperatures. In contrast to Si doping, a strong impact on Ge incorporation is given by the crystal orientation and growth temperature. Ge incorporation is driven by the vapor pressure. Consequently, high vapor pressures of the Ge precursor are required, and incorporation is temperature dependent (Figure 4.6).

On a-plane GaN, the incorporation is about a factor of 50 higher than on the c-plane at identical temperatures, which indicates that bonding to the (0001) Ga face is weak, and stronger on non- or semipolar surfaces.

At present, Si is the only n-type dopant for GaN that is widely applied. Ge still does not play a significant role for devices, but it offers some benefits in comparison to Si for low-resistivity current-spreading layers and enables entirely MOVPE-grown tunnel diode junctions [97]; given this, it could be a replacement for indium-tin-oxide in LED production.

4.3.4.2 p-Type Doping

For p-type doping only Mg, although offering a low electrically active concentration in III-Ns, has been established, due to a lack of suitable alternatives. Mg leads to a relatively deep acceptor state, with a wide range of activation energies of 155–230 meV [100, 101] above the valence band reported. Because of this high activation energy, only a fraction of Mg acceptors are ionized; and typically, doping in the $10^{19}\,cm^{-3}$ range is required to achieve hole concentrations in the $10^{17}\,cm^{-3}$ range at room temperature. One difficulty for Mg doping is self-compensation [102, 103] under MOVPE growth conditions if the layer turns p-type during growth. Thus, growth in MOVPE must be performed under hydrogen carrier gas, with hydrogen passivating the Mg acceptor [104] and thus avoiding self-compensation. Growth must then be followed by annealing, which is best done in a hydrogen-free atmosphere to eliminate hydrogen and thus activate the Mg acceptor [105]. With a maximum hole concentration typically in the upper $10^{17}\,cm^{-3}$ range and low hole mobilities around $10\,cm^2/V\,s$, p-GaN layers show relatively high resistivities and poor current spreading, which limits their application. Therefore, hole concentrations as high as possible are usually targeted. However, with increasing Mg concentration, Mg tends to form Mg-rich inclusions stemming from Mg segregation during growth [106]; these were also identified as inversion domains, lowering the achievable hole concentration and degrading the crystal quality. It should be noted that these inversion domains can be intentionally applied to switch the GaN orientation from Ga-face to N-face, enabling new devices types [107, 108]. Apart from Hall-effect measurements, the activation can be also monitored in photoluminescence where Mg-correlated luminescence is visible [109]. Achieving hole concentrations in excess of $1 \times 10^{18}\,cm^{-3}$ is only reported rarely [110]. Typically, the formation of inversion domains in GaN doped with Mg around $3 \times 10^{19}\,cm^{-3}$ [111, 112] limits the achievable hole concentration. In MBE growth, the situation is quite different, and higher Mg and hole concentrations can be achieved than in MOVPE [113].

In general, in MBE, n-type as well as p-type doping is simpler because of the lower growth temperature and Ga-rich growth conditions as well as a lack of gas-phase prereactions beneficial for Si and Mg concentrations in excess of the limits in MOVPE. Despite these advantages of MBE, MOVPE is the growth technology of choice in industrial applications due to higher growth rates and higher reactor capacity, yielding lower production costs.

4.3.4.3 Semi-Insulating Layers

GaN has a large bandgap energy, but as-grown layers usually exhibit n-type conductivity due to impurities such as Si, O, and intrinsic defects. C as a trace impurity, originating from the metalorganic precursors, partially compensates the background n-type conductivity, but it is also sometimes considered its source, due to its amphoteric character [114]. Although C is always present, the background concentration is not necessarily sufficient to achieve highly resistive material. For HFETs, highly resistive or semi-insulating layers are necessary to avoid carriers leaking through the buffer and allow for complete device pinch-off. For this, the Fermi level should be pinned deep within the bandgap, ideally close to the midgap position. Pinning of the Fermi level implies that even under carrier injection, it is not shifted significantly. This requires doping concentrations of at least 10^{17} cm^{-3}. Such pinning in intrinsically n-type material can be only achieved using deep acceptors. In early studies, Mg or Zn in low concentrations [115, 116] were used to compensate GaN, but the control of material properties is rather difficult, and the achieved resistivities were not sufficient. A well-established dopant in other III-Vs is Fe using bis-cyclopentadienyl-iron (Cp_2Fe), also called ferrocene, as the source. Fe introduces a deep acceptor in the upper half of the bandgap of GaN [117–119]. As an alternative, C as an intrinsic dopant is often used [120]. C on N sites introduces a deep acceptor level in the lower half of the bandgap, whereas on the Ga site, a deep donor in the upper half of the bandgap is introduced. One driver to use C is its advantage of being compatible with Si technology, which is important for the low-cost production of GaN power devices on Si in Si fabs. Fe, although only a dopant in low concentrations, is feared by many Si technologists, and the contamination risk seems too high to introduce Fe-doped GaN into Si fabs. Another benefit of C is its position of the acceptor in the lower half of the bandgap, in contrast to Fe, which is in the upper half. With sufficiently high dopant concentration, the Fermi level pins at or close to the energetic position of the deep level. Therefore, C-doped layers are weakly p-type, while Fe-doped layers are weakly n-type [121]. For excess electrons, as is the case for standard FETs, a p-layer can be expected to have a better blocking behavior than an n-layer because of the slightly higher barrier induced by the difference of the Fermi levels.

Until recently, the typical C doping procedure used intrinsic doping based on a reduced growth temperature, low V/III-ratio, and/or low reactor pressure. However, these conditions are not well suited for high-quality layers that rather demand high growth temperatures and high V/III-ratio. Indeed, reducing the V/III ratio and temperature to introduce C leads to a broadening of the X-ray diffraction (XRD) (0002) reflection full width at half-maximum (FWHM), indicating a lower crystalline quality, and also a splitting of the Θ–2Θ (0002) peak (Figure 4.7). Also, a higher point-defect density, apart from the C acceptor, can be expected at such growth temperatures. It is also indicated that these growth conditions can result in C incorporation not only on N sites as deep acceptor but also on Ga sites as deep donor (N. Jankowski, A. Dadgar, S. Keller et al., manuscript in preparation); this also affects leakage and electrical breakdown, which is inferior for all intrinsically doped samples as compared to the sample doped with C using propane as precursor, as shown in Figure 4.7.

An alternative to intrinsic C doping is the use of gaseous hydrocarbons, such as propane or butane, or liquid compounds, such as pentane, hexane, or CBr_4 as an extrinsic dopant source [122]. They can be used in standard GaN growth conditions where C is usually only incorporated as a deep acceptor. Because standard GaN growth conditions are rather unfavorable for C doping (usually intrinsic $[C] < 10^{17}$ cm^{-3}), relatively high C precursor flows are required for C doping in the 10^{18} to 10^{19} cm^{-3} range. When using propane or butane, an inline gas filter is recommended, because, for example, propane usually is only offered in 3N5 quality or worse; common impurities like O and H_2O can lead to high compensation, without additional purification, because the C doping efficiency is not high and a high propane flow is required.

Another point that has to be taken into account is that C, which is isovalent to Si, has been observed to also induce dislocation climb, leading to tensile stress. This can be a problem for layers grown on silicon substrates where compression during growth is required to avoid cracking of the layers. However, dislocation climb is less pronounced than for Si doping, and achieving high doping concentrations is possible without inducing strong tensile stress.

4.4 Substrates

The preferred choice for substrate is GaN or AlN for homoepitaxy. However, due to the significant mismatch between the binaries, even the use of nitride substrates does not imply that device heterostructures can always be grown pseudomorphically. Unfortunately, GaN and AlN are not

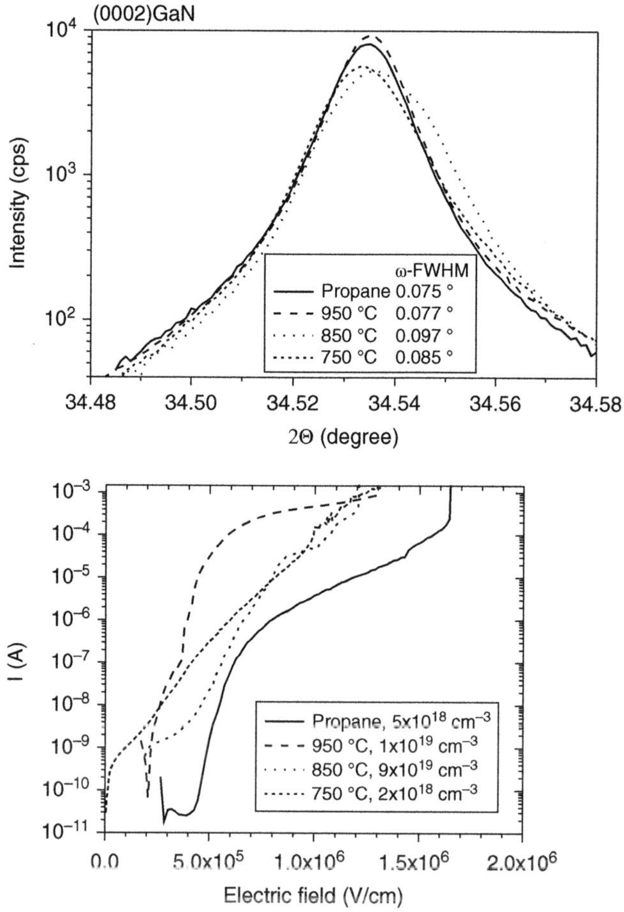

Figure 4.7 *Top: XRD Θ–2Θ (0002) scan of a C-doped GaN layer grown on an undoped GaN buffer. The splitting of the GaN peak (shoulder on the right side) originates in an incoherent growth of the top layer on the buffer structure at lower growth temperatures. The ω-FWHM values stem from an ω-scan of the main peak and also show a deterioration in crystalline quality.*
Bottom: corresponding I–V breakdown measurements showing lower breakdown field strength and higher leakage currents for intrinsically doped samples. The carbon concentration as determined by SIMS is given in the inset of the bottom figure. Source: A. Dadgar, Otto-von-Guericke University, Magdeburg, unpublished results.

readily available as bulk substrates. GaN substrates currently are primarily grown by hydride vapor phase epitaxy (HVPE) as individual layers that are polished [123] or relatively short boules that can be cut into several substrates [124]. Also, crystal growth by the ammonothermal method and the Na flux method is being pursued [125]. Bulk GaN substrates are a must for GaN laser diodes that require a low dislocations density [126]. However, such GaN substrates, due to the not-very-productive manufacturing process, are expensive and unaffordable for cost-sensitive applications like LEDs for lighting applications. In that field, GaN substrates are used only rarely (see www.soraa.com). AlN substrates are manufactured by physical vapor transport (PVT) and currently are available on the open market only in small diameters [127]. They are used for UV LEDs, where their low dislocation density is advantageous [128]. However, despite the high fundamental bandgap of AlN, absorption by impurities, point defects, and complexes can limit transparency at certain wavelengths [129].

Hence, since homoepitaxial growth is difficult due to the lack of affordable substrates, most devices (with the exception of laser diodes, which are all grown on GaN substrates) are grown on foreign substrates. A large variety of substrates has been explored, but there are only three substrates that are relevant for device use. The most widespread is sapphire, which is transparent over the whole wavelength range accessible with nitrides. Consequently, it is the commonly used substrate for all LEDs with diameters currently up to 6" [130], at moderate substrate cost.

However, sapphire has a low heat conductivity that limits its usability for devices with high heat dissipation on small areas. Transistors are thus grown on substrates with superior heat conductivity. For microwave transistors, additionally, a high resistivity is advantageous for low damping of the signals. These properties are offered by SiC, which like AlN is grown at very high temperatures by PVT in diameters of currently up to 6" (http://www.wolfspeed.com/materials/products/sic-substrates), but at high substrate cost. Due to a relatively low in-plane lattice mismatch to Al(Ga)N, growth of device structures on SiC can lead to low initial dislocation densities well below 10^{10} cm^{-2} in the AlN buffer layer [131, 132].

For power-switching transistors, which are needed for power conversion in mass markets like photovoltaics and electrical vehicles, fabrication on large substrate diameters in high-volume fabs known from the silicon industry is desired for low fabrication cost. This is why Si is also finding increasing application, despite the significant challenges posed by lattice and thermal mismatch, the fact that Ga is a good solvent for silicon, which is why contact between Si and liquid Ga has to be strictly avoided [133].

4.4.1 Heteroepitaxy on Foreign Substrates

Heteroepitaxy on these three substrate types, sapphire, SiC, and Si, has some common problems. Due to the lattice mismatch, growth starts by the formation of nuclei. Their size and shape depend on diffusion lengths on the surface and the involved surface energies. While GaN can form relatively large 3D nuclei [134], collecting Ga from a relatively wide surrounding, AlN forms smaller 2D islands. Since these nuclei do not have exactly the same orientation but are rotated (twisted) and/or tilted against each other, their coalescence results in formation of dislocations at the boundaries. Annihilation of dislocations with increasing layer thickness results in tensile strain. Thermal mismatch between the substrate and epitaxial layer stack also results in strain and associated wafer bow or even wafer breakage when the temperature changes: for example, during cooling at the end of a growth process. Thus, strain engineering is an important issue in the growth of III-nitrides.

4.4.1.1 Intrinsic, Thermal, and Mismatch Stresses and Strains

Generally, in crystal growth, strain (lattice deformation) leads to stress (force induced by deformation), which builds up, for example, in a lattice-mismatched system, with increasing layer thickness. This stress leads to a limitation in pseudomorphically achievable layer thickness when the energy built up is sufficient to surmount the energy barrier for misfit dislocation generation or cracking (see also Chapter 2 of this book). Misfit dislocations typically form at the interface of strained heterosystems, and it is quite common in III-Ns under tensile strain for the layer to crack to relax some tension. Furthermore, stress can lead to strong wafer bow, which is a problem during growth and, if present after growth, for processing. The main sources of strain and resulting stress are:

- Island growth and coalescence on lattice-mismatched heterosubstrates
- Edge-type dislocation climb (stress or doping induced)

- Lattice mismatch
- Thermal mismatch

Apart from dislocations and cracking, strain and the correlated stress of a layer on a substrate induce wafer bow, which leads to two limitations in III-N epitaxy. One is that wafer bow can lead to strong temperature inhomogeneities that directly impact the In incorporation into ternary and quaternary layers in the AlGaInN system; the second is wafer bow leading to breakage or hampering device processing. As a rough estimate, In incorporation in such layers changes by around 1.4% in absolute value for a temperature change of ~10 K [135]. Even stronger temperature gradients than 10 K for heavily bowed wafers have been observed [136]. Often, only the temperature in the wafer center is monitored, and the value at the edges is unknown or can be only determined after growth by determining the alloy composition. It has been observed that a substrate temperature change in the wafer center (at constant susceptor temperature) leads to a temperature change with opposite sign at the edge of the wafer. The resulting change in In incorporation shifts the bandgap energy and thus the emission wavelength of light emitters or the channel conductivity of, for example, AlInN-based FETs. By properly balancing strain during growth, one can keep the wafer flat at the critical steps of In-containing layer growth [137]. In addition, susceptor pockets can be slightly curved, which is a common method in the LED industry to accommodate wafer curvature during InGaN growth.

All substrates used in III-N epitaxy are typically thicker than comparable Si SEMI thickness. This accounts for several problems of III-nitride epitaxy:

- Vertical temperature gradients in cold-wall reactors lead to a concave bowing for the bare substrate, which can be reduced using thicker substrates. For 150-mm diameter sapphire, it was also observed that using the standard 675-μm thick (0001) substrates leads to breakage during heating, while a (11$\bar{2}$0) substrate showed extreme bowing but no breakage [137].
- Wafer bow during growth leads to strong lateral composition inhomogeneity [138, 139].
- Residual post-growth wafer bow is too large for processing.

Because of these problems, typical substrate thicknesses today are approximately 1.5 (up to 6) times thicker than standard (SEMI) substrate thicknesses.

4.4.1.2 *Island Growth and Coalescence on Heterosubstrates*

One source of tensile strain originates in high lattice mismatch or high interface energy: for example, due to a valency mismatch, in III-N heteroepitaxy. This leads to nucleation in islands and not as a continuous two-dimensional layer. When looking at the high lattice mismatches in III-N epitaxy on heterosubstrates, it is impossible to grow a continuous layer free from misfit dislocations because, in most cases, lattice mismatch will lead to misfit dislocations for the first monolayer grown, independent of the surface energy. Each misfit dislocation can be seen as a separation of small initial islands (e.g. a monolayer island of ~6 AlN molecules on Si(111) (Figure 4.8) with misfit dislocations in a ~ 4:5 period) and with increasing island size for lower lattice mismatch. It has been recently demonstrated by Mante et al. [140] for AlN on Si grown by MBE that misfit dislocations lead to a twist of around ±3° in the islands for the initial stage of AlN growth on Si. The origins of this twist are 60° misfit dislocations that usually form. They consist of an edge and screw component of which the in-plane screw component leads to a twist of the island (Figure 4.9). Twist varies with the direction of the screw components of all three misfit dislocations surrounding an AlN nucleation island. The expected twist of the island can be calculated from Frank's formula [141] to ±3.31° in good agreement with the findings of MBE-grown material. For AlN on sapphire (~11:10 period) [142], the value is expected to be around ±2° and for AlN on SiC (~63:62 period) [143] around ±0.3°.

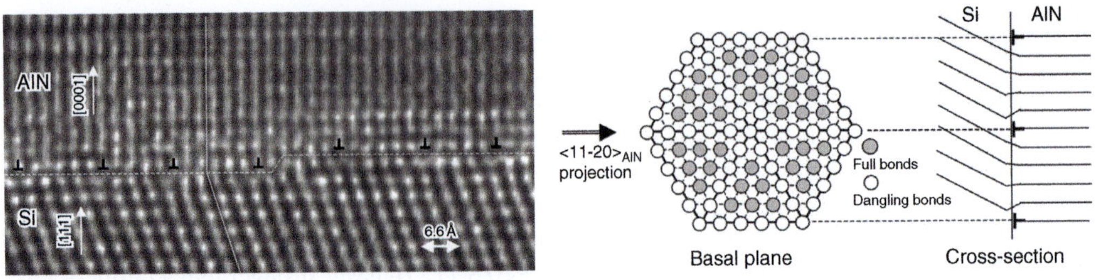

Figure 4.8 TEM of the Si(111)/AlN interface in cross section (left) and the translation into plan view for the interface atomic arrangement (right). Only groups of six AlN molecules are directly bonded to the Si surface. Independent of nonuniformity in island size and shape the misfit dislocation itself is already a source of twist (see also Figure 4.9). Source: Liu et al. 2003 [144]; reprinted with kind permission of American Institute of Physics.

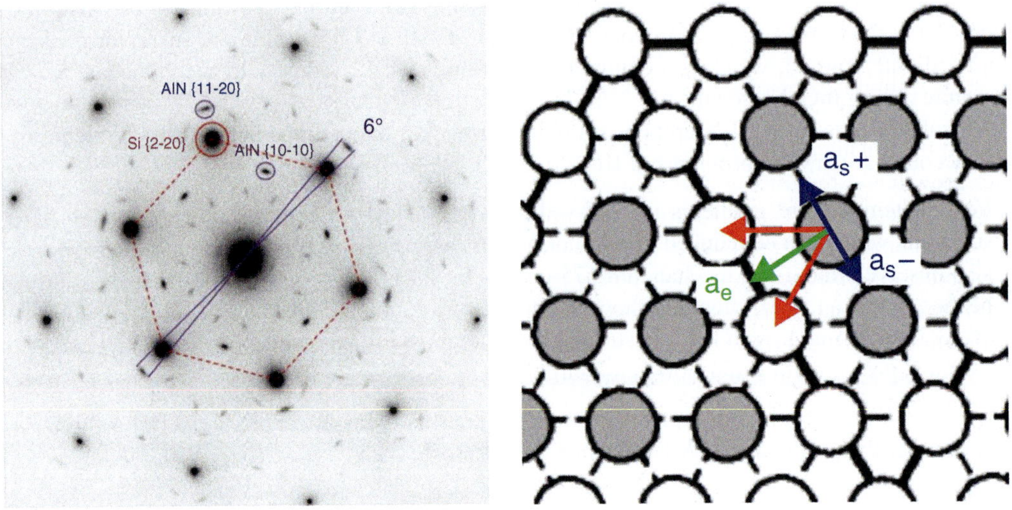

Figure 4.9 Selected-area electron diffraction of AlN on Si epitaxy in the early stage with separated islands and Volmer–Weber growth (left) [140]. A broadening of the in-plane orientation around 6° is observed. This can be understood by the 60° dislocation (red Burger's vector) where the screw component (blue Burger's vectors $a_s\pm$) points in two possible opposite directions in-plane (right, after Mante et al. 2018 [140]). For a single island the three surrounding misfit dislocations result in a twist of ≤ 6.6°. On further growth this twist is averaged when islands increase in size overgrow other islands. Source: Reprinted with kind permission of Philippe Vennéguès, CRHEA France.

Upon further growth, some islands overgrow others, and the twist of the single domains is averaged and strongly reduced. However, strain is introduced even for perfectly aligned islands, when the "misfit gap" is closed by island coalescence of, initially even hydrostatically compressed [145], islands when growth is continued [146, 147]. Usually, such a gap between islands is larger than the atomic distances of the relaxed material. Upon gap closure, local tensile strain is introduced, which sums up to the tensile strain in the whole layer. This gap-closure mechanism is also called the *zipping mechanism*. For thicker layers, some growth domains shrink and disappear, while others increase in size. With a reduction in domain boundaries and domain number, the number of dislocations stemming from misaligned domains, along with the overall dislocation density, is reduced with increasing layer thickness.

4.4.1.3 Edge-Type Dislocation Climb

In addition to tension stemming from island coalescence, edge-type dislocation climb can also introduce tensile strain because the lattice plane, which ends at an edge-type dislocation, is in most cases shortened by climb and thus the crystal volume is reduced. Edge-type dislocation climb can be induced by local strain fields, compression [148], and doping [149]. In the latter cases, it has been always observed that compression is reduced or more general tension increased. In particular, for Si doping, a strong correlation of doping concentration, edge-type dislocation density, and tensile strain has been observed [95, 149]. The generation of tensile strain requires the interaction of dislocations and Si, both of which must be present in sufficient concentration [150]. Such interaction of dopant and dislocations has also been observed for high levels of extrinsic C doping concentrations; however, this is less pronounced than in the case of Si (A. Dadgar et al., Otto-von-Guericke University, Magdeburg, unpublished results).

4.4.1.4 Lattice Mismatch

Lattice mismatch is a problem in all mismatched crystal systems and limits the thickness of pseudomorphically grown layers. If grown pseudomorphically, it induces in-plane strain. When the critical thickness is reached, relaxation relieves all or most of this strain. From wafer-bow measurements, one can indeed determine the composition of pseudomorphically grown ternary alloys if the mismatch is low enough to allow for several tens of nanometers of pseudomorphic layer growth and the thickness is also monitored, e.g. by reflectivity measurements. When not growing pseudomorphically, partially or fully relaxed layers can be applied for strain engineering to influence the in-plane lattice parameter, either to control bow, avoid cracking of layers, or enable the controlled growth of thick lattice-mismatched layers. Here, relaxation should minimize additional threading dislocations and favor in-plane misfit dislocations, to avoid a degradation of active layers.

4.4.1.5 Thermal Mismatch

Because of the relatively high growth temperature and required temperature differences between the layers in MOVPE of III-nitrides, the high thermal mismatch between III-nitrides and the heterosubstrate is the main source of strain and bow. Generally, it has to be considered especially for the growth of In-containing layers due to the temperature-dependent In incorporation. Either wafer bow is optimized by strain engineering to achieve a flat wafer during the growth of In-containing layers, or a curved susceptor is applied, which must then be optimized for the device structure [137].

Thermal mismatch has been a limiting factor in substrate choice since the early days of III-N epitaxy. In particular, GaN-on-Si leads to a strong wafer bow after cooling and, typically, cracking of layers even less than 1 μm in thickness if no strain engineering is applied. On SiC, layer cracking is less problematic but also a limiting factor in overall layer thickness.

4.4.2 GaN Growth on Sapphire

GaN epitaxy on sapphire is the most common approach to grow GaN. The same holds true for the growth of AlN and AlN-based layer structures. Sapphire has a rhombohedral crystal structure (belonging to the hexagonal family), and GaN is not oriented directly with the same orientation on sapphire, but with a 30° rotation, which results in a better lattice match. The crystallographic relationship is GaN$[2\bar{1}\bar{1}0]\|Al_2O_3[1\bar{1}00]$ and GaN$[0001]\|$ $Al_2O_3[0001]$ [151]. In early work, GaN growth was often attempted by gas-phase epitaxy without a dedicated nucleation layer and therefore resulted in poor layer quality [152]. Usually, when heteroepitaxy is performed on a highly lattice-mismatched substrate or a substrate of a different class of compounds, a nucleation layer is a prerequisite to achieve a smooth, continuous layer. The energetic barriers at the heterointerface usually promote island growth instead of layer growth. In particular, at standard growth temperatures, diffusion of adatoms on the heterosubstrate is across long distances, resulting in few nucleation centers. On sapphire, this problem has been solved by two different approaches. First, Amano and Akasaki used a low-temperature grown AlN layer [153]. AlN, also at high temperatures, has low adatom mobility, resulting in a high number of nucleation centers on the surface. Such seeding layers show a polycrystalline texture and only some well-oriented AlN grains [153] that form the buffer when growth continues. From the atomistic side, such an AlN seeding layer often results in a thin mixed Al_2O_3–AlN region. TEM studies have revealed that the first monolayers of AlN on sapphire nucleate at least partially with N polarity. The formation of an AlON layer then not only accommodates the different bonding situation in sapphire and AlN but also leads to a conversion to Al polarity [154]. While this process is possible with the oxygen released from sapphire at high temperatures, it can be further promoted by adding oxygen at the early stages of nucleation [155–157]. AlN seed layers of typical thickness of some 10 nm on sapphire usually are completely Al polar.

The second possibility is to directly grow GaN on sapphire [8, 134]. At high temperatures, however, this does not result in a smooth layer but rather a very rough layer due to the high adatom mobility. At low temperatures, a poorly oriented but continuous GaN film of low crystal quality is grown. Upon heating and a short annealing at growth temperature, GaN recrystallizes and rearranges by diffusion on the surface, and some GaN is lost by desorption. In an ideal process, a high number of GaN islands remains on the surface, and when GaN growth is initiated at high temperatures, they start to grow and coalesce. With increasing density of islands, smaller islands coalesce. By varying the density of the islands, which results in more or fewer coalescence boundaries, one can significantly alter the threading dislocation density because dislocations usually occur at coalescence boundaries of twisted and tilted islands. Dislocation bending can also occur at the inclined interface of growing islands and can help to reduce the dislocation density. The method is limited by a poor coalescence observed for large island distances, resulting in rough layers and, when the island density is too low, new nucleation centers in between the islands. More details on this nucleation layer can be found, for example, in [134] and [158].

When applying an ammonia pretreatment of the sapphire surface, N-face growth can be achieved [159]. N-face GaN opens up the possibility of new device designs, in particular for electronics [160, 161]. Doping of N-face GaN has been investigated in detail in [162], where e.g. steeper Mg doping profiles were demonstrated for growth on the N-face. Although this is a benefit for p-doped structures, a higher incorporation of O makes p-type doping more difficult.

For growth of orientations other than +c or –c, differently oriented sapphire substrates (e.g. m-plane [163] or r-plane [164]) have been used. However, these different orientations have not found their way out of the lab into real applications yet.

Sapphire has the huge benefit of being a transparent substrate in both the visible and UV regions at all wavelengths that can be produced by III-nitride layers. Together with a relatively low price, this makes it the first choice for LEDs. However, its thermal conductivity is low, and heating can be an issue when the device efficiency is low. Also, vertical contacting through the substrate is not possible due to its insulating behavior.

4.4.3 III-N Growth on SiC

SiC is an expensive substrate with good transparency for visible but not for UV light. It has excellent thermal conductivity and is available as conductive and semi-insulating substrates. It is the substrate of choice for high-end power and high-frequency devices but only to a small extent for LEDs. Due to its hexagonal crystalline structure and low lattice mismatch, it is among the best heterosubstrate choices for GaN epitaxy. The epitaxial relationship is GaN[$11\bar{2}0$]||SiC[$11\bar{2}0$] and

GaN[0001]||SiC[0001] [165]. Only its high price limits its usage to high-end applications in radar and power electronics as well as a small number of high-end light-emitting devices.

(Al,Ga)N epitaxy on 6H- or 4H-SiC benefits mainly from the good thermal conductivity of the substrate, making it highly suitable for high-power electronic applications. III-N growth on SiC does not necessarily need low-temperature nucleation due to better lattice matching ($\Delta a/a \sim 1\%$ for AlN/4H-SiC and $\Delta a/a \sim 3.5\%$ for GaN/4H-SiC) than that to sapphire. However, direct growth of GaN on SiC at high temperatures results in rough surface morphologies due to the high adatom mobility. The higher sticking coefficient for Al-containing layers leads to the fact that AlN wetting is the most commonly used growth initiation on SiC [166, 167]. To achieve vertical conduction between epitaxial layers and n-type SiC substrates, Si-doped AlGaN layers have also been used as a wetting layer [168]. It has been demonstrated that the structural and electrical properties as well as wafer bow are correlated with the AlN wetting-layer properties [169, 170]. SiC is the substrate of choice for high-power AlGaN/GaN heterostructure field-effect transistors (HFETs or HEMTs) due its high thermal conductivity of about $3.7\,\mathrm{W\,cm^{-1}\,K^{-1}}$ [171]. However, to fully exploit this, the thermal boundary resistance between the AlN/GaN layer structure and the substrate has to be optimized [172, 173]. Although superior HFET device performance has been reported on 4H-SiC [174], future improvements in view of substrate availability and price competitiveness as well structural properties (i.e. micropipe density) are highly desirable.

4.4.4 GaN Growth on Silicon

Of the three commonly used substrates for GaN epitaxy, silicon has very disadvantageous properties. However, its availability in large diameters compatible with high-volume chip fabrication lines makes it economically very attractive. The low thermal expansion coefficient of Si hinders the simple growth of crack-free layers in excess of 1 μm in thickness. In addition, its high chemical reactivity does not allow for direct growth of GaN-on-Si that would lead to a *meltback etching reaction* [175], which starts with alloying of Ga and Si, leads to deep craters in the Si substrate, and is most likely promoted by ammonia and hydrogen [176]. Nevertheless, using AlN seeding and buffer layers, this reaction can be eliminated. Also, N-rich growth conditions are beneficial [177]. The cracking problem has been solved by strain engineering, as will be explained in more detail later. With solutions to these disadvantageous properties of Si, it offers several advantages such as an unbeatable low price, large substrate diameters, and the possible integration of III-Ns with Si electronics, as well as processing in Si CMOS fabs without significant adaptions: for example, in lithography, as required for transparent GaN-on-sapphire wafers. While the application of Si for LEDs is rather limited, it is the most important substrate for GaN electronics, predicted to soon be a billion-dollar market.

For the growth on Si, usually Si(111) is used because it has a threefold symmetry of the surface atoms best suited for the hexagonal arrangement of III-Ns with the epitaxial relationship GaN[$11\bar{2}0$]||Si[$\bar{1}10$] and GaN[0001]||Si[111] [190]. *c*-Axis oriented GaN growth on Si(110) and Si(001) has been also demonstrated [178–181]. While the growth on Si(110) is straightforward, that on Si(001) is more difficult. In MOVPE, III-Ns grow in the hexagonal phase. However, similar to cubic III-Vs on Si(001), an orientation problem exists. Comparable to the antiphase boundaries formed when, for example, GaAs is grown on Si(001), the hexagonal III-Ns can grow in 90° rotated growth domains that originate in the step structure and surface reconstruction of Si(001) [182]. For the growth of semi- or nonpolar GaN, other Si orientations were investigated, either on planar Si [183, 184] or with etched surface structures [185–187]. With semipolar and nonpolar GaN offering little benefit and significant difficulties, especially the formation of stacking faults [76], which are difficult to reduce [188], direct growth of these orientations is nowadays generally less popular than in the early 2010s.

Si offers a low price, good thermal conductivity, and good electrical conductivity but poor insulation behavior and also low breakdown field strength. The main driver for GaN-on-Si research in the 1990s was LED growth; but around 2000, the first reports on GaN FETs on Si for high-frequency operation came up, and today electronics is the main driver for GaN-on-Si. With the upcoming GaN power converters as the main application of GaN FETs, GaN-on-Si epitaxy has gained in importance, and today one can consider GaN-on-Si in GaN power electronics analogous to what GaN-on-sapphire is in optoelectronics.

4.4.4.1 *Strain Engineering: Principles and Limitations*

The key to growing device-relevant GaN layer thicknesses on Si is to control stress during epitaxy to obtain a well strain-balanced structure after growth that, ideally, is flat. The basic principle in GaN-on-Si epitaxy is to build up compression during growth to counterbalance tensile thermal stress upon cooling to room temperature. This is commonly achieved by first growing AlN, AlGaN, or, more generally, a layer with a smaller in-plane lattice constant than the main device layer: for example, GaN. While this sounds quite simple, the total thickness achievable is strongly influenced by the material quality, in particular the edge-type dislocation density. Due to the structure of screw-type dislocations, they do not reduce strain by dislocation inclination, as is the case for edge-type dislocations. With a high density of edge-type dislocations, compression buildup is usually reduced by edge-dislocation climb. Therefore, it is of utmost importance to grow layers with low edge-type dislocation density if a total layer stack thickness of several micrometers should be achieved.

Here, one limitation of compression buildup is the stiffness of the Si substrate. Si is quite a soft semiconductor at GaN growth temperature, compared to III-nitrides. This is because the energetic barrier to induce slip is low, and compression must be built up during epitaxy to avoid cracks. At growth temperatures above 1000 °C, this can induce slip in the Si substrate, which in turn leads to an even stronger bow [189]. The main problem is that this is an irreversible change of the substrate structure and usually leads to a strongly bowed wafer after growth. To some degree, this deformation can be delayed. High-purity float-zone Si is the worst material choice with regard to plastic deformation. Much better suited in general is Czochralski-grown Si. Here, a high content of n- or p-type dopants and, ideally, also impurities such as oxygen, carbon, and nitrogen can help stiffen the crystal by blocking slip. From the Si manufacturer's viewpoint, a rather poor crystal is the best one can choose for GaN epitaxy, unless thermal properties of the substrate are also of some importance, because they can deteriorate at very high impurity concentrations. Apart from the Si crystal growth method, substrates thicker than SEMI standard are the second possibility and most often the first choice. This is also because SEMI-thick Si substrates bow strongly during epitaxy, independent of their doping concentration and local temperature changes, and layer homogeneity is affected by this wafer bow. Therefore, a thicker substrate eliminates such problems and also increases the compression one can induce in the substrate at only slightly higher cost.

4.4.4.2 *Crack-Free GaN-on-Si*

In early studies, Ishigawa et al. [190] applied an AlGaN *transition layer* between the AlN seed/buffer layer and GaN. This is a simple version for strain engineering, and even without such a transition layer, AlN can act as source of compression in GaN [191]. With a high dislocation density, in particular edge-type dislocations, any compression will be reduced by dislocation climb; the slower this is, the lower the edge-type dislocation density. This has been demonstrated, for example, for the growth of GaN on AlN by introducing an in situ SiN masking layer that reduces the dislocation density [191]. Although it is a kind of separating layer between the AlN and Si, it helped to induce more compression, the thicker it was.

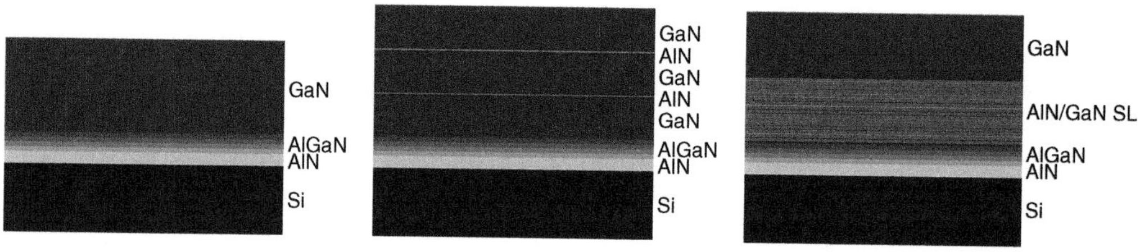

Figure 4.10 *Layer growth scheme of the three main methods to grow crack-free GaN on Si. Graded AlGaN buffer (left), with additional AlN interlayers, typically ~ 5–10 nm in thickness or thicker for AlGaN (middle), and with an AlN/GaN superlattice structure (right).*

Today, there are several established methods for strain engineering to induce compression during growth and thus eliminate cracking. All of them are based on the growth of a material like GaN on one with a lower in-plane lattice constant: for example, AlGaN. Other methods do not play an important role in research and production any longer. The idea of GaN growth on AlGaN is that pseudomorphic or only partially relaxed growth of GaN on a buffer with smaller in-plane lattice constant induces sufficient compression in the GaN layer to compensate for tensile stress upon cooling or tensile stress by dislocation climb. Compression can be achieved by several layer schemes (Figure 4.10):

1. Graded AlN/AlGaN/GaN transition
2. Al(Ga)N interlayers
3. Al(Ga)N/GaN multilayers

In the first case, an originally small in-plane lattice distance of an AlN buffer, typically under tensile strain, on a substrate is transferred to GaN via an AlGaN layer or several AlGaN layers with increasing Ga content. The latter is often grown as one graded layer or, preferentially, by a stepped grading. It is indicated that a stepped grading is superior to a continuous gradient because a sudden change in lattice constant leads to a more pronounced dislocation bending and annihilation at the heterointerface. Here, not small differences but usually a rather high difference in the in-plane lattice constant is preferential to induce high compression in the subsequent layer and to achieve a high dislocation annihilation rate within the AlGaN transition. Thus, a grading of 100% / 80%/ 60%/ 40% / 20% Al content in AlGaN, which is often aimed for, is usually less efficient than one like 100% / ~60% / ~30% Al content in terms of initial compression and possible dislocation annihilation efficiency. A limitation of this method is that the compression induced on the GaN layer will decrease with increasing GaN thickness, mostly due to edge-type dislocation climb, limiting the total thickness of the GaN layer, which is about 2.5 μm without an additional strain-engineering layer in the best case reported so far [192]. Because edge-type dislocation climb is the main factor limiting thickness, the achievable thickness does depend a lot on the quality of the AlN/AlGaN buffer layer. If a low edge-type dislocation density is present, a high GaN thickness can be grown with high compression.

Material quality is also important for a successful application of methods 2 and 3. In method 2, an Al(Ga)N interlayer is grown in such a manner that it relaxes; the subsequently grown GaN layer, ideally, is under high compression. In early work, such layers were mostly grown at low growth temperatures, such as the interlayers described by Amano et al. for the growth on sapphire [193–195]. Low growth temperatures lead to relaxation rather than (micro-)cracking of the Al(Ga)N layer. However, it was observed later that the composition and growth temperature must be optimized for each growth system, most likely due to differences in gas-phase prereactions influencing the quality of this layer. With such relaxed layers, misfit dislocations form at the heterointerface as well as new threading dislocations. While in-plane misfit dislocations are intended, the latter can be minimized by optimizing thickness and composition. The third method, AlN/GaN superlattices, works in a similar way, with partial relaxation of AlN and compression on the GaN layers. AlN/GaN superlattices were investigated early in GaN-on-Si research [196–200]. Nowadays, they are often applied in high-power electronics to achieve thick, highly resistive buffer layers [201–203]. On top of several micrometers of AlN/GaN superlattices, typically a GaN layer < 1 μm in thickness is grown.

The major drawback of Al-rich layers is their limited conductivity, which may, however, be desired for buffer layers in electronics; in addition, they are not trivial to grow in a proper way that does not induce additional threading dislocations. Typically, the dislocation density increases with such layers if they are not grown pseudomorphically, but only within a region of ~ 100 nm above the interlayer if it is well optimized. In the end, it results in a lower or identical dislocation density than before the interlayer for thicker GaN on top. With well-optimized Al(Ga)N interlayers, it is possible to grow a GaN device stack well in excess of 5 μm in thickness and with a final layer more than 4 μm in thickness [189, 204].

As already mentioned, a SiN masking layer is suited to improve the material quality and enable thick GaN-on-Si. Such an in situ SiN masking layer leads to island growth and enables growing larger islands that, after coalescence, show an overall reduced dislocation density [89, 90, 205]. Therefore, especially for optoelectronics, such layers are usually included in the device buffer layer, since lower dislocation density is beneficial for a high radiative recombination efficiency.

4.5 MOVPE Growth Technology

The challenges provided by the growth processes for the III-nitrides also impact the requirements for reactor in situ monitoring and precursors.

4.5.1 Precursors

III-nitrides are grown with NH_3 as the Group-V precursor. Usually, AlGaN is grown under H_2 carrier gas over the whole composition range. As already discussed, In incorporation requires N_2 carrier gas. The trimethyl compounds trimethylaluminum (TMAl), trimethylgallium (TMGa) and trimethylindium (TMIn) are the usual Group-III precursors. However, for the growth of InGaN, often triethylgallium (TEGa) is used [206]. The main reason is that the low In incorporation efficiency results in low growth rates. This implies also that the growth rate of the binary constituent GaN needs to be low but well controlled. TMGa has a much higher vapor pressure than TEGa. While this allows for high growth rates at moderate flows through the bubbler, the control of low growth rates with TMGa requires a dilution line (as used for most dopants). TEGa can be used in a conventional metalorganic line. Additionally, the lower stability of this precursor reduces the tendency for carbon uptake, although at high V/III ratios, this effect does not appear to be relevant [207]. Dopant sources are Cp_2Mg as the only option for p-type, and silane or disilane for n-type. For semi-insulating layers, Cp_2Fe (ferrocene) or the Group-III precursors themselves for intrinsic C doping or extrinsic sources of carbon like propane are used.

4.5.2 Reactors and In Situ Monitoring

For growth of GaN up to very high growth rates, and InGaN and AlGaN at lower growth rates, horizontal (like the AIXTRON planetary reactor [41]) as well as vertical reactors are used in production. The increasing tendency for prereactions with increasing TMAl concentration in the gas phase favors reactors of the vertical type for growth of AlN and AlGaN with high Al content. This results from the short transport length on which prereactions in the gas phase can occur. This allows for higher TMAl concentrations and thus higher growth rates. However, for AlN, high temperatures are also favored, which can lead to thermal problems when gas injection takes place only a short distance away from the substrate. Hence, here the pros and cons of different concepts have to be weighed. While most production reactors currently are used at reduced pressure to cope with gas-phase reactions, growth of InN would benefit from high pressure [208]. However, approaches in this direction have not left the lab yet.

As already discussed, pyrometric control of the temperature is difficult for substrates transparent at the usual infrared wavelength, like 950 nm. This holds for sapphire and SiC. Only Si allows for measurement of the substrate temperature but not the surface temperature of the growing layer by IR pyrometry. Pyrometry at UV wavelengths where GaN absorbs is difficult due to the low signal intensity. Due to the shift of the bandgap to lower energies and thus longer wavelengths, GaN turns opaque at 400 nm after growing around 2 μm thickness (depending on doping). Using this wavelength thus allows measurement of the surface temperature [209]. While this approach can be used for all devices based on a GaN buffer (like blue LEDs or most transistors), it fails, for example, for UV-LEDs that have an AlN or AlGaN buffer transparent at 400 nm even at growth temperature. In this case, one still has to rely on measurement of the temperature of the susceptor underneath the transparent wafer.

As already discussed in Section 4.4.1.2, heteroepitaxy starts with the growth of islands that later coalesce. These processes have a stochastic component, and ex situ assessment is very indirect. Thus, in situ control can greatly help not only in the development of growth processes but especially with their reproducibility. Nakamura, in his early work, used oscillations in the transmission of the radiation from the hot susceptor through the (not continuous or at least rough) GaN layer growing on sapphire to study the nucleation and coalescence process [210] and proposed the usage of reflectometry as the better method.

For example, Figure 4.11a shows the 950-nm reflectance at normal incidence for two GaN growth runs with the same input partial pressure and the same growth temperature at two different reactor pressures. The average value yields information on the material (dependence of reflectance on refractive index) for a flat surface. When the surface becomes rough, the reflected intensity is reduced due to scattering or reflection at inclined facets. This average reflectance is modulated by Fabry–Perot (FP) oscillations, caused by reflection at the substrate/layer interface, and the layer surface, which can easily be translated into growth rate if the refractive index of the layer is known. At low pressure (400 hPa), the reflectance quickly recovers after the initial reduction at the start of the growth and after one oscillation has reached a constant amplitude and average value. This indicates a fast coalescence process. At 800 hPa, the initial drop in reflectance indicating the roughening of the flat sapphire surface by formation of GaN nuclei is much more pronounced, and the recovery of the average reflectivity and the full development of the oscillation amplitude require a much greater layer thickness. This indicates that full coalescence is achieved at a much higher thickness than at 400 hPa. Additionally, it can be seen that the growth rate at high pressure (1.6 μm/h) is around 25% lower than at low pressure due to the increased parasitic reactions in the gas phase.

Figure 4.11b shows the reflectance trace during growth of an LED test structure. Here, it can be seen that the GaN buffer is grown at a much higher rate than the subsequent AlGaN/GaN superlattice. The oscillations from the highly n-type GaN buffer are damped due to absorption, but the average value remains constant, indicating a constant surface roughness.

Combining and comparing different wavelengths allows extraction of even more information. As an example, Figure 4.12 shows the reflectance traces at 405 nm and 280 nm during growth of a heterostructure for a UVB LED. The shorter wavelength shows faster oscillations, allowing the evaluation of layer thickness and growth rate, which can be done for thin layers by fitting a sufficient part of the sine curve stemming from the FP oscillations. Since the AlGaN layer at growth temperature has a bandgap wavelength only slightly below 280 nm, the oscillations are quickly

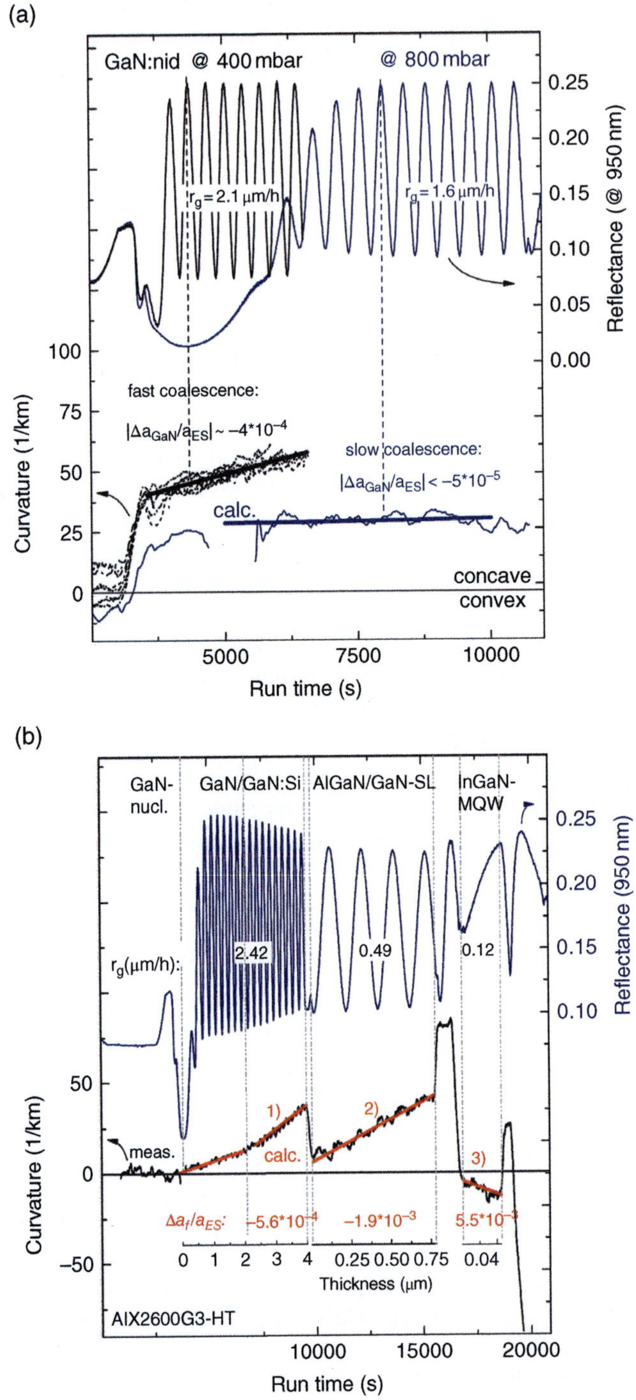

Figure 4.11 *(a) Transients of reflectance (at 950 nm) and curvature during growth of GaN buffer layers on c-plane sapphire at two different reactor pressures. (b) Measured transients of reflectance and curvature during growth of a sapphire-based LED test structure. The curvature change of the different layers (1–3) has been calculated with respect to the underlying layer stack acting as effective substrate. Source: Brunner et al. 2008 [211]; reprinted with kind permission of Elsevier.*

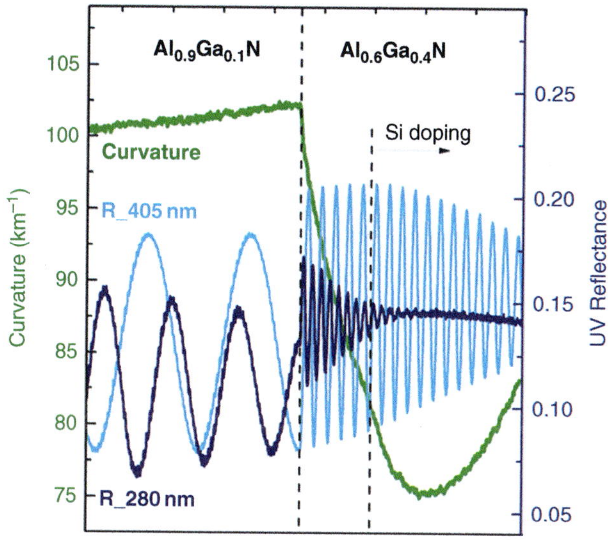

Figure 4.12 *Transients of reflectance at 405 nm (light blue) and 280 nm (dark blue) and curvature (green) during growth of the n-side of a UVB LED heterostructure. First, an AlGaN buffer (x = 0.9) is grown at a low rate and then an* $Al_{0.6}Ga_{0.4}N$ *layer is grown at a high rate, first undoped and then highly Si doped. Source: Adapted from LayTec 2016 [213].*

damped. The decrease of the reflectivity, which is easily seen in the absence of FP oscillations, indicates roughening of the surface. This roughening can hardly be seen in the reflectance at 405 nm. Since the average reflectance at 405 nm is more or less constant, it can be concluded that the roughness seen at 280 nm is on a short length scale. The reflectance at 405 nm also is damped due to the onset of high Si doping but still allows for the assessment of the growth rate over the full AlGaN layer since the bandgap wavelength is much shorter than the monitoring wavelength. Due to the partly complementary information, usage of multiple wavelengths is widespread, and even fully spectroscopic measurements are sometimes used [212].

As discussed earlier, wafer bow needs to be controlled in order to achieve homogeneous device properties. This has led to the introduction of sensors measuring the bow in situ [214] that have become standard in recent years. They are based on the deflection of an array of light beams on a curved surface. As one example of the information gained, Figure 4.11a shows that with the late coalescence at high pressure, a significantly lower concave bow is obtained than with the fast coalescence at low pressure. This confirms that the number and size of GaN grains defined by the nucleation and coalescence process influence the strain evolution of the growing GaN. In Figure 4.11b, the contribution of the individual layers to the bow during growth can be distinguished. This can be exploited to obtain a nearly flat wafer during growth of the InGaN MQW active region (region around 17 500 s growth time). A flat wafer yields uniform temperature over the whole wafer that due to the strong dependence of In incorporation on surface temperature, is a precondition for good uniformity of the emission wavelength. Under the premise of fully pseudomorphic growth, in situ determination of ternary alloy composition based on curvature analysis has been demonstrated [211]. In Figure 4.12, the green curve illustrates the impact of Si doping in the n-AlGaN buffer on the wafer bow. The introduction of tensile stress has been explained by dislocation climb at the growing surface triggered by the presence of Si [215]. Additionally, a reduction of the threading dislocation density introduces tensile strain, since dislocations require additional volume compared to the undisturbed crystal. The reduction in the lattice constant triggered by the presence of Si leads to a reversal of the curvature slope due to overcompensation of compressive strain stemming from the larger lattice constant of the AlGaN buffer with respect to the AlN buffer.

Figure 4.13 shows two other examples of in situ bow analysis. The optimization of an AlGaN buffer layer for a UV LED, on the one hand, aims for a transparent buffer (meaning high Al content); on the other hand, the active region (with lower bandgap but bigger lattice constant) should be grown pseudomorphically on the buffer (meaning the Al content is not too high). The first attempt at a buffer with high Ga content for relatively long wavelength yielded a cracked and thus useless layer, since the stress with respect to the AlN buffer was too large. This cracking can easily be seen in Figure 4.13a. Optimization of composition and growth process yielded the other transient without any cracks, allowing for the deposition of a complete LED structure. In Figure 4.13b, the curvature transients of two wafers in the same growth run are displayed. The lower (red) transient shows a continuous increase of the (negative) curvature over the n-AlGaN

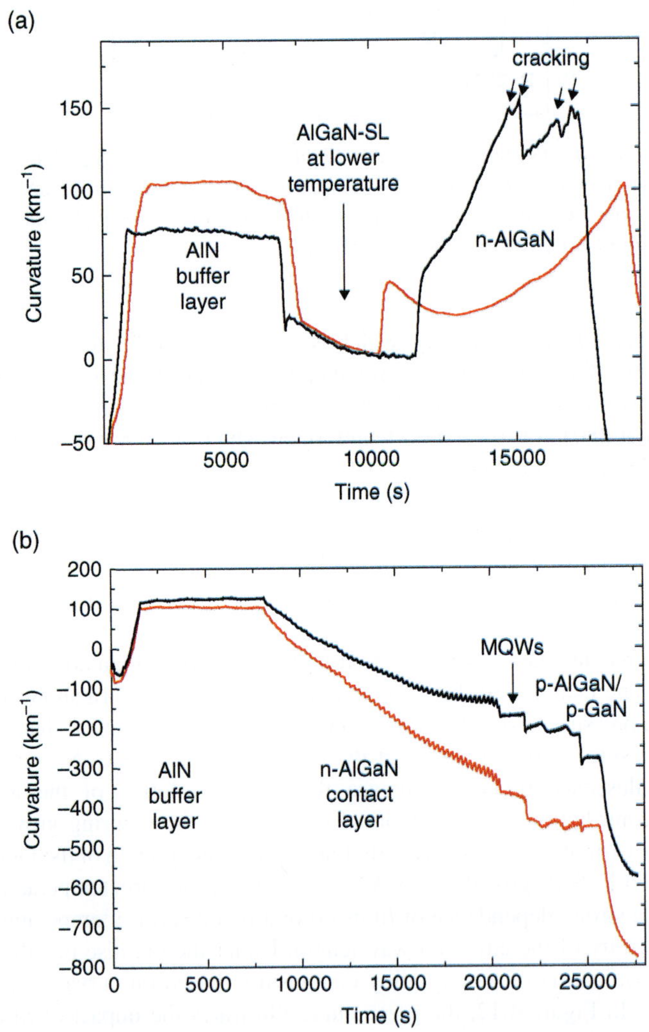

Figure 4.13 *Transients of wafer curvature for (a) different AlGaN buffer layers with (black) and without (red) cracks and (b) the same layer structure on an AlN/sapphire template with low (red) and high (black) dislocation density. Source: A. Knauer, Ferdinand-Braun-Institut, Leibniz-Institut für Höchstfrequenztechnik, unpublished results.*

buffer due to the difference in lattice constants. This trace stems from a layer grown on an AlN/sapphire template with a low threading dislocation density (TDD). The upper (black) transient shows a change in the slope and less curvature on an AlN/sapphire template with higher TDD. The presence of dislocations facilitates the relaxation of the lattice strain and results in less final bow of the complete structure. While low bow facilitates the further processing of the wafer into LEDs, the higher TDD results in reduced output power. So, the curvature trace allows for a prediction of the potential device performance.

As already discussed, patterned sapphire substrates (PSS) are often used for LEDs in order to reduce defect density and increase light extraction. Growth on such patterned wafers of course also has to be optimized to achieve reliable coalescence and, together with a wafer bow, in the ideal case allows the transfer of growth recipes from planar sapphire to PSS without compromising wavelength homogeneity. Figure 4.14 shows one example comparing planar and patterned sapphire grown side by side in a multiwafer reactor at different growth conditions in the first phase of growth.

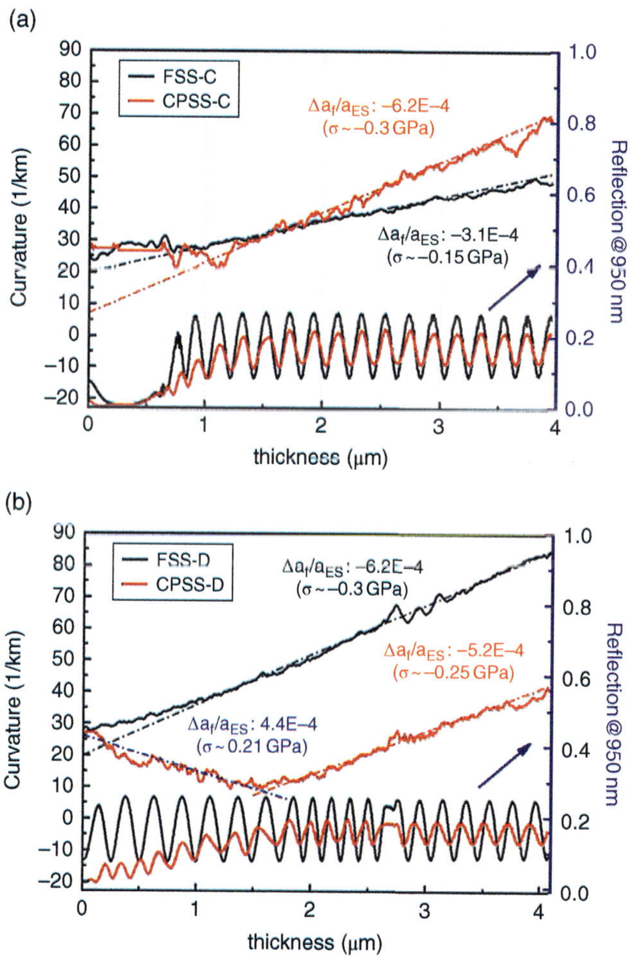

Figure 4.14 *Transients of reflectance and wafer bow on planar (flat, FSS) and patterned (pss with cone shape, CPSS) sapphire substrates with two different growth conditions in the coalescence growth step. (a) 600 hPa for 20 min. (≈0.8 μm thickness) and (b) 400 hPa for 80 min. (≈2.7 μm thickness). Nucleation and growth of the remaining layer were performed under identical growth conditions. Source: Wang et al. [216]; reprinted with kind permission of Elsevier.*

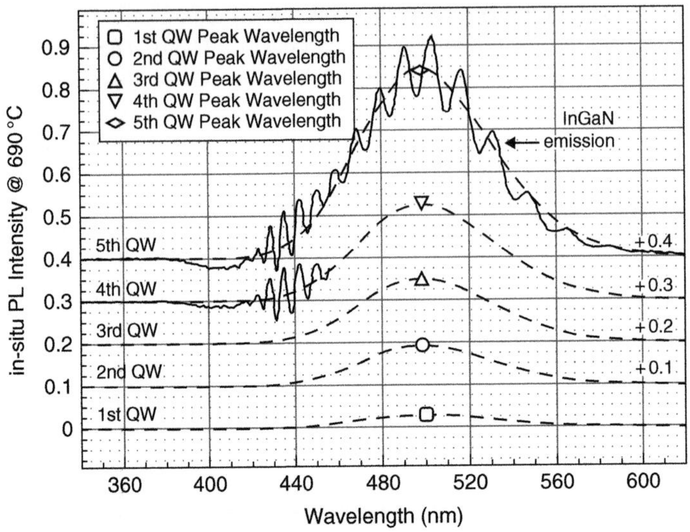

Figure 4.15 *PL spectra taken in situ during the growth of a 5-fold InGaN MQW. The GaN peak from the buffer has been subtracted. The Fabry–Perot oscillations on the raw data can be removed to yield the PL emission that is already visible after the first QW. This allows for optimization of layer homogeneity (from spectra at the center and the edge of a wafer), wafer-to-wafer and run-to-run reproducibility and a precise estimation of the emission wavelength of the final devices. Source: Prall et al. 2015 [217]; reprinted with kind permission of Elsevier.*

These examples illustrate that in situ monitoring of reflectance (at different wavelengths), wafer bow, and (if accessible) the temperature of the growing surface have become indispensable tools in the growth of nitrides during process development and also for monitoring of reproducibility during production. Other tools offering more information are being explored. Despite the high growth temperature, it has been shown that due to the high band offset, III-nitride QWs can show luminescence even at growth temperature [217]. Figure 4.15 shows the development of the photoluminescence (PL) signal of a fivefold QW structure for a 450-nm LED during growth at 700 °C surface temperature. Removing the Fabry–Perot oscillations allows for a precise prediction of the emission wavelength at room temperature by subtracting the offset caused by the smaller bandgap energy at higher temperature.

However, such in situ PL setup is still an expensive tool due to the necessity of a short-wavelength laser source and currently not yet a standard tool to be found in production.

4.6 Economic Importance

As mentioned in the introduction, the blue LEDs that are at the heart of white emitters are the nitride devices of which the largest area is produced, more or less exclusively by MOVPE. Thus, the largest portion of MOVPE reactor capacity worldwide is used for these nitride devices. Other applications, like blue laser diodes and even microwave transistors, are much smaller. Switching power transistors for highly efficient power conversion are currently being developed on multiwafer 6" and 8" systems in order to use existing Si device manufacturing capabilities targeting a potential mass market.

4.6.1 Optoelectronic Devices

Light emitters are today mostly grown on sapphire. Here, low dislocation densities and high light-extraction efficiency are a must. Both can be achieved with a simple method that has been established in today's LED manufacturing. To combine both demands, the substrate is structured (also termed patterned sapphire substrate [PSS]) and growth is performed on small areas and subsequent coalescence as in classical epitaxial lateral overgrowth (ELO). In particular, for low-cost LEDs, replacing incandescent light bulbs, PSS [218, 219] are nowadays routinely applied. They offer GaN growth with a low dislocation density by ELO; and, most importantly, light extraction is enhanced due to inclined surfaces within the structure. This significantly lowers the effort put into techniques like surface texturing, substrate removal, etc. that were applied in the past to extract the majority of light that, otherwise, is trapped within the LED structure.

To manufacture light emitters on Si is, on the one hand, a bad idea, because the Si substrate absorbs a significant amount of light emitted in the structure and also requires strain-engineering layers. For light extraction, the solution is simple because Si can be easily removed from the III-N layer. With such a process, OSRAM Opto Semiconductors demonstrated a comparable quality for GaN-on-Si as for GaN-on-sapphire [220]. The main problem with growth is the quality and thickness of such layer structures. Dislocation densities in the low $10^8\,cm^{-2}$ range and a device stack in the range of 3 micrometers uninterrupted by any strain-engineering layer require well-balanced growth schemes and a precise optimization of seeding and buffer layers. A typical LED thus consists of an AlN seed and buffer followed by graded AlGaN layers and a thin GaN layer. On this, an in situ deposited SiN mask helps to reduce the dislocation density significantly. Upon coalescence after 3-dimensional growth induced by the SiN mask, a strain-engineering layer based on Al(Ga)N is usually required; and if growth is well balanced, the device stack can be grown atop this buffer. For LEDs on Si, but also on sapphire, two things are important: the growth of the MQW structure with good homogeneity and an overall low bow after growth at room temperature. For InGaN layers, a good compositional uniformity requires high temperature uniformity and thus a homogeneous thermal contact of the wafer to the susceptor. This can be achieved by a curved susceptor, optimized for the specific device structure so that during growth of, for example, the InGaN QWs of LED structures, the wafer and susceptor have the same shape and thus are in contact over the whole wafer area. The other approach is to tailor the layer stack such that the wafer is flat when growing the QWs. Low bow after growth is relatively simple to achieve on Si by varying the thickness, composition, and/or amount of the strain engineering layers. On sapphire, the seeding layer has the biggest impact on bow; however, strain engineering is possible, e.g. by introducing relaxed low-In-content layers.

For UV LEDs, the buffer consists of AlN and AlGaN, transparent for the emission wavelengths. Unfortunately, highly Al containing layers due to low Al mobility are more prone to roughening and more difficult to planarize than GaN. Hence, many of the recipes used to obtain low defect density in GaN are difficult to adapt to AlN. Nevertheless, UV LED technology has made significant advances in recent years [20].

GaN laser diodes require much lower dislocation density than LEDs and thus are grown on GaN substrates with low dislocation density. However, in this homoepitaxial approach, strain management is also an important issue. The product of Al content and thickness of AlGaN layers needs to be limited in order to avoid cracking, and the high In content necessary to obtain longer wavelengths beyond 500 nm requires control of strain and strain relaxation.

Not only light emitters, but also light detectors, e.g. solar blind devices [221], are being made from AlGaN. However, while they are commercially available [222], they have not yet found widespread application.

4.6.2 Electronic Devices

Today, most GaN electronic devices are still FETs based on AlGaN/GaN heterostructures. Being power devices with high current and power densities in the range of < 1 A/mm gate width and several tens to hundreds of Watts per cm^2 device area, heat is often a limiting factor for device operation. Therefore, sapphire is usually not applicable due to its low thermal conductivity, and either SiC or Si is used as a substrate. Most devices also require good buffer insulation, either to block high voltages or to promote low losses in high-frequency operation. For such devices, the usual way to achieve semi-insulating behavior is with C doping [223], either by intrinsic doping from the methyl groups in TMGa or with an extrinsic carbon source. This is the preferred approach for high-voltage switching transistors since it is allowed in Si fabs. For μ-wave transistors, Fe doping is standard [224, 225]. When growing on Si, p-type Si substrates enable slightly better blocking behavior in comparison to n-Si. Highly resistive Si has the disadvantage of usually being float zone (FZ), which tends to experience plastic deformation for thinner layers. Highly resistive Czochralsky (CZ) Si is in principal possible but not standard in Si crystal growth. In particular for radio frequency (RF) devices, if SiC is not chosen, such substrates should be preferred to minimize losses. Nevertheless, during growth in an MOVPE environment, highly resistive Si often changes its conductivity, and growth should be optimized to minimize this. Sources for this change are doping and diffusion of precursors used for growth, typically Al and Ga. Nitrogen typically forms a stable nitride on the bare Si wafer that is self-limited in thickness.

For high blocking voltage and normally off operation, vertical transistors are also being developed. One approach is a metal-insulator field effect transistor (MISFET) with the channel formed in a trench on the etched surface of a buried p-GaN layer [226]. When grown on n-type GaN substrate, this resembles a vertical device with the blocking voltage adjusted via thickness and doping of a lowly n-doped GaN drift layer.

In addition to the unipolar FETs, heterostructure bipolar transistors (HBTs) have also been worked on for many years [227]. However, they have not yet found their way into production, most probably due to the challenges of growing abrupt doping profiles on top of a p-layer due to the tendency of Mg to segregate on the surface.

4.7 Conclusions

The III-nitrides have successfully made their way into a number of economically important fields, both in electronics and optoelectronics. This has pushed MOVPE technology with respect to reactor size and also to in situ control of growth, which is indispensable to cope with the challenges of reproducible nucleation on foreign substrates as well as wafer bow. Nevertheless, there is ample room for further improvement of growth procedures for devices emitting in the UV. This partly is due to the high aluminum content required for this spectral region. On the other hand, pushing the limits to ever-longer wavelengths requires high In content, which also will continue to be a topic of research.

References

1. Manasevit, H.M., Erdmann, F.M., and Simpson, W.I. (1971). The use of metalorganics in the preparation of semiconductor materials IV. The nitrides of aluminum and gallium. *Journal of the Electrochemical Society* 118 (11): 1864–1868. doi: 10.1149/1.2407853.
2. Polyakov, A.Y., Shin, M., Skowronski, M. et al. (1997). Growth of GaBN ternary solutions by organometallic vapor phase epitaxy. *Journal of Electronic Materials* 26 (3): 237–242. doi: 10.1007/s11664-997-0157-x.

3. Abid, M., Moudakir, T., Orsal, G. et al. (2012). Distributed Bragg reflectors based on diluted boron-based BAlN alloys for deep ultraviolet optoelectronic applications. *Applied Physics Letters* 100: 051101. doi: 10.1063/1.3679703.
4. Stringfellow, G.B. (1989). *Organometallic Vapor-Phase Epitaxy, Theory and Practice*. San Diego: Academic Press.
5. Stringfellow G.B. (1999). *Organometallic Vapor-Phase Epitaxy, Theory and Practice*, 2e. San Diego: Academic Press.
6. Razeghi, M. (2010). *The MOCVD Challenge: A Survey of GaInAsP-InP and GaInAsP-GaAs for Photonic and Electronic Device Applications*, 2e. CRC Press.
7. Amano, H., Sawaki, N., Akasaki, I. et al. (1986). Metalorganic vapor phase epitaxial growth of a high quality GaN film using an AlN buffer layer. *Applied Physics Letters* 48 (5): 353–355; doi: 10.1063/1.96549.
8. Nakamura, S. (1991). GaN growth using GaN buffer layer. *Japanese Journal of Applied Physics* 30 (10A): L 1705–L 1707. doi: 10.1143/JJAP.30.L1705.
9. Amano, H., Kito, M., Hiramatsu, K. et al. (1989). p-Type conduction in Mg-doped GaN treated with low-energy electron beam irradiation (LEEBI). *Japanese Journal of Applied Physics* 28 (12): L 2112–L 2114.
10. Nakamura, S., Mukai, T., Senoh, M. et al. (1992). Thermal annealing effects on p-type Mg-doped GaN films. *Japanese Journal of Applied Physics* 31 (2B): L 139–L 142.
11. Nakamura, S., Iwasa, N., Senoh, M. et al. (1992). Hole compensation mechanism of p-type GaN films. *Japanese Journal of Applied Physics* 31 (5A): 1258–1266.
12. Nakamura S. and Mukai, T. (1992). High-quality InGaN films grown on GaN Films. *Japanese Journal of Applied Physics* 31 (10B): L 1457–L 1459.
13. Nakamura, S., Mukai, T., and Senoh, M. (1994). Candela-class high-brightness InGaN/AlGaN double-heterostructure blue-light emitting diodes. *Applied Physics Letters* 64 (13): 1687–1689. doi: 10.1063/1.111832.
14. The Nobel Prize. (2014). New light to illuminate the world. Press release (7 October). https://www.nobelprize.org/nobel_prizes/physics/laureates/2014/press.html.
15. Ho, I.H. and Stringfellow G.B. (1996). Solid phase immiscibility in GaInN. *Applied Physics Letters* 69 (18): 2701–2703. doi: 10.1063/1.117683.
16. Ferhat, M. and Bechstedt, F. (2002). First-principles calculations of gap bowing in $In_xGa_{1-x}N$ and $In_xAl_{1-x}N$ alloys: Relation to structural and thermodynamic properties. *Physical Review B* 65 (7): 075213. doi: 10.1103/PhysRevB.65.075213.
17. Rutcrana, P., M. Albrecht, M., and Neugebauer, J. (eds) (2003). *Nitride Semiconductors: Handbook of Materials and Devices*. Weinheim: Wiley-VCH.
18. Morkoç, H. (2013). *Nitride Semiconductor Devices: Fundamentals and Applications*. Weinheim: Wiley-VCH. ISBN 978-3-527-41101-6.
19. Piprek, J. (ed.) (2007). *Nitride Semiconductor Devices: Principles and Simulation*. Weinheim: Wiley-VCH. ISBN 978-3-527-40567-8.
20. Kneissl, M. and Rass, J. (eds.) (2016). *III-Nitride Ultraviolet Emitters; Technology and Applications*. Springer Series in Materials Science, volume 227.
21. Wuertele, M.A., Kneissl, M., Jekel, M. et al. (2011). Application of GaN-based ultraviolet-C light emitting diodes - UV LEDs - for water disinfection. *Water Research* 25: 1481–1489. doi:10.1016/j.watres.2010.11.015.
22. Mehnke, F., Ewald, H., Kneissl, M. et al. (2017). Gas sensing of nitrogen oxide utilizing spectrally pure deep UV LEDs. *IEEE Journal of Selected Topics in Quantum Electronics* 23 (2): 2000108. doi: 10.1109/JSTQE.2016.2597541.
23. Brendel, M., Pertzsch, E. Weyers, M. et al. (2016). Solar- and visible blind AlGaN photodetectors. In: [20].
24. Chow, T.P. and Ghezzo, M. (1996). SiC power devices. *Material Research Society Symposium Proceedings* 423: 69. doi: 10.1557/PROC-423-9.
25. Sze, S.M. (1981). *Physics of Semiconductor Devices*. New York: John Wiley and Sons.
26. Fichtenbaum, N.A., Keller, S., Mishra, U.K. et al. (2008). Impurity incorporation in heteroepitaxial N-face and Ga-face GaN films grown by metalorganic chemical vapor deposition. *Journal of Crystal Growth* 310: 1124–1131. doi: 10.1016/j.jcrysgro.2007.12.051.
27. Keller, S., Speck, J.S., Mishra, U.K. et al. (2008). Properties of N-polar AlGaN/GaN heterostructures and field effect transistors grown by metalorganic chemical vapor deposition. *Journal of Applied Physics* 103: 033708. doi: 10.1063/1.2838214.
28. Scholz, F. (2012). Semipolar GaN grown on foreign substrates: a review. *Semiconductor Science and Technology* 27: 024002. doi: 10-1088/0268-1242/27/2/024002.
29. Kneissl, M. and Wernicke, T. (2013). Optical and structural properties of InGaN light emitters on non- and semipolar GaN. In: *III-Nitride Semiconductors and Their Modern Devices* (ed. B. Gil), 244–279. Series on Semiconductor Science and Technology 18. New York: Oxford University Press.
30. Goldhahn, R., Buchheim, C., and Wenzel, H. (2007). Optical constants of bulk nitrides. In: [19].
31. Schley, R., Goldhahn, R, Feneberg, M. et al. (2009). Influence of strain on the band gap energy of wurtzite InN. *Physica Status Solidi B* 246 (6): 1177–1180. doi: 10.1002/pssb.200880924.
32. Vurgaftman, I., Meyer, J.R., and Ram-Mohan, L.R. (2001). Band parameters for III–V compound semiconductors and their alloys. *Journal of Applied Physics* 89: 5815. doi: 10.1063/1.1368156.

33. Krost, A. and Dadgar, A. (2002). GaN-based devices on Si. *Physica Status Solidi A* 194 (2): 361–375. doi: 10.1002/1521-396X.
34. Levinshtein, M.E., Rumyantsev, S., and Shur, M.S. (eds.) (2001). *Properties of Advanced Semiconductor Materials*. New York: Wiley & Sons.
35. Roder, C., Einfeldt, S., Figge, S.et al. (2005). Temperature dependence of the thermal expansion of GaN. *Physical Review B* 72: 085218. doi: 10.1103/PhysRevB.72.085218.
36. Yim W.M. and Paff, R.J. (1974). Thermal expansion of AlN, sapphire, and silicon. *Journal of Applied Physics* 45: 1456–1457. doi: 10.1063/1.1663432.
37. Roder, C. (2007). Analyse struktureller Eigenschaften von GaN mittels hochauflösender Röntgenbeugung bei variabler Meßtemperatur. PhD dissertation. University Bremen.
38. Madelung, O., Rössler, U., and Schulz, M. (eds.) (2001). Silicon carbide (SiC) thermal expansion. In: *Landolt-Börnstein - Group III Condensed Matter: Group IV Elements, IV-IV and III-V Compounds. Part a - Lattice Properties*. Berlin Heidelberg: Springer. doi: 10.1007/10551045_254.
39. Taylor, A. and Jones, R.M. (1960). In: *Silicon Carbide - A High Temperature Semiconductor* J.R. O'Connor and J. Smiltens, eds), 147. *Proc. of the Conference on Silicon Carbide* (April, 1959, Boston). Oxford: Pergamon Press.
40. Okada, Y. and Tokumaru, Y. (1984). Precise determination of lattice parameter and thermal expansion coefficient of silicon between 300 and 1500 K. *Journal of Applied Physics* 56: 314–320. doi: 10.1063/1.333965.
41. Dauelsberg, M., Yakovlev, E.V., Talalaev, R.A. et al. (2014). On mechanisms governing AlN and AlGaN growth rate and composition in large substrate size planetary MOVPE reactors. *Journal of Crystal Growth* 393: 103–107. doi: 10.1016/j.jcrysgro.2013.08.018.
42. Creighton, J.R., Koleske, D.D., and Mitchell, C.C. (2006). Emissivity-correcting near-UV pyrometry for group-III nitride OMVPE. *Journal of Crystal Growth* 287 (2): 572–576. doi: 10.1016/j.jcrysgro.2005.10.078.
43. Brunner, F., Weyers, M., Kneissl, M. et al. (2008). High-temperature growth of AlN in a production scale reactor. *Physica Status Solidi C* 5 (6): 1799–1801. doi: 10.1002/pssc.200778658.
44. Scholz, F., Härle, V., Hangleiter, A. et al. (1997). Low pressure MOVPE of GaN and GaInNGaN heterostructures. *Journal of Crystal Growth* 170: 321–324. doi: 10.1016/S0022-0248(96)00606-9.
45. Koleske, D.D., Wickenden, A.E., Gorman, R.J. et al. (1998). Growth model for GaN with comparison to structural, optical, and electrical properties. *Journal of Applied Physics* 84: 1998–2010. doi: 10.1063/1.368353.
46. Briot, O., Gil, B., Heuken, M. et al. (2009). MOVPE growth of InN buffer layers on sapphire. *Journal of Crystal Growth* 311: 2787–2790. doi: 10.1016/j.jcrysgro.2009.01.008.
47. Kim, K. S., Razeghi, M., Lim, K. Y. et al. (1997). Determination of the band-gap energy of $Al_{1-}xInxN$ grown by metal–organic chemical-vapor deposition. *Applied Physics Letters* 71 (6): 800–802. doi: 10.1063/1.119650.
48. Kosaki, M., Amano, H., Akasaki, I. et al. (2001). Metalorganic vapor phase epitaxial growth of high-quality AlInN/AlGaN multiple layers on GaN. *Japanese Journal of Applied Physics* 40 (5A): L420–L422. doi: 10.1143/JJAP.40.L420.
49. Yamaguchi, S., Amano, H., Akasaki, I. et al. (2001). Crystal growth of high-quality AlInN/GaN superlattices and of crack-free AlN on GaN: Their possibility of high electron mobility transistor. *Physica Status Solidi A* 188 (2): 895–898. doi: 10.1002/1521-396X.
50. Peng, T., Piprek, J., Schubert, E.F. et al. (1997). Band gap bowing and refractive index spectra of polycrystalline $Al_xIn_{1-x}N$ films deposited by sputtering. *Applied Physics Letters* 71 (17): 2439–2441. doi: 10.1063/1.120112.
51. Walukiewicz, W., Lu, H., Schaff, W.J. et al. (2004). Optical properties and electronic structure of InN and In-rich group III-nitride alloys. *Journal of Crystal Growth* 269 (1): 119–127. doi: 10.1016/j.jcrysgro.2004.05.041.
52. Hums, C., Dadgar, A., Krost, A. et al. (2008). MOVPE growth and characterization of AlInN FET structures on Si(111). *Materials Research Society Symposium Proceedings* 1068: 45.
53. Hums, C., Dadgar, A., Krost, A. et al. (2007). Metalorganic vapor phase epitaxy and properties of AlInN in the whole compositional range. *Applied Physics Letters* 90, 022105. doi: 10.1063/1.2424649. *Erratum* (2007)*: Applied Physics Letters* 91: 139901 (2007). doi: 10.1063/1.2785981.
54. Hums, C. (2012). MOVPE Wachstum und Eigenschaften von AlIn(Ga)N Schichten und Schichtsystemen. PhD thesis. Otto-von-Guericke-Universität Magdeburg. https://core.ac.uk/download/pdf/51448379.pdf.
55. Laskar, M.R., Ghosh, S., Bhattacharya, A. et al. (2011). Anisotropic structural and optical properties of a-plane $(11\bar{2}0)$ AlInN nearly-lattice-matched to GaN. *Applied Physics Letters* 98: 181108. doi: 10.1063/1.3583457.
56. Zeimer, U., Kueller, V., Weyers, M. et al. (2013). High quality AlGaN grown on ELOAlN/sapphire templates. *Journal of Crystal Growth* 377: 32–36. doi: 0.1016/j.jcrysgro.2013.04.041.
57. Knauer, A., Zeimer, U., Kueller, V. et al. (2014). MOVPE growth of $Al_xGa_{1-x}N$ with $x \sim 0.5$ on epitaxial laterally overgrown AlN/sapphire templates for UV-LEDs. *Physica Status Solidi C* 11: 377–380. doi: 10.1002/pssc.201300415.
58. Butté, R., Carlin, J-F., Grandjean, N. et al. (2007). Current status of AlInN layers lattice-matched to GaN for photonics and electronics. *Journal of Physics D (Applied Physics)* 40: 6328–6344. doi:10.1088/0022-3727/40/20/S16.
59. Carlin, J.-F. and Ilegems, M. (2003). High-quality AlInN for high index contrast Bragg mirrors lattice matched to GaN. *Applied Physics Letters* 83 (4): 668–670. doi: 10.1063/1.1596733.

60. Carlin, J.-F., Grandjean, N., Ilegems, M. et al. (2005). Crack-free fully epitaxial nitride microcavity using highly reflective AlInN/GaN Bragg mirrors. *Applied Physics Letters* 86 (3): 31107. doi: 10.1063/1.1849851.
61. Berger, C., Dadgar, A., Krost, A. et al. (2010). Strain profiling of AlInN/GaN distributed Bragg reflectors using in-situ curvature measurements and ex-situ X-ray diffraction. *Materials Science and Engineering: A* 528: 58–64. doi: 10.1016/j.msea.2010.04.053.
62. Berger, C., Dadgar, A., Bläsing, J. et al. (2012). Growth of AlInN/AlGaN distributed Bragg reflectors for high quality microcavities. *Physica Status Solidi C* 9 (5): 1253–1258. doi: 10.1002/pssc.201100132.
63. Berger, C. Dadgar, A., Bläsing, J. et al. (2013). In-situ growth monitoring of AlInN/AlGaN distributed Bragg reflectors for the UV-spectral range. *Journal of Crystal Growth* 370: 87–91. doi: 10.1016/j.jcrysgro.2012.08.046.
64. Matsuoka, T. (1997). Calculation of unstable mixing region in wurtzite $In_{1-x-y}Ga_xAl_yN$. *Applied Physics Letters* 71 (1): 105–106. doi: 10.1063/1.119440.
65. Koukitu, A., Kumagai, Y., and Seki, H. (2000). Thermodynamic analysis of the MOVPE growth of InGaAlN quaternary alloy. *Journal of Crystal Growth* 221 (1–4): 743–750. doi: 10.1016/S0022-0248(00)00810-1.
66. Deibuk, V.G. and Voznyi, A.V. (2005). Thermodynamic stability and redistribution of charges in ternary AlGaN, InGaN, and InAlN alloys. *Semiconductors* 39 (6): 623–628. doi: 10.1134/1.1944849.
67. Karpov, S. Y., Podolskaya, N., Zhmakin, I.A. et al. (2004). Statistical model of ternary group-III nitrides. *Physical Review B* 70: 235203. doi: 10.1103/PhysRevB.70.235203.
68. Gadanecz, A., Dadgar, A., Krost, A. et al. (2007). Thermal stability of metal organic vapor phase epitaxy grown AlInN. *Applied Physics Letters* 90: 221906. doi: 10.1063/1.2743744.
69. Medjdoub, F., Grandjean, N., and Kohn, E. (2006). Can InAlN/GaN be an alternative to high power/high temperature AlGaN/GaN devices? *Proceedings of the International Electron Devices Meeting* (IEDM'06): 1–4.
70. Bolognesi, C.R., Feltin, E., Grandjean, N. et al. (2011). High-speed AlInN/GaN HEMTs on SiC and (111) HR-Si CS. MANTECH Conference, May 16–19, Palm Springs, California, USA.
71. Gaska, R., Gaevski, M. and Shur, M. (2014). Novel AlInN/GaN integrated circuits operating up to 500 °C. 44th European IEEE Solid State Device Research Conference (ESSDERC). doi: 10.1109/ESSDERC.2014.6948778.
72. Sakalauskas, E., Krost, A., and Goldhahn, R. (2010). Dielectric function and optical properties of Al-rich AlInN alloys pseudomorphically grown on GaN. *Journal of Physics D: (Applied Physics)* 43: 365102. doi:10.1088/0022-3727/43/36/365102.
73. Hums, C., Christen, J., Krost, A. et al. (2009). AlInN/GaN based multi quantum well structures–growth and optical properties. *Physica Status Solidi C* 6 (52): S 451–S 454. doi: 10.1002/pssc.200880899.
74. Wang, H., Jiang, D.S., Jahn, U. et al. (2010). Cathodoluminescence study on in composition inhomogeneity of thick InGaN layer. *Thin Solid Films* 518: 5028–5031. doi: 10.1016/j.tsf.2010.03.163.
75. Jeong, H., Lerondel, G., and Jeong, M.S. (2015). Carrier localization in In rich InGaN/GaN multiple quantum wells for green light-emitting diodes. *Scientific Reports* 5: 9373. doi: 10.1038/srep09373.
76. Fischer, A.M., Amano, H., Ponce, F. et al. (2009). Misfit strain relaxation by stacking fault generation in InGaN quantum wells grown on m-plane GaN. *Applied Physics Express* 2: 041002 doi: 10.1143/APEX.2.041002.
77. Dadgar, A., Andreev, Z., Witzigmann, B. et al. (2013). Green to blue polarization compensated c-axis oriented multi-quantum wells by AlGaInN barrier layers. *Applied Physics Letters* 102: 062110. doi: 10.1063/1.4793185.
78. Thon, A. and Kuech, T.F. (1996). High temperature adduct formation of trimethylgallium and ammonia. *Applied Physics Letters* 69: 55–57. doi: 10.1063/1.118117.
79. Sekiguchi, K., Kakimoto, K., Shiraishi, K. et al. (2017). Thermodynamic considerations of the vapor phase reactions in III–nitride metal organic vapor phase epitaxy. *Japanese Journal of Applied Physics* 56: 04CJ04. doi:10.7567/JJAP.56.04CJ04.
80. Mihopoulos, T.G., Gupta, V., and Jensen, K. F. (1998). A reaction-transport model for AlGaN MOVPE growth. *Journal of Crystal Growth* 195 (1–4): 733–739. doi: 10.1016/S0022-0248(98)00649-6.
81. Creighton, J.R., Breiland, W.G., Coltrin, M.E. et al. (2002). Gas-phase nanoparticle formation during AlGaN metalorganic vapor phase epitaxy. *Applied Physics Letters* 81: 2626–2628. doi: 10.1063/1.1510580.
82. Coltrin, M., Creighton, J.R., and Mitchell, C.C. (2006). Modeling the parasitic chemical reactions of AlGaN organometallic vapor-phase epitaxy. *Journal of Crystal Growth* 287: 566–571. doi:10.1016/j.jcrysgro.2005.10.077.
83. Wang, G.T. and Creighton, J.R. (2004). Complex formation between magnesocene ($MgCp_2$) and NH_3: Implications for p-type doping of Group III nitrides and the Mg memory effect. *Journal of Physical Chemistry A* 108 (22): 4873–4877. doi: 10.1021/jp036494e.
84. Ogino, T. and Aoki, M. (1980). Mechanism of yellow luminescence in GaN. *Japanese Journal of Applied Physics* 19 (12): 2395–2405. doi: 10.1143/JJAP.19.2395.
85. Zhang, R. and Kuech, T.F. (1997). Carbon and hydrogen induced yellow luminescence in gallium nitride grown by halide vapor phase epitaxy. *Materials Research Society Symposium Proceedings* 482: 709–714. doi: 10.1557/PROC-482-709.
86. Götz, W., Johnson, N.M., Street, R.A. et al. (1996). Activation of acceptors in Mg-doped GaN grown by metalorganic chemical vapor deposition. *Applied Physics Letters* 68: 667–669. doi: 10.1063/1.116503.

87. Nakamura, S., Grown, T., and Senoh, M. (1992). Si- and Ge-doped GaN films grown with GaN buffer layers. *Japanese Journal of Applied Physics* 31 (9A): 2883–2888. doi: 10.1143/JJAP.31.2883.
88. Murakami, H., Amano, H., Akasaki, I. et al. (1991). Growth of Si-doped $Al_xGa_{1-x}N$ on (0001) sapphire substrate by metalorganic vapor phase epitaxy. *Journal of Crystal Growth* 115 (1–4): 648–651. doi: 10.1016/0022-0248(91)90820-U.
89. Lahrèche, H., Vennégués, P., Beaumont, B. et al. (1999). Growth of high-quality GaN by low-pressure metal-organic vapour phase epitaxy (LP-MOVPE) from 3D islands and lateral overgrowth. *Journal of Crystal Growth* 205 (3): 245–252. doi: 10.1016/S0022-0248(99)00299-7.
90. Tanaka, S., Takeuchi M., and Aoyagi, Y. (2000). Anti-surfactant in III-nitride epitaxy - quantum dot formation and dislocation termination. *Japanese Journal of Applied Physics* 39 (8B): L831–L834. doi: 10.1143/JJAP.39.L831
91. Markurt, T., Stauss, P., Albrecht, M. et al. (2013). Blocking growth by an electrically active subsurface layer: The effect of Si as an antisurfactant in the growth of GaN. *Physical Review Letters* 110: 036103. doi: 10.1103/PhysRevLett.110.036103.
92. Fritze, S. (2012). Wachstumsoptimierung und Charakterisierung von MOVPE-basierten GaN Pufferstrukturen auf Si(111) Substraten. PhD thesis. Otto-von-Guericke-Universität Magdeburg. https://d-nb.info/1058913530/34.
93. Takeda, K., Iwaya, M., and Akasaki, I. (2016). Electrical properties of n-type AlGaN with high Si concentration. *Japanese Journal of Applied Physics* 55: 05FE02. doi: 10.7567/JJAP.55.05FE02.
94. Mehnke, F. Wernicke, T., Kneissl, M. et al. (2013). Highly conductive n-Al*x*Ga_{1-}*x*N layers with aluminum mole fractions above 80%. *Applied Physics Letters* 103: 212109. doi:10.1063/1.4833247.
95. Dadgar, A., Bläsing, J., Diez, A. et al. (2011). Crack-free, highly conducting GaN layers on Si substrates by Ge doping. *Applied Physics Express* 4 (1): 011001. doi: 10.1143/APEX.4.011001.
96. Berger, C., Zettler, T., Strittmatter, A. et al. (2016). Metalorganic chemical vapor phase epitaxy of narrow-band distributed Bragg reflectors realized by GaN:Ge modulation doping. *Journal of Crystal Growth* 440: 6–12. doi: 10.1016/j.jcrysgro.2016.01.027.
97. Neugebauer, S., Dadgar, A., Strittmatter, A. et al. (2017). All metalorganic chemical vapor phase epitaxy of p/n-GaN tunnel junction for blue light emitting diode applications. *Applied Physics Letters* 110 (10): 102104. doi: 10.1063/1.4978268.
98. Fritze, S., Dadgar, A., Krost A. et al. (2012). High Si and Ge n-type doping of GaN–Limits and impact on stress. *Applied Physics Letters* 100 (12): 122104. doi: 10.1063/1.3695172.
99. Young, N.G., Farrell, R.M., Speck J.S. et al. (2016). Germanium doping of GaN by metalorganic chemical vapor deposition for polarization screening applications. *Journal of Crystal Growth* 455: 105–110. doi: 10.1016/j.jcrysgro.2016.09.074.
100. Kim, D.J., Ryu, D.Y., Lee J H. et al. (2000). Thermal activation energies of Mg in GaN:Mg measured by the Hall effect and admittance spectroscopy. *Journal of Applied Physics* 88 (5): 2564–2659. doi: 10.1063/1.1286925.
101. Fischer, S., Wetzel, C., Haller, E.E. et al. (1995). On p-type doping in GaN—acceptor binding energies. *Applied Physics Letters* 67 (9): 1298–1300. doi: 10.1063/1.114403.
102. Obloh, H., Bachem, K.H., Schlotter, P. et al. (1998). Self-compensation in Mg doped p-type GaN grown by MOCVD. *Journal of Crystal Growth* 195 (1–4): 270–273. doi: 10.1016/S0022-0248(98)00578-8.
103. Miceli, G. and Pasquarello, A. (2016). Self-compensation due to point defects in Mg-doped GaN. *Physical Review B* 93 (16): 165207. doi: 10.1103/PhysRevB.93.165207.
104. Neugebauer, J. and Van der Walle, C.G. (1999). Theory of Hydrogen in GaN. In: *Semiconductors and Semimetals*, vol. 61 (ed. H. Nickel), 479–502. New York: Academic Press. doi: 10.1016/S0080-8784(08)62713-1.
105. Nakamura, S., Mukai, T., Senoh, M. et al. (1992). Thermal annealing effects on p-type Mg-doped GaN films. *Japanese Journal of Applied Physics* 31 (2B): L139–L142. doi: 10.1143/JJAP.31.L139.
106. Figge, S., Kröger, R., Hommel, D. et al. (2002). Magnesium segregation and the formation of pyramidal defects in p-GaN. *Applied Physics Letters* 81 (25): 4748–4750. doi: 10.1063/1.1527981.
107. Ramachandran, V., Feenstra, R.M., Greve, D.W. et al. (1999). Inversion of wurtzite GaN(0001) by exposure to magnesium. *Applied Physics Letters* 75 (6): 808–810. doi: 10.1063/1.124520.
108. Wong, M.H., Stacia Keller, S., Mishra, U.K. et al. (2013). N-polar GaN epitaxy and high electron mobility transistors. *Semiconductor Science and Technology* 28 (7): 074009. doi:10.1088/0268-1242/28/7/074009.
109. Callsen, G., Hoffmann, A., Sitar, Z. et al. (2012). Optical signature of Mg-doped GaN: Transfer processes. *Physical Review B* 86 (7): 075207. doi: 10.1103/PhysRevB.86.075207.
110. Nakamura, S., Senoh, M., and Mukai, T. (1991). Highly p-type Mg-doped GaN films grown with GaN buffer layers. *Japanese Journal of Applied Physics* 31 (10A): L1708–L1711. doi: 10.1143/JJAP.30.L1708.
111. Liliental-Weber, Z., Grzegory, I., Porowski S. et al. (1999). Spontaneous ordering in bulk GaN:Mg samples. *Physical Review Letters* 83: 2370–2373. doi: 10.1103/PhysRevLett.83.2370.
112. Vennégués, P., Beaumont, B., Gibart, P. et al. (2000). Pyramidal defects in metalorganic vapor phase epitaxial Mg doped GaN. *Applied Physics Letters* 77 (6): 880–882. doi: 10.1063/1.1306421.
113. Alves, H., Amano, H., Meyer, B.K. et al. (2001). Compensation mechanism in MOCVD and MBE grown GaN:Mg. *Physica B: Condensed Matter* 308–310: 38–41. doi: 10.1016/S0921-4526(01)00663-9.

114. Bogusławski, P., Briggs, E.L., and Bernholc, J. (1996). Amphoteric properties of substitutional carbon impurity in GaN and AlN. *Applied Physics Letters* 69 (2): 233–235. doi: 10.1063/1.117934.
115. Nagai, N., Hiramatsu, K., Sawaki, N. et al. (1998). Hole trap levels in Mg-doped GaN grown by metalorganic vapor phase epitaxy. *Applied Physics Letters* 73 (14): 2024–2026. doi: 10.1063/1.122356.
116. Pankove, J., Berkeyheiser, J., and Miller, E.A. (1974). Properties of Zn-doped GaN. *I. Photoluminescence. Journal of Applied Physics* 45 (3): 1280–1286. doi: 10.1063/1.1663402.
117. Monemar, B. and Lagerstedt, O. (1979). Properties of VPE-grown GaN doped with Al and some iron-group metals. *Journal of Applied Physics* 50 (10): 6480–6491. doi: 10.1063/1.325743.
118. Baur, J., Amano, H., Hiramatsu, K. et al. (1994). Infrared luminescence of residual iron deep level acceptors in gallium nitride (GaN) epitaxial layers. *Applied Physics Letters* 64 (7): 857–859. doi: 10.1063/1.111003.
119. Heitz, R., Hoffmann, A., Meyer, B.K. et al. (1997). Excited states of Fe3+ in GaN. *Physical Review B* 55 (7): 4382–4387. doi: 10.1103/PhysRevB.55.4382.
120. Wright, A.F. (2002). Substitutional and interstitial carbon in wurtzite GaN. *Journal of Applied Physics* 92 (5): 2575–2585. doi: 10.1063/1.1498879.
121. Fariza, A., Dadgar, A., Strittmatter, A. et al. (2017). Leakage currents and Fermi-level shifts in GaN layers upon iron and carbon-doping. *Journal of Applied Physics* 122 (2): 025704. doi: 10.1063/1.4993180.
122. Li, X., Janzén, E., Forsberg, U. et al. (2015). Precursors for carbon doping of GaN in chemical vapor deposition. *Journal of Vacuum Science & Technology B* 33 (2): 021208. doi: 10.1116/1.4914316.
123. Motoki, K., Kumagai, Y., Koukito, A. et al. (2001). Preparation of large freestanding GaN substrates by hydride vapor phase epitaxy using GaAs as a starting substrate. *Japanese Journal of Applied Physics* 40 (2B): L140–L143. doi: 10.1143/JJAP.40.L140.
124. Goubara, S., Matsubara, T., Tadamoto, K. et al. (2017). Bulk GaN substrate with overall dislocation density on the order of 105/cm^2 fabricated by hydride vapor phase epitaxy. *Journal of Crystal Growth* 478: 123–128. doi: 10.1016/j.jcrysgro.2017.08.020.
125. Amano, H. (2013). Progress and prospect of the growth of wide-band-gap Group III nitrides: Development of the growth method for single-crystal bulk GaN. *Japanese Journal of Applied Physics* 52: 050001. doi: 10.7567/JJAP.52.050001.
126. Nakamura, S., Senoh, M., Nagahama, S. et al. (1998). High-power, long-lifetime InGaN/GaN/AlGaN-based laser diodes grown on pure GaN substrates. *Japanese Journal of Applied Physics* 37: L309–L312. doi: 10.1143/JJAP.37.L309.
127. Bickermann, M. (2016). Growth and properties of bulk AlN substrates. In: [20].
128. Grandusky, J. R., Wraback, M., Schowalter, L. et al. (2013). 270 nm pseudomorphic ultraviolet light-emitting diodes with over 60 mW continuous wave output power. *Applied Physics Express* 6 (3): 032101. doi: 10.7567/APEX.6.032101.
129. Hartmann, C., Irmscher, K., Bickermann, M. et al. (2016). Preparation of deep UV transparent AlN substrates with high structural perfection for optoelectronic devices. *Crystal Engineering Communications* 18: 3488–3497. doi: 10.1039/C6CE00622A.
130. Nabulsi, F. (2015). Implications for LEDs of the shift to large-diameter sapphire wafers. *Semiconductor Today* 10 (3): 68–71.
131. Taniyasu, Y., Kasu, M., and Makimoto, T. (2007). Threading dislocations in heteroepitaxial AlN layer grown by MOVPE on SiC (0 0 0 1) substrate. *Journal of Crystal Growth* 298: 310–315. doi: 10.1016/j.jcrysgro.2006.10.032.
132. Okumura, H., Kimoto, T., and Suda, J. (2014). Formation mechanism of threading-dislocation array in AlN layers grown on 6H-SiC (0001) substrates with 3-bilayer-high surface steps. *Applied Physics Letters* 105 (7): 071603; doi: 10.1063/1.4892807.
133. Olesinski, R.W., Kanani, N., and Abbaschian, G.J. (1985). The Ga – Si (gallium-silicon) system. *Bulletin of Alloy Phase Diagrams* 6 (4): 362–364. doi: 10.1007/BF02880523.
134. Koleske, D., Coltrin, M., and Allerman, A. (2004). Understanding GaN nucleation layer evolution on sapphire. *Journal of Crystal Growth* 273: 86–99. doi: 10.1016/j.jcrysgro.2004.08.126.
135. Groh, L., Krost, A., Dadgar, A. et al. (2011). Characterization of AlGaInN layers using X-ray diffraction and fluorescence. *Physica Status Solidi B* 248 (3): 622–626. doi: 10.1002/pssb.201046418.
136. Krost, A., Dadgar, A., Zettler, T. et al. (2005). Simultaneous measurement of wafer curvature and true temperature during metalorganic growth of group-III nitrides on silicon and sapphire. *Physica Status Solidi B* 242 (13): 2570–2574. doi: 10.1002/pssb.200541088.
137. Dadgar, A., Bläsing, J., and Krost, A., (2006). Epitaxy of GaN LEDs on large substrates: Si or sapphire? In: *Advanced LEDs for Solid State Lighting* (eds. C.-H. Hong et al.). *Proceedings of the Asia-Pacific Optical Communications Conf.*, Gwangju, Korea. SPIE Vol. 6355, doi: 10.1117/12.691576.
138. Dadgar, A., Bläsing, J., Krost, A. et al. (2006). Growth of blue GaN LED structures on 150-mm Si(111). *Journal of Crystal Growth* 297 (2): 279–282. doi: 10.1016/j.jcrysgro.2006.09.032.
139. Nishikawa, A., Groh, L., Solari, W. et al. (2013). 200-mm GaN-on-Si based blue light-emitting diode wafer with high emission uniformity. *Japanese Journal of Applied Physics* 52 (8S), 08JB25. doi: 10.7567/JJAP.52.08JB25.

140. Mante, N., Rennesson, S., Frayssinet, E. et al. (2018). Investigation of the mechanisms behind the AlN on Si (111) microstructure. *Journal of Applied Physics* 123: 215701; https://doi.org/10.1063/1.5017550.
141. Frank, F.C. (1950). *Conf. on Plastic Deformation of Crystalline Solids*. Carnegie Inst. of Tech. and Office of Naval Research, 150.
142. O'Connell, J.H., Lee, M.E., Westraadt, J. et al. (2017). Defect characterization of MOCVD grown AlN/AlGaN films on sapphire substrates by TEM and TKD. *Physica B* 535: 293–298. doi: 10.1016/j.physb.2017.08.005.
143. Bai, J., Huang, X., and Dudley, M. (2006). High-resolution TEM observation of AlN grown on on-axis and off-cut SiC substrates. *Materials Science in Semiconductor Processing* 9 (1–3) 180–183. doi: 10.1016/j.mssp.2006.01.055.
144. Liu, R., Ponce, F.A., Dadgar, A. et al. (2003). Atomic arrangement at the AlN/Si (111) interface. *Applied Physics Letters* 83 (5): 860–862. doi: 10.1063/1.1597749.
145. Krost, A., Dadgar, A., Christen, J. et al. (2004). Evolution of stress in GaN heteroepitaxy on AlN/Si(111): from hydrostatic compressive to biaxial tensile. *Applied Physics Letters* 85 (16): 3411–3443. doi: 10.1063/1.1808237.
146. Nix, W.D. and Clemens, B.M. (1999). Crystallite coalescence: A mechanism for intrinsic tensile stresses in thin films. *Journal of Materials Research* 14 (8): 3467–3473. doi: 10.1557/JMR.1999.0468.
147. Sheldon, B.W., Bhandari, A., Redwing, J.M. et al. (2007). Steady-state tensile stresses during the growth of polycrystalline films. *Acta Materialia* 55 (15): 4973–4982. doi: 10.1016/j.actamat.2007.05.008.
148. Romanov, A.E. and Speck, J. S. (2003). Stress relaxation in mismatched layers due to threading dislocation inclination. *Applied Physics Letters* 83 (13): 2569–2571. doi: 10.1063/1.1613360.
149. Dadgar, A., Christen, J., Krost, A. et al. (2004). Strains and stresses in GaN heteroepitaxy - sources and control. In: *Advances in Solid State Physics* 44 (ed. B. Kramer), 313–325. Berlin Heidelberg: Springer. http://www.springer.com/de/book/9783540211488.
150. Brunner, F., Weyers, M., Zettler, J.-T. et al. (2012). Analysis of doping induced wafer bow during GaN:Si growth on sapphire. *Journal of Applied Physics* 112: 033503. doi: 10.1063/1.4739278.
151. Yoshida, S., Misawa, S., and Gonda, S. (1983). Epitaxial growth of GaN/AlN heterostructures. *J. Vacuum Science & Technology B: Microelectronics Processing and Phenomena* 1 (2): 250–253. doi: 10.1116/1.582496.
152. Pankove, J. I. and Miller, E. A. (1973). Electroluminescence in GaN. In: *Luminescence of Crystals, Molecules, and Solutions: Proceedings of the International Conference on Luminescence, August 1972, Leningrad USSR* (eds. J.E. Berkeyheiser et al.). Boston: Springer. doi: 10.1007/978-1-4684-2043-2.
153. Akasaki, I., Amano, H., Sawaki, N. et al. (1989). Effects of AlN buffer layer on crystallographic structure and on electrical and optical properties of GaN and $Ga_{1-x}Al_xN$ ($0<x \leqq 0.4$) films grown on sapphire substrate by MOVPE. *Journal of Crystal Growth* 98 (1–2): 209–219. doi: 10.1016/0022-0248(89)90200-5.
154. Mohn, S., Vennéguès, P., Albrecht, M. et al. (2016). Polarity control in Group-III nitrides beyond pragmatism. *Physical Review Applied* 5 (5): 054004. doi: 10.1103/PhysRevApplied.5.054004.
155. Kuhn, B. and Scholz, F. (2001). An oxygen doped nucleation layer for the growth of high optical quality GaN on sapphire. *Physica Status Solidi A* 188 (2): 629–633. doi: 10.1002/1521-396X(200112)188:2<629::AID-PSSA629>3.0.CO;2-#.
156. Hertkorn, J., Scholz, F., Zweck, J. et al. (2007). Optimization of nucleation and buffer layer growth for improved GaN quality. *Journal of Crystal Growth* 308 (1): 30–36. doi: 10.1016/j.jcrysgro.2007.07.056.
157. Bläsing, J., Krost, A., Scholz, F. et al. (2009). Oxygen induced strain field homogenization in AlN nucleation layers and its impact on GaN grown by metal organic vapor phase epitaxy on sapphire: An X-ray diffraction study. *Journal of Applied Physics* 105 (3): 033504. Doi: 10.1063/1.3074095.
158. Balmer, R.S., Pickering, C., Pidduck, A.J. et al. (2002). Modelling of high temperature optical constants and surface roughness evolution during MOVPE growth of GaN using in-situ spectral reflectometry. *Journal of Crystal Growth* 245 (3–4): 198–206. doi: 10.1016/S0022-0248(02)01685-8.
159. Zúñiga-Pérez, J., Lymperakis, L., Feuillet, G. et al. (2016). Polarity in GaN and ZnO: Theory, measurement, growth, and devices. *Applied Physics Review* 3: 041303. doi: 10.1063/1.4963919.
160. Collazo, R., Dalmau, R.F., Sitar, Z. et al. (2007). Simultaneous growth of a GaN p/n lateral polarity junction by polar selective doping. *Applied Physics Letters* 91 (21): 212103. doi: 10.1063/1.2816893.
161. Keller, S., Speck, J.S., Mishra, U.K. et al. (2008). Properties of N-polar AlGaN/GaN heterostructures and field effect transistors grown by metalorganic chemical vapor deposition. *Journal of Applied Physics* 103 (3): 033708. doi: 10.1063/1.2838214.
162. Fichtenbaum, N.A., Keller, S., Mishra, U.K. et al. (2008). Impurity incorporation in heteroepitaxial N-face and Ga-face GaN films grown by metalorganic chemical vapor deposition. *Journal of Crystal Growth* 310 (6): 1124–1131. doi: 10.1016/j.jcrysgro.2007.12.051.
163. Kappers, M.J., McAleese, C., Humphreys, C.J. et al., (2007). Growth and characterisation of semi-polar (11–21) InGaN/GaN MQW structures. *Journal of Crystal Growth* 300: 155–159. doi: 10.1016/j.jcrysgro.2006.11.008.
164. Brunner, F., Edokam, F., Weyers, M. et al. (2015). Semi-polar (112-2)-GaN templates grown on 100 mm trench-patterned r-plane sapphire. *Physica Status Solidi B* 252 (5): 1189–1194. doi: 10.1002/pssb.201552054.
165. Stemmer, S., Pirouz, P., Ikuhara, Y. et al. (1996). Film/substrate orientation relationship in the AlN/6H-SiC epitaxial system. *Physical Review Letters* 77 (9): 1797–1800. doi: 10.1103/PhysRevLett.77.1797.

166. Weeks, W. Jr., Bremser, M., Davis, R. et al. (1995). GaN thin films deposited via organometallic vapor phase epitaxy on α(6H)–SiC(0001) using high-temperature monocrystalline AlN buffer layers. *Applied Physics Letters* 67: 401–403. doi:10.1063/1.114642.
167. Nishida, T.N. and Kobayashi. N. (2000). Nucleation control in MOVPE of group III-nitrides on SiC substrate. *Journal of Crystal Growth* 221: 297–300.
168. Boeykens, S., Leys, M., Germain, M. et al. (2004). Influence of AlGaN nucleation layers on structural and electrical properties of GaN on 4H-SiC. *Journal of Crystal Growth* 272: 312–317. doi: 10.1016/j.jcrysgro.2004.08.047.
169. Koleske, D.D., Wickenden, A.E., Fatemi, M. et al. (2002). Influence of AlN nucleation layer temperature on GaN electronic properties grown on SiC. *Applied Physics Letters* 80 (23): 4372–4374. doi: 10.1063/1.1484553.
170. Brunner, F., Reentilä, O., Würfl, J. et al. (2009). Strain engineering of AlGaN-GaN HFETs grown on 3 inch 4H-SiC. *Physica Status Solidi C* 6 (52): S1065–S1068. doi:10.1002/pssc.200880984.
171. Sheppard, S.T., Doverspike, K., Jenkins, T.J. et al. (1999). High-power microwave GaN/AlGaN HEMTs on semi-insulating silicon carbide substrates. *IEEE Electron Device Letters* 20 (4): 161–163. doi:10.1109/55.753753.
172. Manoi, A., Pomeroy, J.W., Killat, N. et al. (2010). Benchmarking of thermal boundary resistance in AlGaN/GaN HEMTs on SiC substrates: Implications of the nucleation layer microstructure. *IEEE Electron Device Letters* 31 (12): 1395–1397. doi: 10.1109/LED.2010.2077730.
173. Cho, J., Bozorg-Grayeli, E., Goodson, K.E. et al. (2012). Low thermal resistances at GaN-SiC interfaces for HEMT technology. *IEEE Electron Device Letters* 33 (3): 378–380. doi:10.1109/LED.2011.2181481.
174. Lee, S., Sheppard, S.T., Milligan, J. et al. (2013). Performance and reliability of AlGaN/GaN HEMT on 100-mm SiC substrate with improved epitaxial growth uniformity. In: *CS MANTECH Digest*, New Orleans, Louisiana, USA, 117–120. http://gaas.org/Digests/2013/papers/035.pdf.
175. Ishikawa, H., Jimbo, T, Umeno, M. et al. (1998). Thermal stability of GaN on (111) Si substrate. *Journal of Crystal Growth* 189/190: 178–182. doi: 10.1016/S0022-0248(98)00223-1.
176. Krost A. and Dadgar, A. (2002). GaN-based optoelectronics on Silicon substrates. *Materials Science and Engineering B* 93 (1–3): 77–84. doi: 10.1016/S0921-5107(02)00043-0.
177. Khoury, M., Vennégués, P., and Zúñiga-Pérez, J. (2017). Evolution and prevention of meltback etching: Case study of semipolar GaN growth on patterned silicon substrates. *Journal of Applied Physics* 122: 105108. doi: 10.1063/1.5001914.
178. Contreras, O.E., Krost, A., Ponce, F.A. et al. (2008). Atomic Arrangement at the AlN/Si(110) Interface. *Applied Physics Express* 1 (6): 0611041–0611043. doi: 10.1143/APEX.1.061104.
179. Schulze, F., Dadgar, A., Bläsing, J. et al. (2004). Influence of buffer layers on MOVPE grown GaN on Si(001). *Applied Physics Letters* 84 (23): 4747–4749. doi: 10.1063/1.176021.
180. Schulze, F. Dadgar, A., Krost, A. et al. (2006). Growth of single-domain GaN layers on Si(001) by metalorganic vapor-phase epitaxy. *Journal of Crystal Growth* 289 (2): 485–488. doi: 10.1016/j.jcrysgro.2005.12.073.
181. Dadgar, A., Christen, J., and Krost, A. (2007). Epitaxy of GaN on silicon — impact of symmetry and surface reconstruction. *New Journal of Physics* 9: 389. doi: 10.1088/1367-2630/9/10/389.
182. Gardeniers, J.G.E. and Giling L.J. (1991). Vapour growth of silicon: growth anisotropy and adsorption. *Journal of Crystal Growth* 115 (1–4): 542–550. doi: 10.1016/0022-0248(91)90802-C.
183. Ravash, R., Dadgar, A., Krost, A. et al. (2009). Metal organic vapor phase epitaxy growth of single crystalline GaN on planar Si(211) substrates. *Applied Physics Letters* 95 (24): 242101. doi: 10.1063/1.3272673.
184. Ravash, R., Dadgar, A., Krost, A. et al. (2010). Semipolar single component GaN on planar high index Si(11 h) substrates. *Applied Physics Letters* 97 (14): 142102. doi: 10.1063/1.3492835.
185. Sawaki, N., Honda, Y., Yamaguchi, M. et al. (2009). Growth and properties of semi-polar GaN on a patterned silicon substrate. *Journal of Crystal Growth* 311 (10): 2867–2874. doi: 10.1016/j.jcrysgro.2009.01.032.
186. Sawaki, N. and Honda, Y. (2010). Semi-polar GaN LEDs on Si substrate. *Science China, Technological Sciences* 54 (1): 38–41. doi: 10.1007/s 11431-010-4182-2.
187. Ni, X., Özgür, Ü., Morkoç, H. et al. (2009). Nonpolar m-plane GaN on patterned Si(112) substrates by metalorganic chemical vapor deposition. *Applied Physics Letters* 95 (11): 111102. doi: 10.1063/1.3225157.
188. Dadgar, A., Christen, J., Krost, A. et al. (2011). Eliminating stacking faults in semi-polar GaN by AlN interlayers. *Applied Physics Letters* 99 (2): 021905. doi: 10.1063/1.3610467.
189. Dadgar, A., Christen, J., and Krost, A. (2011). Improving GaN-on-silicon properties for GaN device epitaxy. *Physica Status Solidi C* 8 (5): 1503–1508. doi: 10.1002/pssc.201001137.
190. Ishikawa, H., Jimbo, T., and Umeno, M., (1999). GaN on Si substrate with AlGaN/AlN intermediate layer. *Japanese Journal of Applied Physics* 38 (5A): L492–L494. doi: 10.1143/JJAP.38.L492.
191. A. Dadgar, A., Diez, A., Krost, A. et al. (2003). Reduction of stress at the initial stages of GaN growth on Si(111). *Applied Physics Letters* 82 (1): 28–30. doi: 10.1063/1.1534940.
192. Arslan, E., Ozcelik, S., Ozbay, E. et al. (2008). Buffer optimization for crack-free GaN epitaxial layers grown on Si(1 1 1) substrate by MOCVD. *Journal of Physics D (Applied Physics)* 41 (15): 155317. doi: 10.1088/0022-3727/41/15/155317.
193. Amano, H., Iwaya, M., Figie, J. et al. (1998). Stress and defect control in GaN using low temperature interlayers. *Japanese Journal of Applied Physics* 37 (12 SUPPL. B) L1540–L1542. doi: 10.1143/JJAP.37.L1540.

194. Iwaya, M., Amano H., Akasaki, I. et al. (1998). Reduction of etch pit density in organometallic vapor phase epitaxy-grown GaN on sapphire by insertion of a low-temperature-deposited buffer layer between high-temperature-grown GaN. *Japanese Journal of Applied Physics* 37 (3B) L316–L318. doi: 10.1143/JJAP.37.L316.
195. Amano, H., Wetzel, C., Akasaki, I. et al. (1999). Control of dislocations and stress in AlGaN on sapphire using a low temperature interlayer. *Physica Status Solidi B* 216 (1): 683–689. doi: 0.1002/(SICI)1521-3951(199911)216:1<683::AID-PSSB683>3.0.CO;2-4.
196. Feltin, E., Vennégués, P., Gibart, P. et al. (2001). Crack-free thick GaN layers on silicon (111) by metalorganic vapor phase epitaxy. *Physica Status Solidi A* 188 (2): 531–535. doi: 10.1002/1521-396X(200112)188:2<531:AID-PSSA531>3.0.CO;2-V.
197. Feltin, E., Lahrèche, H., Bouillé, A. et al. (2001). Green InGaN light-emitting diodes grown on silicon (111) by metalorganic vapor phase epitaxy. *Japanese Journal of Applied Physics* 40 (7B): L738–L740. doi: 10.1143/JJAP.40.L738.
198. Feltin, E., Leroux, M., Gibart, P. et al. (2001). Stress control in GaN grown on silicon (111) by metalorganic vapor phase epitaxy. *Applied Physics Letters* 79 (20): 3230–3232. doi: 10.1063/1.1415043.
199. Schenk, H.P.D., Gibart, P., Doghèche, E. et al. (2001). Acoustical and optical gallium nitride waveguides grown on Si(111) by metalorganic vapor phase epitaxy. *Physica Status Solidi A* 188 (2): 537–541.doi: 10.1002/1521-396X(200112)188:2<537::AID-PSSA537>3.0.CO;2-7.
200. Sugahara, T., Lee, J.S., and Kohji Ohtsuka, K. (2004). Role of AlN/GaN multilayer in crack-free GaN layer growth on 5"Ø Si (111) substrate. *Japanese Journal of Applied Physics* 43 (12B): L1595–1597. doi:10.1143/JJAP.43.L1595.
201. Ni, Y., Zhang, B., Liu, Y. et al. (2015). Effect of AlN/GaN superlattice buffer on the strain state in GaN-on-Si(111) system. *Japanese Journal of Applied Physics* 54 (1): 015505. doi: 10.7567/JJAP.54.015505.
202. Iwakami, S., Goto, H., Ohtsuka, K. et al. (2004). AlGaN/GaN heterostructure field-effect transistors (HFETs) on Si substrates for large-current operation. *Japanese Journal of Applied Physics* 43 (7A): L831–L833. doi: 10.1143/JJAP.43.L831.
203. Ubukata, A., Yamazaki, T., Egawa, T. et al. (2007). GaN growth on 150-mm-diameter (1 1 1) Si substrates. *Journal of Crystal Growth* 298: 198–201. doi: 10.1016/j.jcrysgro.2006.10.147.
204. Frayssinet, E., Cordier, Y., Schenk, H.P.D. et al. (2011). Growth of thick GaN layers on 4-in. and 6-in. silicon (111) by metal-organic vapor phase epitaxy. *Physica Status Solidi C* 8 (5): 1479–1482. doi: 10.1002/pssc.201000885.
205. Hageman, P.R., Haffouz, S., Larsen, P.K. et al. (2001). High quality GaN layers on Si(111) substrates: AlN buffer layer optimisation and insertion of a SiN intermediate layer. *Physica Status Solidi A* 188 (2): 523–526. doi: 10.1002/1521-396X(200112)188:2<523::AID-PSSA523>3.0.CO;2-R.
206. Saxler, A., Razeghi, A., Vydyanath, H. R. et al. (1997). Comparison of trimethylgallium and triethylgallium for the growth of GaN. *Applied Physics Letters* 71: 3272. doi: 10.1063/1.120310.
207. Ishibashi, A., Takeishi, H., Mannoh, M. et al. (1996). Residual impurities in GaN/Al_2O_3 grown by metalorganic vapor phase epitaxy. *Journal of Electronic Materials* 25 (5): 799–803. doi.: 10.1007/BF02666639.
208. Alevi, M., Dietz, N., Ferguson, I. et al. (2006). Characterization of InN layers grown by high-pressure chemical vapor deposition. *Applied Physics Letters* 89: 112119. doi: 10.1063/1.2352797.
209. Hoffmann, V., Brunner, F., Zettler, J.-T. et al. (2011). Uniformity of the wafer surface temperature during MOVPE growth of GaN-based laser diode structures on GaN and sapphire substrate. *Journal of Crystal Growth* 315: 5–9. doi: 10.1016/j.jcrysgro.2010.09.048.
210. Nakamura, S. (1991). In situ monitoring of GaN growth using interference effects. *Japanese Journal of Applied Physics* 30 (8): 1620–1627. doi: 10.1143/JJAP.30.1620.
211. Brunner, F., Knauer, A., Zettler, J.-T. et al. (2008). Quantitative analysis of in situ wafer bowing measurements for III-nitride growth on sapphire. *Journal of Crystal Growth* 310 (10): 2432-2438. doi: 10.1016/j.jcrysgro.2008.01.024.
212. Strittmatter, A., Pohl, U.W., Zettler, J.-T. et al. (2004). Optimization of GaN MOVPE growth on patterned Si substrates using spectroscopic in situ reflectance. *Journal of Crystal Growth* 272 (1–4): 76–80. doi:10.1016/j.jcrysgro.2004.08.134
213. LayTec AG. (2016). In-situ metrology solutions for UV LED growth. *LayTec Application Note Nov. 2016.* http://www.laytec.de/uvled/.
214. Dadgar, A. Zettler, T., Krost, A. et al. (2004). In-situ measurements of strains and stresses in GaN heteroepitaxy and its impact on growth temperature. *Journal of Crystal Growth* 272 (1–4): 72–75. doi: 10.1016/j.jcrysgro.2004.08.031.
215. Brunner, F., Mogilatenko, A., Weyers, M. et al. (2013). Stress evolution during $Al_xGa_{1-x}N$/AlN growth on sapphire. *Journal of Crystal Growth* 376: 54–58. doi: 10.1016/j.jcrysgro.2013.04.023.
216. Wang, M.-T., Brunner, F., Weyers, M. et al. (2013). Phase control of semi-polar (11-22) and non-polar (11-20) GaN on cone shaped r-plane patterned sapphire substrates. *Journal of Crystal Growth* 371: 11–16. doi: 10.1016/j.jcrysgro.2012.10.020.
217. Prall, C., Brunner, F., Weyers, M. et al. (2015). In-situ photoluminescence measurements during MOVPE growth of GaN and InGaN MQW structures. *Journal of Crystal Growth* 415: 1–6. doi: 10.1016/j.jcrysgro.2014.12.023

218. Ashby, C., Peake, G., Griego, L. et al. (2000). Low-dislocation-density GaN from a single growth on a textured substrate. *Applied Physics Letters* 77: 3233–323. doi: 10.1063/1.1325394.
219. Yamada, M., Sano M., Mukai, T. et al. (2002). InGaN-based near-ultraviolet and blue-light-emitting diodes with high external quantum efficiency using a patterned sapphire substrate and a mesh electrode. *Japanese Journal of Applied Physics* 41 (12B): L1431–L1433. doi: 10.1143/JJAP.41.L1431.
220. OSRAM OS. (2012). Success in research: first gallium-nitride LED chips on silicon in pilot stage. Press release.
221. Brendel, M., Helbling, M., Weyers, M. et al. (2015). Measurement and simulation of top- and bottom-illuminated solar-blind AlGaN metal-semiconductor-metal photodetectors with high external quantum efficiencies. *Journal of Applied Physics* 118: 244504. doi:10.1063/1.4939283.
222. Park, K.Y., Kwonn, B.J., and Cho, Y.H. (2005). Growth and characteristics of Ni-based Schottky-type $Al_xGa_{1-x}N$ ultraviolet photodetectors with AlGaN/GaN superlattices. *Journal of Applied Physics* 98: 124505. doi: 10.1063/1.2142098.
223. Uren, M.J., Vanmeerbeek, P., Kuball, M. et al. (2015). Electric field reduction in C-doped AlGaN/GaN on Si high electron mobility transistors. *IEEE Electron Device Letters* 36 (8): 826–828. doi: 10.1109/LED.2015.2442293.
224. Heikman, S., Keller, S., DenBaars, S.P. et al. (2002). Growth of Fe doped semi-insulating GaN by metalorganic chemical vapor deposition. *Applied Physics Letters* 81 (3): 439–441. doi: 10.1063/1.1490396.
225. Silvestri, M., Uren, M.J., and Kuball, M. (2013). Iron-induced deep-level acceptor center in GaN/AlGaN high electron mobility transistors: Energy level and cross section. *Applied Physics Letters* 102 (7): 073501. doi: 10.1063/1.4793196.
226. Oka, T., Ina, T., Ueno, Y. et al. (2016). Over 10 A operation with switching characteristics of 1.2 kV-class vertical GaN trench MOSFETs on a bulk GaN substrate. *Proc. 28 ISPSD, June 12–16, 2016, Prague, Czech Republic*. IEEE.
227. Huang, J., Chowdhury, U., Dupuis, R. et al. (2001). Temperature dependent common emitter current gain and collector-emitter offset voltage study in AlGaN/GaN heterojunction bipolar transistors. *IEEE Electron Device Letters* 22 (4) 157–159. doi: 10.1109/55.915594.

5

Metamorphic Growth and Multijunction III-V Solar Cells

N.H. Karam[1], C.M. Fetzer[2], X.-Q. Liu[2], M.A. Steiner[3], and K.L. Schulte[3]
[1]Karamco USA Inc., La Cañada, California, USA
[2]Boeing-Spectrolab, Sylmar, California, USA
[3]National Renewable Energy Laboratory, Golden, Colorado, USA

5.1 Introduction to MOVPE for Multijunction Solar Cells

5.1.1 III-V PV Solar Cell Opportunities and Applications

Metal organic vapor phase epitaxy (MOVPE) has proved to be the primary materials growth technique for low-cost, high-volume III-V multijunction (MJ) solar cells. Figure 5.1 shows the typical applications that drive volume production for MJ solar cells using MOVPE. Today's space satellite industry worldwide predominantly relies on MJ solar cells fabricated using MOVPE to power the satellites [1–4]. Figure 5.1a shows a typical geosynchronous satellite wing showing the MJ solar cells, where the physical size can be appreciated as the arrow points at an individual sitting close to the 8–16 kW satellite. This MJ solar-cell technology also holds promise for near space [5, 6] and low-cost terrestrial concentrator photovoltaics (CPV) [7–11]. Figure 5.1b shows a solar power–enhanced unmanned aerial vehicle (UAV), where integration of ultralight and conformal solar-cell technology into the UAV dramatically increases the flight range of the aircraft, allowing for greater capabilities. Power generation using MJ solar cells for terrestrial use is by far the largest opportunity market, which is driven by low-cost clean energy production. The low cost is typically achieved by concentrating the sunlight by a factor of 500–1000 times using low-cost metal reflectors and glass lenses and shrinking the size of the expensive MJ solar cell by the same factor to bring down the system cost. A good example of a CPV system capable of producing up to 50 kW is shown in Figure 5.1c. Although large-scale CPV has not yet materialized, low-cost and higher-efficiency MJ solar cells will accelerate its market penetration.

Metalorganic Vapor Phase Epitaxy (MOVPE): Growth, Materials Properties, and Applications,
First Edition. Edited by Stuart Irvine and Peter Capper.

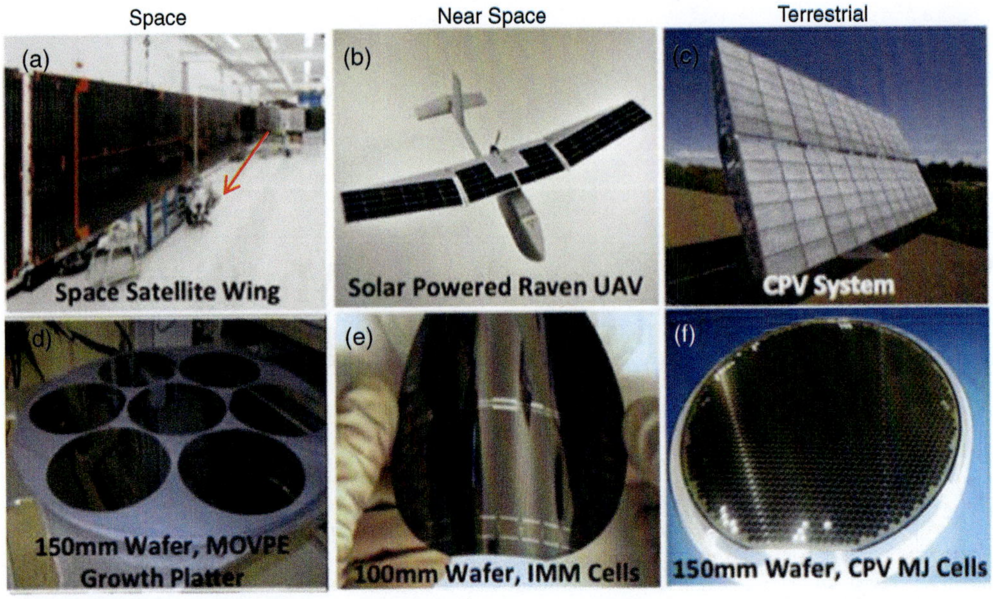

Figure 5.1 *Examples of application of MJ solar cells in space satellites, near space, and terrestrial power generation. (a) Typical geosynchronous space satellite wing populated with MJ solar cells. Source: Image courtesy of Boeing Spectrolab Inc. (b) Raven RQ-11 (manufacturer Aerovironment, Inc.) unmanned aerial vehicle (UAV) with lightweight, flexible solar arrays containing IMM solar cells applied to the wing surface. Source: Image courtesy of MicroLink Devices, Inc. and Air Force Research Labs, Wright-Patterson Air Force Base, Ohio. (c) Typical concentrator photovoltaic system that generates up to 50kW of power. (d) Typical MOVPE platter with five 150 mm MJ solar cells grown on Ge substrates. Source: Image courtesy of Boeing Spectrolab Inc. (e) Processed IMM epitaxial liftoff (ELO) wafer. Source: Image courtesy of MicroLink Devices, Inc. (f) Concentrator photovoltaic MJ solar cells grown by MOVPE and fabricated on 150mm Ge wafers. Source: Image courtesy of Boeing Spectrolab Inc.*

The need for ever-higher photovoltaic cell efficiencies requires incorporation of new multijunction device structures in lattice-matched solar cells, as well as the exploration of subcell bandgap combinations possible only in lattice-mismatched solar cells. Metamorphic III-V semiconductor materials offer access to bandgaps that span key portions of the solar spectrum, enabling new bandgap combinations in multijunction solar cells, and increasing both theoretical and practical efficiency limits for terrestrial and space solar cells [1, 2, 4, 7–12]. Figure 5.2 shows available semiconductor materials and their corresponding lattice constants and bandgaps that are lattice-matched (LM) to conventional substrates (e.g. Si, GaAs, Ge, InP). The MJ solar cell structures available with subcells LM to available substrates are not ideal. Figure 5.3a shows a simplified schematic cross section of a typical series-connected tandem 3 J (three junctions) GaInP/GaAs(In)/Ge solar cell, grown LM to the Ge substrate. Each subcell addresses a portion of the solar spectrum and is series interconnected using an optically transparent and electrically conducting tunnel junction [13]. Here, Ge is a convenient near-LM low-cost substrate that serves as a bottom subcell for the 3 J tandem stack. Only a small amount of In (1%) is needed to achieve the lattice-matching of GaAs to Ge with minimal change to the bandgap of GaAs (Figure 5.2). Although the bulk of today's manufacturing uses variants on this LM structure and the 3 J solar cells are typically twice as efficient as Si, the Ge bottom cell typically generates approximately 60% excess current compared to the top subcells. This limitation means the additional current is lost as heat rather than

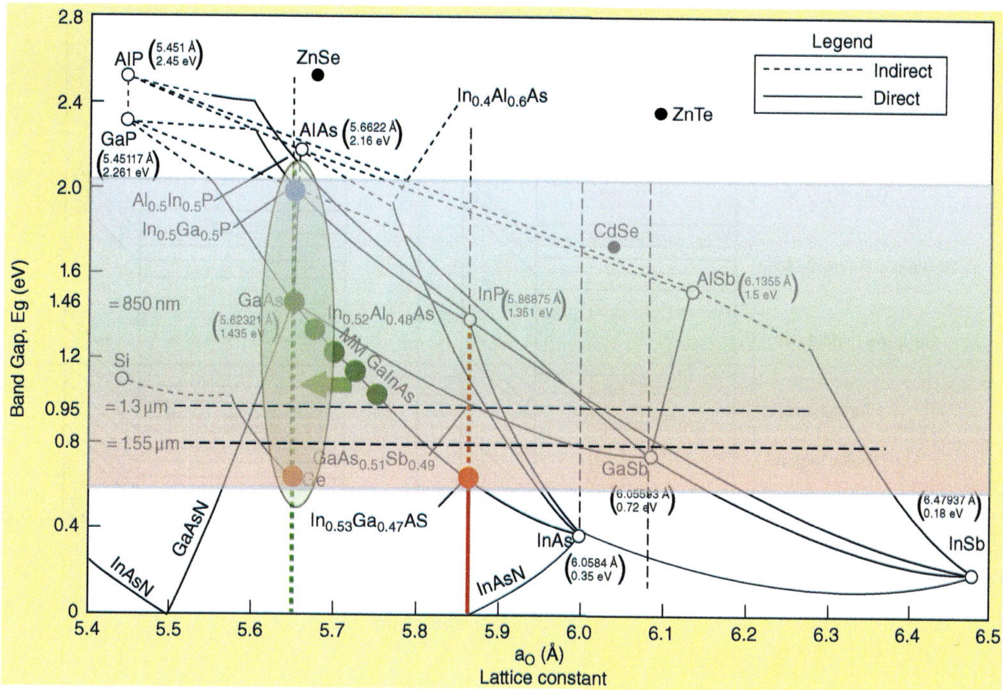

Figure 5.2 Available compound semiconductor materials and their corresponding bandgaps suitable for the fabrication of lattice-matched and metamorphic MJ solar cells.

usable power. In the absence of LM alloys with the appropriate bandgaps, one simple method to increase the efficiency is to evenly distribute the available current among all three subcells by adjusting their bandgaps using different alloy compositions that are lattice-mismatched to the growth substrate lattice constant.

5.1.2 Metamorphic Multijunction Solar Cells

The two commonly used structures for lattice-mismatched MJ solar cells are the upright metamorphic (UMM) structure, shown schematically in Figure 5.3b and addressed in Section 5.2 of this chapter; and the inverted metamorphic (IMM) structure, to be addressed in Section 5.3 of this chapter and shown schematically in Figure 5.3c.

A direct method of accomplishing even distribution of the current in the UMM 3 J solar cell structure is to grow compressive layers to change the lattice constant of the epitaxial films from the Ge substrate to the desired GaInAs alloy composition with the appropriate bandgap. Since solar-cell layers need to be thick relative to the critical thickness for strained layers, to efficiently absorb the sunlight, the resulting epitaxial layers are plastically deformed. In order for a UMM cell film to remain stable during operation, this approach requires near-complete relaxation of the strain. Since the UMM cell structure is grown upright, it faces the challenge of managing the dislocation density, typically $>10^6$-cm^{-2}, in the top two subcells of the multijunction stack as they produce the most power. Although the UMM design is typically a drop-in replacement of the common LM design, the defect density in the top two subcells tends to impact the overall performance of the UMM 3 J cell. The IMM solar-cell structure addresses that issue by growing the cell configuration in an inverted order, Figure 5.3c, where the top two cells, which produce the most power in the stack, are grown lattice-matched to the substrate and hence have a low dislocation density $<10^4\,cm^{-2}$ and potentially

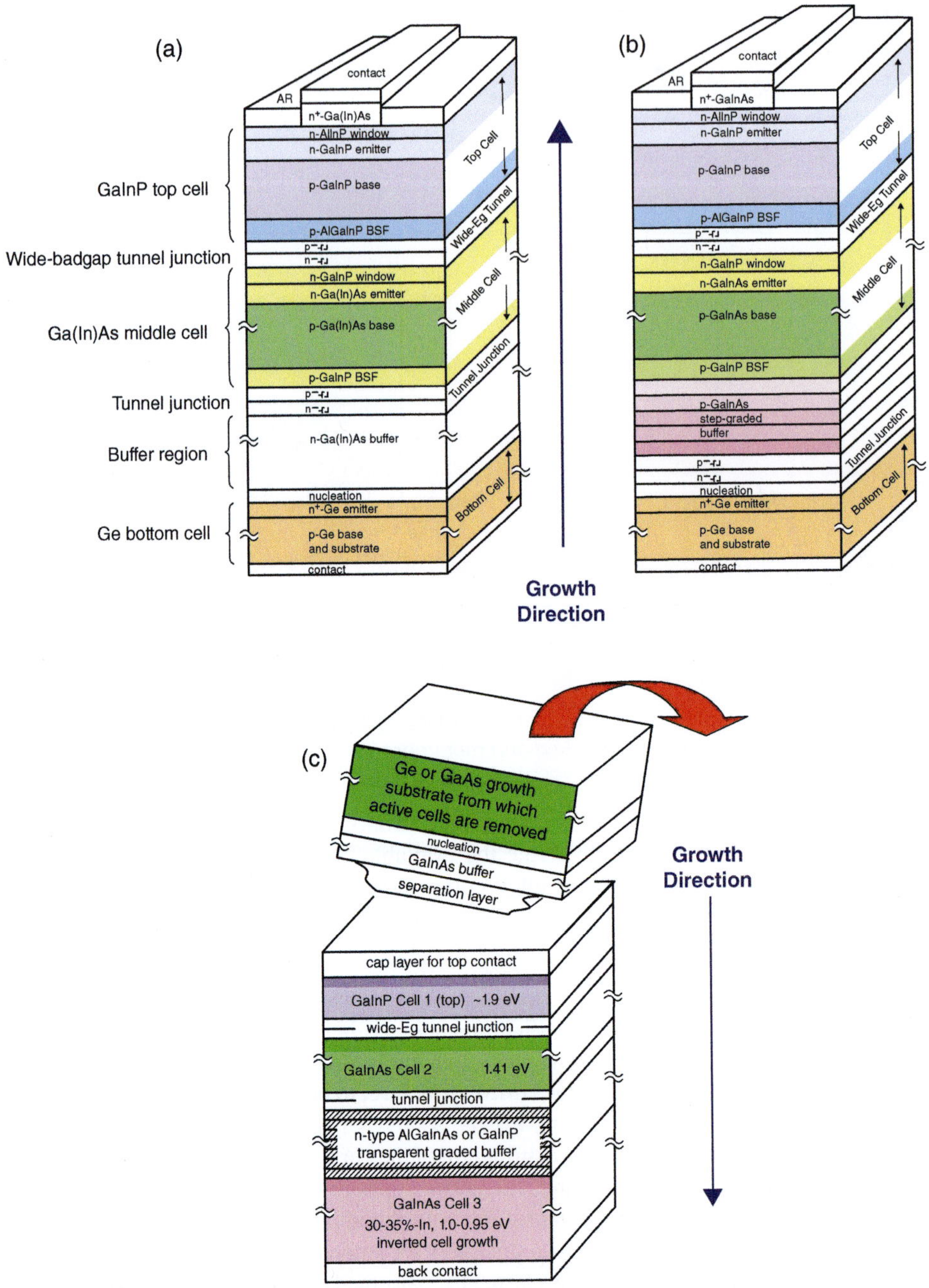

Figure 5.3 *Typical cross-sectional schematic of (a) upright lattice-matched three-junction GaInP/GaAs/Ge solar cell, (b) upright metamorphic three-junction GaInP/InGaAs/Ge solar cell, and (c) inverted metamorphic 3-junction GaInP/GaAs/InGaAs solar cell.*

higher efficiency. The placement of the metamorphic cells last in the IMM growth ensures that defects from these lattice-mismatched layers do not propagate into the high-performing wider-bandgap lattice-matched subcells that produce most of the power in the multijunction stack. The IMM cell is separated from the growth substrate by etching and "lifting off" from the growth substrate, which can be reused, while the active cell structure is mounted on a carrier. This approach will be addressed in Section 5.3 of this chapter.

The management of the defect density in metamorphic active subcell structures in UMM and IMM designs is critical to achieve optimal performance and reliability of the solar cell. The central challenge tempering the bandgap flexibility offered by metamorphic materials is that lattice-mismatched growth typically causes crystal dislocations, introducing energy levels in the bandgap, which enhance deleterious non-radiative, Shockley–Read–Hall (SRH) recombination. A composition-graded buffer (CGB) is used to accommodate misfit and threading dislocations to the greatest extent possible. One way of assessing the quality of the metamorphic cell is by measuring the bandgap-voltage offset (W_{oc}), which is defined as $(E_g/q) - V_{oc} = W_{oc}$ at 1 sun and plotted in Figure 5.4, for a wide range of bandgaps in LM and MM (Al)GaInP and (Al)GaInAs subcells. This bandgap-voltage offset W_{oc} varies only slowly over a wide range of bandgaps, and it can be a helpful simplifying approximation to treat it as a constant to the first order: e.g. at approx. 0.4 V at a fixed solar intensity of 100 mW/cm^2, commonly termed *1 sun*. For a solar cell limited by radiative recombination in the base, for example, the bandgap voltage offset is given by

$$W_{oc} = \left(E_g / q\right) - \left(kT / q\right) \ln\left(qwBN_C N_V / J_{ph}\right), \tag{5.1}$$

The bandgap-voltage offset is a function of the photogenerated current density J_{ph}, radiative recombination coefficient B, solar-cell base thickness w, and density of states in the conduction and valence bands, N_C and N_V. W_{oc} is a slowly varying natural log function with parameters that depend on bandgap and therefore may be approximated as a constant at a given temperature, such

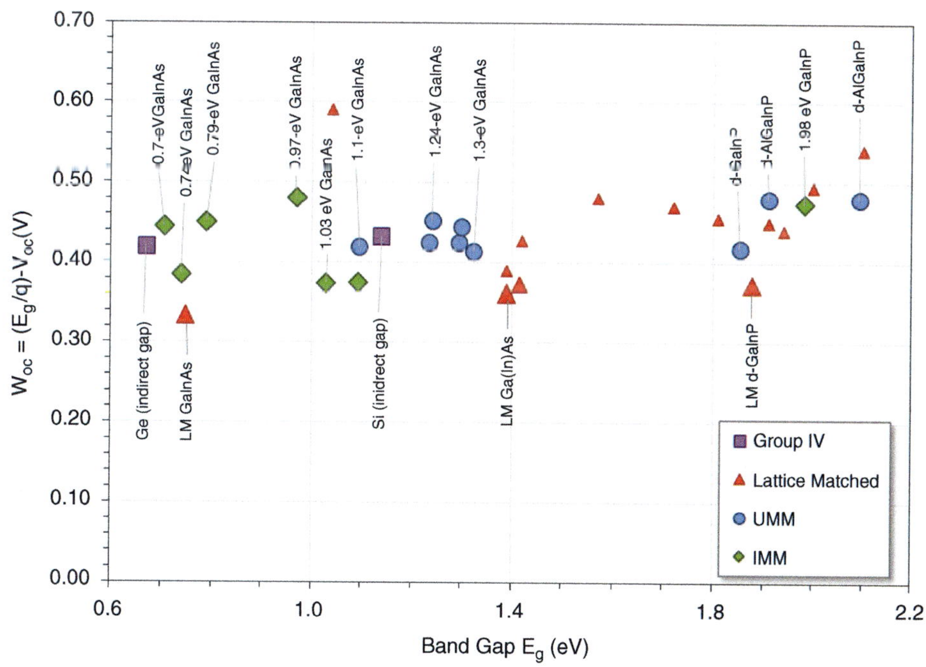

Figure 5.4 *Experimental W_{oc} for a wide range of single-junction LM, UMM, and IMM solar cells with bandgaps, from 0.67 to 2.1 eV, showing that the bandgap–voltage offset, $W_{oc} = (E_g/q) - V_{oc}$, is roughly constant over this range as predicted from theory. Note that W_{oc} for all materials is within 0.55 to 0.35 eV. The lower the W_{oc}, the closer the material is to the radiative limit of 0.33–0.36 eV. The value of this offset approaches the radiative limit for some solar cell materials.*

as room temperature. Deviation from the radiative limit of W_{oc} is a good indicator of the amount of the nonideal mechanism of SRH recombination that is present and, thus, of the semiconductor crystal quality. Figure 5.4 also shows W_{oc} for LM, UMM, and IMM cell structures and will be addressed in the respective sections of this chapter.

5.1.3 Reactor Technology for Metamorphic Epitaxy

While a detailed review of epitaxial reactor technology for MOVPE is beyond the scope of this chapter but is covered in Chapter 12, it is instructive to mention the key aspects of reactor technology pertinent to UMM and IMM solar cell growth. The main commercial low-pressure, As/P MOVPE reactor technology are vertical chambers where wafers are held on a planar rotating graphite susceptor (Figure 5.1d), acting to aid the laminar flow or subplanetary rotation used to planarize the growth. Each reactor type is capable of loading up to seven or eight 150-mm diameter Ge or GaAs wafers [14, 15] necessary for large-area solar cells, reaching 80 cm^2 in area for space applications. Present commercial reactors are capable of ±1% in thickness uniformity by proper selection of the growth conditions [16, 17]. The in situ monitoring becomes a critical component when considering lattice-mismatched growth, in particular the measurement of wafer curvature. One may relate the wafer curvature to the net strain at growth temperature through the use of Stoney's equation [18, 19]. Careful management of the in situ strain of the metamorphic growth layers such as the CGB layers results in complete relaxation while generating a tolerable level of defects and more uniform growth over large-area substrates [20].

5.2 Upright Metamorphic Multijunction (UMM) Solar Cells

5.2.1 Introduction and History of Upright Metamorphic Multijunctions

Investigations of UMM epitaxial III-V solar cells date from early work in MOVPE with the first demonstrated UMM single junctions being GaInAs subcells on GaAs substrate from the 1980s [21, 22]. These simple devices were quickly followed by a tandem UMM AlGaAs/GaInAs on GaAs 2J device in 1985 that employed two CGB layers [23]. It was not until the advent of the GaInP/GaAs dual junction [24] that new opportunities opened up in UMM devices. Specifically, the combination of GaInP and GaInAs opened the door to a single change in lattice constant for the epitaxial layers for devices with more optimal combination of bandgaps to evenly distribute the solar spectrum. Other examples of UMM devices by MOVPE have been explored. Of these, GaInP/GaInAs/Ge 3J devices have achieved the most successful results to date, rising to efficiencies equal to or better than lattice-matched counterparts; they will be the focus of this section [25–27]. The first appearance of UMM tandems based on GaInP/GaInAs was in the late 1990s [28–30]. The UMM concept was the first demonstration of a solar cell device efficiency exceeding 40% and was followed by other demonstrations reaching yet higher performance [31–33]. Today, UMM multijunction solar cells have reached commercialization for terrestrial CPV [34], and 4J UMM devices are presently under qualification for high-efficiency space devices [35, 36]. Expanding upon Figure 5.3, Figure 5.5a shows a general epitaxial stack of a 3J UMM multijunction solar cell. The general components shown are: i) a substrate of either GaAs or Ge, ii) a CGB layer altering the lattice constant, and iii) two or more epitaxial subcells with passivation and interconnecting layers grown at a new lattice constant. Similarly, Figure 5.5b shows a 4J version with a change to the upper subcells to AlGaInP and AlGaInAs utilizing a finer splitting of the solar spectrum for the 4J device.

Figure 5.5c shows a contour map of the efficiency attainable by concentrator cells for independently variable top (GaInP) and middle (GaInAs) subcells of a 3J stack with a Ge bottom subcell. As the subcell bandgaps are tied at a new lattice constant, they form a line on the figure. The lines

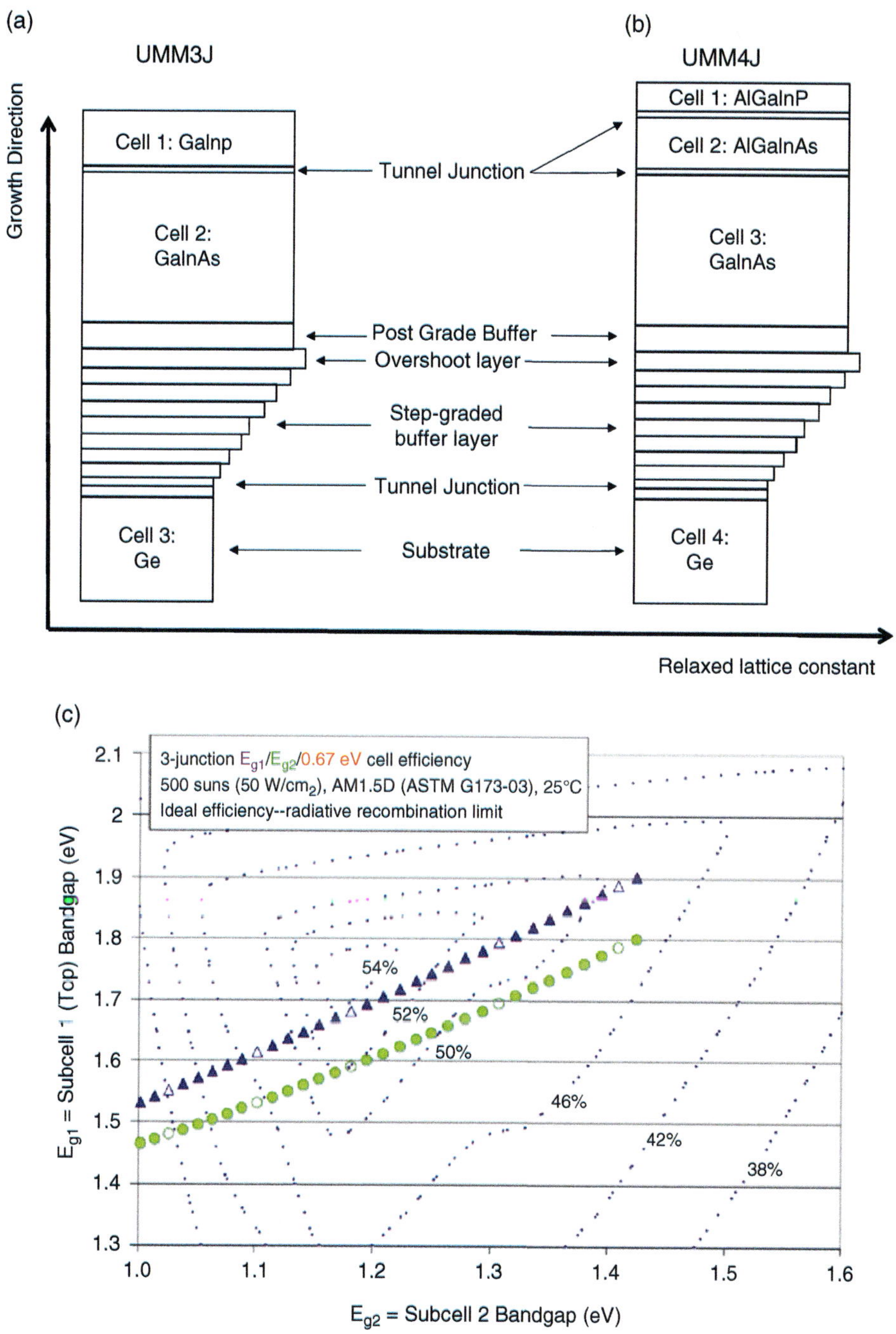

Figure 5.5 *(a) Epitaxial stack cross section of a 3J UMM solar cell (GaInP/GaInAs/Ge); the right-hand side of the figure indicates the relative change in lattice constant for each layer. (b) A similar cross section for a 4J UMM solar cell (AlGaInP/AlGaInAs/GaInAs/Ge). (c) Contour map showing GaInP with partial ordered and disordered track of GaInP/GaInAs/Ge 3J devices at AM1.5d 500 suns.*

in Figure 5.5c are for two potential bandgaps of GaInP, partially $CuPt_B$ ordered and fully disordered GaInP alloys in combination with a GaInAs subcell. Nominal growth conditions of GaInP in such tandem devices leads to a partial $CuPt_B$ ordering of the material and subsequent lowering of the bandgap [37, 38]. Various methods may be employed to increase the bandgap; however, use of triethylantimony (TESb)] as a surfactant provides an efficient method to produce disordered material [39–41]. Also indicated in Figure 5.5b is the combination of $Ga_{1-y}In_yP$ and $Ga_{1-x}In_xAs$ ternary alloys on the nominal bandgap versus lattice constant figure. One may appreciate that as the amount of mismatch between the subcells and the Ge substrate is increased, the bandgaps of the relaxed $Ga_{1-y}In_yP$ and $Ga_{1-x}In_xAs$ alloys decrease together. Either option of partially ordered or disordered $Ga_{1-y}In_yP$ solar cells reaches higher efficiencies than the lattice-matched approaches for the two most-demonstrated combinations utilizing compositions of 8%-In or 17%-In GaInAs [3, 28, 31, 32, 34, 42, 43].

5.2.2 MOVPE Growth Considerations of UMM

The MOVPE growth of UMM solar-cell devices is more complex than growing their lattice-matched counterparts. Epitaxial growth for UMM must include the additional growth of the CBG layers. Care must be taken to account for the addition of indium to the GaAs and GaInP subcells and interlayers to ensure lattice-matching of the epitaxial layers to the CGB final layer. These considerations typically involve more growth time within the reactor beyond the lattice-matched solar cell. A key advantage of the UMM device structure is that the final epitaxial wafer is essentially a drop-in replacement for the fabrication process of similar lattice-matched devices. Similar process steps may be used to produce a final cell from the epitaxial wafer with only small modification. Thus, most of the change required to produce a UMM solar cell over the LM counterpart is contained within the epitaxial layer structure and requires little change to processes downstream within the factory.

5.2.2.1 Compositional-Graded Buffer (CGB) Layers

Forming metamorphic materials leaves residual defects from plastic deformation of the underlying crystal lattice. Plastic deformation and relaxation of the strain from the mismatch between the substrate and the epitaxial layers forms defects such as edge dislocations [44], lattice twist and tilt [45], surface distortion such as cross-hatching [46], and cracking in tensile systems [47]. Such defects are deleterious to solar cells as they form extended deep states that act as nonradiative recombination centers, increasing dark currents and reducing open-circuit voltages [48, 49]. Care must be taken to design the epitaxial layers to relieve the strain within a buffer layer so as to minimize the defects that extend into the active region of the subcells.

A short primer on the method of relaxation of the strain within strained epitaxial layers is requisite to understand the design of the CGB layer. Such a subject has been a topic of deep investigation and is beyond the scope of this chapter. We refer the reader to a few good reviews from which the present discussion is summarized [50–53]. Figure 5.6 shows a diagram of the formation and propagation of dislocation half-loops. Strain relaxation occurs in multiple steps. The process begins when dislocation half-loops are nucleated at the epitaxial-vapor growth surface (Figure 5.6a). Much like heterogeneous nucleation of crystallization, dislocations loops typically occur at defects on the surface where the energy of nucleation is lowered, as the flat surface presents a high energy barrier to the formation of such defects [54]. Once nucleated, the dislocation half-loops propagate along the natural slip planes, typically the {111} planes in III-V materials, until the edge segment intersects the strained epilayer/substrate interface (Figure 5.6b), where the actual strain is relieved. Finally, dislocation half-loops grow and interact in both <110>directions

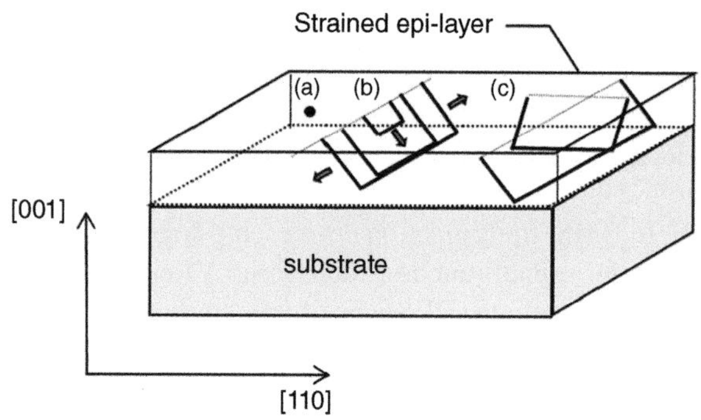

Figure 5.6 *Stages in dislocation half-loop formation: (a) nucleation, (b) propagation to strained interface, (c) relaxation in multiple directions of strain leaving threading segments. Source: Adapted from Tsao 1993 [52], Figure 5.17, p. 184.*

(Figure 5.6c) until the strain energy at the vapor/solid interface is lowered such that the surface energy is reduced below that requisite for additional nucleation.

This simplified approach allows a discussion of a few design criteria for the CGB layer to maximize strain relief with minimal extending threading segments propagating into the epitaxial layers above. Generally, UMM devices rely on architectures of compressive stress to avoid cracking. As surface defects control the nucleation rate, care must be taken to ensure the smoothest possible surfaces by utilizing (i) alloys such as $Ga_{1-x}In_xAs$ with small amounts of strain to avoid surface roughening, and (ii) smoothing dopants such as Te or isoelectronic surfactants like Sb or Bi [53, 55–57]. High growth temperatures and soft materials are employed as well as those with limited phase separation to enhance dislocation loop propagation over nucleation [53]. Small changes in composition in steps are used to trap dislocations in controlled amounts at various interfaces within the buffer [21, 28, 58]. Finally, as the kinetic driving force for relaxation drops as the system approaches 100% relaxation, an overshoot buffer layer is used to force full relaxation in the step grade and provide an interface to help prevent threading dislocations from propagating to the active region of the solar cell [59]. Along with the need for optical transparency to allow light to pass to the lowest subcell, these requirements place a large set of constraints on the growth of UMM solar cells. Further complicating the scenario is the utilization of a Ge substrate to enable a third subcell below the metamorphic epitaxial layers. Such substrates are typically thin (140 μm to 225 μm) compared to GaAs or InP substrates of similar diameter. As such, curvature placed on the substrate from epitaxial strain may produce significant bowing and warping on the wafer during growth, greatly decreasing temperature uniformity across the wafer during growth of critical layers [20, 27, 43]. UMM multijunction devices are typically the result of significant optimization efforts to locate the peak efficiency among a large number of interacting variables.

5.2.2.2 *Passivation Layers*

The CGB layer design and residual defect impact has been the subject of much literature discussion for UMM multijunction solar cells. High-efficiency devices also require multiple passivation and interconnecting layers to be grown with high quality. As shown in Figure 5.3a, these subcell layers include back-surface field layers, window layers, and interconnecting tunnel junction layers, all of which also pose significant design and growth issues. Window layers are typically composed of high-bandgap alloys near the same lattice constant as the subcell. For the GaInP top subcell, an AlInP top layer is employed as an n-type layer to reduce parasitic recombination at the emitter top surface that has a similar range of lattice constants as the top subcell.

5.2.2.3 Tunnel Junction Layers

In multijunction solar cells, tunnel junction layers are required between subcells to allow for series interconnection of multiple diodes. Tunnel junction layers require very high doping levels, typically $>10^{19}\,cm^{-3}$ active dopant atoms for both n-type and p-type materials; optical transparency to allow light to transmit to the lower subcells; and relatively abrupt interfaces to allow tunneling to occur efficiently [60–62]. In UMM multijunction devices, the doping poses a significant challenge. The most common dopants for tunnel junctions are Group VI elements such as Te and Se from diethlytelluride (DETe); diethylselenide (DESe) and hydrogen selenide (H_2Se) for n-type doping; and C from CCl_4 or CBr_4 for p-type doping [62]. While n-type dopants such as Se and Te readily dope materials such as GaAs, higher-bandgap materials such as GaInP have significant resistance to very high Te doping [63–65]. In UMM devices, the situation is further complicated because these layers need to be lattice-matched to the CGB. Typical highly doped layers used in tunnel junctions are of AlGaAs alloys. In UMM devices, indium is used to control the lattice constant of the tunnel junction layer. Organometallic indium sources are known to react with the carbon dopant sources through gas-phase reactions such as between CBr_4 and TMIn, or through vapor-phase etching of the indium by CCl_4. Both of these limitations may reduce the quality of the layer and limit the overall peak doping level and hence peak tunneling currents [66–69]. Most UMM solar-cell devices minimize the use of metamorphic tunnel junction layers by placing these structures in lattice-matched regions, as shown in Figure 5.5a for the lower tunnel junction. In UMM devices, the upper tunnel junction between the GaInP and GaInAs subcells must be metamorphic and is directly encumbered by these limitations. The quality (e.g. doping, abruptness, etc.) of the interconnecting tunnel junction between the epitaxial subcells limits the upper performance of UMM devices under concentration [70].

5.2.3 Growth and Device Results

High-quality UMM device layers are typically characterized for complete relaxation of the epitaxial materials by high-resolution X-ray diffraction (HRXRD), usually by triple-axis diffraction of a zincblende reflection spot. To simultaneously evaluate relaxation and composition, a reciprocal space map (RSM) of an off-axis X-ray diffraction (typically the 115 or 224 spots) is used [71, 72]. Figure 5.7 shows a comparison of 115 RSMs for a series of Ga(In)As-based subcells with compositions ranging from 0–23%In on Ge substrate [42]. Figure 5.7a shows a typical RSM for a strained GaAs layer grown on a Ge substrate. The broadening in the GaAs peak is an indication of the tensile strain in the film. This tensile strain is completely eliminated by adding 1% In, where the InGaAs subcell 2 is perfectly lattice-matched to the Ge substrate in Figure 5.7b. The RSM shows completely vertical diffraction (along the 100% strained direction) along with additional diffraction spots from Pendellosung reflection of the layer. For the composition at 8%-In GaInAs (Figure 5.7c) and the 12%-In GaInAs (Figure 5.7d), up to 23%-In GaInAs used for UMM4J devices showing the buffering techniques described earlier completely relaxes the strain, and the RSM peaks align along the 100% relaxed direction, i.e. back to the origin (e.g. 0,0,0) in reciprocal space.

Figure 5.8 shows the cross-sectional transmission electron microscopy (TEM) of an epitaxial UMM 3 J solar cell. The upper portion including the GaInAs capping layers, GaInP top subcell, and interconnecting tunnel junction layers show no immediate indication of dislocations, whereas the lower figure, including the buffering layers, shows the trapping of the misfit dislocations within the CGB buffer region. Since cross-sectional TEM is useful in observing a small area, to truly evaluate residual threading segments, one uses plan-view techniques. Figure 5.9 shows two different methods to evaluate threading dislocation density (TDD) for the same material. Electron beam induced current (EBIC) and cathodoluminescence (CL) images were taken for the 8%-In

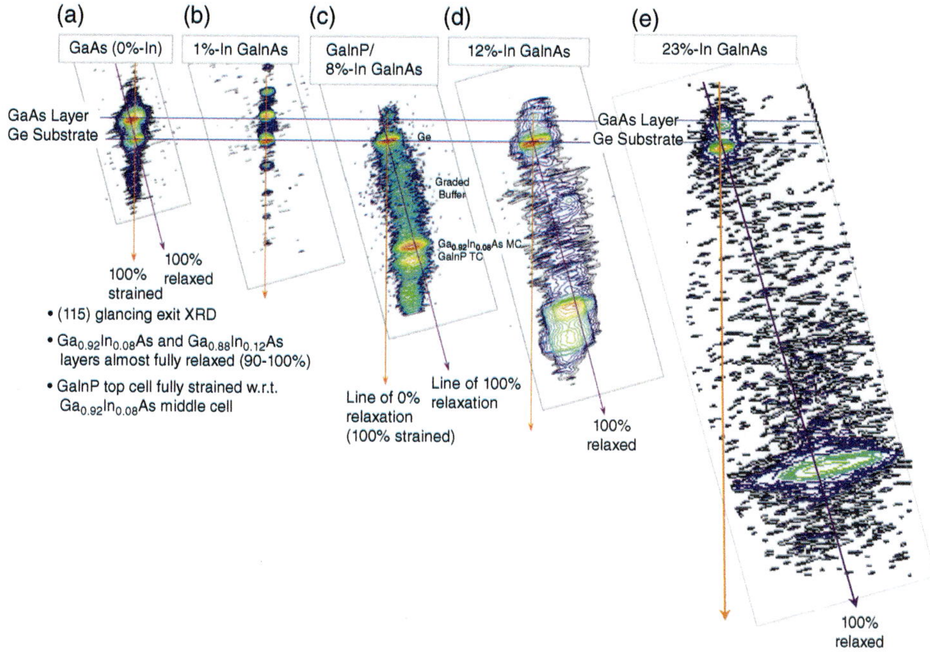

Figure 5.7 *Glancing exit 115 reciprocal space map (RSM) of various compositions of GaInAs subcells grown on Ge substrates: (a) GaAs, (b) 1%-In GaInAs, (c) 8%-In GaInAs, (d) 12%-In GaInAs, (e) 23%-In GaInAs.*

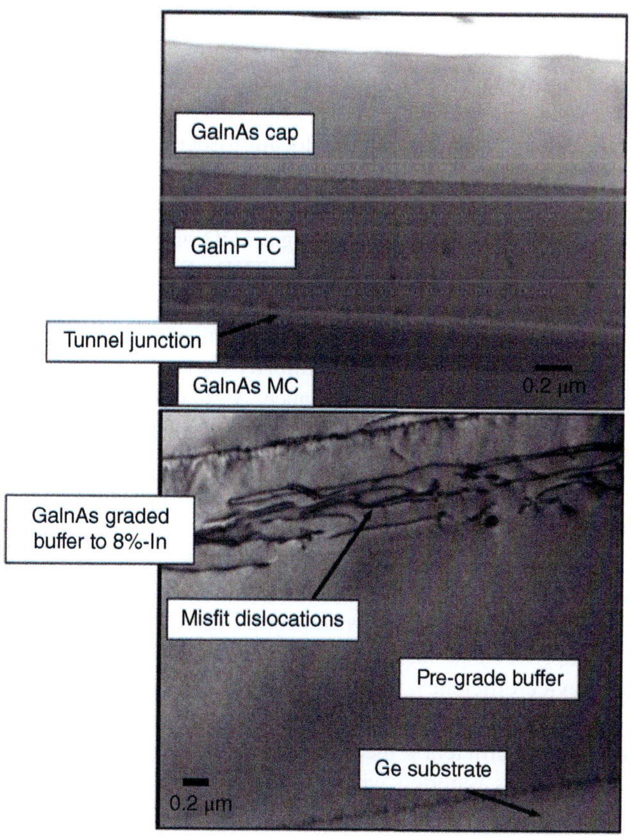

Figure 5.8 *Cross-sectional transmission electron microscopic (TEM) image of a UMM3J device with 8%-In GaInAs middle cell. Two different images are shown, with the upper layers (cap, top cell, and middle cell emitter) positioned above the image of the same region in the graded buffer layer. Note that the observable defects are confined within the buffering region of the device.*

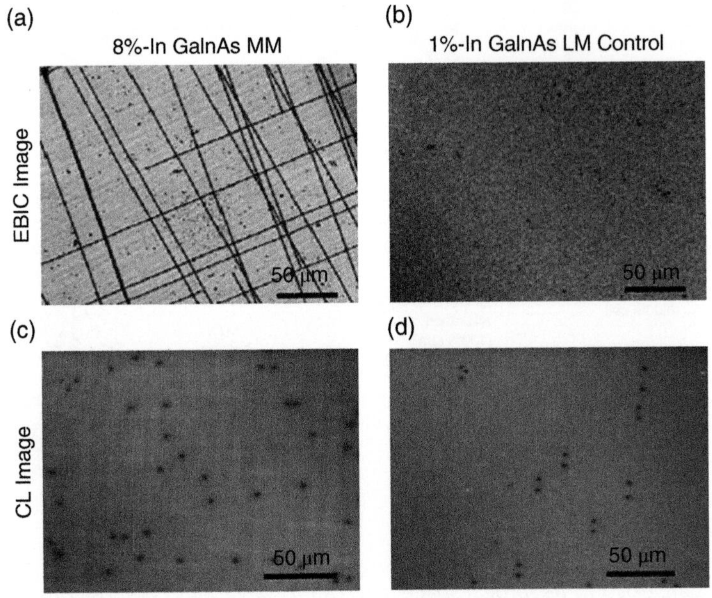

Figure 5.9 *Plan-view electron beam induced current (EBIC) and cathodoluminescence (CL) of lattice-matched (LM) and UMM solar cells, LM (1%-InGaAs 3J) and UMM (8%-In GaInAs 3J), showing threading dislocation density (TDD) of 1e5 cm^{-2} to 2e5 cm^{-2} residual in the middle subcell layers.*

metamorphic subcells (Figures 5.9a and 5.9c) and lattice-matched subcells (Figures 5.9b and 5.9d). Threading dislocations show as dark spots of nonradiative recombination in these images and indicate a near-equal density of threading dislocation segments between the two subcells of TDD of 1 to 2×10^5 cm^{-2} extending into the active regions of the device, similar to the lattice-matched epitaxial layers of Figures 5.9b and 5.9d. As the mismatch increases, the TDD also increases, starting at 8%-In GaInAs subcells showing a TDD of 1×10^5 cm^{-2}, 17%-In GaInAs with a TDD of 1×10^6 cm^{-2}, and 23%-In GaInAs reaching a TDD of 4.4×10^6 cm^{-2} [28, 42, 43].

Figure 5.10 shows the internal quantum efficiency of the series of GaInAs subcells from Figure 5.7 with the addition of one extra at 37%-In GaInAs at 2.4% mismatch. Internal quantum efficiency (IQE) in solar cells is a good measure of the minority carrier lifetime and hence the impact of threading dislocations on the solar-cell performance. All GaInAs devices with base thicknesses of 3 µm reach 100% IQE, indicating that GaInAs subcells exhibit layer quality approaching that of lattice-matched devices. The highest-mismatch device also shows some degradation indicating nonradiative recombination within the cell for 37% indium content on Ge, a mismatch of 2.4%, which appears to be the present upper limit of the state-of-the-art in material quality for such devices. At 1.6% mismatch, the bandgap of 1.1 eV is sufficient to enable 4 J UMM designs in Figure 5.5b. As discussed earlier, the bandgap voltage offset (W_{oc}) for most metamorphic materials reaches limits similar to those of lattice-matched materials, as shown in Figure 5.4 [48, 73].

For all the epitaxial improvements discussed within this chapter, the main objective for UMM solar cells is the final device efficiency. Based on the demonstration of quality CGB layers and subcells, UMM devices have increased significantly in efficiency over the past years. Figure 5.11 shows the historical trend in efficiency measured for AM0 (space) and concentrated AM1.5d (terrestrial concentrator) results reported within the literature for upright metamorphic results. Early 2 J results were quickly superseded by 3 J designs [74]. The highest reported UMM space devices are large-area (73.5 cm^2) 31.5% efficiency averages and a peak of 31.8% AM0 efficiency

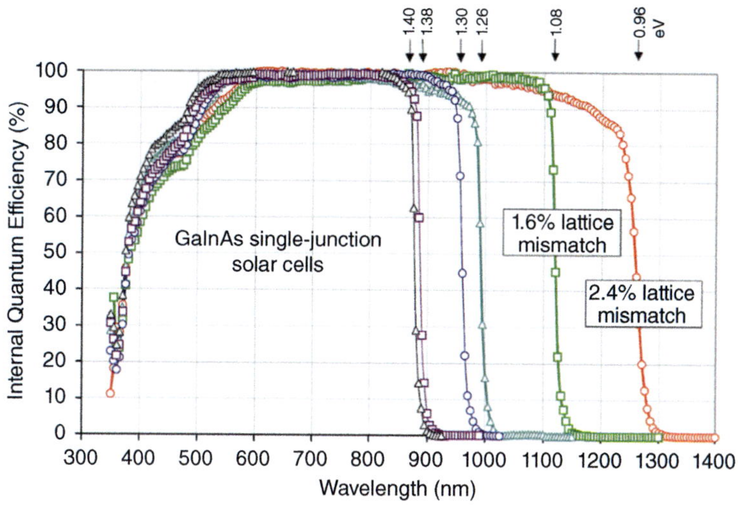

Figure 5.10 *Internal quantum efficiency (IQE) of the series of GaInAs subcells from Figure 5.3 ranging up to 23%-In GaInAs (1.6% mismatch) and as far as 37%-In GaInAs at 2.4% mismatch.*

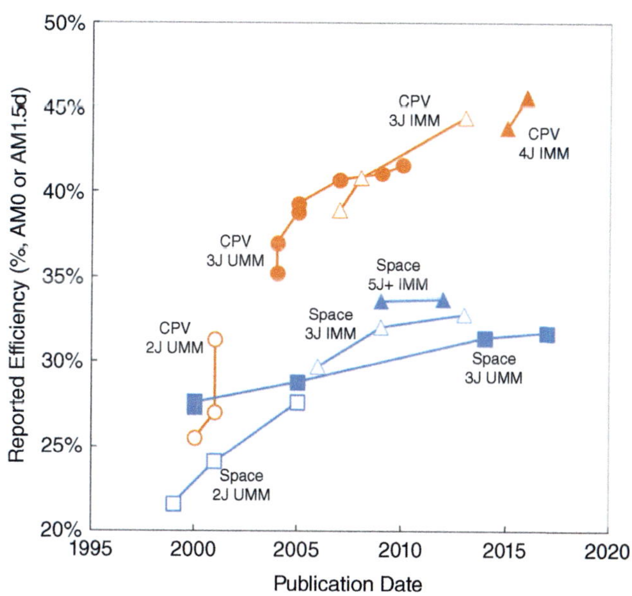

Figure 5.11 *Historical record UMM device efficiency at AM0 (space) and AM1.5d-low AOD concentrated terrestrial spectra.*

for UMM3J devices [36]. The most recent development efforts are exploring UMM4J (Figure 5.5b), have demonstrated 30.7% AM0 efficiency, and are anticipated to exceed the 3J results due to the more optimal utilization of the AM0 solar spectrum [35, 75]. Terrestrial concentrator solar cells fabricated using 3J UMM design reached reported efficiencies 41.6% at 364-sun concentration [73]. There has not been a result of 4J UMM devices for CPV as of yet. Future devices will likely exceed these reports as work continues toward these promising devices as the science of UMM growth by MOVPE advances.

5.2.4 Challenges and Future Outlook

Upright metamorphic multijunction solar cells still have a number of challenges to adoption. UMM4J devices still need to improve their overall efficiency to reach the anticipated 32% AM0 efficiency. Key to reaching that goal is the incorporation of Al in the top AlGaInP subcell of 1.9 eV at a lattice-mismatch of 2.4%. The higher Al content adds substantial oxygen background and has a deleterious impact on the minority carrier lifetime. Such a requirement challenges continual low-oxygen backgrounds within the epitaxial reactors and within metal-organic precursors. As described earlier, another challenge faced by UMM solar cells includes the transparency and peak tunneling current of the interconnecting tunnel junctions. Beyond these epitaxial questions, there continues to be a cost challenge with UMM devices. The UMM solar-cell architecture increases efficiency by inclusion of additional growth layers (such as the CGB layer) not present in lattice-matched designs. For significant market adoption, UMM solar cells require additional work to offset these increased epitaxial costs above the lattice-matched counterparts.

UMM solar cells are a strong potential pathway to achieving satellites' increased power demands beyond the present standard lattice-matched 3 J devices. Upper limitations of lattice-matched solar-cell efficiency is driving the space solar cell market to plan on the power that UMM solar cells provide. Significant market adoption of UMM solar cells appears imminent based on the benefits of increased power, standard cell fabrication, and similarity to heritage lattice-matched devices.

5.3 Inverted Metamorphic Multijunction (IMM) Solar Cells

5.3.1 Introduction and History of Inverted Metamorphic Multijunctions

The UMM architecture described in the previous section potentially suffers from two significant problems. First, because the CGB layer is grown first, both of the upper subcells are lattice-mismatched, with inherently lower material quality compared with lattice-matched versions, as measured by the threading dislocation density. Second, and more important, the bottom subcell is constrained to the Ge bandgap, which limits the design flexibility of the architecture. For a triple-junction cell, the Ge bottom subcell is highly overdriven, generating nearly twice as much photocurrent as the two upper subcells, and contributes a relatively low voltage to the total MJ cell structure. Furthermore, the generally poor quality of Ge solar cells may eventually limit the performance of the device.

The inverted metamorphic multijunction (IMM) solar cell was proposed as a promising architecture to overcome these two challenges [76]. The first modeled result projected a 31.1% efficiency for a triple junction under the global reference spectrum at 1000 W/m^2. The device is grown inverted, with the highest-bandgap upper subcell grown first and the lowest-bandgap bottom cell grown last; the back contact layer is the final layer in the growth. A schematic is shown in Figure 5.12 [77]. A major advantage over the UMM is that inverted growth provides direct access to the back surface of the device, allowing for the implementation of a reflector and/or texturing, which improves photon capture and recycling. The epilayer structure is then reoriented during post-growth processing. First, the back electrical contact is deposited onto the back contact layer and annealed if necessary. Then the epilayer structure is bonded to a secondary handle for mechanical support. The substrate is removed, either by wet chemical etching or by a technique such as epitaxial liftoff (ELO), wherein the substrate (which can cost over \$10 000/m^2) can be reused for another growth, providing another major advantage over UMM. The front metallization is then deposited, the individual devices are isolated, and finally an anti-reflection coating is deposited on the front surface.

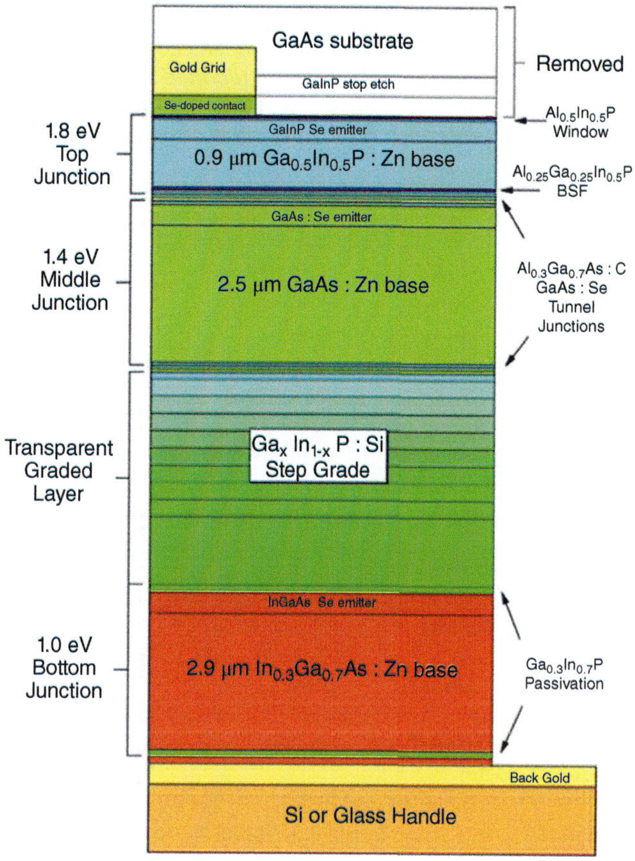

Figure 5.12 Schematic of the first IMM, a GaInP/GaAs/GaInAs 3J cell. The growth direction is from the top to the bottom of the figure. Additional processing layers are shown at the top. Source: Reprinted from Geisz et al. 2007 [77], with the permission of AIP Publishing.

The first demonstrated IMM cell, a triple junction, consisted of lattice-matched GaInP and GaAs subcells, followed by a transparent $Ga_xIn_{1-x}P$ CGB to change the lattice constant, and then a lattice-mismatched, 1.0-eV GaInAs third junction [77]. Growing the device in this way, the highest power-producing GaInP upper cell is grown lattice-matched to the GaAs substrate so that it has the highest possible material quality, with the dislocations that stem from the CGB confined to the lower-power-producing bottom junction(s). More important, the bandgap of the mismatched subcell(s) is unconstrained except by the availability of alloys that are suitable for a CGB and a subcell. Given that there are numerous ternary and quaternary alloy combinations attainable in the III-V materials palette, the cell designer is free to adjust the CGB to achieve whatever lattice constant, and therefore bandgap, is called for by the application. Table 5.1 lists some of the more notable IMM cells demonstrated in the laboratory, as well as the efficiencies.

The extra processing between growth and the final device associated with the IMM cell increases production costs over the UMM cell, but liftoff of the device from the substrate conveys significant advantages to the device. The lifted-off device is thin and needs to be bonded to a secondary handle to maintain structural integrity, but the handle need not be stiff and heavy like the original substrate. Rather, the handle can be flexible and lightweight, made of materials such as plastics. As noted earlier, access to the back surface allows application of a back surface reflector, which improves

Table 5.1 *Notable IMM cells demonstrated in the laboratory. The results are grouped by application. In the second column, the alloys in boldface are grown lattice-mismatched with respect to the substrate. The efficiencies of all space solar cells are reported at one-sun AM0, while those for terrestrial CPV cells are at concentration. One-sun versions of some of these cells have also been demonstrated, but since the main application is for concentrator cells, the efficiencies are listed at concentration.*

Year	Bandgaps(eV)-alloys	Manufacturer	Efficiency	Application	Ref.
2006	1.8-GaInP/1.4-GaAs/**1.0-GaInAs**	NREL	29.7%	Space (AM0)	[78]
2009	1.9-GaInP/1.4-GaAs/**1.0-GaInAs**	SolAero	32.0%	Space	[79]
2013	1.8-GaInP/1.4-GaAs/**1.0-GaInAs**	Sharp	31.5%	Space	[80]
2010	1.9-GaInP/1.4-GaAs/**1.0-GaInAs**	Spectrolab	32.8%	Space	[2]
2009	1.91-GaInP/1.42-GaAs/**1.02-GaInAs/0.70-GaInAs**	SolAero	33.6%	Space	[81]
2012	2.1-AlGaInP/1.7-AlGaAs/1.4-GaAs/**1.1-GaInAs/0.9-GaInAs/0.7-GaInAs**	SolAero	33.7%	Space	[82]
2007	1.8-GaInP/1.4-GaAs/**1.0-GaInAs**	NREL	38.9% (81×)	CPV (AM1.5d)	[77]
2008	1.83-GaInP/**1.34-GaInAs/0.89-GaInAs**	NREL	40.8% (326×)	CPV	[83]
2013	1.8-GaInP/1.4-GaAs/**1.0-GaInAs**	Sharp Solar	44.4% (302×)	CPV	[80]
2015	1.8-GaInP/1.4-GaAs/**1.0-GaInAs/0.74-GaInAs**	NREL	43.8% (327×)	CPV	[84]
2016	1.8-GaInP/1.4-GaAs/**1.0-GaInAs/0.72-GaInAs (+InAsP grade)**	NREL	45.6% (690×)	CPV	[85]

photon recycling and permits thinning of the bottom cell to further boost V_{OC}. Thus, the IMM cell provides unmatched specific power (W/kg), and the extra cost can be justified for applications where minimizing weight is paramount, such as the curved wing of an unmanned aerial vehicle and solar blankets for space applications. For this latter reason, some space photovoltaic manufacturers have been developing 3J and 4J IMM cells as alternatives to the conventional GaInP/InGaAs/Ge cell that presently powers most satellites.

5.3.2 MOVPE Growth Considerations of IMM

In growing the IMM, consideration must be given to the inverted architecture and need to accommodate the more complicated post-growth processing. One cannot simply take a recipe for a UMM device and grow the layers in reverse order, because memory effects in the reactor and the diffusion of dopants can change the structural profile in unexpected ways [86]. The CGB layers also need to be transparent enough to pass the lower-energy light, so as not to reduce the photocurrent generated in the lower-bandgap cells. Here we describe some aspects that need to be considered in the design of the MOVPE growth.

5.3.2.1 Front Contact Layer

Unlike upright devices such as the UMM, where the front contact layer is grown last, in the IMM growth the front contact layer is grown relatively early in the growth and is therefore subject to a lengthy high-temperature anneal over the course of the remainder of the growth. The diffusion of n-type dopants away from the front of this layer can lead to difficulties making ohmic contact during processing. One solution may be to use an alloyed contact like Pd/Ge/Ag and anneal the metals after electron-beam deposition, thus forming an ohmic junction with the higher concentration of dopants deeper in the contact layer. However, by this stage of processing, the substrate will have already been removed and the semiconductor layers secured to the secondary handle, and the sample can be annealed only if the semiconductor–handle bond can withstand the high temperature. Another solution [86] is to grow a thin contact layer of GaInAsN with ~1–4% nitrogen. The lower-bandgap alloy can be more heavily doped; but, more important, the affinity between the nitrogen and the selenium n-type dopant is very high, and the selenium diffusion is considerably reduced.

5.3.2.2 *Tunnel Junctions*

The tunnel junctions in the IMM are grown in the opposite orientation when compared to an upright-grown device. Interconnecting the p-type back layers of one junction with the n-type front layers of the next junction, the p-type layer is grown first, followed by the n-type layer in an IMM consisting of n-on-p cells. This orientation generally reduces the influence of memory effects that are more prominent for dopants like selenium and tellurium than for carbon. Thus, the compensation on the p-side by the residual n-type selenium is avoided, and high carrier concentrations can be obtained in both TJ layers. There is a need, however, to minimize the diffusion of selenium away from the tunnel junction during the remainder of the lengthy growth. A moderately doped n-type AlGaAs diffusion barrier can be effective here, because the diffusivity of selenium in AlGaAs is lower than in GaAs.

5.3.2.3 *Compositionally Graded Buffer Layers*

The glide dynamics of the compositionally graded buffer layers in the IMM are similar to the UMM, but the mismatch of the desired materials is typically much higher in the IMM. Whereas optimal UMM designs rely on a GaInAs cell with a 1.2 eV bandgap, the third junction bandgap of an IMM with GaInP/GaAs top cells is optimized at 1.0 eV. GaInAs cells with 1.2 and 1.0 eV bandgaps have a mismatch relative to GaAs of ~1% and ~2%, respectively. Thus, the 1.0 eV cell requires that double the mismatch strain be relieved, meaning more care has to be taken in the design and growth of the CGB. A thicker CGB is required, since it is not desirable to increase the strain grading rate (%/μm), as this typically leads to increased TDD [51]. The extra thickness increases the growth time, which can potentially exacerbate issues such as dopant diffusion out of the tunnel junctions and lead to increased roughness.

At the National Renewable Energy Laboratory (NREL), a GaInP CGB is utilized to bridge the gap between GaAs and 1.0 eV GaInAs because a low TDD can be maintained in the grade, and the bandgap of this alloy in the desired composition range is well above that of the GaAs cell, avoiding parasitic absorption in the grade [77]. This CGB is grown with a 1%/μm strain grading rate, which yields TDD ~$1 \times 10^6\,cm^{-2}$ [87]. Despite the tendency for mismatched GaInP to phase separate at compositions far from either binary [53], a high-quality grade can be maintained under the proper growth conditions. Through the use of moderately high growth temperatures (which promote dislocation glide) and a high P/III ratio, phase separation can be suppressed, and the surface roughness and TDD can be minimized [87]. The choice of substrate is critical, as mismatched GaInP grown on GaAs miscut toward (111)A exhibits dual variant ordering and a high degree of phase separation. Substrates miscut toward (111)B promote single-variant CuPt ordering, which is believed to help stabilize the material and mitigate phase separation [88, 89].

Ordering provides further benefit through its effects on dislocation glide dynamics. Ordering is metastable, and glide of certain dislocations disorders the material, reducing the energy of the system. Thus, an extra force, equivalent to the energy released by disorder, is placed on these

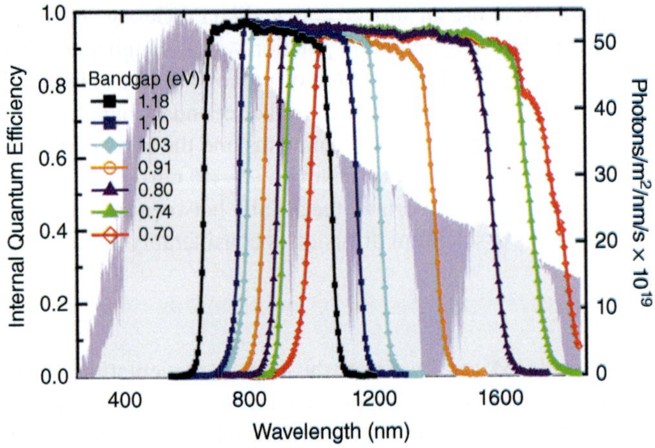

Figure 5.13 *Internal quantum efficiencies for single junction GaInAs cells on CGB layers. The direct solar spectrum appears in gray in the background. Source: France et al. 2016 [85], © IEEE; reprinted with permission.*

dislocations, which increases dislocation glide lengths [90, 91]. In this manner, the crystal grower can select for dislocations on specific glide planes through selection of growth conditions that promote (or inhibit) ordering. Dislocation populations throughout the grade can then be controlled, preventing undesirable dislocation interactions or nucleation of new dislocations near active regions of the device [92]. Through careful design, high-quality GaInP CGBs all the way to InP can be grown with TDD $<3\times10^{6}$ cm^{-2}, as shown in Figure 5.13 [88]. This permits the addition of a fourth junction containing 0.74 eV GaInAs, which has the same lattice constant as InP, to the IMM structure for even higher performance. Efficiency near 46% has been demonstrated with a four-junction IMM CPV structure with two $Ga_xIn_{1-x}P$ CGB stages [85].

$Al_xGa_yIn_{1-x-y}As$ is another common CGB material that also has the requisite transparency for the IMM device [93]. This materials system can be useful when growth on substrates miscut toward (111)A is desired, because the material is less prone to phase separation, at least to $Al_xGa_yIn_{0.30}As$, near the 1.0-eV GaInAs composition. Phase separation can be an issue past this composition, but a combined $Al_xGa_yIn_{1-x-y}As/Ga_xIn_{1-x}P$/InP grade, which employs compositional ranges in each material that are less prone to phase separation, can be employed to reach the 0.74-eV bandgap [94].

The relationship between the CGB and some of the tunnel junctions warrants some consideration. It would seem obvious that a lattice-matched TJ is preferable to a lattice-mismatched (LMM) tunnel junction, and indeed the 3 J IMMs demonstrated by NREL, Sharp, Spectrolab and Emcore (now SolAero) all grow the second tunnel junction immediately after the GaAs second junction but before the CGB. In more-advanced IMM cells with two CGBs, at least one of the TJs must be mismatched [95].

5.3.2.4 Annealing During Growth

Since the high-power-producing upper junctions are grown first in the IMM, there was originally some concern that the thermal annealing during the remainder of the growth would degrade the material quality of those cells. Although we typically find, at a very general level, that the optimum growth temperature for various III-V alloys decreases with decreasing bandgap (e.g. GaInP is grown at >700 °C, GaAs at ~650 °C, GaInAs on InP at ~620 °C), these are still high temperatures where annealing can occur. Moreover, the dislocation glide dynamics in the CGB improve with temperature. In spite of the high temperature and lengthy growth, there has been no compelling evidence of any significant material degradation in the IMM. However, diffusion of dopants may be a concern because of the potential to change the relative carrier concentrations in adjacent layers, for example, and that needs to be considered in the design process.

5.3.3 Growth and Device Results

As noted earlier, the first demonstrated IMM device was grown at NREL and consisted of GaInP and GaAs lattice-matched junctions, followed by a CGB and then a 1.0-eV GaInAs junction. The mismatch was ~1.9% with respect to the substrate, and the threading dislocation density in the $Ga_{0.7}In_{0.3}As$ layer was ~10^6 cm^{-2} as measured by CL and EBIC techniques. The CGB consisted of eight 0.25-μm $Ga_xIn_{1-x}P$ layers, varying in composition from $Ga_{0.51}In_{0.49}P$ ($E_g = 1.81$ eV) to $Ga_{0.22}In_{0.78}P$ ($E_g = 1.52$ eV), followed by a step-back to $Ga_{0.25}In_{0.75}P$. Because all of the layer alloys have bandgaps above 1.42 eV, the grade is transparent to incident light below the bandgap of the GaAs middle junction, which is a key design criterion for a high-performing multijunction solar cell. This first demonstration of an IMM cell achieved an efficiency of 38.9% at a relatively low concentration of 81 suns [77].

Using GaAs for the second junction has the advantage of high material quality, but the bandgap is higher than optimal. The IMM performance was significantly improved by the inclusion of two CGBs: one to enable a second junction at 1.34 eV ($Ga_{0.96}In_{0.04}As$ with 0.3% mismatch), and a second CGB to enable a bottom junction of 0.89 eV ($Ga_{0.63}In_{0.37}As$). This design led to an improved efficiency of 40.8% at 326 suns [83].

Figure 5.14 shows an ion beam image of the cross-section [83], along with TEM dark-field images. The dislocations are largely confined to the two CGB layers, with good material quality in the active layers. The threading dislocation densities were measured by CL to be approximately 1×10^5 cm^{-2} in the first AlGaInP CGB and $2–3 \times 10^6$ cm^{-2} in the second GaInP CGB. This structure successfully demonstrated that two CGBs could be incorporated into a single device, which paved the way to IMM cells with four or more junctions.

3 J IMM cells have also been developed for space applications by companies including Boeing-Spectrolab of Sylmar, California; SolAero (formerly Emcore Photovoltaics) of Albuquerque, New Mexico; Sharp Solar of Japan; Microlink Devices of Niles, Illinois; and others. As of this writing, several IMM-based modules have flown in space for qualification tests.

The major elements of the space IMM architecture are the same as the terrestrial cells, although the bandgaps and junction thicknesses are tuned for the AM0 spectrum, which is more blue-rich and does not have any water or CO_2 absorption bands. The cells tend to be larger, ranging from 1–80 cm^2, and the front metallization is designed for 1367 W/m^2 rather than a terrestrial concentrator incidence of ~1 MW/m^2 at 1000×. Particular attention is also paid to the radiation hardness of the individual subcells, and thus some care must be exercised in interpreting and comparing space cell results. Ionizing radiation tends to degrade the cell performance by reducing the minority carrier diffusion lengths. Cells can be *radiation hardened* by employing strategies that reduce the effects of a decreasing diffusion length. The cells are designed so that the efficiency at the end of life (EOL) still meets some specified operational performance level, but the mitigation strategy may also reduce the beginning of life (BOL) efficiency. For example, the GaAs junction of a space cell might be thinned to improve the radiation hardness. This will lower the BOL efficiency, which would then seem to indicate a lower-performing cell, even though the EOL efficiency remains relatively high and the cell is more commercially valuable.

The first space IMM cell was demonstrated in 2006 with a GaInP/GaAs/GaInAs device [78] and an efficiency of 29.7%. This was followed by a higher-efficiency IMM cell by Emcore (now SolAero) [79], for a 4-cm^2 GaInP/GaAs/GaInAs cell with 32% efficiency. The GaInP junction was fully disordered to give a bandgap of 1.91 eV, more suitable to the AM0 spectrum. Spectrolab followed this

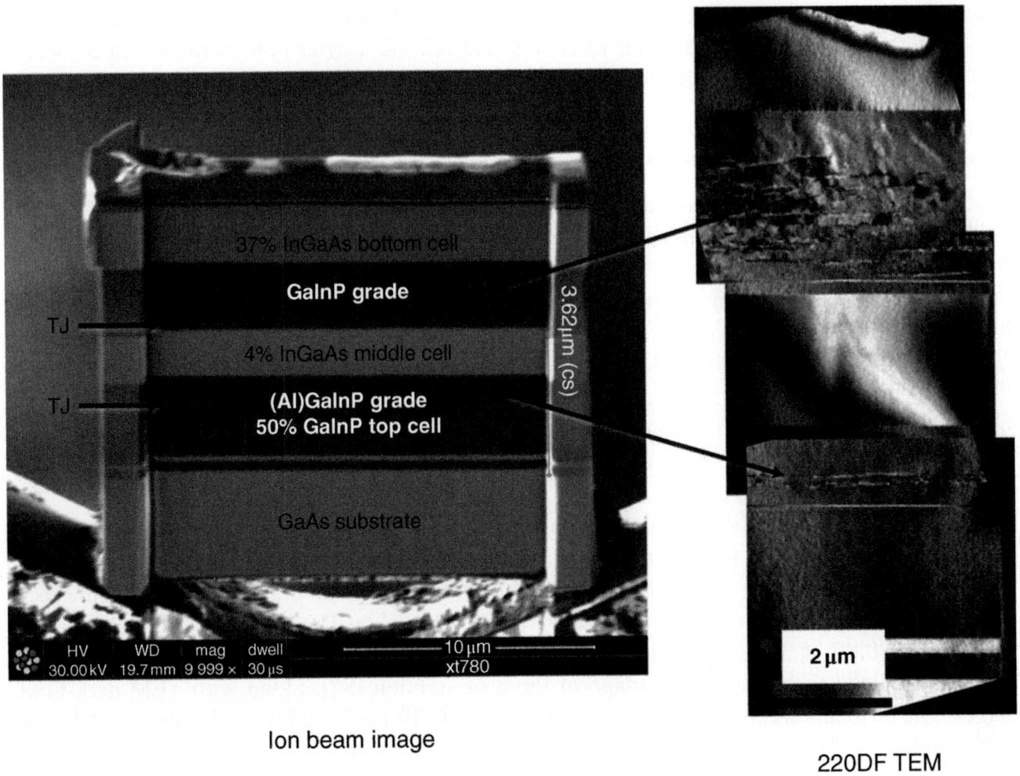

Figure 5.14 Ion beam image and composite 220 dark-field TEM of a FIB cross section of an unprocessed 3J IMM with two CGB layers. Source: Reprinted from Geisz et al. 2008 [83] with the permission of AIP Publishing.

with a 1 cm^2, 32.8% result [9], and Sharp Solar reported a 27 cm^2 with 31.5% BOL efficiency [80]. Sharp also demonstrated a terrestrial concentrator version of the 3J-IMM with an efficiency of 44.4% at 302 suns [80]. This result remains the highest 3J IMM efficiency as of this writing.

Building on the 3J IMM, various manufacturers and laboratories demonstrated a series of 4J devices. The top two junctions included GaInP and GaAs, both lattice-matched to the substrate, followed by a mismatched 1.0-eV $Ga_{0.7}In_{0.3}As$ third junction and a mismatched 0.74-eV $Ga_{0.47}In_{0.53}As$ fourth junction. The second CGB graded the GaInP alloy all the way out to InP, a 3.8% mismatch with respect to the GaAs substrate [84, 88, 96]. This cell demonstrated a peak efficiency of 43.8 ± 2.2 % at 327 suns, with a slightly lower efficiency of 42.9% at the higher concentration of 869 suns [84].

With the same lattice constant as InP, the 0.74-eV $Ga_{0.47}In_{0.53}As$ bottom cell is the lowest-bandgap GaInAs junction that can be grown on a GaInP CGB. The terrestrial solar spectrum does not show a water absorption band until ~1800 nm (0.68 eV), however, so additional performance can be gained by lowering the bottom junction bandgap to generate more current. The CGB can be extended to longer lattice constants by grading with InAsP alloys, as was demonstrated by France et al. in 2016 [85]. The right-most red curve in Figure 5.13 shows an example of a 0.7-eV GaInAs cell on such an extended CGB. The 4J cell using this extended CGB showed an efficiency of 45.6 ± 2.3 % at 690 suns.

For space applications, SolAero presented initial results on a 4J-IMM as early as 2009 with a 33.6% (BOL) [81], 4-cm^2 cell, followed by a 33.9% (BOL) cell in 2010 [97], and a 34.5% (BOL) cell in 2012 [98]. This last cell was not radiation hardened, however.

5.3.4 Challenges and Future Outlook

Increasing efficiency will be key to increased adoption of the IMM cell, and the most straightforward way to do so is to add junctions. Five- and six-junction IMM cells are under development for both terrestrial and space applications. The wavelength range of 350–900 nm spanned by the GaInP and GaAs junctions in the 3 J and 4 J designs are spanned by three junctions in a 5+ junction cell. AlGaInP, with bandgap flexibility of 1.9–2.2 eV, is an obvious choice for the top ~2.1 eV junction, although the potential incorporation of oxygen deep-level traps in the aluminum-containing alloy is a drawback that must be overcome [99–101]. AlGaInP also suffers from the problem that the electron mobility decreases with increasing Al composition, which can lead to an unacceptably high emitter sheet resistance for terrestrial concentrator applications. Increasing the carrier concentration with heavy doping may reduce the sheet resistance but also tends to reduce the minority hole diffusion length, and therefore the photocurrent generation in the emitter. Incorporating a lower-bandgap GaInP emitter is a potential solution that leads to higher emitter photoresponse and lower sheet resistance, but with some loss of voltage.

$Al_{0.23}Ga_{0.77}As$ and $Ga_{0.68}In_{0.32}As_{0.34}P_{0.66}$ [101, 102] are good candidates for the second, ~1.7-eV junction, and GaAs remains the clear choice for the third junction. A study by Heckelman et al. [101] reported on the development of $Al_xGa_{1-x}As$ cells with aluminum compositions as high as 37%. As the second junction, the current is mostly vertical and lateral spreading is largely insignificant, so the low electron mobility in the n-type Al-containing emitter is not an issue here. Indeed, the emitter of the AlGaAs junction can be thinned to a few tens of nanometers to minimize the effects of nonradiative recombination at donor-complex (DX) centers [101]. GaInAsP is another possible choice for the second junction [102]. Compositional control is more challenging because of the nonlinear, temperature-dependent Langmuir adsorption on the Group-V sublattice. The alloy also has a propensity toward phase separation, which can be mitigated with a lower temperature growth that kinetically limits the adatom diffusion along the growth surface.

As in the 4 J cell, the bottom two or three junctions can be fabricated from a stepwise series of GaInAs alloys on CGBs. Another possibility is to design a single CGB layer all the way to the lattice constant of InP (or slightly beyond) and then grow the bottom junctions from various GaInAsP alloys. The composition of the quaternary alloy allows an additional degree of freedom compared to a ternary, and the bandgap can be tuned independently of the lattice constant. Thus, alloys of approximately $Ga_{0.09}In_{0.91}As_{0.18}P_{0.82}$ (1.2 eV), $Ga_{0.23}In_{0.77}As_{0.48}P_{0.52}$ (1.0 eV), and $Ga_{0.47}In_{0.53}As$ (0.74 eV) can all be grown at a fixed lattice constant. Growth of these alloys on a rough, mismatched surface will be a substantial challenge, however, due to surface-driven phase separation. It is anticipated that a 6 J cell should be able to demonstrate an efficiency >50%, at concentrations near 1000 suns [103]. 7+ junction designs, which could potentially incorporate a GaSb junction, are further away from development.

SolAero reported a 6 J IMM for space applications, with an initial result of 33.7% for a 4-cm^2 cell [82]. The lattice-matched upper three subcells consisted of AlGaInP/AlGaAs/GaAs. The lower portion consisted of three CGB layers with three distinct GaInAs junctions. It is anticipated that improvements and optimization will lead to efficiencies exceeding 37%.

5.4 Conclusions

MOVPE has been the preferred growth technique for high-volume manufacturing of MJ solar cells for space, near space, and terrestrial applications. The UMM solar-cell designs provide a near-term opportunity for improved performance efficiency of MJ solar cells, as UMM is a direct drop-in replacement for the LM structure currently in high-volume production. Improved utilization of the solar spectrum for UMM cell designs and optimizing the epitaxial conditions has aided the growth of high-quality alloys for 3 J and 4 J UMM solar cells. Today's UMM subcell materials reach minority carrier lifetime properties comparable to those of lattice-matched subcells. Present designs routinely reach 32% efficiency for large-area space solar cells and in excess of 41.6% efficiency for terrestrial concentrator designs.

The IMM cell provides an alternative to the UMM cell in which the highest-power-producing junctions are lattice-matched to the substrate, ensuring those junctions achieve maximum performance. In order to achieve that, a greater degree of strain must be accommodated, meaning CGBs are thicker and more complicated, and post-growth processing is more complicated due to the inverted structure and the liftoff processes employed. These considerations add to cost, but with that cost come enabling opportunities and unmatched performance efficiency. Furthermore, inverted processing allows bonding of devices to lightweight, flexible handles, making the IMM ideal for applications on nonflat surfaces and an enabler where a high power-to-weight ratio is required, e.g. UAVs and space satellites. IMM space solar cells have already demonstrated 33% and 34% for 3J and 4J devices, and even higher efficiencies are expected for advanced IMM structures. A record terrestrial CPV efficiency of 46% under concentration has already been demonstrated with a 4J IMM device, and 5J and 6J device designs offer the potential to eclipse the 50% barrier in short order. The challenge for IMM is to bring the manufacturing cost down using streamlined and automated growth and fabrication processes.

References

1. Karam, N.H., King, R.R., Haddad, M. et al. (1999). Recent developments in high efficiency Ga0.5In0.5P/GaAs/Ge dual- and triple-junction solar cells: steps to next generation PV cells. *11th International Photovoltaic Science and Engineering Conference*, Sapporo City, Hokkaido, Japan.
2. Boisvert, J., Law, D., King, R. et al. (2010). Development of advanced space solar cells at Spectrolab. *35th IEEE Photovoltaic Specialist Conference*, Honolulu, HI.
3. King, R.R., Fetzer, C.M., Colter. P.C. et al. (2002). High-efficiency space and terrestrial multijunction solar cells through bandgap control in cell structures. *29th IEEE Photovoltaic Specialist Conference*, New Orleans, LA.
4. Edmondson, K.M., Law, D.C., Glenn, G. et al. (2006). Flexible III-V multijunction solar blanket. *4th World Conf. on Photovoltaic Energy Conversion*, Waikoloa, HI.
5. Adams, J.G., Elarde, V.C., Hains, A. et al.(2013). Demonstration of multiple substrate reuses for inverted metamorphic solar cells. *IEEE Journal of Photovoltaics* 3: 899–903.
6. Tatavari, R., Wibowo, A., Elarde, V. et al. (2011). Large-area, epitaxial lift-off, inverted metamoprhic solar cells. *37th Photovoltaic Specialist Conference*, Seattle, WA.
7. Karam, N.H. (2014). Advancements in high efficiency multijunction solar cells for low-cost power generation. *6th World Conference on Photovoltaic Energy Conversion*, Kyoto, Japan.
8. Karam, N.H., Sherif, R.A., and King, R.R. (2007). Multijunction concentrator solar cells: an enabler for low-cost photovoltaic systems. In: *Springer Series in Optical Sciences*, vol. 130. (eds. A. Luque and V. Andreev), 199–220. Berlin: Springer-Verlag.
9. Boisvert, J., Law, D.C., King, R.R. et al.(2010). Development of advanced inverted metamoprhic space solar cells at Spectrolab. *35th IEEE Photovoltaic Specialist Conference*, Honolulu, HI.
10. Cotal. H., Law, D., Karam, N.H. et al. (2010). *Recent Development in High Efficiency Multijunction Solar Cell Technology* (eds. Badescu and Paulescu). New York, Nova Science.
11. Karam, N.H. et al. (2007). Metamorphic concentrator solar cells with over 40% conversion efficiency. *4th International Conference on Solar Concentrators (ICSC-4)*, El Escorial, Spain.
12. King, R.R., Fetzer, C.M., Colter. P.C. et al. (2002). High efficiency space and terrestrial multijunction solar cells through bandgap control in cell structures. *29th IEEE Photovoltaic Specialist Conference*, New Orleans, LA.
13. Bedair, S.M., Lamorte, M.F., and Hauser, J.R. (1979). A two-junction cascade solar-cell structure. *Applied Physics Letters* 34 (1): 38–39.
14. AIX G5 HT-TM MOCVD System (2017). http://www.semiconductor-today.com/news_items/2018/jul/aixtron_190718.shtml.
15. TuboDisk k475i As/P MOCVD System. (2017). https://www.veeco.com/products/turbodisc-k475i-asp-mocvd-system.
16. Bergunde, T., Dauelsberg, M., Kadinski. L. et al.(1997). Heat transfer and mass transport in a multiwafer MOVPE reactor: modelling and experimental studies. *Journal of Crystal Growth* 170: 66–71.

17. Thompson, A.G., Stall, R.A., Kroll, W. et al.(1997). Large scale manufacturing of compound semiconductors by MOVPE. *Journal of Crystal Growth* 170: 92–96.
18. Stoney, G.G. (1909). The tension of metallic films deposited by electrolysis. *Proceedings of the Royal. Society of London, Series A* 82: 172–175.
19. Freund, L.B. and Suresh, S. (2004). *Thin Film Materials; Stress, Defect Formation and Surface Evolution*. Cambridge, U.K.: Cambridge University Press.
20. Fetzer, C.M., Liu, X.Q., Chang, J. et al. (2012). Progress in large area organometallic vapor phase epitaxy for III-V multijunction photovoltaics. *Journal of Crystal Growth* 352: 181–185.
21. Ludowise, M.J., Dietz, W.T., Boettcher, R. et al. (1983). High efficiency (21.4%) $Ga_{0.75}In_{0.25}As$/GaAs(E_g = 1.15 eV) concentrator solar cells and the influence of mismatch on performance. *Applied Physics Letters* 43: 468–470.
22. Yamaguchi, M. and Amano, C. (1985). Efficiency calculations of thin-film GaAs solar cells on Si substrates. *Journal of Applied Physics* 58: 3601.
23. Lewis, C.R., Ford, C.W., Virshup, G.F. et al. (1985). A two-terminal, two-junction monolithic cascade solar cell in a lattice-mismatched system. *18th IEEE Photovoltaic Specialist Conference*, Las Vegas, NV.
24. Olson, J.M., Kurtz, S.R., Kibbler, A.E. et al. (1990). A 27.3% efficient $Ga_{0.5}In_{0.5}P$/GaAs tandem solar cell. *Applied Physics Letters* 56: 623.
25. Yaung, K.N., Vaisman, M., Lang, J. et al. (2016). GaAsP solar cells on GaP/Si with low threading dislocation density. *Applied Physics Letters* 109: 032107.
26. Chiu, P., Wojtczuk, S., Zhang, X. et al. (2011). 42.3% efficient InGaP/GaAs/InGaAs concentrators using bifacial epiGrowth *37th IEEE Photovoltaic Specialist Conference*, Seattle, WA.
27. Fetzer, C.M., Yoon, H., King, R.R. et al. (2005). 1.6/1.1 eV metamorphic GaInP/GaInAs solar cells grown by MOVPE on Ge. *Journal of Crystal Growth* 276: 48–56.
28. Dimroth, F., Schubert, U., Bett, A.W. (2000). 25.5% efficient $Ga_{0.35}In_{0.65}P/Ga_{0.83}In_{0.17}As$ tandem solar cells grown on GaAs substrates. *IEEE Electron Device Letters* 21: 209–211.
29. King, R.R., Haddad, M., Isshiki, T. et al. (2000). Metamorphic GaInP/GaInAs/Ge solar cells. *28th IEEE Photovoltaic Specialist Conference*, Anchorage, AK.
30. Stan, M.A., Aiken, D.J., Sharps, P.R. et al. (2005). *InGaP/InGaAs/Ge high concentration solar cell development at Emcore. 32nd IEEE Photovoltaic Specialist Conference*, Lake Buena Vista, FL.
31. King, R.R., Law, D.C., Edmondson, K.M. et al. (2007). 40% efficient metamorphic GaInP/GaInAs/Ge multijunction solar cells. *Applied Physics Letters* 90: 183516/1–3.
32. Dimroth, F., Guter, W., Schone, J. et al. (2009). Metamorphic GaInP/GaInAs/Ge triple-junction solar cells with >41% efficiency. *34th IEEE Photovoltaic Specialist Conference*, Philadelphia, PA.
33. Guter, W., Schone, J., Philipps, S.P. et al. (2009). Current matched triple junction solar cell reaching 41.1% conversion efficiency under concentrated sunlight. *Applied Physics Letters* 94: 223504/1–3.
34. Al Taher, O., Cravens, R., Pien, P. et al. (2010). Qualification testing of 40% metamorphic CPV solar cells. *35th IEEE Photovoltaic Specialist Conference*, Honolulu, HI.
35. Strobl, G.F.X., Ebel, L., Fuhrmann, D. et al. (2014). Development of lightweight space solar cells with 30% efficiency at end-of-life. *40th IEEE Photovoltaic Specialist Conference*, Denver, CO.
36. Liu, X.Q., Fetzer, C.M., Chiu, P. et al. (2017). Large area multijunction III/V space solar cells over 31% efficiency. *43rd IEEE Photovoltaic Specialist Conference*, Washington, D.C.
37. Bertness, K.A., Kurtz, S.R., Friedman. D.J. et al. (1994). 29.5%-efficient GaInP/GaAs tandem solar cells. *Applied Physics Letters* 65: 989–991.
38. Kurtz, S.R., Olson, J.M., Friedman, D.J. et al. (1994). Ordering and disordering of doped $Ga_{0.5}In_{0.5}P$. *Journal of Electronic Materials* 23: 431–435.
39. Shurtleff, J.K., Lee, R.T., Fetzer, C.M. et al. (1999). Band gap control of GaInP using Sb as a surfactant. *Applied Physics Letters* 75: 1914.
40. Fetzer, C. (2006). Isoelectronic surfactant induced sublattice disordering in optoelectronic devices. US patent 7,126,052, filed Oct. 2, 2002 and issued Oct. 24, 2006.
41. Olson, J.M., McMahon, W.E., Kurtz, S.R. et al. (2006). Effect of Sb on the properties of GaInP top cells. *4th World Conference on Photvoltaic Electric Conversion*, Waikoloa, HI.
42. Fetzer, C.M., King, R.R., Colter, P.C. et al. (2004). High-efficiency metamorphic GaInP/GaInAs/Ge solar cells grown by MOVPE. *Journal of Crystal Growth* 261: 341.
43. Law, D.C., Fetzer, C.M., King, R.R. et al. (2005). Multijunction solar cells with subcell materials highly lattice-mismatched to germanium. *32nd IEEE Photovoltaic Specialist Conference*, Lake Buena Vista, FL.
44. Maree, P.M.J., Barbour, J.C., van der Veen, J.F. et al. (1987). Generation of misfit dislocations in semiconductors. *Journal of Applied Physics* 62: 4414.
45. Ayers, J.E., Ghandhi, S.K., Showalter, L.J. et al. (1991). Crystallographic tilting of heteroepitaxial layers. *Journal of Crystal Growth* 113: 430–440.

46. Chang, K.H., Gilbala, R., Srolovitz, D.J. et al. (1990). Crosshatched surface morphology in III-V semiconductor films. *Journal of Applied Physics* 67: 4093.
47. Matthews, J.W. and Blakeslee, A.E. (1975). Defects in epitaxial multilayers: II. Dislocation pile-ups, threading dislocations, slip lines and cracks. *Journal of Crystal Growth* 29: 273–280.
48. Andre, C.L., Boeckl, J.L., Wilt, D.M. et al. (2004). Impact of dislocations on minority carrier electron and hole lifetimes in GaAs grown on metamorphic SiGe substrates. *Applied Physics Letters* 84: 3447.
49. Yamaguchi, M., Amano, C., Itoh, Y. et al. (1988). Numerical analysis for high-efficiency GaAs solar cells fabricated on Si substrates. *Journal of Applied Physics* 66 (2): 915–919.
50. Tersoff, J.(1993). Dislocations and strain relief in compositionally graded layers. *Applied Physics Letters* 62: 693.
51. Fitzgerald, E.A., Kim, A.Y., Currie, M.T. et al. (1999). Dislocation dynamics in relaxed graded composition semiconductors. *Materials Science and Engineering B* 67: 53–61.
52. Tsao, J.Y. (1993). Relaxation of strain, section 5.3. In: *Materials Fundamentals of Molecular Beam Epitaxy*, 179–197. San Diego, CA: Academic Press.
53. Quitoriano, N.J. and Fitzgerald, E.A. (2007). Relaxed, high-quality InP on GaAs by using InGaAs and InGaP graded buffers to avoid phase separation. *Journal of Applied Physics* 102: 033511.
54. Kemat, S.V. and Hirth, J.P. (1990). Dislocation injection in strained multilayer structures. *Journal of Applied Physics* 67: 6844.
55. Fetzer, C., (2002). Isoelectronic surfactant suppression of threading dislocations in metamorphic epitaxial layers. US patent 7,122,734, filed Oct. 2, 2002 and issued Oct. 17, 2006.
56. Newman, F.D., Stan, M.A., Murray, S.L. et al. (2004). Tellurium surfactant effects in the growth of lattice mismatched $InAs_xP_{1-x}$ by metal organic vapor-phase epitaxy. *Journal of Crystal Growth* 272: 650–657.
57. Valtuena, J.F., Sacedon, A., Alvarez, A.L. et al. (1997). Influence of the surface morphology on the relaxation of low-strained $In_xGa_{1-x}As$ linear buffer layer structures. *Journal of Crystal Growth* 182: 281–291.
58. Dimroth, F., Lanyi, P., and Bett, A.W. (2000). MOVPE grown GaInAs solar cells for GaInP/GaInAs tandem applications. *Journal of Electronic Materials* 29 (1): 42–46.
59. Ebert, C.W., Reynolds, C.L., Wellenius, P. et al. (2011). Optimization of InGaAs metamorphic buffers for triple junction solar cells. *37th IEEE Photovoltaic Specialist Conference*, Seattle, WA.
60. Sze, S.M. (1981). Tunnel devices. In: *Physics of Semiconductor Devices*, 2e, 513–531. New York: John Wiley & Sons.
61. Bedair, S.M., Roberts, J.C., Jung, D. et al. (2000). Analysis of p++-AlGaAs/n++-InGaP tunnel junction for high solar concentration cascade solar cells. *28th IEEE Photovoltaic Specialist Conference*, Anchorage, AK.
62. Bertness, K.A., Friedman, D.J., and Olson, J.M. (1994). Tunnel junction interconnects in GaAs-based multijunction solar cells. *1st World Conferenece on Photovoltaic Energy Conversion*, Waikoloa, HI.
63. Rey-Stolle, I., Garcia, I., Galiana, B. et al. (2007). Improvements in the MOVPE growth of multijunction solar cells for very high concentration. *Journal of Crystal Growth* 298: 762–766.
64. Garcia, I., Rey-Stolle, I., Galiana, B. et al. (2007). Analysis of tellurium as n-type dopant in GaInP: doping, diffusion, memory effect and surfactant properties. *Journal of Crystal Growth* 298: 794–799.
65. Ebert, C., Pulwin, Z., Byrnes, D. et al. (2011). Tellurium doping of InGaP for tunnel junctions applications in triple junction solar cells. *Journal of Crystal Growth* 31: 61–63.
66. Lee, J.S., Kim, I., Choe, B.D. et al. (1994). Carbon doping and growth rate reduction by CCl_4 during metalorganic chemical-vapor deposition of GaAs. *Journal of Applied Physics* 76: 5079–5084.
67. Tateno, K., Kohama, Y., and Amano, C. (1997). Carbon doping and etching in effects of CBr_4 during metalorganic chemical vapor deposition of GaAs and AlAs. *Journal of Crystal Growth* 172: 5–12.
68. Tateno, K. and Amano, C. (1999). Carbon doping and etching in $Ga_xIn_{1-x}As_yP_{1-y}$ on GaAs substrates using CBr_4 by metalorganic chemical vapor deposition. *Journal of Electronic Materials* 28: 63–66.
69. Jung, D., Parker, C.A., Ramdami, J. et al. (1993). AlGaAs/GaInP heterojunction tunnel diode for cascade solar cell application. *Journal of Applied Physics* 74: 2090.
70. Dimroth, F., Beckert, R., Meusel, M. et al. (2001). Metamorphic $Ga_yIn_{1-y}P/Ga_{1-x}In_xAs$ tandem solar cells for space and terrestrial concentrator applications at C>1000 suns. *Progress in Photovoltaics: Research and Applications* 9: 165–178.
71. Fewster, P.F. (1993). X-ray diffraction from low-dimensional structures. *Semiconductor Science and Technology* 8: 1915–1934.
72. Koppensteiner, E., Bauer, G., Kibbel, H. et al. (1994). Investigation of strain-symmetrized and pseudomorphic SimGen superlattices by X-ray reciprocal space mapping. *Journal of Applied Physics* 76: 3489.
73. King, R.R., Bhusari, D., Boca, A. et al. (2011). Band gap-voltage offset and energy production in next-generation multijunction solar cells. *Progress in Photovoltaics: Resources and Applications* 19: 797–812.
74. Liu, X.Q., Fetzer, C.M., Rehder, E. et al. (2012). Organometallic vapor phase epitaxy growth of upright metamorphic multijunction solar cells. *Journal of Crystal Growth* 352: 186–189.

75. Guter, W., Dunzer, F., Ebel, L. et al. (2016). Space solar cells - 3G30 and next generation radiation hard products. *11th European Space Power Conference*, Halkidiki, Greece.
76. Wanlass, M.W., Geisz, J.F., Kurtz, S. et al. (2005). Lattice-mismatched approaches for high performance III-V photovoltaic energy converters. *31st IEEE Photovoltaic Specialist Conference*, Orlando, FL.
77. Geisz, J.F., Kurtz, S.R., Wanlass, M.W. et al. (2007). High-efficiency GaInP/GaAs/InGaAs triple-junction solar cells grown inverted with a metamorphic bottom junction. *Applied Physics Letters* 91: 023502–023505.
78. Wanlass, M.W., Ahrenkiel, P., Albin, D.S. et al. (2006). Monolithic ulta-thin GaInP/GaAs/GaInAs tandem solar cells. *4th World Conference on Photovoltaic Energy Conversion*, Waikoloa, HI.
79. Stan, M., Aiken, D., Cho, B. et al. (2008). Very high efficiency triple junction solar cells grown by MOVPE. *Journal of Crystal Growth* 310 (23): 5204–5208.
80. Takamoto, T., Washio, H., and Juso, H. (2014). Application of InGaP/GaAs/InGaAs triple junction solar cells to space use and concentrator photovoltaic. *40th IEEE Photovoltaic Specialist Conference*, Denver, CO.
81. Stan, M., Aiken, D., Cho, B. et al. (2010). High-efficiency quadruple junction solar cells using OMVPE with inver-eted metamorphic device structures. *Journal of Crystal Growth* 312: 1370–1374.
82. Patel, P., Aiken, D., Chumney, D. et al. (2012). Initial results of the monolithically grown six-junction inverted metamorphic multi-junction solar cell. *38th IEEE Photovoltaic Specicalist Conference*, Austin, TX.
83. Geisz, J.F., Friedman, D.J., Ward, J.S. et al. (2008). 40.8% efficient inverted triple-junction solar cell with two independently metamorphic junctions. *Applied Physics Letters* 93 (12): 123505.
84. France, R.M., Geisz, J.F., Steiner, M.A. et al. (2015). Quadruple junction inverted metamoprhic concentrator devices. *IEEE Journal of Photovoltaics* 5: 432.
85. France, R.M., Geisz, J.F., Garcia, I. et al. (2016). Design flexibility of ultrahigh efficiency four-junction inverted metamoprhic solar cells. *IEEE Journal of Photovoltaics* 6: 578–583.
86. Steiner, M.A., Geisz, J.F., Reedy, R.C. et al. (2008). A direct comparison of inverted and non-inverted growths of GaInP solar cells. *33rd IEEE Photovoltaic Specialist Conference*, San Diego, CA.
87. France, R.M., Geisz, J.F., Steiner, M.A. et al. (2012). Reduction of crosshatch roughness and threading discloation density in metamorphic GaInP buffers and GaInAs solar cells. *Journal of Applied Physics* 111 (10): 103528
88. France, R.M., Garcia, I., McMahon, W.E. et al. (2014). Lattice-mismatched 0.7-eV GaInAs solar cells grown on GaAs using GaInP compositionally graded buffers. *IEEE Journal of Photovoltaics* 4 (1): 190–195.
89. France, R.M., McMahon, W.E., Norman, A.G. et al. (2012). Control of misfit dislocation glide plane distribution during strain relaxation of CuPt-ordered GaInAs and GaInP. *Journal of Applied Physics* 112: 023520.
90. France, R.M., McMahon, W.E., and Guthrey, H.L. (2015). Critical thickness of atomically ordered III-V alloys. *Applied Physics Letters* 107 (15): 151903.
91. McMahon, W.E., Kang, J., France, R.M. et al. (2013). Ordering-enhanced dislocation glide in III-V alloys. *Journal of Applied Physics* 114 (20): 203506.
92. Schulte, K.L., France, R.M., McMahon, W.E. et al. (2017). Reduced dislocation density in $Ga_xIn_{1-x}P$ compositionally graded buffer layers through engineered glide plane switch. *Journal of Crystal Growth* 464: 20–27.
93. Tatavari, R., Hillier, G., Dzankovic, A., *et al.* (2008). Lightweight, low cost GaAs solar cells on 4″ epitaxial liftoff (ELO) wafers. *33rd Photovoltaic Specialist Conference*, San Diego, CA.
94. Schulte, K.L., France, R.M., and Geisz, J.F. (2017). Highly transparent compositionally graded buffers for new metamorphic multijunction solar cell designs. *IEEE Journal of Photovoltaics* 7 (1): 347–353.
95. Garcia, I., Geisz, J.F., France, R.M. et al. (2014). Metamorphic $Ga_{0.76}In_{0.24}As/GaAs_{0.75}Sb_{0.25}$ tunnel junctions grown on GaAs substrates. *Journal of Applied Physics* 116: 074508.
96. France, R.M., Geisz, J.F., Garcia, I., *et al.* (2016). Design flexibility of ultrahigh efficiency four-junction inverted metamorphic solar cells. *IEEE Journal of Photovoltaics* 6: 578–583.
97. Cornfeld, A.B., Aiken, D., Cho, B. et al. (2010). Development of a four sub-cell inverted metamorphic multi-junction (IMM) highly efficient AM0 solar cell. *35th IEEE Photovoltaic Specialist Conference*, Honolulu, HI.
98. Patel, P., Aiken, D., Boca, A. et al. (2012). Experimental results from performance improvement and radiation hardening of inverted metamorphic multijunction solar cells. *IEEE Journal of Photovoltaics* 2: 377–381.
99. Cornfeld, A.B., Patel, P., Spann, J. et al. (2012). Evolution of a 2.05 eV AlGaInP top sub-cell for 5 and 6J-IMM applications. *38th IEEE Photovoltaic Specialist Conference*, Austin, TX.
100. Perl, E.E., Simon, J., Geisz, J.F. et al. (2016). Development of high-bandgap AlGaInP solar cells grown by organometallic vapor-phase epitaxy. *IEEE Journal of Photovoltaics* 6: 770.
101. Hecklemann, S., Lackener, D., Karcher, C. et al. (2015). Investigations on $Al_xGa_{1-x}As$ solar cells grown by MOVPE. *IEEE Journal of Photovoltaics* 5 (1): 446–453.
102. Jain, N., Geisz, J.F., France, R.M. et al. (2017). Enhanced current collection in 1.7 eV GaInAsP solar cells grown on GaAs by metalorganic vapor-phase epitaxy. *IEEE Journal of Photovoltaics* 7 (3): 927–933.
103. Geisz, J.F., Steiner, M.A., Jain, N. et al. (2018). Building a six-junction inverted metamorphic concentrator solar cell. *IEEE Journal of Photovoltaics* 8 (2): 626–632.

6

Quantum Dots

E. Hulicius, A. Hospodková, and M. Zíková
Department of Semiconductors, Institute of Physics, Czech Academy of Sciences, Prague, Czech Republic

6.1 General Introduction to the Topic

6.1.1 Definition and History

This chapter focuses only on quantum dots (QDs) embedded inside a semiconductor structure, like quantum wells (QWs) or partly quantum wires (QWrs). Freestanding QDs with a much larger application field, which can be prepared from the same material, are not covered in this chapter. The reason for this is that we will write about MOVPE QD preparation only (not, for example, about colloidal, metal, or other freestanding QDs). A research report about the market for the whole family of QDs can be found in [1].

QDs are small pieces of semiconductor nanocrystals (mainly) with sizes in the range from a few to tens of nanometers. The first idea of QDs having properties different from those of the bulk material was published by Ekimov and Onushchenko in 1981 [2]. Their QDs were prepared in colloidal solution, not yet QDs embedded in the heterostructure: "The exciton absorption spectrum of microscopic CuCl crystals grown in a transparent dielectric matrix has been studied. The size of the microscopic crystals was varied in a controlled manner from several tens of angstroms to hundreds of angstroms. There is a short-wave shift (of up to 0.1 eV) of the exciton absorption lines, caused by a quantum size effect" [2].

During the early 1990s, embedded QDs were prepared by molecular beam epitaxy (MBE) [3–7] in several laboratories, as described in [8].

MOVPE-grown QDs were reported in 1991 [9, 10], but with little response (12 and 0 citations, respectively). MBE seems to be more suitable for attaining an exact layer thickness down to fractions of a monolayer. QD MOVPE papers published later [11–14] attracted much more interest. An attempt to realize industrial MOVPE-prepared QD-based semiconductor lasers for fiber optic telecommunications appeared during the first years of this century (EU project DOTCOM 2002–05), but without a real impact on industrial production.

Metalorganic Vapor Phase Epitaxy (MOVPE): Growth, Materials Properties, and Applications,
First Edition. Edited by Stuart Irvine and Peter Capper.

QDs exhibit unique electronic and optical properties, intermediate between those of bulk semiconductors and discrete atoms or molecules. The reason for this is that the size of QDs is comparable with the de Broglie wavelength of an electron in a crystal. Electrons (and holes) inside QDs behave differently from those found in a bulk material. The most important difference is that inside the QDs, electrons (and holes) can occupy only discrete energy levels due to strong localization; however, they also can be found outside QDs with nonzero probability. The position of energy levels depends mainly on the smallest dimension of a QD. Thus, electron–hole recombination energy (and, consequently, the wavelength of the emitted radiation) depends on this size, often more than on the bandgap energy of the material, which is determining for the bulk material. A current review of all groups of semiconductor QD applications including the growth and parameters can be found in [15]: "Applications include QD lasers for integrated silicon photonics and quantum computing. QD LEDs for highly efficient solid-state lighting. QDs for intermediate band solar cells and multi-exciton generation. QDs for bio-medical labeling, imaging, targeted drug delivery, sensing, and therapy."

6.1.2 Paradigm of Quantum Dots

Light emission (LED or laser type) from separated atoms in a glass matrix, e.g. neodymium-doped YAG, ruby, etc., or from atoms or molecules in a gas, is nearly monochromatic thanks to these media having sharp energy levels. Semiconductor direct-bandgap materials have much higher energy-conversion efficiency than indirect-bandgap ones, but many possible recombination energies, thus they have broadband or multimode light emission. Electron energy levels in isolated atoms, in bulk semiconductor material, and inside a QD are shown in Figure 6.1. A QD, resembling an artificial atom in its ability to confine the electron (or hole) wave function, emits more monochromatic light than the bulk material.

Reasons for and examples of QD parameters for semiconductor lasers are as follows:

- The emission from quantum levels in QWs, QWrs, or QDs is monochromatic and still very efficient. One of the effects of strong quantum confinement is the discrete energy levels. Another effect is a lower-threshold current density of a QD-based semiconductor laser, when achieving the inversion of population in QDs is needed for stimulated emission, as well as suppression of loss mechanisms such as Auger recombination and intervalence band absorption.
- The photon energy (emitted wavelength) can be changed mainly by the size of QDs, not only by chemical composition. This is similar to quantum-well lasers, but the temperature dependence of these lasers is also much lower than for "classical" two-dimensional QW-based lasers.
- Very small QDs can emit single photons, which is necessary for quantum cryptography.

6.1.3 QD Types

It is worth repeating that only embedded MOVPE (and partly MBE) grown semiconductor QDs will be considered. One can distinguish QD types according to growth modes, growth procedures, materials used, and structures with strain-reducing layers (SRLs), types of interfaces, and spatial arrangement of QDs (uncoordinated, array, and single/individual QDs).

The most common is the self-assembled creation of QDs in the Stranski–Krastanov (SK) growth mode. The Volmer–Weber (VW) growth mode occurs for larger lattice mismatches. The SK mode achieves better size uniformity than can be obtained in the VW growth mode [16]. The Frank–van der Merwe (FM) growth mode occurs when the lattice mismatch is not too high, the AlAs/GaAs system being an example. This growth mode can be used for smooth capping-layer growth on QD structures [17].

A completely different growth procedure for QDs is droplet epitaxy. In this mode, A^{III} element droplets are deposited on the substrate and then crystallized by exposing them to an atom B^{V} flux.

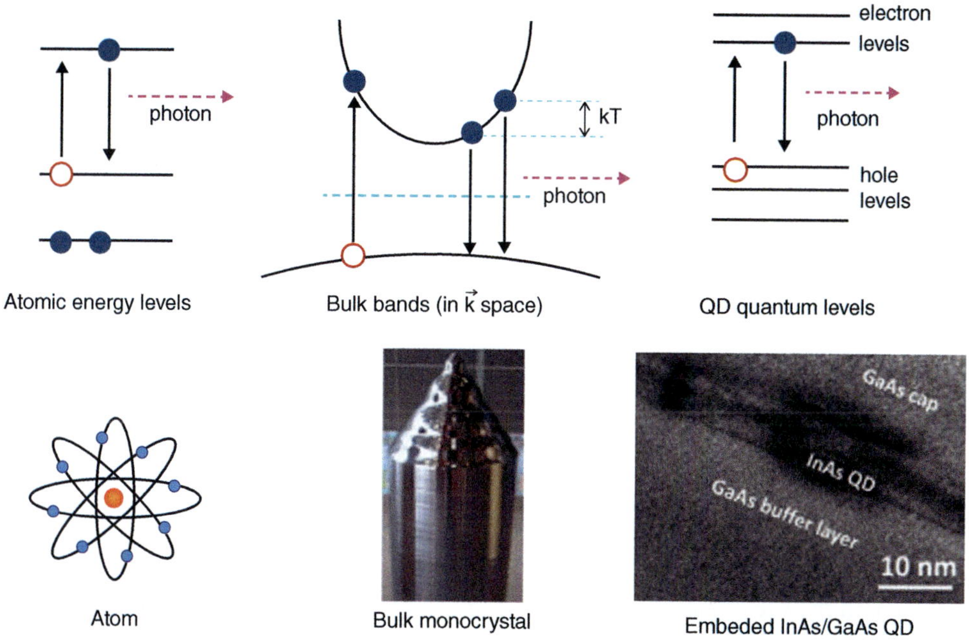

Figure 6.1 *Electron energy levels of atoms, of semiconductor bulk material, and inside a QD. Photons emitted from atoms and QDs are much more monochromatic than from the bulk material.*

This method can be used to grow both lattice-mismatched (e.g. InAs/GaAs) and approximately lattice-matched (e.g. GaAs/AlGaAs) material systems. In contrast to the SK method, there is no wetting layer. Droplet epitaxy is performed at low temperatures, so annealing is usually necessary to improve the optical quality of the dots. The droplet epitaxy process for InAs QDs grown on (100) GaAs is described in [18, 19]. A droplet of In is deposited on the substrate, and the flux of As crystallizes the edges of the droplets, leading to a crater-like structure that is later annealed to form QDs.

QDs prepared by MOVPE preparation techniques using the SK growth mode can be further sorted according to the material. Compound semiconductors with substantially different lattice constants have to be used to reach sufficient tensile strain for self-assembled creation of QDs on the wetting layer. This will be described in detail in the next sections of this chapter. There are plenty of possible combinations of binary compounds [20, 21] (e.g. InAs/GaAs [22], InSb/GaSb [23]) and many more with ternary ones (e.g. InGaAs/AlGaAs [24], InGaAs/GaAs [25, 26], InGaAs/InP [27], InP/InGaP [28, 29], and GaN-based QDs have been also been investigated for a long time [30, 31]. Simple binary QDs are less complicated to prepare, and the energy of the quantum levels can be tuned based on the size and shape of the QDs [32]. Strain produced by the size and shape of QDs is another powerful tool for engineering QD properties, as was recently found [33].

QD types can also be sorted according to interfaces. Type-I, or straddling bandgap, where the edges of the conduction and valence bands of the narrower forbidden gap material are inside of the wider forbidden gap material, is the most frequently used type, or Type-II with staggered bandgap, which was also investigated [34].

Growth of QDs with SRLs adds technological complexity but gives additional degrees of freedom for engineering of device parameters, such as achieving emission at longer wavelengths [34–38]; see also Section 6.3.3.

One way to prepare QDs by MOVPE that is different from the SK growth mode is to use a prepatterned substrate, also known as *size- and site-controlled* QDs, or as *inverted QDs*, since this growth mode is based on the filling of the "nanoholes" [39, 40]. A substrate with lithographically etched pyramids with a surface density of 5×10^8 cm^{-2} was used for QD growth. The surface in the pyramidal pits was Ga-terminated, which eased the decomposition of metalorganic precursors. This determined the place where the growth took place, with the growth-rate anisotropy and capillarity as the main driving forces for the growth. These QDs are usually grown using electron beam or lithographic positioning of the substrate followed by etching of small holes in a predesigned pattern. The QDs can be subsequently grown by MOCVD or MBE and will only grow in these holes.

The fastest triggered single-photon sources to date have been demonstrated using epitaxially grown (mostly by SK mode) semiconductor QDs, which can be conveniently integrated with optical microcavities. Lasers and LEDs based on single QDs are very promising for quantum cryptography, quantum computation, and communications, but are rather difficult to prepare. The most promising for single-photon emitters are single QDs grown on prepatterned substrates [40].

The first demonstration of a single-photon emission from an InGaAs/GaAs QD was made in 2004 by Kapon et al. [41].

Another approach to prepare individual QDs is to create them from two-dimensional electron or hole gases present in remotely doped QWs or semiconductor heterostructures called *lateral QDs*. "The sample surface is coated with a thin layer of resist. A lateral pattern is then defined in the resist by electron beam lithography. This pattern can then be transferred to the electron or hole gas by etching, or by depositing metal electrodes (lift-off process) that allow the application of external voltages between the electron gas and the electrodes. Such QDs are mainly of interest for experiments and applications involving electron or hole transport, i.e., an electrical current" (https://en.wikipedia.org/wiki/Quantum_dot).

The most common QDs for single-photon sources are InAs QDs in a GaAs matrix [42]. These QDs emit in the range from 850–1000 nm and require cryogenic temperatures for operation due to the shallow carrier confinement. These QDs are usually grown in the SK mode by either MBE or MOCVD, although they can also be grown by droplet epitaxy. In the literature, one can read about the first demonstrations of epitaxially prepared QDs for single-photon emission [17], quantum key distribution [43], and electrically pumped single-photon emission [44]; this topic is currently a hot one [45].

6.2 $A^{III}B^{V}$ Materials and Structures

6.2.1 QDs Embedded in the Structure

The three main growth technological procedures used for MOVPE preparation of QDs embedded in the structure are self-assembled Stranski–Krastanov growth mode, formation of QDs in prepatterned inverted pyramids, and droplet epitaxy.

The most widely used procedure is self-assembling of QDs in SK growth mode. This growth mode can be used for highly strained QDs such as InAs on GaAs, InSb on GaSb, or InGaAs on GaP. In this technological procedure, QDs are formed from the material of a strained epitaxial layer; when the system has enough time to find the minimum energy, this means either the growth of the strained layer has to be slow or the growth is interrupted for a few seconds after the epitaxial growth of the strained layer. Using this method, different QD densities can be obtained (10^8–10^{11} cm^{-2}) depending on the amount of deposited strained material and the time of the growth interruption. It is possible to grow stacks of vertically aligned QDs when the separation layer thickness between QDs is sufficiently thin (3–10 nm). In such a case, QDs are formed on the top of the QDs in the previous layer, thanks to the advantageous lattice deformation. An atomic force microscopy (AFM) picture of the surface of a vertically correlated QD structure with the last uncovered QD layer is shown in Figure 6.2a). It nicely demonstrates that the vertical correlation of QDs is high. No surface QDs are formed between hillocks, which are formed above buried QDs in previous layers; the QDs are formed on those hillocks only. A transmission electron microscopy (TEM) example of the cut of a similar structure is shown in Figure 6.2b).

Some applications require precise positioning of QDs. One possible way to control this is SK growth on prepatterned substrates. However, this method requires very precise adjustment of the growth parameters so that QDs are formed only on the prepatterned positions; see [46].

When the QDs and the surrounding barriers do not have sufficient strain for the SK growth mode, such as GaAs QDs with AlGaAs barriers, other alternative procedures for QD formation were developed. One of these is growth on a prepatterned {111}B GaAs substrate using inverted pyramids [47]. This technology takes advantage of the lower growth rate of GaAs on inverted pyramid side walls and the formation of GaAs QDs in the tips of each pyramid pit. A stack of vertically aligned GaAs/AlGaAs QDs can be prepared by this technique, similarly to vertically aligned QDs grown by the SK growth mode. The advantage of this technique is the very precise QD site and photoluminescence (PL) emission energy control; this is why it can be also used for the growth of strained InGaAs/GaAs QDs.

Another widely used method for self-organized QDs preparation, which can be also used for combination of lattice-matched as well as strained materials, is droplet epitaxy. The basic idea of droplet epitaxy originated from Koguchi and his coworkers [48, 49]. In this work, microcrystals of GaAs that were pyramid-shaped and 35 nm in length at the base were grown epitaxially on a ZnSe (001) surface by using As-incorporated growth into Ga droplets for the first time. These microcrystals consisted of (111) facets. The size distribution was controlled within 16% in deviation. The As-incorporated lateral growth of GaAs was not observed.

This technological procedure can be used for both MBE as well as for MOVPE QD structure preparation. The principle of the method is first deposition of metallic nanodroplets of the III-column

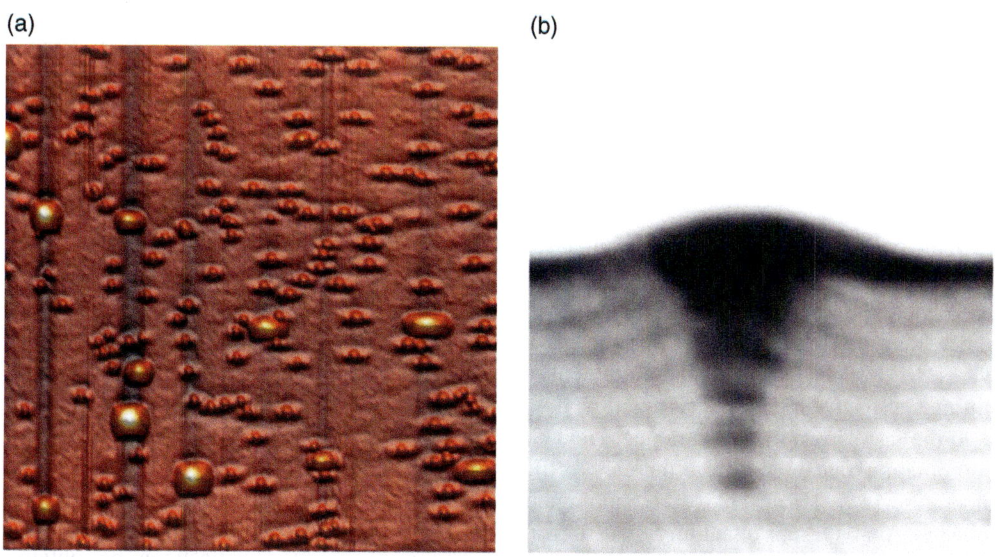

Figure 6.2 AFM picture of a surface with vertically correlated QDs with the last uncovered QD layer (a); TEM picture of the cut of a seven-layer vertical QD structure (b).

elements such as Ga, In, and Al on the substrate surface. During droplet formation, the main driving forces are diffusion, minimization of the surface tension, and Ostwald ripening. The dominance of each of these driving forces depends on technological circumstances. The second step of this technique is the crystallization of the metallic cluster by introduction of the nonmetallic V-column elements into the reactor. It is important to suppress or to control the lateral diffusion of the Group III elements from the cluster. Using this technique, the wetting layer is absent in the structure. A wide range of material combinations can be used for QD structure preparation with this method, such as GaAs/AlGaAs, InGaAs/GaAs, InAs/InP, InSb/GaAs, GaN/AlGaN, and InGaAn/GaN. It can be used for preparation of QD molecules, as in the case when droplet GaAs QDs can serve as a seeding layer for subsequent SK growth of strained InAs QDs [50, 51]. This method is often used for preparation of nitride QD structures grown by MOVPE [52], as well as for narrow-gap QDs (e.g. InSb based) [53].

Submonolayer deposition of 2D islands 1 monolayer (ML) high and a few nanometers wide is considerably less common than conventional self-organized QDs, usually prepared by the SK mechanism. QDs are created when the amount of the deposited material exceeds a critical thickness of the wider-bandgap layer from one to several monolayers. The elastic relaxation due to surface stress discontinuity at the submonolayer island boundaries favors formation of a periodic structure of equal-sized islands and prevents Ostwald ripening of the 2D islands [54, 55]. The existence of nanometer-scale 2D islands has been experimentally confirmed for InAs:GaAs [56] and CdSe:ZnSe [57] systems.

6.2.2 Semiconductor Materials for Embedded QDs

The most common material combination for embedded QDs is InAs QDs in a GaAs matrix. The emission wavelength of these simple QDs is published up to 1250 nm. To further elongate the emission wavelength, different kinds of *strain-reducing layer materials* such as InGaAs, GaAsSb, and AlInAs are used to surround the InAs QDs. Most of the results presented in this chapter are also obtained on this material system.

Useful material combinations are InAs QDs prepared on an InP substrate. To improve the electron localization in QDs, lattice-matched buffers can be used, such as GaAsSb, InGaAs, and AlAsSb [58]. This material system offers the possibility to compensate for the compressive strain of InAs QD by inserting strain-compensating layers with tensile strain (such as GaAs, AlAs, InGaAs, or InGaP) [58]. This enables the growth of very thick active layers, which are necessary for detector or solar cell applications [59].

InP QDs on a GaAs substrate and surrounded by InGaP layers offer visible red luminescence [60]. Surrounding InGaP layers lattice-matched to GaAs are necessary in the structure to obtain sufficient carrier confinement and shorten the emission wavelength. This material system can also take advantage of the strain compensation using InGaP layers with proper composition to achieve a tensile strain.

More attention has also been attracted by GaSb self-assembled QDs recently, because of the type-II band alignment with a GaAs matrix, which brings further potential applications. The large valence-band discontinuity increases the carrier (hole) storage time in GaSb QDs for memory devices [61, 62]. The type-II band alignment provides a wider wavelength tuning range for light-emitting GaSb QD diodes [34]. The GaSb QDs are also useful for the intermediate-band solar cells due to the fast separation of photogenerated carriers in the type-II structure, which decreases the recombination rate of carriers, increases efficiency, and enables a higher solar-cell operation temperature [63]. However, it is technologically difficult to prepare pure GaSb QDs; usually, alloyed GaAsSb QDs or quantum rings are formed, with a reported highest Sb concentration up to 80% being formed during capping [64]. GaSb QDs with the required simple hillock morphology were achieved when InAs QDs were used as the seeding layer; see Figure 6.3a. In this case, strong type-II band alignment of carriers was obtained, as demonstrated in Figures 6.3b–d.

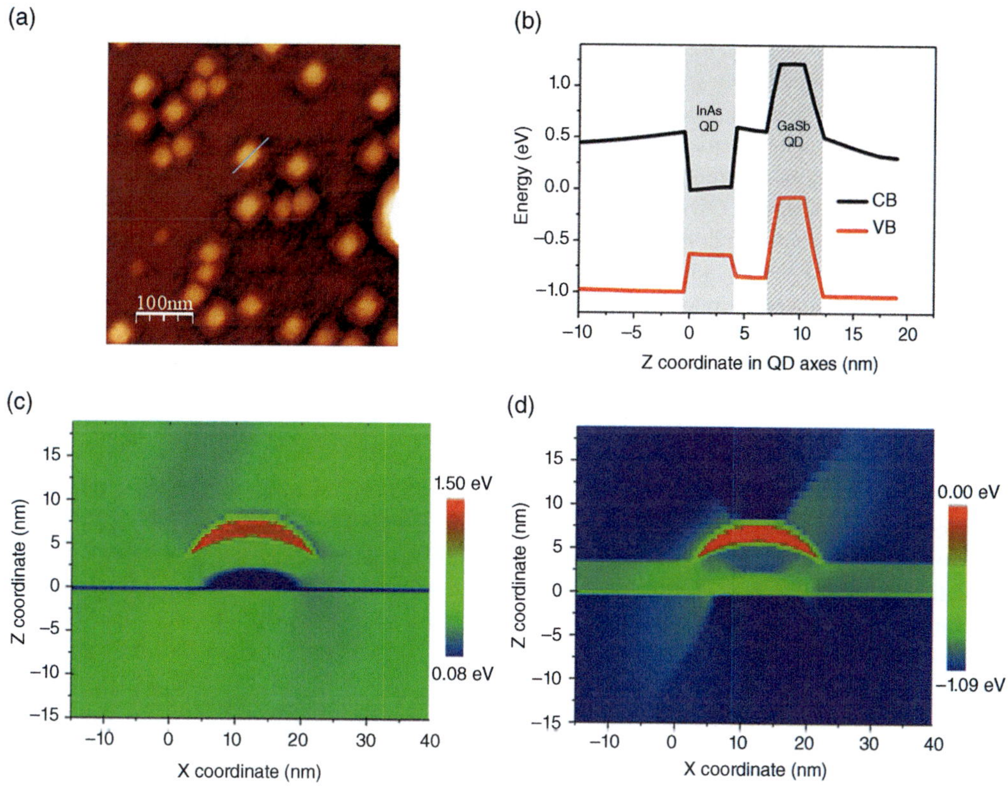

Figure 6.3 *AFM picture of simple hillock morphology of GaSb QDs when InAs QDs were used as the seeding layer (a), calculated band alignment of combined InAs and GaSb QDs along QD axes (b), map of band alignment for vertical cut of combined InAs and GaSb QDs of conduction (c) and valence (d) bands.*

GaN QDs embedded in an AlN matrix are very promising candidates to enable monolithic opto-electronic quantum communication devices via single- and entangled photon emission operating at room temperature. The growth of GaN QDs by MBE has been intensively studied. However, growing GaN QDs by MOVPE is still very challenging, and only a few groups have succeeded with an antisurfactant induced-nucleation or with very low V/III ratios [65–67].

Comparison of ultrathin insertion layers (GaP and GaP/$In_{0.4}Ga_{0.6}P$) on InP self-assembled QDs grown on GaAs (001) substrates using MOVPE was studied in [68]. The effects of insertion layers on the structural and optical properties of InP QDs were investigated. The insertion-layer thickness affects the shape, density, size distribution, and photoluminescence emission of InP QDs. The GaP/InGaP samples are attributed to the best size distribution, photoluminescence intensity, and linewidth.

6.3 Growth Procedures

6.3.1 Comparison of MBE- and MOVPE-Grown QDs

There are few papers about QDs prepared by liquid phase epitaxy (LPE). They were published from 1999 by Krier et al. [69, 70], by Moiseev et al. with possible MIR-region applications [71–73], and more recently in [74]. Only occasionally can we find papers about other epitaxial techniques, e.g. hydride vapor-phase epitaxy [75]. However, the mainstream is focused on MBE and MOVPE technology [76]. Web of Science data (August 2017) show that QD-based MBE research publications prevail over MOVPE ones, as well as in the case of industrial applications; although full data is difficult to obtain, as was expected, MBE still prevails over MOVPE.

Petroff and DenBaars compared MBE and MOCVD growth and the properties of QDs in their paper [8]. This paper and 19 references presented therein show the situation in the early 1990s and concluded "MOCVD grown QDs are similarly uniform in size and show promising optical properties." This is not much different from the current comparison of QDs grown by MBE and MOVPE. For MOVPE growth, one big advantage is the possibility of using reflectance anisotropy spectroscopy (RAS), an in situ technique that allows us to measure monolayer oscillations and changes of the surface during the waiting time and QD creation and to be able to control QD growth, compared to the 1990s, when a comparable in situ technique – *reflection high-energy electron diffraction* (RHEED) was available for MBE only. MBE has better control of layer thickness (including submonoatomic) and interface abruptness than MOVPE. Both are very important for QD growth.

Some of the MBE research and applied activities are described in [77–81]. MOVPE QD structures were studied and used later than MBE, but MOVPE is more suitable for industrial purposes, which makes the MOVPE research also important. The current needs of the industry based on embedded QDs can still be covered by MBE production, but for larger QD devices, production by MOVPE will be necessary. Strong research of MOVPE possibilities to use this technique in the market is underway. A very interesting interview of Philip Ball with Dieter Bimberg and Kang Wang on this subject can be found in [82]. Other reflections on MOVPE QD maturity for industry were presented earlier [83, 84]. In addition, the MOVPE technology producer AIXTRON is very active in EU grant projects and publications, e.g. [85]. Other teams also presented MOVPE-prepared QD device structures recently [86].

The MBE industry offers several types of QD lasers and customized epitaxial structures for distributed Bragg resonators (DBR) mirrors, resonant tunneling diodes (RTDs), and high electron mobility transistors (HEMTS) with QDs nowadays (http://www.qdlaser.com/product04.html, http://www.qdlaser.com/wafer.html).

6.3.2 Growth Parameters

Obtaining reproducible growth of QDs is probably the most difficult task for technologists. The problem is that the size, shape, and density of QDs are very sensitive to many technological parameters, such as growth temperature, growth rate, waiting time for QD formation, QD material precise dosage, and composition of the covering and buffer layers. The strong sensitivity of QD properties to so many growth parameters is the main obstacle for wider industrial application of QDs.

The influence of growth parameters on QD properties will be demonstrated in this chapter on the most common InAs/GaAs QD material system prepared in the SK growth mode. Although the precise growth parameters will be different for other material QDs prepared in the SK growth mode, the growth mechanisms and principles are similar.

6.3.2.1 Growth Temperature

Four AFM images of samples grown at temperatures in the range of 470–510 °C can be found in Figure 6.4 to illustrate the influence of the growth temperature on the QD size and shape. As the growth temperature increases, the size of the QD increases. However, although the influence of the growth temperature on QD size seems to be huge, the influence on the wavelength of emitted light is not so strong. For emission intensity, the growth temperature of the capping layer is more important; see Figure 6.5.

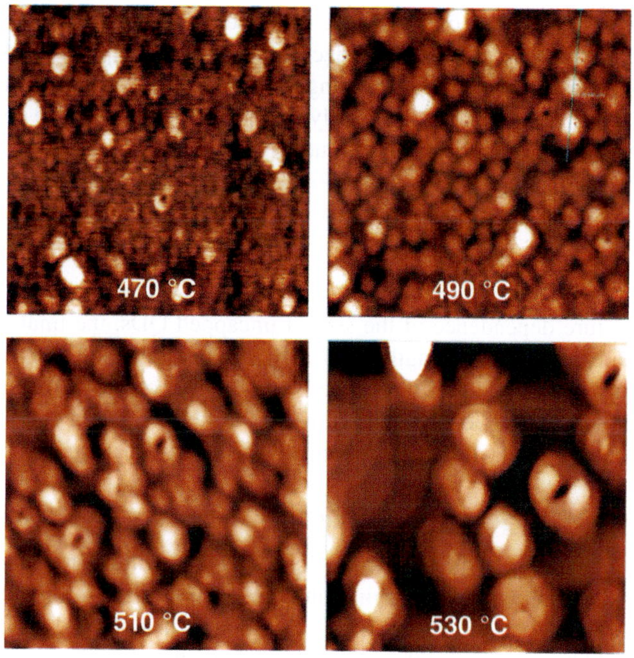

Figure 6.4 *AFM surface images of InAs QDs prepared at different growth temperatures and capped by 3-ML thick GaAs capping to avoid Ostwald ripening during the cooling process. Thin GaAs capping is the cause of quantum ring formation from a few QDs in the figures.*

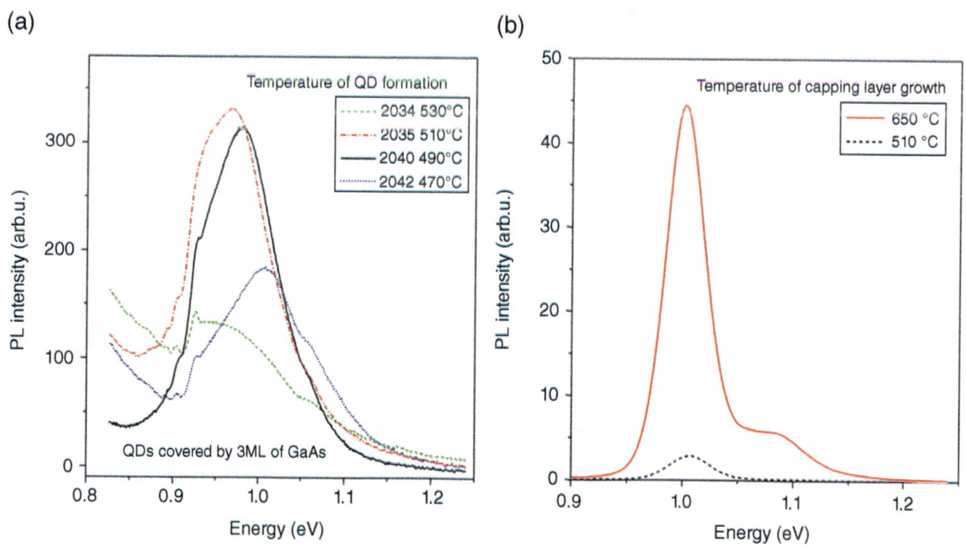

Figure 6.5 *PL of QDs with different growth temperature of InAs layer (a) and of GaAs (b) capping layer.*

6.3.2.2 InAs Dosage

The amount of deposited InAs is very important for successful QD formation. Via the InAs original layer thickness, the QD surface density can be controlled. It was observed by RAS that QD formation starts after the deposition of critical InAs thickness, which is around 1.7 ML of InAs (see the results presented in Section 6.4.1). Above this value, the strain energy is sufficient to start the SK growth mode. To achieve sufficient QD size and density, the required thickness of the original InAs layer should be approximately in the range of 2.0–3.5 ML of InAs. For some single-QD applications, the required QD density is extremely low [87]. In this case, the deposited thickness of InAs is usually even lower than the critical value.

6.3.2.3 Growth Rate of the InAs Layer

Another important parameter is the growth rate of InAs, as demonstrated in Figure 6.6. For the sample where the InAs layer was grown more slowly, the QDs are somewhat bigger. However, similar to the temperature dependence of the size of uncapped QDs, the final height of the QDs after capping and the emission wavelength are barely influenced by the InAs growth rate, but it affects the QD PL intensity. The PL of QDs capped only by 1 nm of GaAs (almost uncapped QDs) and QDs capped by 10 nm of GaAs grown with two different growth rates are compared in Figures 6.7a and b. A slower growth rate increases the final PL intensity of QDs.

6.3.2.4 Growth Interruptions for QD Formation

The deposited InAs layer needs sufficient time to form QDs. The time can be provided either during very slow growth or by the growth interruption after the InAs deposition. The optimal time of growth interruption depends on the growth rate, the amount of deposited InAs, and the growth temperature, and should be adjusted for the precise growth conditions using an in situ method such as reflectance or RAS measurement (see Section 6.4.1).

6.3.2.5 QD Capping Process Growth Rate

The QD properties are strongly influenced by the capping process and its technological parameters. During this technological step, not only are QDs partly dissolved from the not-completely capped QD apex, but the dissolved InAs material is backward-incorporated into the QD's surroundings and forms either hillocks elongated in the [−110] direction or oval rings with a protrusion in the middle; see Figure 6.8. Larger QDs are more influenced by dissolution. Some of the dissolved material from larger QDs is redistributed over the epitaxial surface and incorporated into the surrounding of smaller QDs. When the capping process is slower, or if it is interrupted, the QD dissolution and material redistribution are enhanced. This can be demonstrated on deeper and larger protrusions in the sample with the lowest capping layer growth rate; see the AFM images in Figure 6.8.

It was found that incorporation of the dissolved InAs material is enhanced at lower growth rates, while at higher flow rates, the dissolved In remains as surfactant on the epitaxial surface. This

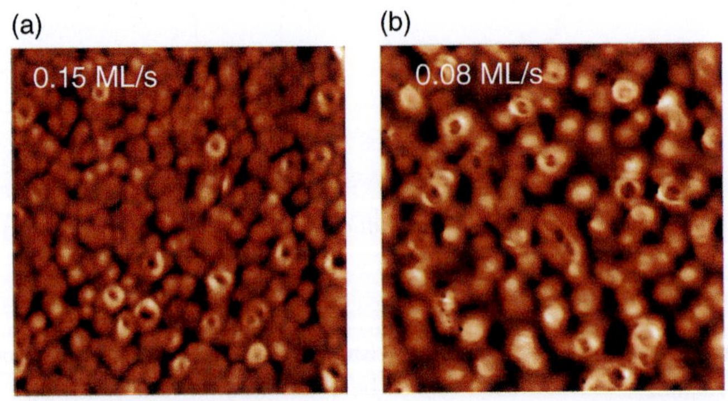

Figure 6.6 AFM surface images of InAs QDs capped by 3 ML of GaAs prepared with different growth rates of InAs layer: 0.15 ML/s (a) and 0.08 ML/s (b). In the case of slower growth, larger QDs are formed.

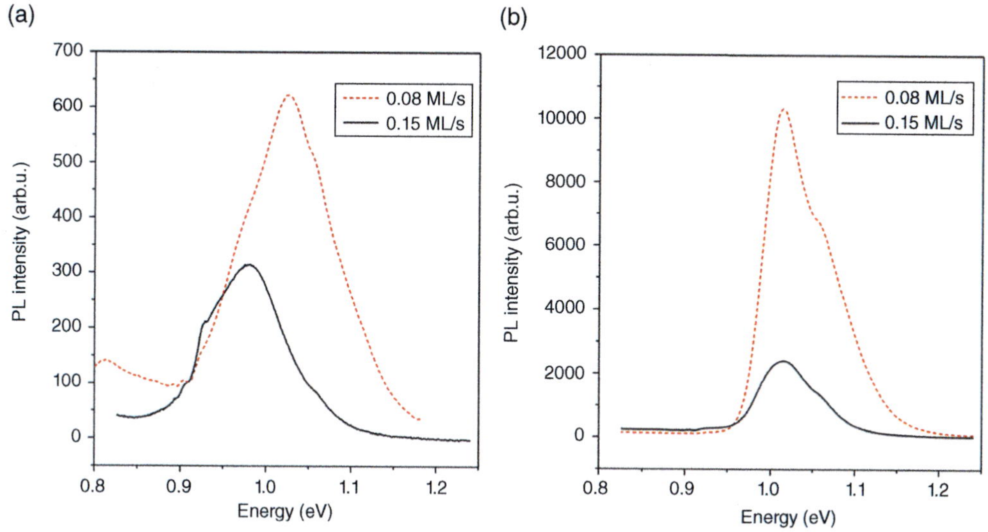

Figure 6.7 PL intensity of QDs with the InAs growth rate 0.08 ML/s for 22 s and 0.15 ML/s for 11 s, with 15 s growth interruption for QD formation capped only by 1 nm of GaAs (almost uncapped) (a) and capped by 10 nm of GaAs (b).

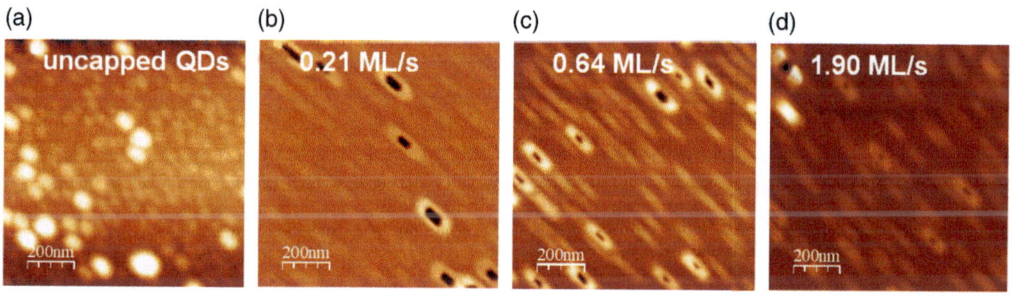

Figure 6.8 AFM images of uncapped QDs (a) and QDs capped by a 10-nm thick GaAs layer grown with different growth rates: 0.21 ML/s (b), 0.64 ML/s (c), and 1.9 ML/s (d). The center of large objects with an uncapped apex is dissolved and redistributed over the epitaxial surface.

redistribution of dissolved InAs material surprisingly has an effect on the PL spectra of QDs capped by layers with different growth rates; see Figure 6.9. QDs capped more quickly have higher emission energy. More details about QD dissolution will be discussed in Section 6.4.1.

6.3.2.6 *Buffer Layer Growth*

However, not only the QD covering layer is important for QD shape and luminescence; the growth rate of the buffer layer is also extremely important. Growth of very smooth buffer layers is important for narrow QD size distribution, as will be demonstrated in the following section.

6.3.3 QD Surrounding Layers

One of the most promising applications of QDs is the creation of a source of light emitting at telecommunication wavelengths 1.30 μm and 1.55 μm (0.95 eV and 0.80 eV), which have advantages for optical signal transmission in silica fibers: 1.3 μm because of the lowest signal dispersion and 1.55 μm because of the lowest attenuation. Although the growth of GaAs by MOVPE is

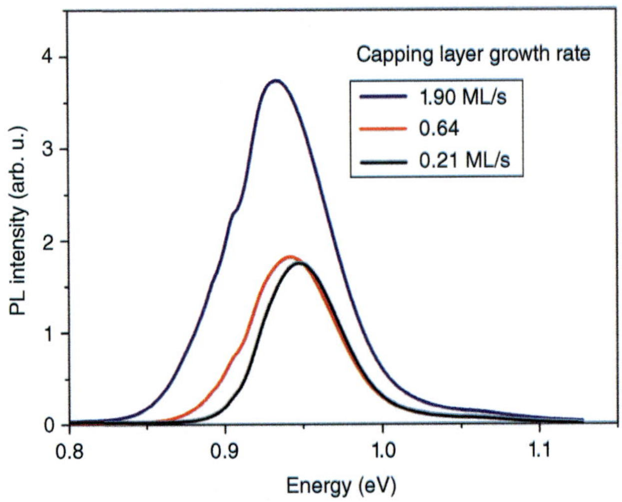

Figure 6.9 *PL spectra of QD samples with different GaAs capping-layer growth rates. Samples with slower QD capping process have lower emission energy due to the incorporation of InAs material from large objects to the QD surrounding.*

well-managed, a GaAs-based structure emitting at such long wavelengths is still missing in the semiconductor industry. Nowadays, InP-based sources of light are used, but they are not very suitable because of their expensive production and strong temperature dependence, and thus shorter lifetime due to degradation. This is why attention was drawn to the GaAs-based structure PL shift to longer wavelengths (lower energy). One possibility is the preparation and use of structures based on InAs QDs grown on a GaAs substrate. Since the emission wavelength of this simple structure is only a little more than 1100 nm, it has to be supplemented by surrounding layers of materials with different parameters that influence the band structure and help shift the PL wavelength to the desired region. Those additional layers can be below the QDs and are called *buffer layers*, or they can be above QDs and called *capping layers*.

The idea of the GaAsSb SRL covering the InAs QDs grown on the GaAs substrate was used in 2004 by Akahane et al. to prepare such structures using an MBE growth technique [88]. In 2005, the structures with InAs QDs and GaAsSb SRL were prepared by MBE and showed a PL up to 1430 nm and room temperature lasing at 1292 nm [89]. In the following year, a study of such structures prepared by MBE continued by reaching a room-temperature emission of light with wavelength of 1600 nm [90], but it was a type-II heterostructure with spatially separated electrons and holes with low-intensity PL. Also in 2006, the first GaAs substrate/InAs QDs/GaAsSb SRL structure was prepared by MOVPE [91]. It led to a decrease of surface energy and a change of surface diffusion kinetics, which resulted in a lower concentration of In atoms in the upper layers and larger QDs with a lower density. As the structures with GaAsSb SRL showed better structure parameters – such as better charge carrier confinement and longer emission wavelength due to the higher QDs – than the InGaAs one [92], the material was studied more intensively from this point.

InAs QDs are mainly prepared on GaAs. The lattice mismatch of those two materials is slightly above 7%, which leads to a formation of a strain field at the interface. This is partly compensated for by a wetting layer, which is a single monolayer of InAs, on which a QD is formed. In addition, the roughness of the buffer layer plays an important role on QD size and homogeneity; see Figure 6.10. The roughness of the GaAs surface is increasing from Figure 6.10a to c. The different roughness was achieved using a different growth rate – for a higher growth rate, the roughness is

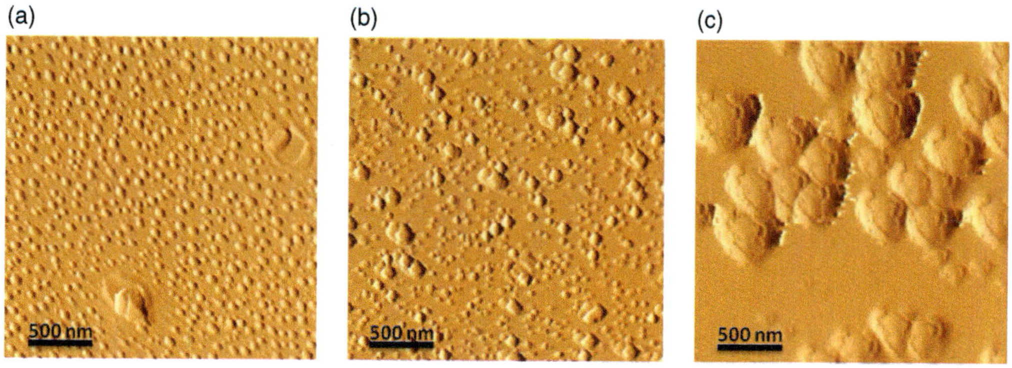

Figure 6.10 AFM images of QDs grown on GaAs substrates with different roughnesses.

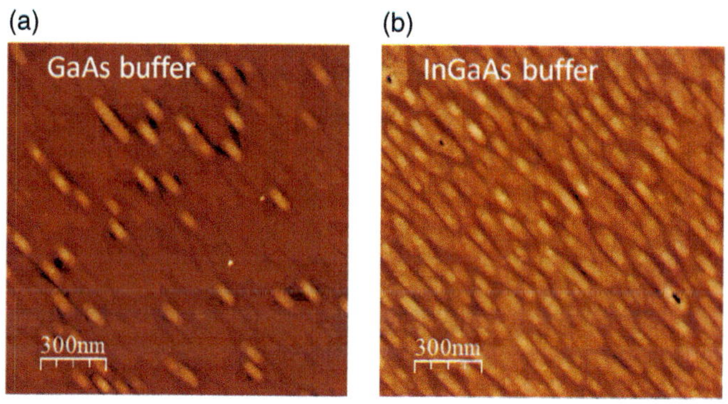

Figure 6.11 QD surface distribution in GaAs (a) and InGaAs buffer layer (b).

higher. As can be seen from the AFM images, for a smoother GaAs underlying buffer layer, the QDs are smaller and more homogeneous in size and distribution. For a rougher GaAs surface, the InAs material tends to coalesce and forms large islands that are no longer QDs.

When InGaAs is used as a buffer layer, it serves as a transition layer from GaAs to InAs and thus helps to reduce the strain in the structure, which then leads to more homogeneous formation of QDs; see Figure 6.11. In this case, the buffer layer is $In_{0.08}Ga_{0.92}As$.

These two samples are also compared based on the PL measurement; see Figure 6.12. When only a GaAs layer is used, there are only a few QDs that show the PL, and the signal intensity is low. When an $In_{0.13}Ga_{0.87}As$ buffer layer was used, the QD density was increased, which resulted in higher PL intensity.

Another type of surrounding layer is a capping layer deposited right over the InAs QDs. This can be GaAs, the same as the buffer layer, or a strain-reducing layer that can influence the strain field in the structure, which is advantageous for the blueshift of the PL emission wavelength. The SRL is usually a few nanometers thick and moderates the change of the lattice constants on the interface. The lattice constant of the SRL lies between the values of lattice constants of GaAs and InAs; thus the InAs QD is not that compressed, as it would be when covered directly with GaAs.

When InAs QDs are covered directly by GaAs, some In atoms can diffuse into the GaAs material, creating a blurred interface; see Figure 6.13.

Moreover, it is difficult to achieve PL of QDs with emission energy smaller than 1 eV when they are covered only by a thick GaAs capping layer; see [93]. The dependence of emission energy on

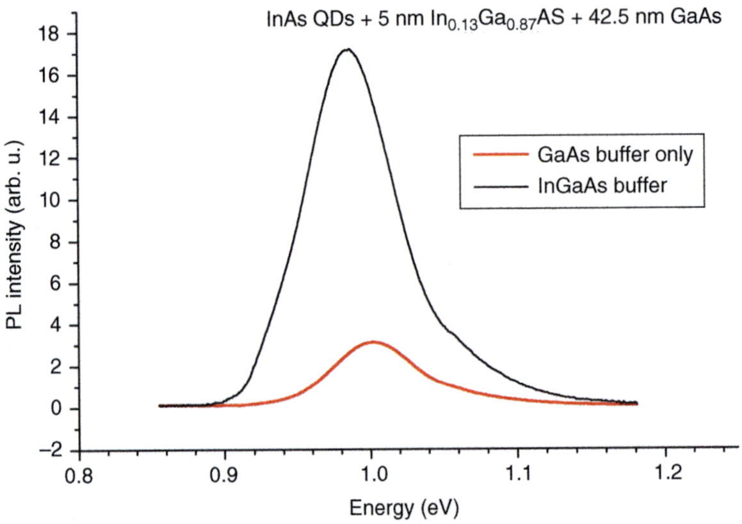

Figure 6.12 Comparison of PL spectra of QDs with GaAs buffer only and a sample with an InGaAs underlying layer.

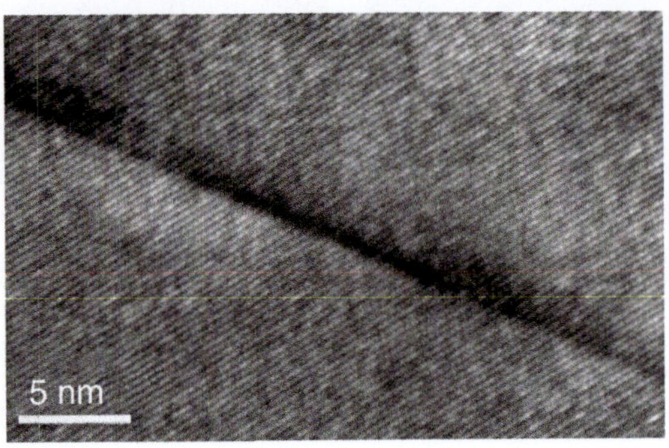

Figure 6.13 HRTEM image of an InAs wetting layer covered by GaAs. A diffusion of In atoms can be seen in the form of a blurred interface.

the thickness of the GaAs capping layer is shown in Figure 6.14a. The shift is caused by increased strain inside the InAs QDs with increased capping layer thickness. The value of the strain was simulated on the top and in the middle of a single QD for different GaAs capping layer thickness, and the results are drawn in Figure 6.14b.

Further improvement of MOVPE-prepared QDs can be achieved by adding the previously mentioned SRL. The aim of the SRL is generally to relax the strain inside the QD (which was later theoretically computed in [94]). There are various consequences of the SRL's presence: the PL maximum is shifted; the QDs are preserved from dissolution, blurring, and changes in shape (which could happen when covered only with a GaAs layer); and the hole localization can be moved to achieve various structures for different applications. There are two commonly used ternary materials for the growth of strain-reducing layers: InGaAs and GaAsSb, which were studied simultaneously during recent years. The mechanism of the strain relaxation is the same for both InGaAs and GaAsSb ternary alloys, but at the same time different from the GaAs-covering mechanism. While GaAs covers the surface with QDs from the bottom upward and levels off the

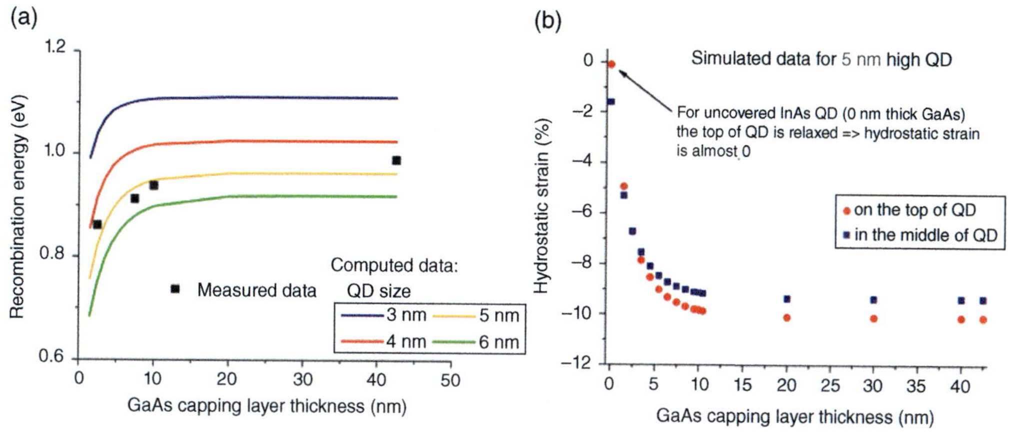

Figure 6.14 *Dependence of recombination energy on the thickness of GaAs capping layer (a), and explanation by simulated hydrostatic strain on the top of and in the middle of a QD (b). Source: Reprinted with permission from Zíková et al. 2017 [93].*

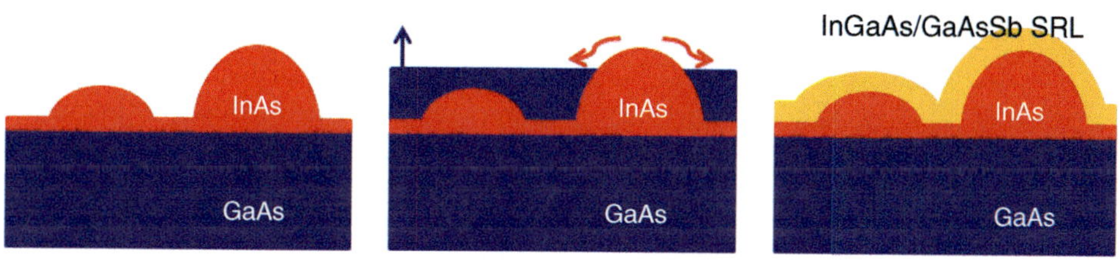

Figure 6.15 *Covering mechanism of different types of covering layers from left: no capping layer, GaAs capping layer, InGaAs/GaAsSb strain-reducing layer.*

surface, both InGaAs and GaAsSb layers copy the topography of the surface; see Figure 6.15. This then enables the preservation of the size and shape of InAs QDs.

For InGaAs SRL, a study of the capping-layer thickness, similar to that carried out for the GaAs CL, was performed. The thickness of the InGaAs was kept at 5 nm, while the thickness of the GaAs was varied. The results for InGaAs with 13% of In are shown in Figure 6.16.

When InGaAs strain-reducing layers are used, it is possible to achieve lower recombination energy. The main problem with the InGaAs SRL is in the weak confinement of electrons in the QD for higher In content [95], due to the lowering of the conduction-band edges, which can then lead to a higher dependency on ambient conditions such as temperature. Another reason for the weak confinement could also be a poor InAs/InGaAs interface, which could lead to delocalization of electrons in the structure. Then, the structure shows PL with low intensity because the recombination rate of electrons and holes decreases. This can be seen in Figure 6.17 from PL spectra measured for different InGaAs compositions. There, the PL energy is decreased for higher In content, but from 13% of In, the PL intensity steadily decreases. Concerning the PL maximum shift and intensity, the most suitable composition of the InGaAs layer would be around 13% or slightly higher.

For an InGaAs SRL, the shift could be also caused by effectively enlarged InAs QDs. For a covering layer, the preferred deposition site for atoms depends on the SRL composition, and large atoms like In and Sb may accumulate in the vicinity of the QD apex. The QD is relaxed such that the lattice constant of InAs is highest on the apex, which matches the higher content of In or Sb

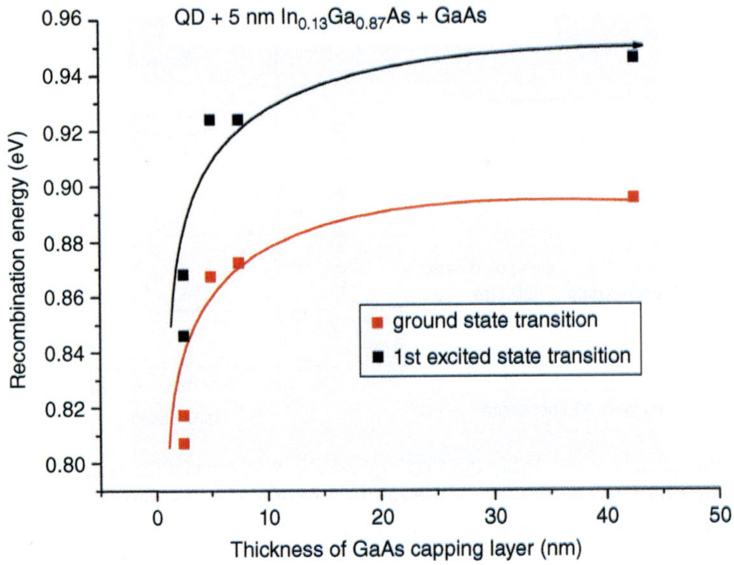

Figure 6.16 Dependence of recombination energy of a structure with InGaAs strain-reducing layer on the thickness of GaAs capping layer.

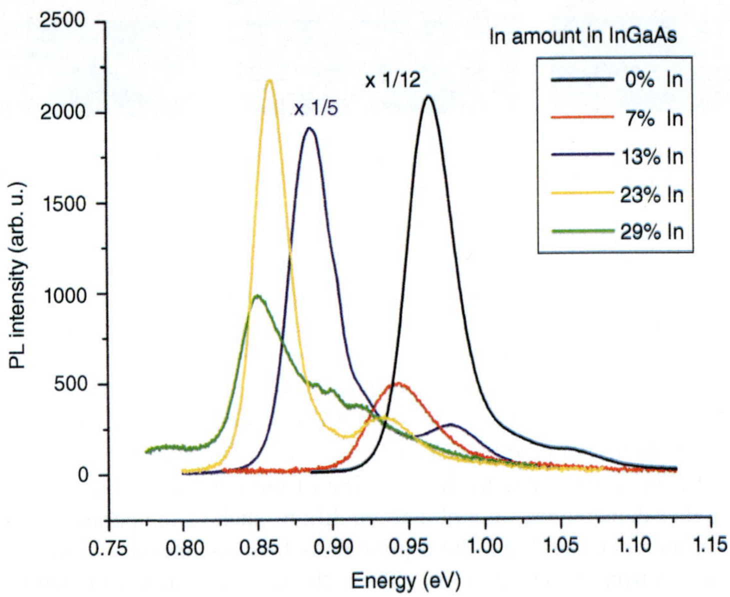

Figure 6.17 PL spectra of InAs QDs covered by InGaAs SRLs with different In content.

atoms in InGaAs or GaAsSb SRL, respectively. On the QD "sides," the lattice constant of InAs is smaller due to the compressive strain; this fits better to the SRL with lower content of In or Sb atoms. In the case of an InGaAs SRL, the surrounding In atoms can then effectively enlarge the QD size, which further influences the recombination energy. For the GaAsSb layer, the InAs QD size is well conserved by the SRL. In addition, the interface of InAs/GaAsSb is sharp, which was also revealed by the high-resolution transmission electron microscopy (HRTEM) image in Figure 6.18.

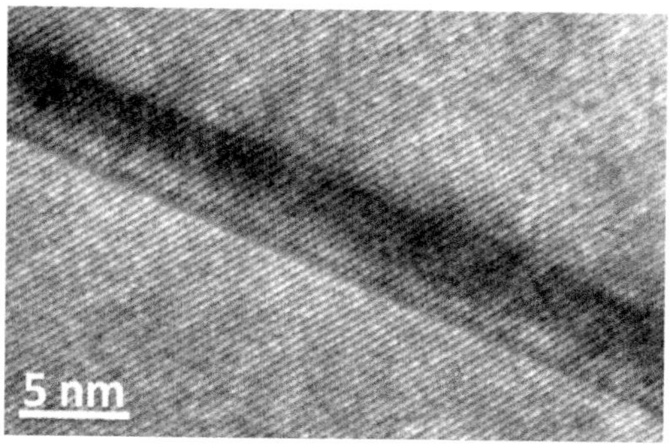

Figure 6.18 HRTEM image of InAs wetting layer covered by GaAsSb SRL that suppresses In atom diffusion, thus creating a sharp InAs/GaAsSb interface.

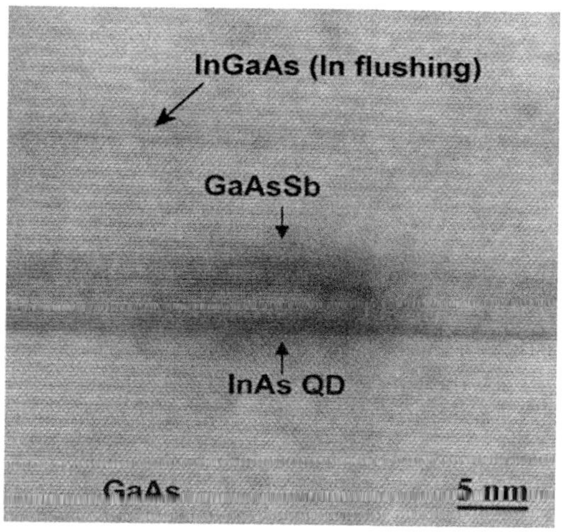

Figure 6.19 HRTEM image of quasicombined GaAsSb+InAs QDs self-organized during GaAsSb capping of InAs QD. Source: Reprinted with permission from Nadtochiy et al. 2017 [132].

As was shown earlier, a GaAsSb SRL covers the surface with InAs QDs equally and preserves their height and shape. Sb incorporation is enhanced in the InAs QD surroundings, which could lead to the quasicombined InAs + GaAsSb QD formation, as can be seen in the HRTEM image in Figure 6.19. The In flushing was performed to eliminate the rest of the In atoms floating on the surface during the GaAs growth.

For a GaAsSb layer, it is possible to obtain a heterostructure of type I or type II by adjusting the composition. The transition from a type-I to type-II heterostructure occurs for approximately 14% of Sb [89, 96], and this value depends mainly on the QD size and SRL thickness. For an InGaAs layer, the type-II heterostructure is not considered at all, because even when it is theoretically possible to reach a type-II heterostructure for approximately 45% of In in InGaN, when a QD is present, the In content would be so high that the structure would deteriorate and would contain a high number of dislocations.

Both types of heterostructures can be used for different applications. For the type I, electrons and holes are localized in the same area, which is in this case a QD; see Figure 6.20a. Their electron and hole wavefunctions have good overlap, and the radiative recombination rate is high. This means that the structure shows PL with high intensity, and thus it can be used as a source of

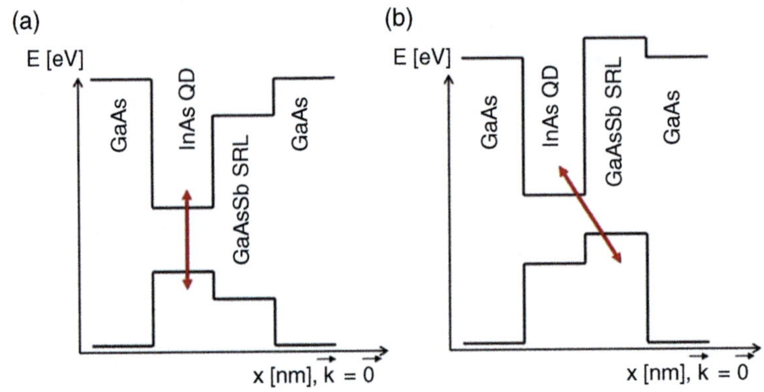

Figure 6.20 *Schematic drawing to explain type-I (a) and type-II (b) heterostructures.*

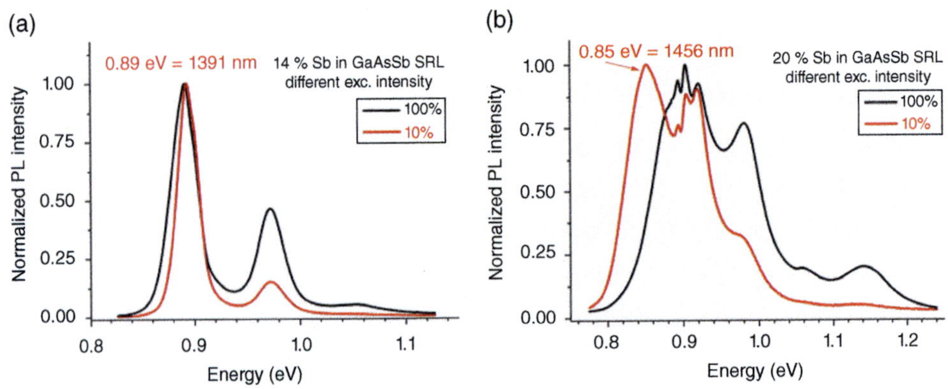

Figure 6.21 *PL spectra of a type-I and type-II heterostructure.*

light: for example, in telecommunications. In a type-II heterostructure (see Figure 6.20b), charge carriers are spatially separated from each other: in this case, electrons in QDs and holes in SRL. Then, the wave-function overlap is lower and the recombination rate is lower. On the other hand, the charge carriers can be quickly separated, which would be advantageous for applications in detectors or solar cells.

It is possible to distinguish those two types based on the PL measurement with different excitation intensity. When a heterostructure is of type I (Figure 6.21a), the PL wavelength peaks remain at the same energy, in this case 1391 nm for the ground state. When a heterostructure is of type II, the PL maximum peaks shift to lower energy for lower excitation intensity; see Figure 6.21b. This mechanism is usually explained by electrons and holes creating a dipole and a triangular-like potential on the interface. For higher excitation intensity the potential becomes steeper, and the energy levels shift to higher values, which then increase the recombination energy. For more details, see [97, 98].

The PL results obtained on samples with InAs QD covered by GaAsSb SRL with various composition are shown in Figure 6.22. On increasing Sb content, the recombination energy decreases, but the PL intensity decreases for Sb contents above 12%. This is caused by the transition to the type-II heterostructure with decreased wave-function overlap and thus lower radiative recombination rate. For light-emitting applications, the ideal composition would be around 12%, which is the value when the PL wavelength is shifted to lower energies and the PL intensity is still

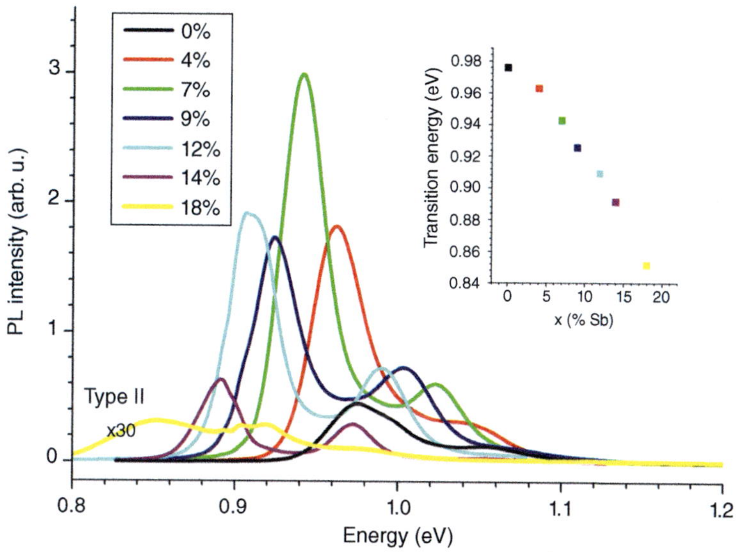

Figure 6.22 *PL spectra of a structure with InAs QDs covered by GaAsSb SRL with various Sb contents.*

intense. The higher energy peak that is present for all spectra is a first excited state of QDs. With a GaAsSb SRL, long-wavelength PL from a type-I heterostructure was achieved in 2013: 1371 nm for an appropriate QD size, and 14% of Sb in GaAsSb [34].

6.4 In Situ Measurements

It is always important for any epitaxial process to use an in situ method that can provide information about the growth and processes on the epitaxial surface. In the case of QDs, the growth procedure is so complex that usage of an in situ method is extremely important. The most-often-used methods are reflectance measurement and reflectance anisotropy spectroscopy.

6.4.1 Reflectance Anisotropy Spectroscopy of QD Growth

This optical method can provide information about processes taking place on a 1-ML thick epitaxial surface during epitaxy. It is based on measuring the difference of reflectance in two crystallographic directions: [110] and [–110] [99]. The sample surface has anisotropic reconstruction during the growth due to the dangling bonds. The surface sensitivity of the RAS signal is then achieved by subtracting the value of reflectance in two crystallographic directions, [110] and [–110], which eliminates the contribution of the isotropic bulk reflection. The output signal is described as

$$\mathrm{Re}\left\{\frac{\Delta r(E)}{r(E)}\right\} = \mathrm{Re}\left\{\frac{r_{[\bar{1}10]}(E) - r_{[110]}(E)}{1/2\left(r_{[\bar{1}10]}(E) + r_{[110]}(E)\right)}\right\}$$

where r is the reflectance with corresponding direction in square brackets. There are two possible modes in which the RAS data can be collected. First, the spectroscopic mode is used to obtain information about the RAS spectrum, which has a characteristic shape fingerprint for different

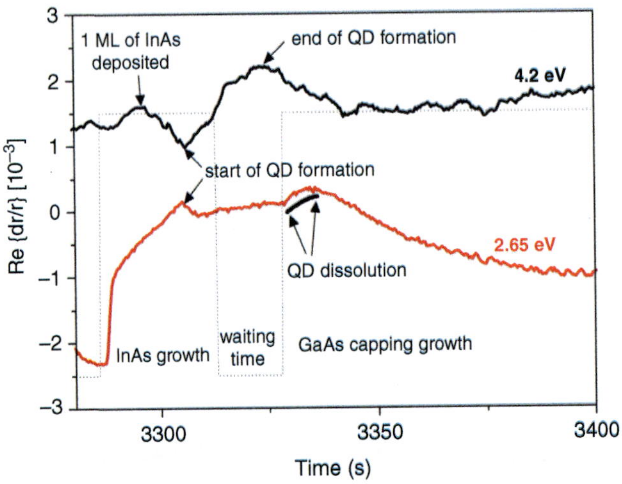

Figure 6.23 *Reflectance anisotropy signal measured at two different photon energies: 4.2 eV and 2.65 eV. Different phases of QD growth procedures are marked by arrows.*

surface reconstructions or composition of the surface layer. This measurement mode has very low time resolution (several tens of seconds), but the color plot consisting of these spectra measured during epitaxy can elucidate epitaxial mechanisms and different types of atom-incorporation mechanisms, and can be used for an immediate check of the process's reproducibility.

Second, the time-resolution mode measures the reflectance anisotropy for a chosen photon energy. The time-resolved mode must be used for monitoring of fast processes such as QD formation. It is necessary to choose the appropriate energy with strongest RAS changes during the monitored process. For QD formation monitoring, two energies are usually used: 2.65 eV and 4.2 eV [100] (Figure 6.23). Both energies are suitable to monitor different parts of the structure growth. The maximum RAS intensity at 4.2 eV is achieved when 1 ML of InAs is deposited [101].

The RAS signal at 4.2 eV is more sensitive to InAs thickness than at 2.65 eV. This is useful for monitoring and estimating the minimal InAs amount, when QDs start to form, as well as finding the moment when QD formation is completed. On the other hand, the start of QD formation can also be monitored at 2.65 eV. This energy is very sensitive to the presence of In on the surface, since it turns the orientation of As dimers (absorption energy 2.65 eV) from the [110] to the [−110] direction. This is why it is more suitable for observation of the dissolution of QDs during the capping process. Different discussed moments during the QD structure growth are indicated in the RAS signals in Figure 6.23.

The example of optimization of InAs amount estimation from RAS signal at 4.2 eV is shown in Figure 6.24a. During the waiting time, QD formation can be observed on the RAS signal for the sample with 1.8 ML of deposited InAs, while the 1.5-ML InAs is not sufficient for QD formation, as published in [102]. RAS observations were also confirmed by PL spectra of both samples in Figure 6.24b, where the sample with 1.5-ML InAs has very weak QD PL with a higher energy maximum (smaller QDs).

Waiting-time optimization is important to achieve a QD structure with high PL intensity. *Waiting time* is the growth interruption after InAs deposition under an arsine partial pressure, to prevent desorption of As atoms from the surface. During this time, QDs are formed. The RAS signal in time-resolved mode at 4.2 eV of two samples prepared with different WTs of 14 s (blue line) and 30 s (red line) are compared in Figure 6.25a. The QD formation during waiting time is characterized by an increase of the RAS signal, since the QD formation is accompanied by formation of a 1-ML

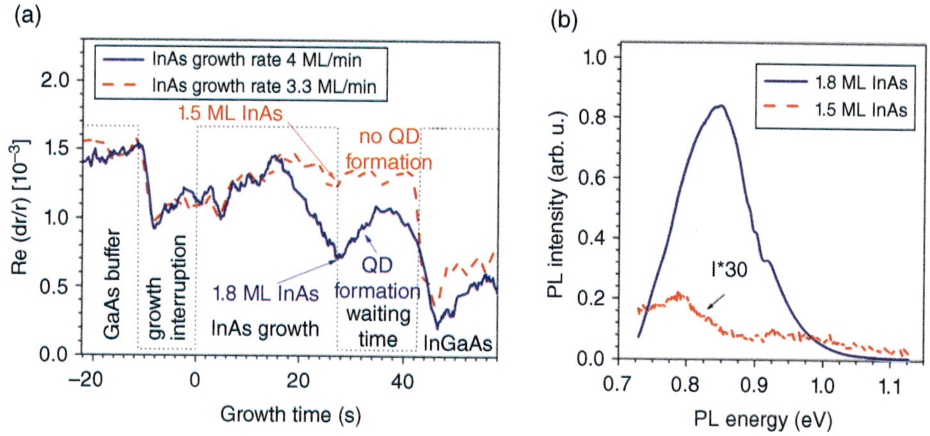

Figure 6.24 *Comparison of RAS at 4.2 eV of two samples with sufficient and subcritical amount of deposited InAs (a). PL spectra of the same samples (b). Source: Reprinted with permission from Zíková et al. 2016 [102].*

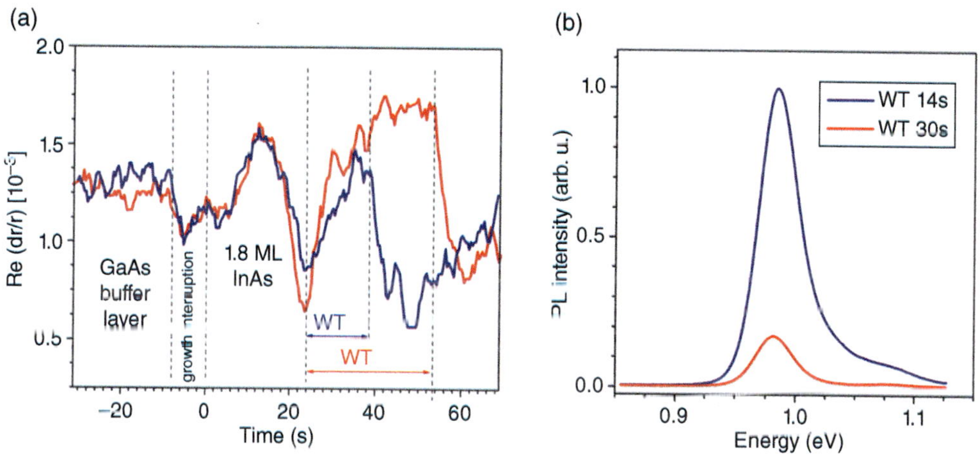

Figure 6.25 *Comparison of RAS at 4.2 eV of two samples with different waiting times – optimal (14 s, blue line) and prolonged (30 s, red line) (a); PL of the same samples (b). Source: Reprinted with permission from Zíková et al. 2016 [102].*

thick InAs wetting layer. When the wetting layer as well as QD formation is completed, the RAS signal remains at the same value. According to the RAS signal, the time necessary for QD formation is 14 s (blue line). For longer waiting times, Ostwald ripening takes place, when large objects with dislocations and without PL are formed, as in the case of the sample with a 30 s waiting time (red line for RAS as well as PL spectrum) in Figures 6.25a and b.

Considerably lower PL intensity for longer-than-optimal waiting time occurs due to the Ostwald ripening problem, during which dislocations are created; see Figure 6.26.

With the help of the color plot mode, the presence of different types of atoms on the epitaxial surface can be recognized, so it is very beneficial when the time resolution is not crucial. Different compositions of epitaxial layer can have different fingerprints; see Figure 6.27. The shape of the RAS spectrum can be also influenced by the partial pressure of precursors, namely by the V/III ratio.

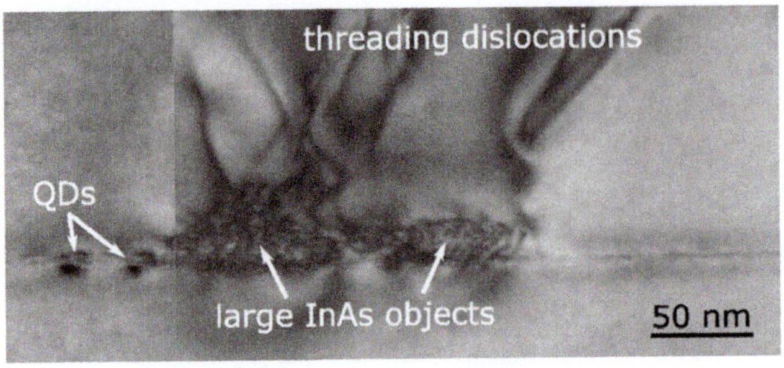

Figure 6.26 *TEM images of two InAs QDs and of large objects formed due to Ostwald ripening. Threading dislocations originate in partly relaxed large objects.*

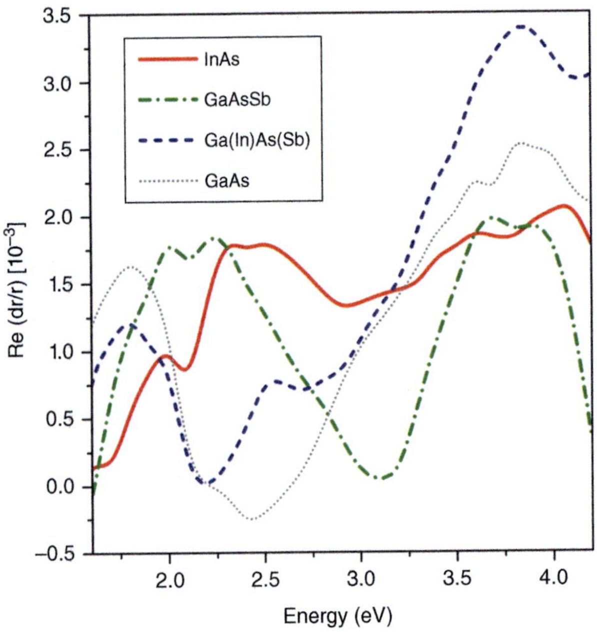

Figure 6.27 *RAS spectra of epitaxial layers with different compositions. Each material has its characteristic spectrum: the GaAsSb spectrum with higher Sb content has a minimum around 3.0 eV and two maxima near 2 eV and 3.8 eV. InAs has a characteristic maximum near 2.6 eV; this local maximum can be observed also for InGaAs layers. GaAs with 4 × 4 surface reconstruction has a minimum near 2.5 eV. Source: Reprinted with permission from Nadtochiy et al. 2017 [132].*

In Figure 6.28, an example is shown of how QD dissolution during the capping process can be observed by RAS in the spectroscopic mode measurement imaged in the color plot. In this case, the color plots were used to optimize the GaAsSb composition, with the aim to suppress QD dissolution during the capping process.

The TEM image of the sample with GaAsSb capping layer with lowest Sb content also confirmed the presence of a nonintentional In(Ga)As layer after GaAsSb deposition. During the growth interruption, other nonintentional QDs were formed; see Figure 6.29.

The RAS can also be used for optimization of the In-flushing steps that are necessary for the growth of multiple QD structures to prevent transfer of In atoms from the lower to upper QD layers, due to the surfactant ability of large In atoms [101, 103].

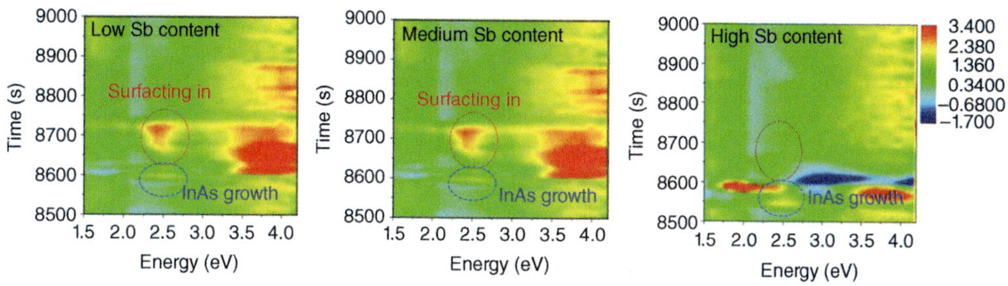

Figure 6.28 *RAS color plots of the growth of QDs capped by GaAsSb layer. It can be seen that high Sb content in GaAsSb SRL (shown by a deep blue minimum at 3 eV) can prevent the dissolution of QDs. For lower Sb content, a fingerprint spectrum of nonintentional InGaAs growth can be recognized after GaAsSb deposition (the red maxima around 2.5 eV and 3.8 eV).*

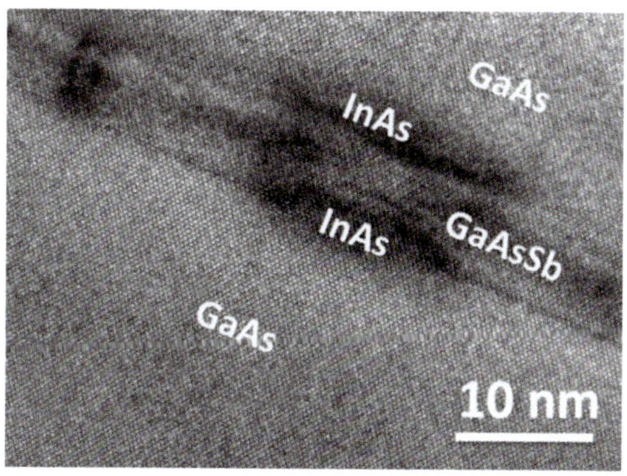

Figure 6.29 *TEM images of a QD capped by GaAsSb layer with low Sb content at the beginning of the capping process. The dissolved amount of InAs during QD capping was so large that another unintentional QD layer was formed during the growth interruption after the GaAsSb growth. Source: Reprinted with permission from Zíková et al. 2017 [93].*

6.4.2 Other Supporting In Situ Measurements

For structure growth, it is advantageous to measure temperature, if possible in situ and continuously. The best way is to measure surface temperature and, if possible, the whole surface temperature homogeneity. MOVPE systems are usually equipped with thermocouples, but true temperature optical surface measurements, based on emissivity corrected pyrometry, are better.

Measurements of internal ambient quality (hydrogen, nitrogen, precursors, residual vacuum, etc.) are necessary as well, but this is standard for newer MOVPE equipment.

For the growth of structures with QDs, it is also crucial to measure layer thicknesses. The best and the most exact is RAS, as previously described. If the machine is not equipped with RAS or the material is not suitable for RAS (discussed earlier), growers have to measure the time of growth and extrapolate thin layer thickness (less than 1 nm) from the growth speed determined from the thick layer (around 1 μm) growth time.

Other in situ measurement that can be useful for QD growth observation, like in situ scanning tunneling microscopy (STM) (see Richter et al. [104, 105]), are still too exotic. The first in situ STM measurements showed Ostwald ripening of InAs QDs grown on Si-doped (001) GaAs by MOVPE [92]. During an annealing step in the reactor under an arsenic overpressure immediately after the QD growth at 475 °C, the change in QD density and size distribution could be observed directly in a sequence of in situ STM images over 50 min. The density of InAs QDs decreased reciprocally during the annealing step, which agrees well with Ostwald ripening limited by indium incorporation/decomposition.

Direct QD in situ measurements, apart from RAS, are ellipsometry and reflectance measurements [106]. These measurements are also described by Steimetz [107]. The overgrowth of InAs islands with a GaAs capping layer has been investigated by optical in situ measurements under various growth conditions. The status of the surface during and after the capping-layer growth can be determined from RAS and spectroscopic ellipsometry (SE) measurements during growth. This correlation between the optical data and the surface morphology was confirmed by ex situ AFM measurements.

The influence of different As-precursors – TMAs (trimethylarsine), TBAs (tertiarybutylarsine), and AsH_3 (arsine) – on the formation of InAs QDs and their evolution was studied by applying RAS and spectroscopic ellipsometry (SE). Hydrogen radicals at the surface are found to enhance the ripening and coalescence processes. The use of TBAs instead of AsH_3 can partly reduce coalescence. The level of the GaAs covering layer during the overgrowth of InAs QDs was determined by RAS measurements. Temperature-dependent In segregation effects were observed in situ; see [108].

6.5 Structure Characterization

6.5.1 Optical: Photo-, Magnetophoto-, Electro-luminescence, and Spin Detection

The luminescence of the sample provides crucial information about QD and QW energy levels, which is very important for a structure preparation that should serve as a source of light with emission wavelength at a required value, which is 1.3 mm or 1.55 mm for GaAs-based structures. The simplest and widely used method is PL. Simple InAs/GaAs QDs emit on a wavelength range of 1150–1250 nm, depending on the QD size. Possible ways to influence the PL include changing the QD size, covering the QDs with various materials with different band structures, or just changing the composition of the QDs or QD surrounding material, as was discussed in Section 6.3.3. PL gives us information about QD size, which strongly affects the ground (s-like) state electron–hole transition, and also about the QD aspect ratio (QD height/lateral dimension ratio), which influences the energy difference between the ground (s-like) and first excited (p-like) QD state transition [109, 110]. PL measured under different excitation intensities can also be used for identification of type-I and type-II band alignment in GaAsSb capped InAs QDs (see Figure 6.21), as was described in Section 6.3.3.

The magneto-photoluminescence (photoluminescence measured in very high magnetic fields), can give us information about the elongation of QDs buried in the structure, especially about elongation in lateral directions [110] and [–110], since in the strong magnetic field the degenerate p-like QD state became separated; see Figure 6.30 and [109].

When the QD structure is prepared for laser or LED application, it is necessary to characterize the structure by electroluminescence (EL). Due to the different excitation mechanism and higher carrier density in the case of EL, the excited state transitions of QDs are enhanced, and the long-wavelength EL at 1300 nm is much more difficult to obtain. Very strong enhancement of excited-state transitions was observed for structures where an InGaAs SRL was used for QD covering, in comparison to a simple GaAs capping layer or to structures with a GaAsSb SRL. The comparison of PL and EL spectra for InGaAs SRL and GaAsSb SRL can be found in Figure 6.31.

For a GaAsSb SRL, electrons are better confined in QDs, and the excited-state EL enhancement with increasing injection current is weak (Figure 6.32), which is very a strong advantage over the InGaAs SRL.

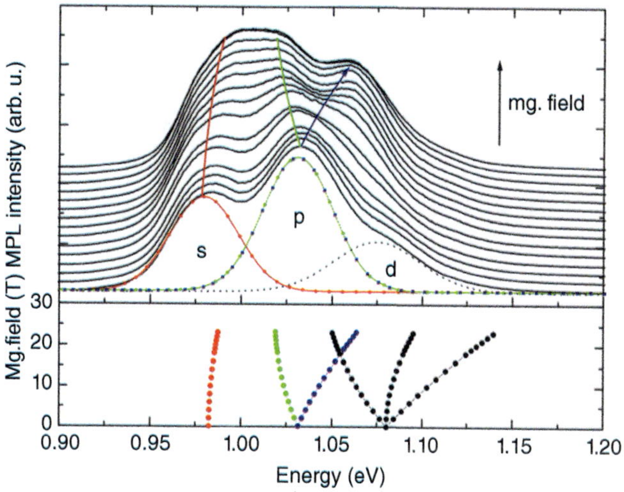

Figure 6.30 Magnetophotoluminescence spectra of our QD sample. Source: Reprinted with permission from Hospodková et al. 2006 [109].

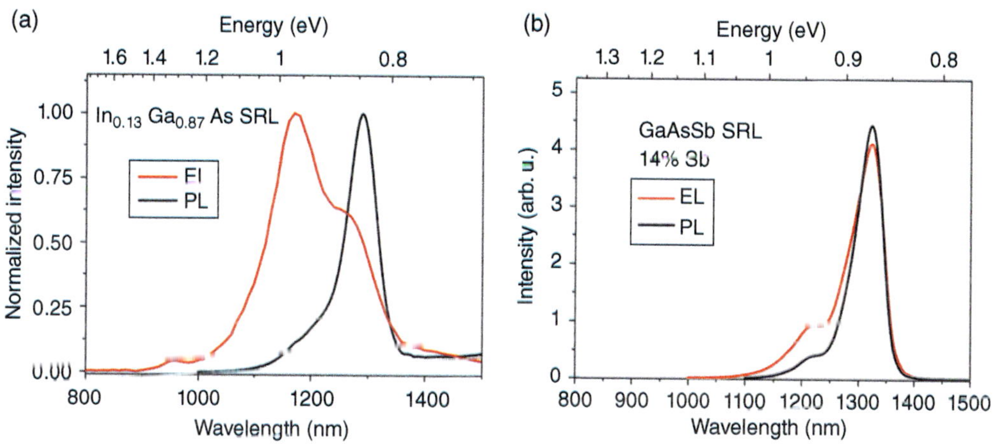

Figure 6.31 Comparison of EL and PL spectra of QD sample with InGaAs (a) and GaAsSb (b) SRLs.

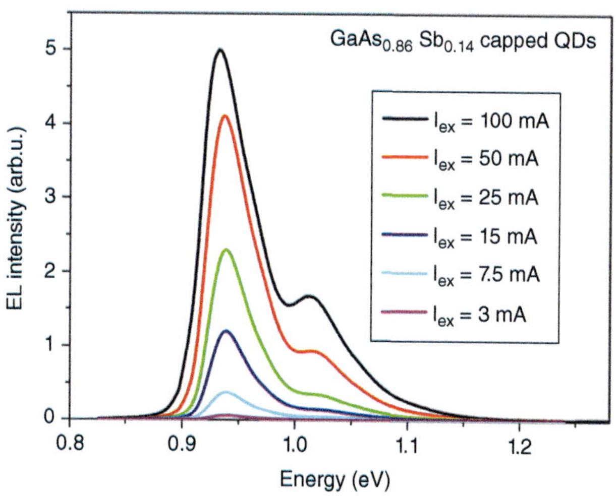

Figure 6.32 EL spectra of QD sample GaAsSb SRL measured with different injection currents.

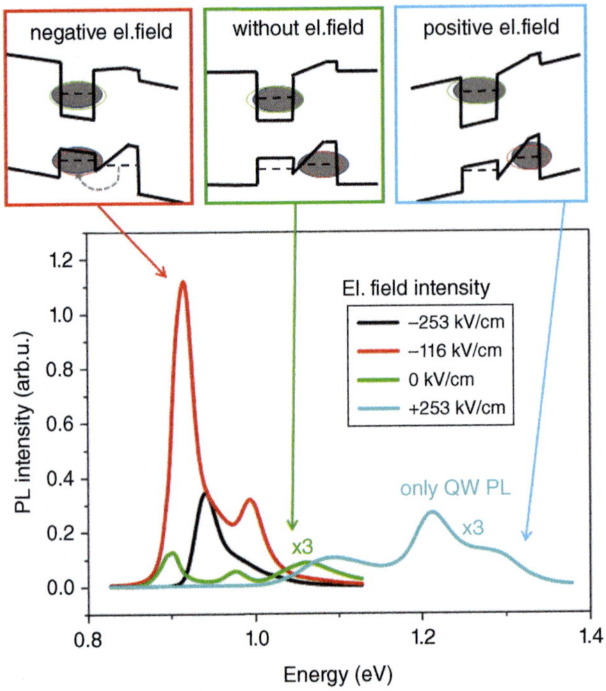

Figure 6.33 PL spectra of samples with QD covered by GaAsSb SRL embedded in PIN structures with different built-in electric field determined by the thickness of the intrinsic region. Acceptor and donor concentrations in p- and n-type layers were the same for all measured samples. Source: Reprinted with permission from Hospodková et al. 2016 [111].

When QDs are optimized for electroluminescent or detector applications, which require a structure with a p-n junction and a built-in electric field, the structure growth has to be optimized together with p- and n-type surrounding layers, since the presence of a built-in electric field may considerably influence the electron–hole wave function overlap and, consequently, the QD PL and photocurrent (PC) properties; see [111]. In Figure 6.33, the influence of an electric field on PL of QD structures is demonstrated. The presented PL spectra were measured on QD structures with a GaAsSb SRL prepared with the same technological procedure but differing in intensity and orientation of the built-in electric field. It can be noticed that proper orientation of the electric field can increase the electron–hole wave function overlap and enhance the PL intensity. On the other hand, the opposite electric field orientation can switch off the QD PL, and only PL of the wetting layer combined with a GaAsSb SRL was observed.

In the last decade, considerable effort has been devoted to development of QDs for quantum information processing applications. The ability to sequentially initialize, manipulate, and read out the state of a qubit, such as an electron spin in a QD, is required. For such applications, new optical methods for detection of coherent spin dynamics of a single electron in a QD were developed [112].

6.5.2 Microscopies – AFM, TEM, XSTM, BEEM/BEES

AFM is usually used to measure the density and distribution of QDs. It is a useful ex situ method to check the surface morphology at different stages of the QD structure growth when the growth is stopped. It can help to elucidate processes like QD formation, dissolution (see Figure 6.8 in Section 6.3.2), or incorporation of dissolved material, or the probability of QD stacking in MQD

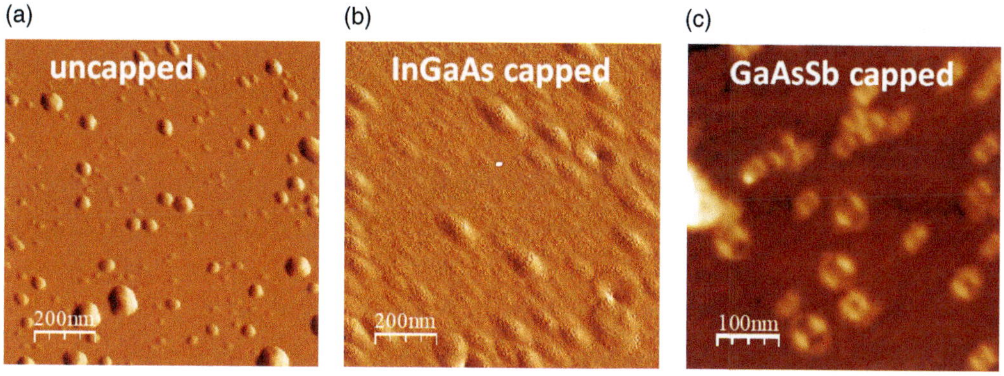

Figure 6.34 *AFM surface image of sample with uncapped QDs (a), QDs capped by 10 nm of InGaAs (b), QDs capped by 10 nm of GaAsSb (c).*

structures (see Figure 6.2 in Section 6.2.1). AFM can also be used to determine the shape of QDs after capping by layers with different compositions and different growth parameters of the capping layers. Different morphologies of the capped QDs were observed, such as quantum rings, elongated hillocks when capping by GaAs or InGaAs SRL (Figure 6.34b), and double peak elongated hillocks when capped by GaAsSb (Figure 6.34c). However, when considering the AFM images, it has to be kept in mind that the surface morphology may be changed after the growth stops during the cooling time.

To obtain information about the size and shape of QDs embedded in a structure, other measurement methods are necessary, such as cross-sectional scanning tunneling microscopy (XSTM) and TEM or, better, HRTEM. Using these methods, we can observe the final shape of embedded QDs and composition gradients inside QDs as well as in the surrounding layers resulting from complex processes, which take place during QD capping. XSTM and HRTEM can prepare images with atomic resolution. The disadvantage of both methods is the rather complicated sample preparation.

In the case of XSTM, the sample has to be cleaved in a high vacuum, and the strain in the QD structure usually makes it very difficult to obtain nicely planar cleavage in the QD surrounding, so many tries are required to obtain the required cleaved surface quality. Despite this obstacle, many nice XSTM studies were published on capped QDs [113–115].

The preparation of samples for HRTEM is even more complicated, since the structure slice has to be thinned down to electron transparency using cutting and mechanical polishing followed by Ar + ion milling. Preparing one sample for the measurement takes more time than the measurement itself. For illustration, some HRTEM images are shown in Sections 6.3.3 and 6.4.1. The area with a higher concentration of large atoms is seen as a dark region in the image; however, it can also be depicted as a light area, depending on the thickness of the sample; for more information, see [116].

QDs buried beneath an Au/GaAs interface were studied for the first time by Rubin et al. [117] using the imaging and spectroscopic models of ballistic electron emission microscopy/spectroscopy (BEEM/BEES). The BEEM images show enhanced current flow through each QD. Spectra taken with the tip positioned on a dot show shifted current thresholds when compared with the off-dot spectra, which are essentially the same as those of Au on bulk GaAs. Shifts in Gamma and L conduction-band thresholds are attributed to strain in the GaAs cap layer. The fine structure below the Gamma threshold is consistent with resonant tunneling through zero-dimensional states within the QDs.

We studied QDs with an image of elliptical shape by BEEM/BEES; see Wallachová et al. [118]. A scheme of BEEM/BEES equipment and complex QD structure scheme can be found in

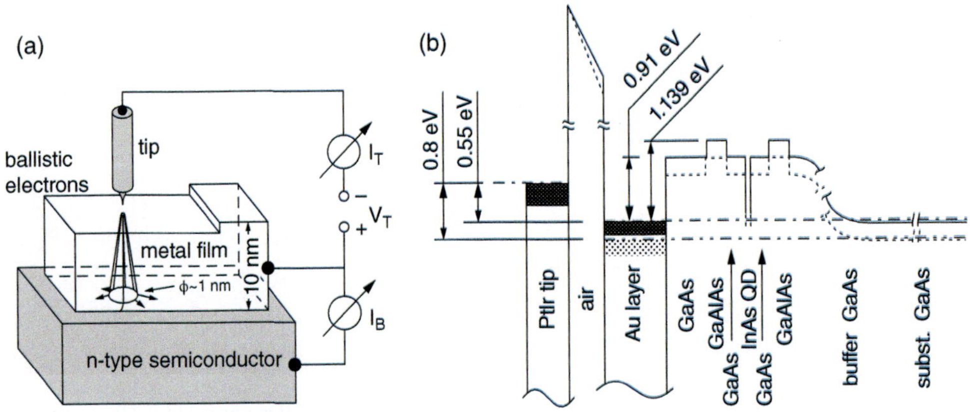

Figure 6.35 *Scheme of BEEM/BEES equipment (a) and QD structure scheme (b). Source: Reproduced from Walachová et al. 2007 [118], with the permission of AIP Publishing.*

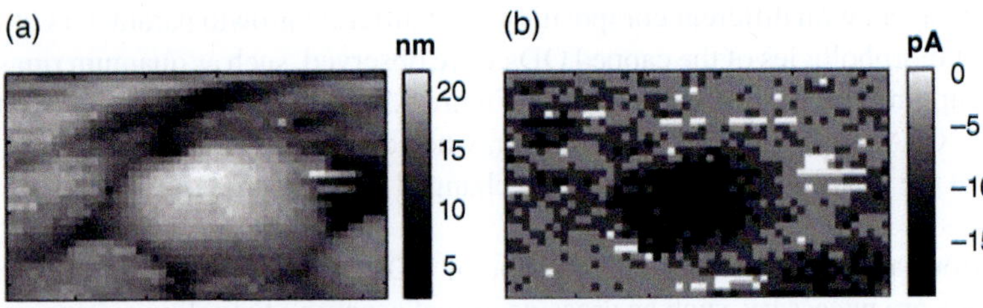

Figure 6.36 *Picture of QD obtained by STM (a) and by BEEM (b). Source: Reproduced from Walachová et al. 2007 [118], with the permission of AIP Publishing.*

Figure 6.35. Ballistic current–voltage characteristics through the QD outside the QD are compared in the voltage range of 0.55–2 V. Examples of ballistic characteristics and their derivatives for the voltage range from 0.55 to 0.8 V are given in [118]. In a detailed study, measurements with 1-mV steps in the energy range from 0.55 to 0.63 V were presented.

In Figure 6.36, pictures of a QD obtained by STM (a) and by BEEM (b) can be found. The difference between STM and BEEM currents is, remarkably, of three orders of magnitude.

Good agreement of the BEES spectrum [119] and theoretical calculations made by Bimberg et al. [120] can be found. This method is how to "directly" measure quantum levels in QDs. A recent review of BEEM/BEES was published in [121].

6.5.3 Electrical: Photocurrent, Capacitance Measurements

PC measurements are a necessary characterization method for QD structures prepared for solar cells or photodetector applications. InAs multiple-QD structures can be very promising for application in solar cells when they are surrounded by buffer and covering layers of a suitable type. To increase the extracted PC from QD structures and to suppress the recombination rate of photogenerated electrons and holes in InAs QDs, it is advantageous to cover them with a GaAsSb SRL with a type-II band alignment. In this structure, the incident light is absorbed predominantly in QD excited states with a better overlap of the electron and hole wave functions. With fast relaxation of carriers to the ground state, the electrons and the holes are quickly spatially separated (electrons to InAs QDs and holes to GaAsSb SRL). The extracted PC can be further increased using composition

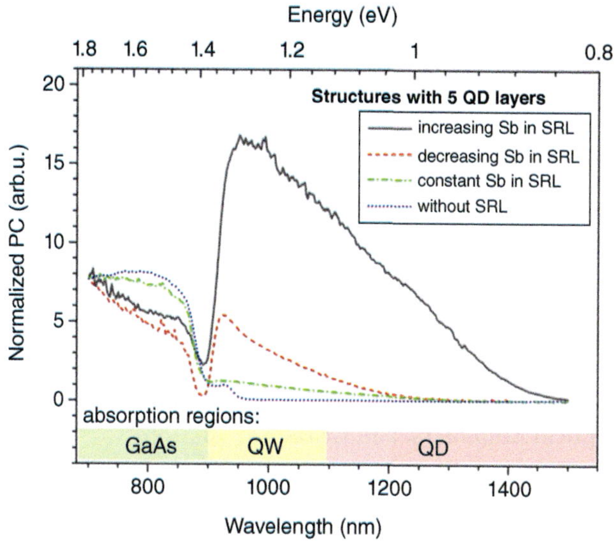

Figure 6.37 *Normalized photocurrent measurement of QDs covered by GaAsSb SRL with different composition GaAsSb layer profile. PC measured on simple InAs QD structure is shown for comparison. Significant PC enhancement in QW as well as QD absorption region for structures with GaAsSb is evident.*

profiling of the surrounding layers (Figure 6.37) or using InGaAs buffer layers, which can increase the QD density (see Section 6.3.3) and improve electron extraction [122].

For memory and detector or photovoltaic applications, the escape rate of carriers from QDs is an important characteristic, which has to be measured. For escape-rate measurement, capacitance spectroscopy and deep-level transient spectroscopy (DLTS) are used. An example of time resolved observation of electron escape from QDs as a function of temperature and electric field by DLTS and capacitance–voltage measurements can be found in [123].

6.6 Applications

As mentioned in Section 6.3.1, the MBE industry offers QD lasers for 1240–1310 nm wavelength, for high-temperature 1300 nm wavelength lasers, and for 3-inch GaAs high-quality QD wafers for DBR mirrors, RTDs, and HEMTs with QDs; see [124]. It is clear that the MOVPE industry is able to offer the same, but customers' needs are not great enough for them to switch from MBE to MOVPE.

MOVPE QD applications can be divided into the real running and potential ones.

Free QDs can be used in many more applications fields than the MOVPE-embedded ones that are our focus in this chapter. Embedded MOVPE-prepared QDs are currently used for prototypes of semiconductor lasers, optical amplifiers, LEDs, and photodetectors.

Potential fields of QD application are promising for optical applications due to their high extinction coefficient. They can operate like a single-electron transistor and show the Coulomb blockade effect. QDs have also been suggested as implementations of qubits for quantum information processing. QD technology is relevant to solid-state quantum computation. Current through a QD can be controlled, and thus precise measurements of the spin and other properties can be implemented. Several interwoven QDs, plus a way of performing operations, quantum calculations, and the computers that would perform them might be possible.

Tuning the size of QDs is attractive for many potential applications. For instance, larger QDs have a greater spectrum-shift toward the red, compared to smaller dots, and exhibit less-pronounced quantum properties.

InAs QDs embedded in an AlGaAs matrix have been produced by MOVPE with a partial capping and annealing technique to achieve controllable QD energy levels that could be useful for solar-cell applications. The resultant spool-shaped QDs are around 5 nm in height and have a lognormal diameter distribution, which is observed by TEM in the range from 5–15 nm. Two photoluminescence peaks associated with QD emission are attributed to the ground and the first excited-state transitions. The luminescence peak width is correlated with the distribution of QD diameters through the diameter-dependent QD energy levels.

Examples of the general research leading to the real applications of QDs and some specific solutions of growth research of QD devices are described in [84, 125–138].

6.6.1 QD Lasers, Optical Amplifiers, and LEDs

The great advantage of QD semiconductor lasers in general, not just diode lasers, is not only the small volume of the active laser zone, where the inversion of carrier population has to be achieved, but also the different density of states compared to QWs. This is why QD types of lasers have lower temperature dependence. The emitted wavelength can also be controlled based on the size and shape of the QD, not only the active-zone composition, as in the case with classical double-heterostructure laser materials.

As described in Section 6.1.3 and in the related literature [40–45], single-photon lasers based on single/individual QDs are very promising for quantum cryptography and quantum computation. These QDs can be prepared using either MBE or MOCVD. It may be that this type of QDs will be the most promising of all the QD types described here.

There are many publications dealing with QD lasers and optical amplifiers, which have nearly the same structure. Here are a few of the most relevant and actual papers:

- Highly efficient wavelength conversion observed by nondegenerate four-wave mixing in an InAs/GaAs QD Fabry–Perot laser. A maximum normalized conversion efficiency of –3 dB, along with a high 35-dB optical SNR is presented in [139].
- 40-GHz QD mode-locked lasers providing low-jitter optical and electrical microwave signals are being investigated. 160-Gb/s data transmission based on differential quadrature phase-shift keying and optical time-division multiplexing is demonstrated using a packaged module, described in [140, 141].
- The static properties and large-signal modulation capabilities of directly modulated p-doped QD distributed-feedback lasers are presented in [142]. A broad gain spectrum, which is typical for QD material, and emission wavelengths from 1290–1310 nm are covered by the transversal and longitudinal single-mode lasers fabricated from the same wafer. Thus, these lasers are ideal devices for on-chip wavelength division multiplexing within the original band according to the IEEE802.3ba standard. 10-Gb/s data transmission across 30 km of single mode fiber was demonstrated. The maximum error-free data rate was found to be 15 Gb/s.
- The dynamic properties of 1.31-μm InAs/GaAs and 1.55-μm InAs/InP QD Fabry–Perot lasers with the main focus on the increase of their large-signal modulation capabilities are reported in [143]. A GaAs-based edge-emitter structure incorporating a standard p-doped active region with ten QD layers enables 15 Gbit/s data transmission at 70 °C upon direct modulation. 80-Gbit/s on-off keying and 80-GBd (160-Gbit/s) differential quadrature phase-shift keying data transmission based on optical time-division multiplexing are demonstrated using a packaged 40-GHz module.
- The potential impact of QD lasers in communication links by characterizing their long-delay optical feedback responses as well as the role of the lasing states on the multimode dynamics of

InAs/GaAs QD Fabry–Perot devices sharing the same design were studied in [144]. The results unveil that the excited-state laser shows a much larger sensitivity to optical feedback, with a more complex route to chaos, and a first destabilization point occurring at lower feedback strengths than for a comparable ground-state laser, which remains almost unaffected.

- The vast majority of QDs used for LEDs, OLEDs, AMOLEDs, etc. are freestanding QDs (secondary embedded = cores covered by wider-bandgap material) prepared by technologies other than MOVPE. QDs for many types of LED screens are a solution that is feasible and suitable for wet-processing techniques. The two major fabrication techniques for QD-LED are called *phase separation* and *contact-printing* [145], but MOVPE-prepared LEDs are also currently being used. The current-injected spectrum of flat-topped LEDs using self-assembled SK InAs/InP QDs grown by selective-area low-pressure MOVPE with a double-capping procedure is described in [146]. By increasing the first capping layer thickness of three QDs layers, the output intensity of each layer has become equal, and a more than 500 nm spectrum width with flat-topped spectrum shape was obtained.
- Green and red LEDs based on InGaN/GaN QDs grown by controlling the process of the growth-interruption method using MOVPE are presented in [147]. It was found that the three-step growth-interruption method and the underlying InGaN/GaN superlattice structure are beneficial for achieving greater indium incorporation in InGaN QDs. As a result, green and red LEDs with electroluminescence peak energies of 2.28 eV at 20 mA and 1.70 eV at 80 mA, respectively, were demonstrated. The EL emission energy blue shift of the green QD LEDs is 140 meV as the injection current increases from 5 to 50 mA, while that of the red LED is 70 meV as injection current increases from 75 to 100 mA.
- High internal efficiency and high-temperature stability UV LEDs at 308 nm were achieved using high-density ($2.5\times10^{9}\,cm^{-2}$) GaN/AlN QDs grown by MOVPE. Photoluminescence shows the characteristic behaviors of QDs: nearly constant linewidth and emission energy, and a linear dependence of the intensity with varying excitation power. More significantly, the radiative recombination was found to be dominant from 15 to 300 K, with a high internal quantum efficiency of 62% even at room temperature; see [148]
- InGaN QDs formed via spinodal and binodal decomposition are used as an optically active region for LEDs. It is shown that the emission wavelength of the electroluminescence can be shifted from blue to green by adjusting the deposition time of the InGaN QD layer. A first approach toward a monolithic multicolour-emitting diode is presented. A strong difference of the EL on the InGaN QD stacking sequence is observed and is attributed to both low hole concentration and mobility; see [149].

6.6.2 QD Detectors, FETs, Photovoltaics, and Memories

Detectors for thermal imaging in the long-wavelength infrared region are dominated by the HgCdTe and QW photodetector technologies, both of which need cooling systems. Some alternative technologies are currently being developed: type-II superlattices and QD infrared photodetectors; see [150]. QD infrared photodetector properties and preparation are described in general by Razeghi in [151].

The design and fabrication of (Al)GaAs(Sb)/InAs tensile-strained QD-based detector material for thermal infrared imaging applications were presented in [152]. The detection is based on transitions between confined dot states and continuum states in a type-II band lineup, and we therefore refer to it as a *dot-to-bulk infrared photodetector* with expected benefits including long carrier lifetime due to the type-II band alignment, suppressed Shockley–Read–Hall generation–recombination due to the relatively large-bandgap matrix material, and inhibited Auger recombination processes due to the tensile strain and epitaxial simplicity. Multiple (Al)GaAs(Sb) QD layers on InAs substrates at different QD nominal thicknesses, compositions, doping conditions, and multilayer periods were prepared by MOVPE.

A high-temperature operating mid-infrared InAs/GaAs QD photodetector is reported in [153]. This detector covers a wide detection spectrum range from 3 μm to 6 μm. A large photoresponsivity of 6.4 A/W at a low bias voltage of 0.5 V and a high peak-specific photodetectivity D* of 6.0×10^7 cmHz$^{1/2}$/W were obtained at a high operating temperature of 230 K.

InAs QDs embedded in an InGaAs/InAlAs QW structure on InP substrate for mid-infrared detectors and focal plane arrays were presented in [154]. This combined dot–well structure has weak dot confinement of carriers, and as a result, the device behavior differs significantly from the one in more common dot systems with stronger confinement. The detectors operate at relatively high temperature (150 K) and have high quantum efficiency over 50%.

QD structures with type-II band alignment to the surrounding matrix material have been proposed as a III/V material approach to realize small-bandgap device structures suitable for photon detection and imaging in the long-wavelength infrared region, and are described in [155]. A review of alternative design approaches is presented there, and a choice of matrix material is discussed in terms of band alignment and its effect on the photoresponse. Photodiodes were fabricated consisting of 10 layers of $In_{0.5}Ga_{0.5}Sb$ QDs grown on InAs (001) substrates by MOVPE. The photoresponse and dark current were measured in single-pixel devices as a function of temperature in the range 20–230 K. These type-II InGaSb QDs can potentially be a competitive sensor material for the long-wavelength infrared region detection.

More than 20 years of research of QDs has been applied in field-effect transistors (FETs). Transformation of a simple quantum wire into a coupled QD array is described in [156]. A clear change in transport properties is observed, with changes in the barrier height at low temperatures. The experimental results are consistent with the theory of one-dimensional sub-bands and the Coulomb blockade of single-electron tunneling. Currently, MOVPE-prepared QDs strongly compete with Si QDs [157], graphene QDs [158], and CdSe/ZnS core/shell QDs [159] based FETs. Si-based FETs reach a fundamental limit and will no longer be a material of choice for sustainable continuation of Moore's law, due to low carrier mobility and chemical limitations. At this size, silicon dioxide is no longer an option due to charge leakage and bias-voltage variability. Other semiconductors such as GaAs and InGaAs have much higher mobility than silicon, thus increasing device performance. The use of a high-kappa ZnS/ZnMgS/ZnS heteroepitaxial tunneling layer in InGaAs FETs with QDs as the gate material is described in [160]. Fabrication of this semiconductor structure is promising in terms of reducing threshold-voltage variation by replacing silicon and hafnium oxides as a gate insulator. The II-VI tunneling insulator is grown on an InGaAs substrate using ultraviolet-irradiated MOVPE. GeO (x) -Ge cladded QDs are self-assembled to form the gate material.

A transistor and memory operation of a new AlGaAs/InGaAs FET in a tetrahedral-shaped recess on the (111)B GaAs substrate was investigated at a temperature up to 120 K; see [161]. This type of FET memory has a channel on the (111)A side surfaces of the recess and a single floating QD gate at the bottom. The charge in the floating QD gate can effectively modulate the channel current. They found a clear hysteresis in the current–voltage (I–V) characteristics with an abrupt increase and decrease in the current at the subthreshold gate bias region. Random telegraph signals with constant amplitude of about 70 nA were also observed in the memory-retention characteristics. These phenomena were considered to be attributed to the current modulation by hole charging/discharging in the QD.

The status (until 2013) of the storage times in self-organized QDs, surveying a variety of heterostructures advantageous for strong electron and/or hole confinement, is described by Bimberg et al. [162]. Experimental data for the electronic properties, such as localization energies and capture cross sections, are listed. Based on the theory of thermal emission of carriers from

QDs, we extrapolate the values for materials that would increase the storage time at room temperature to more than millions of years. For electron storage, GaSb/AlSb, GaN/AlN, and InAs/AlSb are proposed. For hole storage, GaSb/$Al_{0.9}Ga_{0.1}As$, GaSb/GaP, and GaSb/AlP are promising candidates. The description and parameters of one of these structures can be found in [163]. Fast read-out of two to six charges per dot from the ground and first excited state in a QD-based memory is demonstrated using a two-dimensional electron gas. Single-shot measurements on modulation-doped FET structures with embedded InAs/GaAs QDs show read-out times as short as 3 ns. At low temperature (T=4.2 K), this read-out time is still limited by the parasitic signal of the setup and the device structure. Faster read-out times and a larger read-out signal are expected for an improved setup and device structure.

The general description of QD memories is given in [164]. MOVPE-prepared QDs compete with many other QD materials and growth technologies. A few recent examples: MoS_2 QDs were synthesized by a facile solvothermal method "Quantum conductance in MoS_2 QDs-based nonvolatile resistive memory device" in [165]; the resistive memory devices with the Ti/quantum dot/Ti/QD/ITO structure were prepared by a spin-casting method. CdSe-ZnS core-shell QDs of ~5.2 nm in size in [166].

Nevertheless, we can conclude, as written in [162] that material systems based on self-organized QDs with the room-temperature storage time of electrons and holes can be shifted further, possibly up to millions and billions of years. These are GaAs/AlSb, GaN/AlN, and InAs/AlSb for electron storage and GaSb/$Al_{0.9}Ga_{0.1}As$, GaSb/GaP, and GaSb/AlP for hole storage. It can be concluded that although growth of the materials proposed will be challenging, a nonvolatile memory based on self-organized QDs is feasible.

As with many other applications, freestanding QDs secondarily embedded into the solar-cell structures or systems prevail over the MOVPE (or MBE) grown ones. An internet search shows around 5000 citations for *QD solar-cell* references, and the Web of Science shows more than 9000 times.

When we add the word *MOVPE* to *QD solar cell*, these numbers decrease to 9 and 20 citations, respectively. Nevertheless, MOVPE QDs incorporated into active solar cell structures can increase the efficiency of photovoltaics, which is of course why they are currently prepared and studied in research. There are optoelectronic devices based on novel types of active region – quantum well–dot hybrid nanostructures. This hybrid type of the active region can be described as a quantum well, which has an ultradense array of narrow-gap In-rich regions with sizes of 20–30 nm, which serve as localization centers of charge carriers. Such structures can be formed spontaneously during the MOVPE deposition of $In_xGa_{1-x}As$ ($0.3<x<0.5$) on GaAs substrate. The optimal average thickness and composition of this layer to achieve maximal PL intensity and PC in QWD structures were determined and characteristics of edge-emitting lasers based on 5-QWD layers were described in [132].

Other possible applications are InAs QDs for enhanced InGaAs space solar cells. For this field, an increase of energy-conversion efficiency is very important. InAs QDs in the InGaAs cell will provide sub-gap absorption and thus improve its short-circuit current. This cell could then be integrated into the three-cell stack to achieve a space solar cell whose efficiency exceeds current state-of-the-art standards. A theoretical estimate [167] predicts that an InGaAlP(1.95 eV)/InGaAsP(1.35 eV)/InGaAs(1.2 eV) triple-junction cell incorporating QDs to improve the bottom cell current would have an efficiency exceeding 40%. In addition, it was theoretically estimated that the use of QD structures might hold other cell benefits such as improved temperature coefficients and better radiation tolerance, which are important, especially for utilization in space.

Embedded nanostructures such as QDs have been studied for many applications including enhanced miniband absorption in intermediate-band solar cells and current matching in multijunction cells; see [168]. Furthermore, solar cells with QDs have shown improved radiation hardness and temperature tolerance due to the presence of QDs.

InAs/GaAs QD cells were grown by MOVPE, fabricated, and processed by epitaxial lift-off, creating thin and flexible devices that exhibit enhanced sub-GaAs bandgap current collection.

For sufficient capture of light, it is necessary to increase the thickness of the structure, but this increase generates internal strain. Thus, it is necessary to compensate for this strain. GaP tensile strain-compensation layers were introduced into GaAs solar cells enhanced with a five-layer stack of InAs QDs; see [169]. One-sun air mass zero-illuminated current–voltage curves show that the strain-compensation results in improved conversion efficiency and reduced dark current. The strain-compensated QD solar cell shows a slight increase in short-circuit current compared to a baseline GaAs cell, due to sub-GaAs bandgap absorption by the InAs QD.

Recent and older MOVPE QD photovoltaic research results – AIIIBV QD structures for intermediate-band solar cells, multijunction solar cell spectral tuning with QDs, and increasing the quantum efficiency of InAs/GaAs QD arrays – can be found in [134, 170, 171], respectively.

6.7 Summary

MOVPE-grown QDs have been prepared and studied by many MOVPE technological teams for more than 25 years, but their real current application is still limited. The situation is comparable with QWrs. In comparison with QWs, neither of these structures is mature. It is an open question whether QDs will ever have an impact on our lives comparable to QWs (internet, laser diodes, QW-based devices in computers, mobile phones, etc.). It may happen that improvement of the final chip/device parameters in comparison with QW-based ones will not be sufficient to introduce new growth procedures and device constructions.

Materials that can be used for QD structures come from many semiconductor families, such as arsenide, antimonide, phosphide, telluride, and nitride binaries and ternaries (and quaternaries). QD structure material choice (including substrate/wafer materials such as GaAs, Si, sapphire, perovskite, etc.) can drive the emitted wavelength, influence temperature and structure stability, and, of course, affect the cost of devices.

Complex internal QD device structures like QD size and shape (including layout, in the case of QD arrays), SRLs and spacers (materials and thicknesses), QD layer multiplicity for stacking-layer structure, material and thickness of the capping layer, and the system of buffer layers can affect QD device parameters and their possible applications.

MOVPE growth parameters such as growth temperature, material type of precursors (which can be different for the same element), reactor pressure, precursor flow ratio, total pressure and flow, complex purity of materials, precursors, and reactor setup can strongly influence the structure and device parameters, as well.

6.8 Future Perspectives

Applications:

- QD-based single-photon or -electron devices. Mainly single-photon QD lasers for quantum cryptography.
- QD MIR photomultipliers can be more useful than other QD photomultiplier wavelengths applications.
- QD long-life memories are very promising.
- Telecom lasers, very probably for 1.3 μm, maybe also for 1.55 μm.
- We are not sure about the prospects of QD solar cells; this is at a very early stage of research, with many exciting possibilities.

Research:

- Extension of QD device wavelengths to the far-MIR and to the UV using different materials as well as size and shape of QDs.
- Improvement of structure parameters over standard ones.
- The cost of devices has to be lowered, as usual, and thus research into low-cost and mature (Si or GaAs wafers based?) technology must be improved.

Acknowledgment

The authors acknowledge support from the MSMT NPU project no. LO1603–ASTRANIT.

References

1. WiseGuy Research Consultants Pvt Ltd. (2017). Project Portfolio Quantum Dots: 2017 Global Market Expected to Grow at CAGR 11.39% and Forecast to 2019, Quantum Dots. *Market Research Report* 2017. Press release.
2. Ekimov, A.I. and Onushchenko, A.A. (1981). Quantum size effect in three-dimensional microscopic semiconductor crystals. *JETP Letters* 34: 345–349.
3. Leonard, D., Krishnamurthy, M., Reaves, C.M. et al. (1993). Direct formation of quantum-sized dots from uniform coherent islands of InGaAs on GaAs surfaces. *Applied Physics Letters* 63: 3203–3205.
4. Moisson, J.M., Houzay, F., Barthe, F. et al. (1994). Self-organized growth of regular nanometer-scale InAs dots on GaAs. *Applied Physics Letters* 64: 196–198.
5. Krishnamurthy, M., Drucker, J.S., and Venables, J.A. (1991). Microstructural evolution during the heteroepitaxy of Ge on vicinal Si(100). *Journal of Applied Physics* 69: 6461–6471.
6. Eaglesham, D.J. and Cerullo, M. (1990). Dislocation-free Stranski-Krastanow growth of Ge on Si(100). *Physical Review Letters* 64: 1943–1946.
7. Grundmann, M., Christen, J., Ledentsov, N.N. et al. (1995). Ultranarrow luminescence lines from single quantum dots. *Physical Review Letters* 74: 4043–4046.
8. Petroff, P.M. and DenBaars, S.P. (1994). MBE and MOCVD growth and properties of self-assembling quantum dot arrays in III-V semiconductor structures. *Superlattices and Microstructures* 15: 15–21.
9. Moseley, A.J., Thompson, J., Kightley, P. et al. (1991). CODE: a novel single step MOVPE technique for the fabrication of low-dimensional devices, quantum wires and quantum dots. *Journal of Crystal Growth* 108: 203–218.
10. Galeuchet, Y.D., Rothuizen, H., and Roentgen P. (1991). MOVPE on patterned substrates: a new fabrication method for nanometer structure devices. *Microlectronic Engineering* 15: 667–670.
11. Hirayama, H., Matsunaga, K., Asada, M., and Suematsu, Y. (1994). Lasing action of $Ga_{0.67}In_{0.33}As$/GaInAsP/InP tensile-strained quantum-box laser. *Electronics Letters* 30: 142–143.
12. Carlsson, N., Junno, T., Montelius, L. et al. (1998). Growth of self-assembled InAs and $InAs_xP_{1-x}$ dots on InP by metalorganic vapour phase epitaxy. *Journal of Crystal Growth* 191: 347–356.
13. Schlenker, D., Miyamoto, T., Chen, Z. et al. (2000). Growth of highly strained GaInAs/GaAs quantum wells for 1.2 mu m wavelength lasers. *Journal of Crystal Growth* 209: 27–36.
14. Mukai, K. and Sugawara, M. (1999). Suppression of temperature sensitivity of interband emission energy in 1.3-mu m-region by an InGaAs overgrowth on self-assembled InGaAs GaAs quantum dots. *Applied Physics Letters* 74: 3963–3965.
15. Zhou, W.D. and Coleman, J.J. (2016). Semiconductor quantum dots. *Current Opinion in Solid State and Material Science* 20: 352–360.
16. Michler, P. (2009). *Single Semiconductor Quantum Dots*. Book Series: Nanoscience and Technology. Berlin: Springer Verlag.
17. Michler, P., Kiraz, A., Becher, C. et al. (2000). A quantum dot single-photon turnstile device. *Science* 290: 2282–2285.
18. Mano, T., Watanabe, K., Tsukamoto, S. et al. (2000). Fabrication of InGaAs quantum dots on GaAs(001) by droplet epitaxy. *Journal of Crystal Growth* 209: 504–508.
19. Surnina, M.A., Akchurin, R.Kh., Marmalyuk, A.A. et al. (2016). Growing InAs/GaAs quantum dots by droplet epitaxy under MOVPE conditions. *Technical Physics Letters* 42: 747–749.

20. Glaser, E.R., Bennett, B.R., Shanabrook, B.V. et al. (1996). Photoluminescence studies of self-assembled InSb, GaSb, and AlSb quantum dot heterostructures. *Applied Physics Letters* 68: 3614–3616.
21. Norman, A.G., Mason, N.J., Fisher, M.J. et al. (1997). Structural and optical characterisation of MOVPE self-assembled InSb quantum dots in InAs and GaSb matrices. *Institute of Physics Conference Series* 157: 353–356.
22. Brandt, O., Tapfer, L., Ploog, K. et al. (1991). InAs quantum dots in a single-crystal GaAs matrix. *Physical Review B* 44: 8043–8053.
23. Bennett, B.R., Thibado, P.M., Twigg, M.E. et al. (1996). Self-assembled InSb and GaSb quantum dots on GaAs(001). *Journal of Vacuum Science and Technology B* 14: 2195–2198.
24. Nötzel, R., Temmyo, J., and Tamamura, T. (1994). Self-organization of box-like microstructures on GaAs (311)B surfaces by metalorganic vapor-phase epitaxy. *Japanese Journal of Applied Physics* 33: L275–L278.
25. Warburton, R.J., Nicholas, R.J., Howard, L.K., and Emeny, M.T. (1991). Intraband and interband magnetooptics of p-type $In_{0.18}Ga_{0.82}As$/GaAs quantum-wells. *Physical Review B* 43: 14124–14133.
26. Oshinowo, J., Nishioka, M., Ishida, S., and Arakawa, Y. (1994). Highly uniform InGaAs/GaAs quantum dots (~15 nm) by metalorganic chemical-vapor-deposition. *Applied Physics Letters* 65: 1421–1423.
27. Ahopelto, J., Lipsanen, H., Sopanen, M. et al. (1994). Selective growth of InGaAs on nanoscale InP islands. *Applied Physics Letters* 65: 1662–1664.
28. DenBaars, S.P., Reaves, C.M., Bressler-Hill, V. et al. (1994). Formation of coherently strained self-assembled InP quantum islands on InGaP/GaAs(001). *Journal of Crystal Growth* 145: 721–727.
29. Carlsson, N., Seifert, W., Petersson, A. et al. (1994). Study of the 2-dimensional/3-dimensional growth mode transition in metalorganic vapor-phase epitaxy of GaInP/InP quantum-sized structures. *Applied Physics Letters* 65: 3093–3095.
30. Damilano, B., Grandjean, N., Semond, F. et al. (1999). From visible to white light emission by GaN quantum dots on Si(111) substrate. *Applied Physics Letters* 75: 962–964.
31. Gogneau, N., Jalabert, D., Monroy, E. et al. (2003). Structure of GaN quantum dots grown under "modified Stranski-Krastanow" conditions on AlN. *Journal of Applied Physics* 94: 2254–2261.
32. Hospodková, A., Křápek, V., Mates, T. et al. (2007). Lateral-shape of InAs/GaAs quantum dots in vertically correlated structures. *Journal of Crystal Growth* 298: 570–573.
33. Roberts, T.S., Stevens, B.J., Clarke, E. et al. (2017). Strain balancing of metal-organic vapour phase epitaxy InAs/GaAs quantum dot lasers. *IEEE Journal of Selected Topics in Quantum Electronics* 23: 1901208.
34. Hospodková, A., Zíková, M., Pangrác, J. et al. (2013). *Type I-type II band alignment of GaAsSb/InAs/GaAs quantum dot heterostructure influenced by dot size and strain reducing layer composition. Journal of Physics D (Applied Physics)* 46: 095103.
35. Zíková, M., Hospodková, A., Pangrác, J. et al. (2015). *MOVPE prepared InAs/GaAs quantum dots covered by GaAsSb layer with long wavelength emission at 1.8 μm. Journal of Crystal Growth* 414: 167–171.
36. Hospodková, A., Pangrác, J., Zíková, M. et al. (2014). *Effect of the lower and upper interfaces on the quality of InAs/GaAs quantum dots. Applied Surface Science* 301: 173–177.
37. Hospodková, A., Oswald, J., Pangrác, J. et al. (2013). *Combined vertically correlated InAs and GaAsSb quantum dots separated by triangular GaAsSb barrier. Journal of Applied Physics* 114: 174305.
38. Hospodková, A., Zíková, M., Pangrác, J. et al. (2013). *Graded GaAsSb strain reducing layers covering InAs/GaAs quantum dots. Journal of Crystal Growth* 370: 303–306.
39. Mohan, A., Gallo, P., M. Felici, M. et al. (2010). Record-low inhomogeneous broadening of site-controlled quantum dots for nanophotonics. *Small* 6: 1268–1272.
40. Michler, P., Kiraz, A., Becher, C. et al. (2001). A quantum dot single photon source. *Advances in Solid State Physics* 41: 3–14.
41. Baier, M.H., Pelucchi, E., Kapon, E. et al. (2004). Single photon emission from site-controlled pyramidal quantum dots. *Applied Physics Letters* 84: 648–650.
42. Petroff, P.M. (2011). Semiconductor self-assembled quantum dots: Present status and future trends. *Advanced Materials* 23: 2372–2376.
43. Waks, E., Inoue, K., Santori, C. et al. (2002). Secure communication: Quantum cryptography with a photon turnstile. *Nature* 420: 762.
44. Yuan, Z.L., Kardynal, B.E., Stevenson, R.M. et al. (2002). Electrically driven single-photon source. *Science* 295: 102–105.
45. Kim, H., Choi, J., Lingley, Z. et al. (2017). Selective growth of strained (In)GaAs quantum dots on GaAs substrates employing diblock copolymer lithography nanopatterning. *Journal of Crystal Growth* 465: 48–54.
46. Zhong, Z.Y., Halilovic, A., Mühlberger, M. et al. (2003). Positioning of self-assembled Ge islands on stripe-patterned Si(001) substrates. *Journal of Applied Physics* 93: 6258–6264.
47. Rigal, B., Jarlov, C., Rudra, A. et al. (2015). Site-controlled InGaAs/GaAs pyramidal quantum dots grown by MOVPE on patterned substrates using triethylgallium. *Journal of Crystal Growth* 414: 187–191.
48. Koguchi, N., Takahashi, S., and Chitkow, T. (1991). New MBE growth method for InSb quantum well boxes. *Journal of Crystal Growth* 111: 688–692.

49. Sanguinetti, S. and Koguchi, N. (2013). Droplet epitaxy of nanostructures. In: *Molecular Beam Epitaxy from Research to Mass Production* (ed. M. Henini), 95–111. Amsterdam: Elsevier Science.
50. Lee, J.H., Wang, Zh.M., Liang, B.L. et al. (2006). Size and density control of InAs quantum dot ensembles on self-assembled nanostructures templates. *Semiconductor Science and Technology* 21: 1547–1551.
51. Liang, B.L., Dorogan, V.G., Mazur, Yu.I. et al. (2009). InAs quantum dot clusters grown on gas droplet templates: Surface morphologies and optical properties. *Journal of Nanoscience and Nanotechnology* 9: 3320–3324.
52. Rice, J.H., Oliver, R.A., Robinson, J.W. et al. (2004). InGaN quantum dots grown by MOVPE via a droplet epitaxy route. *Physica E* 21: 546–550.
53. Shusterman, S., Raizman, A., and Paltiel, Y. (2009). Narrow gap nano-dots growth by droplets heteroepitaxial mode. *Infrared Physics and Technology* 52: 229–234.
54. Marchenko, V.I. (1981). Possible structures and phase transitions on the surface of crystals. *JETP Letters* 33: 381–383.
55. Bimberg, D., Heinrichsdorff, F., Ledentsov, N.N. et al. (2000). Self-organized growth of semiconductor nanostructures for novel light emitters. *Applied Surface Science* 159: 1–7.
56. Wang, P.D., Torres, C.M.S. et al. (1994). Optical characterization of submonolayer and monolayer InAs structures grown in a GaAs matrix on (100) and high-index surfaces. *Applied Physics Letters* 64: 1526–1528.
57. Ledentsov, N.N., Krestnikov, I.L., Maximov, M.V. et al. (1996). Ground state exciton lasing in CdSe submonolayers inserted in a ZnSe matrix. *Applied Physics Letters* 69: 1343–1345.
58. Simmonds, P.J., Sun, M., Laghumavarapu, R.B. et al. (2014). Improved quantum dot stacking for intermediate band solar cells using strain compensation. *Nanotechnology* 25 (44): 25445402.
59. Alonso-Álvarez, D., Taboada, A.G., Ripalda, J.M. et al. (2008). Carrier recombination effects in strain compensated quantum dot stacks embedded in solar cells. *Applied Physics Letters* 93: 123114.
60. Walter, G., Elkow, J., N. Holonyak, N. Jr. et al. (2004). Visible spectrum (645 nm) transverse electric field laser operation of InP quantum dots coupled to tensile strained $In_{0.46}Ga_{0.54}P$ quantum wells. *Applied Physics Letters* 84: 666–668.
61. Marent, A., Geller, M., Schliwa, A. et al. (2007). 10^6 years extrapolated hole storage time in GaSb/AlAs quantum dots. *Applied Physics Letters* 91: 242109.
62. Geller, M., Marent, A., Nowozin, T. et al. (2008). A write time of 6 ns for quantum dot-based memory structures. *Applied Physics Letters* 92: 092108.
63. Laghumavarapu, R.B., Moscho, A., Khoshakhlagh, A. et al. (2007). GaSb/GaAs type II quantum dot solar cells for enhanced infrared spectral response. *Applied Physics Letters* 90: 173125.
64. Timm, R., Eisele, H., Lenz, A. et al. (2004). Structure and intermixing of GaSb/GaAs quantum dots. *Applied Physics Letters* 85: 5890.
65. Bellmann, K., Tabataba-Vakili, F., Wernicke, T. et al. (2015). Desorption induced GaN quantum dots on (0001) AlN by MOVPE. *Physica Status Solidi RRL* 9 (9): 526–529.
66. Bürger, M., Lindner, J.K.N., Reuter, D. et al. (2015). Investigation of cubic GaN quantum dots grown by the Stranski-Krastanov process. *Physica Status Solidi C* 12 (4–5): 452–455.
67. Zhuravlev, K.S., Aleksandrov, I.A., and Holtz P.-O. (2010). Linear polarized photoluminescence from GaN quantum dots imbedded in AlN matrix. *Physica Status Solidi C* 7 (7–8): 2227–2229.
68. Han, S.S., Higo, A., Yunpeng, W. et al. (2013). Effect of GaP and GaP/InGaP insertion layers on the structural and optical properties of InP quantum dots grown by metal-organic vapor phase epitaxy. *Microelectronic Engineering* 112: 143–148.
69. Krier, A., Labadi, Z., and Hammiche, A. (1999). InAsSbP quantum dots grown by liquid phase epitaxy. *Journal of Physics D Applied Physics* 32: 2587–2589.
70. Krier, A., Huang, X.L., and Labadi, Z. (2007). Liquid phase epitaxy of quantum wells and quantum dots. In: *Liquid Phase Epitaxy of Electronic, Optical and Optoelectronic Materials* (eds. P. Capper and M. Mauk), 227–257. John Wiley & Sons.
71. Moiseev, K.D., Parkhomenko, Ya.A., Ankudinov, A.V. et al. (2007). InSb/InAs quantum dots grown by liquid phase epitaxy. *Technical Physics Letters* 33: 295–298.
72. Moiseev, K.D., Parkhomenko, Ya.A., Mikhailova, M.P. et al. (2008). Self-assembled InSb/InAs quantum dots for the mid-infrared spectral range 3-4 μm. In: *Narrow Gap Semiconductors*, Springer Proceedings in Physics 119: 125–127. Dordrecht, Netherlands: Springer.
73. Moiseev, K.D., Dement'ev, P., Romanov, V., and Mikhailova, M.P. (2011). Quantum dots and quantum dashes in the narrow-gap InSb/InAsSbP system. *Journal of Crystal Growth* 318: 379–384.
74. Maronchuk, I.I., Sanikovitch, D.D., Cherkashin, A.S. et al. (2017). Improvement of growing of Ge QDs by the method of liquid phase epitaxy. *Journal of Physics Conference Series* 794: 012012.
75. Li, F., Lee, S.H., You, J.H. et al. (2010). UV photovoltaic cells fabricated utilizing GaN nanorod/Si heterostructures. *Journal of Crystal Growth* 312: 2320–2323.
76. Wikipedia. (2019). Quantum dot. https://en.wikipedia.org/wiki/Quantum_dot.

77. Bimberg, D., Grundmann, M., Ledentsov, N.N. et al. (1995). Self-organization processes in MBE-grown quantum dot structures. *Thin Solid Films* 267: 32–36.
78. Krebs, R., Deubert, S., Reithmaier, J.P., and Forchel, A. (2003). Improved performance of MBE grown quantum-dot lasers with asymmetric dots in a well design emitting near 1.3 μm. *Journal of Crystal Growth* 251: 742–747.
79. Pchelyakov, O.P., Nikiforov, A.I., Shanetsky, B.Z. et al. (2008). MBE growth of ultra small coherent Ge quantum dots in silicon for applications in nanoelectronics. *Journal of Physics C (Solid State Physics)* 69: 669–672.
80. Liu, A.Y., Zhang, Ch., Snyder, A. et al. (2014). MBE growth of P-doped 1.3 μm InAs quantum dot lasers on silicon. *Journal of Vacuum Science and Technology* 32: 02C108.
81. Manna, S., Aluguri, R., Katiyar, A. et al. (2013). MBE-grown Si and $Si_{1-x}Ge_x$ quantum dots embedded within epitaxial Gd_2O_3 on Si(111) substrate for floating gate memory device. *Nanotechnology* 24: 505709.
82. Ball, P., Bimberg, D., and Wang, K. (2017). Quantum engineering of matter from the laboratory to the market: an interview with Dieter Bimberg and Kang Wang. *Nature Science Review* 4: 210–212. https://academic.oup.com/nsr/article-lookup/doi/10.1093/nsr/nww067.
83. Bimberg, D. (2014). Fundamentals. In: *Self-Organized Quantum Dots for Memories: Electronic Properties and Carrier Dynamics* (T. Nowozin), 5–24. Springer Theses – Recognizing Outstanding PhD Research.
84. Bimberg, D. and Pohl, U.W. (2011). Quantum dots: promises and accomplishments. *Materials Today* 14: 388–397.
85. Briot, O., Ruffenach, S., Moret, M. et al. (2009). Growth of InN films and nanostructures by MOVPE. *Journal of Crystal Growth* 311: 2761–2766.
86. Karomi, I., Smowton, P.M., Shutts, S. et al. (2015). InAsP quantum dot lasers grown by MOVPE. *Optical Express* 23: 27282–27291.
87. Torelly, G., Jakomin, R., Pinto, L.D. et al. (2016). Early nucleation stages of low density InAs quantum dots nucleation on GaAs by MOVPE. *Journal of Crystal Growth* 434: 47–54.
88. Akahane, K., Yamamoto, N., and Ohtani, N. (2004). Long-wavelength light emission from InAs quantum dots covered by GaAsSb grown on GaAs substrates. *Physica E* 21 (2–4): 295–299.
89. Liu, H.Y., Steer, M.J., Badcock, T.J. et al. (2005). Long-wavelength light emission and lasing from InAs/GaAs quantum dots covered by a GaAsSb strain-reducing layer. *Applied Physics Letters* 86 (14): 143108.
90. Liu, H.Y., Steer, M.J., Badcock, T.J. et al. (2006). Room-temperature 1.6 μm light emission from InAs/GaAs quantum dots with a thin GaAsSb cap layer. *Journal of Applied Physics* 99: 046104.
91. R. Timm, R., Eisele, H., Lenz, A. et al. (2006). Structure of InAs/GaAs quantum dots grown with Sb surfactant. *Physica E* 32 (1–2): 25–28.
92. Ulloa, J.M. Çelebi, C. Koenraad P.M. et al. (2007). Atomic scale study of the impact of the strain and composition of the capping layer on the formation of InAs quantum dots. *Journal of Applied Physics* 101 (8): 081707.
93. Zíková M., Hospodková A., Pangrác J. et al. (2017). Comparison of MOVPE grown GaAs, InGaAs and GaAsSb covering layers for different InAs/GaAs quantum dot applications. *Journal of Crystal Growth* 464: 59–63.
94. Haxha, V., Drouzas, I., Ulloa, J.M. et al. (2010). Control of Strain in GaSbAs/InAs/GaAs Quantum Dots. *Journal of Physics: Conference Series* 245: 012065.
95. Hospodková, A., Pangrác, J., Oswald, J. et al. (2011). Influence of strain reducing layers on electroluminescence and photoluminescence of InAs/GaAs QD structures. *Journal of Crystal Growth* 315 (1): 110–113.
96. Liu, H.Y. and Hopkinson, M. (2003). Tuning the structural and optical properties of 1.3-μm InAs/GaAs quantum dots by a combined InAlAs and GaAs strained buffer layer. *Applied Physics Letters* 82 (21): 3644–3646.
97. Jo, M., Sato, M., Miyamura, S. et al. (2012). Origin of the blueshift of photoluminescence in a type-II heterostructure. *Nanoscale Research Letters* 7: 654.
98. Ledentsov, N.N., Bohrer, J., Beer, M. et al. (1995). Radiative states in type-II GaSb/GaAs quantum wells. *Physical Review B* 52 (19): 14058–14066.
99. Zettler, J.-T., Haberland, K., Zorn, M. et al. (1998). Real-time monitoring of MOVPE device growth by reflectance anisotropy spectroscopy and related optical techniques. *Journal of Crystal Growth* 195: 151–162.
100. Hospodková, A., Vyskočil, J., Pangrác, J. et al. (2010). Surface processes during growth of InAs/GaAs quantum dot structures monitored by reflectance anisotropy spectroscopy. *Surface Science* 604: 318–321.
101. Steimetz, E., Wehnert, T., Kirmse, H. et al. (2000). Optimizing the growth procedure for InAs quantum dot stacks by optical in situ techniques. *Journal of Crystal Growth* 221: 592–598.
102. Zíková, M., Hospodková, A., Pangrác J. et al. (2016). The metalorganic vapour phase epitaxy growth of AIIIBV heterostructures observed by reflection anisotropy spectroscopy. *Acta Physica Polonica A* 129: A-75–A-78.
103. Hospodková, A. (2012). Capping of InAs/GaAs quantum dots for GaAs based lasers quantum dots – A variety of new applications (ed. A. Al-Ahmadi), 27–46. Rijeka: InTech open access. DOI: 10.5772/35632.
104. Esser, N., ReschEsser, U., Pristovsek, M. et al. (1996). Scanning-tunneling-microscopy study of InP(001) surfaces prepared by UHV decapping of metal-organic vapor-phase-epitaxy-grown samples. *Physical Review B* 53 (20): 13257–13259.
105. Kremzow, R., Pristovsek, M., and Kneissl, M. (2008). Ripening of InAs quantum dots on GaAs (001) investigated with in situ scanning tunneling microscopy in metal-organic vapor phase epitaxy. *Journal of Crystal Growth* 310 (23): 4751–4753.

106. Steimetz, E., Wehnert, T., Haberland, K. et al. (1998). GaAs cap layer growth and In-segregation effects on self-assembled InAs quantum dots monitored by optical techniques. *Journal of Crystal Growth* 195 (1–4): 530–539.
107. Steimetz, E., Richter, W., Schienle, F. et al. (1998). The effect of different group V precursors on the evolution of quantum dots monitored by optical in situ measurements. *Japanese Journal of Applied Physics* 37 (3B): 1483–1486.
108. Massoudi, I., Habchi, M.M., Rebey, A. et al. (2012). Study of surface roughness using spectral reflectance measurements recorded during the MOVPE of InAs/GaAs heterostructures. *Physica E* 44 (7–8): 1282–1287.
109. Hospodková, A., Křápek, V., Kuldová, K. et al. (2006). Photoluminescence and magnetophotoluminescence of vertically stacked InAs/GaAs quantum dot structures. *Physica E* 36: 106–113.
110. Křápek, V., Kuldová, K., Oswald, J. et al. (2006). Elongation of InAs/GaAs quantum dots from magnetophotoluminescence measurements. *Applied Physics Letters* 89: 153108 1–3.
111. Hospodková, A., Oswald, J., Pangrác, J. et al. (2016). Growth and properties of the MOVPE GaAs/InAs/GaAsSb quantum dot structures. *Physica B* 480: 14–22.
112. Mikkelsen, M.H., Berezovsky, J., Stoltz, N.G. et al. (2007). Optically detected coherent spin dynamics of a single electron in a quantum dot. *Nature Physics* 3 (11): 770–773.
113. Offermans, P., Koenraad, P.M., Wolter, J.H. et al. (2005). Formation of InAs quantum dots and wetting layers in GaAs and AlAs analyzed by cross-sectional scanning tunneling microscopy. *Physica E* 26: 236–240.
114. Koenraad, P.M., Bruls, D.M., Davies, J.H. et al. (2003). Composition profiling at the atomic scale in III–V nanostructures by cross-sectional STM. *Physica E* 17: 526–532.
115. Bozkurt, M., Ulloa, J.M., and Koenraad, P.M. (2011). An atomic scale study on the effect of Sb during capping of MBE grown III–V semiconductor QDs. *Semiconductor Science and Technology* 26: 064007.
116. Beanland, R. (2005). Dark field transmission electron microscope images of III-V quantum dot structures. *Ultramicroscopy* 102: 115–125.
117. Rubin, M.E., MedeirosRibeiro, G., O'Shea, J.J. et al. (1996). Imaging and spectroscopy elf single InAs self-assembled quantum dots using ballistic electron emission microscopy. *Physical Review Letters* 77 (26): 5268–5271.
118. Walachova, J., Zelinka, J., Malina, V. et al. (2007). Study of InAs quantum dots in AlGaAs/GaAs heterostructure by ballistic electron emission microscopy/spectroscopy. *Applied Physics Letters* 91 (4): 042110.
119. Walachova, J., Zelinka, J., Malina, V. et al. (2008). Ballistic electron emission spectroscopy/microscopy of self-assembled InAs quantum dots of different sizes embedded in GaAs/AlGaAs heterostructure. *Applied Physics Letters* 92 (1): 012101.
120. Stier, O., Grundmann, M., and Bimberg, D. (1999). Electronic and optical properties of strained quantum dots modeled by 8-band k · p theory. *Physical Review B* 59: 5688–5701.
121. Bell, L.D. (2016). Ballistic electron emission microscopy and spectroscopy: Recent results and related techniques. *Journal of Vacuum Science and Technology B* 34 (4): 040801.
122. Vyskočil J., Hospodková, A., Petříček, O. et al. (2017). GaAsSb/InAs/(In)GaAs type II quantum dots for solar cell applications. *Journal of Crystal Growth* 464: 64–68.
123. Kapteyn, C.M.A., Heinrichsdorff, F.O., Stier, O. et al. (1999). Electron escape from InAs quantum dots. *Physical Review B* 60: 14265–14268.
124. QD Laser. (2019). http://www.qdlaser.com.
125. Bimberg, D., Arsenijevic, D., Schmeckebie, H. et al. (2012). Nanophotonics for datacom and telecom applications. *Proceedings of the SPIE - The International Society for Optical Engineering* 8265: 82650E.
126. Gustafsson, O., Karim, A., Berggren, J. et al. (2012). Photoluminescence and photoresponse from InSb/InAs-based quantum dot structures. *Optical Express* 20: 21264–21271.
127. Bimberg, D., Arsenijevic, D., Larisch, G. et al. (2014). Green nanophotonics for future datacom and Ethernet networks. *Proceedings of the SPIE – The International Society for Optical Engineering* 9134: 913402.
128. Kaptan, Y., Roehm, A., Herzog, B. et al. (2014). Stability of quantum-dot excited-state laser emission under simultaneous ground-state perturbation. *Applied Physics Letters* 105: 191105.
129. Arsenijevic, D., Kleinert, M., and Bimberg, D. (2014). Breakthroughs in photonics 2013: Passive mode-locking of quantum-dot lasers. *IEEE Photonics Journal* 6: 0700306.
130. Xie, H., Prioli, R., Torelly, G. et al. (2017). Correlation between size distribution and luminescence properties of spool-shaped InAs quantum dots. *Semiconductor Science and Technology* 32: 055013.
131. Gocalinska, A., Manganaro, M., Juska, G. et al. (2014). InP-AlInAs "strain free" early stages heteroepitaxy leading to nanostructure formation by MOVPE. Book Series: 26th International Conference on Indium Phosphide and Related Materials, May 11–15, Montpellier, France. IEEE.
132. Nadtochiy, A.M., Maximov, M.V., Mintairov, S.A. et al. (2017). Light-emitting and photovoltaic devices based on quantum well-dots hybrid nanostructures. *Proceedings of the SPIE – The International Society for Optical Engineering* 10114: 101140Y.
133. Ben Naceur, H., Mzoughi, T., Moussa, I. et al. (2010). Properties of InAs grown on misoriented GaAs substrates by atmospheric pressure metal-organic vapor phase epitaxy. *Nuclear Instruments & Methods in Physics Research Section B (Beam Interactions with Materials and Atoms)* 268: 236–240.

134. Vyskočil, J., Gladkov, P., Petříček, O. et al. (2015). Growth and properties of A(III)B(V) QD structures for intermediate band solar cells. *Journal of Crystal Growth* 414: 172–176.
135. Bimberg, D. (2014). Self-organized quantum dots for memories electronic properties and carrier dynamics. introduction and fundamentals. In: *Self-Organized Quantum Dots For Memories: Electronic Properties And Carrier Dynamics* (T. Nowozin), 1–5. Springer Theses –Recognizing Outstanding PhD Research.
136. Sakaki, H. (2007). Roles of quantum nanostructures on the evolution and future advances of electronic and photonic devices. *Technical Digest - International Electron Devices Meeting,* 9–16. 10.1109/IEDM.2007.4418851.
137. Higo, A., Thomas, C., Kiba, T. et al. (2016). The effect of asymmetric barriers of GaAs quantum nanodisks light emitting diode. 16th Int. Conf. on Nanotechnology (IEEE-NANO), Aug. 22–25, Sendai, Japan. IEEE.
138. Iwane, Y., Kawashima, F., Hirooka, M. et al. (2012). InAs/InP QDs grown by selective MOVPE growth using double-cap procedure for broadband LED improved p-cladding layer. *Physica Status Solidi C* 9: 210–213.
139. Huang, H., Schires, K., Raghunathan, R. et al. (2015). Highly efficient wavelength conversion in InAs/GaAs quantum dot lasers. IEEE Photonics Conf., Oct. 04–08, Reston, VA, USA. IEEE.
140. Arsenijevic, D., Schmeckebier, H., Kleinert, M. et al. (2015). Quantum-dot mode-locked lasers for microwave-signal generation and 160 Gb/s optical communication. IEEE Photonics Conf., Oct. 04–08, Reston, VA, USA. IEEE.
141. Schmeckebier, H., Zeghuzi, A., Arsenijevic, D. et al. (2016). 40 GBd D(Q)PSK and OOK amplification using O-band quantum-dot semiconductor optical amplifiers. 18th Int. Conf. on Transparent Optical Networks (ICTON), July 10–14, Trento, Italy. IEEE.
142. Stubenrauch, M., Stracke, G., Arsenijevic, D. et al. (2014). 15 Gb/s index-coupled distributed-feedback lasers based on 1.3 μm InGaAs quantum dots. *Applied Physics Letters* 105: 011103.
143. Arsenijevic, D. and Bimberg, D. (2016). Quantum-dot lasers for 35 Gbit/s pulse-amplitude modulation and 160 Gbit/s differential quadrature phase-shift keying. *Proceedings of the SPIE - The International Society for Optical Engineering* 9892: 98920S.
144. Huang, H., Arsenijevic, D., Schires, K. et al. (2016). Multimode optical feedback dynamics of InAs/GaAs quantum-dot lasers emitting on different lasing states. *AIP Advances* 6: 125114.
145. Coe-Sullivan, S., Steckel, J.S., Kim, L.A. et al. (2005). Method for fabrication of saturated RGB quantum dot light emitting devices. *Proceedings of the SPIE – The International Society for Optical Engineering* 5739: 108–115.
146. Yoshikawa, S., Yamauchi, M., Yamamoto, Y., and Shimomura Y. (2013). Current injected spectrum change in flat-topped InAs/InP QD arrayed waveguide LED with different QD heights. *Physica Status Solidi C* 10: 1438–1441.
147. Lv, W., Wang, L., Wang, J. et al. (2013). Green and red light-emitting diodes based on multilayer InGaN/GaN dots grown by growth interruption method. *Japanese Journal of Applied Physics* 52: 08JG13.
148. Yang, W., Li, J., Zhang, Y. et al. (2014). High density GaN/AlN quantum dots for deep UV LED with high quantum efficiency and temperature stability. *Scientific Reports* 4: 5166.
149. Tessarek, C., Figge, S., and Hommel, D. (2014). Colour and multicolour tuning of InGaN quantum dot based light-emitting diodes. *Journal of Physics D (Applied Physics)* 47: 055108.
150. Krishna, S., Gunapala, S.D., Bandara, S.V. et al. (2017). Quantum dot based infrared focal plane arrays, *Proceedings of the IEEE* 95: 1838–1852.
151. Razeghi, M. (2010). Quantum dot infrared photodetectors. In: *Technology of Quantum Devices*, 395–423. New York: Springer.
152. Gustafsson, O., Berggren, J., Ekenberg, U. et al. (2011). Long-wavelength infrared quantum-dot based interband photodetectors. *Infrared Physics and Technology* 54: 287–291.
153. Vaillancourt, J., Vasinajindakaw, P., Lu, X.J. et al. (2011). Thermo-electrically cooled shortwave infrared and long-wave infrared dual-band quantum-dot photodetector. *Proceedings of the SPIE – The International Society for Optical Engineering* 8012: UNSP 80120O.
154. Tsao, S., Myzaferi, A., and Razeghi, M. (2010). High performance quantum dot-quantum well infrared focal plane arrays. *Proceedings of the SPIE – The International Society for Optical Engineering* 7605: 76050J.
155. Gustafsson, O., Karim, A., Asplund, C. et al. (2013). A performance assessment of type-II interband $In_{0.5}Ga_{0.5}Sb$ QD photodetectors. *Infrared Physics and Technology* 61: 319–324.
156. Kufer, D. and Konstantatos, G. (2016). Photo-FETs: Phototransistors enabled by 2D and OD nanomaterials. *ACS Photonics* 3: 2197–2210.
157. Shankar, R.R., Hasaneen, V.E.S., Heller, E.K. et al. (2017). Floating gate nonvolatile memory using individually cladded monodispersed quantum dots. *IEEE Transactions on VLSI Systems* 25: 1774–1781.
158. Mao, S., Pu, H.H., Chang, J.B. et al. (2017). Ultrasensitive detection of orthophosphate ions with reduced graphene oxide/ferritin field-effect transistor sensors. *Environmental-Science Nano* 4: 856–863.
159. Huang, Y., Zang, H., Chen, J.-S. et al. (2016). Hybrid quantum dot-tin disulfide field-effect transistors with improved photocurrent and spectral responsivity. *Applied Physics Letters* 108: 123502.
160. Suarez, E., Chan, P.-Y., Gogna, M. et al. (2015). Fabrication and simulation of InGaAs field-effect transistors with II-VI tunneling insulators. *Journal of Electronic Materials* 44: 3064–3068.

161. Shima, M., Sakuma, Y., Futatsugi, T. et al. (2000). Tetrahedral-shaped recess (111)A facet channel AlGaAs/InGaAs heterojunction field-effect transistor with an InGaAs floating quantum dot gate. *IEEE Transactions on Electron Devices* 47 (11): 2054–2060.
162. Nowozin, T., Bimberg, D., Daqrouq, K. et al. (2013). Materials for future quantum dot-based memories. *Journal of Nanomaterials* 2013, 215613.
163. Nowozin, T., Beckel, A., Bimberg, D. et al. (2014). 3 ns single-shot read-out in a quantum dot-based memory structure. *Applied Physics Letters* 104 (5): 053111.
164. Darma, Y. and Rusydi, A. (2014). Quantum dot based memory devices: Current status and future prospect by simulation perspective. *AIP Conference Proceedings* 1586: 20–23.
165. Wang, D., Ji, F., Chen, X. et al. (2017). Quantum conductance in MoS_2 quantum dots-based nonvolatile resistive memory device. *Applied Physics Letters* 110 (9): 093501.
166. Kannan, V., Kim, H.-S., and Park, H.-C. (2016). Non-volatile resistive memory device fabricated from CdSe quantum dot embedded in thermally grown In_2O_3 nanostructure by oblique angle deposition. *Physics Letters A* 380: 3743–3747.
167. Raffaelle, R.P., Sinharoy, S., King, C.W. et al. (2005). InAs quantum dot development for enhanced InGaAs space solar cells. *Materials Research Society Symposium Proceedings* 851: 195–203.
168. Bennett, M.F., Bittner, Z.S., Forbes, D.V. et al. (2013). Alpha radiation effects on n-i-p quantum dot epitaxial lift-off solar cells. 39th Photovoltaic Specialists Conf. (PVSC), Jun 16–21, Tampa, FL, USA, 2784–2789. IEEE.
169. Hubbard, S.M., Cress, C.D., Bailey, C.G. et al. (2008). Effect of strain compensation on quantum dot enhanced GaAs solar cells. *Applied Physics Letters* 92: 123512.
170. Kalyuzhnyy, N.A., Mintairov, S.A., Salii, R.A. et al. (2016). Increasing the quantum efficiency of InAs/GaAs QD arrays for solar cells grown by MOVPE without using strain-balance technology. *Progress in Photovoltaics* 24: 1261–1271.
171. Raffaelle, R.P., Sinharoy, S., Wilt, D.M. et al. (2006). Multi-junction solar cell spectral tuning with quantum dots. 4th World Conf. on Photovoltaic Energy Conversion, May 07–12, Waikoloa, HI, USA, 162–166. IEEE.

7

III-V Nanowires and Related Nanostructures: From Nitrides to Antimonides

H.J. Joyce
Department of Engineering, Cambridge University, UK

7.1 Introduction to Nanowires and Related Nanostructures

The versatility of MOVPE is apparent when considering the diverse range of functional semiconductor nanostructures that can be grown. Through MOVPE, semiconductor nanomaterials have been grown in a wide variety shapes and sizes, and spanning a vast range of materials including III-V, II-VI, II-V, and IV-V-VI semiconductors. This chapter focuses on a subset of III-V nanostructures that are commonly known as *nanowires*, and on related free-standing nanostructures such as nanorods and nanopillars.

Semiconductor nanowires are driving new technologies and innovations across the boundaries of electronics, optics and photonics, chemistry, fundamental physics, and biology. These nanowires are quasi-one-dimensional crystals with narrow diameters (d) of a few tens of nanometers, and with lengths (L) at least an order of magnitude longer than the diameter (Figure 7.1a). Nanowires are commonly several hundreds of nanometers to several micrometers in length. Nanorods and nanopillars are similar to nanowires but usually feature smaller length-to-diameter aspect ratios (Figure 7.1b). The nanowire geometry lends itself to three different types of heterostructure or doped homostructure: nanowire-substrate, axial, and radial. Axial structures feature junctions along the axis of the nanowire (Figure 7.1c), whereas radial structures feature junctions between the core and a shell (Figure 7.1d), or even between multiple shells. Through sophisticated MOVPE techniques, it is possible to achieve branched nanowires in the form of kinked V-shapes, Y-shapes, zig-zags, and nanotrees [1–3]; interconnected nanowire structures such as X-shapes and hash-shapes [4]; and other shapes such as nanotubes [5, 6], nanorings [7], nanofins/nanosheets [8], and nanocubes [9].

Metalorganic Vapor Phase Epitaxy (MOVPE): Growth, Materials Properties, and Applications,
First Edition. Edited by Stuart Irvine and Peter Capper.

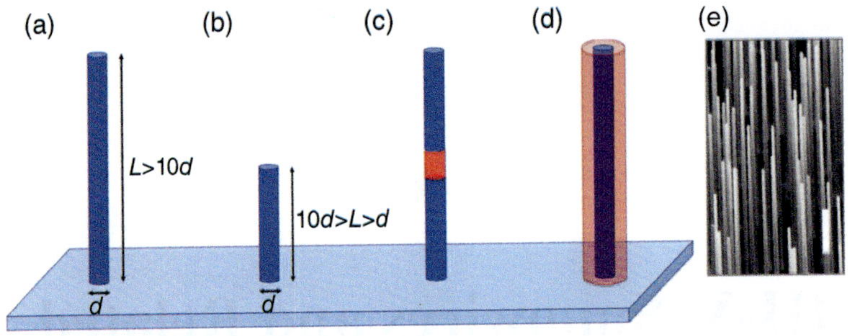

Figure 7.1 (a–d) Schematic diagram of freestanding semiconductor nanostructures achievable by MOVPE: (a) freestanding nanowire on a substrate with diameter d of a few tens of nanometers and length L; (b) nanorod with diameter d and length L; (c) axial heterostructure nanowire composed of two different materials; (d) radial, or core-shell, heterostructure nanowire composed of a core of one material and a shell of another material. Colors represent different materials. (e) SEM image of InAs nanowires: scale bar is 1 μm. Image (e) source: Reproduced with permission from Joyce et al. 2010 [13]; copyright American Chemical Society 2010.

The freestanding nature and unique quasi-one-dimensional geometry confer nanowires with several unique properties and advantages over conventional planar semiconductors. First, the inherently narrow diameter of nanowires affords their crystal lattice with the ability to expand (or contract) radially to accommodate significant amounts of strain. This property enables dislocation-free heteroepitaxy of highly lattice-mismatched materials within axially heterostructured nanowires (Figure 7.1c) and on lattice-mismatched and thermal expansion coefficient-mismatched substrates, with layer thicknesses and lattice mismatch ranges far beyond the limits of conventional planar heterostructures. Through this property, III-V nanowires offer a means to achieve direct epitaxy of III-V materials on Si, which has long been a goal of the III-V community, and is anticipated to enable the monolithic integration of optoelectronic and high-speed III-V devices with standard Si microelectronics. Strain relaxation in radially heterostructured nanowires, although less efficient than in axial heterostructure nanowires, also surpasses that of planar structures [10–12].

Other advantages of nanowires include their high surface area-to-volume ratios, the possibility of engineering quantum confinement in the two radial dimensions within nanowires, reduced and more efficient consumption of precursor materials, their waveguiding properties, enhanced light trapping in arrays of nanowires, enhanced out-coupling of emitted light, mechanical flexibility, and the ability to tune the crystal structure in ways not possible with planar materials. These unique properties are propelling new applications as diverse as nanowire-based single junction [14, 15] and multijunction solar cells [16], single-photon sources [17–19], light-emitting diodes (LEDs) [20], photodetectors [21, 22], nanoscale lasers [23, 24], and wrap-gate transistors [25]. Nanowires are also a powerful platform for the study of fundamental physics. For example, MOVPE-grown InSb nanowires [26] have been employed in semiconductor–superconductor hybrid devices to investigate the existence of theoretically predicted Majorana quasiparticles, which are of significance for topological quantum computing.

MOVPE of nanowires is by definition a *bottom-up* fabrication process, as the nanowires self-assemble from the controlled reaction of precursor materials. It is therefore distinct from *top-down* fabrication processes whereby nanostructures are successively patterned and etched from the parent material. As a bottom-up process, nanowire fabrication by MOVPE fully harnesses the just-mentioned benefits of nanowires, such as dislocation-free strain relaxation during growth, and the reduced consumption of precursor materials. It also avoids the shortcomings of top-down nanowire fabrication, which is inherently wasteful and in which the quality of the nanowires is dependent on the etch damage to the surfaces and the properties of the parent material, which is in turn limited by the constraints of conventional planar growth. As a caveat, it should be noted that in some cases, the fabrication of nanostructures by MOVPE may involve some top-down steps, such as the pregrowth lithographic patterning of masks that is performed prior to selective area epitaxy of nanowires, or the postgrowth selective etching of the cores of core-shell nanowires to form nanotubes [5, 6].

Two growth modes occur in nanowire growth: axial growth and radial growth. *Axial growth* is the term describing the anisotropic elongating growth of nanowires. *Radial growth*, also known as *conformal* or *lateral growth*, is the epitaxial growth of material on the side-facets of the nanowire. The interplay between axial and radial growth modes is used to tailor imaginative and new nanowire structures, including core-shell heterostructures.

The two dominant categories of nanowire growth by MOVPE are (i) particle-assisted growth and (ii) selective-area epitaxy (SAE). In both cases, anisotropic nanowire growth is favored because the growth rate at the tip of the nanowire is greater than the growth rate at other locations, including on the side-facets of the nanowire and on the substrate. This chapter describes the fundamental processes underlying the different methods of nanowire growth, and how these techniques can be tailored to achieve exquisite control over nanowire composition, doping, crystal structure, and morphology. It then gives examples of how the nanowire geometry is currently being exploited to achieve devices with new and improved functionalities, including LEDs and solar cells.

7.2 Geometric and Crystallographic Properties of III-V Nanowires

7.2.1 Crystal Phase

One of the features that distinguishes nanowires from their bulk and planar counterparts is their crystal phase. In nanowires, the crystal very often adopts a metastable phase rather than the stable phase that exists in bulk crystals. For III-V materials (excluding III-nitrides), the cubic zincblende (ZB) crystal structure is the thermodynamically stable phase in the bulk, yet nanowires have presented not only ZB structure [27], but also hexagonal wurtzite (WZ) structures [13], twinned [28] and periodically twinned [29, 30] ZB structures, mixed WZ/ZB structures [30], and even 4H crystal structures [31]. For III-nitrides, the WZ crystal structure is stable in the bulk, but nanowires have been grown with ZB inclusions [32]. Figure 7.2 shows transmission electron microscope (TEM) images of InAs nanowires with ZB and WZ crystal structures.

The difference between ZB and WZ structures lies in their stacking sequence along the cubic [111]/hexagonal [0001] direction, where ZB lattices follow an ABCABC stacking sequence (Figure 7.2c) and WZ lattices follow an ABAB stacking sequence (Figure 7.2f). Different crystal structures have different bandstructures, and consequently different electrical and optical properties. The ability to control the crystal structure within nanowires therefore provides a new paradigm for the exploration of novel optoelectronic and electronic devices. GaP is a powerful example. In the bulk form, GaP has a ZB crystal structure with indirect bandgap, which makes it unsuitable for light-emitting devices. GaP nanowires, on the other hand, can be grown with WZ structure, which has been shown to confer a direct bandgap allowing efficient emission at 594 nm [33]. These WZ direct-bandgap GaP nanowires, along with WZ AlGaP alloys, therefore have potential for light-emitting devices that bridge the so-called *green gap* in which there is a lack of traditional planar semiconductors that efficiently emit green light [33].

The WZ crystal structure in III-V nanowires (excluding III-nitrides) is possible because of their [111] growth direction, coupled with the relatively low surface energy of WZ side-facets, and the very small difference in lattice energies between ZB and WZ crystal structures. In nanowire epitaxy, the crystal structure is determined when each layer nucleates on the (111) growth facet:

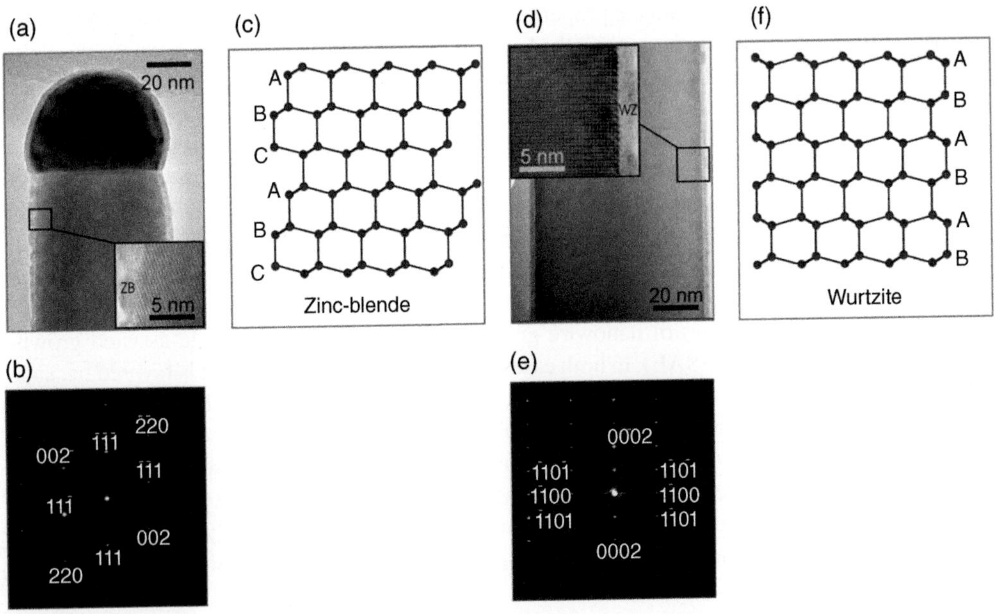

Figure 7.2 III-V nanowires with (a–c) zincblende and (d–f) wurtzite crystal structures. (a) High-resolution TEM image of a zincblende InAs nanowire taken along the <1–10> zone axis with (b) selective-area diffraction pattern and (c) schematic diagram of the characteristic ABC zincblende stacking sequence. (d) High-resolution TEM image of a wurtzite InAs nanowire taken along the <1–10> zone axis with (e) selective-area diffraction pattern and (f) schematic diagram of the stacking sequence. Parts (a, b, d, and e) source: Adapted and reproduced with permission from Joyce et al. 2010 [13]; copyright American Chemical Society, 2010.

the nucleus adopts either a WZ or ZB stacking sequence relative to the pre-existing crystal and then expands via step flow growth [34]. WZ growth occurs when the lower surface energy of nuclei forming WZ side-facets outweighs the influence of the lower lattice energy of a nucleus joining a ZB stacking sequence [13, 29]. The supersaturation of the growth system and, in particle-assisted growth, the surface energy of the particle, also affect the energy required for nucleation of WZ and ZB layers.

Growth conditions such as growth temperature and V/III ratio affect the relative surface energies of WZ and ZB facets, and hence determine the likelihood of WZ or ZB growth. Higher growth temperatures, for example, favor WZ formation in GaAs [13], GaP [33], InAs [13], and InP [35, 36] nanowires. Temperature and V/III ratio can be tailored very effectively to select for either purely WZ or purely ZB nanowire growth [13], to create crystal phase heterostructures with WZ/ZB interfaces [37], and to achieve precisely positioned single twin planes and stacking faults [38].

7.2.2 Growth Direction, Morphology, and Side-Facets

Most commonly, nanowires elongate along the energetically preferred cubic <111>/hexagonal <0001> axes that lie perpendicular to the close-packed {111}/{0001} planes. A nanowire's shape is determined not only by its growth direction, but also by the side-facets it adopts. [111]-oriented ZB nanowires frequently adopt the shape of hexagonal prisms with six {1–10} or six {11–2} side-facets, as shown in Figures 7.3a and b. The {1–10} side-facets are non-polar and are commonly seen in nanowires [39] and nanosheets [8] grown by SAE. The {11–2} planes, commonly seen in particle-assisted nanowires [40], are polar and can be subdivided into three {112}A facets

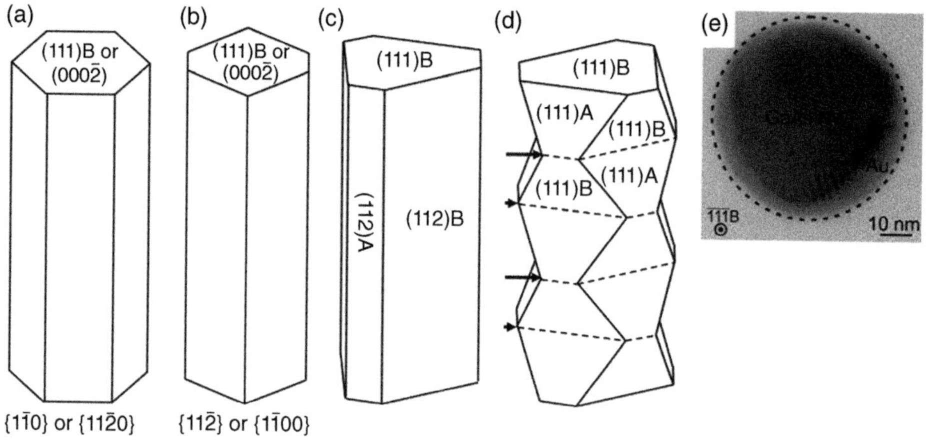

Figure 7.3 *Schematic illustration of morphologies exhibited by [111]B-oriented nanowires: (a) six {1–10}/{11–20} side-facets, (b) six {11–2}/{1–100} side-facets, (c) three dominant {112}B side-facets and three smaller {112}A side-facets, (d) alternating {111}A and {111}B side-facets with twin planes indicated by the arrows and dotted lines. (e) TEM image of the interface between an Au particle and the underlying GaAs nanowire tip, which adopts a rounded cross-section in the shape of a Reuleaux triangle. Sources: Parts (a–d) adapted from Joyce et al. 2011 [40]; part (e) reproduced from Jiang et al. 2014 [45], copyright American Chemical Society 2014.*

and three {112}B facets. Under high V/III ratios, the {112}B facets are more stable and feature a slower growth rate than the {112}A facets. Consequently, the {112}A facets grow out at high V/III ratios, leaving dominant {112}B facets together with a triangular cross section, as shown in Figure 7.3c [41]. ZB nanowires can also feature alternating {111}A/{111}B microfaceted side-facets, where a twin plane perpendicular to the growth axis occurs wherever the facets switch along the axis from {111}A to {111}B and vice versa [42], as shown in Figure 7.3d.

WZ nanowires commonly elongate along the energetically preferred [0001] or [000–1] directions (c-axes), which are equivalent to the [111]A or [111]B directions, respectively. These WZ nanowires adopt the shape of a hexagonal prism with six side-facets that are nonpolar {11–20} (a-plane) or {1–100} (m-plane) planes, as shown in Figures 7.3a and b. At the growing tip of the hexagonal prism, [0001]-oriented (Ga-polar) WZ GaN nanowires sometimes present a pyramidal shape dominated by semi-polar {1–101} planes that are stabilized under high V/III ratios.

For particle-assisted III-V nanowires (excluding nitrides), the energetically preferred direction for axial growth is usually the Group V-polar <111>B direction [40]. Often, (111)B substrates are employed so that epitaxial nanowires grow vertically from the substrate. Particle-assisted III-nitride nanowires are an exception, as they tend to grow along <11–20>(a-axis) or <1–100>(m-axis) directions and exhibit cross sections in the shape of isosceles triangles. The <11–20>-oriented nanowires are bounded by one (0001) polar side-facet and two semipolar {1–101} side-facets [43]. The <1–100>-oriented wires are bounded by one {0001} polar side-facet and two semi-polar {11–22} side-facets [44].

In particle-assisted growth, the shape of the nanowire is also influenced by how the rounded particle wets the tip of the nanowire. The TEM image in Figure 7.3e shows that immediately below the particle, the ZB nanowire tip adopts a cross-section in the shape of a Reuleaux triangle with rounded edges [45]. Later, as radial growth proceeds, the Reuleaux triangle converts to a sharply facetted hexagonal or triangular shape consistent with Figure 7.3b or c [45].

7.3 Particle-Assisted MOVPE of Nanowires

The particle-assisted method – also known as the *particle-seeded method* – is a widely studied and utilized means of achieving nanowires by MOVPE because of its simplicity, reproducibility, and controllability. This method is widely applicable for a number of semiconductor material systems, including III-Vs [40], II-VIs such as ZnO [46] and CdTe [47], II-Vs such as Zn_3As_2 [48], and even IV-V-Vs such as $Ge_1Sb_2Te_4$ [49]. In the particle-assisted method, anisotropic nanowire growth occurs because the growth rate is strongly enhanced at the interface between a nanoscale particle and the semiconductor crystal. The salient feature of particle-assisted growth is the particle present at the growing tip of each nanowire. This particle features a truncated spherical shape with its flat side contacting the nanowire, as illustrated in the scanning electron microscope (SEM) and TEM images in Figure 7.4. The particle's diameter is usually several tens of nanometers and is comparable to the diameter of the nanowire.

The most commonly used particle is Au, but the particle-assisted process is not unique to Au particles; MOVPE of III-V nanowires has been achieved with a variety of metallic particles including Ni for GaN nanowires [50–52], Fe for GaN and GaAs nanowires [53, 54]; Cu for InP and InAs nanowires [55, 56]; Ag for GaAs nanowires [57]; Pd for GaAs and InAs nanowires [58, 59]; Sn for GaAs and GaSb [60, 61]; Ga for GaN [62]; and In for InP, InAs, and InSb nanowires [31, 63–65]. The seed particle can be a foreign element that is not consumed into the growing nanowire, such as Au, or may be an element that comprises the nanowire itself, such as In for InAs nanowires.

By the early 1980s, particle-assisted *whisker* growth in MOVPE was a well-known phenomenon [66, 67]. The whiskers achieved were typically micrometers in diameter, rather than true nanowires, and were often obtained unintentionally when planar growth was the desired objective. Deliberate efforts toward the MOVPE of ultrathin nanowires with nanoscale diameters began in earnest in the 1990s with work performed by Hiruma and colleagues [28, 68]. Their initial experiments revealed that GaAs and InAs nanowires grew exclusively on the exposed regions of the GaAs substrate defined within a patterned SiO_x mask [28, 68]. It was soon discovered that the fluorocarbon plasma process used to etch the SiO_x mask was introducing Au contaminants on the substrate surface, and nanoscale clusters of Au atoms were acting as seeds for nanowire growth [69]. Well-controlled nanowire growth was thereafter achieved by deliberately depositing Au particles on the substrate prior to MOVPE [69, 70]. The demonstration of such tightly defined nanoscale diameters sparked worldwide interest in particle-assisted MOVPE of nanowires, and the name of these structures evolved from *needle-shaped microcrystals*, *quantum wire crystals*, and *whiskers* and converged on *nanowires*.

7.3.1 The Phase of the Particle

The particle-assisted method encompasses two possible growth mechanisms: the vapor-liquid-solid (VLS) and vapor-solid-solid (VSS) growth models. These mechanisms are so-called because of the phases involved. In both mechanisms, the precursors are supplied in the *vapor* phase, and the growing nanowire is in the *solid* phase. The physical state of the particle is *liquid* for VLS and *solid* for VSS growth.

Hiruma and colleagues proposed that the Au particle-assisted MOVPE of nanowires proceeded via a VLS growth mechanism [69], akin to the mechanism originally proposed by Wagner and Ellis in 1964 for the chemical vapor deposition of Si whiskers [74]. Consider the following simple and intuitive explanation for the case of MOVPE of GaAs nanowires by VLS, together with the schematic diagram in Figure 7.5. Initially, the Au particle is solid. Ga and As are supplied from the *vapor* phase through the decomposition of trimethylgallium (TMGa) and arsine (AsH_3), respectively. At growth temperature, the metallic Au particle alloys with supplied Ga, eventually forming

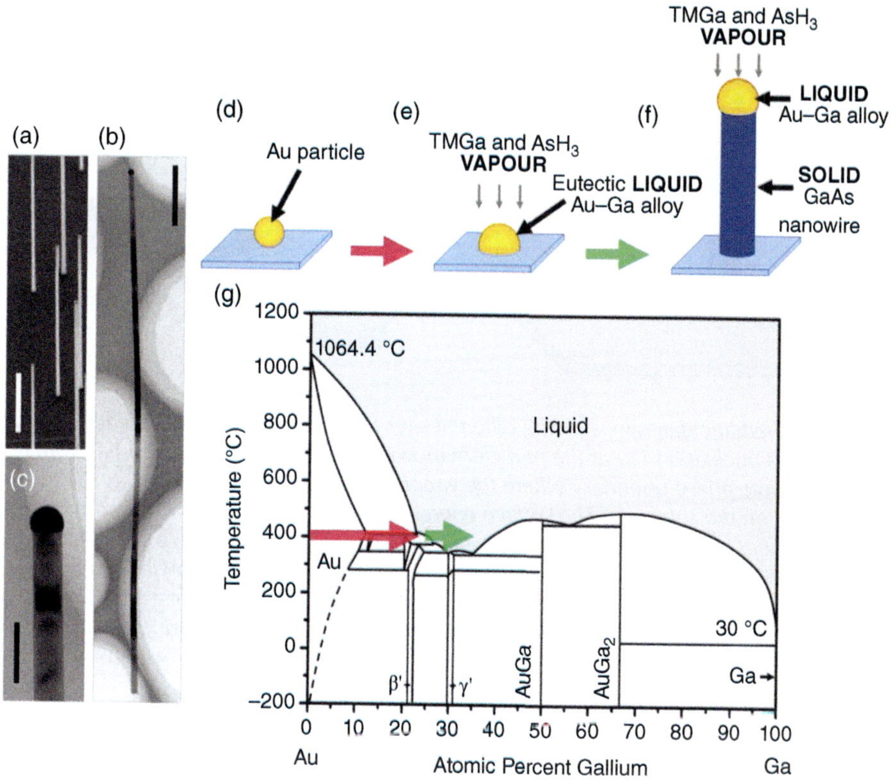

Figure 7.4 Au particle-assisted MOVPE of GaAs nanowires. (a) SEM image of nanowires on the substrate with substrate tilted at 40° to the electron beam, and (b, c) TEM images of GaAs nanowires, illustrating an entire nanowire in (b) and the particle at the nanowire tip in (c). (d–g) Schematic illustration of the VLS growth of GaAs nanowires using Au particles to drive nanowire growth. The simplified Au-Ga binary phase diagram (g) illustrates the various phases the Au-Ga particle can adopt under different temperatures. The arrows indicate the phase transitions occurring in the particle as its Ga content increases at a constant growth temperature of 400 °C. Initially, the Au particle is solid (d). At the growth temperature and with the supply of Ga from TMGa in the vapor, the Au particle alloys with the Ga until the liquidus line is crossed, yielding an Au-Ga eutectic liquid (e). As the Ga atomic percentage in the particle increases, the particle becomes supersaturated with Ga. The supersaturation of Ga in the particle coupled with the supersaturation of As species in the vapor phase leads to the crystallization of GaAs. As crystallization continues at the newly formed particle/GaAs interface, an elongated GaAs nanowire grows (f). Sources: Phase diagram (g) produced from data presented in references [71–73]. Images (a–c) reproduced with permission from [27]; copyright American Chemical Society, 2007.

a *liquid* eutectic Au-Ga droplet. With further supply of Ga, the liquid Au-Ga droplet becomes supersaturated with Ga, and *solid* GaAs is deposited at the droplet/semiconductor interface. As growth continues at the droplet/semiconductor interface, the crystal elongates to form a nanowire with diameter defined by the interface with the Au-Ga droplet. The annotated Au-Ga binary phase diagram in Figure 7.5 illustrates how the composition and phase of the Au-Ga nanoparticle may change during the VLS growth process. The Au-assisted VLS process for other III-V materials is analogous, insofar as the Group III elements (Al, Ga, and In) feature eutectic points in their binary phase diagrams with Au and readily dissolve into the Au particle under typical growth conditions.

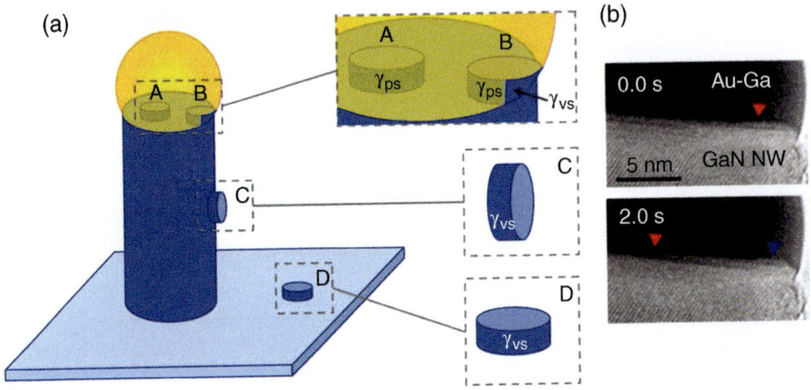

Figure 7.5 *(a) Schematic diagram showing different sites for nucleation in the presence of a particle, with heterogeneous nucleation (A) at the particle/nanowire interface without contacting with the vapor; (B) at the triple-phase boundary where the vapor, particle, and nanowire meet; (C) on the side-facet; and (D) on the substrate. The surface energies at the vapor-nucleus (γ_{vs}) and particle/nucleus (γ_{ps}) interfaces are shown for a hypothetical cylindrical nucleus shape. (b) ETEM images of nucleation and step-flow growth at the interface between a growing GaN nanowire and the Au-Ga alloy particle driving growth. The red arrow indicates the growing edge of the first step, and the blue arrow indicates the growing edge of the next step, 2 s later. Source: Adapted and reproduced with permission from Gamalski et al. 2016 [92]; copyright American Chemical Society 2016.*

In contrast, the Group V elements N, P, and As all have very low solubility in Au at typical growth temperatures and are thought to reach the growth front by diffusion along the particle/nanowire interface [75]. Antimony (Sb) is an exception amongst Group V elements, as it alloys with Au at typical growth temperatures.

In the mid-2000s, the VLS process was called into question when Persson [76] and Dick [77] presented compelling evidence that solid Au-alloy particles, rather than liquid particles, seed III-V nanowire growth. The VSS mechanism was proposed as a likely alternative to the VLS mechanism. Modeling showed that diffusion of Group III species through a solid Au-Ga particle was sufficient to maintain the observed GaAs nanowire growth rates [76]. More recently, in situ studies of nanowire growth by environmental transmission electron microscopy (ETEM) have shown that the particle driving growth can be liquid [78, 79] or solid [80]. In most cases, it is difficult to ascertain the physical phase of the particle, and hence whether a VLS or VSS mechanism is at work, so *particle-assisted* is used as a generic descriptor.

7.3.2 The Role of the Particle

The role of the particle – that is, how the particle promotes nanowire growth – has been the topic of much debate. A number of possible mechanisms have been proposed to explain the enhancement of growth rate brought about by the particle, with the most likely being (i) that the particle acts as a catalyst, (ii) that compared to the substrate, the particle has a higher sticking coefficient and hence acts as a sink for growth species arriving from the vapor, or (iii) that nucleation at the particle/nanowire interface is preferable over nucleation directly on the substrate or on the nanowire sidewalls. These processes are not mutually exclusive, and a combination of these factors may contribute to particle-assisted growth.

The first possibility is that the particle is acting as a catalyst for the chemical reactions of precursor decomposition and nanowire growth. Catalysts are substances that increase the reaction rate by lowering the activation energy barrier, without being consumed in the reaction. Like true

Table 7.1 *Typical growth temperatures used to promote axial nanowire growth via Au particle-assisted epitaxy and SA-MOVPE.*

Nanowire material	Typical temperatures for Au particle-assisted epitaxy	Typical temperatures for SA-MOVPE of nanowires
GaAs	375–550 °C [27, 37]	700–750 °C [111, 129, 133, 134]
InAs	400–500 °C [13, 77]	540–580 °C [130]
GaP	400–580 °C [37, 42]	770–790 °C [39]
InP	400–510 °C [35]	600–730 °C [36, 131]
GaN	780–1000 °C [44, 50]	1050–1125 °C [126, 127]

catalysts, foreign particles such as Au are not consumed in particle-assisted growth. Several groups have reported that metallic particles lower activation energies for nanowire growth [81–83] by catalyzing the pyrolysis of hydride or metalorganic species such as PH_3 [81], AsH_3 [83], NH_3 [44, 51], or TMGa [82]. It is notable that the growth temperatures used for particle-assisted MOVPE of arsenides, phosphides, and antimonide nanowires lie below 600 °C (see Table 7.1) and are usually at least 100 °C lower than temperatures used for planar growth. At these low temperatures, crystal growth is kinetically limited, and the decomposition of precursors is a rate-limiting step [84]. This is precisely the temperature range in which catalysis by the particle would be expected to enhance the nanowire growth rate above planar growth. However, a number of other studies have concluded that the metallic particles do not appreciably lower activation energies compared with particle-free growth [85–89]. Therefore, catalytic behavior is not considered a universal phenomenon or a necessary condition for particle-assisted growth, although it can play a significant role and can be exploited to achieve growth at lower temperatures and with a greater variety of precursors.

The second factor is that the particle may have a higher accommodation coefficient for vapor-phase reaction species than the other surfaces. Thus, the particle acts as a favorable collection site, or sink, and locally increases concentration of reactants to enhance the growth rate [87]. Liquids tend to have higher accommodation coefficients than solids, so this effect is considered to be more applicable to the VLS mechanism than the VSS mechanism.

The third mechanism, that of preferential nucleation at the particle/nanowire interface, has been proposed as the overarching mechanism of nanowire growth [90]. Here, the Gibbs free energy of nucleation predicts the thermodynamically preferable sites for growth. The Gibbs free energy of nucleation is the energetic barrier for nucleation and is minimized for nuclei with low surface energies γ and under high supersaturation growth conditions. Possible sites for nucleation are schematically illustrated in Figure 7.5a. Nucleation at positions A and B, both at the particle/nanowire interface, would contribute to nanowire growth. On the other hand, nucleation at position C or D, on the substrate and nanowire side-facets, respectively, would not contribute to nanowire growth. The nuclei in positions C and D have exposed edges in contact with the vapor, with surface energy γ_{vs}. At position A, the edges of the nucleus are in contact with the particle with interface energy γ_{ps}. At a given supersaturation, nucleation at position A will be favored over positions C and D, provided γ_{ps} is lower than γ_{vs}. The condition $\gamma_{ps} < \gamma_{vs}$ is met for particles that effectively wet the semiconductor surface, such as Au particles [91]. Consider also the triple-phase boundary where the vapor, particle, and nanowire meet, as labelled B in Figure 7.5. Similar to position A, some nucleus edges at B are in contact with the particle, although part of the nucleus contacts the vapor directly. At position B, the supersaturation is high because of the direct contact with the vapor, and the nucleus can adjust its shape to minimize its total interface energy [90]. Therefore, the lowest nucleation barrier may be expected at position B, so that nucleation occurs preferentially at the triple-phase boundary and growth occurs fastest at the particle/nanowire interface. Nuclei that form at position A or B then expand by step-flow growth across the particle/nanowire interface, and each layer elongates the nanowire [34, 92]. The ETEM images in Figure 7.5b show such behavior at the interface between a growing GaN nanowire and the Au-Ga alloy particle driving growth.

7.3.3 Axial and Radial Growth Modes

In particle-assisted growth, axial growth receives contributions from reaction species that directly impinge on the particle, and from species that have adsorbed on the substrate and on the nanowire side-facets that diffuse along these surfaces toward the particle. These mass-transport pathways are illustrated in Figure 7.6a. Desorption also takes place but is in most cases minimal at the low temperatures used for particle-assisted growth.

Surface diffusion is significant for Group III species arising from metalorganic precursors. These have long diffusion lengths, λ, of the order of micrometers [93], and so may be adsorbed a considerable distance from the particle and still contribute to axial growth. If nanowires are spaced within a diffusion length of one another, these nanowires compete for diffusing Group III species, so that their axial growth rates are reduced [93]. Diffusion lengths of several hundred nanometers have also been reported for Sb species arising from TMSb [9]. In contrast, Group V species from hydride precursors typically have shorter diffusion lengths and are thought to incorporate within a few tens of nanometers of their point of adsorption.

Unlike axial growth, radial growth does not involve the particle and occurs via the vapor-solid growth mode rather than the VLS or VSS growth modes. Radial growth receives contributions from species that adsorb on the substrate and nanowire side-facets and then diffuse and incorporate on the side-facets. As these diffusing species also contribute to axial growth, radial growth competes with axial growth. Therefore, when axial growth is desired, growth conditions are chosen to minimize parasitic radial growth.

Due to radial growth, particle-assisted nanowires frequently adopt tapered morphologies, with wider bases than tips, as shown in Figure 7.6b. Two factors are at play in the development of tapering. First, nanowire bases are grown first, and are therefore exposed to reactants for longer than the

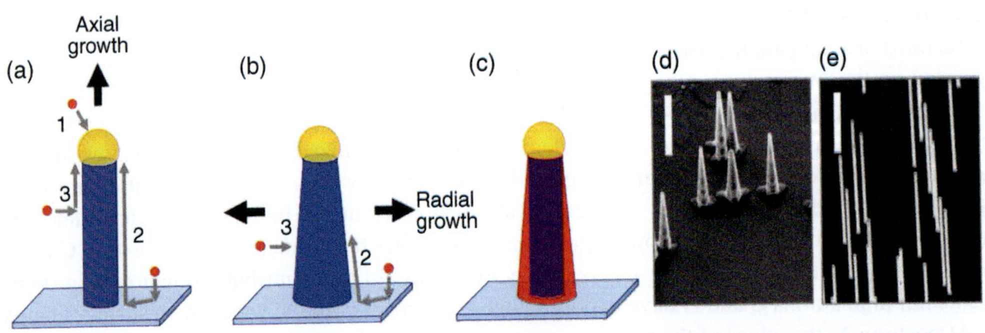

Figure 7.6 *Nanowire growth modes and resulting structures. (a, b) Schematic illustration of the mass-transport processes in particle-assisted growth leading to (a) axial growth and (b) radial growth. The contributions include (1) species that directly impinge on the particle, (2) species that adsorb on the substrate and then diffuse toward the particle, and (3) species that adsorb on the nanowire side-facets and then diffuse towards the particle. (c) Schematic illustration of an InGaAs ternary nanowire that spontaneously grows with an In-rich shell and Ga-rich core, owing to different incorporation efficiencies of In (red) and Ga (blue) into the axial and radial growth modes. (d, e) SEM images of (d) tapered GaAs nanowires grown at a high temperature of 500 °C and (e) minimally tapered GaAs nanowires grown at a low temperature of 375 °C. Source: SEM images in (d) and (e) reproduced with permission from Joyce et al. 2011 [40].*

tips. Second, reaction species that are collected on the substrate must diffuse past the base of the nanowire before they can reach the nanowire tip. As these species diffuse upward, they are consumed into radial growth, with the bases seeing the highest concentration of species and therefore experiencing a higher radial growth rate. Tapered morphologies can be useful and have been exploited, for example, in single-photon emitters for efficient out-coupling of emitted photons [17, 18]. In other cases, tapered morphologies are undesirable, such as when the nanowire is used as a resonant cavity in a nanowire laser and the modes must be well-confined along the nanowire [24].

Parasitic radial growth is often associated with poorer optical and electrical properties than particle-assisted growth [27, 94]. Carbon impurities that arise from the decomposition of metalorganic precursors are more readily incorporated into radial growth than into particle-assisted growth [95]. Radial growth also leads to compositional nonuniformity when growing doped, ternary, and quaternary nanowires, because different elements incorporate at different rates into radial growth and axial growth. To demonstrate this phenomenon, consider the Au-assisted growth of InGaAs nanowires, in which In species have higher decomposition efficiencies [96], longer diffusion lengths [93], and higher affinity for the Au particle [97] compared to Ga species. Therefore, In and Ga incorporate at different rates by the particle-assisted and vapor-solid routes. As a result, a Ga-rich core forms by particle-assisted growth, surrounded by an In-rich shell that forms by radial growth [98], as illustrated in Figure 7.6. Similarly, dopants incorporate with different efficiencies through particle-assisted and radial growth [95], leading to nonuniform doping throughout the nanowire. Parasitic radial growth is also problematic when growing axial heterostructures and p-n junctions, as any unintentional shell structure that grows radially around the lower section of the nanowire can short-circuit the device [14].

The key to obtaining compositionally uniform nanowires is therefore to eliminate parasitic radial growth. Low temperatures are often employed to minimize radial growth, which is effective because radial growth is kinetically limited [27]. In addition, diffusion lengths decrease with decreasing growth temperature, so low temperatures reduce the flux of species diffusing from the substrate [27]. The SEM images in Figures 7.6d and e compare the degree of radial growth experienced at high and low growth temperatures, respectively. Low V/III ratios also minimize growth rates on the side-facets and increase the diffusion length of Group III species on these side-facets, which increases the likelihood that Group III species will reach the tip of the nanowire and contribute to axial growth [35, 96, 99]. In situ etching using HCl gas is another effective way to prevent parasitic radial overgrowth while axial growth proceeds [100].

On the other hand, the radial growth mode is required when growing the shells of radial core-shell heterojunctions or homojunctions, as illustrated in Figure 7.1d. To enhance radial growth in particle-assisted MOVPE, the key is to increase the growth temperature, which shifts growth from the kinetically limited regime to the mass transport-limited growth regime [40]. As axial growth does not cease entirely under these conditions, a short axial segment grows as shell growth proceeds.

Achieving axial heterostructures via the particle-assisted mechanism requires careful consideration of how the particle alloys with and wets the different materials that form the heterostructure. During the growth of the first segment, the particle forms an alloy with at least one element of that segment (e.g. In) and therefore behaves as a reservoir for that element even after its gas-phase precursor (e.g. TMIn) is removed. When the gases are switched to grow the next segment of the axial heterostructure, the particle may continue to release the previous element, resulting in a graded interface rather than an atomically sharp interface [97]. Growth interruptions are required to deplete the particle of the initial element to achieve sharp interfaces. A further challenge occurs when the interfacial energy of the particle with the initial segment is lower than with the next segment. In this case, the particle will slide off the tip of the nanowire and onto the side-facet of the initial segment to maintain the lower energy interface, resulting in a change of growth direction [2].

7.3.4 Self-Assisted Growth

A special case of particle-assisted growth is *self-assisted* or *self-seeded* growth; here the particle comprises one of the elements – usually the Group III element – of the growing nanowire itself. For example, a particle (or droplet) of liquid In drives the self-assisted MOVPE of InP [63, 66], InAs [64], InSb [31, 101], InAsSb [102, 103], InAsP [65], or InPSb [104] nanowires. Similarly, a droplet of liquid Ga drives the self-assisted MOVPE of GaAs whiskers [67], GaN [62], and GaSb [105] nanowires.

To achieve self-assisted growth, the growth conditions must allow the formation and maintenance of the liquid droplet. To achieve self-assisted growth with Ga or In droplets, relatively low growth temperatures and low V/III ratios are chosen. Under these growth conditions, the rate of pyrolysis of the Group V precursor is low and the vapor pressure of liquid Ga or In droplets is low [66, 67], which favors the formation of In or Ga droplets.

The most important advantage of self-assisted growth is that it avoids the need for foreign particles. Atoms of foreign particles, such as Au or Sn, have been observed to incorporate as impurities or dopants into the growing nanowire [106, 107], which can affect or even degrade the electronic or optoelectronic properties of the nanowire. Au particles, in particular, are incompatible with Si-based microelectronics, as Au forms deep levels in Si. Therefore, self-assisted growth is more appropriate when the nanowires are to be integrated with Si or directly grown on Si substrates. The self-assisted growth mechanism is also useful for achieving high Sb content in InAsSb nanowires [102, 103]. The disadvantage of the self-assisted growth mechanism is that the range of growth conditions is restricted to those that maintain liquid droplets. This narrows the scope for tuning the growth parameters and thereby controlling nanowire properties such as the crystal phase and doping.

7.4 Selective-Area MOVPE of Nanowires and Nanostructures

Selective-area epitaxy (SAE), also known as selective-area growth (SAG), has found widespread application in the growth of III-V nanowires. SAE of nanowires requires that the substrate is partially masked by a dielectric layer, such as SiO_x or Si_xN_y, chosen such that the nucleation of material on the mask is unfavorable. The underlying principle in SAE is that the nucleation of semiconductor material occurs selectively in apertures in the mask where the underlying substrate is exposed. Under appropriate growth conditions chosen to promote axial growth and reduce radial overgrowth, the growth occurs in a highly anisotropic fashion to form structures such as nanowires. The process is schematically illustrated in Figures 7.7a–c, and an example of a grown nanowire sample is shown in Figure 7.7e. The apertures in the mask can be lithographically defined prior to growth, with pitch a and diameter d_0, to control the position and diameter of the nanowires (Figure 7.7d).

Research into selective-area-MOVPE (SA-MOVPE) of ordered arrays of III-V nanowires was initiated in the late 1990s [108] and early 2000s [109, 110] as part of efforts to develop two-dimensional photonic crystals via bottom-up fabrication processes. These structures were termed *pillars*, *micropillars*, or *nanopillars* and featured heights of the order of a micrometer and sub-micrometer radial dimensions. The term *nanowire* became common for such SA-MOVPE-grown structures in the mid-2000s as efforts toward increasing the length-to-diameter aspect ratios and reducing the diameter to below 100nm intensified [111]. Since then, SA-MOVPE has been extended to a wide variety of nanostructures spanning arsenides, phosphides, and nitrides [112], antimonides [113, 114] and including axial nanowire heterostructures [115], radial nanowire heterostructures [5, 21, 116, 117], ternary nanowires [118], nanowire-substrate heterostructures [119–121], and nanosheets [8].

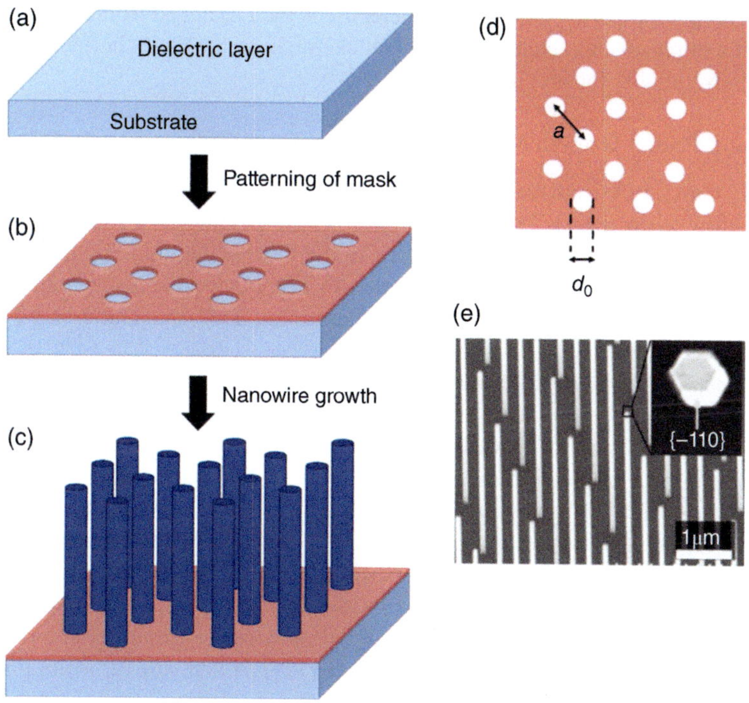

Figure 7.7 Schematic diagram illustrating SA MOVPE of nanowires. (a) A dielectric layer is used as a mask on the substrate, (b) apertures are created in the dielectric mask via lithography and etching, and (c) nanowires are grown by MOVPE in the mask apertures. (d) The diameter and pitch of the apertures are denoted d_0 and a, respectively. (e) SEM image of GaAs nanowires grown via SA-MOVPE. The inset shows a nanowire's hexagonal cross section bounded by six {1–10} side-facets. Source: Image reproduced with permission from Ikejiri et al. 2007 [125].

One major advantage of SA-MOVPE is that it avoids the foreign metal impurities used in particle-assisted growth (with the exception of self-assisted growth). The most commonly used particle, Au, is incompatible with CMOS and other Si-based processes, as Au forms deep levels in Si. SA-MOVPE, on the other hand, has shown particular promise for the integration of III-V nanowires with Si substrates [25, 120, 122].

7.4.1 The Role of the Mask

In SA-MOVPE, the selectivity for growth in the apertures originates from the preferential decomposition and reaction of the metalorganic precursors on the semiconductor substrate, compared to the dielectric mask. By using metalorganic precursors, MOVPE achieves a greater selectivity for growth in the apertures than that achieved with elemental fluxes as in molecular beam epitaxy [123]. The diffusion lengths of adsorbed metalorganic species on dielectric surfaces are significant, so these species readily diffuse to the apertures where they complete pyrolysis and contribute to nanowire growth [39].

The dielectric mask (Figure 7.7a) may be deposited by a number of means, including plasma-enhanced chemical vapor deposition and radio-frequency sputtering. In the case of Si substrates, thermal oxidation is often used to form the SiO_2 layer as the resultant film is relatively tolerant to temperature changes during MOVPE. E-beam lithography is frequently used to expose the apertures in the resist [39], although nanoimprint lithography [124] and other high-throughput lithography techniques are also suitable. The dielectric is then removed from the apertures by wet chemical etching or reactive ion etching (RIE). Wet chemical etching with buffered hydrogen fluoride is particularly suitable for SiO_x masks.

7.4.2 Axial and Radial Growth Modes

In SA-MOVPE, GaAs, InAs, and GaP nanowires all grow preferentially along the [111]B direction and therefore grow vertically on (111)B substrates. These nanowires feature a hexagonal cross section with six {1–10} side-facets, as illustrated in the inset of Figure 7.7e. InP nanowires also exhibit {1–10} side-facets, but their preferential growth direction is [111]A. GaN nanowires grown by SA-MOVPE elongate in the [0001] or [000–1] direction (*c*-axis) and feature six {1–100} side-facets (*m*-plane) [126–128].

The mass-transport processes that lead to axial growth are illustrated in Figure 7.8a. Reaction species that directly impinge on the aperture and, at later stages of growth, on the nanowire tip, decompose, and are deposited to drive axial nanowire growth (1). Axial nanowire growth also receives contributions from reaction species that adsorb on the mask and are transported to the mask apertures via surface diffusion (2). Third, axial growth may receive contributions from reaction species that reach the mask and then desorb, return to the boundary layer in the vapor phase, and diffuse toward the aperture via vapor-phase diffusion (3). As the nanowire grows, species that adsorb on the nanowire side-facets also diffuse toward the tip (4) and contribute to axial growth (Figure 7.8b). If process (4) is dominant, then the axial growth rate is inversely proportional to the nanowire diameter d, as observed for GaAs nanowires [129]. If pathway (2) is dominant, then the axial growth rate is inversely proportional to the square of d, as observed for InAs nanowires [130].

Mass-transport pathways (2), (3), and (4) also contribute to radial growth, as illustrated in Figure 7.8c. Radial growth causes the diameter of the nanowires to expand beyond the diameter of the aperture.

SA-MOVPE uses substantially higher growth temperatures than used for particle-assisted growth. A comparison of typical growth temperatures, although not exhaustive, is given in Table 7.1. SA-MOVPE takes place in the mass-transport limited-growth regime, in which the

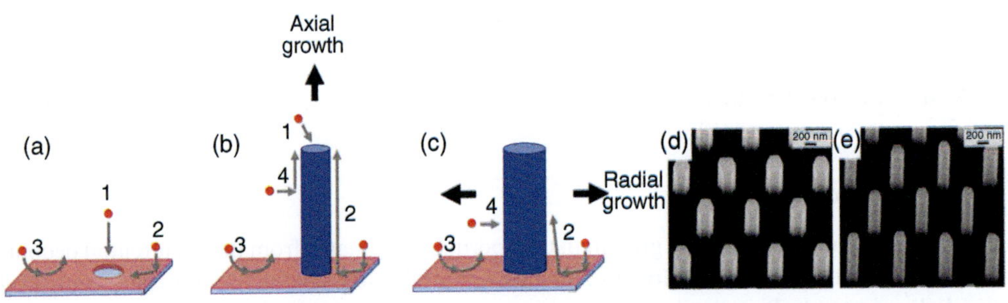

Figure 7.8 *Nanowire growth modes and resulting structures. (a–c) Schematic illustration of the mass-transport processes in selective-area epitaxy leading to (a) growth in the aperture, (b) axial growth, and (c) radial growth. The contributions include (1) species that directly impinge on the aperture or nanowire tip; (2) species that adsorb on the mask and then diffuse on the mask surface; (3) species that desorb from the mask, return to the boundary layer, and diffuse in the vapor phase toward the aperture or nanowire; and (4) species that adsorb on the nanowire side-facets and then diffuse. (d, e) SEM images of (d) InP nanowires featuring substantial lateral overgrowth grown at a high tertiarybutylphosphine (TBP) partial pressure of 5.5×10^{-3} atm and (e) narrower InP nanowires grown under identical conditions to (d) but with lower TBP partial pressure of 5.5×10^{-4} atm. Source: SEM images in (d) and (e) reproduced with permission from Mohan et al. 2005 [131].*

growth rate is limited by the diffusion of precursors, rather than by temperature [39]. At the relatively high temperatures used for SA-MOVPE, desorption of precursors from the substrate and nanowire side-facets is also significant.

In SA-MOVPE, the axial and radial growth modes can be isolated by choosing an appropriate growth temperature and appropriate V/III ratio. At high temperatures [39] and low V/III ratios [125, 131, 132], the growth rate of the top facet (either (111)A or (111)B) becomes significantly faster than that of the side-facets (often {1–10}). Therefore, these conditions suppress radial growth and promote axial growth. The SEM images in Figures 7.8d and e illustrate the effect of V/III ratio, with the nanowires grown at higher V/III ratio (Figure 7.8d) featuring more significant lateral overgrowth. Pulsed growth, in which NH_3 flow is periodically switched on and off, has been employed to yield low effective V/III ratios and thereby achieve GaN nanowires with minimal lateral overgrowth [127]. Lower temperatures and higher V/III ratios are used to promote shell growth and thereby create radial heterostructures or radial p-n junctions [39].

7.5 Alternative Techniques for MOVPE of Nanowires

Although particle-assisted growth and selective-area epitaxy are the dominant methods for MOVPE of nanowires, other strategies are used. One alternative strategy is a hybrid between the particle-assisted and SAE techniques. Here, nanowire growth is driven by Au particles placed in the apertures of a dielectric mask [124, 135]. An advantage of this hybrid approach is that the mask locks the particle in place and hence preserves the shape and position of the particle [124, 136]. A selection of other promising strategies is outlined next.

GaN nanowires have been achieved by MOVPE using a particle-free self-assembly technique [137]. In essence, the technique is a special form of selective-area epitaxy in which the SiN_x mask is deposited in situ during MOVPE. Before the introduction of the Group III precursor, SiH_4 and NH_3 are introduced at around 1000 °C to cover the substrate with a thin layer (~2 nm thick) of SiN_x [137]. Localized holes that form spontaneously in the SiN_x layer are thought to act as favorable sites for the growth of nanoscale GaN seeds. After seed formation, SiH_4 injection favors longitudinal growth of nanowires. The SiH_4 flow can be turned off after growing sufficiently long wires (~5 μm), and the wires continue growing longitudinally. This technique has been used to grow the core-shell LEDs illustrated in Figure 7.9a [20].

Another noteworthy method for achieving nanowires is van der Waals heteroepitaxy with two-dimensional atomic-layered materials, such as graphene [138] or monolayer MoS_2 [139], acting as the substrate. No particle or mask is needed [140]. Instead, seeds of III-V material nucleate on two-dimensional materials, presumably at vacancies or point defects on the two-dimensional material. These seeds grow along in the energetically preferable [111]B direction to form nanowires. The case of InAs nanowires on graphene is particularly interesting because the distance of the nearest InAs primitive lattices along the zincblende <110>direction is 4.28 Å, which is almost equal to the distance of the nearest-neighbor carbon honeycomb lattices along the $<1\bar{2}10>$in the graphene layer of 4.26 Å. Therefore, nearly lattice-matched growth of [111]B-oriented InAs nanowires occurs on graphene surfaces [138, 139].

7.6 Novel Applications of Nanowires

Nanowires exhibit many unique properties that are now being exploited to achieve devices that do not merely mimic conventional planar devices, but instead achieve entirely new functionalities with performance that surpasses that of conventional planar devices. Several of these properties

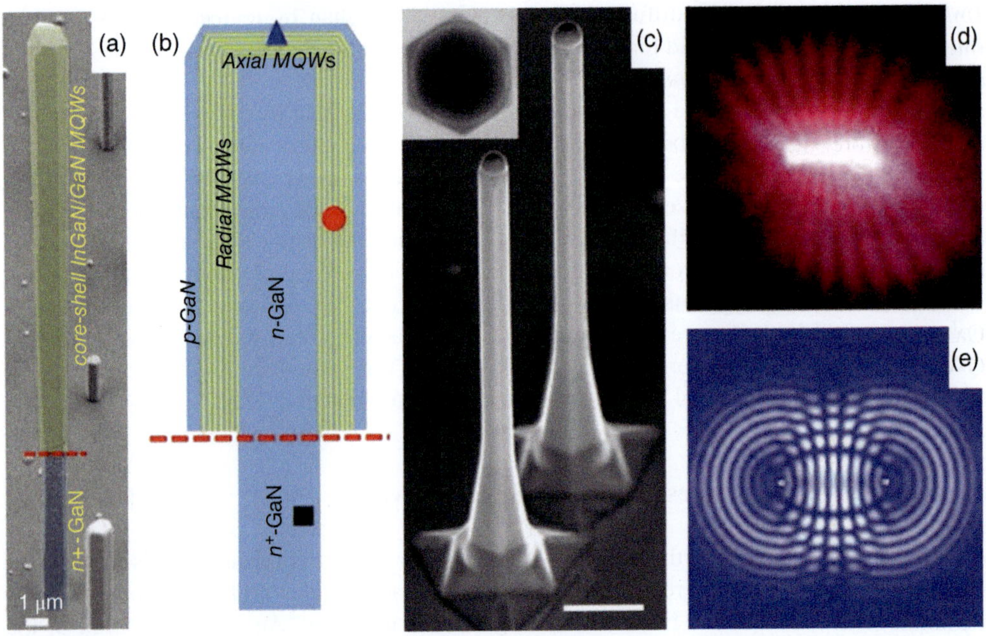

Figure 7.9 *(a) SEM image and (b) schematic illustration of a core-shell nanowire with n⁺-GaN core, multiple radial InGaN quantum wells, and outer p-GaN shell designed as the active element of an LED. (c) SEM image of GaAs/AlGaAs core-shell nanowires capable of lasing action at room temperature. Scale bar is 100 nm. (d) Optical image of a GaAs/AlGaAs core-shell nanowire showing emission from the nanowire above the lasing threshold with a distinct interference pattern. (e) Simulated emission-intensity pattern for a lasing nanowire showing similar interference pattern to (d). Sources: (a, b) Reproduced with permission from [20]; copyright American Chemical Society 2015. Images (c–e) reproduced from [24]; copyright Springer Nature 2013.*

have already been described in detail, such as efficient strain relaxation (Section 7.1) and the ability to achieve crystallographic structures not possible in conventional bulk materials (Section 7.2). Other properties enabling new devices include:

- *Large surface area-to-volume ratio*: The large surface area-to-volume ratio of nanowires means their electronic properties are particularly sensitive to surfaces. This property is exploited in surround-gate nanowire field effect transistors (FETs), as the gate can be wrapped around the entire surface of the nanowire to provide the best possible electrostatic gate control over the conducting channel in the nanowire [141, 142]. A further example of surface properties dominating the electrical properties of nanowires is the long-lived negative photoconductivity observed in InAs nanowires. This effect occurs due to surface states that trap hot photoexcited electrons and underpins optically writable and electronically readable memory devices based on InAs nanowires [22].
- *One-dimensionality:* By virtue of their one-dimensionality, nanowires provide a unique platform for the investigation of fundamental physics and for the engineering of new quantum electronic devices. Quantized conductance, the hallmark of truly ballistic one-dimensional transport, has been observed in MOVPE-grown InSb nanowires [143]. Nanowires with strong spin-orbit coupling (such as InSb, InAs) that are proximity coupled to conventional *s*-wave superconductors (such as Al) provide an ideal test bed for the investigation of topological superconducting phases together with Majorana zero modes [4, 26].

- *Unique interfacial geometry of core-shell nanowires:* Core-shell nanowires feature a large interfacial area between core and shell that far exceeds the footprint of the nanowire on the substrate. In solar cells, the core-shell geometry decouples the directions of light absorption and charge-carrier separation: light is absorbed vertically, whereas carriers are separated radially. Therefore, the dimensions for absorption and carrier collection can be optimized independently, which is not possible in conventional planar devices. For LEDs, the core-shell geometry allows quantum wells (QWs) to be grown as concentric shells on the side-facets of the nanowires, which significantly increases the area of the emitting QWs relative to planar architectures. By maximizing the emissive area, the geometry allows efficient light emission while effectively reducing the current density through the QWs. Furthermore, *c*-axis oriented GaN nanorods spontaneously form nonpolar *m*-plane side-facets upon which can be grown QWs that are minimally affected by polarization fields [144]. This mitigates the quantum confined Stark effect that limits radiative recombination in conventional planar LEDs. Figures 7.9a and b illustrate a GaN/InGaN radially heterostructured nanowire designed for mechanically flexible LEDs.
- *Mechanical flexibility and elasticity*: Nanowires are highly flexible and can withstand high stresses without undergoing plastic deformation or fracture [145]. This property has enabled the development of flexible LEDs consisting of vertically aligned III-nitride nanowires embedded in a flexible polymer matrix [20]. Piezoelectric nanowire-based devices also benefit from this flexibility [146].
- *Waveguiding, light trapping, and outcoupling:* Similar to optical fibers, nanowires with approximately cylindrical shape behave as waveguides. Despite their subwavelength diameters, nanowires support the propagation of guided modes along their axes due to their high refractive indices. A nanowire will behave as a Fabry–Pérot cavity when guided modes reflect from its end facets, and hence integrates a semiconductor gain medium with a resonant cavity, as required for lasing [23, 24]. At the wavelengths corresponding to guided modes, nanowires exhibit absorption cross sections much larger than their physical cross sections. In the context of solar cells and photodetectors, this means incident light is concentrated into the nanowire volume resulting in enhanced absorption of light. Therefore, nanowires of only a few micrometers in length achieve high light-absorption efficiency with a small volume of semiconductor material, which represents significant cost savings for solar cells and photodetectors. The nanowire pitch and diameter can be selected so that resonant absorption matches the solar spectrum optimally [14]. Furthermore, the intrinsic antireflection properties of nanowire arrays are advantageous for enhancing light extraction from nanowire-based LEDs and single-photon emitters [17].

7.7 Concluding Remarks

MOVPE, combined with nanowire growth methods such as particle-assisted epitaxy and selective-area epitaxy, offers enormous flexibility and high accuracy in the fabrication of complex nanowire structures. This chapter has focused on III-V nanowires, but it should be noted that MOVPE is much more widely applicable. MOVPE of CdTe nanowires and ZnO nanostructures, for example, are discussed in Chapters 10 and 11, respectively. MOVPE-grown semiconductor nanowires are driving a host of new applications ranging from photovoltaics to topological quantum computing, and the number and impact of these devices will continue to grow in the years to come.

References

1. Dick, K.A., Deppert, K., Larsson, M.W. et al. (2004). Synthesis of branched "nanotrees" by controlled seeding of multiple branching events. *Nature Materials* 3 (6): 380–384.
2. Paladugu, M., Zou, J., Auchterlonie, G.J. et al. (2007). Evolution of InAs branches in InAs/GaAs nanowire heterostructures. *Applied Physics Letters* 91 (13): 133115.
3. Wang, J., Plissard, S.R., Verheijen, M.A. et al. (2013). Reversible switching of InP nanowire growth direction by catalyst engineering. *Nano Letters* 13 (8): 3802–3806.
4. Gazibegovic, S., Car, D., Zhang, H. et al. (2017). Epitaxy of advanced nanowire quantum devices. *Nature* 548: 434–438.
5. Noborisaka, J., Motohisa, J., Hara, S., and Fukui, T. (2005). Fabrication and characterization of freestanding GaAs/AlGaAs core-shell nanowires and AlGaAs nanotubes by using selective-area metalorganic vapor phase epitaxy. *Applied Physics Letters* 87 (9): 093109.
6. Mohan, P., Motohisa, J., and Fukui, T. (2006). Realization of conductive InAs nanotubes based on lattice-mismatched InP/InAs core-shell nanowires. *Applied Physics Letters* 88 (1): 13110.
7. Paladugu, M., Zou, J., Guo, Y.-N. et al. (2009). Formation of hierarchical InAs nanoring/GaAs nanowire heterostructures. *Angewandte Chemie - International Edition* 48 (4): 780–783.
8. Chi, C., Chang, C., Hu, S. et al. (2013). Twin-free GaAs nanosheets by selective area growth: Implications for defect-free nanostructures. *Nano Letters* 13 (6): 2506–2515.
9. Plissard, S.R., Slapak, D.R., Verheijen, M.A. et al. (2012). From InSb nanowires to nanocubes: Looking for the sweet spot. *Nano Letters* 12 (4): 1794–1798.
10. Pendyala, C., Jasinski, J.B., Kim, J.H. et al. (2012). Nanowires as semi-rigid substrates for growth of thick, $In_xGa_{1-x}N$ ($x > 0.4$) epi-layers without phase segregation for photoelectrochemical water splitting. *Nanoscale* 4 (20): 6269–6275.
11. Kim, J.H., Ko, Y.H., Cho, J.H. et al. (2014). Toward highly radiative white light emitting nanostructures: a new approach to dislocation-eliminated GaN/InGaN core-shell nanostructures with a negligible polarization field. *Nanoscale* 6 (23): 14213–14220.
12. Wallentin, J., Jacobsson, D., Osterhoff, M. et al. (2017). Bending and twisting lattice tilt in strained core-shell nanowires revealed by nanofocused X-ray diffraction. *Nano Letters* 17 (7): 4143–4150.
13. Joyce, H.J., Wong-Leung, J., Gao, Q. et al. (2010). Phase perfection in zincblende and wurtzite III-V nanowires using basic growth parameters. *Nano Letters* 10 (3): 908–915.
14. Wallentin, J., Anttu, N., Asoli, D. et al. (2013). InP nanowire array solar cells achieving 13.8% efficiency by exceeding the ray optics limit. *Science* 339 (6123): 1057–1060.
15. Yao, M., Huang, N., Cong, S. et al. (2014). GaAs nanowire array solar cells with axial p-i-n junctions. *Nano Letters* 14 (6): 3293–3303.
16. Yao, M., Cong, S., Arab, S. et al. (2015). Tandem solar cells using GaAs nanowires on Si: Design, fabrication, and observation of voltage addition. *Nano Letters* 15 (11): 7217–7224.
17. Reimer, M.E., Bulgarini, G., Akopian, N. et al. (2012). Bright single-photon sources in bottom-up tailored nanowires. *Nature Communications* 3: 737.
18. Bulgarini, G., Reimer, M.E., Bouwes Bavinck, M. et al. (2014). Nanowire waveguides launching single photons in a Gaussian mode for ideal fiber coupling. *Nano Letters* 14 (7): 4102–4106.
19. Holmes, M.J., Choi, K., Kako, S. et al. (2014). Room-temperature triggered single photon emission from a III-nitride site-controlled nanowire quantum dot. *Nano Letters* 14 (2): 982–986.
20. Dai, X., Messanvi, A., Zhang, H. et al. (2015). Flexible light-emitting diodes based on vertical nitride nanowires. *Nano Letters* 15 (10): 6958–6964.
21. Farrell, A.C., Senanayake, P., Meng, X. et al. (2017). Diode characteristics approaching bulk limits in GaAs nanowire array photodetectors. *Nano Letters* 17 (4): 2420–2425.
22. Alexander-Webber, J.A., Groschner, C.K., Sagade, A.A. et al. (2017). Engineering the photoresponse of InAs nanowires. *ACS Applied Materials and Interfaces* 9: 43993–44000.
23. Gradečak, S., Qian, F., Li, Y. et al. (2005). GaN nanowire lasers with low lasing thresholds. *Applied Physics Letters* 87 (17): 173111.
24. Saxena, D., Mokkapati, S., Parkinson, P. et al. (2013). Optically pumped room-temperature GaAs nanowire lasers. *Nature Photonics* 7 (12): 963–968.
25. Tomioka, K., Yoshimura, M., and Fukui, T. (2012). A III-V nanowire channel on silicon for high-performance vertical transistors. *Nature* 488 (7410): 189–192.
26. Mourik, V., Zuo, K., Frolov, S.M. et al. (2012). Signatures of Majorana fermions in hybrid superconductor-semiconductor nanowire devices. *Science* 336 (6084): 1003–1007.

27. Joyce, H.J., Gao, Q., Tan, H.H. et al. (2007). Twin-free uniform epitaxial GaAs nanowires growth by a two-temperature process. *Nano Letters* 7 (4): 921–926.
28. Hiruma, K., Katsuyama, T., Ogawa, K. et al. (1991). Quantum size microcrystals using organometallic vapor phase epitaxy. *Applied Physics Letters* 59 (4): 431–433.
29. Algra, R.E., Verheijen, M.A., Borgstrom, M.T. et al. (2008). Twinning superlattices in indium phosphide nanowires. *Nature* 456 (7220): 369–372.
30. Caroff, P., Dick, K.A., Johansson, J. et al. (2009). Controlled polytypic and twin-plane superlattices in III-V nanowires. *Nature Nanotechnology* 4 (1): 50–55.
31. Kriegner, D., Panse, C., Mandl, B. et al. (2011). Unit cell structure of crystal polytypes in InAs and InSb nanowires. *Nano Letters* 11 (4): 1483–1489.
32. Yun, S.-H., Ra, Y.-H., Lee, Y.-M. et al. (2010). Growth of hexagonal and cubic InN nanowires using MOCVD with different growth temperatures. *Journal of Crystal Growth* 312 (15): 2201–2205.
33. Assali, S., Zardo, I., Plissard, S. et al. (2013). Direct band gap wurtzite gallium phosphide nanowires. *Nano Letters* 13 (4): 1559–1563.
34. Jacobsson, D., Panciera, F., Tersoff, J. et al. (2016). Interface dynamics and crystal phase switching in GaAs nanowires. *Nature* 531 (7594): 317–322.
35. Paiman, S., Gao, Q., Joyce, H.J. et al. (2010). Growth temperature and V/III ratio effects on the morphology and crystal structure of InP nanowires. *Journal of Physics D: Applied Physics* 43 (44): 445402.
36. Gao, Q., Saxena, D., Wang, F. et al. (2014). Selective-area epitaxy of pure wurtzite InP nanowires: high quantum efficiency and room-temperature lasing. *Nano Letters* 14 (9): 5206–5211.
37. Lehmann, S., Wallentin, J., Jacobsson, D. et al. (2013). A general approach for sharp crystal phase switching in InAs, GaAs, InP, and GaP nanowires using only group V flow. *Nano Letters* 13 (9): 4099–4105.
38. Dick, K.A., Thelander, C., Samuelson, L., and Caroff, P. (2010). Crystal phase engineering in single InAs nanowires. *Nano Letters* 10 (9): 3494–3499.
39. Tomioka, K., Ikejiri, K., Tanaka, T. et al. (2011). Selective-area growth of III-V nanowires and their applications. *Journal of Materials Research* 26 (17): 2127–2141.
40. Joyce, H.J., Gao, Q., Tan, H.H. et al. (2011). III – V semiconductor nanowires for optoelectronic device applications. *Progress in Quantum Electronics* 35 (2–3): 23–75
41. Zou, J., Paladugu, M., Wang, H. et al. (2007). Growth mechanism of truncated triangular III-V nanowires. *Small* 3 (3): 389–393.
42. Johansson, J., Karlsson, L.S., Svensson et al. (2006). Structural properties of <111>B-oriented III-V nanowires. *Nature Materials* 5 (7): 574–580.
43. Qian, F., Li, Y., Gradečak, S. et al. (2004). Gallium nitride-based nanowire radial heterostructures for nanophotonics. *Nano Letters* 4 (10): 1975–1979.
44. Kuykendall, T.R., Altoe, M.V., Ogletree, D.F., and Aloni, S. (2014). Catalyst-directed crystallographic orientation control of GaN nanowire growth. *Nano Letters* 14 (12): 6767–6773.
45. Jiang, N., Wong-Leung, J., Joyce, H.J. et al. (2014). Understanding the true shape of Au-catalyzed GaAs nanowires. *Nano Letters* 14 (10): 5865–5872.
46. Park, W. Il, Lee, C.H., Chae, J.H. et al. (2009). Ultrafine ZnO nanowire electronic device arrays fabricated by selective metal-organic chemical vapor deposition. *Small* 5 (2): 181–184.
47. Di Carlo, V., Prete, P., Dubrovskii, V.G. et al. (2017). CdTe nanowires by Au-catalyzed metalorganic vapor phase epitaxy. *Nano Letters* 17 (7): 4075–4082.
48. Burgess, T., Caroff, P., Wang, Y. et al. (2015). Zn_3As_2 nanowires and nanoplatelets: Highly efficient infrared emission and photodetection by an earth abundant material. *Nano Letters* 15 (1): 378–385.
49. Hardtdegen, H., Mikulics, M., Rieß, S. et al. (2015). Modern chemical synthesis methods towards low-dimensional phase change structures in the Ge–Sb–Te material system. *Progress in Crystal Growth and Characterization of Materials* 61 (2–4): 27–45.
50. Kuykendall, T., Pauzauskie, P., Lee, S. et al. (2003). Metalorganic chemical vapor deposition route to GaN nanowires with triangular cross sections. *Nano Letters* 3 (8): 1063–1066.
51. Weng, X., Burke, R.A., and Redwing, J.M. (2009). The nature of catalyst particles and growth mechanisms of GaN nanowires grown by Ni-assisted metal-organic chemical vapor deposition. *Nanotechnology* 20 (8): 85610.
52. Maliakkal, C.B., Hatui, N., Bapat, R.D. et al. (2016). The mechanism of Ni-assisted GaN nanowire growth. *Nano Letters* 16 (12): 7632–7638.
53. Kuykendall, T., Pauzauskie, P.J., Zhang, Y. et al. (2004). Crystallographic alignment of high-density gallium nitride nanowire arrays. *Nature Materials* 3 (8): 524–528.
54. Regolin, I., Khorenko, V., Prost, W. et al. (2007). GaAs whiskers grown by metal-organic vapor-phase epitaxy using Fe nanoparticles. *Journal of Applied Physics* 101 (5): 054318.

55. Hillerich, K., Ghidini, D.S., Dick et al. (2013). Cu particle seeded InP-InAs axial nanowire heterostructures. *Physica Status Solidi - Rapid Research Letters* 7 (10): 850–854.
56. Hillerich, K., Dick, K.A., Messing, M.E. et al. (2012). Simultaneous growth mechanisms for Cu-seeded InP nanowires. *Nano Research* 5 (5): 297–306.
57. Lindberg, C., Whiticar, A., Dick, K.A. et al. (2016). Silver as seed-particle material for GaAs nanowires - Dictating crystal phase and growth direction by substrate orientation. *Nano Letters* 16 (4): 2181–2188.
58. Hallberg, R.T., Lehmann, S., Messing, M.E., and Dick, K.A. (2016). Palladium seeded GaAs nanowires. *Journal of Materials Research* 31 (2): 175–185.
59. Xu, H., Wang, Y., Guo, Y. et al. (2012). Defect-free <110>zincblende structured InAs nanowires catalyzed by palladium. *Nano Letters* 12 (11): 5744–5749.
60. Sun, R., Vainorius, N., Jacobsson, D. et al. (2016). Sn-seeded GaAs nanowires grown by MOVPE. *Nanotechnology* 27 (21): 215603.
61. Zamani, R.R., Ghalamestani, S.G., Niu, J. et al. (2017). Polarity and growth directions in Sn-seeded GaSb nanowires. *Nanoscale* 9 (9): 3159–3168.
62. Tessarek, C., Bashouti, M., Heilmann, M. et al. (2013). Controlling morphology and optical properties of self-catalyzed, mask-free GaN rods and nanorods by metal-organic vapor phase epitaxy. *Journal of Applied Physics* 114 (14): 144304.
63. Novotny, C.J. and Yu, P.K.L. (2005). Vertically aligned, catalyst-free InP nanowires grown by metalorganic chemical vapor deposition. *Applied Physics Letters* 87 (20): 203111.
64. Mandl, B., Stangl, J., Hilner, E. et al. (2010). Growth mechanism of self-catalyzed group III-V nanowires. *Nano Letters* 10 (11): 4443–4449.
65. Park, J.H., Pozuelo, M., Setiawan, B.P.D., and Chung, C.H. (2016). Self-catalyzed growth and characterization of In(As)P nanowires on InP(111)B using metal-organic chemical vapor deposition. *Nanoscale Research Letters* 11 (1): 208.
66. Hsu, C.C., Cohen, R.M., and Stringfellow, G.B. (1983). OMVPE growth of InP using TMIn. *Journal of Crystal Growth* 63 (1): 8–12.
67. Stringfellow, G.B. (1984). Thermodynamic aspects of OMVPE. *Journal of Crystal Growth* 70 (1–2): 133–139.
68. Yazawa, M., Koguchi, M., and Hiruma, K. (1991). Heteroepitaxial ultrafine wire-like growth of InAs on GaAs substrates. *Applied Physics Letters* 58, 1080–1082.
69. Yazawa, M., Koguchi, M., Muto, A. et al. (1992). Effect of one monolayer of surface gold atoms on the epitaxial growth of InAs nanowhiskers. *Applied Physics Letters* 61 (17): 2051–2053.
70. Hiruma, K., Yazawa, M., Haraguchi, K. et al. (1993). GaAs free-standing quantum-size wires. *Journal of Applied Physics* 74 (5): 3162.
71. Cooke, C.J. and Hume-Rothery, W. (1966). The equilibrium diagram of the system of gold–gallium. *Journal of the Less-Common Metals* 10 (1): 42–51.
72. Elliott, R.P. and Shunk, F.A. (1981). The Au–Ga (gold–gallium) system. *Bulletin of Alloy Phase Diagrams* 2 (3): 356–358.
73. Jendrzejczyk-Handzlik, D. (2017). Thermodynamic study and re-optimization of the Au-Ga binary system. *Journal of Phase Equilibria and Diffusion* 38 (3): 305–318.
74. Wagner, R.S. and Ellis, W.C. (1964). Vapour-liquid-solid mechanism of single crystal growth. *Applied Physics Letters* 4 (5): 89–90.
75. Dick, K.A. (2008). A review of nanowire growth promoted by alloys and non-alloying elements with emphasis on Au-assisted III–V nanowires. *Progress in Crystal Growth and Characterization of Materials* 54 (3–4): 138–173.
76. Persson, A.I., Larsson, M.W., Stenstrom, S. et al. (2004). Solid-phase diffusion mechanism for GaAs nanowire growth. *Nature Materials* 3 (10): 677–681.
77. Dick, K.A., Deppert, K., Mårtensson, T. et al. (2005). Failure of the vapor–liquid–solid mechanism in Au-assisted MOVPE growth of InAs nanowires. *Nano Letters* 5 (4): 761–764.
78. Chou, Y.C., Hillerich, K., Tersoff, J. et al. (2014). Atomic-scale variability and control of III-V nanowire growth kinetics. *Science* 343 (6168): 281–284.
79. Gamalski, A.D., Tersoff, J., Sharma, R. et al. (2010). Formation of metastable liquid catalyst during subeutectic growth of germanium nanowires. *Nano Letters* 10 (8): 2972–2976.
80. Wen, C.Y., Reuter, M.C., Tersoff, J. et al. (2010). Structure, growth kinetics, and ledge flow during vapor-solid-solid growth of copper-catalyzed silicon nanowires. *Nano Letters* 10 (2): 514–519.
81. Verheijen, M.A., Immink, G., de Smet, T. et al. (2006). Growth kinetics of heterostructured GaP-GaAs nanowires. *Journal of the American Chemical Society* 128 (4): 1353–1359.
82. Borgstrom, M.T., Immink, G., Ketelaars, B. et al. (2007). Synergetic nanowire growth. *Nature Nanotechnology* 2 (9): 541–544.
83. Dayeh, S.A., Yu, E.T., and Wang, D. (2007). Growth of InAs Nanowires on SiO_2 substrates: Nucleation, evolution, and the role of Au nanoparticles. *Journal of Physical Chemistry C* 111 (36): 13331–13336.

84. Reep, D.H. and Ghandhi, S (1983). Deposition of GaAs epitaxial layers by organometallic CVD. *Journal of The Electrochemical Society* 130 (3): 675.
85. Samuelson, L., Thelander, C., Björk, M.T. et al. (2004). Semiconductor nanowires for 0D and 1D physics and applications. *Physica E: Low-dimensional Systems and Nanostructures* 25 (2–3): 313–318.
86. Paiano, P., Prete, P., Lovergine, N., and Mancini, A.M. (2006). Size and shape control of GaAs nanowires grown by metalorganic vapor phase epitaxy using tertiarybutylarsine. *Journal of Applied Physics* 100 (9): 94305.
87. Johansson, J., Wacaser, B.A., Dick, K.A., and Seifert, W. (2006). Growth related aspects of epitaxial nanowires. *Nanotechnology* 17 (11): S355–S361.
88. Soci, C., Bao, X.-Y., Aplin, D.P.R., and Wang, D. (2008). A systematic study on the growth of GaAs nanowires by metal-organic chemical vapor deposition. *Nano Letters* 8 (12): 4275–4282.
89. Borgström, M., Deppert, K., Samuelson, L., and Seifert, W. (2004). Size- and shape-controlled GaAs nano-whiskers grown by MOVPE: a growth study. *Journal of Crystal Growth* 260 (1–2): 18–22.
90. Wacaser, B.A., Dick, K.A., Johansson, J. et al. (2009). Preferential interface nucleation: An expansion of the VLS growth mechanism for nanowires. *Advanced Materials* 21 (2): 153–165.
91. Hilner, E., Mikkelsen, A., Eriksson, J. et al. (2006). Au wetting and nanoparticle stability on GaAs(111)B. *Applied Physics Letters* 89 (25): 251912.
92. Gamalski, A.D., Tersoff, J., and Stach, E.A. (2016). Atomic resolution in situ imaging of a double-bilayer multistep growth mode in gallium nitride nanowires. *Nano Letters* 16 (4): 2283–2288.
93. Kim, Y., Joyce, H.J., Gao, Q. et al. (2006). Influence of nanowire density on the shape and optical properties of ternary InGaAs nanowires. *Nano Letters* 6 (4): 599–604.
94. Bolinsson, J., Ek, M., Trägårdh, J. et al. (2015). GaAs/AlGaAs heterostructure nanowires studied by cathodoluminescence. *Nano Research* 7 (4): 473–490.
95. Thelander, C., Dick, K.A., Borgström, M.T. et al. (2010). The electrical and structural properties of n-type InAs nanowires grown from metal-organic precursors. *Nanotechnology* 21 (20): 205703.
96. Joyce, H.J., Gao, Q., Wong-Leung, J. et al. (2011). Tailoring GaAs, InAs, and InGaAs nanowires for optoelectronic device applications. *IEEE Journal on Selected Topics in Quantum Electronics* 17 (4): 766–778.
97. Paladugu, M., Zou, J., Guo, Y.-N. et al. (2008). Nature of heterointerfaces in GaAs/InAs and InAs/GaAs axial nanowire heterostructures. *Applied Physics Letters* 93 (10): 101911.
98. Guo, Y.-N., Xu, H.-Y., Auchterlonie, G.J. et al. (2013). Phase separation induced by Au catalysts in ternary InGaAs nanowires. *Nano Letters* 13 (2): 643–650.
99. Joyce, H.J., Gao, Q., Tan, H.H. et al. (2008). High purity GaAs nanowires free of planar defects: Growth and characterization. *Advanced Functional Materials* 18 (23): 3794–3800.
100. Borgström, M.T., Wallentin, J., Trägårdh, J. et al. (2010). In situ etching for total control over axial and radial nanowire growth. *Nano Research* 3 (4): 264–270.
101. Pozuelo, M., Zhou, H., Lin, S. et al. (2011). Self-catalyzed growth of InP/InSb axial nanowire heterostructures. *Journal of Crystal Growth* 329 (1): 6–11.
102. Du, W.N., Yang, X.G., Wang, X.Y. et al. (2014). The self-seeded growth of InAsSb nanowires on silicon by metal-organic vapor phase epitaxy. *Journal of Crystal Growth* 396: 33–37.
103. Namazi, L., Ghalamestani, S.G., Lehmann, S. et al. (2017). Direct nucleation, morphology and compositional tuning of $InAs_{1-x}Sb_x$ nanowires on InAs (111) B substrates. *Nanotechnology* 28 (16): 165601.
104. Ngo, C., Zhou, H., Mecklenburg, M. et al. (2011). Effect of precursor flux on compositional evolution in $InP_{1-}xSbx$ nanowires grown via self-catalyzed vapor-liquid-solid process. *Journal of Crystal Growth* 336 (1): 14–19.
105. Weng, X., Burke, R.A., Dickey, E.C., and Redwing, J.M. (2010). Effect of reactor pressure on catalyst composition and growth of GaSb nanowires. *Journal of Crystal Growth* 312 (4): 514–519.
106. Perea, D.E., Allen, J.E., May, S.J. et al. (2006). Three-dimensional nanoscale compositional mapping of semiconductor nanowires. *Nano Letters* 6 (2): 181–185.
107. Sun, R., Jacobsson, D., Chen, I.J. et al. (2015). Sn-seeded GaAs nanowires as self-assembled radial p-n junctions. *Nano Letters* 15 (6): 3757–3762.
108. Hamano, T., Hirayama, H., and Aoyagi, Y. (1997). New technique for fabrication of two- dimensional photonic bandgap crystals by selective epitaxy. *Japanese Journal of Applied Physics* 36: L286–288.
109. Takeda, J., Akabori, M., Motohisa, J., and Fukui, T. (2002). Formation of $Al_xGa_{1-x}As$ periodic array of micro-hexagonal pillars and air holes by selective area MOVPE. *Applied Surface Science* 190: 236–241.
110. Akabori, M., Takeda, J., and Motohisa, J. (2003). InGaAs nano-pillar array formation on partially masked InP(111)B by selective area metal–organic vapor phase epitaxial growth for two-dimensional photonic crystal application. *Nanotechnology* 14 (10): 1071–1074.
111. Motohisa, J., Noborisaka, J., Takeda, J. et al. (2004). Catalyst-free selective-area MOVPE of semiconductor nanowires on (111)B oriented substrates. *Journal of Crystal Growth* 272 (1–4): 180–185.
112. Lin, Y.T., Yeh, T.W., and Dapkus, P.D. (2012). Mechanism of selective area growth of GaN nanorods by pulsed mode metalorganic chemical vapor deposition. *Nanotechnology* 23 (46): 465601.

113. Farrell, A.C., Lee, W.J., Senanayake, P. et al. (2015). High-quality InAsSb nanowires grown by catalyst-free selective-area metal-organic chemical vapor deposition. *Nano Letters* 15 (10): 6614–6619.
114. Ji, X., Yang, X., Du, W. et al. (2016). Selective-area MOCVD growth and carrier-transport-type control of InAs(Sb)/GaSb core-shell nanowires. *Nano Letters* 16 (12): 7580–7587.
115. Ren, D., Farrell, A.C., and Huffaker, D.L. (2018). Axial InAs(Sb) inserts in selective-area InAsP nanowires on InP for optoelectronics beyond 2.5 μm. *Optical Materials Express* 8 (4): 1075–1081.
116. Motohisa, J., Takeda, J., Inari, M. et al. (2004). Growth of GaAs/AlGaAs hexagonal pillars on GaAs (111)B surfaces by selective-area MOVPE. *Physica E: Low-dimensional Systems and Nanostructures* 23 (3–4): 298–304.
117. Ishizaka, F., Hiraya, Y., Tomioka, K. et al. (2017). Growth of all-wurtzite InP/AlInP core – multishell nanowire array. *Nano Letters* 17 (3): 1350–1355.
118. Sato, T., Motohisa, J., Noborisaka, J. et al. (2008). Growth of InGaAs nanowires by selective-area metalorganic vapor phase epitaxy. *Journal of Crystal Growth* 310 (7–9): 2359–2364.
119. Yoshimura, M., Tomioka, K., Hiruma, K. et al. (2011). Lattice-mismatched InGaAs nanowires formed on GaAs(111) B by selective-area MOVPE. *Journal of Crystal Growth* 315 (1): 148–151.
120. Tomioka, K., Tanaka, T., Hara, S. et al. (2011). III – V nanowires on Si substrate: selective-area growth and device applications. *IEEE Journal of Selected Topics in Quantum Electronics* 17 (4): 1112–1129.
121. Tomioka, K., Izhizaka, F., and Fukui, T. (2015). Selective-area growth of InAs nanowires on Ge and vertical transistor application. *Nano Letters* 15 (11): 7253–7257.
122. Lee, W., Kim, H., Farrell, A.C. et al. (2016). Nanopillar array band-edge laser cavities on silicon-on-insulator for monolithic integrated light sources. *Applied Physics Letters* 108 (8): 081108.
123. Dick, K.A. and Caroff, P. (2014). Metal-seeded growth of III – V semiconductor nanowires : towards gold-free synthesis. *Nanoscale* 6 (6): 3006–3021.
124. Otnes, G., Heurlin, M., Graczyk, M. et al. (2016). Strategies to obtain pattern fidelity in nanowire growth from large-area surfaces patterned using nanoimprint lithography. *Nano Research* 9 (10): 2852–2861.
125. Ikejiri, K., Noborisaka, J., Hara, S. et al. (2007). Mechanism of catalyst-free growth of GaAs nanowires by selective area MOVPE. *Journal of Crystal Growth* 298: 616–619.
126. Hersee, S.D., Sun, X., and Wang, X. (2006). The controlled growth of GaN nanowires. *Nano Letters* 6 (8): 1808–1811.
127. Lin, Y.-T., Yeh, T.-W., Nakajima, Y., and Dapkus, P.D. (2014). Catalyst-free GaN nanorods synthesized by selective area growth. *Advanced Functional Materials* 24 (21): 3162–3171.
128. Wang, X., Hartmann, J., Mandl, M. et al. (2014). Growth kinetics and mass transport mechanisms of GaN columns by selective area metal organic vapor phase epitaxy. *Journal of Applied Physics* 115 (16): 163104.
129. Ikejiri, K., Sato, T., Yoshida, H. et al. (2008). Growth characteristics of GaAs nanowires obtained by selective area metal-organic vapor-phase epitaxy. *Nanotechnology* 19 (26): 265604.
130. Tomioka, K., Mohan, P., Noborisaka, J. et al. (2007). Growth of highly uniform InAs nanowire arrays by selective-area MOVPE. *Journal of Crystal Growth* 298: 644–647.
131. Mohan, P., Motohisa, J., and Fukui, T. (2005). Controlled growth of highly uniform, axial/radial direction-defined, individually addressable InP nanowire arrays. *Nanotechnology* 16 (12): 2903–2907.
132. Choi, K., Arita, M., and Arakawa, Y. (2012). Selective-area growth of thin GaN nanowires by MOCVD. *Journal of Crystal Growth* 357 (1): 58–61.
133. Shapiro, J.N., Lin, A., Wong, P.S. et al. (2010). InGaAs heterostructure formation in catalyst-free GaAs nanopillars by selective-area metal-organic vapor phase epitaxy. *Applied Physics Letters* 97 (24): 243102.
134. Bassett, K.P., Mohseni, P.K., Li, X. et al. (2015). Evolution of GaAs nanowire geometry in selective area epitaxy. *Applied Physics Letters* 106 (13): 133102.
135. Berg, A., Lenrick, F., Vainorius, N. et al. (2015). Growth parameter design for homogeneous material composition in ternary $Ga_xIn_{1-x}P$ nanowires. *Nanotechnology* 26 (43): 435601.
136. Calahorra, Y., Kelrich, A., Cohen, S., and Ritter, D. (2017). Catalyst shape engineering for anisotropic cross-sectioned nanowire growth. *Scientific Reports* 7: 40891.
137. Koester, R., Hwang, J.S., Durand, C. et al. (2010). Self-assembled growth of catalyst-free GaN wires by metal-organic vapor phase epitaxy. *Nanotechnology* 21 (1): 15602.
138. Hong, Y.J., Lee, W.H., Wu, Y. et al. (2012). Van der Waals epitaxy of InAs nanowires vertically aligned on single-layer graphene. *Nano Letters* 12 (3): 1431–1436.
139. Mohseni, P.K., Behnam, A., Wood, J.D. et al. (2013). $In_xGa_{1-x}As$ nanowire growth on graphene: Van der Waals epitaxy induced phase segregation. *Nano Letters* 13 (3): 1153–1161.
140. Choi, J.E., Yoo, J., Lee, D. et al. (2018). Crystal-phase intergradation in InAs nanostructures grown by van der Waals heteroepitaxy on graphene. *Applied Physics Letters* 112 (14): 142101.
141. Tomioka, K., Yoshimura, M., and Fukui, T. (2012). A III-V nanowire channel on silicon for high-performance vertical transistors. *Nature* 488 (7410): 189–192.

142. Carrad, D.J., Burke, A.M., Lyttleton, R.W. et al. (2014). Electron-beam patterning of polymer electrolyte films to make multiple nanoscale gates for nanowire transistors. *Nano Letters* 14 (1): 94–100.
143. Van Weperen, I., Plissard, S.R., Bakkers, E.P.A.M. et al. (2013). Quantized conductance in an InSb nanowire. *Nano Letters* 13 (2): 387–391.
144. Yeh, T.W., Lin, Y.T., Stewart, L.S. et al. (2012). InGaN/GaN multiple quantum wells grown on nonpolar facets of vertical GaN nanorod arrays. *Nano Letters* 12 (6): 3257–3262.
145. Wang, Y.-B., Wang, L.-F., Joyce, H.J. et al. (2011). Super deformability and Young's modulus of GaAs nanowires. *Advanced Materials* 23 (11): 1356–1360.
146. Huang, C., Song, J., Lee, W. et al. (2010). GaN nanowire arrays for high-output nanogenerators. *Journal of the American Chemical Society* 132 (25): 4766–4771.

8

Monolithic III/V integration on (001) Si substrate

B. Kunert[1] *and K. Volz*[2]

[1]*imec, Leuven, Belgium*

[2]*Material Sciences Center and Faculty of Physics, Philipps-Universität Marburg, Germany*

8.1 Introduction

Silicon (Si) is the most important semiconductor in our current information and communication technology. It is widely available, easy to fabricate and process, and has good mechanical and thermal properties. Used in combination with silicon dioxide (SiO_2) for metal oxide semiconductors (MOSs), Si has been the backbone of the impressive development of integrated circuits (ICs) and electronic device scaling over the last several decades.

Just as Si-based devices dominate the current micro- and nanoelectronics industry, III/V compound semiconductors are the material of choice to realize optoelectronic devices. Many III/V materials have a direct bandgap and hence strong photon emission and absorption efficiency in comparison to the indirect semiconductor Si. Therefore, light-emitting semiconductor devices such as LEDs and lasers are mainly made of III/V semiconductors. In addition, detectors, sensors, and highly efficient multijunction solar cells are also based on III/V materials due to their optoelectronic properties. The direct band structure implies higher electron mobility in comparison to Si; therefore, III/Vs are used for high-electron-mobility transistors (HEMTs) and heterojunction bipolar transistors (HBTs) in radio frequency (RF) applications. Figure 8.1 shows the bandgap of most III/V and some group IV semiconductors versus the lattice constant. Also considering the epitaxial growth of mixed III/V alloys as well as strained heterostructures, the large variety of different III/V compound material systems enables the design and fabrication of optoelectronic devices covering a very large application spectrum.

Metalorganic Vapor Phase Epitaxy (MOVPE): Growth, Materials Properties, and Applications,
First Edition. Edited by Stuart Irvine and Peter Capper.

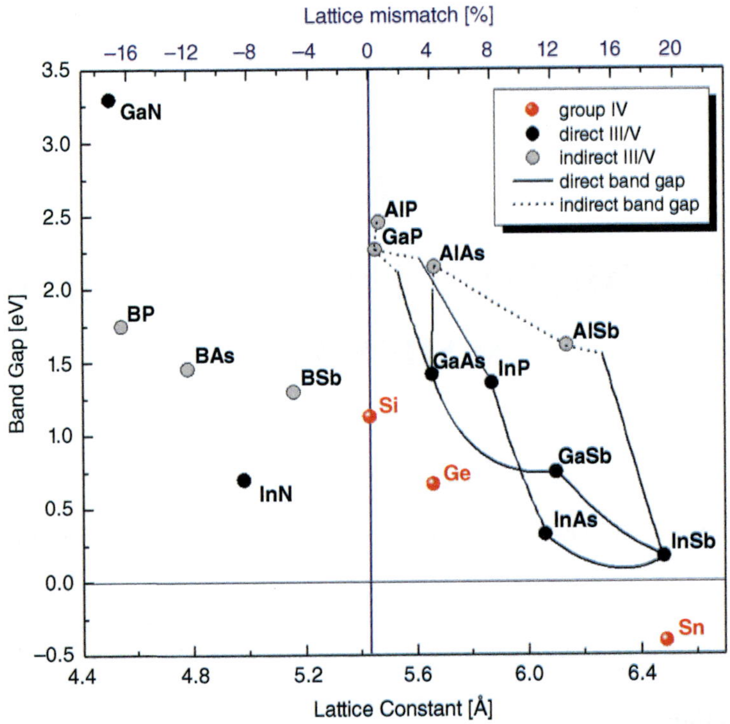

Figure 8.1 *Bandgap versus lattice constant of most binary III/V material systems and Si, Ge, and Sn. The top x-axis indicates the lattice mismatch toward Si. The red and grey data points indicate semiconductors with an indirect bandgap, whereas the black data points belong to direct gap semiconductors.*

The marriage of III/V compound materials with Si-based ICs would open up a wide field of new device functionalities that would simultaneously benefit from the excellent optoelectronic properties of III/Vs and the very mature and cost-efficient Si-based IC process technology.

The integration of III/Vs is discussed for various applications. For example, III/V is considered to replace Si as a channel material in future technology nodes based on Fin field-effect transistors (FinFETs), Gate-all-around nanowire FETs (GAA-NWFETs), or even tunnel field-effect transistors (TFETs) [1–4]. One of the key device components still missing in silicon photonics is a monolithically integrated laser diode. A cost-efficient integration scheme of a light source on Si substrates would clearly boost the use of silicon photonics and launch the option of optical data transmission even within IC chips [5–7]. However, adding any kind of III/V-based devices, such as HEMT, HBT, detector, modulator, etc., to Si-based electronics could lead to cost-efficient multifunctional ICs. Besides the merging of III/V- and Si-based functionalities, the use of Si as a cheap substrate with a large diameter is also very attractive. It would be clearly advantageous if III/V device fabrication could profit from the process and fabrication techniques that have been well established in the Si industry.

In addition to the monolithic deposition of III/V on Si, wafer-, chip-, and die-bonding are also heavily explored to combine III/V devices with Si-based electronics [8]. These techniques have to face their own challenges and represent a very interesting and fast-growing research field. Still, the successful monolithic growth of III/V layers directly on Si substrates is the "silver bullet" of a scalable and cost-efficient integration approach for III/V device components. Obviously, success inevitably implies a very low defect density in the integrated active III/V layer for good device performance.

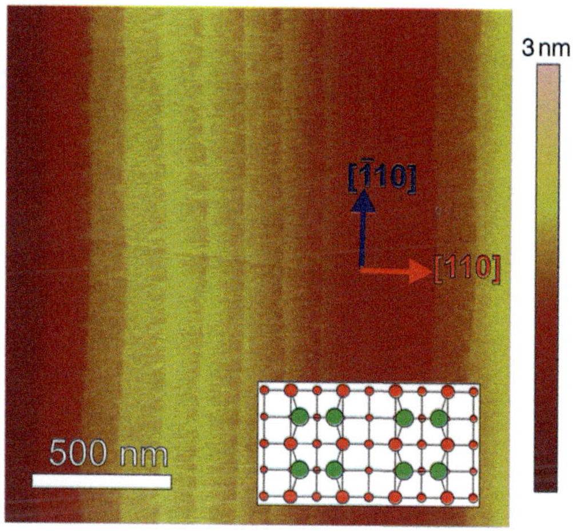

Figure 8.4 Atomic force micrograph of a Si(001) surface after the annealing procedure described in the text. The different steps on the surface are clearly visible. Source: Reprinted from Beyer et al. 2012 [40] with the permission of AIP Publishing. A sketch of the surface reconstruction of the major domain is shown in the inset.

8.2.2 Interface Formation in the Presence of Impurities and MO Precursors

The Si surface, especially if it is not completely hydrogenated, is extremely reactive to contamination (carbon, oxygen) as well as to species that are present in the reactor. The latter may be molecules/atoms, which intentionally arrive from the gas phase at the surface in order to nucleate the III/V semiconductor; or species, which desorb from reactor walls, where they have been adsorbed during previous epitaxy steps. Residual contamination of the Si surface can result in three-dimensional growth and stacking fault formation in the subsequent III/V growth [42]. As explained in more detail later, the III/V nucleation on Si is very sensitive to the very first Si surface structure. A group V-dominated nucleation generally leads to superior III/V layer morphology and fewer defects [40, 43, 44]. The adsorption behavior of tertiarybutyl phosphine (TBP), the phosphorous precursor most frequently used for low-temperature nucleation of GaP on Si, has been studied using a combination of STM of vacuum-transferred samples and density-functional theory (DFT) [45]. It was shown that the TBP adsorbs on the surface at reactive sites, still carrying its tertiarybutyl group at the low temperatures used for the initial nucleation steps. The arrangement of multiple adsorption was shown to be dominated by the steric repulsion of the tertiarybutyl group. Neighboring defect sites are stabilized by the adsorbed TBP, resulting in an anisotropic agglomeration perpendicular to the Si surface dimer rows as the coverage increases. Figure 8.5 shows this arrangement with increasing coverage. The tertiarybutyl group will accordingly only be removed in the Ga precursor pulse following the group V pulse.

Gallium-dominated nucleation schemes are more problematic due to the well-known melt-back etching of Si by Ga [44, 46]. Experiments where a Si surface has been exposed to triethylgallium (TEGa) using TEGa partial pressures and temperatures as they are typically used during the GaP nucleation show that instead of forming a monolayer of Ga on the surface, Ga clusters in pyramidal etch structures in the Si are formed [47]. In Figure 8.6, a cross-sectional scanning transmission electron microscopy (STEM) micrograph is shown, which clearly depicts that the Ga, which is liquid at the growth temperature, has dissolved the Si and etched on {111} lattice planes into the Si substrates.

Hence, Ga-dominated nucleation conditions are highly unfavorable for controlled and defect-free III/V nucleation on Si (001) substrates. It was also shown that – on top of the pyramidal Ga droplets – the polarity of the GaP is reversed with respect to the main polarity. These findings have important

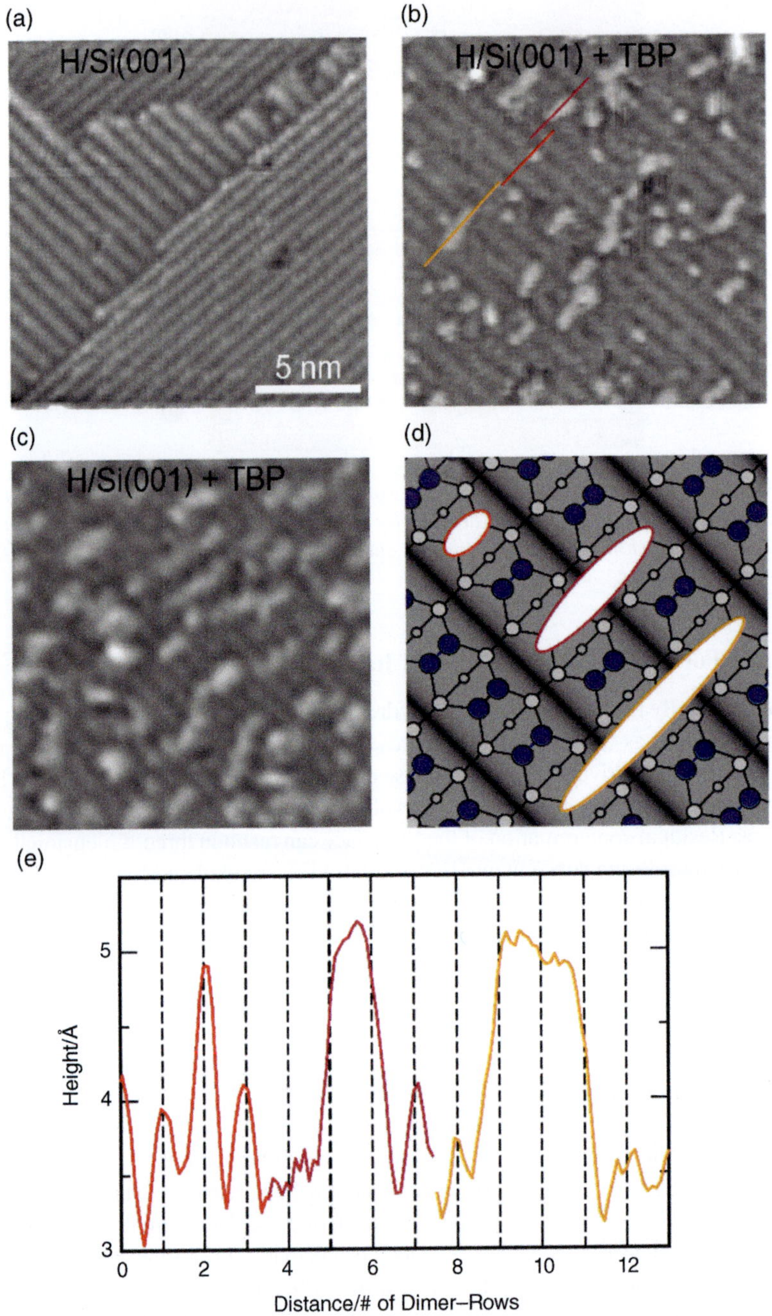

__Figure 8.5__ Scanning tunneling microscopy of Si(001) surface after different TBP exposures after vacuum-transfer from an MOVPE machine. Source: Reproduced from Stegmüller et al. 2016 [45] with permission from Wiley, copyright (2016). (a) Hydrogenated Si surface without TBP coverage after the annealing procedure described in the text. (b) and (c) TBP precoverage. (d) Schematics of the alignment of the TBP molecules with respect to the Si dimer formation on the surface and analysis of the TBP chains with different lengths (e).

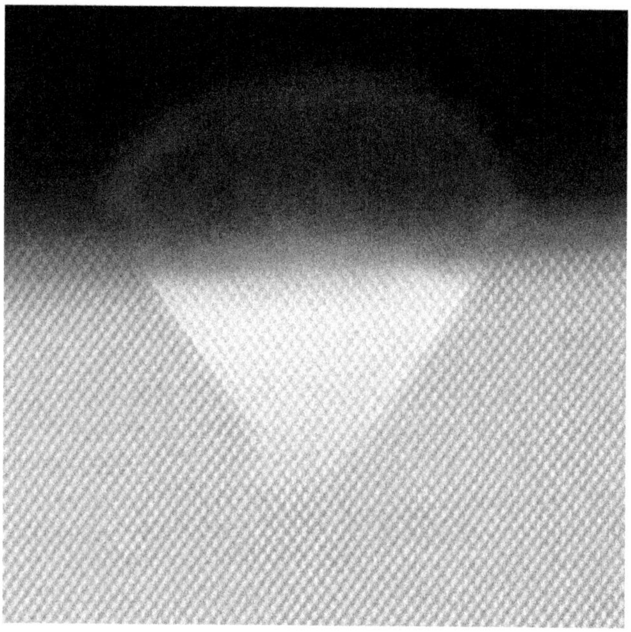

Figure 8.6 *Cross-sectional high-angle annular dark-field STEM image of a Si(001) substrate, which has been exposed to TEGa (triethyl gallium) at 450 °C after the oxide has been removed. The pyramidal region, where the Ga melt-back etched the Si, is clearly visible from the bright contrast of the Ga compared to the Si substrate. Source: After Werner et al. 2014 [47].*

implications, if a one-chamber configuration is used for III/V on Si heteroepitaxy. During high-temperatures steps, reactor residuals will desorb from the walls and alter the nucleation conditions of the III/V semiconductor compared to a clean reactor environment. As shown in [48] by a correlation of RAS with X ray photoelectron spectroscopy (XPS), Ga residuals from the reactor also result in a change in polarity. This has of course important implications on the control of antiphase domain-free III/V heteroepitaxy, which cannot be established if, locally on the Si surface, the polarity switches due to Ga residuals, either from nonoptimized nucleation or from the reactor wall.

Of course, the influence of other reactor residuals resulting from device epitaxy has to be investigated in detail. Early studies on the influence of As on Si(001) surfaces in an UHV environment have shown that As can bond either additively or replacively to the dimerized Si surface, resulting in an inverted polarity of GaAs grown on these surfaces with respect to GaAs grown on clean Si surfaces [49, 50]. In an MOVPE environment, it has been shown that AsH_3 as well as tertiarybutyl arsine (TBAs) are capable of etching Si(001) at elevated temperatures, which can also be exploited to reduce deoxidation temperatures of Si substrates by about 100 °C [51–53]. As in the previously mentioned case for GaAs growth on As-precovered Si(001), As residuals on the Si(001) surface were also shown to alter the crystal polarity in subsequent GaP growth [54].

All these studies show that a well-controlled III/V on Si nucleation process might not be achievable any longer in the (uncontrollable) presence of reactor residuals. Hence, a two-chamber epitaxy cluster is preferential, especially if thicker III/V layers are needed for devices.

8.2.3 Atomic III/V on Si Interface Structure

The exact atomic structure of III/V on Si (001) heterointerfaces and hence the band alignment between the III/V semiconductor and the Si substrate are still open questions today. Oversimplified models of III/V on Si (001) interfaces are shown in Figures 8.7a and b [55]. Here, the Si bulk has been simply cut at the (001) plane, and the Si surface atoms form bonds solely to P (a) or Ga (b) at the interface.

As explained in detail in [56, 57], this should be an energetically highly unfavorable scenario, as each III–Si and V–Si bond lacks or adds $q/4$ charge. The charge-carrier density resulting from such configurations is in the range of 3×10^{14} carriers/cm^2. Electric fields of the order of 4×10^7 V/cm across the interface would be the consequence [56, 57]. Such high charged-carrier concentrations have never been measured using electrical characterization. However, in [58], it has been shown by a comparison of XPS with RAS in situ data that GaP/Si (001) heterointerfaces with solely Si–P bonds can form kinetically controlled under certain growth conditions, although they are thermodynamically highly unstable. A large number of interface states has been shown to be the result of such an atomic arrangement [59].

Usually, an electric field of the order of several MV/cm should lead to an atomic rearrangement during growth to form a *charge-neutral interface* [56, 57]. These interface configurations should have an equal number of III–Si and V–Si bonds, respectively. The "first" charge-neutral interfaces have been suggested in [56] and are shown schematically in Figures 8.7c and d. The interface shown in Figure 8.7c comprises one monolayer of atomic intermixing by removing half of the Si atoms to the first III/V layer and replacing them by Ga and P atoms, respectively. The finite electric dipole, which is still present for this interface configuration, can be suppressed, if two monolayers of intermixing are present, as schematically shown in Figure 8.7d.

Using atomically resolved high-angle annular dark-field (HAADF) imaging in STEM, more complex interface configurations have also been experimentally motivated [60].

Figure 8.8a shows a cross-sectional view through a GaP/Si interface together with the quantification of the composition (Figure 8.8b). From these images, it can be concluded that there is a

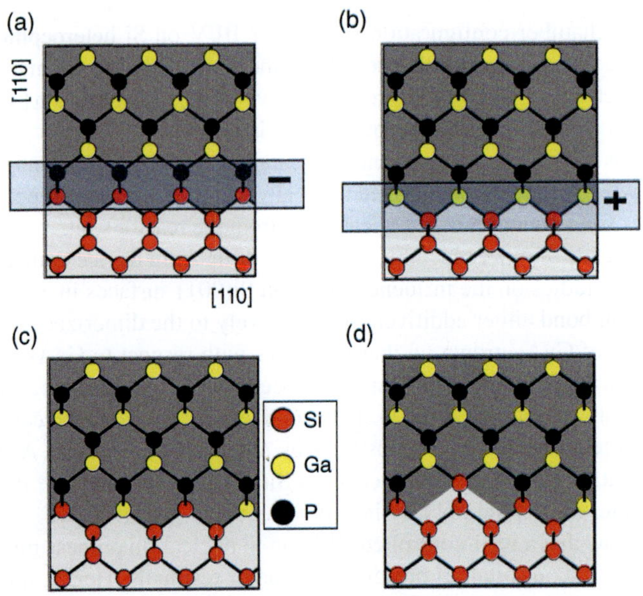

Figure 8.7 Schematics of different atomic arrangements at GaP/Si (001) interfaces: (a) and (b) are not charge neutral. In c) the simplest, one-monolayer-wide charge-neutral GaP/Si(001) interface structure is shown. There is, however, still a dipole shift. Interfaces without dipole shift are at least two monolayers wide, as exemplarily shown in (d). Source: Reprinted from Volz et al. 2015 [55], with permission from Elsevier.

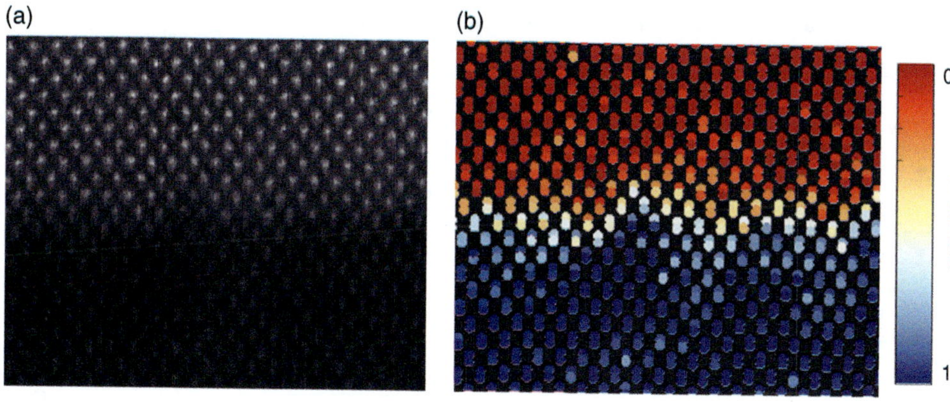

Figure 8.8 Atomically resolved HAADF STEM cross-section of a GaP/Si (001) interface (a) and the quantification of the Si content across the structure (b). The pyramidal interface structure can be clearly seen. Source: Adapted with permission from Beyer 2016 [60]; copyright American Chemical Society.

pyramidal morphology at the interface, with the sidewalls of the Si pyramids lying on {111} and {112} lattice planes, respectively. This morphology is a result of kinetic as well as thermodynamic driving forces. The different adatom kinetics of Ga and P, respectively, with Ga being significantly more mobile than P, results in a kinetically driven intermixing of the interface, with termination on the thermodynamically most stable {112} lattice planes [60]. This result is also in accordance with earlier studies, where growth of III/Vs on (110) or (211) Si substrates has been proposed [56, 57, 61, 62] to avoid interface reconstruction due to charge-distribution effects.

8.2.4 Antiphase Domains

As mentioned earlier, the crystal structure of III/V semiconductors (with the exemption of pure nitrides) and Si is similar in the sense that both have a cubic crystal structure, consisting of two face-centered cubic (fcc) sublattices, which are shifted by one-quarter of the body diagonal in the <111>direction. In the diamond structure, in which Si crystallizes, both sublattices are occupied with Si; in the cubic zincblende structure of the III/V semiconductors, one of the sublattices is occupied by the group III and the other by the group V atom. This results in Si being nonpolar and III/Vs being polar semiconductors, for which the polarity is determined by the electronegativity difference of the atoms on both fcc sublattices. The differences in crystal structure, however, result in challenges at the III/V on Si (001) interface, if the Si surface prior to III/V growth was covered with monoatomically high steps. This situation is schematically shown in Figure 8.9.

Wherever an odd-number high step is present on the Si(001) surface, an antiphase boundary (APB) is formed in the III/V semiconductor. These APBs can be located at different lattice planes, as schematically shown in Figure 8.9. APBs enclose antiphase domains (APDs). The two III/V crystalline regions forming an APB have different polarity. An APB consists of wrong III–III or V–V bonds, because the (001) planes of the III/V semiconductor get out of registry when grown on Si(001) surfaces, which exhibit odd-number high steps. Depending on the exact atomic structure of APBs, they can add a significant amount of charge to the crystal. The effective charge per wrong, homopolar bond amounts to +/– $q/2$ for wrong III–III and wrong V–V bonds, respectively, with q being the electron charge. The charged-carrier type can be p- or n-type, respectively, depending on which wrong homopolar bonds are formed. Only if the APBs kink to higher indexed planes than {110} can they self-annihilate, and only then is the subsequent III/V layer free of these defects.

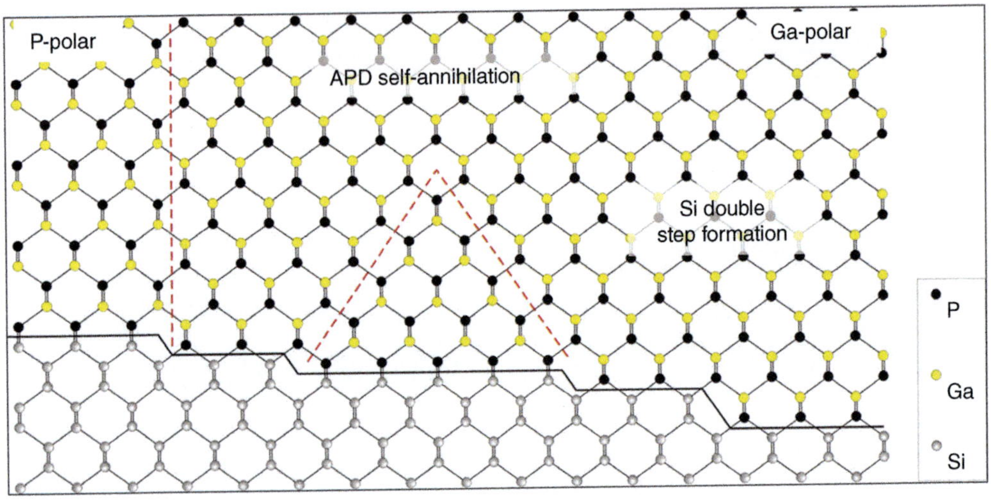

Figure 8.9 Schematic cross-section through a GaP/Si (001) interface in [110] projection. The Si atoms are depicted in gray, and Ga and P in yellow and black, respectively. Wherever monoatomic steps on the Si surface exist, antiphase boundaries (APBs) form in the III/V semiconductor. They can lie on different lattice planes and separate antiphase domains (APDs) of different polarity from each other. Cubic zincblende semiconductors are called Ga(P)-polar, if the Ga(P) is the upper atom in the III/V "dumbbell." When double-layer high steps are formed on the Si-surface, no antiphase disorder is expected in the III/V material. Source: Reprinted from Volz et al. 2015 [55] with permission from Elsevier.

The situation is different for growth on {111}-oriented silicon. It is shown in [32] that the {111} sidewalls of V-shaped trenches have the same polarity in any <110>direction. As depicted in Figure 8.20, the {111} surface of the trench can even have monoatomic steps, because those steps do not lead to the formation of APDs on this surface. Growing in these trenches allows the retention of the preferred (001) orientation of the Si substrate and nevertheless the suppression of antiphase disorder [63, 64].

8.2.5 III/V Growth on Si(001)

III/V deposition on Si typically consists of two growth steps: a low-temperature nucleation to establish a defined heterointerface, and a high-temperature overgrowth to achieve better crystal quality. In addition to the challenge of combining polar and nonpolar semiconductors, the lattice mismatch will also have a considerable impact on finding the optimal growth conditions. Optimized GaP/Si nucleation and APD-free growth are based on a low-temperature, flow-rate-modulated (FME) seed layer to suppress the formation of extended defects and a high-temperature overgrowth step to annihilate the APDs. In the following, both steps will be addressed separately.

The GaP nucleation on Si(001) takes place in the form of islands containing a large density of stacking faults and twins, if FME is not used in the nucleation phase [44, 65–70]. The nucleation temperature has an important influence on island density and size, with the trend toward smaller island size and larger island density the lower the temperature [71]. Heterogeneous nucleation from the vapor onto the substrate results in an incubation time for the initial stages of GaP nucleation on Si (001). If the nucleation conditions are changed from continuous growth mode to pulsed growth mode (FME), the stacking fault density can be minimized [44, 70, 72]. In FME growth mode, the group V and group III precursors are fed to the reactor alternatingly. The disappearance of the stacking faults is correlated with two-dimensional nucleation, which is observed in FME growth mode.

A good quality of the thin nucleation layers was only achieved if the GaP nucleation temperature was low (around 400–450 °C) [44]. Hence, all-liquid precursors, like TBP, which have a lower decomposition temperature compared to the hydrides, have mainly been used [44, 73]. Higher nucleation temperatures result, independent of the growth mode, again in stacking faults and twins [44]. It should be mentioned that these defects, as they have been mainly observed on the {111} A-planes, can only be observed in the Ga-polar GaP direction [43]. As mentioned in the previous paragraph, it is important to start the nucleation with a P-precoverage to avoid melt-back etching of Si by Ga as well as to control the Ga-offer to avoid droplet formation. The polarity of the GaP layer is set within the thin nucleation layer. It was shown that the polarity is fixed by the Si dimer orientation and the atom (group III or group V) prevalently binding to the Si surface first [40].

APBs are the greatest challenge in GaP/Si heteroepitaxy – aside from nucleation defects like stacking faults and twins – at least if exactly oriented Si (001) substrates are used. Typically, the III/V overgrowth conditions are tuned in a way that APBs self-annihilate after a minimum III/V layer thickness. There are several methods to detect APDs with different lateral and depth resolution in the III/V layers. On the surface of the III/V layer, APDs exceeding a certain size and intersecting the surface can be detected by anisotropic etching [24]. They can also be seen in AFM images of the surface as dips, if the GaP layer was heated to high temperatures around 675 °C after growth [74], because material evaporates more easily at the position of these defects than from the perfect III/V surface. Low-energy electron microscopy (LEEM) can be applied to tackle the fraction of the III/V surface, which is covered by APDs, as well as to resolve their spatial arrangement [75]. It has also been shown that RAS can be applied to quantify the APD content on the III/V surface [76–78]. RAS, however, does not give any information about the shape of the APDs. All these techniques have in common that they only can be used to address APDs intersecting the surface. However, the APB shape and depth distribution in the bulk is important, because the APBs can be charged, and this charge may alter the electric characteristics of device layers. X-ray diffraction can be used to gain information on the APD content in the entire III/V layer stack, if appropriate reflections are chosen [24, 79–81]. However, this technique does not yield information on the spatial arrangement of the APBs. TEM has been widely applied to investigate the lateral distribution of APBs and also to research the APB structure at high resolution. Dark-field imaging with specific reflections, being sensitive to the polarity of the material, can be applied in cross-sectional as well as in plan-view geometry [82–84]. High-resolution TEM techniques, such as phase-contrast TEM in combination with image simulation [85] as well as aberration-corrected high-angle annular dark-field imaging in a scanning TEM [86], have been used to determine the composition as well as the shape of an APB at an atomic scale.

APBs in III/V on Si(001) heteroepitaxy form, in theory, if surface steps consisting of an uneven number of monolayers are present on the Si surface [44, 57, 85] (see also Figure 8.9). As double steps are not stable on the exact Si(001) surface, one has to optimize the III/V overgrowth parameters with respect to minimizing the APD density on the III/V layer by finding growth conditions that confine the APBs to {111} or higher-index planes so that they do not penetrate into the device structures. It has also been shown experimentally that an APB arises at each monolayer-high step on the Si(001) surface [44]. AFM images of exact Si(001) surfaces with different unintentional miscut orientations and treated under different conditions are shown in Figures 8.10a and b.

A terrace structure of alternating smooth and triangular steps having different sizes depending on the preconditions can be clearly seen. The structure of the APBs in GaP grown on these Si surfaces exactly reflects the previous Si surface structure (Figures 8.10c and d). In this investigation, the growth conditions were intentionally tuned so that the APBs propagate on {110} planes only throughout the entire sample in order to be able to correlate their arrangement with the step structure of the underlying Si surface. From these investigations, it can be concluded that APDs will nucleate at – or close to – the positions of uneven monolayer high steps on the Si(001) surface.

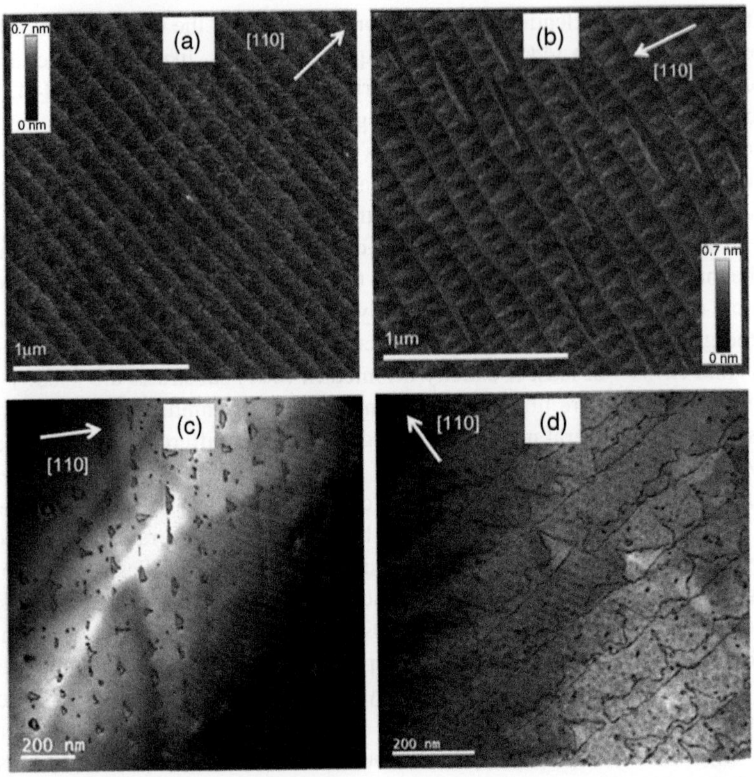

Figure 8.10 *(a and b) AFM images depicting the morphologies of Si-buffer layers on exact Si(001) substrates with different unintentional misorientation directions and differently annealed. (c and d) Plan-view TEM dark-field micrographs, which show the position of APBs in the GaP layers grown on these Si buffers with dark contrast. (c) corresponds to (a), and (d) to (b). Source: Reprinted from Volz et al. 2011 [44] with permission from Elsevier.*

The size of the APD base, i.e. on the (001) Si plane, is determined by the size of the monolayer high terrace on the Si substrate. This is important, because this size, together with the lattice planes on which the APBs lie, will determine the final size of the APD.

There are various studies in the literature investigating the MOVPE growth conditions to minimize the APD volume and study the dependence of APD formation and shape on the growth conditions [40, 43, 44, 55, 73, 74]. One of the decisive parameters for the lattice plane on which APBs lie is the GaP overgrowth temperature. If overgrowth takes place at low temperatures, around 450 °C, the APBs penetrate straight – on {110} planes – through the GaP layer [44, 74]. If the overgrowth temperature is increased to values around 600 °C and above, it was shown [43] that the APBs kink in the Ga-polar GaP direction to {111} and {112} planes. In the P-polar direction, the APBs remain on the {110} planes, irrespective of the overgrowth temperature. This geometry can be concluded from Figures 8.11a, b, and c, where TEM dark-field images of the same GaP/Si (001) layer are displayed in different viewing directions: plan view, perpendicular to the Si-step direction, and along the Si step direction of the substrate, respectively.

Comparison of the micrographs confirms that the APD shape is anisotropic, with APBs lying on {110} planes in the P-polar GaP direction and on {111} and {112} planes in the Ga-polar direction (from [43]). As a result, the APD shape is rather complex and schematically depicted in Figure 8.11d. As mentioned earlier, this shape can only be reversed if either the polarity of the GaP or the Si-surface reconstruction direction is reversed [40].

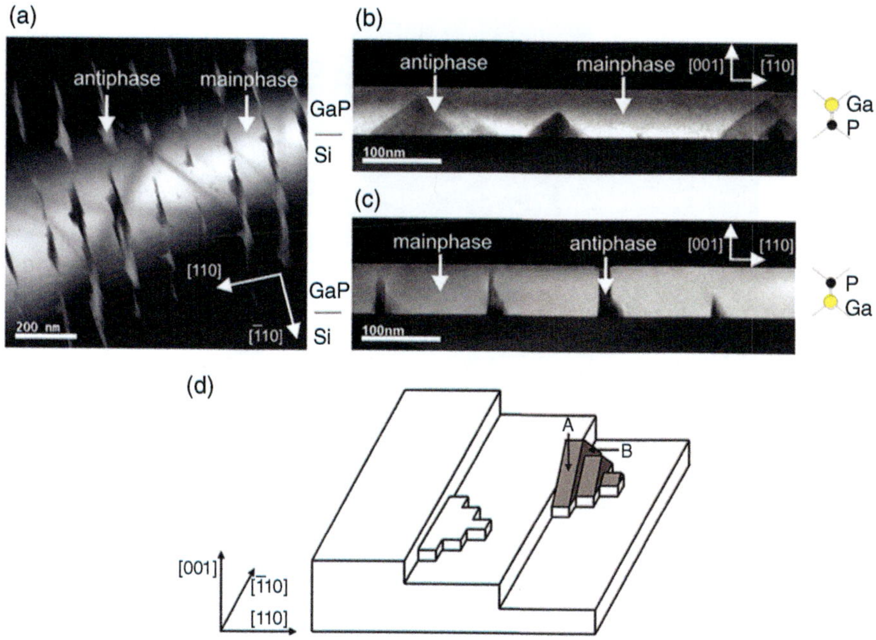

Figure 8.11 Plan-view TEM (020) dark-field micrograph of GaP on Si in the [001] direction showing the antiphase domain distribution along steps in the Si buffer (a), (002) dark-field micrographs in [110] cross-section (b), and [−110] cross-section (c), indicating anisotropic APD shape. (d) Model for APD geometry relative to steps on the silicon substrate, depicting the APBs lying on {110} planes (marked by B) and self-annihilating APBs lying on {112} planes (marked by A). Source: Reproduced from Beyer et al. 2011 [43] with the permission of AIP Publishing.

It is important to determine on which lattice planes the APBs lie and what their atomic arrangement is, because this will determine whether they supply charged-carriers to the III/V layer. As can be seen from Figure 8.9, APDs on {111} planes consist of wrong bonds only, supplying a significant number of charged carriers to the layers. APDs on {112} and {110} planes can be macroscopically charge neutral in the sense that they still consist of wrong homopolar III–III and V–V bonds, respectively, but these bonds alternate, as schematically shown in Figure 8.9 for a {110} boundary. It has been shown by atomically resolved scanning TEM that an APB on a {110} plane can consist either of alternating Ga–Ga and P–P wrong bonds, respectively, or of {111} segments, which align in a way that the APB is charge-neutral again on a larger length scale [87]. After careful optimization of the growth conditions, an APD-free GaP layer can be achieved even on exactly oriented Si(001) substrates after about 50–100 nm III/V layer thickness. The surfaces of these templates can be perfect, as shown in Figure 8.12.

Of course, if offcut Si(001) substrates are used, the overgrowth conditions are not as critical, because on these substrates, double-layer Si surface steps are stable and, hence, the APD density can be minimized to start with [73].

8.3 Heteroepitaxy of Bulk Layers on Si

Heteroepitaxial growth of mismatched material leads to the build-up of strain energy. If elastic relaxation does not ensure the full strain release, as observed for low-dimensional nanostructures, plastic relaxation will induce misfit and threading dislocations once the critical layer thickness t_{crit}

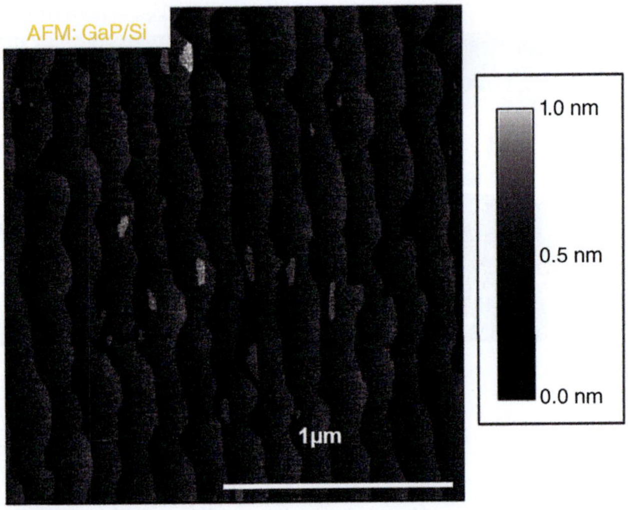

Figure 8.12 *AFM image of a (001)GaP surface of a thin GaP film on exact Si(001) template grown by MOVPE. Source: Reprinted from Volz et al. 2015 [55] with permission from Elsevier.*

is exceeded. Most III/V devices require a reasonable bulk-layer thickness beyond t_{crit}. Hence, the control, manipulation, and restriction of strain-induced defects such as misfit and threading dislocations or planar defects is the main topic of this section.

Misfit and *threading dislocations* (MDs and TDs) are line defects with open and wrong bonds, introducing a strong local distortion and strain field as well as accumulating charge. Driven by the strain-field interaction, dislocations can move in stressed crystals and repel or attract each other. It was reported in several publications and textbooks that these dislocations are electrically very active and induce undesired recombination paths. They act as carrier recombination centers and radically reduce the minority-carrier diffusion length. On the other hand, misfit defects can also cause the generation of carriers, which increases the reverse bias leakage current. All this is crucial for good device performance of solar cells, detectors, imagers, HBTs, etc. In addition, a threading dislocation in the region of high carrier concentration and radiative recombination processes initiates an additional network of defects, *dark line defects* (DLDs), which causes unavoidably the degradation of laser diodes. This so-called recombination enhanced dislocation climb (<100>DLDs) and glide (<110>DLDs) mechanism is so pronounced that only a very low threading dislocation density (TDD) in the active region permits a sufficient laser diode lifetime. However, it should be noted that the impact of DLDs also depends on the III/V material system and is less destructive for InP-based laser structures.

Planar defects such as stacking faults and twins show a less-destructive effect on device performance than misfit defects but still induce a negative impact due to the perturbation of the crystal periodicity. Antiphase boundaries are also planar defects with highly charged interfaces caused by wrong bonds and have again a strong effect on the device performance. For more details about defect characteristics and the influence on device performance, please see the following references: [88–101].

The overall objective in the heteroepitaxy of III/V semiconductors on Si is to confine and clearly separate the area of strain release and defect formation from the active region of the final III/V device. Achieving a low defect density is therefore the main criteria that determines the success of a III/V integration approach. Another important aspect is the total integration cost and/or compatibility to current Si-based complementary metal-oxide semiconductor (CMOS) processes.

This section is subdivided into three main topics that cover the most important cases of III/V heteroepitaxy on Si: lattice-matched growth on Si, metamorphic growth on blanket Si, and selective area growth (SAG) on Si. For lattice-matched growth, we discuss GaP-based III/V alloys with a lattice parameter close to that of Si. The metamorphic growth of lattice-mismatched III/Vs on blanket Si substrates was investigated in heteroepitaxy from the beginning. Many textbooks and overview articles [102–104] discuss this topic in detail. Chapter 5 concerns the metamorphic growth of solar cells, which is facing comparable challenges concerning the management of defect formation; hence this section rather focuses on recent laser integration approaches based on metamorphic growth. SAG for the integration of III/V on patterned Si wafers has gained a lot of attention in the last couple of years. The topic of SAG of nanostructures is also addressed in Chapter 7, so the main focus here is the application of SAG for III/V heteroepitaxy on Si. The latest achievements and device applications will be described in depth, in particular for growth in trenches, but some integration concepts beyond stripes will also be introduced.

8.3.1 Lattice-Matched Growth on Si

The lattice constant of GaP is only slightly larger than that of Si; hence pseudomorphic growth of 50–100-nm thick GaP on Si is possible without nucleation of strain-induced defects. As discussed in the previous paragraph, an antiphase-disorder-free GaP surface can be achieved on (001) offcut as well as on exactly oriented Si substrate, which can serve as a perfect template for further III/V deposition. Adding B or N to GaP decreases its lattice constant and allows lattice-matched growth of thick III/V layers on Si substrate. Figure 8.13 shows a comparison of the lattice constant of Si,

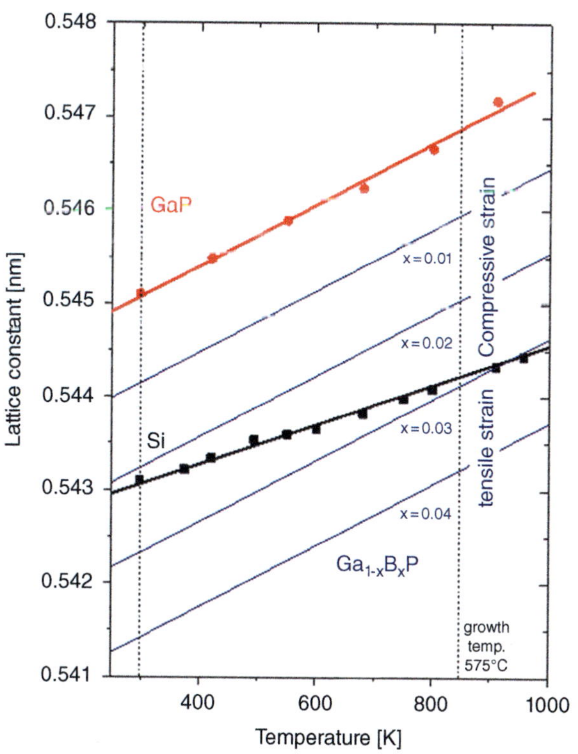

Linear thermal expansion coefficient [$\times 10^{-6}$ K^{-1}]	
Si	2.9
Ge	5.9
GaP	4.7
GaAs	5.7
AlAs	5.2
InP	4.6
GaSb	7.8
AlSb	4.0
InAs	4.5
InSb	5.4

temperature range: ~300K

Figure 8.13 *The graph shows the lattice constants of Si, GaP, and BGaP of different B concentrations as a function of temperature. Source: Reprinted from Kunert et al. 2008 [105] with permission from Elsevier. The table contains the linear thermal expansion coefficients selected from [106] of most binary III/V material systems around 300 K.*

GaP, and ternary BGaP alloys with different B content as a function of the temperature between 300 and 1000 K. It becomes obvious that the lattice mismatch between GaP and Si increases with rising temperature due to the difference in the thermal expansion coefficient. This correlation counts for all III/V material systems deposited on Si, because their thermal expansion coefficients are always larger than that of Si. Figure 8.13 also contains a table with the linear thermal expansion coefficient around 300 K of most binary III/V compound materials. In order to attain a lattice-matched condition of BGaP to Si at room temperature, a B concentration of about 2% is necessary, whereas the lattice adjustment requires 3% B at a growth temperature of 575 °C. As strain-induced relaxation defects are mainly nucleated during growth, it is essential to adjust the lattice constant of bulk layers at the corresponding growth temperature rather than at room temperature. As a consequence, cooling down to room temperature induces a tensile strain in the III/V layer, which can lead to wafer bowing or even crack formation.

In addition to BGaP, different alloys can be grown lattice-matched or pseudomorphically strained to Si. Dilute nitrides such as GaNP and GaNAsP with a low (up to 30%) As-concentration were explored for tandem III/V-Si solar cell applications [107, 108] and photoelectrochemical devices for water splitting [109–111]. The quaternary alloy $GaNAs_{>75\%}P$ with an As-concentration of more than 75% reveals a pronounced direct bandgap due to the high As content and the N-induced band coupling known in dilute nitrides [112, 113]. Laser devices with compressively strained $GaN_{>6\%}As_{>75\%}P$ quantum well (QW) structures as an active material for light emission and BGaP/BGaAsP-based separate confinement heterostructures (SCH) were integrated on (001) exact Si substrates without nucleation of misfit defects. Electrically pumped laser operation at an emission wavelength of 800–900 nm was demonstrated at up to ~150 K, emphasizing the potential of this device concept [114]. Since the laser diode is based on novel dilute nitrides and borides, which are highly metastable alloys and complex to grow [115–120], further development is necessary to improve the material quality, conductivity, and carrier confinement to progress in device performance. Nevertheless, this lattice-matched laser integration concept is unique among all other approaches, because the nucleation of strain-release defects is not necessary and hence the fundamental limitation of device lifetime is removed. Other alloys, such as GaInNP [121] and, lately, Bi-containing layers such as GaPBi [122], also hold the potential for pseudomorphic growth on Si substrate for optoelectronic applications.

8.3.2 Metamorphic Growth on Blanket Si

Each heteroepitaxial growth starts with a pseudomorphically strained epitaxial layer that is tetragonally distorted with an out-of-plane lattice constant different from the relaxed lattice constant of the heterolayer. With increasing layer thickness, the strain energy rises until it becomes energetically more favorable to introduce misfit dislocations to release the strain. At this point, the critical layer thickness t_{crit} is reached, which might be only a few monolayers thick, depending on the lattice mismatch. Pre-existing threading dislocations start to bend over to increase the length of misfit dislocation at the heterointerface. Other relaxation mechanisms are the homo- and heterogeneous nucleation of dislocation half-loops at the surface. If these half-loops glide down to the heterointerface, misfit-dislocation segments are formed again between the two associated threading dislocation "arms" to release the strain. Dislocation multiplication is another path to increase the plastic relaxation. It is a quite complex mechanism and can occur at pinned dislocations or result from an interaction of two or more dislocations. Depending on the lattice mismatch and growth conditions, mainly 60° misfit dislocations (less than 2% mismatch), a mixture of 60° and 90° (Lomer) misfit dislocations or mostly 90° misfit dislocations are nucleated (more than 6% mismatch). A three-dimensional growth mode (Volmer–Weber growth mode), likely for highly mismatched materials, can promote the injection of 90° misfit dislocations at the island boundaries.

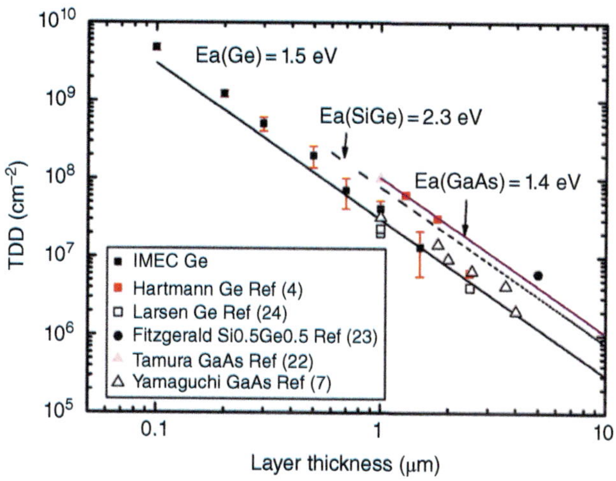

Figure 8.14 *Threading dislocation density (TDD) as a function of the layer thickness of Ge, SiGe, and GaAs deposited on Si substrates. Source: Reprinted from Wang et al. 2009 [125] with the permission of AIP Publishing.*

In the heteroepitaxial growth of a uniform buffer with constant composition on Si substrates, the threading dislocation density decreases with increasing layer thickness. This dependence was reported for various heteroepitaxial material systems on Si. Threading dislocations glide, move, and react with other threading dislocations, leading to coalescence or annihilation during growth. This causes a strong reduction in defect density with increasing layer thickness. Raising the growth temperature and/or adding an annealing treatment can further reduce the defect density [123, 124] until the TDD saturates at a minimum value [90].

Figure 8.14 shows a log log graph illustrating a linear dependence of the defect density in Ge, SiGe, and GaAs on Si substrate as a function of the layer thickness. It becomes obvious that independent from the material system, a layer thickness of several micrometers is necessary to achieve a defect density below 10^6 cm^{-2}.

The use of a graded metamorphic buffer allows for better control of defect formation [126–128]. In a graded buffer, the composition and hence the lattice constant are continuously changed with the distance from the Si substrate surface. A slow increase in lattice mismatch leads to the nucleation of a limited number of *glissile* threading dislocations. An efficient glide of these mobile threading dislocations can increase the misfit dislocation segments to promote the strain release per threading dislocation and hence minimize the nucleation of additional threading dislocations. The increasing strain energy in slowly graded buffers can reinforce the glide and initiate threading dislocation bending. Growth conditions and gradient design should be optimized to achieve maximal dislocation glide velocity and length together with minimal dislocation nucleation. With a slow composition gradient and a thick layer bridging a relatively low lattice mismatch, very low defect densities of 10^4 to 10^6 cm^{-2} were achieved [129]. However, utilizing a faster change in lattice constant per μm layer thickness (misfit %/μm) by inserting a more "aggressive" gradient, in order to reduce the total layer thickness, can clearly increase the nucleation of defects and induce phase separation. Any compositional fluctuations, surface roughness, existing dislocations, or other defects should be avoided, because they can prohibit threading-dislocation glide. If the threading-dislocation glide is prevented, then higher levels of strain build up in the layer and hence induce the nucleation of a large number of additional defects.

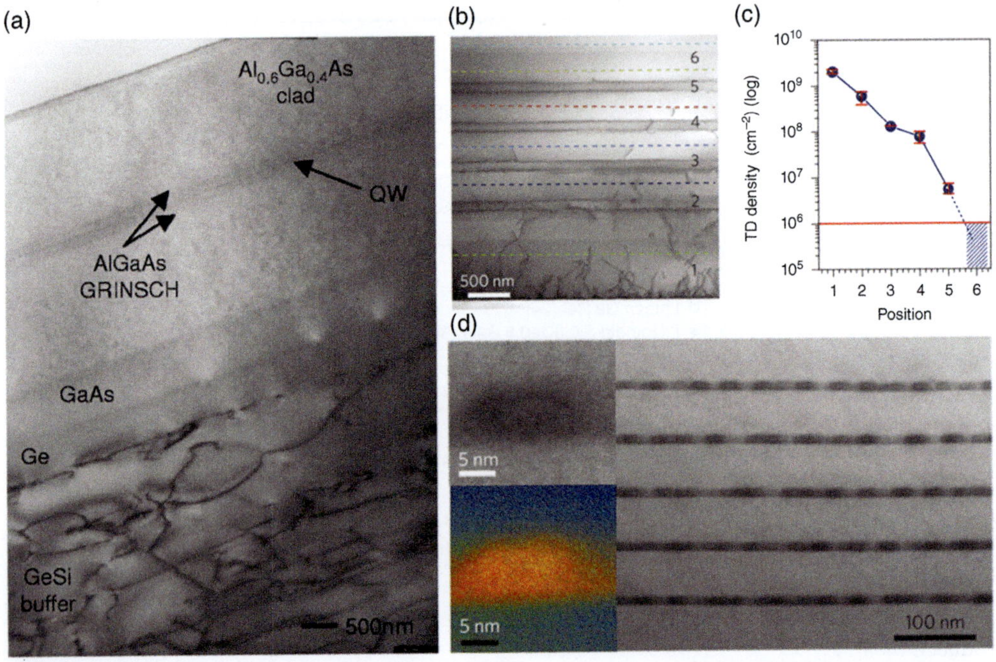

Figure 8.15 (a) A TEM cross-sectional image of an AlGaAs/GaAs laser diode deposited on a SiGe graded buffer on Si substrate. Source: Reprinted from Groenert et al. 2003 [130] with the permission of AIP Publishing. (b) BF-STEM image of an $In_{0.18}Ga_{0.82}As$/GaAs SLS on Si. (c) The defect density measured at the position indicated in (b). (d) BF-STEM image of a 5x InAs/GaAs dot-in-well (DWELL) active region. Source for (b), (c), and (d): Reprinted by permission from Macmillan Publishers Ltd from [131]. The two inserts are a BF-STEM micrograph and a DF-STEM image in false color of a single QD. The typical dimension of QD is ~20 nm in width and ~5 nm in height.

The design of the composition gradient can be limited to the III/V compound material. However, a common approach to bridge the lattice mismatch between Si and III/V is the use of a thick, relaxed Ge layer or a SiGe graded buffer to achieve the lattice constant of Ge before the transition to the III/V deposition. Figure 8.15a shows a TEM of an AlGaAs/GaAs-based laser diode integration on a SiGe graded buffer on Si.

Embedding a strained layer superlattice (SLS) as a defect filter layer in mismatched heteroepitaxy also enhances the defect reduction and has been reported for very different heteroepitaxial material systems [132–135]. Threading dislocations bend over at the interfaces and can serpentine back and forth between the different superlattice interfaces, which increases the chance of annihilation or coalescence reactions with other dislocations. The SLS consists of alternating strained layers and, if possible, even with different strain signs. It is important that each layer is thick enough to bend threading dislocations but not so thick that it initiates additional misfit defects. Gourley et al. [136] even suggested that a built-in strain is not necessary, and a difference in elastic stiffness is sufficient to bend threading dislocations. Figure 8.15a is a BF-STEM image of an $In_{0.18}Ga_{0.82}As$/GaAs SLS on Si. Figure 8.15c shows the defect density determined at the position indicated in Figure 8.15b. The defect density in this SLS was further reduced by applying an in situ annealing treatment. After a total III/V layer thickness of about 3 μm, a defect-density range below 10^6 cm^{-2} is reached [131]. The thickness comparison with Figure 8.14 emphasizes the efficiency of this mechanism to additionally reduce the defect density with a SLS.

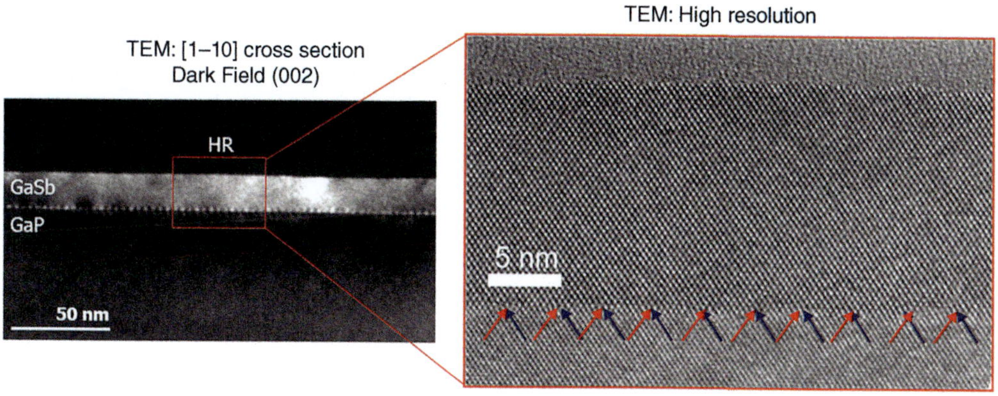

Figure 8.16 *Left: a (002) DF-TEM image of a GaSb layer deposited on a GaP/Si template. The Si substrate is not shown due to the magnification of the TEM micrograph. Right: a HR-TEM image revealing the periodic arrangement of additional (111) Si lattice planes, indicated by the red and blue arrows, typical for a 90° misfit dislocation.*

The use of a quantum dot (QD) defect-filter layer to reduce the defect density is a continuation of the SLS. The surrounding strain field of a three-dimensional QD is much larger than the one introduced by a SLS. Therefore, the stronger Peach–Koehler force can bend propagating threading dislocations more efficiently in order to reduce the defect density [137, 138]. This effect was observed for GaN QDs [139] and for InGaAs as well as for InAs QDs in GaAs [140, 141].

The lattice mismatch between III/Sb and Si is >12% (see Figure 8.1), which leads to a large force to form 90° misfit dislocations. In addition, III-Sb-based compound materials are relatively soft with reduced energy for defect nucleation. It was reported in the literature that under optimized growth conditions, a periodic interfacial misfit (IMF) array can be achieved to release the strain, which is a two-dimensional network of 90° misfit dislocations at the III/Sb-Si interface [142, 143]. A comparable periodic misfit array was even observed for the growth of GaSb on GaAs substrates with less the 8% mismatch [144]. Monolithic heteroepitaxy is clearly simplified by applying a GaP/Si template, since the annihilation of antiphase disorder and the strain relief by misfit formation is located at different interfaces [145, 146]. Figure 8.16 contains a TEM investigation of a thin GaSb layer deposited by MOVPE on a GaP/Si template, illustrating a typical IMF array. The Si substrate is not shown due to the magnification of the TEM images. The periodic alignment of 90° misfit defects along a 200-nm long segment is already visible in the DF-TEM image in Figure 8.16 due to the induced strain fields. In the HR-TEM image, arrows indicate the two additional inserted (111) planes typical for a 90° Lomer dislocation. The growth of highly mismatched materials is a very interesting approach since the IMF array offers the possibility to achieve a low defect density within a clearly reduced layer thickness. To date, reported defect densities are in the range of 10^5 to 10^6 cm^{-2}.

A review of the literature shows that most reported values for defect densities in V/III on Si are in the range of 10^6 to 10^7 cm^{-2}. Some authors stated a defect density in the range of 10^5 to 10^6 cm^{-2} for very thick and/or optimized buffers. Very few groups determined a defect density in the range of ~10^4 cm^{-2} but have not provided any information about device integration and performance as of yet. The use of graded buffers and defect filter layers can clearly lead to a reduction in total layer thickness for a given defect density. However, defect densities of $\leq 10^6$ cm^{-2} are still difficult to reach within an acceptable layer thickness. Certain challenges are correlated to the growth of thick III/V layers on Si substrate: the difference in the thermal expansion coefficient between III/Vs and Si leads to the build-up of tensile strain upon cooling from growth temperature to room temperature. The induced strain leads to wafer bowing, which can reach critical values especially for large-scale wafers. This complicates or even hinders further wafer handling and processing, e.g. advanced lithography fails due to alignment problems. Crack formation is another important risk, which has to be controlled [147, 148]. Furthermore, the surface morphology of metamorphic graded buffers most of the time shows a cross-hatch pattern caused by a growth-rate fluctuation due to strain fields, which can be detrimental for realizing abrupt device interfaces.

8.3.2.1 Device Applications

Integrating a III/V solar cell on a Si substrate is very attractive to reduce the substrate cost and to profit from large-scale wafers. In addition, theoretical calculations predict a high efficiency of 37–44% under AM 1.5G (1–500 suns) illumination [149] for a tandem solar cell based on a Si- and a III/V-junction with a bandgap between 1.6–1.8 eV [150]. A defect density in the range of $\leq 10^6$ cm^{-2} should be sufficient to avoid degradation of the minority carrier diffusion length in solar cell applications [151, 152]. Hence, much effort was spent on optimizing a GaPAs graded buffer to realize a III/V junction at 1.7 eV for a tandem solar cell on Si [150, 153, 154]. This integration approach also profits from the use of the GaP/Si template, as discussed in the previous paragraph. More general details about III/V solar cells are provided in Chapter 5 and recent overview articles [102, 155, 156].

The successful integration of III/V-based detectors and imagers on Si also relies on an appropriate minority carrier diffusion length as well as on a low dark current, which are only guaranteed by a low defect density in the active region. III/V-based MOSFETs and HEMTs on blanket Si were also explored [104, 146, 157–159]. Due to the stringent requirements concerning MOSFET scaling and the transition from planar to three-dimensional device architectures such as FinFET and GAA-FET in the current and future technology nodes, significant challenges are still linked to the device fabrication process. Some examples are developing the gate stack for III/Vs, reducing contact resistance, and avoiding leakage current paths. Therefore, it is not always straightforward to separate the impact of bulk crystal defects on device performance from other process-related factors. Considering all III/V device options of interest integrated on Si, a laser diode is most sensitive to crystal defects and requires a very low defect density for an appropriate device lifetime. The target threshold defect density is still under discussion and also relies on the device architecture. Hence, the focus in the remainder of this section will be on laser integration.

The attempt to integrate III/V lasers on Si goes back to the 1980s. At that time, the first AlGaAs/GaAs laser diodes on Si were demonstrated with a threshold current comparable to the same laser structure grown on a GaAs substrate [160]. Even room-temperature continuous wave (cw) operation was reported at the emission wavelength of 840 nm. The laser structure was grown by MBE on a 2–2.5-μm thick GaAs buffer. An InP-based laser diode with an emission at ~1.55 μm was deposited on a heteroepitaxial buffer of 2 μm GaAs plus 13 μm InP (also partly containing a SLS). Electrically pumped cw operation of more than 2000 h [161] and later of even 7000 h at 50 °C [162] was reported. In the latter case, an etch-pit defect density of 5×10^6 cm^{-2} was determined. Groenert et al. [130] used a 10-μm thick graded SiGe buffer to achieve the lattice constant of Ge with a final defect density of 2×10^6 cm^{-2} before the integration of an AlGaAs/GaAs laser diode. A TEM micrograph of this structure is shown in Figure 8.15a. A laser diode lifetime in the range of just 15 min was observed for cw operation at room temperature, although the threshold current was identical to that of the same device structure deposited on GaAs substrate. At this point, it should be mentioned that the device failure characteristic in InP- and GaAs-based multiquantum well (MQW) structures is not the same for a comparable defect density. In past years, offcut (001) Si substrates were mainly utilized, but GaP/Si (001) templates simplify laser integration on exactly oriented substrates [163], which are standard in Si technology.

Infrared laser diodes based on GaSb were also realized on Si substrates with a relatively thin buffer layer of about 1 μm. Rodriduez et al. [164] reported an electrically pumped laser based on a GaInAsSb/AlGaAsSb MQW structure with an emission wavelength of 2.25 μm at room temperature. The emission wavelength was also shifted to 1.55 μm by depositing a GaInSb/AlGaAsSb MQW structure [165]. These first laser devices operated under pulse condition, but the first cw operation at room temperature was later demonstrated at emission wavelengths of 2 μm and 1.6 μm [166, 167]. Since no detailed study about the GaSb-based laser lifetime has been published yet, it is very likely that these devices still suffer from the destructive impact of the residual defect density in the active region.

All these examples based on different material systems demonstrate very clearly the lifetime limitation due to the high TDD. It is possible to realize very low threshold currents and good output powers comparable to devices based on lattice-matched growth on III/V substrates even with a defect density $\gg 10^5$ cm^{-2}. But realizing a sufficient device reliability on Si is the biggest remaining challenge in heteroepitaxy.

The use of QDs as an optical gain medium in laser diodes has recently led to very promising device results on Si substrates. Layers containing QDs as an active material in laser devices induce several general advantages. The three-dimensional confinement of carriers leads to a near-discrete density of states. Very low threshold currents and excellent temperature invariance during operation were demonstrated, which is an advantage over the QW laser. However, the most important characteristic is that a QD laser is less sensitive to defects [168, 169] in comparison to its QW counterpart. This enhanced tolerance to defects is caused by the strong strain field induced by the QD, which inhibits the in-plane movement of threading dislocations. Threading dislocations are pinned or pushed away from the QDs. In addition, the strong carrier localization inside the QDs reduces the interaction of radial recombination processes with defects, which is the typical driving force for recombination enhanced DLD formation and dislocation climb. It could also be possible that the formation of this additional network of defects is constrained inside a QD, and hence DLDs expand less efficiently over the full active region.

Liu et al. [169] reported InAs/GaAs QD laser operation at 30 °C for more than 2700 h and extrapolated a mean time to failure (MTTF) of 4600 h. The defect density of this device structure was fairly high, in the range of 2×10^8 cm^{-2}. Therefore, the demonstrated device lifetime is quite impressive and emphasizes the tolerance of QD-based laser diodes to defects. This first impressive reliability result has since been improved upon by a QD laser integrated on a GaAs buffer with a defect density in the range of $<10^6$ cm^{-2} [131]. Chen et al. observed a cw laser operation at room temperature of 3100 h and calculated a MTTF of 100 158 h. The low defect density in the buffer was achieved using a SLS, as discussed for Figures 8.15b and c. Figure 8.15d shows the active material of this QD laser consisting of a 5-times InAs/GaAs dot-in-well (DWELL) layer stack. The emission wavelength was 1.3 μm. The best results to date were recently announced in joined publications from UCSB and Intel [170]. An extrapolated lifetime of over 10 million hours was obtained from an GaAs-based QD laser diode on (001) Si substrate with a defect density of 7×10^6 cm^{-2} (35 °C aging at about twice the threshold current).

It seems that for a QD laser diode, a defect density in the range of 10^5–10^6 cm^{-2} is not an obstacle to realize devices with excellent reliability. Lasers based on QWs are much more sensitive to the defect density, which results in an evidently limited lifetime for a comparable defect density. Due to the promising potential of III/V QD lasers, research activity has increased in recent years, exploring different buffer approaches and device architectures [138, 171–174]. Reliability tests have to be extended to more application-relevant operating conditions to substantiate the impressive potential of QD lasers.

Remaining challenges for the successful integration of III/V QD lasers are wafer bowing and the risk of crack formation due to the difference in thermal expansion coefficients. The deposition of strain-compensation layers on the back side of the Si substrate, such as silicon nitride, can reduce wafer bowing. The threat of cracking depends a lot on the buffer material and design, the strain evolution inside the buffer, the wafer offcut, as well as the process conditions. It can be strongly reduced by optimizing each parameter. Another option to reduce the risk of cracking is growth on patterned wafers on the micrometer scale [175].

The final success criterion is compatibility with the mature CMOS technology. Laser diodes in the micrometer scale are huge structures in comparison to Si-based device components. This difference in device size complicates cointegration. Efficient heat dissipation during laser operation is one of the key challenges that has to be solved. A well-defined circuit design as well as an adjusted total IC process flow has to be developed for the successful integration of QD lasers in Si microelectronics.

8.3.3 Selective-Area Growth (SAG) on Si

Epitaxial growth on prepatterned substrates, where a dielectric layer is used to mask specific regions of the surface, is especially attractive if high selectivity can be achieved. This means that growth takes place only in the uncovered windows, and no parasitic deposition occurs on the dielectric mask. This SAG or selective-area epitaxy (SAE) can be easily realized using MOVPE and HVPE and has also been demonstrated by MBE. A high selectivity is achieved if the absorbed material on the mask surface either diffuses to the exposed crystal surface or desorbs again. These conditions are easily achieved over a large process parameter window with chemical sources as used as in MOVPE, HVPE, and CBE or atomic hydrogen-assisted MBE.

Selectivity in MOVPE obviously depends on the III/V material system, the choice of MO precursors, and the growth conditions, but also on the pattern size, dimension, and nature of the dielectric mask. Local changes in the mask coverage can have an impact on the effective surface temperature. In addition, parasitic nucleation centers such as impurities, macrosteps, and large defects on the mask surface can degrade selectivity. Loading effects due to material diffusion from the mask to the crystal surface as well as out of the pattern opening onto the mask have a very strong impact on the growth rate. Moreover, the incorporation efficiency for the deposition of alloyed semiconductors is also affected by the ad/desorption rate of the different precursors on the mask. Generally, deposition in constrained areas leads to characteristic facet formation as a function of the growth conditions and crystal polarity, especially when growing out of mask openings.

Altogether, SAG adds a certain complexity to MOVPE but also introduces new advantages for heterointegration. The selective growth of mismatched materials in highly confined structures offers a very novel method to reduce the crystal defect density. In the literature, this method is called either *necking* of threading dislocation [176–178] or *aspect ratio trapping* (ART) [179, 180], but both describe the same fundamental mechanism. Crystal defects are trapped, blocked, or terminated inside a highly confined region, whereas the top or overgrown part becomes defect free. A sufficient aspect ratio (AR) is important to achieve a pronounced defect reduction and is defined by the pattern layer height h divided by the opening width w ($AR = h/w$). Another advantage of SAG is that the difference in thermal expansion coefficient between Si and III/V has a minor impact due to the growth in restricted openings. Hence, wafer bowing and crack formation in the III/V layer are less of an issue in comparison to the growth on blanket Si wafers.

8.3.3.1 SAG in Trenches

Deposition in trenches is particularly interesting because a sufficient confinement in one direction leads to a clear defect reduction, whereas a certain trench length provides enough material volume for various device applications. As the standard alignment of most Si-based microelectronics is along the <110>direction, the trench pattern is accordingly arranged to ensure the best CMOS compatibility. Nevertheless, other trench orientations are in principle possible.

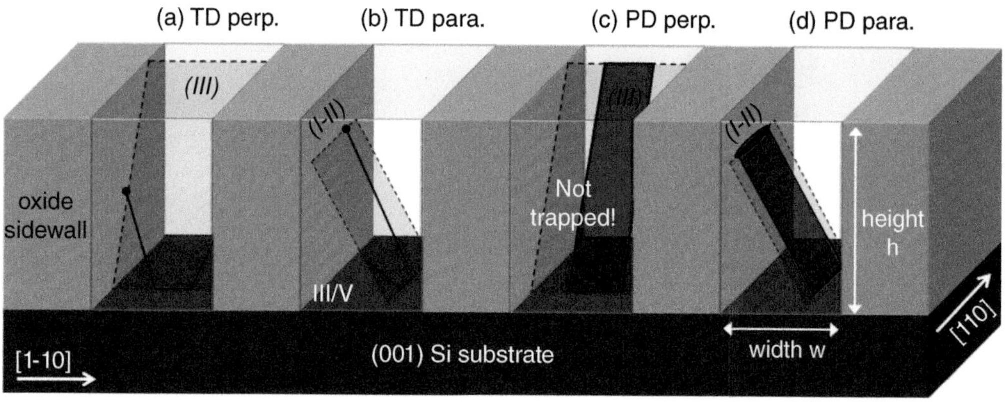

Figure 8.17 *Sketch of aspect ratio trapping to reduce defects. (a) and (b) consider a threading dislocation (TD) lying in a {111} slip plane perpendicular and parallel to the trench orientation. (c) and (d) depict a planar defect (PD) such as a stacking fault or twin in a {111} plane perpendicular and parallel to the trench orientation. The aspect ratio is defined by the trench height over the trench width (AR=h/w).*

Figure 8.17 shows a sketch of the deposition of a mismatched III/V material in narrow oxide trenches on Si substrate. The difference in lattice constant inevitably demands strain release via defect formation. If a dislocation half-loop nucleated during growth is based on 60° misfit dislocations, the slip plane is a {111} plane. As the dislocation half-loop nucleated at the growth surface glides down toward the III/V-Si interface to form and extend the misfit segment, the two TD arms will hit the sidewalls and hence be trapped. This is very obvious in the case where the slip plane is perpendicular to the oxide sidewalls and trench orientation, respectively; see Figure 8.17a. If the slip plane is parallel to the trench orientation, it will also hit the sidewalls due to the inclined angle of 54.7° if the AR of the trench is high enough (AR>1.43). As a result, the two TD ends are trapped again by the oxide pattern

The trapping of planar defects (PDs) is only possible for defects along a {111} crystal plane parallel to the trench orientation; see Figure 8.17d. Since each {111} plane will hit the sidewalls for a sufficient high AR, all PD nucleated close to the trench bottom are blocked. If the PD lies in a {111} plane perpendicular to the trench orientation, no blocking in the growth direction occurs, and the PD penetrates the full III/V layer, as shown in Figure 8.17c.

In contrast, a two-dimensional mask pattern such as holes could ensure the trapping of all PDs but leads to a very limited III/V template. The method of lateral overgrowth to increase the III/V volume will be discussed later in this section. Since both kinds of PDs cannot be trapped in trenches, their generation has to be avoided from the very beginning. In contrast to PDs, MDs and TDs have to be generated in order to release the strain. If all misfit defects lie in {111} planes, full trapping in both directions is possible. The differentiation of PDs and misfit defects is often ignored in the literature regarding ART, incorrectly concluding that full defect trapping is not possible. However, if we can circumvent the formation of PDs in a plane perpendicular to the trench orientation and fully control the misfit defects gliding along {111} planes, ART can lead to the integration of an entirely defect-free active III/V material on Si.

Selective epitaxy of compound semiconductors on Si was demonstrated in various publications exploring very different pattern structures. Mostly, epitaxial III/V growth is divided into a low-temperature seed deposition followed by a higher-temperature overgrowth comparable to the deposition on blanket Si wafers, but not limited to this approach. Special care has to be taken concerning the Si surface preparation, which is more challenging in narrow openings. The quality

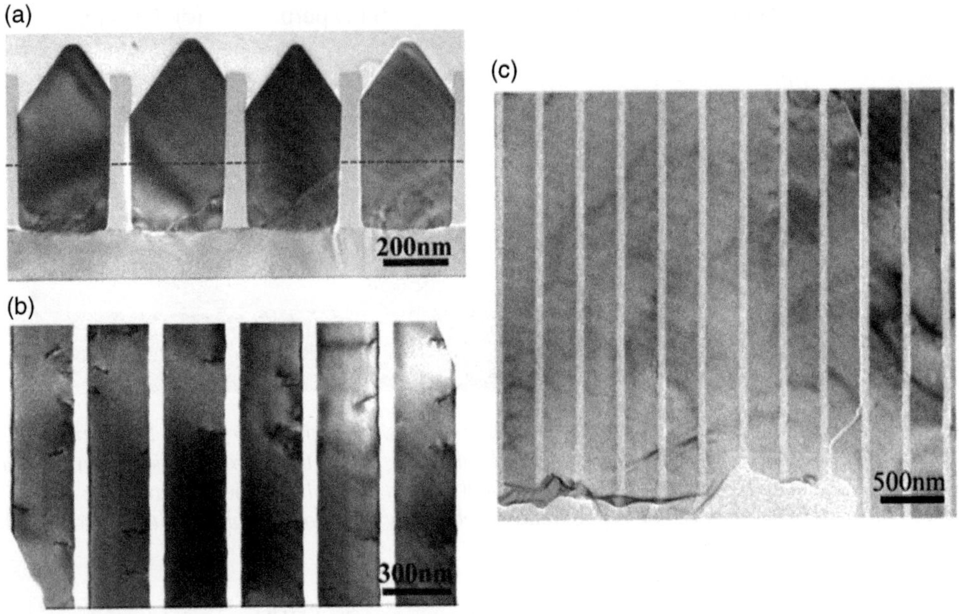

Figure 8.18 *(a) Cross-section TEM images showing that all TDs are trapped by the sidewalls below the dashed line. (b) A plan-view TEM image without the Si substrates, which contains both the TD-trapping region and the top defect-free region. Thinning down the sample further leads to image (c), containing mainly the defect-free GaAs region. The right side is a thicker sample area with still some TD present. Source: Reproduced from Li et al. 2007 [180] with the permission of AIP Publishing.*

of the dielectric pattern such as the sidewall roughness and any possible remaining contamination also play an important role in achieving good selectivity and low planar defect density.

The efficiency of ART for GaAs on (001) Si was demonstrated conclusively in [180]. GaAs was deposited inside oxide trenches with an AR > 1.6. Figure 8.18a is a TEM image across the trench orientation revealing clearly the effect of defect trapping at the sidewalls: the III/V material above the dashed line is free of threading dislocations. Figure 8.18b is a plan-view image including both the defect-trapping and the defect-free regions. It becomes obvious how the threading dislocations end at the oxide sidewalls. The plane-view image in Figure 8.18c contains the defect-free region, verifying the absence of threading dislocations, although the right TEM sample part is slightly thicker and hence still holds some defects.

SAG starting with an exact (001) Si surface bears the risk of forming APD. Cipro et al. [181] demonstrated that APBs can also be trapped by the sidewalls; they observed a clear decrease in antiphase disorder with rising AR. Figure 8.19 contains four top-view scanning electron microscopy (SEM) images of GaAs deposited inside oxide trenches on (001) Si with different width and AR. For an AR of 1.3, no APDs were observed in the top GaAs surface. Nevertheless, it would be beneficial to avoid the formation of antiphase disorder in the first place.

A high-temperature treatment of the exactly orientated (001) Si surface to form double-layer steps is very likely restricted due to a limitation in the temperature budget for most highly confined pattern structures or due to the risk of Si reflow. The use of offcut substrates introduces additional challenges, because the crystal quality and surface morphology depend strongly on the offcut angle and trench orientation [182]. An interesting approach to simplify double-step formation on the group-IV surface was reported by Wang et al. [183]. The flat (001) Si trench bottom was etched

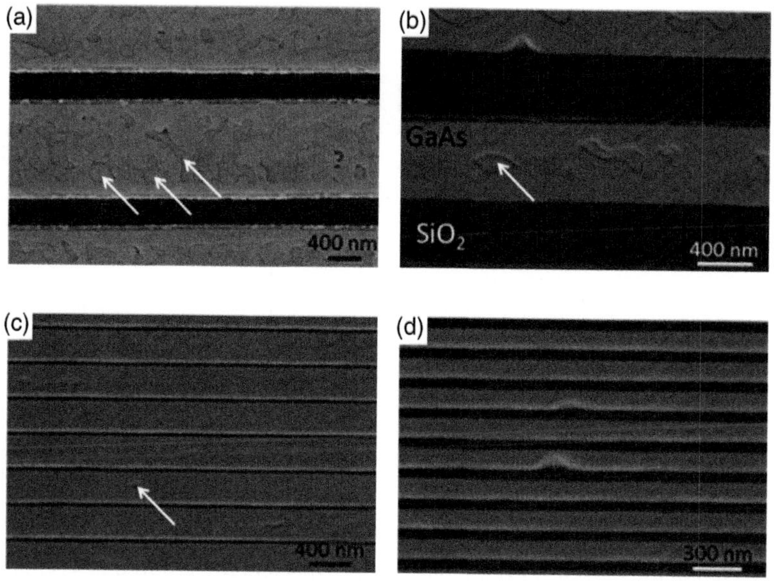

Figure 8.19 *Four top-view scanning electron microscopy (SEM) images of GaAs deposited in 180-nm deep oxide trenches with different widths and AR: (a) 1000 nm, AR=0.18; (b) 570 nm, AR=0.32; (c) 300 nm, AR=0.6; (d) 140 nm, AR=1.3. The top GaAs surface is flat and reveals a (001) surface. Antiphase domains are exemplarily indicated by white arrows. Source: Reproduced from Cipro et al. 2014 [181] with the permission of AIP Publishing.*

in a HCl vapor phase to form a slightly concave Si surface, which consists of different facets. This marginally curved surface is then overgrown with a thin Ge layer. Double-atomic steps are more stable on Ge surfaces and can be induced at lower temperatures [184, 185] as compared to Si. It is important to realize a rounded Ge layer to simulate an offcut Ge surface, although with a changing angle. An adapted thermal treatment after Ge deposition initiates the double-step formation to enable the nucleation of APD-free III/V layers.

An ingenious method to prevent any antiphase disorder in SAG on (001) Si is the preparation of a V-shaped or V-grooved trench bottom with two {111} facets. This is easily achieved for trenches along <110>directions by applying a wet etch – with potassium hydroxide (KOH), ammonium hydroxide (NH_4OH), or tetramethylammonium hydroxide (TMAH)). The great advantage is that a monoatomic surface step on {111} Si does not induce an antiphase boundary, and thus III/V can be deposited inside V-shaped trenches with a low risk of antiphase disorder formation [32]. Figure 8.20 (left) is a corresponding illustration of the III/V-Si interface. This formation of a V-grooved trench bottom is mainly attractive for trench widths in the submicrometer scale, as the additional triangle-shaped volume can be easily filled again during III/V deposition. If the trench width is very large, an overgrowth of a large V-groove might lead to new challenges.

A patterned wafer with Si and SiO_2 regions can be fabricated with the shallow trench isolation (STI) process typically used in CMOS technology. Si is partly replaced by oxide, and the remaining Si regions are then wet-etched to create trenches. The position of the trench bottom can be easily adjusted with the etch time. Depending on the process flow, the V-shaped Si surface is still surrounded by the sidewalls or even expanded under the oxide (underetched), as shown in Figure 8.20 (right). A SiO_2 pattern can also be realized by other processes such as oxidation of the Si substrate surface followed by lithography and etching.

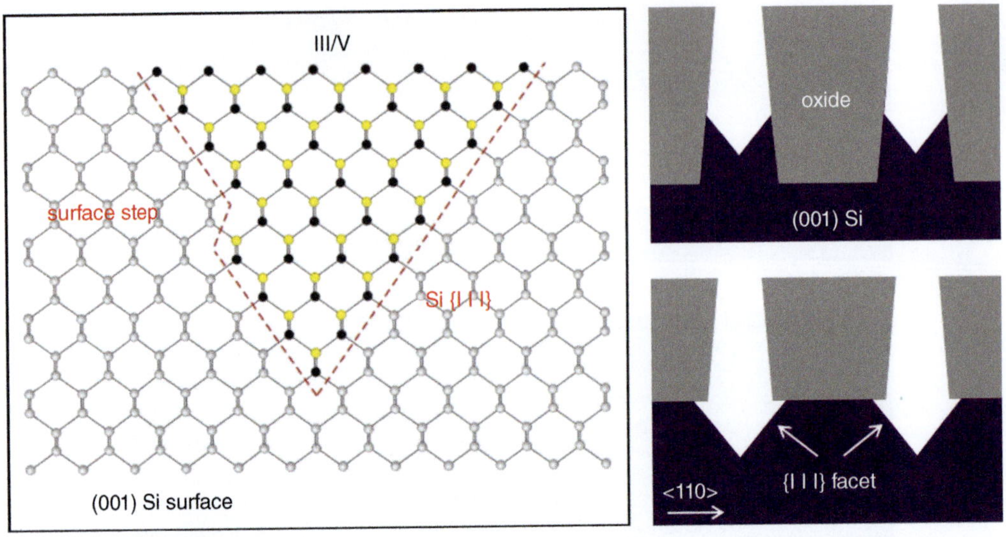

Figure 8.20 *Sketch of the III/V-Si {111} interface (left). The etched V-grooved can be surrounded by the oxide or even expanded under the oxide (right).*

The integration of III/V on a {111} Si surface has already been investigated some time ago. Krost et al. report in [187, 188] that InP forms a heavily twinned region of 25 nm thickness with microtwins parallel to the {111} Si interface. Hence, the lattice mismatch between InP and Si is accommodated via partial dislocations within this highly twinned region, whereas the InP material above this zone is free of planar defects and threading dislocations. A comparable relaxation mechanism was observed for the deposition of InP in V-shaped trenches [32, 189]. An example is shown in Figure 8.21b, where the twinned InP layer along the {111} Si facets is indicated by the white dashed lines. The InP region on top in Figure 8.21b is strain-released and free of defects. The formation of a highly twinned region to release the strain was also reported for GaAs [190], although the formation energy for stacking faults is much higher than for InP [191]. Figures 8.22a and b show clearly the restricted twinned region on top of the {111} Si facets, whereas the GaAs above this layer is defect free. However, some TDs are still nucleated, as revealed by the TEM investigation along the trench; see Figure 8.22c.

The strain release in GaAs on Si can also be achieved using only misfit and threading dislocation formation along the III/V-Si surface [192–194]. An array of 60° misfit dislocation can be precisely defined in a TEM study applying an inverse fast Fourier transformation (IFFT), as shown in Figure 8.23. Some threading dislocations are still nucleated and trapped at the sidewall [194]. The formation of PDs or a twinned region, as discussed before, is not necessary and should be completely suppressed from the beginning.

Finding very different relaxation mechanisms for III/V on {111} facets in the literature reveals that the chosen MOVPE growth conditions will determine which defect formation is energetically preferred. In particular, the effective growth temperature, growth rate, and V/III ratio for the deposition of the III/V seed layer play an important role; but the {111} surface preparation and III/V-precursor switching sequence as well as the polarity may also have an impact on the relaxation mechanisms.

If the strain release in both crystallographic directions can be fully ensured by the formation of interfacial twinned layers along the two {111} planes, only a low sidewall height is necessary to block this twinned region. Hence, a very low aspect ratio should clearly improve the defect density in mismatched material, if no threading dislocations are initiated. Ismail et al. reported in 1991 [195] a pronounced defect reduction in merging GaAs films deposited on a sawtooth-patterned Si substrate

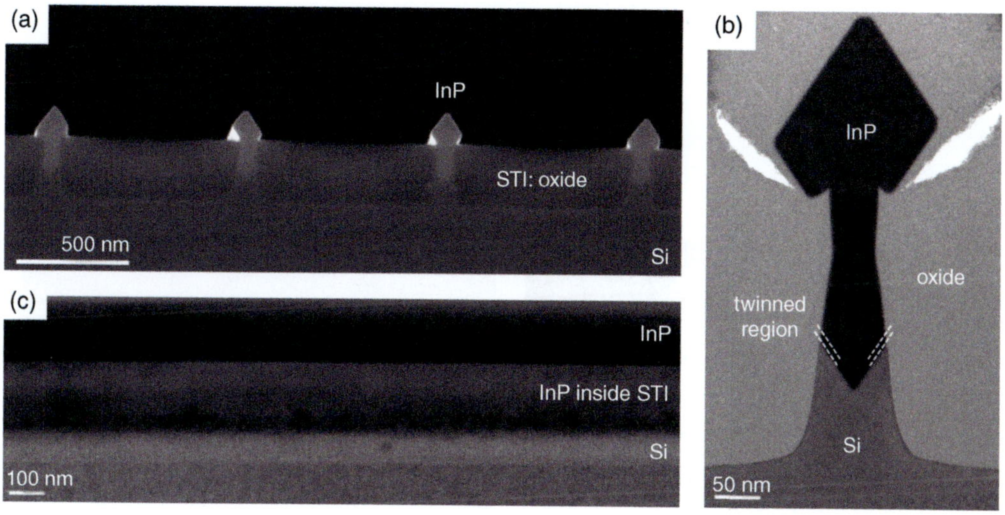

Figure 8.21 (a) A cross-sectional SEM image of InP deposited in V-grooved Si trenches of ~50 nm width. The InP grown out of the trench forms a diamond-shaped top surface with {111} facets. (b) and (c) TEM images across and along a 50-nm wide trench with InP. The twinned III/V region along the {111} plane is indicated by white dashed lines in (b). InP on top of this twinned region does not show any additional defects. The TEM image in (c) along the trench also indicates a defect reduction going from the trench bottom to the top III/V material. Source: Reproduced from Waldron et al. 2016 [186] with permission from Elsevier.

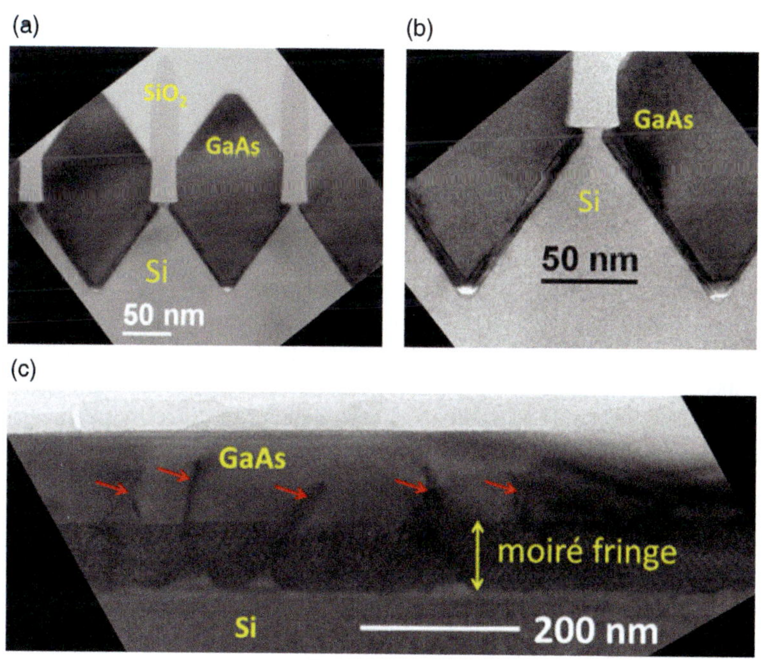

Figure 8.22 TEM images of GaAs deposited in ~90-nm wide trenches. (a) and (b) TEM images across the trenches with different magnifications. A clearly restricted twinned region is visible along the {111} Si facets. (c) A TEM image along the trench. The moiré fringe is caused by the interference of the GaAs layer and the V-shaped Si bottom. Although the main strain release is caused via the twin formation, some TDs are nucleated (indicated by the red arrows), which are trapped at the oxide sidewalls. Source: Reproduced from Li et al. 2015 [190] with the permission of AIP Publishing.

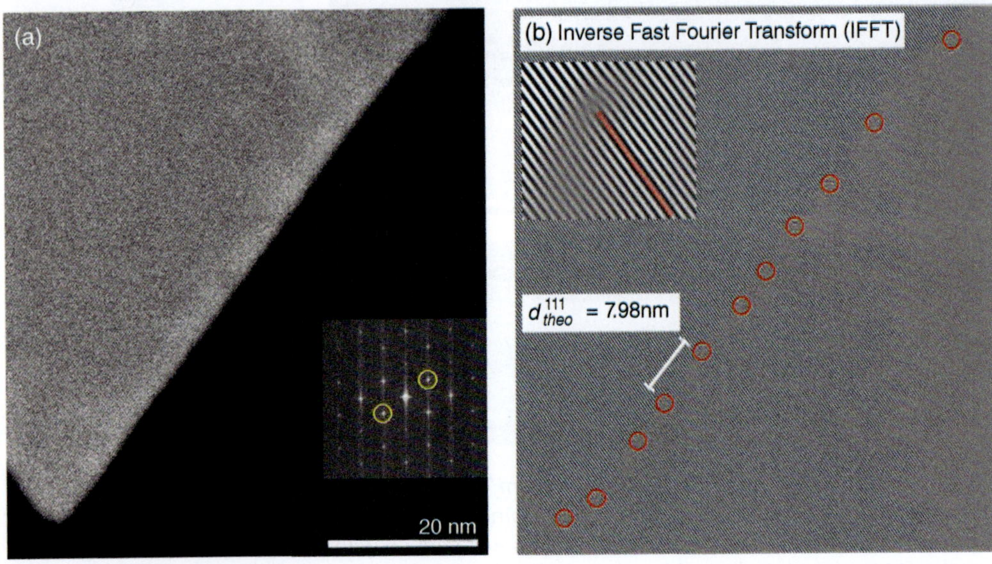

Figure 8.23 (a) A high-magnification STEM image of GaAs deposited in a V-shaped Si trench. (b) The corresponding inverse fast Fourier transformation (IFFT) showing the {111} lattice planes perpendicular to the GaAs/Si interface. The insert in (a) is the associated FFT pattern. The inset in (b) is a magnification of a 60° misfit dislocation core emphasizing in red the additional lattice plane in Si. The red circles in (b) indicate the core position of the 60° misfit array along the GaAs/Si interface. Source: Reproduced from Guo, W. et al. 2017 [193] with the permission of AIP Publishing.

with {111} V-grooves separated by a thin oxide layer. Li et al. [190] recently introduced an approach where only a "tiara"-like edge at the V-grooved Si rim is used to confine the highly twinned GaAs region. This tiara edge is formed by a Si undercut. A two-dimensional merged GaAs layer was achieved after the removal of the oxide sidewalls and a GaAs regrowth step. The TEM images in Figures 8.24a and b show a coalesced GaAs layer and PD trapping at the Si undercut. This is a novel approach to deposit antiphase disorder-free GaAs on (001) Si substrates without the necessity of a specific Si surface bake and to achieve a reduced defect density within a low layer thickness at the same time. An investigation applying electron channeling contrast imaging (ECCI) of such a merged GaAs on a V-groove Si template reveals defect densities of 7×10^7 cm^{-2} and 2×10^7 cm^{-2} for threading dislocations and stacking faults, respectively [172]. The first InAs QD microdisk lasers [174] and electrically pumped InAs QD lasers [172] were also demonstrated on these GaAs templates. The near future will show whether the strain relaxation can be completely ensured by the twinned interfacial layer along the {111} Si facets. Any formation of threading dislocations and PD in a plane perpendicular to the trench orientation has to be avoided to further reduce the defect density.

Looking at the integration examples in Figure 8.18, Figure 8.21, and Figure 8.22 shows that a triangular-shaped top III/V surface is easily achieved in SAG, but a flat (001) III/V surface is also possible: see Figure 8.19. The top surface has a strong impact on any device integration and can be manipulated by the growth conditions. The manipulation of the III/V ridge formation will be discussed later in more detail.

8.3.3.1.1 Applications of SAG in Trenches

The lattice constants of InP and related heterostructures are very interesting for various devices. Very different integration approaches based on SAG are explored, depositing InP directly on Si or using GaAs as an intermediate seed layer or buffer between Si and InP. The strong interest in

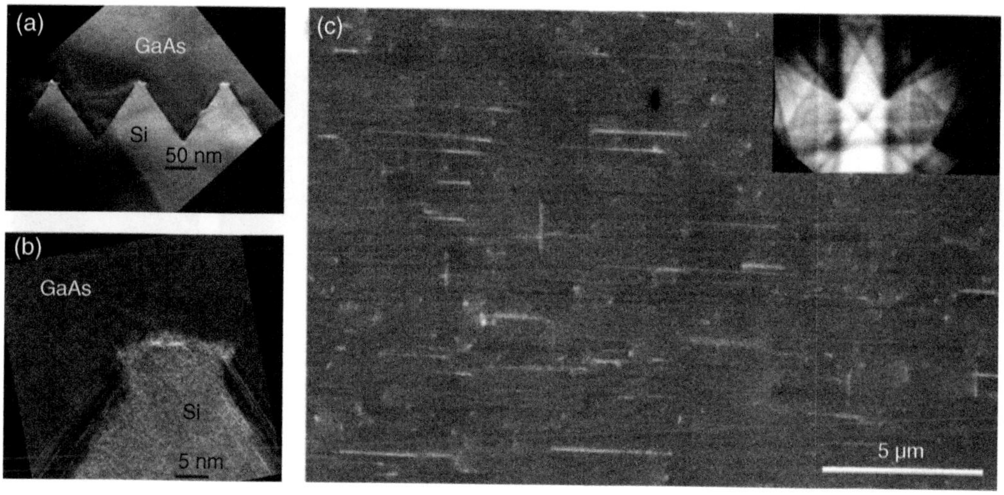

Figure 8.24 *(a) and (b) TEM images in cross-section showing the merged GaAs film from the tiara-shaped undercut of the Si edge. Source: Reproduced from Li et al. 2015 [190] with the permission of AIP Publishing. (c) A plan-view ECCI image. Source: Reproduced from Norman et al. 2017 [172] with the permission of The Optical Society. The bright dashes and points indicate the presence of stacking faults and threading dislocations in the plan-view image. The inset is an electron channeling pattern corresponding to the applied imaging conditions.*

replacing Si as a channel material with III/V in future CMOS technology nodes has especially boosted research on the growth of InGaAs lattice matched to InP in narrow trenches. Waldron et al. reported the first InGaAs FinFET device on 300-mm Si substrate with a STI pattern fabricated in a CMOS-compatible R&D pilot line [196]. InP was deposited in narrow trenches to achieve a high-quality InP buffer. The InP top surface evolved to a triangular-shaped ridge, as shown in Figure 8.21a for the applied MOVPE growth conditions, which was planarized by chemical mechanical polishing (CMP) before the regrowth of the InGaAs channel. The precise device process flow is described in [196] or [186]. Figure 8.25a shows a dark-field STEM image of the complete FinFET device stack on a 300-mm Si substrate. The usage of a selective etch process enables the InP buffer removal under the InGaAs channel region and leads to horizontal GAA-NWFETs [197]. Figures 8.25b and c are some examples of different horizontal nanowire shapes that illustrate the scalability and different shapes of III/V FETs in SEG. The growth of InGaAs channel material for a FinFET was also explored directly on a GaAs buffer, as shown in Figure 8.25d. A GaAs buffer has the advantage over InP of being less conductive and hence leading to lower buffer leakage. On the other hand, the growth of InGaAs on top of a GaAs buffer induces new challenges such as an anisotropic strain relaxation behavior, as reported in [193].

Due to the potential of SEG in narrow trenches for III/V channel integration on Si, various institutes and companies have explored different integration routes and channel material systems [199–203].

The InP buffer is also explored for optoelectronic applications, since InP allows for the integration of heterostructures based on InGaAs or InGaAsP QWs for light emission at telecommunication wavelengths of 1310 nm and 1550 nm. Wang et al. [204] reported the first distributed feedback (DFB) laser array based on SEG of InP in 500-nm wide trenches. The defect-reduction mechanism in this integration approach is based on forming the twinned interfacial layer along the {111} facets; therefore, deposition in 500-nm wide trenches with an aspect ratio of just 0.6 should still lead to a clear defect reduction. In this approach, growth in wider trenches is preferred to achieve enough III/V volume for wave guidance. The developed top triangular shape of the InP was planarized by

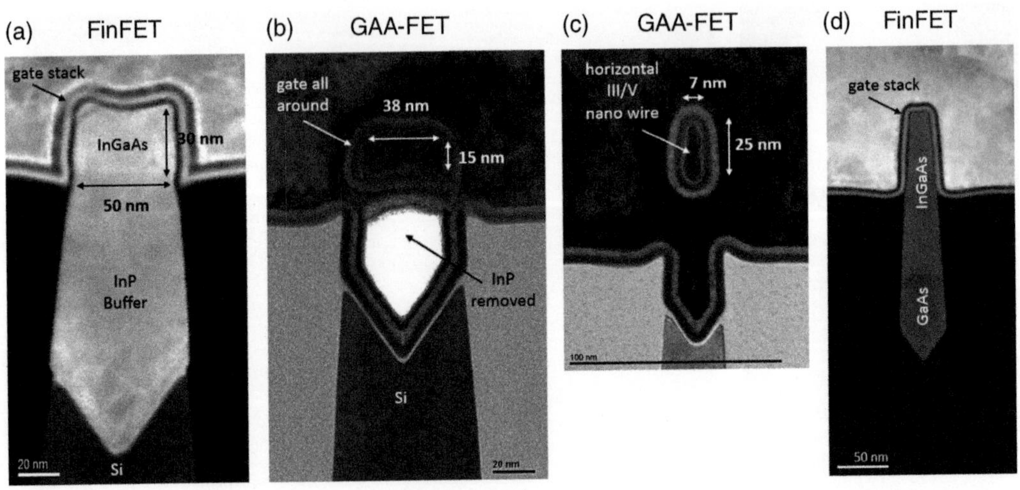

Figure 8.25 *(a) A TEM image of an InGaAs FinFET deposited on an InP buffer in narrow trenches. Source: Adapted from Waldron et al. 2014 [196]. (b) and (c) TEM micrographs illustrating the scalability and different InGaAs nanowire shapes after InP removal. Source: Adapted from Waldron et al. 2016 [197] and Zhou et al. 2016 [198]. (d) A TEM image of an InGaAs FinFET grown on a GaAs buffer. A multilayer gate stack is deposited around the Fin- and nanowire channels.*

CMP before the DFB and output grating fabrication. The STI oxide and the Si substrate were partly removed to reveal a 200-μm long, 500-nm wide free-standing waveguide without absorption losses in the substrate. A schematic of the DFB laser is shown in the top image of Figure 8.26. Optical pumped-laser operation around 925 nm at room temperature proves that this kind of nanoscaled laser structure provides sufficient optical gain to realize a laser diode. The planarized (001) InP buffer surface in Figure 8.26 was also used for the regrowth of an InGaAs/InP heterostructure to shift the emission wavelength toward 1310 nm [205].

Other research groups have investigated the growth of InGaAs or InGaAsP QWs directly on the InP buffer without a planarization step. This is especially challenging if the InP reveals a triangular-shaped or multifaceted surface, because the growth rate and In incorporation may change as a function of the surface facet. The bottom part of Figure 8.26 ((a) and (b)), contains an example from [206]; other approaches are reported in [207, 208]. For more details, see the corresponding references. Overall, the final device application requires very good control of the MQW deposition and hence emission wavelength from each single QW. In addition, all present heterostructures should ensure an efficient carrier capture in the active QWs.

8.3.3.2 III/V Nanoridge Engineering

The evolved ridge shape is defined by the growth-rate hierarchy on the different crystal facets, also taking into account the crystal polarity. This kind of dependence in homoepitaxial SEG of InP and GaAs was reported in the literature for different deposition techniques. Since the growth-rate hierarchy on the different crystal facets can be manipulated by the applied growth conditions, very different ridge shapes can be achieved [202, 209–211]. If the growth rates are known for the different facets and chosen deposition conditions, e.g. the growth-rate polar diagram [212], a Wulff construction allows for simulating the final ridge shape [210].

Recently, it was reported that the ridge shape of GaAs can be nicely engineered for heteroepitaxial deposition in nanoscaled trenches on (001) Si substrate [213]. Figure 8.27 summarizes four cross-sectional SEM images showing GaAs nanoridges with very different shapes achieved by

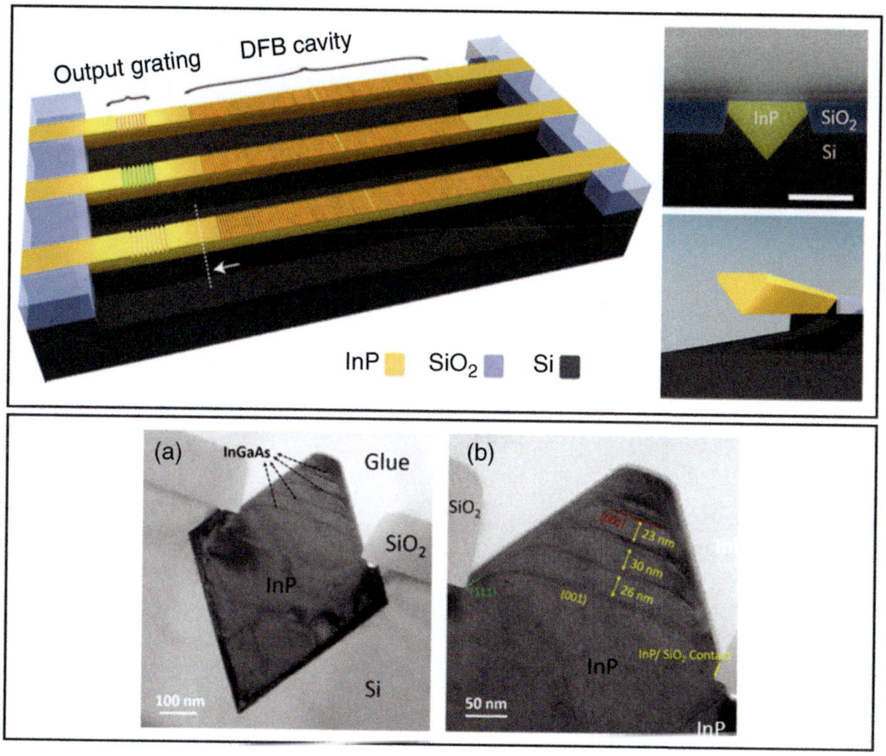

Figure 8.26 The top image is a schematic of the monolithic integration approach of an InP DFB laser on patterned Si substrates. Source: Reprinted by permission from Macmillan Publishers Ltd. from Wang et al. 2015 [204]. In the bottom, (a) and (b), are two TEM images showing an example of InGaAs growth on a multifaceted InP nanoridge. Source: Reproduced from Han et al. 2016 [206] with the permission of AIP Publishing.

applying different MOVPE growth conditions. The ridge shape is always defined and constrained by the facets with the lowest growth rates, whereas the facets with the highest growth rate disappear with rising ridge size. More details about the correlation of the growth-rate hierarchy and shape engineering of GaAs can be found in [210, 213].

Optimized growth parameters within a stable growth window ensure a very uniform shape formation independent of the trench size. Figure 8.28 shows an example of a box-shaped GaAs nanoridge deposited in 20-nm and 100-nm wide trenches realized in one growth experiment. The basic nanoridge shape is the same for 20-nm and 100-nm wide trenches, although with different ridge dimensions, which depend on the trench width, oxide environment, and loading effects [213]. The tilted top-view image in Figure 8.28c clearly emphasizes the very uniform nanoridge shape also along the pattern line.

This III/V integration approach ensures efficient defect trapping due to the deposition in trenches with a high aspect ratio. Pronounced growth out of the trench together with a well-controlled nanoridge engineering leads to increased III/V volume with high crystal quality, which is essential for many device applications. The box-shaped GaAs nanoridge of Figure 8.28b in particular provides a broadened (001) top facet, which is suitable for the integration of pseudomorphically strained MQWs as an optical gain medium. At the same time, the increased nanoridge volume enables efficient wave guidance along the nanoridge.

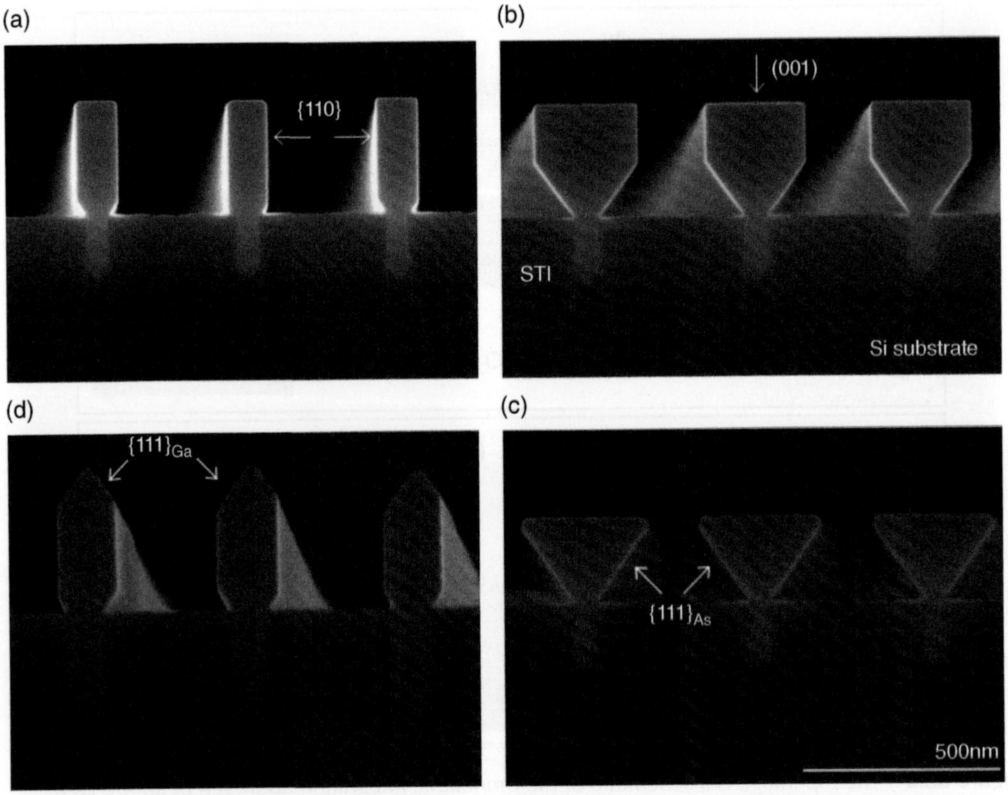

Figure 8.27 Cross-sectional SEM images of four different growth experiments applying specific MOVPE growth conditions to engineer the GaAs nanoridge shape. The different crystal facets are indicated by white arrows. $\{111\}_{Ga}$ and $\{111\}_{As}$ are the facets {111} A and {111} B, respectively. Source: Reproduced from Kunert et al. 2016 [213] with the permission of the Electrochemical Society.

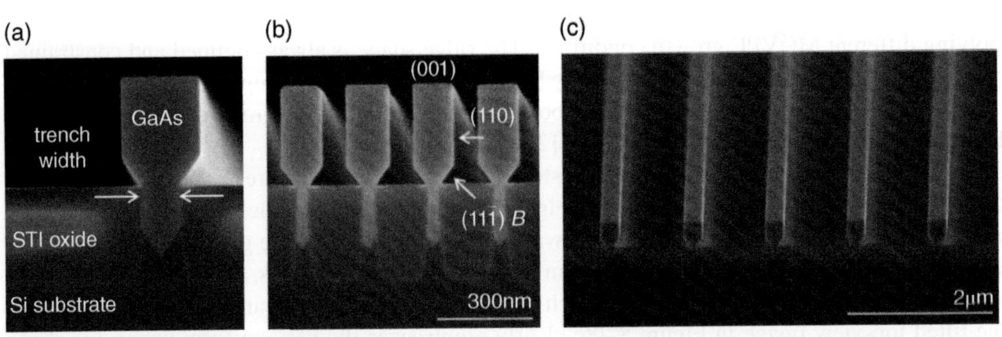

Figure 8.28 (a) and (b) SEM images in cross-section of GaAs nanoridge grown in 100-nm and 20-nm wide trenches. (c) A tilted top view SEM image of a nanoridge array on top of 100-nm wide trenches. Source: Reproduced from Kunert et al. 2016 [194] with the permission of AIP Publishing.

A growth experiment depositing a 7× InGaAs/GaAs MQW on GaAs nanoridge structures reveals the complex growth behavior in SAG. Figure 8.29 contains the corresponding HAADF-STEM images from [213]. The growth time for the InGaAs QW and the GaAs layer on top of the QWs were kept constant. The lighter contrast reveals the InGaAs material, which is mainly grown

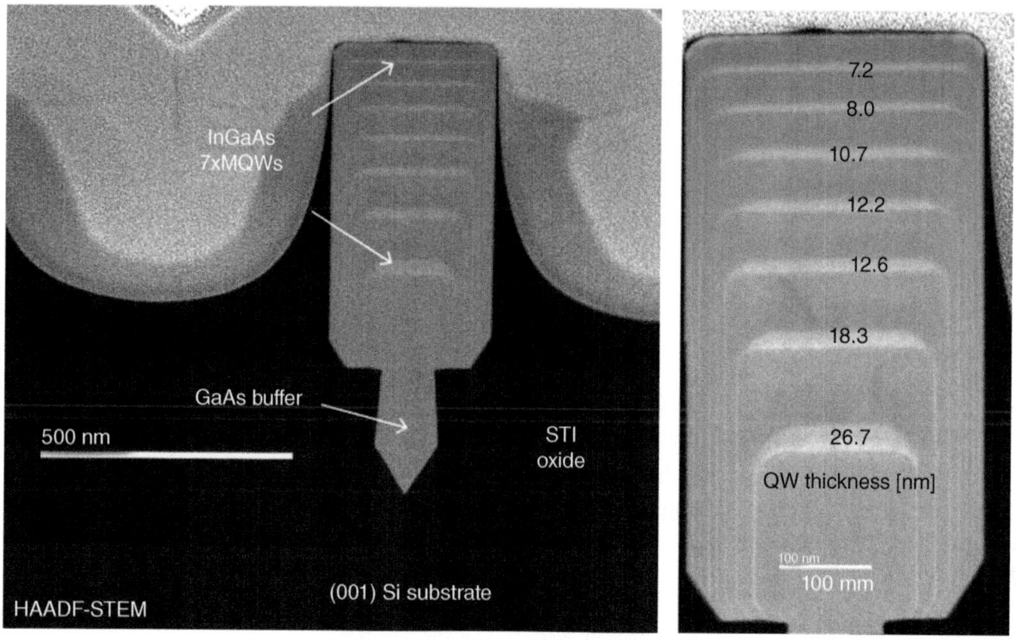

Figure 8.29 *HAADF-STEM images of a 7× MQW stack deposited in a 100-nm wide trench. The InGaAs layers are visible with the light contrast. In the right image, with larger magnification, the InGaAs layer thickness on the (001) facet is indicated in nanometers for each QW. Source: Reproduced from Kunert et al. 2016 [213] with the permission of the Electrochemical Society.*

on the (001) plane; but a thin layer is also deposited on the two {110} side facets. In the same way, GaAs is also grown on the top and side nanoridge facets. It is obvious that the layer thicknesses of the QWs as well as of the GaAs barrier decreases with increasing box size.

In SAG, the III/V material supply to the growth surface is given by the direct precursor transport through the gas phase and by loading effects on the mask. As the growth surface area changes, the first contribution will rise as the nanoridge surface increases. The contribution from the loading effect is defined by the ad- and desorption rate of fully and/or partly decomposed precursors on the dielectric mask as well as by the diffusion length toward the growth surface. All these factors are governed by the applied growth conditions and should not change a lot over the growth time. However, if the direct material transport to an increased III/V surface leads to a depletion in the gas phase, the total precursor amount adsorbed on the mask surface could decrease. Hence, the contribution via the loading effect would drop as the growth surface increases. Figure 8.29 is a typical example for a growth window, where a relatively constant material supply for a given deposition time is distributed over the full box surface. As the nanoridge size increases, the QW and GaAs barrier thicknesses on the (001) and {110} facets decrease, but the total amount of material is roughly constant. The strong interplay between nanoridge size and (001) growth rate as well as the deposition on all exposed nanoridge facets complicates device integration on pattern wafers but can be controlled after a detailed analysis.

Yuting et al. [214] reported the first optically pumped laser device based on this GaAs nanoridge approach implementing a InGaAs/GaAs MQW structure with about 20% In [194]. The QW growth time was adjusted for each QW to achieve a comparable layer thickness. In order to reduce carrier losses due to surface recombination, an InGaP cap layer with a large bandgap was grown, lattice-matched to GaAs around the nanoridge. Figure 8.30 contains a layer sketch, a slightly tilted

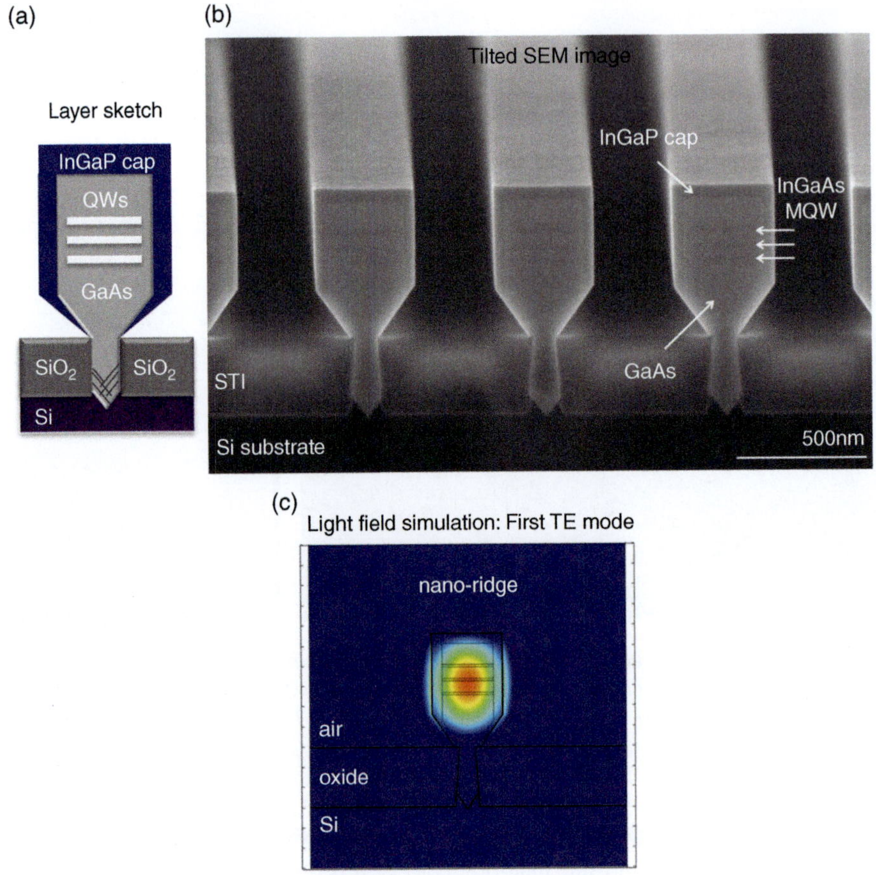

Figure 8.30 *(a) A sketch of the nanoridge laser. (b) The tilted SEM images of a nanoridge laser array emphasizes the uniform and reproducible ridge shape. (c) A light-field simulation of the first TE mode, given a confinement factor of 8.8% whereas the losses into the Si substrate is less than <5 dB/cm. Reproduced from Shi et al. 2016 [214] with the permission of The Optical Society.*

cross-sectional SEM image, and a light-field simulation of the first transverse electric (TE) mode in such a nanolaser device structure. Optically pumped lasing operation at room temperature underlines the potential of this integration approach, where light losses toward the Si substrate are negligible [214], which clearly simplifies the integration flow.

The application of a GaAs buffer inside narrow trenches is also interesting beyond optoelectronic applications and was explored for tunnel diodes on Si substrate [215] and FinFET devices to reduce buffer leakage, as discussed before. However, the integration of GaSb and InAs has also gained a lot of interest [216, 217]. Figure 8.31 shows some examples from [218]. The merging of GaSb from adjacent trenches as discussed for GaAs in Figure 8.24 was also recently reported in [219]. Achieving a high crystal quality inside a trench could enable a top-down etch approach to fabricate highly scaled vertical GAA-NWFETs and TFETs based on SAG [220, 221]. Controlling the nanoridge formation outside the trenches or realizing a high-quality coalesced GaSb film has promise for various optoelectronic application in the infrared region.

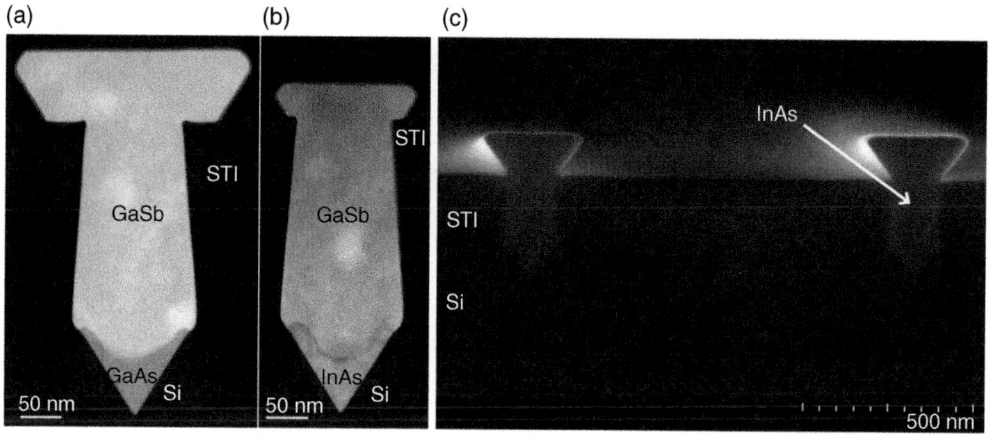

Figure 8.31 (a) and (b) Cross-sectional HAADF-STEM images of GaSb integrated on a GaAs and InAs seed layer. (c) A cross-sectional SEM image showing an InAs nanoridge with a flat (001) surface. Source: Mols et al. 2017 [218].

8.3.3.3 SAG on Various Patterns

The defect-trapping approach via SAG in confined structures was extended to many other geometries. One interesting approach is the application of nanotube templates that define the dimension and orientation of nanowire crystals [222, 223]. The fabrication of a SiO_2 nanotube template is illustrated in Figure 8.32f and described in [222]. The III/V material is selectively nucleated on the exposed Si surface inside the nanotube. Defect trapping is very efficient due to the circular confinement. In addition, the very small Si surface or pattern opening enhances III/V growth from a single nucleation point, which prevents merging of separate crystallites and allows elastic expansion from a single seed. This growth mechanism is comparable to the growth of free-standing nanowires. If strain release is ensured completely via elastic relaxation, no misfit dislocation defects are nucleated. The application of this *template-assisted selective epitaxy* (TASE) enables the integration of vertical nanowires on differently oriented Si substrates including (001) orientation, as shown in Figures 8.32a–d. III/V facet formation depends strongly on the Si substrate orientation and applied growth conditions and can change once the deposited III/V material extends outside the oxide template [224]. Very different material systems and heterostructures based on InAs, GaAs, and InSb were explored for a CMOS compatible III/V integration [225].

TASE is also used to realize lateral III/V structures with a horizontal arrangement of the constricted template. This can lead to more complex and versatile device configurations such as horizontal nanowires, III/V cross junctions, as well as 3D-stacked nanowires [226]. Figure 8.33 contains an illustration of the horizontal template process and the InGaAs growth as described in [226]. The top Si film of a SOI wafer is patterned to define the device dimension. After a thin oxide layer deposition, the Si is etched back inside the confined structures, and a {111} Si surface is revealed for the III/V nucleation. Successful InGaAs growth by MOVPE as illustrated in Figures 8.33c and d leads toward novel device-integration schemes and allows for cointegrating III/V and Si on the same platform [226].

The integration of InGaAs, InAs, GaAs, and GaSb as well as related devices, such as n-FETs, p-FETs, and TFET, on (001) Si were already demonstrated [227–229]. The best crystal quality

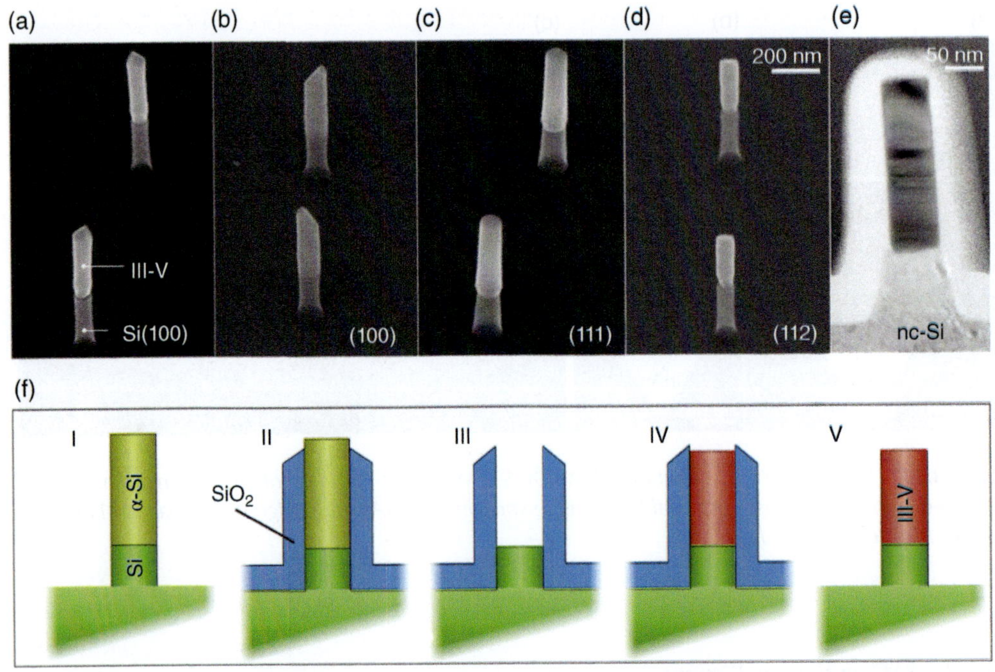

Figure 8.32 (a)–(d) Tilted SEM image of vertical InAs nanowires deposited in oxide nanotubes fabricated on differently oriented Si substrates. (e) A TEM image of a GaAs nanowire. (f) An illustration of the main fabrications steps of the nanotube template. Source: Reprinted from Borg et al. 2014 [222] with permission from the American Chemical Society.

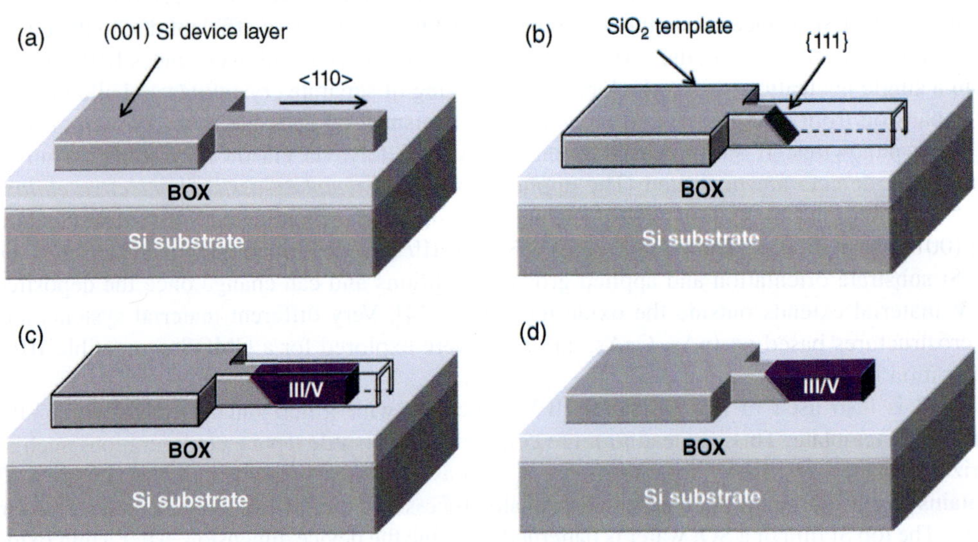

Figure 8.33 A sketch of the template formation on a (001) SOI substrate. (a) The Si film on top of the buried oxide (BOX) is processed into a device layer with a specific ridge to determine the dimensions of the template. (b) A thin (~30 nm) SiO_2 layer is deposited around this Si device layer. The SiO_2 cap is removed at the end of the device ridge, and the Si is etched back to the target length with a defined {111} crystal plane. (c) InGaAs is deposited into the horizontal oxide channel by MOVPE, nucleating at the {111} Si facet. (d) The SiO_2 layer is completely removed, and the III/V film is accessible for further device processing. Source: After Schmid et al. 2015 [226].

is ensured when III/V growth is initiated by one nucleation point. This is most likely achieved in a very narrow template. However, this highly confined region can be followed by extended growth inside a broadened template to achieve a larger III/V area [228], which enables the integration of III/V devices requiring larger dimensions. In addition, the TASE method is not restricted to a monocrystalline Si seed surface. The application of polycrystalline Si as a nucleation surface was used to fabricate multilayer structures, but the final III/V crystal orientation is then arbitrary [226].

8.3.3.4 Lateral Overgrowth

Epitaxial lateral overgrowth (ELOG) is well known for GaN deposition on foreign substrates, but it is also investigated for the integration of III/V on Si [230–234]. The deposition starts with SAG on a patterned wafer, but growth proceeds laterally over the dielectric mask. Defects are either completely trapped inside the constrained pattern if the AR is sufficient or can only propagate through the open area, whereas the overgrown material on the mask is defect free. The mask pattern is fabricated either directly on the Si substrate or on top of a III/V seed layer, which still contains defects. Figures 8.34a, b, and c illustrates these three cases. Since all planar and most misfit dislocation defects continue along the <111>directions in zincblende semiconductors, propagating defects through the mask openings increases the defective material volume on top of the mask (Figure 8.34b); hence, efficient defect trapping inside the mask pattern is clearly preferable. An optional coalescence of the III/V material originating from different open areas leads to large III/V templates, which is very useful for device integration. However, achieving a defect-free coalescence boundary is very difficult. Finally, realizing an active III/V layer on an oxide can be very beneficial for specific device functionalities due to optional electrical isolation.

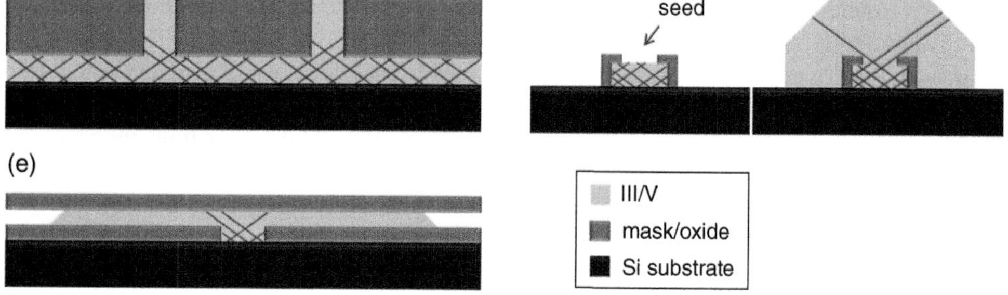

Figure 8.34 *Sketch of different approaches investigated for lateral overgrowth. ELOG starts with SAG in a confined area, but growth proceeds laterally over the dielectric mask. The efficiency of defect trapping depends strongly on the design and confinement of the patterned substrate. A possible merging of III/V layers from adjacent openings can lead to a coherent film if the quality of the coalescence boundary is well controlled.*

Various III/V material systems were investigated, exploring very different Si substrate patterns such as holes [235, 236] and trenches [195, 237–241], applying MOVPE as well as HVPE. As the dimensions of the pattern have decreased to a submicrometer scale in recent years, the names *nanoepitaxy* and *nanopattern* are often used in the literature.

In an interesting laser-integration approach, the dielectric mask is used to block defects inside narrow trenches and to contain a Si waveguide structure at the same time [242]. The principle of defect reduction is shown in Figure 8.34c. An InP seed layer deposited on a blanket Si substrate is planarized and polished by CMP before the deposition of a thick SiO_2 layer. A narrow trench pattern leads to efficient trapping of all remaining defects from the seed layer during InP regrowth. It is reported in [243] that a smooth InP surface morphology was also achieved in the area of InP coalescence from adjacent openings, which enabled the integration of InGaAsP/InP MQWs for light emission. The novelty of this integration approach lies in the fabrication of a Si/SiO_2 waveguide inside the mask oxide, which is close to the MQW structure, acting as an optical gain medium. As the light field of the MQW structure penetrates the waveguide, a monolithic evanescently coupled Si laser (MECSL) might be realized.

A slightly modified ELOG approach is based on the application of a restricted III/V seed mesa surrounded by an oxide, as shown in Figure 8.34d. The regrown III/V layer will first laterally extend over the top mask layer and then expand vertically downward and finally cover the Si substrate. The growth of InP and InGaP from an InP seed mesa by HVPE was investigated in [244, 245].

Lateral overgrowth was also reported in a geometrically confined cavity. This *confined epitaxial overgrowth* (CLEO) was explored based on very different mask approaches that led to a final III/V layer directly on Si [246] or on SiO_2 [247]. Figure 8.34e illustrates the approach used in [247]. An oxide cavity is fabricated, applying a sacrificial layer, which restricts the height during lateral overgrowth. The III/V nucleation starts from a confined Si surface where defects are efficiently trapped and extends laterally in between the oxide cavity. The first reported InGaAs MOSFET devices [247] underline the potential of this approach in which III/V integration on an isolator holds a clear advantage to realize fully depleted MOSFETs. Also, using an additional lateral restriction, as introduced for TASE, Wirths et al. reported very recently a first GaAs-based microdisk laser with a diameter of about 3.1 μm [248] based on CLEO.

The lateral overgrowth of III/V offers an interesting opportunity for a CMOS-compatible III/V integration approach, although based on a complex fabrication flow. The epitaxial growth control as well as the dielectric mask quality and purity are very crucial for the III/V material quality of the active region. Defect densities down to the range of 10^5 cm^{-2} are reported for different material systems. Nevertheless, the number of publications about complete device integration is still very limited to date [249].

The growth in very confined mask structures holds the clear advantage of efficient defect trapping. Nevertheless, the smaller the template dimensions, the more challenging is the dielectric pattern fabrication, in order to achieve a high oxide surface quality. At the same time, the Si surface pretreatment might also be more difficult. Growth conditions have to be carefully adjusted to ensure a uniform filling of highly constricted templates and to prevent stacking faults and twin formation [250].

One aspect that has not been addressed yet is interdiffusion at the III/V-Si interface. Si atoms can diffuse into the III/V material and vice versa for group-III and -V atoms, leading to unintentional doping of the Si and the heterolayer. The total impact of this interdiffusion depends a lot on the total device architecture and functionality, the process flow, and the total thermal budget. Generally, it can be concluded that, especially in SAG, much more attention must be given to this aspect. The layer and device dimensions are very small and more easily infiltrated by diffusing atoms, whereas a micrometer-thick metamorphic buffer can prevent the influence of interdiffusion on the active device layers.

Overview of different integration approaches of III/V on (001) Si.

Approach	Advantages	Challenges	Application/devices
Blanket layer growth	• Freedom in the choice of the fina buffer lattice parameter	• Wafer bowing • Crack formation • Relatively high precursor consumption • CMOS compatibility	• Good CMOS compatibility is only feasible for localized growth on patterned wafers • Device size and placement are very crucial for cointegration with nanoelectronics
Lattice-matched growth	• No misfit defect formation, hence no thick buffer required • Reduced risk of crack formation	• Includes many novel and metastable material systems such as dilute nitrides and borides → still difficult to improve material quality and to achieve good device performance	• GaNAsP laser diodes • GaN(As)P/Si tandem solar cells and photoelectrochemical devices
Metamorphic growth	• Integration of a defect filter layer allows for a clear decrease of the total buffer thickness	• Strong wafer bowing • High risk of crack formation	• QD laser diodes: very good lifetimes were observed because QD lasers are less sensitive to defects in comparison to MQW lasers
Interfacial misfit array in III/Sb	• No thick buffer required • Reduced risk of crack formation	• Ensuring full relaxation via 90° misfit dislocations at the III/Sb-Si interface	• Infrared MQW laser diodes
Selective area growth	• Efficient defect trapping • Less impact of wafer bowing and crack formation • Small footprint → toward nanoscaled devices • Good CMOS compatibility	• Additional process steps necessary to fabricate patterned wafers • Device architecture needs to meet the smaller III/V footprint • Interdiffusion can affect device performance	• High CMOS compatibility in comparison to blanket wafer growth • Limited to nanosized devices
SAG in trenches	• *Used STI templates are directly compatible with CMOS cointegration*	• *Growth control of heterostructures* • *Device architecture*	• *First demonstration of FETs and optical pumped-laser diodes*
TASE	• *Highly scaled devices* • *Device stacking possible*	• *Very complex pattern fabrication* • *Risk of stacking fault formation*	• *Cointegration of III/V- and IV-channels*
ELOG	• *Growth of large active III/V areas from small openings*	• *Defect-free merging*	• *Merged layers enable the integration of large-size devices*

8.4 Conclusions

The strong interest in recent decades in integrating III/V-based devices on Si substrates and the increasing demand to add new functionalities for Si-based ICs have led to tremendous research efforts in the field of monolithic III/V heteroepitaxy on Si. Although this book chapter covers only a selection of topics, it is obvious that many different integration approaches exist, all having their own advantages and disadvantages. A summary of the different heteroepitaxial approaches discussed in this chapter, including their pros and cons, is shown in Table 8.1.

The future will show which integration approach will be most successful or if multiple concepts will enter the market. Two key features will decide whether a III/V device concept on Si will go into commercial production: the defect density in the active region, which determines the device lifetime, and the feasibility of the integration flow, which has a direct impact on production costs. In particular, for cointegration with Si-based microelectronics, CMOS compatibility must be fully ensured.

The requirements concerning the integration flow and production costs might be more relaxed if the integration of III/V on Si leads to new device functionalities. However, if III/V is considered as a replacement for Si-based devices, the demands with respect to integration and production costs are very high and difficult to meet due to the already well-established Si technology. For example, the current FinFET technology is so highly scaled that III/V has to meet similar scaling requirements to be relevant for CMOS. This necessitates many additional process and flow developments, so the potential gain in device performance might be washed out by the increasing R&D costs.

The assessment of the different approaches summarized in Table 8.1 is based on the current development status (as of 2017) and on the personal opinion of the authors. However, research and development can open up new possibilities that are not yet known, especially in this fast-evolving research field of heteroepitaxy. Therefore, we should always be open-minded concerning each research route and curious about the future of III/V on Si.

References

1. Del Alamo, J.A., Antoniadis, D.A., Lin, J. et al. (2016). Nanometer-scale III-V MOSFETs. *IEEE Journal of the Electron Devices Society* 4 (5): 205–214.
2. Kuhn, K.J. (2012). Considerations for ultimate CMOS scaling. *IEEE Transactions on Electron Devices* 59 (7): 1813–1828.
3. Del Alamo, J.A. (2011). Nanometre-scale electronics with III–V compound semiconductors. *Nature* 479 (7373): 317–323.
4. Riel, H., Wernersson, L.-E., Hong, M., and del Alamo, J.A. (2014). III–V compound semiconductor transistors–from planar to nanowire structures. *MRS Bulletin*, 39 (8): 668–677.
5. Thomson, D. et al. (2016). Roadmap on silicon photonics. *Journal of Optics* 18 (7): 73003.
6. Wang, Z. et al. (2017). Novel light source integration approaches for silicon photonics. *Laser Photonics Review* 11 (4): 1700063.
7. Lourdudoss, S. (2012). Heteroepitaxy and selective area heteroepitaxy for silicon photonics. *Current Opinion in Solid State & Materials Science* 16 (2): 91–99.
8. Ren, B., Hou, Y., and Liang, Y. (2016). Research progress of III–V laser bonding to Si. *Journal of Semiconductors* 37 (12): 124001.
9. Nishinaga, T. and Kuech, T.F. (2015). Eds., Handbook of Crystal Growth. Elsevier.
10. Riedl, T. and Lindner, J.K.N. (2017). Heteroepitaxy of III–V Zinc Blende Semiconductors on Nanopatterned Substrates. In: *Nanoscaled Film Layers*, chapter 4. InTech Publishers.
11. Frost, T. et al. (2014). Monolithic electrically injected nanowire array edge-emitting laser on (001) silicon. *Nano Letters* 14 (8): 4535–4541.
12. Hazari, A. et al. (2017). 1.3μm optical interconnect on silicon: A monolithic III-nitride nanowire photonic integrated circuit. *IEEE Journal of Quantum Electronics* 9197: 1–1.

13. Liu, J., Kimerling, L.C., and Michel, J. (2012). Monolithic Ge-on-Si lasers for large-scale electronic – photonic integration. *Semiconductor Science and Technology* 27: 94006.
14. Wirths, S. et al. (2015). Lasing in direct-bandgap GeSn alloy grown on Si. *Nature Photonics* 9 (2): 88–92.
15. Grydlik, M. et al. (2016). Lasing from glassy Ge quantum dots in crystalline Si. *ACS Photonics* 3 (2): 298–303.
16. Grydlik, M. et al. (2016). Laser level scheme of self-interstitials in epitaxial Ge dots encapsulated in Si. *Nano Letters* 16 (11): 6802–6807.
17. Hamers, R.J., Tromp, R.M., and Demuth, J.E. (1986). Scanning tunneling microscopy of Si (001). *Physical Review B* 34 (8): 5343–5357.
18. Tromp, R.M., Hamers, R.J., and Demuth, J.E. (1985). Si (001) dimer structure observed with scanning tunneling microscopy, *Physical Review Letters* 55 (12): 1303–1306.
19. Wolkow, R.A. (1992). Direct observation of an increase in buckled dimers on Si (001) at low temperature. *Physical Review Letters* 68 (17): 2636–2639.
20. Baski, A.A., Erwin, S.C., and Whitman, L.J. (1997). The structure of silicon surfaces from (001) to (111). *Surface Science* 392 (1–3): 69–85.
21. Chadi, D.J. (1987). Stabilities of single-layer and bilayer steps on Si (001) surfaces. *Physical Review Letters* 59 (15): 1691–1694.
22. Aspnes, D.E. and Ihm, J. (1986). Biatomic steps on (001) silicon surfaces. *Physical Review Letters* 57 (24): 3054–3057.
23. Pehlke, E. and Tersoff, J. (1991). Phase diagram of vicinal Si (001) surfaces. *Physical Review Letters* 67 (10): 1290–1293.
24. Fang, S.F. et al. (1990). Gallium arsenide and other compound semiconductors on silicon. *Journal of Applied Physics* 68 (7): R31.
25. Dürr, M. and Höfer, U. (2006). Dissociative adsorption of molecular hydrogen on silicon surfaces. *Surface Science Reports* 61 (12): 465–526.
26. van Dijken, S., Zandvliet, H.J.W., and Poelsema, B. (1998). Energetics and structure of the stable and unstable biatomic step edges of Si (100). *Surface Review and Letters* 5 (1): 15–20.
27. Laracuente, A.R. and Whitman, L.J. (2003). Step structure and surface morphology of hydrogen-terminated silicon: (001) to (114). *Surface Science* 545 (1–2): 70–84.
28. Aoyama, T., Goto, K., Yamazaki, T., and Ito, T. (1995). Silicon (001) surface after annealing in hydrogen ambient. *Journal of Vacuum Science and Technology A* 14 (5): 2909.
29. Chabal, Y.J. (1986). High-resolution infrared spectroscopy of adsorbates on semiconductor surfaces: Hydrogen on Si (100) and Ge (100). *Surface Science* 168 (1–3): 594–608.
30. Boland, J.J. (1990). Structure of the H-saturated Si (100) surface. *Physical Review Letters* 65(26), 3325–3328.
31. Telieps, W. and Bauer, E. (1985). The (7×7) ↔ (1×1) phase transition on Si (111). *Surface Science* 162 (1–3): 163–168.
32. Paladugu, M. et al. (2012). Site selective integration of III–V materials on Si for nanoscale logic and photonic devices. *Crystal Growth and Design* 12 (10): 4696–4702.
33. Ogawa, H., Arai, T., Yanagisawa, M. et al. (2002). Dry cleaning technology for removal of silicon native oxide employing hot NH_3/NF_3 exposure. *Japanese Journal of Applied Physics* 41 (Part 1 (8)), 5349–5358.
34. Tang J., Ingle N., and Yang, D. (2009). Smooth siconi etch for silicon-containing films. US Patent.US20110151674.
35. Nishino, H., Hayasaka, N., and Okano, H. (1993). Damage-free selective etching of Si native oxides using NH_3/NF_3 and SF_6/H_2O down-flow etching. *Journal of Applied Physics* 74 (2): 1345–1348.
36. Kunert, B., Németh, I., Reinhard, S. et al. (2008). Si (001) surface preparation for the antiphase domain free heteroepitaxial growth of GaP on Si substrate. *Thin Solid Films* 517 (1): 140–143.
37. Brückner, S., Döscher, H., Kleinschmidt, P. et al. (2012). Anomalous double-layer step formation on Si (100) in hydrogen process ambient. *Physical Review B* 86 (19): 195310.
38. Brückner, S., Döscher, H., Kleinschmidt, P., and Hannappel, T. (2011). In situ investigation of hydrogen interacting with Si (100). *Applied Physics Letters* 98 (21): 211909.
39. Komeda, T. and Kumagai, Y. (1998). Si (001) surface variation with annealing in ambient H_2. *Physical Review B* 58 (3): 1385–1391.
40. Beyer, A. et al. (2012). GaP heteroepitaxy on Si (001) : Correlation of Si-surface structure, GaP growth conditions, and Si-III/V interface structure. *Journal of Applied Physics* 111 (8): 83534.
41. Brückner, S., Kleinschmidt, P., Supplie, O. et al. (2013). Domain-sensitive in situ observation of layer-by-layer removal at Si (100) in H_2 ambient. *New Journal of Physics* 15 (11): 113049.
42. Bachmann, K.J., Rossow, U.,Sukidi, N. et al. (1996). Heteroepitaxy of GaP on Si (100). *Journal of Vacuum Science and Technology B* 14 (4): 3019–3029.
43. Beyer, A., Németh, I., Liebich, S. et al. (2011). Influence of crystal polarity on crystal defects in GaP grown on exact Si (001). *Journal of Applied Physics* 109 (8): 83529.

44. Volz, K. et al. (2011). GaP-nucleation on exact Si (001) substrates for III/V device integration. *Journal of Crystal Growth* 315 (1): 37–47.
45. Stegmüller, A. et al. (2016). Surface chemistry of *tert-* butylphosphine (TBP) on Si (001) in the nucleation phase of thin-film growth. *Chemistry - A European Journal* 22 (42): 14920–14928.
46. Yamane, K., Kobayashi, T., Furukawa, Y. et al. (2009). Growth of pit-free GaP on Si by suppression of a surface reaction at an initial growth stage. *Journal of Crystal Growth* 311 (3): 794–797.
47. Werner, K., Beyer, A., Oelerich, J.O. et al. (2014). Structural characteristics of gallium metal deposited on Si (001) by MOCVD, *Journal of Crystal Growth* 405: 102–109.
48. Supplie, O. et al. (2015). Formation of GaP/Si (100) heterointerfaces in the presence of inherent reactor residuals. *ACS Appled Materials and Interfaces*, 7 (18): 9323–9327.
49. Bringans, R.D., Biegelsen, D.K., and Swartz, L.-E. (1991). Atomic-step rearrangement on Si (100) by interaction with arsenic and the implication for GaAs-on-Si epitaxy. *Physical Review B* 44 (7): 3054–3063.
50. Kipp, L., Biegelsen, D.K., Northrup, J.E. et al. (1996). Reflectance difference spectroscopy: Experiment and theory for the model system Si (001):As and application to Si (001). *Physical Review Letters* 76 (15): 2810–2813.
51. McMahon, W.E., Batyrev, I.G., Hannappel, T. et al. (2006). 5-7-5 line defects on As/Si(100): A general stress-relief mechanism for V/IV surfaces. *Physical Review B* 74 (3): 33304.
52. Bork, T., McMahon, W.E., Olson, J.M., and Hannappel, T. (2007). Surface science studies including low-temperature RDS on MOCVD-prepared, As-terminated Si(100) surfaces. *Journal of Crystal Growth* 298: 54–58.
53. Hannappel, T., McMahon, W.E., and Olson, J.M. (2004). An RDS, LEED, and STM study of MOCVD-prepared Si (100) surfaces. *Journal of Crystal Growth* 272 (1–4): 24–29.
54. Supplie, O. et al. (2015). *In situ* controlled heteroepitaxy of single-domain GaP on As-modified Si (100). *APL Materials* 3 (12): 126110.
55. Volz, K., Stolz, W., Dadgar, A., and Krost, A. (2015). Growth of III/Vs on silicon: nitrides, phosphides, arsenides and antimonides. In: *Handbook of Crystal Growth*, 1249–1300.
56. Harrison, W.A., Kraut, E.A., Waldrop, J.R., and Grant, R.W. (1987). Polar heterojunction interfaces. *Physical Review B* 18 (8): 4402–4410.
57. Kroemer, H. (1987). Polar-on-nonpolar epitaxy. *Journal of Crystal Growth* 81 (1–4): 193–204.
58. Supplie, O. et al. (2014). Atomic scale analysis of the GaP/Si (100) heterointerface by *in situ* reflection anisotropy spectroscopy and *ab initio* density functional theory. *Physical Review B* 90 (23): 235301.
59. Romanyuk, O., Hannappel, T., and Grosse, F. (2013). Atomic and electronic structure of GaP/Si (111), GaP/Si (110), and GaP/Si (113) interfaces and superlattices studied by density functional theory. *Physical Review B* 88 (11): 115312.
60. Beyer, A. et al. (2016). Pyramidal structure formation at the interface between III/V semiconductors and silicon. *Chemistry of Materials* 28 (10): 3265–3275.
61. Kroemer, H., Polasko, K.J., and Wright, S.C. (1980). On the (110) orientation as the preferred orientation for the molecular beam epitaxial growth of GaAs on Ge, GaP on Si, and similar zincblende-on-diamond systems. *Applied Physics Letters* 36 (9): 763–765.
62. Wright, S.L., Kroemer, H., and Inada, M. (1984). Molecular beam epitaxial growth of GaP on Si. *Journal of Applied Physics* 55 (8): 2916–2927.
63. Sprung, K.R., Wilke, K., Heymann, G. et al. (1993). GaAs single-domain growth on exact (100) Si substrate. *Applied Physics Letters* 62 (21): 2711–2712.
64. Grundmann, M., Krost, A., Bimberg, D. et al. (1992). Maskless growth of InP stripes on patterned Si(001): Defect reduction and improvement of optical properties. *Applied Physics Letters* 60 (26): 3292.
65. Ernst, F. and Pirouz, P. (1989). The formation mechanism of planar defects in compound semiconductors grown epitaxially on {100} silicon substrates. *Journal of Materials Research* 4 (4): 834–842.
66. Ernst, F. and Pirouz, P. (1988). Formation of planar defects in the epitaxial growth of GaP on Si substrate by metal organic chemical-vapor deposition. *Journal of Applied Physics* 64 (9): 4526–4530.
67. Soga, T., Jimbo, T., and Umeno, M. (1995). Initial stage of heteroepitaxial growth by metalorganic chemical vapor deposition. *Journal of Crystal Growth* 163 (1–2): 165–170.
68. Soga, T., Jimbo, T., and Umeno, M. (1994). Epitaxial growth of a two-dimensional structure of GaP on a Si substrate by metalorganic chemical vapor deposition. *Applied Surface Science* 82–83: 64–69.
69. Soga, T., George, T., Suzuki, T. et al. (1991). Transmission electron microscopy characterization of the initial stage of epitaxial growth of GaP on Si by low-pressure metalorganic chemical vapor deposition. *Applied Physics Letters* 58 (19): 2108.
70. Takagi, Y., Yonezu, H., Samonji, K. et al. (1998). Generation and suppression process of crystalline defects in GaP layers grown on misoriented Si (100) substrates. *Journal of Crystal Growth* 187 (1): 42–50.
71. Narayanan, V.,Mahajan, S., Sukidi, N. et al. (2000). Orientation mediated self-assembled gallium phosphide islands grown on silicon. *Philosophical Magazine A* 80 (3): 555–572.

72. Grassman, T.J. et al. (2009). Control and elimination of nucleation-related defects in GaP/Si (001) heteroepitaxy. *Applied Physics Letters* 94 (23): 232106.
73. Grassman, T.J. et al. (2013). Nucleation-related defect-free GaP/Si (100) heteroepitaxy via metal-organic chemical vapor deposition. *Applied Physics Letters* 102 (14): 142102.
74. Németh, I., Kunert, B., Stolz, W., and Volz, K. (2008). Heteroepitaxy of GaP on Si: Correlation of morphology, anti-phase-domain structure and MOVPE growth conditions. *Journal of Crystal Growth* 310 (7–9): 1595–1601.
75. Döscher, H., Borkenhagen, B., Lilienkamp, G. et al. (2011). III–V on silicon: Observation of gallium phosphide anti-phase disorder by low-energy electron microscopy. *Surface Science* 605 (15–16): L38–L41.
76. Döscher, H., Hannappel, T., Kunert, B. et al. (2008). In situ verification of single-domain III-V on Si (100) growth via metal-organic vapor phase epitaxy. *Applied Physics Letters* 93 (17): 172110.
77. Döscher, H. et al. (2010). In situ antiphase domain quantification applied on heteroepitaxial GaP growth on Si(100). *Journal of Vacuum Science and Technology B* 28 (4): C5H1–C5H6.
78. Döscher, H. et al. (2011). Indirect in situ characterization of Si(100) substrates at the initial stage of III–V heteroepitaxy. *Journal of Crystal Growth* 315 (1): 16–21.
79. Létoublon, A. et al. (2011). X-ray study of antiphase domains and their stability in MBE grown GaP on Si. *Journal of Crystal Growth* 323 (1 SI): 409–412.
80. Dixit, V.K. et al. (2006). Studies on MOVPE growth of GaP epitaxial layer on Si (001) substrate and effects of annealing. *Journal of Crystal Growth* 293 (1): 5–13.
81. Guo, W. et al. (2012). Thermodynamic evolution of antiphase boundaries in GaP/Si epilayers evidenced by advanced X-ray scattering. *Applied Surface Science* 258 (7): 2808–2815.
82. Németh, I., Kunert, B., Stolz, W., and Volz, K. (2008). Ways to quantitatively detect antiphase disorder in GaP films grown on Si(001) by transmission electron microscopy. *Journal of Crystal Growth* 310: 4763–4767.
83. Kuan, T.S. and Chang, C.-A. (1983). Electron microscope studies of a Ge–GaAs superlattice grown by molecular beam epitaxy. *Journal of Applied Physics* 54 (8): 4408.
84. Pond, R.C., Gowers, J.P., Holt, D.B. et al. (1984). A general treatment of antiphase domain formation and identification at polar-nonpolar semiconductor interfaces. *MRS Proceedings* 25: 273.
85. Narayanan, V., Mahajan, S., Bachmann, K.J. et al. (2002). Antiphase boundaries in GaP layers grown on (001) Si by chemical beam epitaxy. *Acta Materialia*, 50 (6): 1275–1287.
86. Beyer, A. et al. Resolving the intrinsic atomic structure of interfaces in III/V semiconductors on silicon. Submitted for publication.
87. Beyer, A. et al. (2013). Atomic structure of (110) anti-phase boundaries in GaP on Si(001). *Applied Physics Letters* 103 (3): 32107.
88. Ueda, O. and Pearton, S.J. (2013). *Materials and Reliability Handbook for Semiconductor Optical and Electron Devices*. New York: Springer-Verlag.
89. Holt, D.B. and Yacobi, B.G. (2007). *Extended defects in Semiconductors Electronic Properties, Device Effects and Structures*. Cambridge University Press.
90. Ayers, J.E. (2007). *Heteroepitaxy Of Semiconductors Theory, Growth and Characterization*. Taylor & Francis Group.
91. Mahajan, S. (2000). Defects in semiconductors and their effects. *Acta Materialia* 48: 137–149.
92. Waters, R.G. (1991). Diode laser degradation mechanisms: a review. *Progress in Quantum Electronics* 15 (3): 153.
93. Petroff, P. and Hartman, R.L. (1973). Defect structure introduced during operation of heterojunction GaAs lasers. *Applied Physics Letters* 23 (8): 469–471.
94. Petroff, P. and Hartman, R.L. (1974). Rapid degradation phenomenon in heterojunction GaAlAs–GaAs lasers. *Journal of Applied Physics* 45 (9): 3899–3903.
95. Ueda, O. (1982). Degradation of I II-V opto-electronic devices. *Journal of the Electrochemical Society* 135 (1): 11–22.
96. Hasegawa, Y., Egawa, T., Jimbo, T., and Umeno, M. (1995). Influences of dark line defects on characteristics of AlGaAs/GaAs quantum well lasers grown on Si substrates. *Japanese Journal of Applied Physics* 34 (6R): 2994–2999.
97. Hutchinson, P.W. and Dobson, P.S. (1980). Climb asymmetry in degraded gallium arsenide lasers. *Philosophical Magazine A* 41 (4): 601–614.
98. O'Hara, S., Hutchinson, P.W., and Dobson, P.S. (1977). The origin of dislocation climb during laser operation. *Applied Physics Letters* 30 (8): 368–371.
99. Akiyama, M., Kawarada, Y., and Kaminishi, K. (1984). Growth of single domain GaAs layer on (100)-oriented Si substrate by MOCVD. *Japanese Journal of Applied Physics* 23 (Part 2, 11): L843–L845.
100. Brammertz, G. et al. (2006). Low-temperature photoluminescence study of thin epitaxial GaAs films on Ge substrates. *Journal of Applied Physics* 99: 9.
101. Alcotte, R. et al. (2016). Epitaxial growth of antiphase boundary free GaAs layer on 300 mm Si (001) substrate by metalorganic chemical vapour deposition with high mobility. *APL Materials* 4: 4.

102. France, R.M., Dimroth, F., Grassman, T.J., and King, R.R. (2016). Metamorphic epitaxy for multijunction solar cells. *MRS Bulletin*, 41 (3): 202–209.
103. Tournie, E., Cerutti, L., Rodriguez, J.B. et al. (2016). Metamorphic III-V semiconductor lasers grown on silicon. *MRS Bulletin*, 41 (3): 218–223.
104. Lee, K.E. and Fitzgerald, E.A. (2016). Metamorphic transistors: Building blocks for hetero-integrated circuits. *MRS Bulletin*, 41 (3): 210–217.
105. Kunert, B., Zinnkann, S., Volz, K., and Stolz, W. (2008). Monolithic integration of Ga (NAsP)/(BGa)P multi-quantum well structures on (001) silicon substrate by MOVPE. *Journal of Crystal Growth* 310 (23): 4776–4779.
106. Landolt-Börnstein (2002). *Group IV Elements, IV-IV and III-V Compounds*. Springer Berlin Heidelberg.
107. Geisz, J.F., Olson, J.M., Friedman, D.J. et al. (2005). Lattice-matched GaNPAs-on-silicon tandem solar cells. In: *Conference Record of the Thirty-first IEEE Photovoltaic Specialists Conference*, 695–698.
108. Yamane, K., Sato, K., Sekiguchi, H. et al. (2017). Doping control of GaAsPN alloys by molecular beam epitaxy for monolithic III-V/Si tandem solar cells. *Journal of Crystal Growth* 473: 55–59.
109. Deutsch, T.G., Koval, C.A., and Turner, J.A. (2006). III-V nitride epilayers for photoelectrochemical water splitting: GaPN and GaAsPN. *Journal of Physical Chemistry B* 110 (50): 25297–25307.
110. Döscher, H. et al. (2012). Epitaxial III-V films and surfaces for photoelectrocatalysis. *Chemphyschem* 13 (12): 2899–2909.
111. Kargar, A. et al. (2017). GaP/GaNP heterojunctions for efficient solar-driven water oxidation. *Small* 13 (21): 1–6.
112. Kunert, B., Volz, K., Koch, J., and Stolz, W. (2006). Direct-band-gap Ga (NAsP) -material system pseudomorphically grown on GaP substrate. *Applied Physics Letters* 88 (18): 182108.
113. Koukourakis, N. et al. (2012). High room-temperature optical gain in Ga (NAsP)/Si heterostructures. *Applied Physics Letters* 100 (9): 1–4.
114. Liebich, S. et al. (2011). Laser operation of Ga (NAsP) lattice-matched to (001) silicon substrate. *Applied Physics Letters* 99 (7): 2011–2014.
115. Geisz, J.F. et al. (2001). Epitaxial growth of BGaAs and BGaInAs by MOCVD. *Journal of Crystal Growth* 225 (2–4): 372–376.
116. Sommer, N. et al. (2013). Growth of (BGa)As, (BGa)P, (BGa)(AsP) and (BGaIn)P by MOVPE. *Journal of Crystal Growth* 370, 191–196.
117. Leibiger, G., Krahmer, C., Bauer, J. et al. (2004). Solar cells with (BGaIn)As and (InGa) NAs) as absorption layers. *Journal of Crystal Growth* 272 (1–4): 732–738.
118. Kunert, B., Volz, K., Koch, J., and Stolz, W. (2007). MOVPE growth conditions of the novel direct band gap, diluted nitride Ga(NAsP) material system pseudomorphically strained on GaP-substrate. *Journal of Crystal Growth* 298, 121–125.
119. Ayse, E., (Ed.) (2008). *Dilute III-V Nitride Semiconductors and Material Systems*. Springer Berlin Heidelberg.
120. Ostheim, L., Klar, P.J. Liebich, S. et al. (2016). Interplay of boron localized states and electron transport in $B_xGa_{1-x}As_{0.11}P_{0.89}$:Te. *Semiconductor Science and Technology* 31 (7): 07LT01.
121. Odnoblyudov, V.A. and Tu, C.W. (2006). Growth and fabrication of InGaNP-based yellow-red light emitting diodes. *Applied Physics Letters* 89 (19): 2004–2007.
122. Nattermann, L., Beyer, A., Ludewig, P. et al. (2017). MOVPE growth of Ga(PBi) on GaP and GaP on Si with Bi fractions up to 8%. *Journal of Crystal Growth* 463, 151–155.
123. Yamaguchi, M. (1991). Dislocation density reduction in heteroepitaxial III-V compound films on Si substrates for optical devices. *Journal of Materials Research* 6 (2): 376–384.
124. Ayers, J.E. (1992). Post-growth thermal annealing of GaAs on Si (001) grown by organometallic vapor phase epitaxy. *Journal of Crystal Growth* 125 (1–2): 329–335.
125. Wang, G. et al. (2009). A model of threading dislocation density in strain-relaxed Ge and GaAs epitaxial films on Si(100). *Applied Physics Letters* 94: 10.
126. Mori, M.J., Boles, S.T., and Fitzgerald, E.A. (2010). Comparison of compressive and tensile relaxed composition-graded GaAsP and (Al) InGaP substrates. *Journal of Vacuum Science and Technology A Vacuum, Surfaces, Films* 28 (2): 182–188.
127. Lee, K.E. and Fitzgerald, E.A. (2010). High-quality metamorphic compositionally graded InGaAs buffers. *Journal of Crystal Growth* 312 (2): 250–257.
128. Mols, Y., Kunert, B., Gaudin, G. et al. (2016). Study toward integration of $In_{0.53}Ga_{0.47}As$ on 300 mm Si for CMOS sub-7 nm node: Development of thin graded $In_xGa_{1-x}As$ buffers on GaAs. *Journal of Crystal Growth* 452: 244–247.
129. Richardson, C.J.K. and Lee, M.L. (2016). Metamorphic epitaxial materials. *MRS Bulletin* 41 (3): 193–198.
130. Groenert, M.E. et al. (2003). Monolithic integration of room-temperature cw GaAs/AlGaAs lasers on Si substrates via relaxed graded GeSi buffer layers. *Journal of Applied Physics* 93 (1): 362–367.
131. Chen, S. et al. (2016). Electrically pumped continuous-wave III–V quantum dot lasers on silicon III–V quantum dot lasers on silicon. *Nature Photonics* 10 (5): 307–311.

132. Qian, W., Skowronski, M., and Kaspi, R. (1997). Dislocation density reduction in GaSb films grown on GaAs substrates by molecular beam epitaxy. *Journal of the Electrochemical Society* 144 (4): 1430–1434.
133. Nozawa, K. and Horikoshi, Y. (1992). Low threading dislocation density GaAs on Si (100) with InGaAs/GaAs strained-layer superlattice grown by migration-enhanced epitaxy. *Journal of Electronic Materials* 21 (6): 641–645.
134. Sakai, S., Soga, T., Takeyasu, M., and Umeno, M. (1986). Room-temperature laser operation of AlGaAs/GaAs double heterostructures fabricated on Si substrates by metalorganic chemical vapor deposition. *Applied Physics Letters* 48 (6): 413.
135. Tang, M. et al. (2016). Optimizations of defect filter layers for 1.3-μm InAs/GaAs quantum-dot lasers monolithically grown on Si substrates. *IEEE Journal of Selected Topics in Quantum Electronics* 22 (6): 1900207.
136. Gourley, P.L., Drummond, T.J., and Doyle, B.L. (1986). Dislocation filtering in semiconductor superlattices with lattice-matched and lattice-mismatched layer materials. *Applied Physics Letters* 49 (17): 1101–1103.
137. Yang, J., Bhattacharya, P., and Mi, Z. (2007). High-performance $In_{0.5}Ga_{0.5}As$/GaAs quantum-dot lasers on silicon with multiple-layer quantum-dot dislocation filters. *IEEE Transactions on Electron Devices* 54 (11): 2849–2855.
138. Bhattacharya, P., Mi, Z., Yang, J. et al. (2009). Quantum dot lasers: From promise to high-performance devices. *Journal of Crystal Growth* 311 (7): 1625–1631.
139. Huang, D., Reshchikov, M.A., Yun, F. et al. (2002). Defect reduction with quantum dots in GaN grown on sapphire substrates by molecular beam epitaxy. *Applied Physics Letters* 80 (2): 216–218.
140. Mi, B.Z., Yang, J., Bhattacharya, P., and Qin, G. (2009). High-performance quantum dot lasers and integrated optoelectronics on Si. *Proceedings of the IEEE* 97 (7): 1239–1249.
141. Lacombe, D., Ponchet, A., Gerard, J.M., and Cabrol, O. (1997). Structural study of InAs quantum boxes grown by molecular beam epitaxy on a (001) GaAs-on-Si substrate. *Applied Physics Letters* 70 (18): 2398–2400.
142. Huang, S.H., Balakrishnan, G., Khoshakhlagh, A. et al. (2008). Simultaneous interfacial misfit array formation and antiphase domain suppression on miscut silicon substrate. *Applied Physics Letters* 93: 7.
143. Kim, Y.H., Noh, Y.K., Kim, M.D. et al. (2010). Transmission electron microscopy study of the initial growth stage of GaSb grown on Si(001) substrate by molecular beam epitaxy method. *Thin Solid Films* 518 (8): 2280–2284.
144. Huang, S.H., Balakrishnan, G., Khoshakhlagh, A. et al. (2006). Strain relief by periodic misfit arrays for low defect density GaSb on GaAs. *Applied Physics Letters* 88 (13): 131911.
145. Wang, Y., Ruterana, P., Chen, J. et al. (2012). Strain relief and growth optimization of GaSb on GaP by molecular beam epitaxy. *Journal of Physics Condensed Matter* 24 (33): 335802.
146. Desplanque, L. et al. (2012). Monolithic integration of high electron mobility InAs-based heterostructure on exact (001) silicon using a GaSb/GaP accommodation layer. *Applied Physics Letters* 101 (14): 142111.
147. Murray, R.T., Kiely, C.J., and Hopkinson, M. (2000). General characteristics of crack arrays in epilayers grown under tensile strain. *Semiconductor Science and Technology* 15 (4): 325–330.
148. Yang, V.K., Groenert, M., Leitz, C.W. et al. (2003). Crack formation in GaAs heteroepitaxial films on Si and SiGe virtual substrates. *Journal of Applied Physics* 93 (7): 3859–3865.
149. Geisz, J.F. and Friedman, D.J. (2002). III N V semiconductors for solar photovoltaic applications. *Semiconductor Science and Technology* 17 (8): 769–777.
150. Yaung, K.N., Vaisman, M., Lang, J.,and Lee, M.L. (2016). GaAsP solar cells on GaP/Si with low threading dislocation density. *Applied Physics Letters* 109: 3.
151. Yamaguchi, M. and Amano, C. (1985). Efficiency calculations of thin-film GaAs solar cells on Si substrates. *Journal of Applied Physics* 58 (9): 3601–3606.
152. Andre, C.L., Wilt, D.M., Pitera, A.J. et al. (2005). Impact of dislocation densities on n +/p and p +/n junction GaAs diodes and solar cells on SiGe virtual substrates. *Journal of Applied Physics* 98: 1.
153. Grassman, T.J., Chmielewski, D.J., Carnevale, S.D. et al. (2016). $GaAs_{0.75}P_{0.25}$/Si dual-junction solar cells grown by MBE and MOCVD. *IEEE Journal of Photovoltaics*, 6 (1): 326–331.
154. Vaisman, M. et al. (2017). 15.3%-efficient GaAsP solar cells on GaP/Si templates. *ACS Energy Letters* acsenergylett.7b00538.
155. Jain, N. and Hudait, M.K. (2014). III–V Multijunction solar cell integration with silicon: Present status, challenges and future outlook. *Energy Harvesting Systems* 1 (3–4): 121–145.
156. Yamaguchi, M., Lee, K.-H., Araki, K. et al. (2016). Potential and activities of III-V/Si tandem solar cells. *ECS Journal of Solid State Science and Technology* 5 (2): Q68–Q73.
157. Zhou, X., Li, Q., Member, S. et al. (2012). 30-nm inverted $In_{0.53}Ga_{0.47}As$ MOSHEMTs on Si substrate grown by MOCVD with regrown source/drain. *IEEE Electron Device Letters* 33 (10): 1384–1386.
158. Li, Q., Zhou, X., Tang, C.W., and Lau, K.M. (2013). Material and device characteristics of metamorphic $In_{0.53}Ga_{0.47}As$ MOSHEMTs grown on GaAs and Si substrates by MOCVD. *IEEE Transactions on Electron Devices* 60 (12): 4112–4118.
159. Huang, M.L. et al. (2016). High performance $In_{0.}$53Ga0.47As FinFETs fabricated on 300 mm Si substrate. *Symposium on VLSI Technology*, 16–17.

160. Chen, H.Z., Ghaffari, A., Wang, H. et al. (1987). Continuous-wave operation of extremely low-threshold GaAs/AlGaAs broad-area injection lasers on (100) Si substrates at room temperature. *Optical Letters* 12 (10): 812–813.
161. Sugo, M., Mori, H., Sakai, Y., and Itoh, Y. (1992). Stable cw operation at room temperature of a 1.5-μm wavelength multiple quantum well laser on a Si substrate. *Applied Physics Letters* 60 (4): 472–473.
162. Yamada, T.,Tachikawa, M.,and Sasaki, T. (1997). 7000 h continuous wave operation of multiple quantum well laser on Si at 50°C. *Applied Physics Letters* 70: 1614.
163. Jung, D., Song, Y., Lee, M. et al. (2014). InGaAs/GaAs quantum well lasers grown on exact GaP/Si(001). *Electronics Letters* 50 (17): 1226–1227.
164. Rodriguez, J.B., Cerutti, L., Grech, P., and Tourní, E. (2009). Room-temperature operation of a 2.25 μm electrically pumped laser fabricated on a silicon substrate. *Applied Physics Letters* 94 (6): 10–12.
165. Cerutti, L., Rodriguez, J.B., and Tournie, E. (2010). GaSb-based laser, monolithically grown on silicon substrate, emitting at 1.55 μm at room temperature. *IEEE Photonics Technology Letters* 22 (8): 553–555.
166. Reboul, J.R., Cerutti, L., Rodriguez, J.B. et al. (2011). Continuous-wave operation above room temperature of GaSb-based laser diodes grown on Si. *Applied Physics Letters* 99 (12): 2009–2012.
167. Castellano, A. et al. (2017). Room-temperature continuous-wave operation in the telecom wavelength range of GaSb-based lasers monolithically grown on Si. *APL Photonics* 61301 (2): 61301.
168. Kazi, Z.I., Egawa, T., Umeno, M.,and Jimbo, T. (2001). Growth of $In_xGa_{1-x}As$ quantum dots by metal-organic chemical vapor deposition on Si substrates and in GaAs-based lasers. *Journal of Applied Physics* 90 (11): 5463–5468.
169. Liu, A.Y., Herrick, R.W., Ueda, O. et al. (2015). Reliability of InAs/GaAs quantum dot lasers epitaxially grown on silicon. *IEEE Journal of Selected Topics in Quantum Electronics* 21 (6): 1–8.
170. Jung, D., Herrick, R., Norman, J. et al. (2018). Impact of threading dislocation density on the lifetime of InAs quantum dot lasers on Si, *Applied Physics Letters* 112 (15): 153507
171. Liu, A.Y. et al. (2017). Electrically pumped continuous-wave 13 μm quantum-dot lasers epitaxially grown on on-axis (001) GaP/Si. *Optical Letters* 42 (2): 338–341.
172. Norman, J. et al. (2017). Electrically pumped continuous wave quantum dot lasers epitaxially grown on patterned, on-axis (001)Si. *Optical Express*, 25 (4): 3927.
173. Liao, M. et al. (2017). Monolithically integrated electrically pumped continuous-wave III-V quantum dot light sources on silicon. *IEEE Journal of Selected Topics in Quantum Electronics* 23 (6): 1–1.
174. Li, Q. et al. (2016). 1.3 μm InAs quantum-dot micro-disk lasers on V-groove patterned and unpatterned (001) silicon. *Optical Express*, 24 (18): 21038–21045.
175. Scaccabarozzi, A. et al. (2016). Integration of InGaP/GaAs/Ge triple-junction solar cells on deeply patterned silicon substrates. *Progress in Photovoltaics: Research and Applications* 24: 1368–1377.
176. Fitzgerald, E.A., Chand, N., Laboratories, T.B.,and Hill, M. (1991). Epitaxial necking in GaAs grown on pre-patterned Si substrates. *Journal of Electronic Materials* 20 (10): 839–853.
177. Krost, A., Grundmann, M., and Bimberg, D. (1992). InP on patterned Si(001): defect reduction by application of the necking mechanism. *Journal of Crystal Growth* 124 (1-4): 207–212.
178. Langdo, T.A., Leitz, C.W., Currie, M.T. et al. (2000). High quality Ge on Si by epitaxial necking. *Applied Physics Letters* 76 (25): 3700–3702.
179. Park, J.S., Bai, J., Curtin, M. et al. (2007). Defect reduction of selective Ge epitaxy in trenches on Si (001) substrates using aspect ratio trapping. *Applied Physics Letters* 90 (5): 43–46.
180. Li, J.Z. et al. (2007). Defect reduction of GaAs epitaxy on Si(001) using selective aspect ratio trapping. *Applied Physics Letters* 91 (2): 1–4.
181. Cipro, R. et al. (2014). Low defect InGaAs quantum well selectively grown by metal organic chemical vapor deposition on Si(100) 300 mm wafers for next generation non planar devices. *Applied Physics Letters* 104 (26): 0–5.
182. Wang, G. et al. (2011). Growth of high quality InP layers in STI trenches on miscut Si(001) substrates. *Journal of Crystal Growth* 315 (1): 32–36.
183. Wang, G. et al. (2010). Selective area growth of high quality InP on Si(001) substrates. *Applied Physics Letters* 97 (12): 1–4.
184. Zandvliet, H.J.W. (2003). The Ge (001) surface. *Physics Reports*, 388 (1): 1–40.
185. McMahon, W. and Olson, J. (1999). Atomic-resolution study of steps and ridges on arsine-exposed vicinal Ge (100). *Physical Review B* 60 (4): 2480–2487.
186. Waldron, N. et al. (2016). Replacement fin processing for III-V on Si: From FinFets to nanowires. *Solid State Electronics* 115: 81–91.
187. Krost, A., Heinrichsdorff, F., Bimberg, D., and Cerva, H. (1994). InP on Si(111): Accommodation of lattice mismatch and structural properties. *Applied Physics Letters* 64 (6): 769–771.
188. Krost, A., Schnabel, R.F., Heinrichsdorff, F. et al. (1994). Defect reduction in GaAs and InP grown on planar Si(111) and on patterned Si(001) substrates. *Journal of Crystal Growth* 145 (1–4): 314–320.

189. Merckling, C. et al. (2014). Heteroepitaxy of InP on Si (001) by selective-area metal organic vapor-phase epitaxy in sub-50 nm width trenches: The role of the nucleation layer and the recess engineering. *Journal of Applied Physics* 115: 2.
190. Li, Q., Ng, K.W., and Lau, K.M. (2015). Growing antiphase-domain-free GaAs thin films out of highly ordered planar nanowire arrays on exact (001)silicon. *Applied Physics Letters* 106 (7): 2–6.
191. Takeuchi, S., Suzuki, K., Maeda, K., and Iwanaga, H. (1985). Stacking-fault energy of II-VI compounds. *Philosophical Magazine A* 50 (2): 171–178.
192. Orzali, T. et al. (2015). GaAs on Si epitaxy by aspect ratio trapping: Analysis and reduction of defects propagating along the trench direction. *Journal of Applied Physics* 118: 10.
193. Guo, W. et al. (2017). Anisotropic relaxation behavior of InGaAs/GaAs selectively grown in narrow trenches on (001) Si substrates. *Journal of Applied Physics* 122 (2): 25303.
194. Kunert, B. et al. (2016). III/V nano ridge structures for optical applications on patterned 300 mm silicon substrate. *Applied Physics Letters* 109: 9.
195. Ismail, K., Legoues, F., Karam, N.H. et al. (1991). High-quality GaAs on sawtooth-patterned Si substrates. *Applied Physics Letters* 59 (19): 2418–2420.
196. Waldron, N. et al. (2014). An InGaAs/InP quantum well finfet using the replacement fin process integrated in an RMG flow on 300 mm Si substrates. *Digest Technical Papers - Symposium on VLSI Technology*, 32–33.
197. Waldron, N. et al. (2016). Gate-all-around InGaAs nanowire FETS with peak transconductance of 2200 S/m at 50 nm Lg using a replacement Fin RMG flow. *Technical Digest - International Electron Devices Meeting IEDM*, 3111–3114.
198. Zhou, X. et al. (2016). Scalability of InGaAs gate-all-around FET integrated on 300 mm Si platform: Demonstration of channel width down to 7 nm and Lg down to 36 nm. *Digest Technical Papers - Symposium on VLSI Technology*, 166–167.
199. Doornbos, G. et al. (2016). High-performance InAs Gate-All-around nanowire MOSFETs on 300 mm Si substrates. *Journal of the Electron Devices Society* 4 (5): 253–259.
200. Sun, X. et al. (2017). T3-4 high performance and low leakage current InGaAs-on-silicon FinFETs with 20 nm gate length. *2017 IEEE Symposium on VLSI Technology*, Kyoto, T40–T41. DOI: 10.23919/VLSIT.2017.7998193.
201. Vert, A., Orzali, T., Satyavolu, P. et al. (2016). Integration of InP and InGaAs on 300 mm Si wafers using chemical mechanical planarization. *ECS Journal of Solid State Science and Technology* 5 (9): P478–P482.
202. Lee, S.-M. et al. (2015). Effects of growth temperature on surface morphology of InP grown on patterned Si(001) substrates. *Journal of Crystal Growth* 416 (1): 113–117.
203. Hsu, C.-W., Chen, Y.-F., and Su, Y.-K. (2011). Nano epitaxial growth of GaAs on Si(001). *Applied Physics Letters* 99 (13): 133115.
204. Wang, Z. et al. (2015). Room-temperature InP distributed feedback laser array directly grown on silicon. *Nature Photonics* 9 (12): 837–842.
205. Tian, B. et al. (2017). Room temperature O-band DFB laser array directly grown on (001) silicon. *Nano Letters* 17 (1): 559–564.
206. Han, Y., Li, Q., and Lau, K.M. (2016). Highly ordered horizontal indium gallium arsenide/indium phosphide multi quantum-well in wire structure on (001) silicon substrates. *Journal of Applied Physics* 120: 24.
207. Li, S. et al. (2016). Ridge InGaAs/InP multi-quantum-well selective growth in nanoscale trenches on Si (001) substrate. *Applied Physics Letters* 108 (2): 1–5.
208. Megalini, L. et al. (2017). 1550-nm InGaAsP multi-quantum-well structures selectively grown on v-groove-patterned SOI substrates. *Applied Physics Letters* 111 (3): 32105.
209. Fahed, M. et al. (2015). Impact of P/In flux ratio and epilayer thickness on faceting for nanoscale selective area growth of InP by molecular beam epitaxy. *Nanotechnology* 26 (29): 295301.
210. Gil-Lafon, E., Napierala, J., Castelluci, D. et al. (2001). Selective growth of GaAs by HVPE: Keys for accurate control of the growth morphologies. *Journal of Crystal Growth* 222 (3): 482–496.
211. Sun, Y.T., Rodríguez Messmer, E., Söderström, D. et al. (2001). Temporally resolved selective area growth of InP in the openings off-oriented from 110 direction. *Journal of Crystal Growth* 225 (1): 9–15.
212. Jones, S.H., Salinas, L.S., Jones, J.R., and Mayer, K. (1995). Crystallographic orientation dependence of the growth rate for GaAs low pressure organometallic vapor phase epitaxy. *Journal of Electronic Materials* 24 (1): 5–14.
213. Kunert, B., Guo, W., Mols, Y. et al. (2016). Integration of III/V hetero-structures by selective area growth on Si for nano- and optoelectronic. *ECS Transactions* 75 (8): 409–419.
214. Shi, Y., Wang, Z., Van Campenhout, J. et al. (2016). Optical pumped InGaAs/GaAs nano-ridge laser epitaxially grown on a standard 300-mm Si wafer. *Optica* 4 (12): 1468–1473.
215 Li, Q., Han, Y., Lu, X, and Lau, K.M. (2016). GaAs-InGaAs-GaAs fin-array tunnel diodes on (001) Si substrates with room temperature peak-to-valley current ratio of 5.4. *IEEE Electron Device Letters* 37 (1): 24–27.

216. Hsu, C.W., Chen, Y.F., and Su, Y.K. (2012). Dislocation reduction of InAs nanofins prepared on Si substrate using metal-organic vapor-phase epitaxy. *Nanoscale Research Letters* 7 (1): 642.
217. Orzali, T. et al. (2016). Epitaxial growth of GaSb and InAs fins on 300 mm Si (001) by aspect ratio trapping, *Journal of Applied Physics* 120 (8): 1–6.
218. Mols, Y. et al. (2017). Low-defect density GaSb and InAs selective area growth for vertical (T) FET on 300 mm Si. in *MRS 2017 Fall Meeting*.
219. Li, Q., Lai, B.,and Lau, K.M. (2017). Epitaxial growth of GaSb on V-grooved Si(001) substrates with an ultrathin GaAs stress relaxing layer. *Applied Physics Letters* 111 (17): 172103.
220. Lu, W. et al. (2017). Alcohol-based digital etch for III – V vertical nanowires with sub-10 nm diameter. *IEEE Electron Device Letters* 38 (5): 548–551.
221. Zhao, X. and Del Alamo, J.A. (2014). Nanometer-scale vertical-sidewall reactive ion etching of InGaAs for 3-D III-V MOSFETs. *IEEE Electron Device Letters* 35 (5): 521–523.
222. Borg, M. et al. (2014). Vertical III–V nanowire device integration on Si(100). *Nano Letters* 14 (4): 1914–1920.
223. Das Kanungo, P. et al. (2013). Selective area growth of III–V nanowires and their heterostructures on silicon in a nanotube template: toward monolithic integration of nano-devices. *Nanotechnology* 24 (22): 225304.
224. Borg, M., Schmid, H., Moselund, K.E. et al. (2015). Mechanisms of template-assisted selective epitaxy of InAs nanowires on Si. *Journal of Applied Physics* 117: 14.
225. Cutaia, D., Moselund, K.E., Member, S., and Member, M.B. (2015). Vertical InAs-Si gate-all-around tunnel FETs integrated on Si using selective epitaxy in, *IEEE Journal of the Electron Devices Society* 3 (3): 176–183.
226. Schmid, H. et al. (2015). Template-assisted selective epitaxy of III-V nanoscale devices for co-planar heterogeneous integration with Si. *Applied Physics Letters* 106: 23.
227. Cutaia, D., Moselund, K.E., Schmid, H. et al. (2016). Complementary III-V heterojunction lateral NW tunnel FET technology on Si. *IEEE Symposium on VLSI Technology* Honolulu, 1–2.
228. Schmid, H. et al. (2017). Monolithic integration of multiple III-V semiconductors on Si for MOSFETs and TFETs. *Technical Digest - International Electron Devices Meeting IEDM*, 3.6.1–3.6.4.
229. Borg, M. et al. (2017). High-mobility GaSb nanostructures cointegrated with InAs on Si. *ACS Nano*, 11 (3): 2554–2560.
230. Ujie, Y. and Nishinaga, T. (1989). Epitaxial lateral overgrowth of GaAs on a Si. *Japanese Journal of Applied Physics* 28: 337.
231. Naritsuka, S., Nishinaga, T.,Tachikawa, M., and Mori, H. (1995). InP layer grown on (001) silicon substrate by epitaxial lateral overgrowth. *Japanese Journal of Applied Physics* 34 (11): L1432–L1435.
232. Nishinaga, T. (2002). Microchannel epitaxy: An overview. *Journal of Crystal Growth* 237–239 (1–4 II), 1410–1417.
233. Zytkiewicz, Z.R. (2002). Laterally overgrown structures as substrates for lattice mismatched epitaxy. *Thin Solid Films* 412 (1–2): 64–75.
234. Hersee, S.D. et al. (2002). Nanoheteroepitaxy for the integration of highly mismatched semiconductor materials. *IEEE Journal of Quantum Electronics* 38 (8): 1017–1028.
235. Hsu, C.-W., Chen, Y.-F., and Su, Y.-K. (2012). Nanoepitaxy of GaAs on a Si(001) substrate using a round-hole nano-patterned SiO_2 mask. *Nanotechnology* 23 (49): 495306.
236. Nakamura, Y., Miwa, T., and Ichikawa, M. (2011). Nanocontact heteroepitaxy of thin GaSb and AlGaSb films on Si substrates using ultrahigh-density nanodot seeds. *Nanotechnology* 22 (26): 265301.
237. Li, J.Z., Bai, J., Major, C. et al. (2008). Defect reduction of GaAs/Si epitaxy by aspect ratio trapping. *Journal of Applied Physics* 103 (10): 2006–2009.
238. He, Y., Wang, J., Hu, H. et al. (2015). Coalescence of GaAs on (001) Si nano-trenches based on three-stage epitaxial lateral overgrowth. *Applied Physics Letters* 106, 20.
239. Junesand, C. et al. (2014). Defect reduction in heteroepitaxial InP on Si by epitaxial lateral overgrowth. *Materials Express* 4 (1): 41–53.
240. Junesand, C. et al. (2013). Study of planar defect filtering in InP grown on Si by epitaxial lateral overgrowth. *Optical Materials Express* 3 (11): 1960–1973.
241. Julian, N.H., Mages, P.A., Zhang, C., and Bowers, J.E. (2014). Improvements in epitaxial lateral overgrowth of InP by MOVPE. *Journal of Crystal Growth* 402: 234–242.
242. Wang, Z., Junesand, C., Metaferia, W. et al. (2012). III-Vs on Si for photonic applications - A monolithic approach. *Material Science and Engineering B Solid-State Materials and Advanced Technology* 177 (17): 1551–1557.
243. Kataria, H. et al. (2013). Toward a monolithically integrated III-V laser on silicon: optimization of multi-quantum well growth on InP on Si. *Semiconductor Science and Technology* 28 (9): 094008.
244. Sun, Y.-T., Kataria, H., Metaferia, W., and Lourdudoss, S. (2014). Realization of an atomically abrupt InP/Si heterojunction via corrugated epitaxial lateral overgrowth. *CrystEngComm* 16 (34): 7889.
245. Omanakuttan, G., Stergiakis, S., Sahgal, A. et al. (2017). Epitaxial lateral overgrowth of $Ga_x In_{1-x}P$ toward direct $Ga_xIn_{1-x}P$/Si heterojunction. *Physica Status Solidi* 1600631 (3): 1600631.

246. Parillaud, O., Gil-Lafon, E., Gérard, B. et al. (1996). High quality InP on Si by conformal growth. *Applied Physics Letters* 68 (19): 2654–2656.
247. Czornomaz, L. et al. (2015). Confined epitaxial lateral overgrowth (CELO) : A novel concept for scalable integration of CMOS-compatible InGaAs-on-insulator MOSFETs on large-area Si substrates. *Digest Technical Papers - Symposium on VLSI Technology,* 172–T173.
248. Wirths, S. et al. (2017). *Room Temperature Lasing from Monolithically Integrated GaAs Microdisks on Si*, in CLEO/Europe-EQEC.
249. Kazi, Z.I., Thilakan, P., Egawa, T. et al. (2001). Realization of GaAs/AlGaAs lasers on Si substrates using epitaxial lateral overgrowth by metalorganic chemical vapor deposition. *Japanese Journal of Applied Physics* 40: 4903–4906.
250. Knoedler, M. et al. (2017). Observation of twin-free GaAs nanowire growth using template-assisted selective epitaxy, *Crystal Growth and Design* 17: 6297–6302.

9

MOVPE Growth of Cadmium Mercury Telluride and Applications

C.D. Maxey[1], P. Capper[2], and I.M. Baker[1]
[1]Leonardo MW Ltd, Southampton, UK
[2]Ex-Leonardo MW Ltd, Southampton, UK

9.1 Requirement for Epitaxy

Mercury cadmium telluride (MCT, $Cd_xHg_{1-x}Te$) remains the preeminent material for many infrared (IR) applications due to its high sensitivity over a wide spectral range (0.8–20 μm) [1]. The key component in any IR system is the focal plane array sensor responsible for converting the incident IR photons into an electrical signal that can be transferred into a readout circuit and subsequent image processing. Therefore, as system performance improvements have been sought, great emphasis has been placed on improving the performance of the infrared focal plane arrays (IRFPAs). Such improvements range from improved sensitivity and resolution (both spatial and spectral) to lower manufacturing and operating costs.

For these reasons, work over the last 35 years has concentrated on developing epitaxial growth techniques such as liquid phase (LPE), metal-organic vapor phase (MOVPE), and molecular beam epitaxy (MBE) and moving away from the use of bulk-grown MCT. This materials system is extremely challenging, and it presents many different problems to each growth technique.

Epitaxial techniques not only remove many of the hazards associated with the high-pressure processes required for conventional bulk growth but also improve diode performance routes by offering increased area and improved crystalline quality, alloy composition, and dopant and thickness control, and by enabling the use of semiconductor manufacturing processing methods. The ability to control composition, thickness, and dopants within the growth facilitates device-engineering opportunities such that heterostructure diode designs can be realized. For example, these can improve sensitivity by use of electron avalanche photodiodes (eAPD), spectral resolution through multispectral detector device designs, and spatial resolution through confined sensitive area pixels and larger-format arrays.

Metalorganic Vapor Phase Epitaxy (MOVPE): Growth, Materials Properties, and Applications,
First Edition. Edited by Stuart Irvine and Peter Capper.

Ultimately, lower manufacturing costs arise through the use of large-area growth on cheaper substrate materials, and lower operating costs arise from higher operating temperature devices using heterostructure device designs. However, achieving all of these advantages in a single array has remained a formidable challenge. The various material technologies available offer different potential advantages in several combinations of these criteria. At Leonardo MW Ltd., MOVPE technology is now used for all the IRFPA applications sensor formats; see Sections 9.13–9.17.

9.2 History

Research into the use of MOVPE for MCT began in the early 1980s in many centers around the world following the success that the process achieved in the growth of a variety of III/V compounds. This early work has been reviewed previously [2–4]. In this chapter, the current status of MOVPE growth will be described, along with the key factors that need to be considered when using this growth technique. Many of the key parameters are inherently linked, which can make deconvoluting the effect of any process change very complex.

A key breakthrough in the use of MOVPE as a growth technique for MCT production came with the invention of the interdiffused multilayer process (IMP) in the early 1980s at the Royal Signals and Radar Establishment (RSRE) [5]. This process takes advantage of the high diffusion coefficients of HgTe and CdTe to enable MCT to be grown by depositing alternating layers of these binary compounds sufficiently thin that after relatively short times, they interdiffuse to form homogeneous MCT (see Figure 9.1). As such, it enables the growth of each binary to be controlled and optimized separately for key parameters such as uniformity and alkyl ratio. It also removed the problem of controlling low Cd alkyl fluxes that were needed to maintain direct alloy growth (DAG), which inherently resulted in depletion issues in horizontal reactor design configurations. In addition, it meant that control of the Cd alloy fraction no longer required changes to the growth conditions as only changes to the relative thicknesses of the binary layers were needed. Of course, the control of the interface between the binary IMP pairs becomes very important in this growth method.

Equation (9.1) shows how, when using IMP, the relative thicknesses within the IMP-grown MCT are related to the ratio of growth rates and growth times. As such, with the stable binary growth rates, only changes in the relative binary growth times are required to control the composition, x, within a structure.

$$x = t_{\mathrm{CdTe}} / \left(t_{\mathrm{CdTe}} + t_{\mathrm{HgTe}} \right) = \left[\frac{\left(\tau_{\mathrm{CdTe}} / \tau_{\mathrm{HgTe}} \right)}{\left(G_{\mathrm{CdTe}} / G_{\mathrm{HgTe}} \right)} + 1 \right]^{-1} \tag{9.1}$$

where t is the binary-IMP thickness, τ is the binary-IMP growth time, and G is the binary growth rate.

This means that once a growth reactor is established, only the growth conditions of the CdTe and HgTe need to be maintained in order to grow complex alloy composition profiles in heterostructure device designs for any spectral range. This avoids having to operate MCT production from dedicated machines set up for the different IR spectral bands or device types with all the associated increased overheads, etc. This growth flexibility is an important aspect of MOVPE growth using the IMP technique.

The use of the IMP technique also has important ramifications for introducing and controlling dopants (see Section 9.9), as conditions within the HgTe and CdTe can be adjusted to improve the probability of correctly incorporating the dopant atoms. The position of any dopant-defined junction location can be modified with respect to the heterointerfaces purely by the choice of the IMP cycle during which the dopant switching is initiated or terminated, which is a major strength of the process.

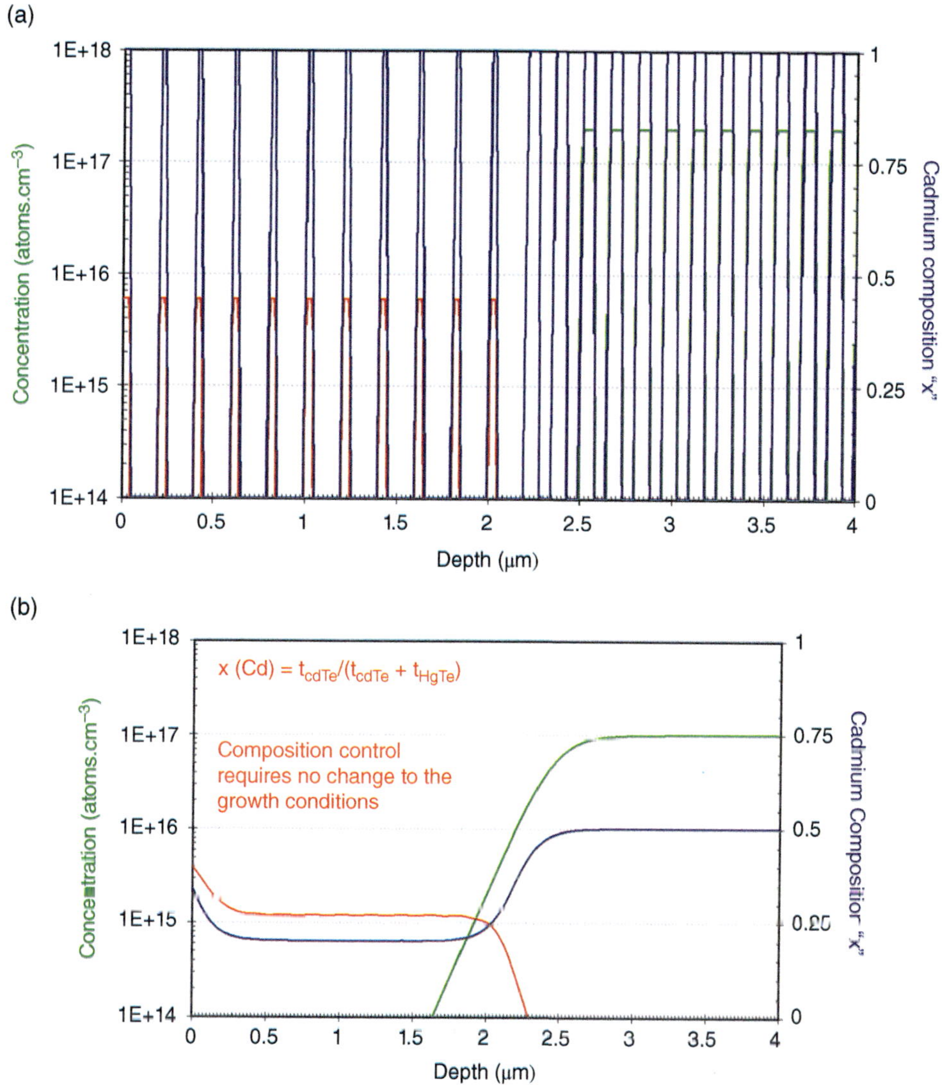

Figure 9.1 (a) Schematic of doped IMP growth of MCT growth before interdiffusion. (b) Schematic of doped IMP growth of MCT layers after interdiffusion. (Red = intentional donor impurity concentration; green = intentional acceptor impurity concentration.)

MOVPE has become the epitaxial growth technique of choice for third-generation MCT-based IR devices in the UK and Poland (at Vigo Systems S.A.), whereas in the United States, Germany, and France, MBE is favored.

9.3 Substrate Choices

In all epitaxial processes, by definition, growth is controlled by the choice of substrate crystal. Therefore, the substrate is the key component in determining growth quality. MOVPE employs a higher growth temperature for MCT growth than equivalent MBE systems, but in both cases, growth will be dominated by control over the initial nucleation conditions employed. In this section, the options for material orientation and type are described.

9.3.1 Orientation

Growth orientation plays a very strong role in determining nucleation conditions, growth rates, and dopant incorporation efficiency due to the different step energies involved [6]. (111)B orientations enable faster growth rates and generate layers with suppressed macrodefect densities. However, on (111)B, the CdTe growing within IMP layers will generate twin boundaries, and it was shown that these resulting microtwins exhibit donor-like activity [7]. In addition, the As dopant incorporation rates were found to be suppressed by over an order of magnitude when AsH_3 was used as the doping source [8–10]; conversely, iodine incorporation is enhanced by several orders of magnitude [11].

Conversely, the lowest growth rates are obtained on substrates with (100) orientation. However, it is necessary to grow on substrates that are misoriented from the (100) plane by a few degrees to suppress the formation of pyramid-shaped macrodefects (known as hillocks). Section 9.10 describes the macrodefect types in more detail. The choice of orientation becomes a balance between the doping, the macrodefect control and the background morphologies that can be achieved in the reactor system and in the UK the preferred orientation for MOVPE growth is still misoriented (100). However, the (111)B orientation is used by Polish workers for their optical immersion high operating temperature device fabrication as this gives lower macrodefect densities in their process. The background roughness and residual background doping concentrations, however, are shown to be significantly worse on (111)B orientations so work has continued on trying to optimize the growth on (100) orientation as this gives improved device performance [12–18].

Alternatives like (211)B and (552)B have also been explored to try to achieve the compromise of both well-behaved growth, dopant incorporation, and macrodefects while still suppressing microtwin formation [13, 19, 20]. It was found that As incorporation was slightly reduced on these orientations but less severely than on (111)B. However, the iodine dopant incorporation was still found to be several orders higher on the (211)B orientations, which although not an insurmountable issue, makes the reproducible control of low doping concentrations much more difficult. This would require either more extensive dilution systems in the gas train, lower vapor pressure sources, or low alkyl source temperatures [11, 21]. On CdZnTe substrates, (211)B was the preferred orientation for device-quality material in the late 1990s in the United States; but this was largely due to the issue of hillock density suppression, which is easier on this orientation, and the availability arising from this being the preferred MBE growth orientation.

9.3.2 Substrate Material

CdZnTe (CZT) should be the ideal substrate material for achieving lattice-matched growth by using a 2–4% ZnTe alloy fraction (depending on the target MCT composition). However, there are many problems associated with CZT substrates. The bulk-growth processes limit the size of commercially available substrates to 7×7.5 cm, and the difficulties of controlling precipitation and twinning effects in the growth result in a very high substrate cost per cm^2 that are ~50–100 times higher than alternatives like GaAs. The mechanical properties of CZT mean that is it very fragile, and this leads to an increased risk of losses during the handling required for substrate preparation, growth, and processing. While these issues are all common to both MBE and MOVPE, there are additional difficulties in MOVPE growth that make it a poor substrate choice.

First, the very tight Zn-alloy composition uniformity control required for true lattice matching in a LW device structure growth cannot be achieved in the preferred growth orientation using the bulk-growth processes. This is because bulk CZT growth is naturally oriented to the <111>

orientation, but segregation of the Zn along the boule length results in nonuniform zinc concentrations within substrates that are cut from the (100) planes.

Secondly, the MOVPE growth temperature is higher than that used in MBE. Consequently, this enhances the release of any residual Cu impurities from within the Te precipitates in the CZT substrates. The subsequent accumulation of Cu within the growing MCT layer leads to large run-to-run variations in the background residual doping because Cu is an active dopant species. This affects the ability to reproducibly control the low n-type properties required in many device designs. These complications, along with the high cost per unit area and fragility issues of the CZT substrates, forced the MOVPE growth community to seek alternative substrates.

In all cases, MCT growth onto alternative substrates requires intermediary buffers layers to manage the lattice mismatch, defect generation, and autodoping issues. These are usually produced using either ZnTe or CdTe.

Silicon is considered the ultimate substrate solution primarily due to its low cost, large wafer areas, and thermal matching to the readout integrated circuit (ROIC), but this option is very difficult to realize in an MOVPE system due to the problem of removing the thermal oxide in situ and maintaining the preferred (100)-type orientations. Various approaches have been reported using either high-temperature oxide desorption annealing [22], or seed layers of Ge [23, 24], GaAs [20, 27, 28], or ZnTe/CdTe [25, 26]. The use of GaAs on Si substrates came closest to being able to demonstrate IR detector performance; but the quality of the films achievable was still poorer than the bulk GaAs substrates [20, 27], and the distribution of the device performance characteristics was found to mirror this limitation [28].

GaAs has proven to be the most useful substrate for MOVPE growth of MCT. It is readily available in wafer sizes up to 150 mm diameter and is only marginally more expensive than Si. Studies of CdTe nucleation have been carried out to determine the thickness of CdTe required to both optimize the crystal quality and minimize the level of gallium out-diffusion, which can otherwise lead to autodoping of the subsequent MCT layers [13, 29, 30]. Once the technologically challenging issues of controlling the growth process were established, it has been shown that low defect density, device-quality MCT wafers can be produced [20].

In order to be able to use GaAs substrates for FPA production, differential thermal expansion between the GaAs substrate and the Si readout multiplexer chip also has to be addressed, but once resolved it enables the route for producing large-area FPAs and will be discussed in Sections 9.13–9.17.

9.4 Reactor Design

MOVPE reactor designs for MCT growth evolved from successful III/V systems and encompassed both vertical and horizontal concept configurations. The majority of work has been performed on horizontal systems similar to the schematic shown in Figure 9.2. Fundamentally, the growth process involves the introduction of controlled concentrations of either volatile metal-organic compounds (known as *alkyls*) or metal hydrides in a carrier gas (usually H_2) into a growth reactor chamber containing a substrate that is mounted on a temperature-controlled susceptor (usually made from high-purity graphite or SiC). The susceptor design normally has an associated rotation mechanism to improve the deposition uniformity when large-area substrates are used. The gas concentrations are controlled by the relative carrier gas flows (by the use of mass flow controllers [MFCs]), the alkyl source temperature (by the use of thermostatic baths), and the pressure within the bubblers (by mass flow pressure controllers [MFPCs]). Improvements in vapor-pressure control have also been reported by using Piezocon controllers that use direct ultrasonic precursor concentration sensors [13].

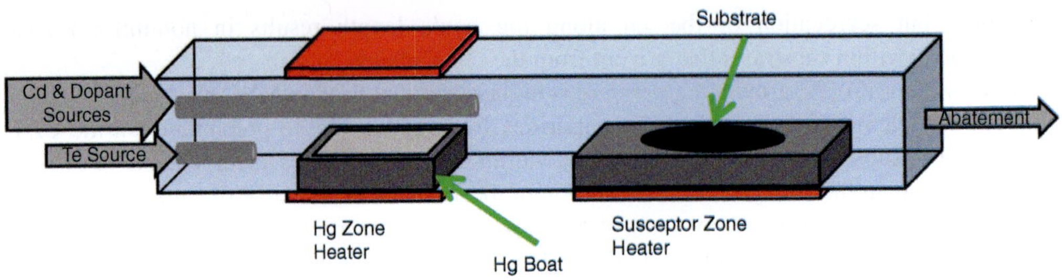

Figure 9.2 *Schematic of MOVPE cell for MCT growth.*

The main complication in moving from the III/V compounds to Hg-based II/VI compounds that has to be addressed in the reactor design arises from the high Hg overpressure control that is required during the MCT growth due to phase-diagram considerations. Metal-organic Hg compounds are highly toxic and, due to their relatively low vapor pressure/pyrolysis temperature characteristics, cannot provide this Hg overpressure. The solution to this problem has been to supply Hg from an elemental source heated to the required temperature to yield the overpressures required. This, in turn, then exerts many limitations on the reactor design, as it is imperative to avoid Hg condensation between the source and the growth zones.

The presence of Hg in the gas phase has a strong influence on the already-complex chemistry of the pyrolysis processes [31]. In addition, it was found that the chemical interactions between Hg and di-methyl cadmium (DMCd) can lead to heterogeneous gas-phase reactions that have to be avoided, as the resulting dust particles lead to macrodefect generation on the growing wafer [7, 32]. Many commercial reactor designs in the 1990s suffered from extremely complex quartz component designs that were, due to the limitations in engineering quartz components, inherently both fragile and nonreproducible. In addition, the cleanliness of all the components within the reactor needs to be maintained to a very high level to avoid either unwanted autocatalytic reactions [33] or impurity crosscontamination of the MCT wafers.

Initially many reactor designs were only intended for laboratory-level R&D investigations using substrate samples with dimensions <30 mm. Scaling up these concepts into reactors that could grow on wafers with 3- or 4-inch diameter introduced further challenges in terms of controlling flow, concentration, temperature uniformity, and chemical interactions adequately to achieve sufficiently uniform growth rates, dopant uniformity, and morphology quality. Needless to say, the details of such custom-designed systems have remained strictly proprietary. Relatively few commercial MOVPE equipment manufacturers exist, and currently only AIXTRON offers an AX200 variant reactor cell design for use with Hg-based compounds. Although the reactor geometry in the system is limited to handling 2-inch (50-mm) wafers, it is used for device production at Vigo, Poland [34].

9.4.1 Process Abatement Systems

Fundamentally, all MOVPE growth systems have to deal with the waste byproducts from the process. In the growth of MCT, these are a highly toxic combination of unreacted alkyls, Hg vapor, and decomposition products. Therefore, the exhaust-gas abatement system is a crucial reactor design requirement.

The key to abatement systems is the choice of filter media and rigorous adherence to routine maintenance schedules, ideally coupled with exhaust-gas concentration monitoring. Commercial MCT systems tend to deal with the exhaust using a combination of particle and activated carbon traps. However, as the reactor cell sizes are increased, the maintenance frequency of these systems increases; therefore, to minimize this requirement, additional filtration stages and gas-scrubbing systems can be introduced to increase the efficiency of the abatement system. These include multistage particulate filtering along with high-temperature alkyl-cracking furnaces. The final stage of filtration is often activated carbon filtering and should ideally be custom made to use carbon that is treated for optimum uptake of Hg vapor and alkyls to ensure no toxic gas emissions into the atmosphere. As Hg vapor is the highest-concentration source material in the MOVPE process, it can also be easily routinely monitored in the exhaust gas line with commercial equipment to ensure the activated carbon filters are functioning correctly. The trapped toxic material within the abatement then has to be contained and disposed of via approved waste-management companies. Clearly, adherence to legislative protocols when dealing with toxic materials is imperative, and these may differ around the world.

9.5 Process Parameters

Optimization of the initial nucleation conditions is essential to achieving good-quality epitaxy [13, 20]. It has been shown that Te or Cd flux exposure can switch the orientation of the epitaxial layers from (100) to (111). The MOVPE nucleation is usually performed under conditions to help promote 2D nucleation before then optimizing for more commercially viable growth rates. Fundamentally, all growth conditions are constrained to occur under conditions that are consistent with the pressure–temperature phase diagrams for MCT, and control of temperature and vapor pressure are critical throughout the process. To be practical, IMP growth of MCT requires the growth temperature to be the same for both the CdTe and HgTe phases. The process parameters are required to make growth both uniform and consistent with practical IMP growth-period control. For device heterostructure design to be realized, growth rates need to be consistent with the generation of sufficiently sharp heterointerfaces while still balancing the need to minimize growth defects that can arise from growing too rapidly.

Even with these limitations, the parameter space of temperature, flow, alkyl concentration, and reactor design inputs still affords a very large degree of freedom. Studies have shown the influence of many of the parameters on the morphology [13]. However, any literature claims for limits on specific parameters may well only apply to a particular cell design. Consequently, MCT layers produced by the MOVPE process are highly dependent on the reactor design and process used in their production.

9.6 Metalorganic Sources

After establishing that Hg had to be provided by a heated elemental Hg bath, the choice for the Te and Cd alkyls alkyl sources is also quite extensive. The options can be found in a more detailed review article [35]. Fundamentally, the most important primary requirement for a good metal organic source for the matrix elements is that the thermal stability of the alkyl has to be balanced with adequate volatility such that the concentrations in the growth zone of the cell design are sufficient to enable reasonable growth rates.

The cadmium alkyl of choice is di-methyl cadmium (DMCd), as it undergoes decomposition in the temperature range 200–420 °C. Alternative sterically hindered equivalent compounds [36] were considered as a method of reducing the reactivity of the interaction between DMCd and Hg or the other alkyls in the reactor but never demonstrated significant advantages.

Initial studies used di-ethyl telluride (DETe), but the preferred Te source has now become di-*iso* propyl telluride (DiPTe). This allowed reduction in the HgTe growth temp from ~420 °C to ~350 °C and is more consistent with the DMCd pyrolysis temperature. The highly reactive nature of DMCd meant that chemical interactions with DiPTe allow CdTe growth to occur at much lower temperatures.

Other Te sources have been investigated for MCT growth. Di-tertiary-*butyl* telluride (DtBTe) was reported for growth of MCT by the DAG technique at very low temperatures (~275 °C) [37], and allyl-*iso*propyl telluride was considered at QinetiQ [38]. Although di-methyl telluride was used in CdTe growth at Rockwell [39], it was too stable for use in IMP growth of MCT. Methyl allyl telluride (MATe) and allyl-*iso*propyl telluride (AiPTe) were sources that were developed in the hope of offering lower growth temperatures on the basis that the asymmetric ligand groups would yield the higher volatility of di-methyl telluride or di-iso propyl telluride but with a lower thermal stability of di-allyl telluride. However, using MATe, it was not possible to grow CdHgTe or HgTe due to the preferred reaction route of forming the more thermally stable di-allyl telluride compounds [40]. MATe can still be used for CdTe growth, but concerns have been expressed over the stability of this source at room temperature, and no new Te sources have supplanted DiPTe as the most useful Te precursor.

A crucial aspect to all alkyl compounds that are used in the MOVPE process is the purity of the source material supplied, as any unwanted organic compounds or electrically active impurities could disrupt the growth processes and electrical properties, respectively. Hence, the suppliers play a key role in the success of the final product. Adduct-purification processes were developed that significantly reduced the contaminant concentrations [35].

9.7 Uniformity

A key property of MCT epitaxial layers needed for FPA production is the uniformity of the composition and thickness both within the depth of the device structure and laterally across the area of the substrate. Clearly, rotation mechanisms can significantly reduce the nonuniformity that would arise from temperature, flow, and concentration uniformities within the growth zone, but these can generally only eliminate the effects of linearly changing growth rates.

FTIR mapping techniques are the most commonly used methods for evaluating growth uniformity. The layer thickness can be determined by analysis of the Fabry–Perot interference fringes using literature values of the refractive indices. The 295 K cut-off wavelength can be used to deduce the alloy composition so that the IMP growth rates can be deduced. The composition map can then predict the uniformity of the device cut-off wavelength at the operating temperature so that preprocessing screening can be implemented. Typically, the characterization of 3-inch diameter wafers can be performed at between 2- to 5-mm spatial resolutions using a 125-μm beam spot size. Of course, higher spatial resolutions can be achieved, but this increases the analysis time. Figure 9.3a shows the predicted 80 K cut-off wavelength maps for a typical MW device layer. Much of the apparent edge nonuniformity is due to the spatial resolution and interpolation effects used in the generation of the map. The reproducibility of the growth uniformity that is relevant to the devices is evaluated such that the statistics are calculated from the data points associated with the area from which the devices are to be made. The standard deviation of the cut-off wavelength data for the layer shown in Figure 9.3 was 0.04 μm. Figure 9.3b shows the thickness uniformity of the same hetero-epitaxial layer and has a standard deviation of 0.09 μm. This uniformity is sufficient to obtain 30 potential IRFPA sites with 640×512 pixels on a 16-μm pitch or 17 sites with 1024×768 pixels on a 12-μm pitch from a 3-inch diameter wafer.

Control limits for different wavebands are set by the relationship between the Cd alloy and device cut-off wavelength. For LW device bands, the composition target window is much tighter than for the MW device bands (±0.01 for MW compared with ±0.002 for LW). Figure 9.4 shows the predicted composition uniformity of a LW heterostructure wafer that demonstrates that MOVPE can produce the required uniformity control required for LW IRFPA production. The standard deviation of the composition of this wafer was 0.0004 over the device area, such that all the die sites are within 0.228 ± 0.001.

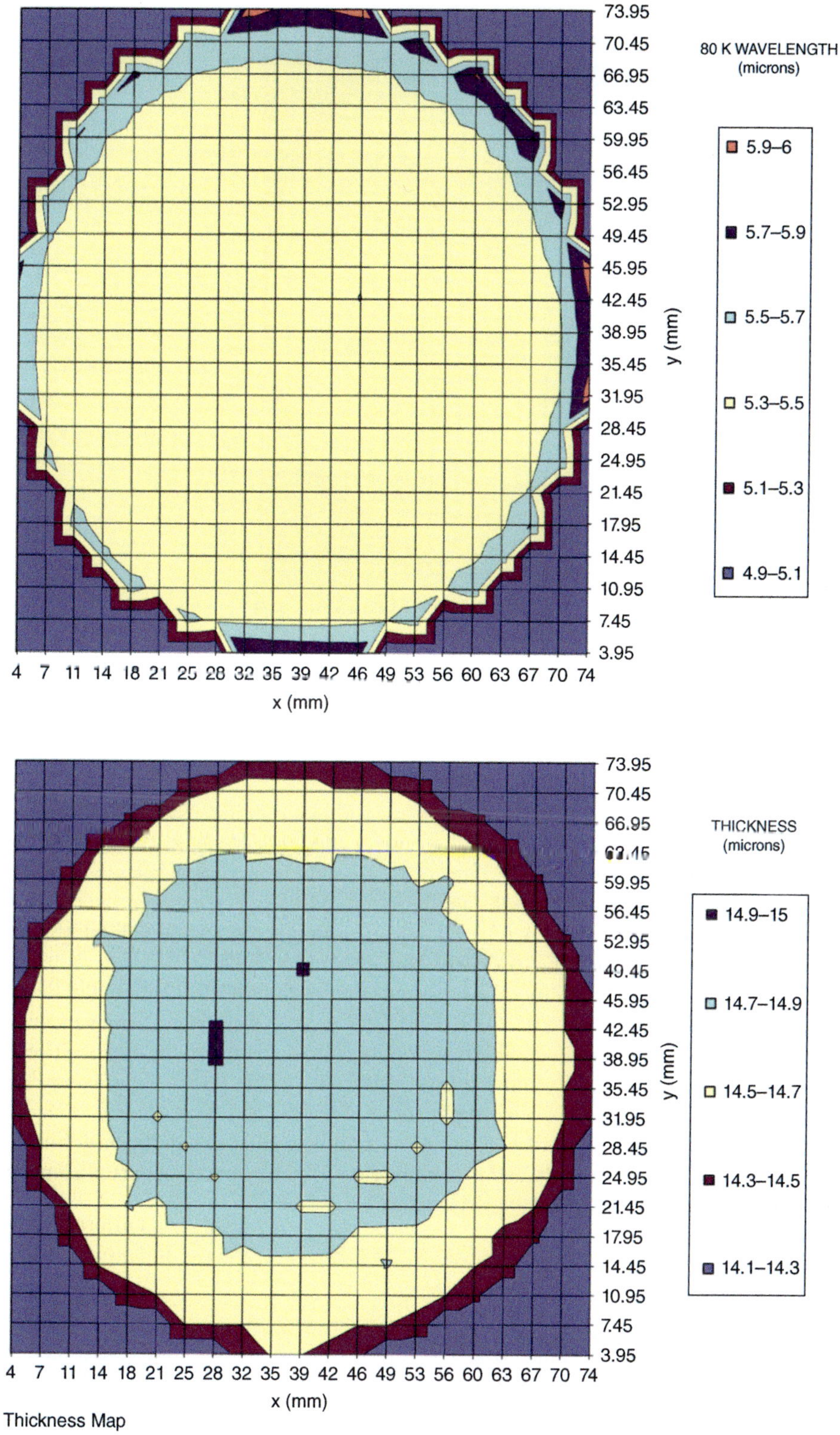

Figure 9.3 *Uniformity maps (3.5 mm spatial resolution) of predicted device cut-off wavelength and thickness of a heterostructure device wafer grown on a 3″ diameter GaAs substrates.*

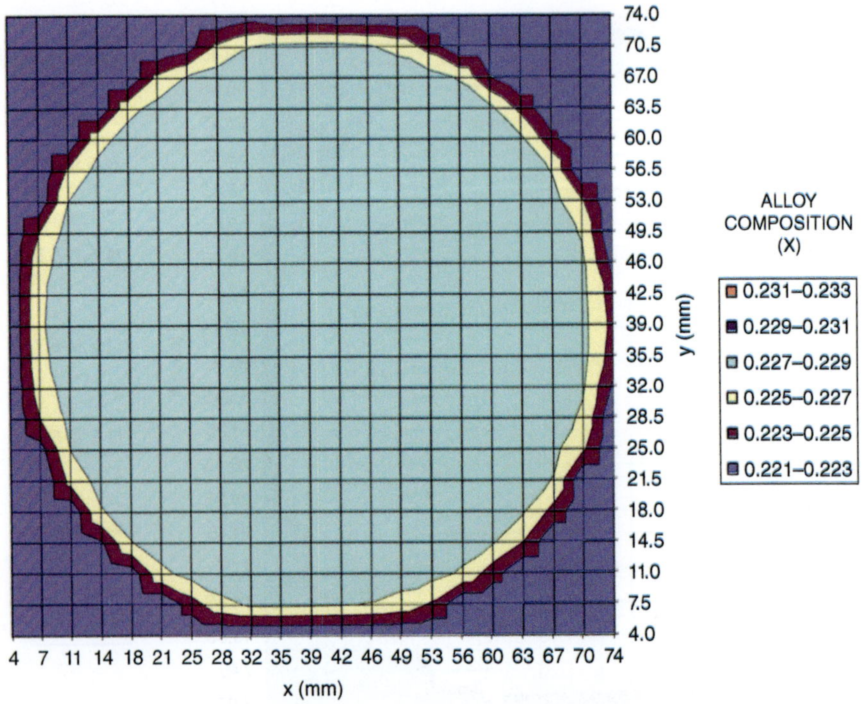

Figure 9.4 *Composition uniformity map from a LW device heterostructure.*

9.8 Reproducibility

In addition to the requirements for uniformity control within a run, it is essential to achieve good reproducibility of MCT heterostructure growth over extended time periods. The MOVPE process needs to be able to demonstrate a stable ratio of CdTe and HgTe growth rates within the IMP growth. Figure 9.5 shows that the growth-rate data from MWIR layers grown over a 12-month period under similar IMP conditions is reproducible enough that the critical CdTe:HgTe growth-rate ratio had a standard deviation that was only 2% of the mean value.

Figure 9.6 shows the reproducibility of the median composition over a 12-month period during which the target window was 0.301 ± 0.011. The MOVPE process was able to achieve a 96% yield to this target specification. The reproducibility of the standard deviation of the composition is seen to range between 0.0008–0.0033. Even the poorest uniformity values do not significantly limit the yield of potential die sites for MW device applications. Over the last 10 years since this data was generated, reactor design changes at Leonardo MW have been able to significantly improve the reproducibility of both the uniformity and growth rate stability by removing design elements that had been intentionally flexible during the R&D phase.

9.9 Doping

In order to realize the potential of improving device performance by using fully doped heterostructure designs, suitable dopants have to be selected. The choices and limitations of available intentional dopants for MCT have been described previously [4, 21]. Currently, the most commonly used dopants for MOVPE growth are arsenic (As) as the p-type dopant and iodine (I) as the n-type dopant. These

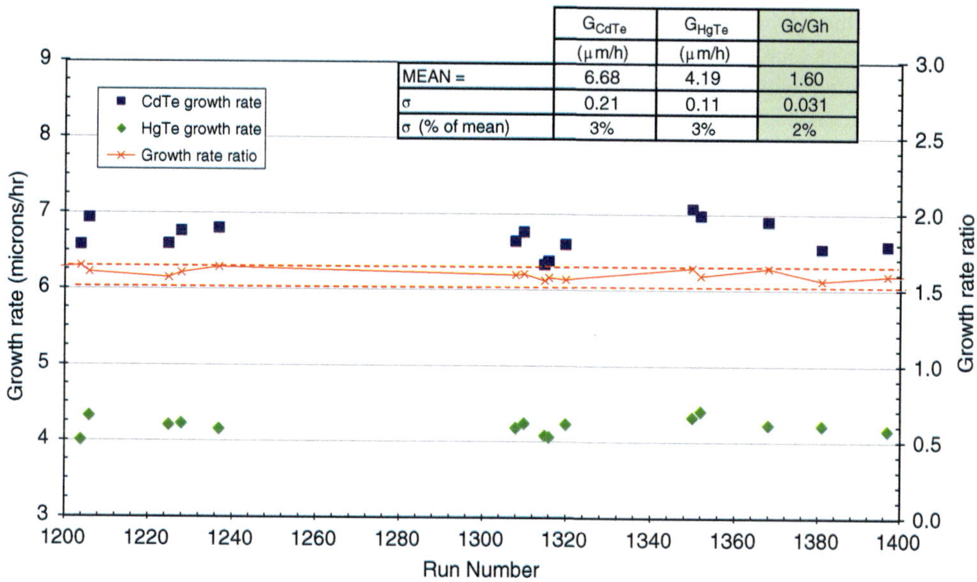

Figure 9.5 *Growth-rate reproducibility for a MW MCT heterostructure design grown on GaAs over a 12-month period. Source: Reproduced from Mercury Cadmium Telluride: Growth, Properties and Applications (eds. P. Capper and J. Garland), John Wiley & Sons Ltd, 2011, Chapter 6.*

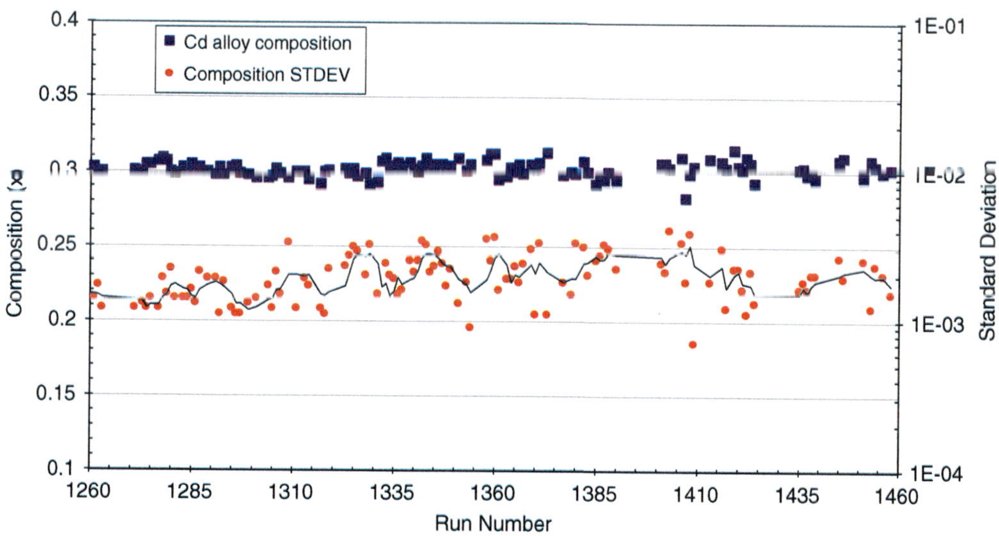

Figure 9.6 *Compositional targeting and uniformity reproducibility for MW MCT heterostructures grown over a 12-month period. Source: Reproduced from Mercury Cadmium Telluride: Growth, Properties and Applications (eds. P. Capper and J. Garland), John Wiley & Sons Ltd, 2011, Chapter 6.*

have demonstrated the key characteristics of being stable during growth and subsequent annealing processes, yielding good electrical transport characteristics from Hall and lifetime studies and without affecting the MCT growth quality. In order to achieve this, the dopant atoms have to be incorporated (or, alternatively, transferred by anneal processes) onto the Te lattice sites. When growing using the

IMP growth method, it has been found that dopant species often only incorporate in the CdTe growth cycles. The dopant sources are thermally stable at the growth temperature but will decompose via chemical interaction with DMCd. Consequently, it is possible to control the II/VI ratio during CdTe growth to ensure metal-rich conditions under which the correct lattice placement of the dopants is promoted.

Sources of the chosen dopants are typically either metal organic (MO) or hydride molecule forms. Good doping sources need to have minimal interaction with the other MO-source materials being used in the growth process. Such problems can lead to issues with significant in-system or run-to-run memory characteristics. Indium from many common indium alkyl sources was found to exhibit both of these problems in earlier In-doping studies. Although tri-isopropylindium was eventually identified as a potential In alkyl source, it required modification to the growth reactor design and has not been used since by other workers [41].

The metal-organic form of the compound used to deliver the dopant into the reactor has to fulfill different criteria from the Te and Cd alkyls. Typically, it is necessary to achieve control of the dopant vapor concentration within the growth reactor in the range 1 ppb to 100 ppm (depending on the incorporation efficiency within the growth conditions), so high-vapor-pressure sources are usually best avoided even if double-dilution dopant control channels are available on the reactor gas lines. Although ethyl iodide (EtI) [4, 13] and iso-propyl iodide (IPI) [42] have successfully shown donor doping capability, a study using iso-butyl iodide (IBI) from the family of iodide compounds showed that improved control of very low donor concentrations could be achieved by using the lower-volatility IBI. Doping control from mid $\times 10^{14}$ – mid $\times 10^{17}$ cm^{-3} has been achieved without excessive cooling of the source or using high-dilution-factor channels [43].

In the case of As doping, a wider variety of source materials have been investigated, including arsine (AsH_3), phenylarsine, and tertiarybutylarsine. Consideration was given to minimizing the probability of As–C or As–H bond incorporation, as these were theoretically expected to be associated with Shockley–Read–Hall (SRH) traps [44]. The favored As-doping source has become *tris*-dimethylaminoarsine (tDMAAs), which avoids As–H and As–C bonds, and this has demonstrated doping control from mid $\times 10^{14}$ – mid $\times 10^{17}$. Lifetime studies of MCT layers grown using tDMAAs showed that the SRH trap densities could be reduced sufficiently such that the lifetimes were dominated by fundamental Auger recombination mechanisms [4, 17, 45, 46]. However, recent results [10] have shown that p-type layers produced using AsH_3 have been able to show radiative-limited lifetimes equivalent to using tDMAAs. These results are summarized, along with historical data, in Figure 9.7.

One limitation of tDMAAs is that it has been observed that attempts to dope any higher than mid $\times 10^{17}$ cm^{-3} on a (100) growth orientation result in an interaction with the DMCd such that the CdTe growth rate is observed to accelerate, leading to a loss of control of the composition and higher defect densities [45, 46].

Once all calibrations have been established, complex heterostructures can be produced with full control of dopants and composition, as is required for third-generation IRFPA devices. Figure 9.8 shows an example of a LW/MW dual-band device structure grown at Leonardo MW that shows control of low- and highly doped iodine and arsenic regions within a complex composition profile.

9.10 Defects

The size and density of macrodefects (known as *hillocks*) in epitaxial layers control the potential size and probability of nonoperative pixel cluster defects in the IRFPA. Figure 9.9 shows Nomarski contrast micrographs of typical hillock defects observed on misoriented (100) substrates. Figure 9.9b shows how sometimes the top of the hillock develops a polycrystalline decoration.

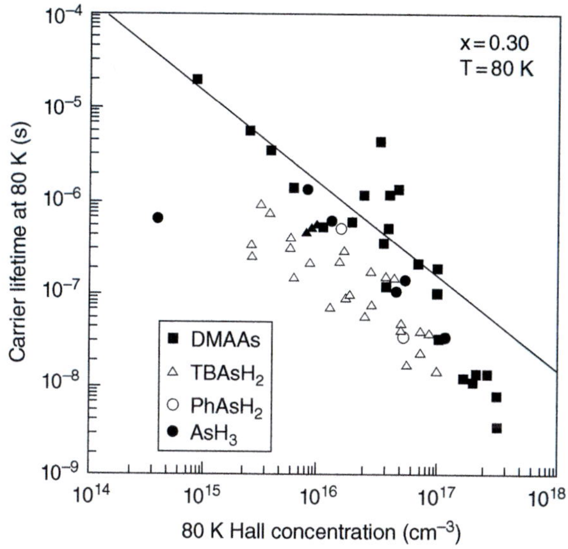

Figure 9.7 *Comparison of 80 K lifetimes in As-doped $Cd_xHg_{1-x}Te$ for $x > 0.30$. Source: Madejczyk et al. 2005 [10].*

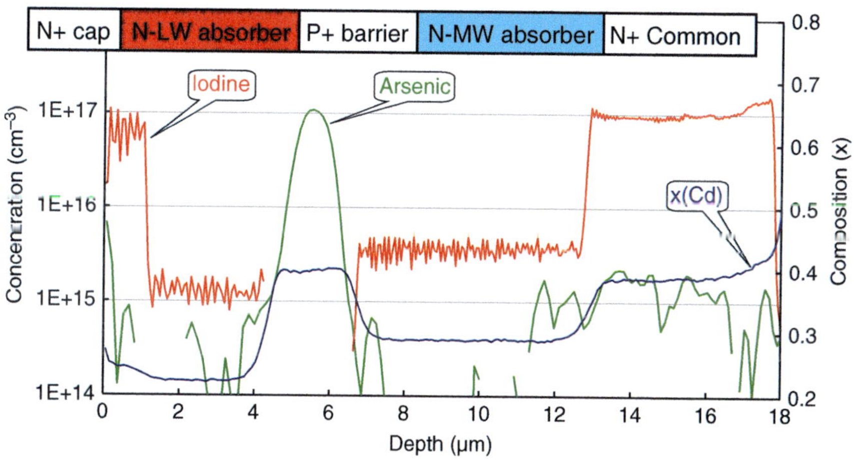

Figure 9.8 *SIMS profile from a dual-band device heterostructure. Source: Reproduced from Mercury Cadmium Telluride: Growth, Properties and Applications (eds. P. Capper and J. Garland), John Wiley & Sons Ltd, 2011, Chapter 6.*

It has been suggested that there may be a single origin of hillock macrodefects based on MO-induced etching of the substrate surface prior to nucleation commencing [47]. However, the appearance of hillocks is more universal and the substrate exposure time too short to make this single mechanism the most likely origin. It is considered more likely that the fundamental disparity in the growth rates between the various orientations (discussed in Section 9.3.1) means that any perturbation of the surface perfection that offers a growth face closer to a (111) orientation will grow faster than the (100) surface growth plane. As growth proceeds, the facets then develop, and the height of the hillock defect feature increases. This would also explain why macrodefects have higher Cd content [7] and are less prevalent on the (111) and higher-order orientations.

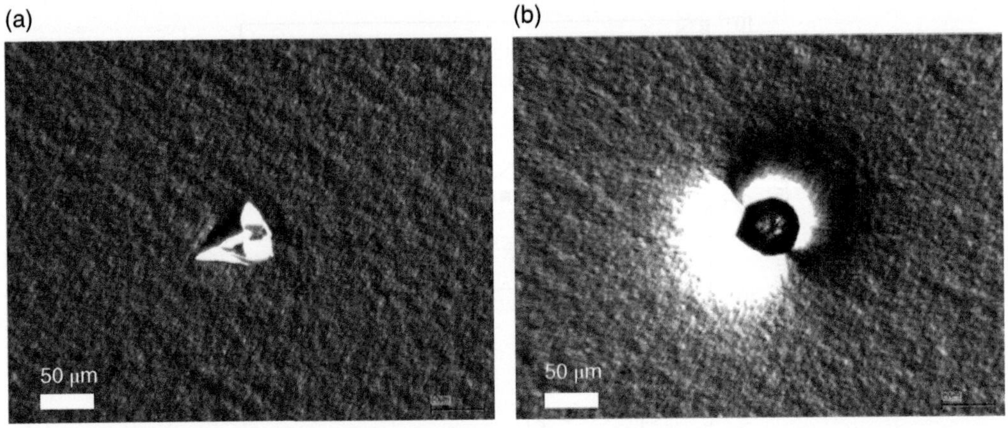

Figure 9.9 (a) A typical macrodefect defect found on 4° off (100) orientation substrates. (b) A polycrystalline decorated macrodefect found on 4° off (100) orientation substrates.

The surfactant role of adatoms of either Na or K has been discussed as a way of facilitating the nucleation of CdTe onto GaAs [13, 48]. Some papers describe significant reductions in defect levels but still have much higher defect densities than 2D FPA arrays require, which suggests that although this may be beneficial in some instances, there is no single factor that eliminates the macrodefect problem. It is instead the careful consideration of many aspects of the substrate treatment and growth nucleation that is crucial.

Statistical data of the macrodefect surface density nature is rarely found in the literature for any epitaxial material system, although there have been one-off claims of low defect density or occasional examples of deviant behavior. As hillocks have a large impact on the growth yield of material suitable for large-area 2D arrays using MOVPE, it is essential to optimize the reactor integrity and growth-nucleation processes in order to minimize macrodefect densities. Large-scale defects that can arise from sources such as poor substrate processing and growth deposit debris from the reactor should be suppressed by careful control of the reactor cleanliness, substrate preparation quality control, and working environment control.

With ever-higher FPA specification requirements, e.g. larger die sizes and reduced pixel pitches, the density of normal hillocks, which are typically ~ 50 × 75 μm in size, needs to continue to be driven down. A figure of between 0–1 cm^{-1} is typically required in order to ensure > 50% of all sites are suitable for IRFPA manufacture. However, larger macrodefects (>100 × 100 μm) can sometimes occur that have a larger impact on the number of defective pixels in an array, and these cannot be tolerated for any pixel pitch designs below 25 μm. Consequently, the allowable density of these features drops to < 0.5 cm^{-1} in order to achieve the same yield target of potential die sites.

Data from one of Leonardo MW's MOVPE reactors is shown in Figure 9.10; the defect densities were achieved over a 12-month growth period on 3″ diameter GaAs substrates. On average, ~65–70% of wafers had hillock densities below 1 cm^{-2} and macrodefect densities below 0.15 cm^{-2}. Steady improvements have been achieved over the years since this data was gathered by improving the various aspects of growth-process reproducibility, as discussed in Section 9.8.

The primary microscale defects that are considered to have the highest impact on device performance are misfit and threading dislocations. With the 14% lattice mismatch between the CdTe and GaAs, the dislocation density is expected to be much higher than for a lattice-matched system. Etch pit density (EPD) assessment on the (100) orientation has proven to be difficult to quantify, due to the indistinct nature of the etch pits that are achieved, but these are estimated to be in the 10^6–10^7 cm^{-2} range. Conventional understanding and theory dictate that such high densities would

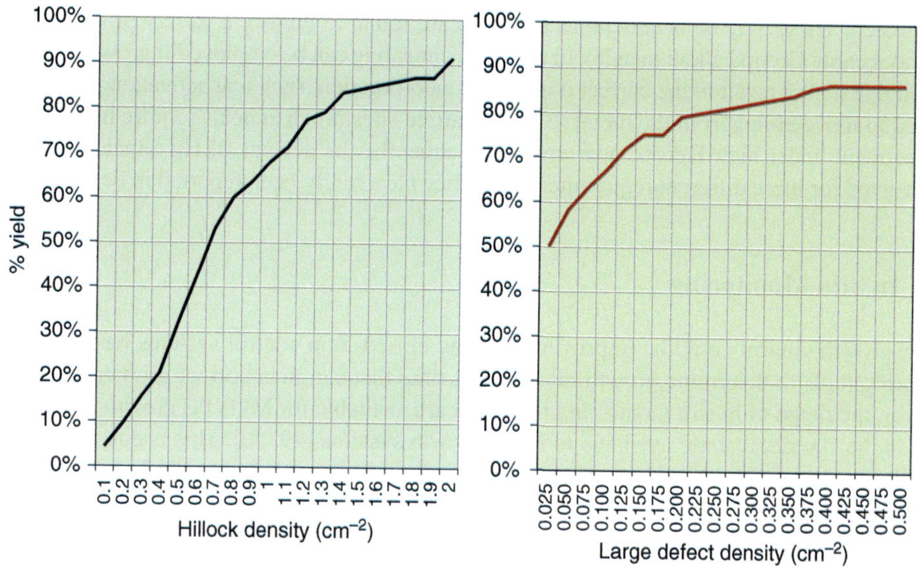

Figure 9.10 *Examples of the cumulative yield plots of hillock defect densities and the subset of larger defects (>100 μm × 100 μm). Source: Reproduced from Mercury Cadmium Telluride: Growth, Properties and Applications (eds. P. Capper and J. Garland), John Wiley & Sons Ltd, 2011, Chapter 6.*

be expected to prohibit any possibility of achieving high-operability LW or LW/MW devices. However, at Leonardo, these types of FPAs have been successfully manufactured in MOVPE-grown material, as will be described in the application sections of this chapter. The reason for this apparent disparity is still being studied, but it is thought possible that the chemical etching used in the standard EPD methodology is revealing features other than threading dislocations.

9.11 Annealing

The properties of MCT epitaxial layers can be controlled by the annealing procedures that are employed after growth. This is due to the extensive range of native defects that can be formed in the MCT alloy. The most commonly considered defects are metal-site vacancies (often referred to as *Hg vacancies*), Te antisites, Hg interstitials, as well as many complex defects. The growth conditions within the phase diagram predict significant metal vacancies are present during growth under normal MOVPE growth conditions. Anneals are therefore performed, for example, if fully doped heterostructure device layers are grown, to remove these metal vacancies so that the electrical properties of the wafers are dominated by the dopants that have been intentionally incorporated during growth.

In situ annealing [10] procedures have been incorporated into the growth reactor process to reduce metal vacancies; however, this is not compatible with production processes due to the extensive time that the anneal processes take. Another novel approach that has been adopted in Poland has been to introduce a very thin transitional layer of Cd between the IMP layers in order to create excess metal. This was found to remove the requirement for Hg-vacancy annealing or dopant-activation anneal steps from the wafer-processing sequence [34].

Ex situ annealing methods are used to heat-treat wafers (with controlled Hg overpressures) to re-equilibrate or resite native defects within the phase diagram and can be performed with multiple wafers/batch. These may be beneficial as they also are found to remove a variety of mobile atoms (most noticeably Group I alkali metals). In addition, annealing can be employed if growth conditions are used that do not result in the correct dopant site incorporation, such that activation processes are required to move atoms from the metal to the Te lattice sites. Such anneals require high-temperature treatment under high Hg overpressures. Consequently, the designs of Hg annealing furnaces have evolved for these different requirements, but they have rarely been reported in detail.

9.12 In Situ Monitoring

Ideally, growth control could be improved by using closed-loop control systems based on in situ monitoring tools to determine growth rates in real time that can then correct growth in real time. However, the range of in situ monitoring tools that are available for MOVPE growth is less extensive than the choices that are available to MBE growth systems [49, 50]. This is due to the lack of a vacuum environment, high Hg concentrations, and more restricted accessibility to the growth zone. Laser interferometry is the most commonly reported technique that has been found to yield useful information on growth rates and film roughening. CdTe and HgTe have very high absorption coefficients to standard laser wavelengths. However, with IMP-grown MCT, the large refractive index change between CdTe and HgTe combined with the fact that very thin layers are being deposited enables a constant interferogram to be produced. Key information such as growth rates (which allow the alloy composition to be monitored) and roughening factors of the growing epitaxial film (which allow the morphology to be monitored) can be ascertained. Such systems are now commercially available (www.laytec.de) but require optical access holes in the growth reactor design to prevent growth deposits in the reactor from attenuating the signal, and therefore the effect of any such reactor modification has to be carefully reviewed to ensure that the changes are not detrimental to the growth process. If these criteria can be met, then with a suitable choice of laser wavelengths sufficient to give good resolution of growth rates, growth rates within a growth run may be monitored and used in a dynamic feedback configuration.

Workers in Poland developed an in-situ alkyl concentration tool based on IR absorption. This has been used to determine arrival rates, flow-switching characteristics, and concentrations within a reactor cell during growth. Such information is important for gaining understanding of the growth cell design. However, there are limitations reported on the use of this system once growth has led to deposition within the chamber [34].

9.13 Background for Applications of MOVPE MCT

9.13.1 Introduction to Infrared Imaging and Atmospheric Windows

The Earth's atmosphere is conveniently transparent in the visible region but severely filters much of the radiation at longer wavelengths. Figure 9.11 shows the quantum spectra for solar and thermal radiation at the Earth's surface for specific conditions. The solar spectrum is scaled to a lux equivalent of 17 500, which is typical for a bright sunlit day. The thermal spectrum is the natural photon emission by objects around us and depends on the temperature and the spectral emissivity of the materials in the scene. *Passive thermal imaging* is the term used to describe imaging of this radiation. Figure 9.11 shows the photon emittance of a body at a temperature of 20 °C, peaking around 10 μm. After atmospheric absorption, the solar and infrared spectrum is divided into many windows that have names or letters as shown. The thermal infrared is particularly compromised by water-vapor absorption, and only two windows are used for imaging: the medium-wave infrared

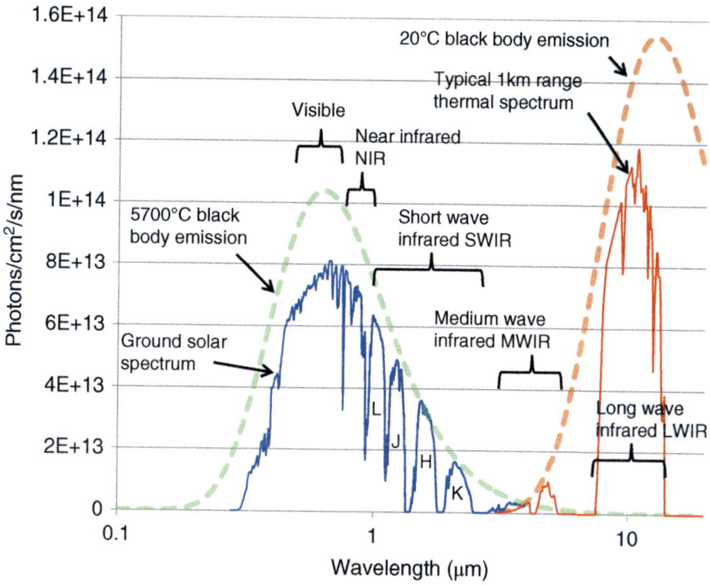

Figure 9.11 Solar and thermal quantum spectra at the Earth's surface with associated band definitions. The solar data corresponds to a bright sunlit day of 17 500 lux. Source: Reproduced by permission from Springer Nature, Handbook of Electronic and Photonic Materials, S. Kasap and P. Capper (eds), 2017, Chapter 34.

band (MWIR), between 3.2 and 5.0 μm, with a carbon dioxide notch around 4.3 μm; and the long-wave infrared band (LWIR), from 7.7 to 14 μm. Shortwave infrared (SWIR) is used by astronomers and in active imaging where the scene is illuminated by an infrared source, such as a neodymium-YAG laser at 1.55 μm. MCT is uniquely sensitive to the whole solar and thermal infrared range from 0.4 to beyond 14 μm and can be tuned to match the atmospheric windows.

9.13.2 MCT Infrared Detector Market in the Modern Era

The LWIR band is often chosen for imaging in poor conditions with fog and smoke. The infrared detector market is dominated by uncooled cameras that are sensitive to heat and use ferroelectrics such as barium strontium titanate, or microbolometers such as vanadium oxide (VOx) or amorphous silicon (α-Si). Uncooled cameras have a relatively short range and tend to need fast optics to achieve useful sensitivity with a typical focal ratio of F/1.2. For longer ranges, high-performance, cooled cameras based on MCT have so much sensitivity in reserve that they can use focal ratios around F/4, and it is straightforward and affordable to use long telephoto optical designs. They can also have fast frames rates, with some cameras able to run at more than 1000 frames per second. LWIR MCT systems are favored by the military, particularly in the United States.

At first sight, the MW window appears to be compromised by photon flux, some two orders lower than the LW window. However, it is much easier to make 2D detectors for this wavelength, and the photon-flux disadvantage can be offset by staring architectures that enable a longer integration (or stare) time. The pixel size can be proportionately smaller, with wavelength and typical sizes range from 8–15 μm. Array formats can be much larger than uncooled arrays, and HD 1024 × 768 and 1280 × 768 formats are used in top-end cameras. The high-performance thermal imaging market is currently shared between two material systems: indium antimonide (InSb), which has sensitivity to 5.5 μm, and MCT. InSb has most of the cooled thermal imaging market

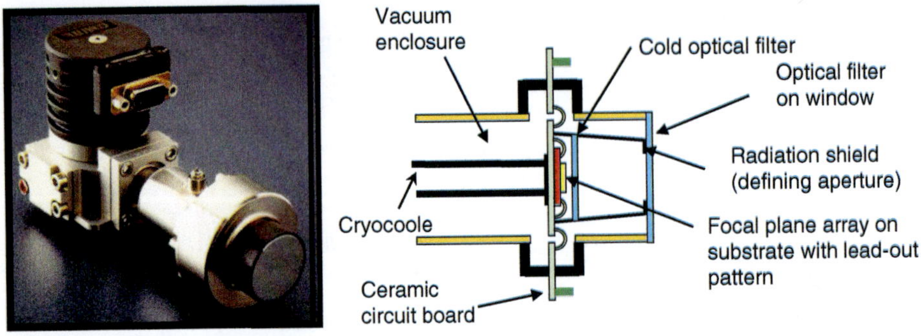

Figure 9.12 *Typical integrated detector–cooler assembly (IDCA). Source: Reproduced by permission from Springer Nature, Handbook of Electronic and Photonic Materials, S. Kasap and P. Capper (eds), 2017, Chapter 34.*

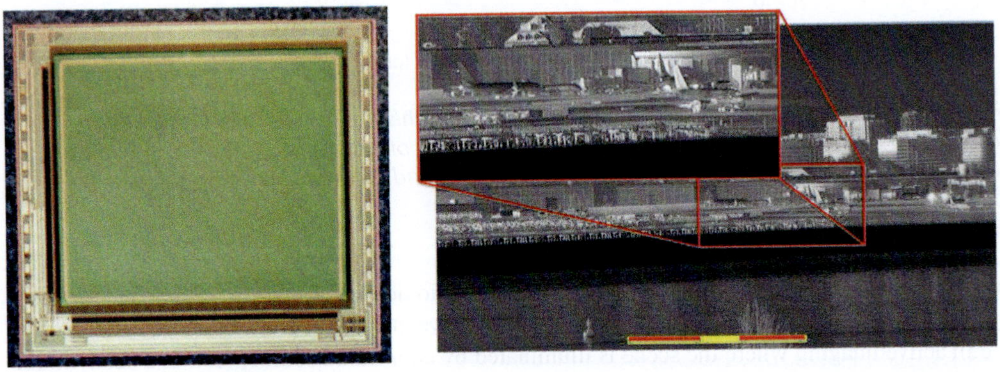

Figure 9.13 *Typical focal plane array and thermal image.*

because it was the developed earliest. MCT has some key discriminators that may make it the material of choice for future third-generation developments. MCT can be operated at a much higher temperature than InSb, so cameras can be smaller and have lower power consumption.

Modern high-performance thermal imaging cameras utilize miniature cryocoolers to cool the arrays and suppress dark current. A typical integrated detector-cooler assembly (IDCA) is illustrated in Figure 9.12.

MOVPE infrared detectors, in conjunction with mesa-pixel isolation, provide near-ideal imaging quality with high quantum efficiency and low crosstalk. As a consequence, the device technology is amenable to large arrays with small pixels, as illustrated in Figure 9.13, which shows a typical image for a 1280 × 1024 array called the SLX-SuperHawk with 8 μm pixels.

The SW window is rarely used for passive thermal imaging because of low flux, but active imaging using a source such as infrared illuminators, lasers, or even nightglow is of continuing interest and is the main band used in astronomy. Other materials can be used for SW detectors, but products such as InGaAs tend to have high noise beyond 1.7 μm due to defects arising from the lattice mismatch with the InP substrate. MCT has almost-ideal solid-state properties for electron avalanche photodiode devices (eAPDs). The high mobility ratio of MCT results in single-carrier multiplication with low noise figures, and such devices are exploited in photon-starved applications, such as long-range imaging and astronomy.

There are emerging technologies for the higher-performance market based on quantum effects in III-V materials such as Type II superlattices, barrier photoconductors (so-called nBn devices), and multiple quantum well (MQW) detectors (usually using AlGaAs/GaAs technology). These systems are claimed to have higher ultimate performance than MCT, better reliability, and lower manufacturing costs. Recent developments in wafer-scale growth of MCT, however, have set new standards. Also, the main drawback of III-V materials is the absence of a good surface-passivation process (MCT can use interdiffused CdTe), and this may limit how close these systems can get to ultimate performance.

9.14 Manufacturing Technology for MOVPE Photodiode Arrays

Infrared detector technology must ensure high quantum efficiency and minimal excess noise so that a condition called *background limited performance* (BLIP) is achieved [51]. Here, the signal-to-noise is determined by the photon statistics and not device-related noise. The physics of photovoltaic detectors is described in [52], and the fundamental limits are discussed in [53–55]. MOVPE allows for a step change in excess noise. The P-N junction can be placed in wider-bandgap material so that junction-related leakage currents (such as trap-assisted tunneling) are strongly suppressed. The absorber volume can be minimized to reduce thermally generated leakage currents. The absorber can be bounded by high-carrier-concentration regions that confine the absorber volume and suppress 1/f noise on the photocurrent (a problem with most homojunction technologies). The contact regions can be heavily doped to avoid non-ohmic contacts that contribute to 1/f noise and carrier injection. Absorbers can have graded bandgaps to promote field-assisted diffusion to the P-N junction and enhance quantum efficiency. In summary, the flexibility of MOVPE growth allows devices to be designed specifically for BLIP performance.

9.14.1 Mesa Heterojunction Devices (MHJ)

The main advantage of MOVPE material is that doping and thicknesses can be fully optimized. For heterostructures that are grown in situ, it is necessary to completely isolate the junction, forming a *mesa device*. The device structure is illustrated in Figure 9.14 and is sometimes called a *double-layer heterojunction device* (DLHJ). Reviews of MOVPE device fabrication at Leonardo MW Ltd is given by Hipwood et al. [56, 57] and Baker et al. [58].

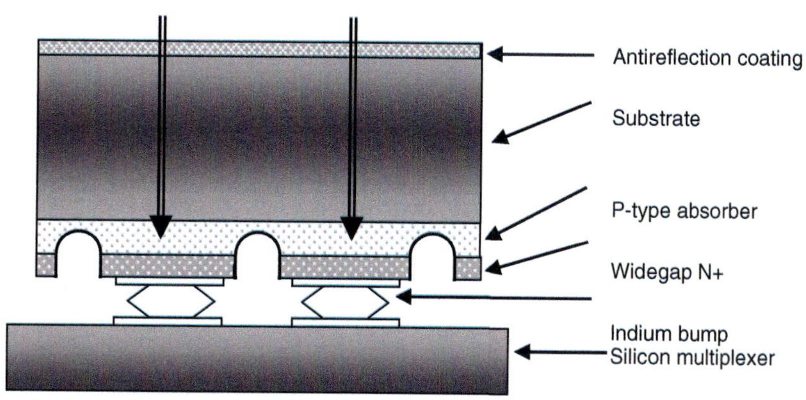

Figure 9.14 Schematic of mesa heterojunction (MHJ).

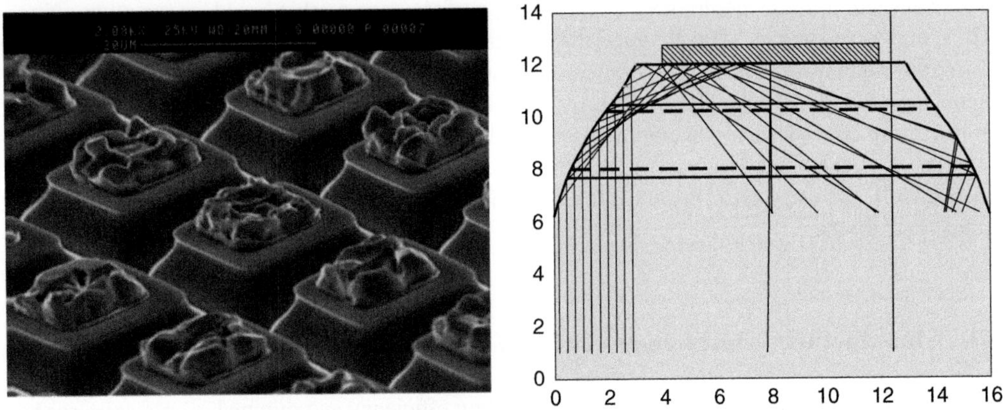

Figure 9.15 16-μm pitch mesa MOVPE device and optical modelling. Source: Reproduced by permission from Springer Nature, Handbook of Electronic and Photonic Materials, S. Kasap and P. Capper (eds), 2017, Chapter 34.

Each pixel is electrically isolated by an etched mesa slot that extends through the absorber to eliminate lateral collection that can limit the MTF in LPE-based processes. The sidewalls are then coated with an electron-beam evaporated CdTe layer that is interdiffused at high temperature. The widening of the bandgap around the edges of the absorber effectively separates carriers from surface states and minimizes junction currents where the junction intercepts the sidewall. Because the absorber can have minimal volume in this process, the dark currents are usually at least an order less than LPE homojunction technology.

With properly shaped mesa sidewalls, the cone can become an optical concentrator with photons trapped within the mesa cone. MHJs can therefore have low optical crosstalk, which is especially important in small pixels. Photons that are not absorbed can contribute to the background of other pixels and can result in images that lack contrast. Figure 9.15 shows a mesa heterojunction array with a 16-μm pitch and the associated ray-tracing analysis. The sidewall reflections have skewed paths through the absorber effectively, lengthening the absorber by about $\times 3$ so the absorber thickness can be minimized. A short absorber results in high collection efficiency and low modulation by traps and dislocations. Hence, 1/f noise on the photocurrent is minimal compared with lateral collection diodes used in planar and via-hole device structures.

9.14.2 Wafer-Scale Processing

There is a dual purpose in developing vapor phase epitaxy technologies. First, it opens up the possibility of growing complex multilayer materials for a whole range of new device structures; and secondly, it allows wafer-scale growth on lower-cost substrates than the CdZnTe used in LPE. Figure 9.16 illustrates the process.

A number of device types, including SW, MW, LW, e-APD, and two-color arrays, have been fabricated using MOVPE growth on low-cost 75-mm GaAs [58]. The GaAs substrate is removed after bump bonding to avoid thermal expansion stress and to eliminate optical crosstalk due to internal reflections.

9.15 Advanced MCT Technologies

Third generation or *GEN III* is a commonly used term to describe advanced infrared detectors. The definition of a GEN III detector can differ between nations, but the general guideline is any detector that offers an imaging advantage over conventional first- and second-generation systems.

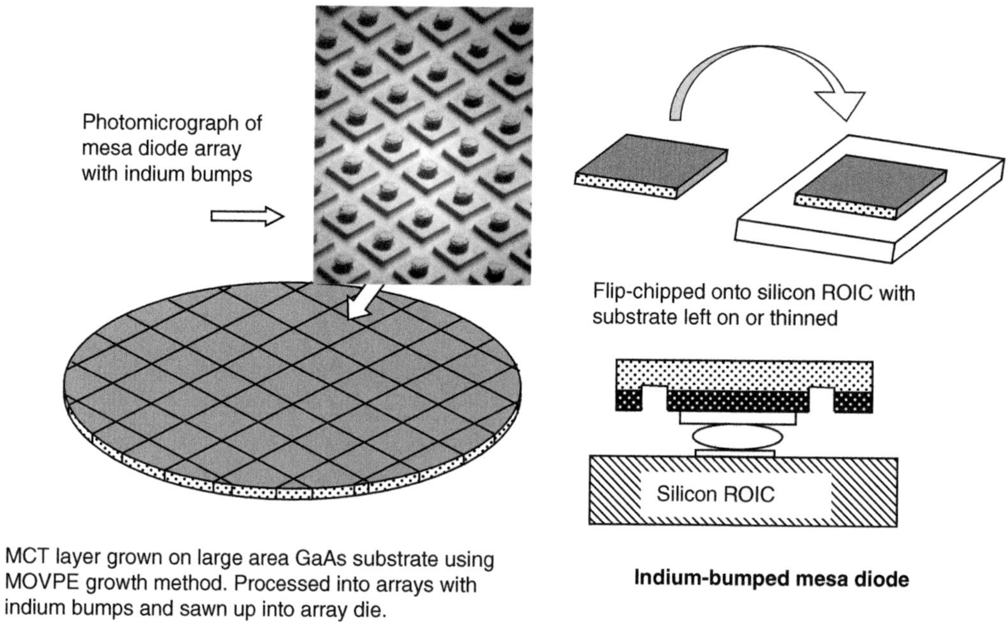

Figure 9.16 *Wafer-scale growth. Source: Reproduced by permission from Springer Nature, Handbook of Electronic and Photonic Materials, S. Kasap and P. Capper (eds), 2017, Chapter 34.*

Commonly agreed-on examples are, for instance, megapixel arrays with high density, dual-color, or even multispectral arrays; higher operating temperature; fast readout rates; very low NETD due to pixel-level signal compression; and retina-level signal processing. Some of the research progress is described here in more detail.

9.15.1 Small-Pixel Technology

Electro-optic systems strive to reduce pixel size because it has a direct effect on the cost, size, weight, and power consumption of the overall camera. Alternatively, in conjunction with faster optics, resolution can be improved, and this is a useful retrofit path for existing cameras. Faster optics also increase the ratio of photocurrent to dark current, allowing higher operating temperatures. Clearly, the limitation of small pixel sizes is set by the diffraction limit of the optics (for instance, Driggers [59]). McEwen et al. [60] proposed that the pixel size should satisfy Sparrow's "Undulation Condition" [61], setting a minimum value of 0.5 λ.Fno., where λ is the mean photon wavelength that depends on the scene temperature and atmospheric conditions. This means a pixel size of 8 μm is viable in long-range imaging systems.

Pixel sizes in the 12–20 μm range for MWIR systems are available in the marketplace with good maturity from mesa heterojunctions [62]. McEwen et al. [60] describe mesa heterojunctions with a pixel size of 8 μm and implemented in a 1280×1024 pixel array called SLX-SuperHawk. MOVPE together with mesa delineation allows small pixel sizes while maintaining very low optical cross-talk. In consequence, near-ideal MTF was measured using an F/2.8 optic and 3.5–5.0 μm spectral range [60]. A photomicrograph of the 8 μm pitch array is shown in Figure 9.17.

9.15.2 Higher Operating Temperature (HOT) Device Structures

Operating focal-plane arrays at higher temperatures is a key modern requirement and has far-reaching consequences for the size, weight, power, and cost of future thermal imaging systems. The power consumption of high-performance thermal imaging cameras is often dominated by the

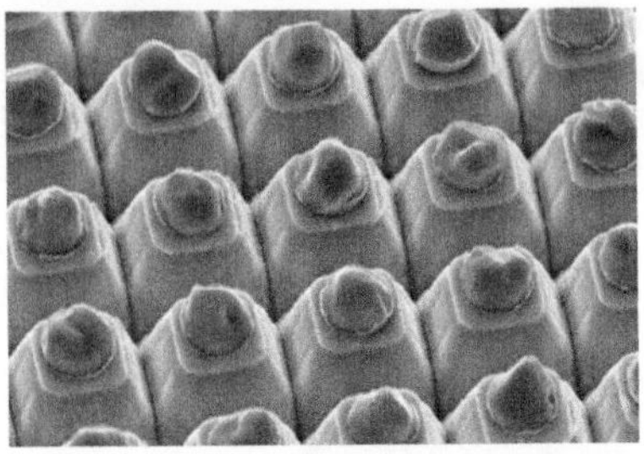

Figure 9.17 *Photomicrograph of an 8 μm pitch MOVPE array.*

Figure 9.18 *Imaging at 210 K operating temperature. Source: Courtesy Leonardo Ltd.*

Stirling cooling engine; operating a small cooler at 150 K instead of 80 K approximately halves the power as well as reduces the cooldown time. A satisfactory operating temperature clearly depends on optics, scene temperature, waveband, and signal processing. Making comparisons from the literature is therefore difficult, but rationalizing the data to a typical practical camera specification of F4, 290 K background, 3.5–5.0 μm waveband, and 50 mK NETD provides a rough guide to the status quo. Epitaxial InSb can be operated up to 100 K with stretch potential to 120 K. LPE processes, nBn barrier devices, Type II superlattices, and MBE MCT processes using planar device structures currently operate around 130–150 K. MOVPE processes using mesa cones can operate at 160 K with stretch potential to 200 K. The main reason MOVPE processes can be better is due to geometry. The absorber volume can be minimized, and reflections within the mesa cone effectively create long path lengths to maintain quantum efficiency (QE).

The best reported HOT imaging is described by Pillans et al. [63] using mesa heterodiodes in MOVPE with lightly doped P-type absorbers; good-quality imaging has been achieved at temperatures up to 210 K, as shown in Figure 9.18. The array was a 16-μm pixel Hawk with 3-5 μm bandpass and operated with a focal ratio of F4.

9.15.3 Two-Color Array Technology

Thermal emission from scenes with a mixture of materials can have spectral variations analogous to color in the visible. Realtime multicolor sensors are still in development, but a good stepping stone is the dual-waveband detector (DWB) with simultaneous readout of flux levels in two

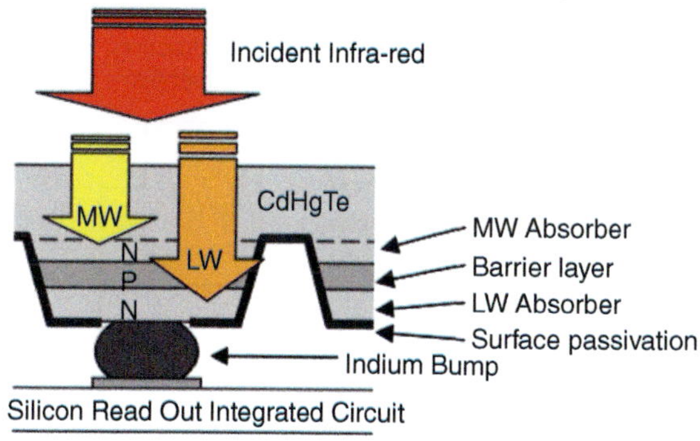

Figure 9.19 *MW/LW bias-selectable architecture for MOVPE. Source: Reproduced by permission from Springer Nature, Handbook of Electronic and Photonic Materials, S. Kasap and P. Capper (eds), 2017, Chapter 34.*

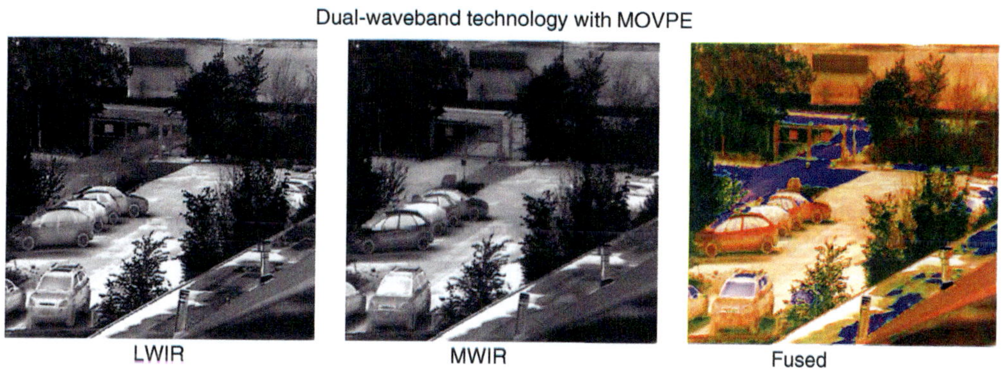

Figure 9.20 *Imaging from CONDOR bispectral detector (F2, 4 ms/0.4 ms integration time). Source: Courtesy Leonardo MW Ltd.*

separate infrared bands (usually MWIR and LWIR). DWBs incorporate both wavebands into a single detector, giving the user the ability to switch between the two as circumstances require. This clearly offers a cost advantage over the need for two separate cameras. Image fusion can be used to highlight areas of interest by exposing spectral differences.

MOVPE is particularly suited to a device structure called a *bias-selectable detector* [64] that needs only one bump and is compatible with small pixel sizes. Figure 9.19 shows a schematic structure. The simplest device is an n^--P- n^- structure, i.e. wide-bandgap P-type material surrounded by narrow bandgap n-type material (capitalized letters are wider bandgap). The two n-type regions are the MWIR and LWIR absorbers, with appropriately chosen bandgaps. The intermediate wide-gap P-type region prevents transistor action between the two absorbers. The two junctions are stacked vertically so that the radiation travels through the MW absorber before the LW absorber. The band to be detected is selected by the polarity of the applied bias, and the ROIC has two separate integration capacitors so the integration times can be adjacent before reading out both bands.

Figure 9.20 shows typical imaging performance from a CONDOR bispectral camera with wavebands of 3–5 μm and 8–10 μm on alternate frames. The focal ratio was F2, and integration periods were 4 ms and 0.4 ms, respectively. Radiometric performance in both bands was similar to that of cameras dedicated to each band [62]. Spectral crosstalk due to unabsorbed MWIR radiation was measured as 0.2%.

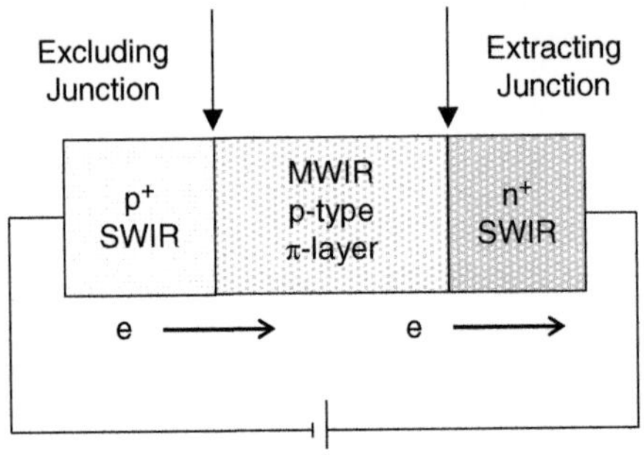

Figure 9.21 *Schematic of nonequilibrium HOT device structure. Source: Reproduced by permission from Springer Nature, Handbook of Electronic and Photonic Materials, S. Kasap and P. Capper (eds), 2017, Chapter 34.*

9.15.4 Nonequilibrium Device Structures

MOVPE is well suited to advanced device concepts that require low doping. Operation at close to room temperature has been proposed primarily as a result of research in the UK in the 1980s and 1990s [65]. The possibility of operating detectors at temperatures near room temperature was proposed by Elliott and Ashley [66]. The basic structure is $Pp^{-}N$, as illustrated in Figure 9.21.

By biasing the extracting contact into reverse bias all, of the intrinsically generated minority carriers in the lightly doped p-layer can be removed, and operation is said to be in *nonequilibrium.* To preserve space-charge neutrality, the majority carrier concentration drops to the background dopant concentration, and Auger generation is effectively suppressed. Remaining components are associated with S–R centers and possibly injection from the contact and surface regions. The realization of HOT LW arrays will depend on the suppression of high 1/f noise associated with the reverse-bias operation. Nevertheless, the device concept is a route toward high-performance infrared detectors with minimal cooling.

More recent work in Poland [18] has extended the three-layer structure of Figure 9.21 into a 10-layer structure with barrier layers to suppress dark current from Schottky–Read–Hall and trap-assisted tunneling mechanisms. It is reported that LWIR diodes in nonequilibrium mode have a D* of $10^{11}\,cm\,Hz^{1/2}/W$, a response time of 120 ps, and an extraction ratio of 2.3.

9.16 MOVPE MCT for Scientific Applications

Scientific applications place emphasis on sensitivity to individual photons; the largest commercial market is astronomy, both ground-based and space-based. Infrared detectors based on MBE MCT have an established presence and achieve very low dark current over a wide spectral range from 0.4 to over 13 μm. In recent years, there has been more interest in avalanche photodiode arrays (APDs) to increase sensitivity and allow faster readout speeds. The first application was adaptive optics, where high frame-rate NIR/SW wavefront sensors were required in conjunction with deformable mirrors to correct for atmospheric distortion. Attention has turned to dark background astronomic imaging, where avalanche gain can reduce exposure times and improve the science return of large telescopes.

9.16.1 Linear-Mode Avalanche Photodiode Arrays (LmAPDs) in MOVPE

This concept dates back to the 1980s [67], but the first LmAPD arrays were only produced in the early 2000s (for instance, Beck et al. [68]). A number of centers developed LmAPDs for laser-gated imaging (LGI) or burst-illumination LIDAR (BIL), now well established for long-range target identification [69]. Conventional detectors need a signal of around 200 photons to achieve a useful signal-to-noise. This is due to the photon responsivity, which is typically only 5 μV per photon; many photons are required to overcome the noise floor of the ROIC. APDs amplify the photon responsivity at the pixel, so individual photons could be resolved with enough gain.

The solid-state mechanisms for avalanche gain in MCT have been well described: three recommended papers are Rothman et al. [70], Beck et al. [68], and Kinch and Baker [71] and the references therein. Figure 9.22 shows a potential-energy schematic for a typical photodiode illustrating the single-carrier, cascade-like gain mechanism, together with the history-dependent behavior that leads to the observed low noise figures. The heavy-hole mass is ~ $0.55m_0$ across the entire MCT composition range, resulting in low hole mobility values, and a significant degree of optical phonon scattering. Thus, the heavy hole acquires energy very inefficiently, and readily loses what it does to optical phonons. The effective mass of the electron, on the other hand, is very small, $m_e^*/m_0 \sim 7 \times 10^{-2} E_g$ [8], where E_g is the bandgap in eV. Its mobility is high, particularly at low temperatures, and scattering by optical phonons is weak, which results in significant energy gain at even modest applied fields. The conduction band of MCT is also devoid of any low-lying secondary minima, which allows for large electron energy excursions deep into the band, and hence a high probability of impact ionization, with the generation of electron–hole pairs. In MCT, this is an electron gain mechanism, so the absorber must be p-type, which dictates an n-on-p structure.

MOVPE allows the absorber, junction region, and multiplication region to be independently optimized. This is important for controlling the peak fields around the junction to allow high voltages without incurring excessive dark current. Figure 9.23 shows two device structures employed for e-APDs depending on the required spectral range.

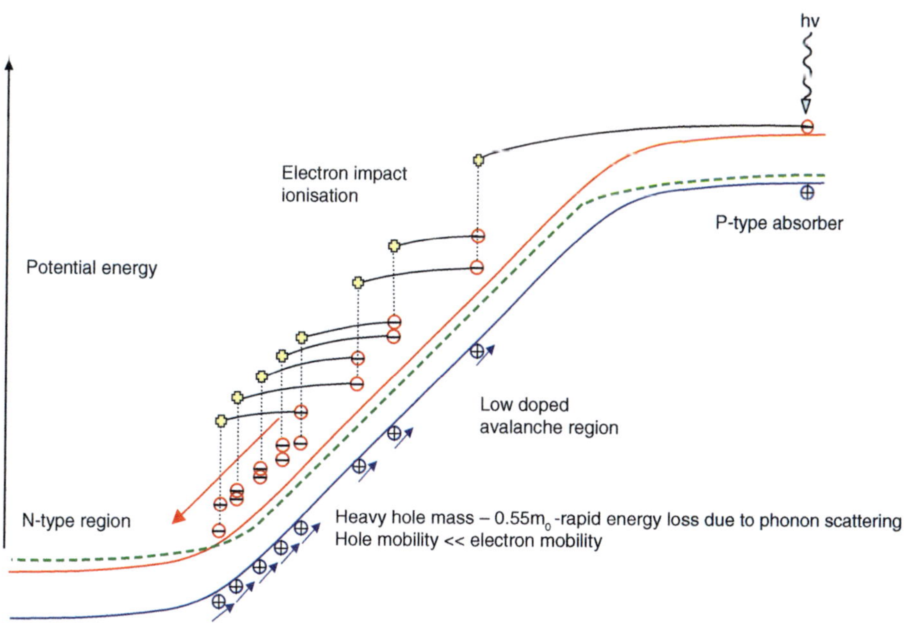

Figure 9.22 *Potential-energy schematic – illustrating history-dependent avalanche gain in MCT. Source: Reproduced by permission from Springer Nature, Handbook of Electronic and Photonic Materials, S. Kasap and P. Capper (eds), 2017, Chapter 34.*

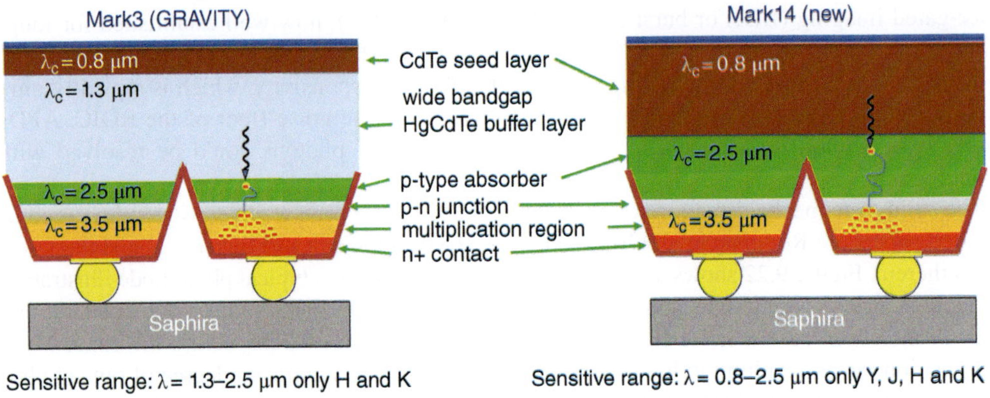

Figure 9.23 *Schematics of two MOVPE e-APD device structures. Source: Courtesy Leonardo MW Ltd.*

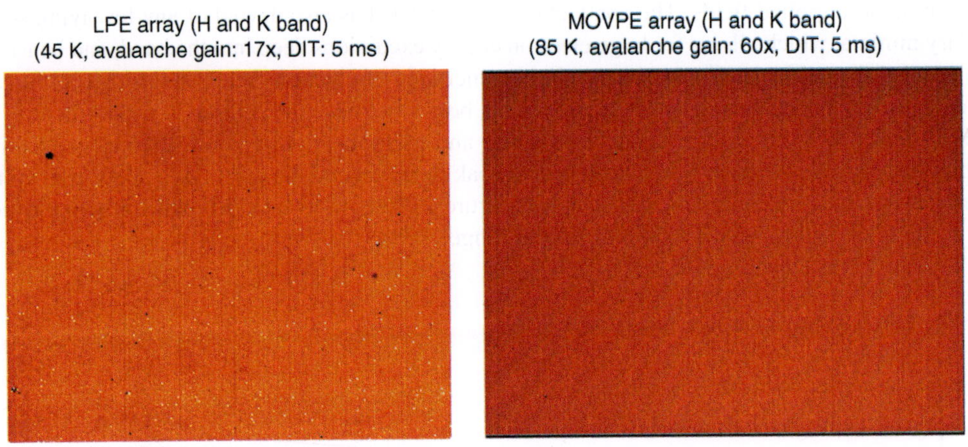

Figure 9.24 *Comparison of an LPE and MOVPE e-APD device. Source: Courtesy Gert Finger ESO.*

It is often quoted that VPE processes cannot achieve the crystalline quality of higher-temperature liquid-phase processes. However, the ability to profile the bandgap through the device (*bandgap engineering*) more than compensates for this effect. Figure 9.24 shows the comparison of a state-of-the-art LPE e-APD with an MOVPE e-APD. The LPE array is on the limit of useful operation due to cosmetics (defective pixels), whereas the MOVPE array has no defective pixels.

Finger et al. [72] and Atkinson et al. [73] summarize the status of LmAPDs using a device called SAPHIRA, following evaluation activity at the European Southern Observatory (ESO) and University of Hawaii, respectively. The excess noise due to the avalanche process is defined by the figure of merit – noise figure. Values of 1.3 at 90 K are typical at gains up to × 400 (noise increase of 1.14 over Poisson statistics) and near-unity at 60 K with gains up to × 600. The avalanche gain versus bias voltage is illustrated in Figure 9.25. In effect, LmAPDs in MOVPE produce noise-free avalanche gain.

MOVPE APDs have set new standards in sensitivity for astronomy. Subelectron read noise is routinely obtained using Fowler sampling, as shown in Figure 9.26 from research at ESO (G. Finger, 2017, to be published). The record read noise of 0.14 electrons RMS compares well with minimum values around 3 electrons RMS from conventional detectors.

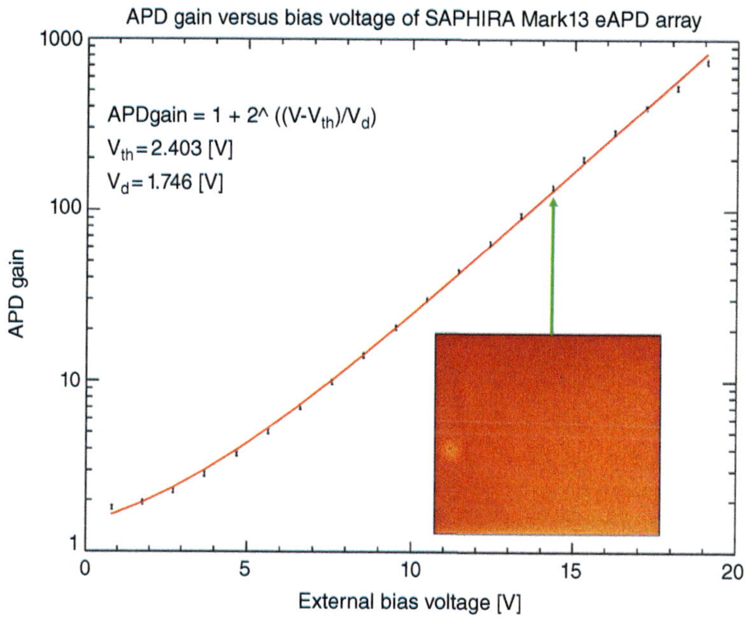

Figure 9.25 *Avalanche gain versus bias voltage of a SAPHIRA eAPD array. Source: Courtesy Gert Finger, ESO.*

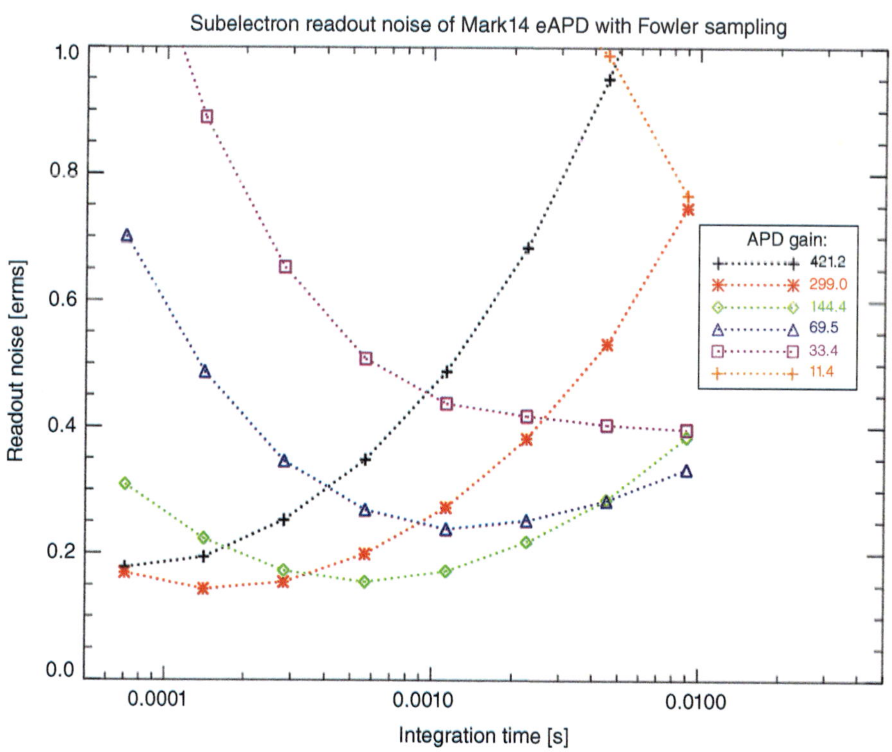

Figure 9.26 *Readout noise with Fowler sampling for Leonardo e-APD. Source: Courtesy Gert Finger, ESO.*

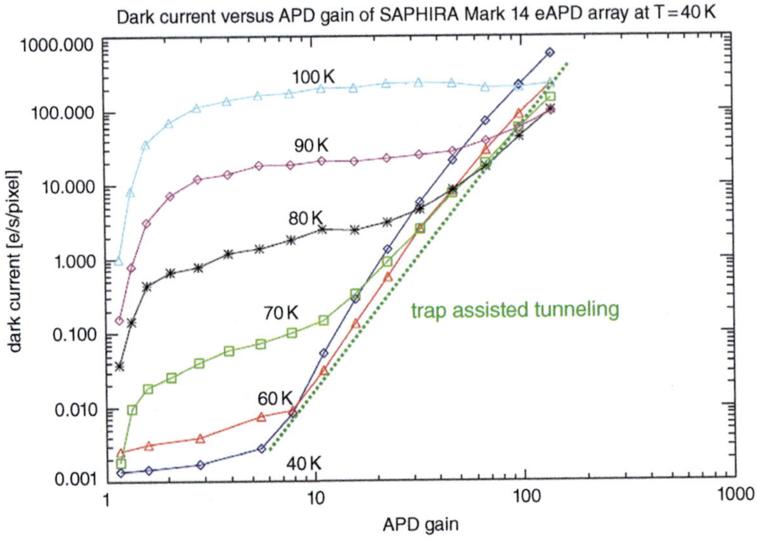

Figure 9.27 *Dark current as a function of avalanche gain and temperature. Source: Courtesy Dani Atkinson (UH) and Gert Finger, (ESO).*

The ultimate benefit of e-APDs arrays is in low-background astronomy where there is potential for shorter observation times, especially in photon-starved applications, such as spectroscopy. In large telescope installations, this has the potential to improve the science return and economics. Key to low-background applications is dark current. The University of Hawaii and ESO have measured dark current to the limit of the instrumentation, and the work is reported in Atkinson (D. Atkinson, 2017, to be published) and Finger. Exposures of up to five days were used to collect enough dark electrons at the lowest temperatures. Figure 9.27 shows the result.

The dark current at low gain (bias voltage) follows a classic diffusion current trend down to 40 K, where stray photons are thought to take over at a level of 4 photons per hour. At higher bias voltages, tunnel current via a trap takes over.

The results allow two conclusions:

1. The flexible bandgap and doping engineering of MOVPE enable far higher performance than other crystal growth systems where the crystal quality may be better but the ability to switch off dark current and noise mechanisms is reduced.
2. The dark current and avalanche gain noise figure of MOVPE matches or possibly exceeds the performance of any other MCT growth system in the most demanding applications.

9.17 Conclusions and Future Trends for MOVPE MCT Arrays

From the mid-to-late 1990s, the MOVPE technique was largely rejected by many countries, and all research efforts concentrated on using MBE for third-generation MCT device technologies. However, it has been shown that MOVPE processes can be used to produce high-quality, large-area IRFPA devices for many applications. In addition, MOVPE offers the advantage of lower overall maintenance costs for reactor systems compared to MBE reactor systems. As a consequence, there has been a renewed interest in the MOVPE process due to the combined benefits of low surface macrodefect densities and the ability to produce large-area growth on low-cost substrates with both n- and p-type doping control. Most importantly, this means the

MOVPE process has the flexibility to grow many different device designs covering a wide spectral range from 0.8–17 μm in a single reactor system.

The field of high-performance thermal imaging requires arrays with larger physical size, smaller pixels, and higher sensitivity. Wafer-scale MOVPE allows this step with a cost similar to current half-TV formats. For small handheld cameras, the emphasis is on low power consumption to provide long battery life, which is a strong function of the operating temperature. There is still considerable potential to raise the operating temperature of MOVPE thermal imaging arrays by geometry reduction and nonequilibrium operation. In poor imaging conditions and difficult scenes, the ability to perform spectral discrimination is paramount, and bispectral and multispectral devices will become essential within the range of 3.5–14 μm. MOVPE has produced outstanding MW/LW arrays and has the flexibility to make many bispectral combinations.

In the scientific field, the spectral response is not limited to the atmospheric windows, and sensors are needed from 0.4 to >14 μm. Such applications are often photon starved, requiring low dark current and avalanche gain. MOVPE has achieved useful gain at the lowest astronomical backgrounds, providing better spectrometer performance, shorter integration periods, and ultimately greater science return from large telescope assets. The minimal absorber volume provides immunity from high-energy protons, making MOVPE devices a good candidate for deep-space missions.

The ultimate performance potential of MCT will ensure that it is the material of choice for all high-performance infrared systems.

Definitions

- LWIR: LW is short for *long wave* and is the term for the transparent atmospheric window between the wavelengths 7.5 μm and 14 μm (sometimes called the *thermal band*). The edges of this window are influenced by the water vapor content and atmospheric conditions.
- MWIR: MW is short for *medium wave* and is the term for the transparent atmospheric window between the wavelengths 3.3 μm and 5 μm. A CO_2 absorption band around 4.25 μm divides the band in two, with better atmospheric transmission in the 3.3–4.2 μm band.
- SWIR: SW stands for *short wave* and is often used for wavelengths between 1.0 μm and 3 μm. The atmosphere is transparent in relatively narrow bands within the SW region. The most common is 2.0–2.25 μm, although 1.5 μm is also important as it is the wavelength for eye-safe lasers.
- ROIC: ROIC stands for *readout integrated circuit* and is commonly used to describe the silicon chip on which the detector material is mounted. The role of the ROIC is to integrate the signal, perform some signal processing, and read out the array. Another term is *multiplexer* or *mux*.
- IDCA: IDCA stands for *integrated detector–cooler assembly* and is a commonly used infrared detector scheme in which the detector is mounted directly on the cold finger of a cryocooler (often based on Stirling cycle engines). The detector and cold finger are then enclosed in a vacuum vessel with a transparent window and optical baffles.
- Narrow bandgap: Refers to a semiconductor with a forbidden energy gap of less than about 0.7 eV, making it suitable for detection in the infrared wavebands.
- Photoconductive: A photoconductive device uses the change in resistance of a slab of semiconductor to measure the extra electron–hole pairs created by the absorption of photons.
- Photovoltaic: A photovoltaic device utilizes a p-n junction in a semiconductor to separate electron–hole pairs (created by the absorption of photons) to generate a voltage.
- Homojunction: When the composition (or bandgap) of the semiconductor is the same on each side of the p-n junction, it is known as a homojunction.
- Heterojunction: When the composition (or bandgap) of the semiconductor is different on each side of the p-n junction, it is known as a heterojunction.

- BLIP: BLIP stands for *background limited performance* and describes the best signal-to-noise performance that can be achieved with a detector. In this condition, the only noise stems from the random arrival rate of photons and is therefore at a theoretical minimum.
- NETD: The *noise equivalent temperature difference* is a measure of sensitivity for a multiplexed infrared detector and is the change in scene temperature that produces a signal equivalent to the RMS noise level of the detector.

References

1. Norton P. (2002). MCT infrared detectors. *Opto-Electronics Review* 10 (3):159–174.
2. Mullin, J.B. et al. (1990). MOVPE growth of narrow and wide gap II/VI materials. *Journal of Crystal Growth* 101: 1.
3. Irvine, S.J.C., et al. (1991). MOVPE growth of MCT. *Semiconductor Science and Technology* 6: C15.
4. Mitra, P., Case, F.C., and Reine M.B. (1998). Progress in MOVPE of MCT for advanced infrared detectors. *Journal of Electronic Materials* 27 (6): 510–520.
5. Irvine, S.J.C., Tunnicliffe, J., and Mullin, B. (1984). The growth of highly uniform cadmium mercury telluride by a new MOVPE technique. *Materials Letters* 2 (4b): 305.
6. Snyder, D.W., Mahajan. S., Ko, E.I., and Sides, P.J. (1991). Effect of substrate misorientation on surface morphology of homoepitaxial CdTe films grown by OMVPE. *Applied Physics Letters* 58: 848.
7. Capper, P., Maxey, C., Whiffin, P., and Easton, B. (1989). Substrate orientation effects in $Cd_xHg_{1-x}Te$ grown by MOVPE. *Journal of Crystal Growth* 96: 519.
8. Capper, P., Maxey, C., Whiffin, P., and Easton, B. (1989). Incorporation and activation of Group V elements in MOVPE–grown CdHgTe. *Journal of Crystal Growth* 97: 833–844.
9. Svob, L., Cheze, I., Lusson, A. et al. (1998). Crystallographic orientation dependence of As incorporation in MOVPE-grown CdTe and corresponding acceptor electrical state activation. *Journal of Crystal Growth* 184/185: 459–464.
10. Madejczyk, P., Piotrowski, A., Gawron, W. et al. (2005). Growth and properties of MOCVD MCT epilayers on GaAs substrates. *Opto-Electronics Review* 13 (3): 239–251.
11. Mitra, P., Case, F.C., Reine, M.B. et al. (1997). Doping in MOVPE of MCT: orientation effects and growth of high performance IR photodiodes. *Journal of Crystal Growth* 170: 542–548.
12. Kebłowski, A., Kopytko, M., Młynarczyk, K. et al. (2017). Investigation of hillocks formation on (100) MCT layers grown by MOCVD on GaAs epi-ready substrates. *Infrared Physics and Technology* 84: 87–93.
13. Madejczyk, P., Piotrowski, A., Kłos, K. et al. (2009). Surface smoothness improvement of MCT layers grown by MOCVD. *Bulletin of the Polish Academy of Technical Sciences* 57 (2): 139–146
14. Madejczyk, P., Gawron, W., Martyniuk, P. et al. (2013). MOCVD grown MCT device structure for ambient temperature LWIR detectors. *Semiconductor Science and Technology* 28: 105017.
15. Piotrowski, A., Madejczyk, P., Gawron, W. et al. (2007). Progress in MOCVD growth of MCT heterostructures for uncooled infrared photodetectors. *Infrared Physics and Technology* 49 (3): 173–182.
16. Piotrowski, A. and Piotrowski, J. (2012). Uncooled infrared detectors in Poland, history and recent progress. *Procedia Engineering* 47: 1506–1512.
17. Madejczyk, P., Piotrowski, A., Kłos, K. et al. (2010). Control of acceptor doping in MOCVD MCT epilayers. *Opto-Electronics Review* 18 (3): 271–6.
18. Martynuik, P., Gawron, W., Stepien, D., et al. (2015). Status of long-wave Auger suppressed MCT detectors operating >200 K, *Opto-Electronics Review* 23 (4): 278.
19. Mitra, P., Case, F.C., Glass, H.L. et al. (2001). MCT growth on (552) oriented CdZnTe by metalorganic vapor phase Epitaxy. *Journal of Electronic Materials* 30 (6): 779–784.
20. Maxey, C.D., Fitzmaurice, J.C., Lau, H.W. et al. (2006). Current status of large-area MOVPE growth of MCT device heterostructures for infrared focal plane arrays. *Journal of Electronic Materials* 35: 6.
21. Maxey, C.D. (1993). Doping studies in MOVPE-grown CdHgTe. *Semiconductor Science and Technology* 8: S183.
22. Maruyama, K., Nishino, H., Okomoto, T. et al. (1996). Growth of (111) MCT on (100) Si by MOVPE using metalorganic tellurium adsorption and annealing. *Journal of Electronic Materials* 25 (8): 1353–1357.
23. Wang, W. and Bhat, I. (1995). Growth of high quality CdTe and ZnTe on Si substrates using OMVPE. *Journal of Electronic Materials* 24 (5): 451–455.
24. Rao, S., Shintri, S., and Bhat, I. (2009). MOVPE Growth of (211)B CdTe on Si using Ge interface layer. *Journal of Electronic Materials* 38 (8): 1618–1623
25. Hall, D., Buckle, L., Gordon, N.T. et al. (2004). High-performance long-wavelength HgCdTe infrared detectors grown on silicon substrates. *Applied Physics Letters* 85 (11): 2113.

26. Hails, J., Keir, A., Graham, A. et al. (2007). Influence of the Si substrate on defect formation in MCT grown on II/VI buffered Si using a combined MBE/MOVPE technique. *Journal of Electronic Materials* 36 (8): 864.
27. Irvine, S.J.C., Edwall D.D., Bubulac, L.O. et al. (1992). Study of microinhomogeneities in midwave infrared mercury cadmium telluride grown by metalorganic chemical vapor deposition-interdiffused multilayer process onto GaAs and GaAs/Si substrates. *Journal of Vacuum Science and Technology B* 10 (4): 1392.
28. Jones, C.L., Hipwood, L.G., Shaw, C.J. et al. (2006). High-performance MW and LW IRFPAs made from MCT grown by MOVPE. *Proceedings of the SPIE - The International Society for Optical Engineering* 6206: 620610.
29. Tennant, W.E., Cockrum, C.A., Gilpin, J.B. et al. (1992). Key issues in MCT-based focal plane arrays: An industry perspective. *Journal of Vacuum Science and Technology B* 10: 1359–1369.
30. Nishino, H., Murakami, S., Saito, T. et al. (1995). Dislocation profiles in MCT (100) on GaAs (100) grown by metalorganic chemical vapor deposition. *Journal of Electronic Materials* 24 (5): 533.
31. Hails, J. (1994). Identification of the volatile decomposition products produced in the deposition of (Hg,Cd)Te by MOVPE. *Advanced Materials for Optics and Electronics* 3: 151.
32. Suh, S., Kim, S., Kim, H., and Song, J. (2002). Control of hillock formation during MOVPE growth of MCT device by suppressing the pre-reaction of Cd precursor with Hg. *Journal of Crystal Growth* 236: 119–124.
33. Czerniak, M. and Easton, B. (1984). An investigation of the pyrolysis of dimethyl cadmium and diethyltelluride by in-situ gas sampling and analysis. *Journal of Crystal Growth* 68: 128–135.
34. Piotrowski, A. and Klos, K. (2007). Metalorganic chemical vapor deposition of MCT fully doped heterostructures without post growth anneal for uncooled MWIR and LWIR detectors. *Journal of Electronic Materials* 36 (8): 1052.
35. Jones, A. (1993). Metalorganic precursors for vapor phase epitaxy. *Journal of Crystal Growth* 1291: 728–773.
36. Malik, M. and O'Brien, P. (1994). Mixed alkyl zinc or cadmium complexes with dialkyl thio- or selenocarbamates: Precursors for cadmium chalcogenides. *Advanced Materials for Optics and Electronics* 3: 171.
37. Yasuda, K., Hatano, H., Ferid, T. et al. (1994). Low temperature growth of (100) MCT layers with DtBTe in metalorganic vapor phase epitaxy. *Journal of Electronic Materials* 24: 1093.
38. Hails, J., Cole-Hamilton, D., Stevenson, J., and Bell, W. (2000). Allyl-iso-propyltelluride, a new MOVEP precursor for CdTe, HgTe and (Hg,Cd)Te. *Journal of Crystal Growth* 214/215: 45–49.
39. Edwall, D., Chen, J.S., Bajaj, J. et al. (1990). MOCVD $Hg_{1-x}Cd_xTe$/GaAs for IR detectors. *Semiconductor Science and Technology* 5: S221–S224.
40. Hails, J., Cole-Hamilton, D., and Bell, W. (1994). Methyl(allyl)telluride as a Te precursor for growth of (Hg,Cd)Te by metalorganic vapor phase epitaxy. *Journal of Crystal Growth* 145: 74.
41. Irvine, S., Bajaj, J., and Bubulac, L. (1993). MOVPE growth of MCT For LWIR detectors. *Journal of Electronic Materials* 22: 859.
42. Song, J., Kim, J., Park, M. et al. (1998). Iodine and arsenic doping of (100) MCT/GaAs grown by metal organic vapor phase epitaxy using IPI and tDMAAs. *Journal of Crystal Growth* 184 (5): 1232–1236.
43. Maxey, C.D., Jones, C.L., Metcalfe, N.E. et al. (1996). MOVPE growth of improved non-equilibrium MCT device structures for near ambient temperature heterodyne detectors. *Journal of Electronic Materials* 25: 8–1276.
44. Clerjaud, B., Cote, D., Svob, L. et al. (1993). Hydrogen-acceptor pairing in CdTe epitaxial layers grown by OMVPE. *Solid State Communications* 85: 167.
45. Mitra, P., Tyan, Y.L., Case, F.C. et al. (1996). Improved arsenic doping in metalorganic chemical vapor deposition of MCT and in-situ growth of high performance long wavelength infrared photodiodes. *Journal of Electronic Materials* 25: 1328–1335.
46. Maxey, C.D., Jones, C.L., Metcalfe, N.E. et al. (1997). MOVPE growth of improved non-equilibrium MCT device structures for near ambient temperature heterodyne detectors. *Proceedings of the SPIE - The International Society for Optical Engineering* 3122: 453–464.
47. Hails, J.E., Cole-Hamilton, D.J., and Giess, J. (1998). The origin of hillocks in (Hg, Cd)Te grown by MOVPE. *Journal of Electronic Materials* 27 (6): 33.
48. Suh, S., Song, J., and Moon, S. (1996). Metalorganic vapor phase epitaxial growth of hillock free (100) MCT/GaAs with good electrical properties. *Journal of Crystal Growth* 159: 1132–1135.
49. Capper, P. (2017). *Handbook of Electronic and Photonic Materials* (eds. S. Kasap and P. Capper), ch 15.3.3, 357. New York: Springer.
50. Irvine, S. and Bajaj, J. (1993). In situ characterization techniques for monitoring and control of VPE growth of $Hg_{1-x}Cd_xTe$. *Semiconductor Science and Technology* 8: 860.
51. Gordon N.T. and Baker I.M. (2001). In: *Infrared Detectors and Emitters: Materials and Devices* (eds. Capper and Elliott), ch 2, 23–42. Kluwer Academic Publishers.
52. Reine, M.B. (2001). In: *Infrared Detectors and Emitters: Materials and Devices*, (eds. Capper and Elliott), ch 12. Kluwer Academic Publishers.
53. Tennant, W.E., Lee, D., Zandian, M. et al. (2008). MBE MCT technology: A very general solution to IR detection, described by Rule 07, a very convenient heuristic. *Journal of Electronic Materials* 37: 1406.

54. Tennant, W.E. (2010). Rule 07 revisited: Still a good heuristic predictor of p/n MCT photodiode performance? *Journal of Electronic Materials* 39: 1030.
55. Kinch, M.A. (2007). *Fundamentals of Infrared Detector Materials*. SPIE Press.
56. Hipwood, L.G., Jones, C.L., Walker, D. et al. (2007). Affordable high-performance LW IRFPAs made from HgCdTe grown by MOVPE. *Proceedings of the SPIE - The International Society for Optical Engineering* 6542: 65420I.
57. Hipwood, L.G., Baker, I.M., Jones, C.L. et al. (2008). LW IRFPAs made from MCT grown by MOVPE for use in multispectral imaging. *Proceedings of the SPIE - The International Society for Optical Engineering* 6940: 69400G.
58. Baker, I.M., Maxey, C.D., Hipwood, L. et al. (2012). Developments in MOVPE MCT arrays for passive and active infrared imaging. *Proceedings of the SPIE - The International Society for Optical Engineering* 8542: 85421A.
59. Driggers, R. et al. (2012). Infrared detector size – how low should you go? *Optical Engineering* 51 (6): 063202.
60. McEwen, K., Jeckells, D., Bains, S., and Weller, H. (2015). Developments in reduced pixel geometries with MOVPE grown MCT arrays. *Proceedings of the SPIE - The International Society for Optical Engineering* 9451: 94512D.
61. Sparrow, C.M. (1916). On spectroscopic resolving power. *Astrophysics Journal* 44: 76–87.
62. Thorne, P., Weller, H., Hipwood, L.G. et al. (2012). 12 µm pixel pitch development for 3-side buttable megapixel MW FPAs. *Proceedings of the SPIE - The International Society for Optical Engineering* 8353: 83532J.
63. Pillans, L., Ash, R.M, Hipwood, L., and Knowles, P. (2012). MWIR mercury cadmium telluride detectors for high operating temperatures. *Proceedings of the SPIE - The International Society for Optical Engineering* 8353: 83532W.
64. Casselman, T.N., Chapman, G.R., Kosai, K. et al. (1991). *US Workshop on Physics and Chemistry of MCT and other II-IV compounds*, Dallas, TX, Oct.
65. Elliott, C.T., Gordon, N.T., and White, A.M. (1999). Towards background-limited, room-temperature, infrared photon detectors in the 3–13 µm wavelength range. *Applied Physics Letters* 74: 2881.
66. Elliott, C.T. and Ashley, T. (1985). Nonequilibrium devices for infra-red detection, *Electronics Letters* 21: 451.
67. Elliott, C.T., Gordon, N.T., Hall, R.S., and Crimes, G.J. (1990), Reverse breakdown in long wavelength lateral collection $Cd_xHg_{1-x}Te$ diodes. *Journal of Vacuum Science and Technology A* 8: 1251–1253.
68. Beck, J.D., Scritchfield, R., Mitra, P. et al. (2011). Advanced photon counting techniques. *SPIE Conference Series* 8033.
69. Baker, I.M., Duncan, S., and Copley, J. (2004). Low noise laser gated imaging system for long range target identification. *Proceedings of the SPIE - The International Society for Optical Engineering* 5406: 133–144.
70. Rothman, J., Mollard, L., Bosson, S. et al. (2012). Short-wave infrared MCT avalanche photodiodes. *Journal of Electronic Materials* 41 (10): 2928–2936.
71. Kinch, M.A. and Baker, I.M. (2011). MCT electron avalanche photodiodes. In: *Mercury Cadmium Telluride - Growth, Properties and Applications* (eds. P. Capper and J. Garland), ch 21. Wiley UK.
72. Finger, G., Baker, I.M., Alvarez, D. et al. (2013). NIR MCT avalanche photodiode arrays for wavefront sensing and fringe tracking. *Optical and IR Telescope Instrumentation and Detectors*, Scientific Detector Workshop, Florence Italy.
73. Atkinson, D., Hall, D., Baranec, C. et al. (2014). Observatory deployment and characterization of SAPHIRA MCT APD arrays. *Proceedings of the SPIE - The International Society for Optical Engineering* 9154: 915419.

10

Cadmium Telluride and Related II-VI Materials

G. Kartopu and S.J.C. Irvine
Centre for Solar Energy Research, College of Engineering, Swansea University, OpTIC Centre, St. Asaph, UK

10.1 Introduction and Historical Background

Cadmium telluride is a II-VI semiconductor material with a room-temperature bandgap of 1.45 eV. This places it in the ideal range for solar-absorber materials for efficient photovoltaic (PV) solar cells and the relatively high atomic weight of Cd and Te enable efficient X-ray and gamma-ray detectors. These applications will be described in greater detail in this chapter.

The origins of CdTe MOVPE go back to the early work of Manasevit and Simpson [1]. In this paper, the authors were attempting to reproduce their earlier successes with III-V semiconductors, and the approach was essentially the same. An organometallic Group II precursor was reacted over a hot substrate with a Group VI hydride. This worked well for ZnSe and ZnS but not for CdTe. The hydride H_2Te is unstable and was not available commercially, so alkyl alternatives were used. Successful heteroepitaxy of (111) CdTe onto (0001) sapphire (Al_2O_3) was demonstrated, with dimethyl cadmium (DMCd) and dimethyl telluride (DMTe), and was probably the first successful demonstration of epitaxial growth with all organometallic sources. Later, both ZnSe and ZnS were grown from all organometallic sources as the preferred approach to achieve better control over uniformity of the epitaxial layers [2].

Anyone who has worked with chalcogenide organometallics will know that the smell is very strong and extremely unpleasant. The organometallic molecule is also an effective way of absorbing these elements into the body through the respiratory system. Each of the chalcogenides has its own characteristic odor, and for tellurium it can be described as a "garlic" odor. Therefore, extreme caution must be taken to ensure that all surfaces exposed to tellurium organometallics are not handled with bare hands and that there is sufficient fume extraction and filtration to avoid inhaling the

Metalorganic Vapor Phase Epitaxy (MOVPE): Growth, Materials Properties, and Applications,
First Edition. Edited by Stuart Irvine and Peter Capper.

fumes. As with all chalcogenides, the nose is very sensitive to tellurium organometallics and so can act as an early warning of unwanted contamination. With good "housekeeping" and by following strict procedures, these precursors can be handled safely. An advantage of using all organometallic sources is that the reactor exhaust abatement is actually easier, where disposable graphite filters are extremely effective [3]. This avoids the need for chemical scrubbers and pyrolysis abatement that is commonly used in III-V MOVPE.

Since the early work using DMTe, there has been a lot of research on seeking alternative tellurium precursors that decompose at lower temperature (less than 500 °C) to obtain better control of the electrical properties of the CdTe films. A motivation for lower-temperature growth has also been the growth of the infrared detector alloy mercury cadmium telluride (MCT); see Chapter 9. The alternative precursors to achieve growth temperature below 300 °C are discussed in Section 10.4; alternative methods to achieve low-temperature growth involve energy-assisted decomposition, where the homolyses of the organic ligands is assisted by either UV irradiation or plasma-assisted MOCVD. Both topics are covered later in this chapter.

Homoepitaxy has been achieved using CdTe and (Cd,Zn)Te (CZT) substrates. However, the issue of obtaining large-area substrates has proved to be difficult and expensive. This has led to research on high-quality single-crystal layers of CdTe onto more readily available substrates such as Si, GaAs, and sapphire. These substrates are also of interest in device applications such as infrared detector arrays that need to be connected to a silicon multiplexer with the minimum of bond strain during cooling to operational temperature. The search for alternative substrates goes from near-lattice-matched in the case of InSb to 14% lattice mismatch in the case of GaAs and even 20% lattice mismatch for Si. The energy bandgap versus lattice parameter plot of II-VI compounds and alloys is reproduced in Figure 10.1 from [4].

From Figure 10.1, it can be seen that CdTe has a large lattice parameter of ~0.65 nm for the cubic zincblende structure. The only substrate with a similar lattice parameter is InSb, which will be discussed in greater detail in Section 10.3. One of the applications of CdTe is as a buffer for the growth of epitaxial MCT, where the CdTe layer is used to reduce the strain due to the large lattice mismatch and provide a near lattice-matched substrate for the MCT layers. The very large lattice

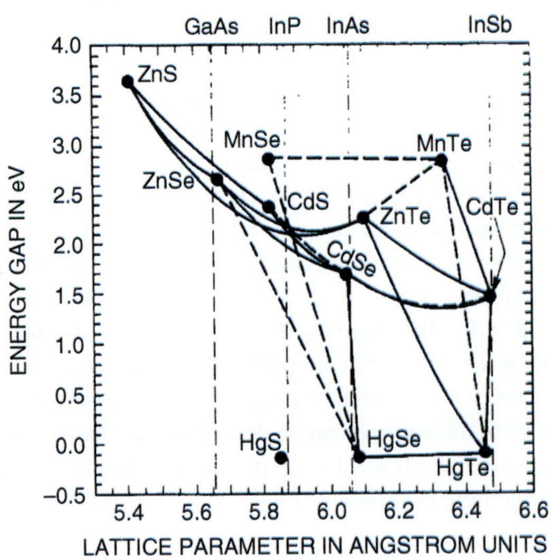

Figure 10.1 *Bandgap energy and lattice parameters for II-VI compound and alloys with lattice parameters of commonly used substrates indicated. Source: Reprinted from Kisker 1989 [4] with permission from Elsevier.*

mismatch with GaAs and Si (0.54 nm) can be reduced with an intermediate lattice parameter compound such as ZnTe. For ZnSe and related alloys, there is a good match with GaAs, and this has been a popular substrate for electroluminescent devices structures. The upsurge in GaN in the 1990s has left these materials as largely of historical interest, but current research interests remain in quantum dots of CdS and CdSe.

10.2 CdTe Homoepitaxy

A lot of the interest in MOVPE of CdTe has been directed toward the remarkable flexibility with heteroepitaxy on alternative substrates, and there was relatively little work on homoepitaxy. One of the challenges with CdTe substrates was to produce a stoichiometric and oxide-free surface for epitaxial growth. Surface preparation is normally through the use of a dilute Br/methanol solution that is used for chemomechanical polishing of the surface to remove damage. This leaves a Te-rich surface that will naturally form TeO_2 during a prolonged exposure to air. However, this can be readily removed by heating the substrate to over 400 °C under a stream of hydrogen prior to growth [5].

Substrate orientation has proved to be very important in controlling surface morphology and defects. The (111)A surface (Cd-terminated) produces a rough pyramid-like surface, while the (111)B (Te-terminated surface) produces a very smooth surface, but it is highly twinned. On-orientation (100) produces a rough surface with pyramidal hillocks; but with vicinal surfaces, where the substrates are off-cut toward the (110) planes, smooth surfaces can be produced that are also twin free [6].

The crystalline quality of CdTe homoepitaxial layers is not superior to the nonlattice-matched comparisons of sapphire and GaAs. For example, Sochinskii et al. [7] found that Rutherford backscattering (RBS) channeling was lowest for sapphire; and low-temperature photoluminescence (PL) for CdTe on CdTe(111) substrates was dominated by the donor–acceptor pair band ($D^0 - A^0$) at 843 nm. This was attributed to an increase in acceptor concentration associated with Cd vacancies compared with CdTe grown onto sapphire substrates. Details of nonlattice-matched CdTe heteroepitaxy will be given in the next section.

10.3 CdTe Heteroepitaxy

The limited size and availability of CdTe substrates has stimulated the search for alternative substrates that are more readily available. There are also considerations of the optical, electrical, and thermal properties of alternative substrates where they become a functional part of the application. For example, infrared detectors may require the substrate to be a suitable infrared window. Sapphire is highly transparent in the 3–5 µm infrared window but not in the 8–14 µm window where both GaAs and Si are both highly transparent. This chapter is concerned with the MOVPE of CdTe onto different substrates to provide a good crystalline and morphological surface for subsequent MCT epitaxy; see Chapter 9.

10.3.1 InSb

InSb substrates have the closest lattice match to CdTe with a lattice parameter of 0.6479 nm compared with 0.6482 nm for CdTe. They are also more readily available in larger substrate sizes than for CdTe and with excellent crystal quality and purity. The first demonstration of the growth of CdTe on InSb was by Farrow et al. [8] using molecular beam epitaxy (MBE). The CdTe films were lattice-matched with misfit being taken up with a tetragonal strain in the 1-µm thick CdTe.

This was followed by the first MOVPE growth of CdTe on InSb in 1984 by Hoke et al. [9], where (110) and (211) InSb substrates were used. The optimum growth temperature of 440 °C was somewhat higher than that for the earlier MBE work, and specular CdTe films were reported on both orientations that were used for subsequent growth of MCT epitaxial layers. A concern with InSb substrates is the noncongruent evaporation of In and Sb, which can leave the surface rich in In and can leave In droplets on the surface. This will lead to macrodefects in the epitaxial layer and diffusion of In (an n-type dopant) into the CdTe layer. Growth of epitaxial CdTe onto InSb (100) was demonstrated by Taskar et al. at temperatures similar to the work of Hoke et al. [10]. Again, excellent crystal quality was demonstrated, and the PL was superior to that of homoepitaxy on CdTe substrates. Growth of CdTe epitaxy onto InSb (111)B substrates gave specular surfaces in the temperature range 350–375 °C with good near band-edge PL emission up to a growth temperature of 410 °C [11]. This early success was not followed up due to the relative instability of InSb compared with that of sapphire and GaAs and problems with segregation of In into the CdTe films. There was also a realization that CdTe had the remarkable property that large lattice mismatches could be accommodated, more so than for III-V epitaxial layers.

10.3.2 Sapphire

Deposition of CdTe onto sapphire (0001) substrates goes back to the original work by Manasevit and Simpson and became important because the sapphire substrate is robust and is an excellent window for midwave infrared. A detailed study by Zuniga-Perez et al. on the structural characterization of CdTe deposited onto (0001) sapphire has shown that the precursor ratio of DMCd to diisopropyl tellurium (DIPTe) is critical to obtain a single-crystal layer of CdTe [12]. As the temperature is increased toward 400 °C, the II/VI precursor ratio is less critical in obtaining a single-crystal film but does affect polarity and therefore twinning in the CdTe films. This is illustrated in the work by Zuniga-Perez et al. [12], where they varied the II/VI precursor ratio at a substrate temperature of 414 °C. The micrographs of the surfaces are shown in Figure 10.2. At the lower II/VI ratio, the surface has shallow pyramidal facets and becomes more disorganized around the stoichiometric ratio in Figure 10.2b. This occurs due to 60° rotational twins in the (111) plane. At higher II/VI ratios the surface morphology is more faceted but has a lower twin content and narrower X-ray rocking curve width. This surface morphology is consistent with (111)B face morphology of CdTe on CdTe (111)B substrates.

Figure 10.2 *SEM images of CdTe on (0001) sapphire substrates at 414 °C where the II/VI ratio was varied from (a) 0.67, (b) 1.05, and (c) 1.98. The scale marker is 2.5 μm. Source: Reprinted from Zuniga-Perez et al. 2004 [12] with permission from Elsevier.*

It was observed by Glass et al. that the microtwin content can be detected by shining a light at the rotating substrate where bright reflections are seen every 60°, rather than the expected 120° rotation, and was verified by X-ray diffraction [13]. These authors also showed that the microtwin content increased with film thickness but could be largely removed by annealing for 20 min at 650 °C. Glass et al. [13] used the Manasevit combination of DMCd and DMTe that enabled higher-temperature growth (around 460 °C) than for DIPTe. This process is the basis of the Rockwell PACE-1 process for producing high-quality CdTe substrates for Liquid Phase Epitaxy (LPE) growth of MCT in the 3–5 μm infrared window.

10.3.3 GaAs

GaAs has been the most widely studied alternative substrate for CdTe epitaxy and has produced high-quality CdTe despite the large lattice mismatch (14%), as indicated in Figure 10.1. In a study on the effect of GaAs substrate orientation on the orientation of the CdTe layer, it was found that the in situ heat treatment, to remove surface oxide, and the growth temperature both contributed to the CdTe film orientation [14]. The presence of residual CdTe in the reactor from previous growth runs also has an influence. For a 10-min anneal of the GaAs substrate at 580 °C, with residual CdTe in the reactor, the orientation is (100) for growth temperatures below 400 °C and (111) for growth temperatures above 400 °C.

A detailed study of the surface chemistry of GaAs substrates prior to MOVPE of CdTe by Ekawa et al. has revealed that annealing for 5 min at 500 °C, under a flow of H_2 in the MOVPE reactor, prior to growth, restores the stoichiometry of the GaAs surface and also leaves a monolayer of Te on the surface, sublimed from previously deposited CdTe [15].

For the (100)//(100) orientation relationship, it has been shown by high-resolution TEM that the mismatch is taken up by a regular array of misfit dislocations, spaced approximately 3 nm apart [16]. The Matthews criterion for strain relaxation with misfit dislocations, with such a large lattice mismatch, occurs within one monolayer, which enables the formation of the misfit dislocations within the first monolayer. Between the misfit dislocations, the lattice planes are continuous through the interface. However, this does not mean the crystal structure above the interface is free of dislocations; and threading dislocations are a common feature, as in all heteroepitaxy.

The lowest crystal-defect density is achieved for growth on the GaAs (100) surface, but "hillocks" are formed for thick films, producing a poor substrate for subsequent MCT growth. As with the growth onto CdTe (100) substrates, the best crystalline structure and smoothest morphology are achieved for misoriented substrates, just a few degrees off the (100) plane. This promotes step-flow growth that is the preferred mode for CdTe MOVPE and avoids defect formation and additional strain at the coalescence of the islands in 2D growth. This will also help to organize the misfit dislocation where there may otherwise be a discontinuity of the periodicity of the misfit dislocations. A detailed study of the effects of substrate tilt was made by Hamilton et al. using a dome-polished GaAs (100) substrate [17]. Misalignment of the crystal surface toward the <111>A, <011>, and <111>B all suppressed the hillock formation. Misalignment in the <111>B direction showed a reduction in the X-ray rocking curve width and was considered to be the best crystal direction for misalignment to yield good surface morphology and good crystal structure. In this work, a thin nucleation layer of ZnTe was used and produces an intermediate lattice parameter to "step" the lattice mismatch (see Figure 10.1). The ZnTe layer also has the benefit of ensuring a (100) orientation and avoiding the (111) growth described earlier. A comparative study of CdTe (100) growth onto GaAs (100) with and without a thin ZnTe buffer clearly showed via both RBS channeling and double-crystal X-ray rocking curve width that the crystalline quality in the CdTe layer is improved with the presence of the ZnTe buffer [18]. More recent work on ZnTe-buffered CdTe has shown that there is an optimum thickness of the ZnTe film of around 30 nm.

This is considerably thinner than in the earlier work and shows how critical the initial growth can be in determining the crystalline quality of the CdTe layer [19].

10.3.4 Silicon

The intention to grow CdTe-based epitaxial layers on crystalline Si stems from the need to increase dimension and cost efficiency of optoelectronic devices such as imaging arrays. For example, using single-crystal substrates, such as (Cd,Zn)Te, excellent quality detectors and X-ray/gamma-ray imaging arrays can be fabricated [20], but since those substrates are expensive and small, devices must be put into a mosaic array to increase the surface area.

Heteroepitaxial CdTe films can be grown on Si substrates despite the 19% lattice mismatch. It is common to both MOVPE and MBE techniques that use of a 211-oriented Si wafer, usually with a few degrees of misorientation cut surface, facilitates the epitaxial crystal growth. The (211)Si substrate is typically prepared by dilute HF acid etching to remove the native oxide, and then an anneal is performed with arsenic (As) overpressure to retain the surface crystallinity and promote Cd bonding to As first during CdTe growth, which enhances the interface abruptness. Additionally, Ge and ZnTe interlayer films with several tens to hundreds of nanometer thickness are shown to improve the crystal quality for thick CdTe epilayers (Figure 10.3) by observing XRD rocking curves and Nomarski microscopy coupled with Everson etch [21].

Heteroepitaxial CdTe films can then be used to template other epitaxial films, such as MCT and CZT, for various detector applications, as well as photovoltaics. For example, for a tandem solar cell fabricated by forming an epitaxial CZT top cell (1.8 eV bandgap) on a (211)Si bottom cell (1.1 eV bandgap), an usually high open-circuit potential of 1.75 V was demonstrated, displaying ~17% power conversion efficiency [22]. However, further work would be needed to develop this technology for it to be cost-competitive with other technologies including polycrystalline thin-film CdTe photovoltaic solar devices; these technologies are described in the following sections.

10.4 Low-Temperature Growth and Alternative Precursors

The low-temperature growth of CdTe was originally stimulated by the growth of the infrared detector alloy MCT; see Chapter 9. The original work on MOVPE of CdTe by Manasevit [1] used the Te precursor DMTe, which requires temperatures over 500 °C. A study of the independent pyrolysis of Cd and Te precursors by Irvine and Mullin [23] showed that the limitation on growth temperature was the Te precursor and not the DMCd. In this study, DETe was used to enable growth around 400 °C. Lower-temperature growth of CdTe and MCT was achieved using alternative Te precursors with increasing molecular weight of the alkyl ligands. Hoke et al. [24] explained the trend in stability as the delocalization of electron density over the organic ligand that resulted in the stability trend for homolytic fission as follows:

$$R_2Te \text{ where } R = Me > Et > i-Pr > t-Bu > allyl$$

This work led to the popular use of disopropyltelluride (DIPTe) where it was possible to grow CdTe in the range 320–400 °C, as first reported by Hoke and Lemonias [25]. Work by Berrigan et al. [26] was able to take this further where growth of CdTe using DIPTe at a substrate temperature as low as 300 °C was demonstrated.

Developing precursors with ever-decreasing ligand stability risks the precursor not being sufficiently stable in the bubbler. The best balance could be found by using mixed-ligand organometallics

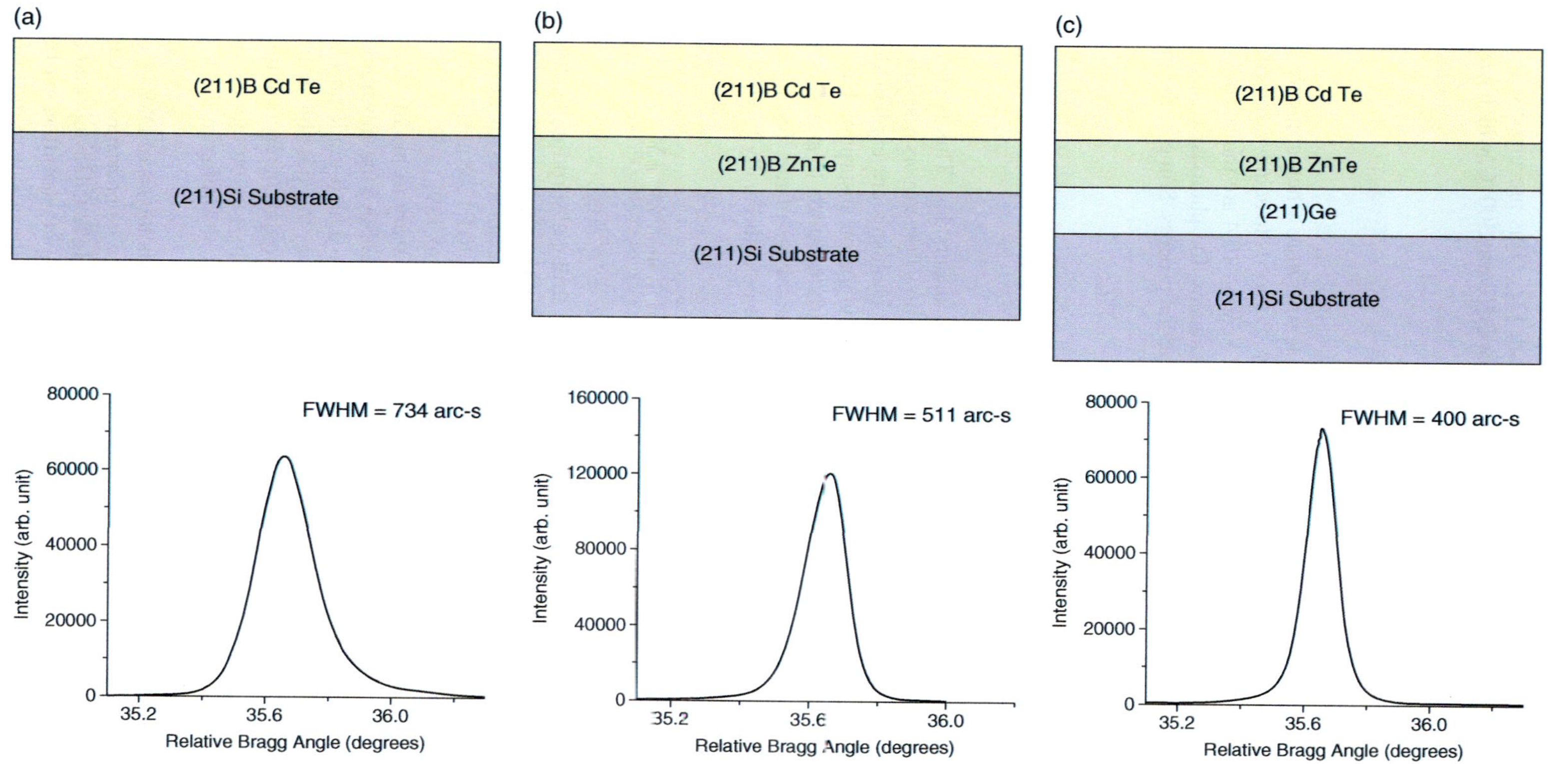

Figure 10.3 *Layer structures and XRD (422) rocking-curve scans of 2-μm MOVPE CdTe films deposited (a) directly on Si (no misorientation cut), (b) using 200-nm ZnTe interlayer, and (c) using 200-nm ZnTe/300-nm Ge interlayers. CdTe film grown using ZnTe/Ge intermediate layers displays the lowest full width at half-maximum (FWHM), indicating the best crystal quality with the CdTe film. Source: Reprinted from Rao et al. 2011 [21] with permission from Springer.*

such as methylallytelluride. One of the most-optimized precursors for low-temperature growth of Te compounds is allylisopropyltelluride, reported by Hails et al. [27]. This precursor was shown to decompose at 200 °C, but the minimum growth temperature for CdTe was 300 °C, which was reported to be limited by the stability of DMCd.

10.5 Photoassisted MOVPE

Photoassisted MOVPE provides a lowering of the growth temperature when compared with the normal thermal reactions of the precursors. Therefore, the motivation is to achieve epitaxial growth at temperatures below 300 °C. The precursors discussed in Section 10.4 are the starting point for any energy-assisted-MOVPE process; however, the mechanisms for photoassistance can vary widely and, effectively, insert into different stages in the reaction process. Some of the more commonly used photoassisted processes are as follows:

1. Ultraviolet (UV) photolysis of the gas-phase precursor.
2. UV photosensitization, where an intermediate UV absorber produces a reactive excited state.
3. Surface photocatalysis with bandgap radiation to release charge at the surface, to aid decomposition of the precursor adsorbate.

The first mechanism is the most common and the easiest to understand. This is also the most appropriate for II-VI MOVPE, where both the Group II and Group VI precursors normally undergo photolysis for accessible wavelengths >200 nm. The basic mechanisms are explained in Irvine et al. [28]. For the Group II alkyls, there is a broad electronic-vibration absorption band between 200 and 260 nm with a long-wave absorption tail associated with transition to a dissociative electronic state. Any photon energy in this band would lead to photolysis of the first ligand, which results in an unstable Me(M)* radical that would rapidly decompose thermally to yield the metal in the reaction sequence shown here:

$$\mathrm{Me_2}(\mathrm{M}) + h\nu(200-260\,\mathrm{nm}) \rightarrow \mathrm{Me}(\mathrm{M})^* + \mathrm{Me}^* \rightarrow 2\mathrm{Me}^* + \mathrm{M}$$

where M = Zn, Cd, Hg, etc.

For the Group VI alkyls, the spectrum is more complex, but strong absorption, leading to photolysis, occurs in the same waveband region. The broad nature of the absorption spectra has enabled a variety of different UV sources to be used to bring about photolysis of the precursors, including both low- and high-pressure Hg lamps, Xenon lamps, and KrF (248 nm) excimer lasers. A more-detailed analysis of reaction mechanisms is given by Liu et al. [29].

To grow epitaxial layers using photoassisted MOVPE, it is necessary to modify the MOVPE reactor cell to allow the UV radiation to illuminate the reaction zone above the substrate. This can be achieved in both horizontal and vertical cell configurations where a high-grade quartz window is provided to allow good transmission down to 200 nm. This may be possible using with a hydroxyl-radical-free grade of quartz. In some cases, the wall of the quartz reaction tube can be used, if it is of sufficiently high-grade material. In addition to transmission through the quartz, the inner surface must be kept free of reaction products that are highly absorbing in the UV. This is achieved through flushing pure H_2 or N_2 over the surface.

Successful epitaxial growth of CdTe has been achieved using UV photolysis of DMCd and DIPTe using a KrF excimer laser in the temperature range 100–150 °C [30]. Low-temperature growth of CdTe and HgTe using photoassisted MOVPE has enabled the growth of HgTe-CdTe superlattices in the range 182–240 °C by Ahlgren et al. [31]. Higher temperatures cause rapid interdiffusion of the superlattice, and a growth temperature below 200 °C is required to preserve the quality of the interfaces.

The photoassisted mechanisms 2 and 3 both relate to promoting surface-reaction mechanisms as opposed to gas-phase photolysis. The latter can produce very high gas-phase supersaturation that, at low temperature, can lead to homogeneous gas-phase nucleation. This results in particulates that can introduce defects into the layer and reduce the supply of the precursors (or products) to the surface. The UV photosensitization can work well if an appropriate photosensitizer is available. This was used effectively for HgTe, where adsorbed Hg atoms provided the sensitizer for photoassisted decomposition of DETe [32].

The third approach of surface photocatalysis does not rely on absorption of the adsorbate, which will have a relatively low cross section compared with the vapor, but by the substrate itself. This does make the process substrate selective, but it has been shown to work very well for wide-bandgap semiconductors such as ZnSe [33]. This mechanism has a decreasing impact for narrower-bandgap semiconductors, possibly due to a reduction in the charge energy available for surface catalysis. However, surface photoassisted growth of CdTe was demonstrated using a UV laser to produce a selective pattern of CdTe on a GaAs substrate at growth temperatures in the range 250–300 °C [34]. The laser source was a frequency-doubled argon-ion laser with a wavelength of 257 nm producing relatively weak direct photolysis. This approach was used to demonstrate patterning of the CdTe epitaxial layer by projecting the UV excitation beam through a mask to replicate a mesa pattern on the substrate, with feature sizes ranging from 500 down to 50 μm.

10.6 Plasma-Assisted MOVPE

Plasma-assisted MOVPE has the same objective of photoassisted MOVPE in achieving high growth rates at temperatures <300 °C. For example, a growth temperature of 150 °C produced radiation-detector quality, iodine-doped CdTe [35]. Plasma-assisted MOVPE/MOCVD has been used for a variety of II-VI and III-V materials. In the work by Lu et al., doped ZnO was grown at 250 °C using precursors that would normally require growth temperatures >500 °C [36]. The objective with II-VI semiconductors for low-temperature growth is to achieve low defect density, which can be important when doping the semiconductors to avoid deep-level complex formation or compensation. Most plasma-assisted MOCVD systems are a combination between a vacuum deposition system, with remote plasma, and an MOCVD system with vapor sources directed toward the substrate. The design of the system requires collisions between plasma-generated ions and the precursors in the vapor but must avoid high-energy ions reaching the substrate. Niraula et al. [37] used a 60 W radio frequency (RF) remote plasma to generate hydrogen radicals to achieve low-temperature heteroepitaxy of CdTe on GaAs from the precursors DMCd and DETe in the temperature range 150–250 °C. Iodine doping produced CdTe layers up to 1×10^{18} cm^{-3} donor concentration at a growth temperature of 150 °C. This is well below the expected epitaxial growth temperature for CdTe and shows the potential for high-quality epitaxial layers to be grown at low temperatures using plasma assistance.

10.7 Polycrystalline MOCVD

In contrast to MOVPE, which utilizes lattice-matching substrates and low growth rates, MOCVD takes place at higher rates and usually on substrates having little lattice match to the intended growth layer, leading to a polycrystalline film structure. In this case, the crystals can have a variety of textures and sizes depending on the growth conditions. Furthermore, since stringent requirements with epitaxial layers, such as wafer abruptness, low density of crystal defects, and thickness uniformity, are no longer valid, MOCVD can take advantage of using inexpensive substrates,

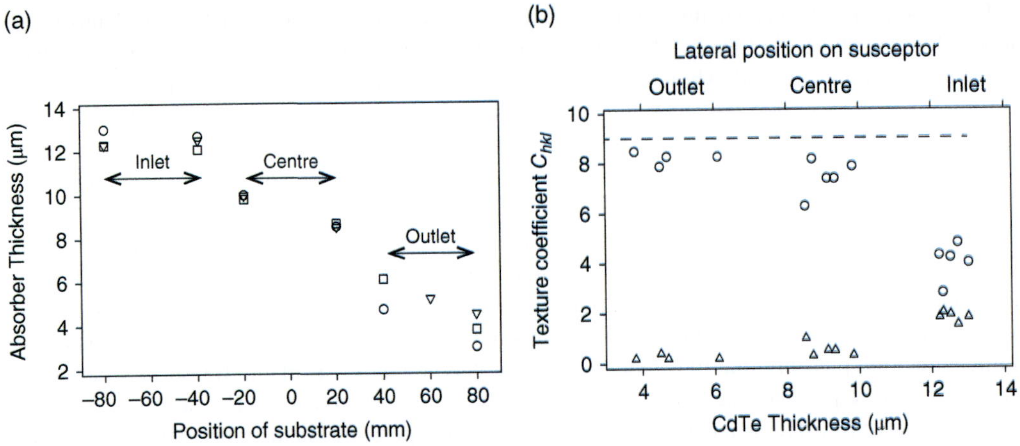

Figure 10.4 *(a) Thickness of polycrystalline CdTe films grown using a horizontal MOCVD reactor as a function of the substrate position along the graphite susceptor. Data for the three different growth runs is plotted. (b) Texture coefficients of the (111) (o) and (422) (Δ) reflections as a function of the absorber thickness. The dashed line represent values for a completely oriented sample. Source: Reprinted from Zoppi 2005 [39] with permission from Durham E-Theses.*

higher growth rates/throughput, and simplified reactor geometries (instead of a rotating-substrate assembly with shower-head precursor delivery as with MOVPE).

MOCVD has been developed and most commonly used for the growth of III-V and II-VI compound semiconductors for optoelectronic applications. Notable II-VI examples are CdTe and its alloys (MCT, CZT) for photodetectors and, more recently, photovoltaic solar cells.

Many single-wafer MOCVD research systems utilize a horizontal tube reactor with a static substrate, by which the film growth usually takes place nonuniformly across the length of the susceptor due to variations in surface temperature and precursor depletion effects [38]. These lead to films with a *wedge profile*, typically thicker on the inlet side and thinner on the outlet/exhaust side, or thinning from the center toward the edges, as well as variations in film crystallinity and composition. Data for polycrystalline MOCVD CdTe films, shown in Figure 10.4a, illustrates such a thickness distribution and film texture across the ~16 cm length of the susceptor, with the thickness gradient attributed mainly to the depletion effect since temperature control was found to be tight [39]. Thinner sections of the film are highly 111-oriented in the out-of-plane direction; however, the texture is lost, with nearly random crystal orientation visible at the thicker sections (Figure 10.4b).

More recently, interest in MOCVD has been extended to oxides, superconductors, metals, etc. for other electronic applications. Such systems, however, are less established and require more research toward fundamentals of deposition compared to compound semiconductors. For the latter, MOCVD technology reached sufficient maturity, enabling researchers to focus more on the development of technological applications.

10.8 In Situ Monitoring of CdTe

In situ optical monitoring of the growth of CdTe layers and related II-VI materials such as MCT, ZnTe, and CdS has proved to be a valuable tool in understanding the kinetic mechanisms for MOVPE growth and growth optimization. In a survey of in situ monitoring techniques,

Irvine and Bajaj [40] classified the techniques according to the operating pressure in the growth chamber. At one extreme, with MBE, it is possible to use electron-beam techniques that are not applicable at higher chamber pressures. For MOVPE, the range of techniques depends on photons, normally around the visible wavelengths, but in situ X-ray techniques have also been used [41] where grazing-incidence X-ray scattering was used to determine the surface structure. For optical wavelengths, it is possible to use techniques that are surface sensitive or bulk sensitive where reflection anisotropy spectroscopy (RAS) is sensitive to the surface reconstruction [42]. Spectroscopic ellipsometry (SE) [43] is sensitive to surface and bulk, whereas laser reflectance (LR) is sensitive to the bulk of the film. In its simplest form, a single-wavelength laser-reflectance technique is used to monitor the Fabry–Perot interference as the CdTe layer is growing. Both RAS and SE use polarization information from the reflected beam, whereas LR just looks at the intensity.

The advantage of LR is that it is less sensitive to the optical quality of the reaction chamber window than SE, which needs the polarization to be preserved and is more suited to routine monitoring purposes [44]. The first reported used of laser-reflectance monitoring of CdTe MOVPE was by Zinck et al. [45] where they measured CdTe growth rate in situ during laser-assisted deposition. LR monitoring of multiple layers during the HgTe/CdTe multilayer IMP process (described in more detail in Chapter 9) was reported by Bajaj et al. [46]. The laser used in this monitoring, as with many of the earlier works on LR monitoring, was the 633 nm HeNe laser. The absorption in the HgTe layers ensured that only the top few IMP layers influenced the laser reflectance signal, and the modelling of the reflectance took account of the interdiffusion occurring during growth, in addition to the multilayer contribution. This gave a continuous monitoring of the composition of the film as it was growing.

One of the useful benefits of monitoring the growth rate in situ was to change growth parameters, such as temperature and organometallic concentration, in real time to monitor the effect on the growth kinetics. This technique was first used to gain insights into photoassisted MOVPE of CdTe and later used to understand both the low-temperature kinetic regime of growth temperature and higher-temperature desorption processes [47].

10.8.1 Mechanisms for Laser Reflectance (LR) Monitoring

The reflectance intensity can be calculated from the complex refractive index:

$$N = n - ik \tag{10.1}$$

where n is the refractive index and k the extinction coefficient. Each layer will be characterized by a value for n and k, depending on the reflectance wavelength and the temperature of the growing film. The reflectance intensity for any stack of thin films can be calculated using the matrix method by Macleod [48], where the reflectance intensity can be calculated from the complex conjugate of the reflectance amplitudes:

$$R = rr^{*} \tag{10.2}$$

$$r = {(N_0 - Y)}\big/{(N_0 + Y)} \tag{10.3}$$

N_0 is the complex refractive index of the incident medium, and $Y = B/C$, which is given by the matrix

$$\begin{bmatrix} B \\ C \end{bmatrix} = \begin{bmatrix} \cos\delta_1 & i\sin\delta_1 \\ iN_1\sin\delta_1 & \cos\delta_1 \end{bmatrix} \begin{bmatrix} 1 \\ N_s \end{bmatrix} \tag{10.4}$$

δ_1 is a phase factor that, for normal incidence, is as follows:

$$\delta_1 = {2\pi N_1 d}/{\lambda} \tag{10.5}$$

N_1 and N_s are the complex refractive indices of the epilayer and substrate, respectively; d is the epilayer thickness (which is the variable to be determined), and λ is the laser wavelength. This matrix approach can be generalized to any stack of layers. In practice, an incident wavelength is chosen that is absorbing in the growing layer, so the interference oscillations will attenuate before the next layer is grown. For CdTe, a red laser around 630 nm will give five or six oscillations, which is equivalent to a thickness of approximately 0.7 μm. To monitor thicker layers, a near-infrared laser around 950 nm could be used.

It is not unusual for laser interferometers to combine a number of lasers to provide this flexibility. This is illustrated in Figure 10.5, which shows laser reflectance monitoring of a polycrystalline CdTe layer with simultaneous reflectance of green (532 nm), red (670 nm) and infrared (905 nm) lasers. The film is roughening as it grows. The green and red lasers are absorbing in the CdTe film and show significant attenuation of the oscillations, while the infrared laser continues to monitor the growth for more than 1 μm thickness. The more rapid attenuation of the Fabry–Perot oscillations at the shorter wavelengths can be useful when monitoring very thin films and multiple films so only the growing film (and its interface with the underlying film) is contributing to the interference. The shorter wavelength also has an advantage in being more sensitive to small changes in thickness, where a half-oscillation at 532 nm wavelength corresponds to a thinner layer than the longer wavelengths at 670 nm and 905 nm where the thickness for half an oscillation is given by

$$d_\pi = {\lambda}/{4n} \tag{10.6}$$

The other characteristic that is measured is the rate of roughness, which can be determined from the rate of decrease in the reflectance of mean reflectance or *offset* reflectance. It can be seen from Figure 10.5 that the shorter wavelength decreases more rapidly than the longer wavelength

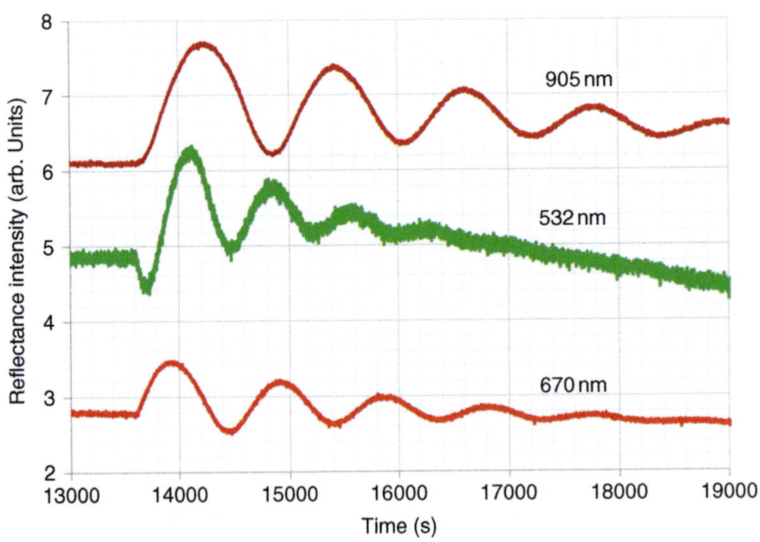

Figure 10.5 *Example of triple laser-reflectance monitoring of a CdTe polycrystalline layer for a PV CdS/CdTe structure. The laser wavelengths 532 nm, 670 nm, and 905 nm are indicated.*

and is expected to be more sensitive to the rate of surface roughening. Equations (10.2)–(10.5) assume smooth layers with smooth interfaces, so the roughening is not taken into account. This is a good approximation for an epitaxial layer, but for polycrystalline CdTe, this approach needs to be modified to take account of the roughening. A method for taking account of the roughening to enable accurate measure of thickness and in situ growth rates was proposed by Irvine et al. [49]. This uses a correction factor (F_{rs}) that in turn provides a measure of the roughening using the following equation:

$$R = R_{\mathrm{m}}\left(1 + F_{\mathrm{rs}}\right) \tag{10.7}$$

where R is the theoretical reflectance, calculated from Eq. (10.2), and R_m is the measured reflectance from a roughening layer. This was used to correlate the rate of roughening with different growth conditions for CdTe/CdS polycrystalline layers. The layers were smoother for Te-rich precursor ratios, while the rate of roughening increased for Cd-rich precursor ratios. It was also found that the PV device series resistance decreased for Cd-rich precursor ratios, hence linking film properties to the growth conditions.

10.9 MOCVD of CdTe for Planar Solar Cells

Among several thin-film deposition techniques used for CdTe deposition, MOCVD is an attractive method as it offers greater flexibility for doping and alloying and can operate at atmospheric pressure. While MOVPE is commonly associated with epitaxial III-V multijunction solar cells, MOCVD can produce relatively inexpensive, polycrystalline CdTe solar-cell structures with promising device performance. MOCVD CdTe is utilized as the active *absorber* layer within a multi-layer polycrystalline thin-film stack on a glass substrate, as exemplified by Figure 10.6. This is the *superstrate structure*, in which solar irradiation enters the cell via the glass substrate before passing through a transparent front contact (TCO) and then a semiconducting *window* layer. The window layer is typically made of CdS- or a CdZnS-type wide-bandgap n-type compound.

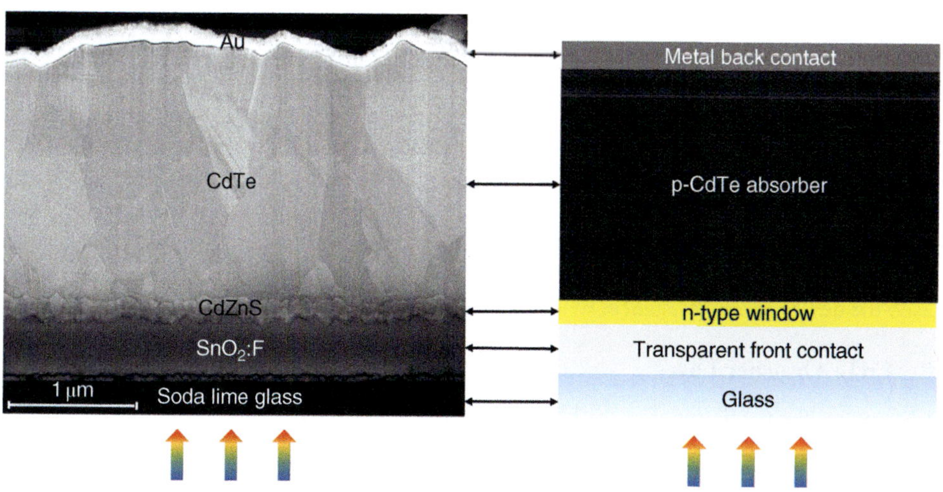

Figure 10.6 *A cross-sectional SEM image of a superstrate CdTe solar cell structure employing polycrystalline MOCVD CdTe absorber and CdZnS window layers, thermally evaporated gold (Au) back-contact, and a commercial* SnO_2*:F (FTO)/glass front contact. Source: Reprinted with the kind permission of Dr. Sachit Grover, First Solar.*

A sufficiently p-type doped CdTe absorber completes the semiconductor p-n junction. A metal back-contact material having a high work function (typically gold or nickel) that is comparable to the electron affinity of CdTe is deposited on the CdTe back-surface to finish the solar-cell fabrication. The photoexcited carriers produced within the absorber are dissociated and carried toward the front and back electrodes, via diffusion and/or drift, under the potential difference (*built-in voltage*) produced by the junction field. In the following sections, typical device layers and key processes utilized for MOCVD CdTe solar cells will be described in detail.

10.9.1 CdS and CdZnS Window Layers

Historically, CdS has been the preferred n-type window layer in CdTe and $CuInGaSe_2$ (CIGS) type chalcogenide thin-film solar cells. It is considered that CdS forms a relatively low-defect-density interface to these absorbers. In the case of CdTe, for example, an ultrathin (~50 nm) $CdTe_{1-x}S_x$ alloy transition layer is usually formed between CdS and CdTe films, via S and Te interdiffusion, thereby reducing the lattice mismatch (10%) between them. Despite such benefits, the relatively narrow bandgap (2.4 eV) of CdS is a limiting factor to the PV performance since CdS does not contribute to the current generation directly, leaving the absorber layer to act as the "workhorse." Thus, photons with energies above the CdS bandgap are usually wasted, as they generally cannot reach the absorber. This is traditionally resolved in part by thinning the CdS layer to below 100 nm for it to transmit more photons in the blue/near UV region. Although some pinholes may appear in the CdS layer upon extreme thinning, a maximum power conversion efficiency (PCE) of 16.5% was demonstrated through this approach for a CdS/CdTe solar cell [50].

As an alternative to the CdS thinning approach, alloying of CdS with, for example, Zn to widen the bandgap beyond 2.4 eV was also shown to enhance the cell current and hence the PCE. The bandgap of $Cd_{1-x}Zn_xS$ (CZS) is expected to vary via the relationship

$$E_g(x) = E_g(\mathrm{CdS}) + \left[E_g(\mathrm{ZnS}) - E_g(\mathrm{CdS}) - b\right]x + bx^2 \tag{10.8}$$

where E_g(CdS) = 2.42 eV, E_g(ZnS) = 3.54 eV, and the band-bowing parameter b = 0.91 eV [51]. Although limited success was achieved in controlling the CZS alloy composition by the wet-deposition methods [51, 52] that are conveniently used for CdS synthesis, MOCVD was shown to deliver excellent compositional tunability from CdS to ZnS (Figure 10.7a) [53].

Series of CZS thin films grown using ditertiarybutylsulphide (DtBS), dimethylcadmium (DMCd), and diethylzinc (DEZn) organometallic precursors and hydrogen carrier gas were also evaluated in CdTe solar cells (Figure 10.7b, Figure 10.8) [53]. Although nearly all CZS compositions outperformed pure CdS performance, x = 0.70 was determined to be the optimum, providing a 37% maximum boost in PCE (η). In addition to the 17% enhancement to the current density (J_{sc}) due to improved spectral utilization, the open-circuit voltage (V_{oc}) and fill factor (FF) were also improved. At above x = 0.70, the CZS film is believed to become more insulating as these compositions increased the device series resistance, which adversely affects mainly the FF and hence the PCE.

10.9.2 CdTe Absorber Layer

In the literature, the most common method for depositing the CdTe polycrystalline layer for thin-film solar cells is physical vapor deposition, including close-space sublimation (CSS) and vapor transport deposition (VTD), carried out in low/moderate vacuum conditions [54, 55]. In these methods, CdTe vaporized from a powder source (usually kept at ≥600 °C) is deposited as a thin film onto a heated substrate (usually at ≥500 °C). For CSS, the substrate is held close to the source in the same deposition chamber, while for VTD the vapor is transported to a heated translation stage to achieve large-area deposition. As it is difficult to achieve controlled (in situ) doping of

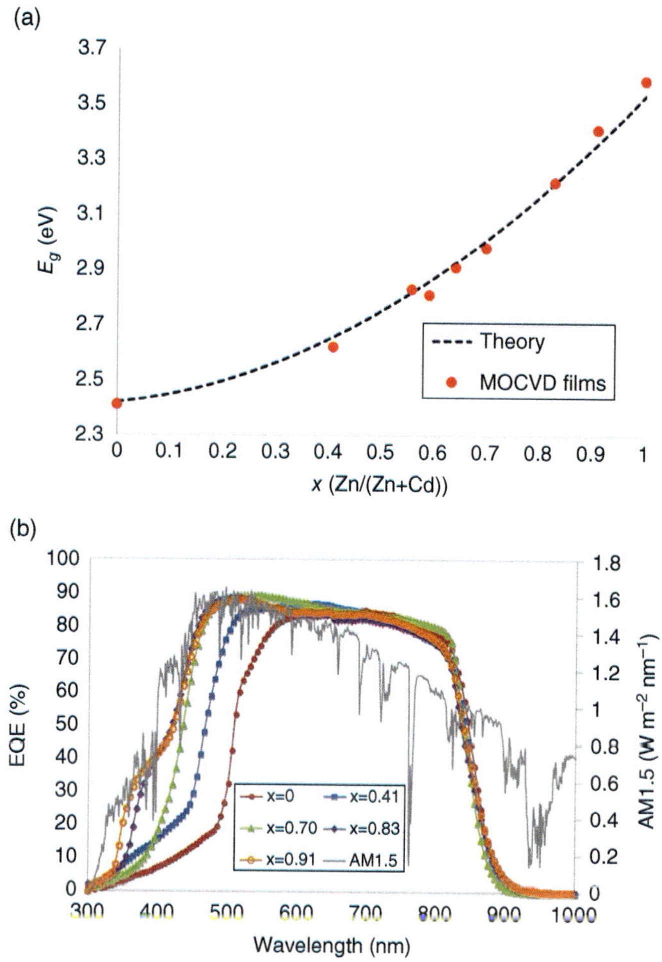

Figure 10.7 *(a) Compositional control and (b) device performance (spectral response) of MOCVD* $Cd_{1-x}Zn_xS$ *alloy films. In (a), composition determined from optical gap measurements are compared to the theoretical change in bandgap with Zn alloying (Eq. (10.7)). In (b), external quantum efficiency data shows a better spectral match of device response to solar irradiation (AM1.5) in the blue/UV region with increasing Zn concentration. Source: Reprinted from Kartopu et al. 2014 [53] with permission from Wiley.*

CdTe film by these methods, traditionally a small amount of oxygen is used in the processing gas during growth in addition to copper (Cu) indiffusion through the back-contact surface of CdTe introduced by a postdeposition anneal. However, these steps necessitate chemical etching of the CdTe surface (prior to Cu deposition) and annealing (after Cu deposition), which should be carefully controlled. Surface etching is performed to remove oxide compounds and to yield a Te-rich surface. Formation of a Cu_xTe interlayer then helps to obtain an ohmic contact. The Cu doping is usually introduced by depositing an ultrathin (up to a few nm) Cu film or a copper-containing compound (such as ZnTe:Cu) followed by the heat treatment. The highest doping (i.e. acceptor density) achievable with ex situ Cu doping is known to be $\sim 1 \times 10^{15}\,cm^{-3}$ [56], which indicates the difficulty with p-type doping of CdTe. It is widely accepted that CdTe produces *anti-sites* (usually near mid-bandgap), which works against the intentional doping through a *self-compensation* process.

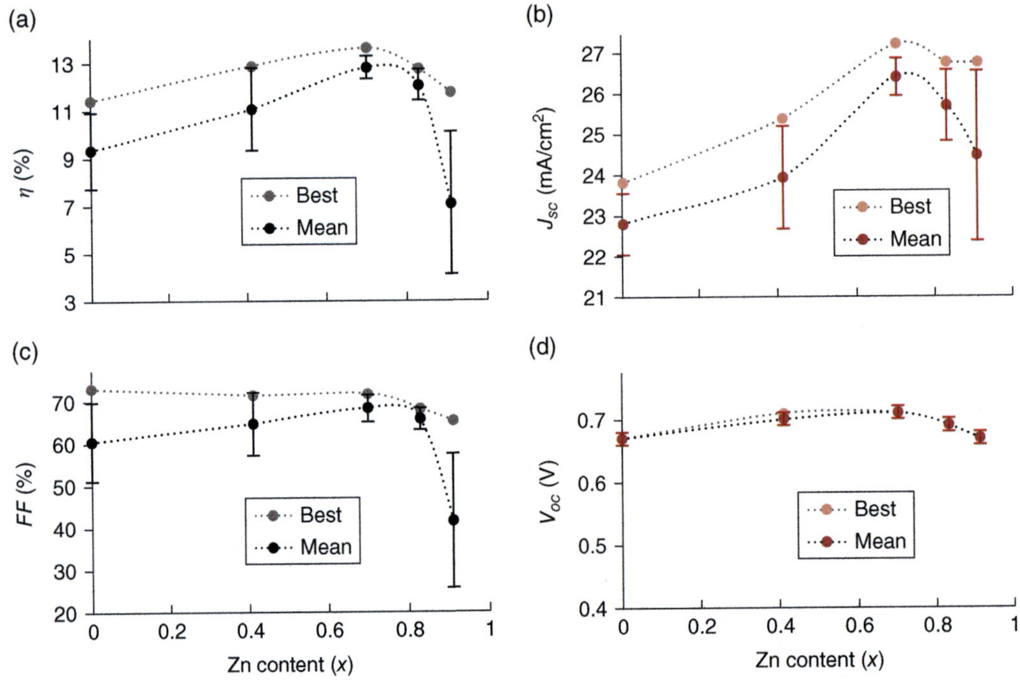

Figure 10.8 *Solar simulator current–voltage performance data summarized for MOCVD CZS/CdTe solar cells as a function of the Zn content incorporated to the CZS layer: the best and mean cell efficiency (a) goes through an optimum near the composition* $Cd_{0.3}Zn_{0.7}S$*, with current density (b) being the main contributor. At this composition, fill factor (c) and open-circuit potential (d) parameters are also near their maxima. Source: Reprinted from Kartopu et al. 2014 [53] with permission from Wiley.*

Therefore, even at high dopant ion concentrations (≥1 × 10^{16} atom/cm^{-3}), the achievable activation ratio, i.e. (net) carrier density/dopant atom concentration, usually remains below 1%.

An alternative approach to deposit the CdTe absorber layer is by MOCVD, which relies on chemical synthesis of metalorganic compounds on a heated substrate. While it is relatively easy and reliable to introduce dopants to CdTe film via MOCVD, the growth temperature is limited by the highest achievable growth rate. For CdTe deposition, this temperature is reached around 400 °C in most cases (Figure 10.9). As a drawback, the grain sizes obtained in a typical as-grown MOCVD CdTe film are usually small (≤1 μm) compared to high-temperature deposition methods (CSS, VTD), in which case they can exceed several μm.

The flexibility imparted by MOCVD toward CdTe doping is demonstrated in the case of in situ arsenic (As) incorporation, using diisopropyltelluride, dimethylcadmium, and tris(dimethylamino) arsenic precursors [58]. In this process, to facilitate As substitution on Te sites, the II/VI precursor ratio, [Cd]/[Te], is kept above 1. The highest carrier density achievable using As doping is determined to be ~1 × 10^{16} cm^{-3}, which is an order of magnitude higher than ex situ Cu doping, despite the still-low activation ratios (~1%) obtained with As. Another basic difference of MOCVD from physical vapor-deposition methods is that a carrier gas such as hydrogen (H_2) must be used. Since the deposition atmosphere is rich in the carrier gas, H atoms can become incorporated into the CdTe lattice. The role of the processing environment on the dopant activation remains an area to explore, since even a low-temperature postdeposition anneal in air for MOCVD CdTe:As films leads to higher (by two to three times) activation rates (Figure 10.10) [59].

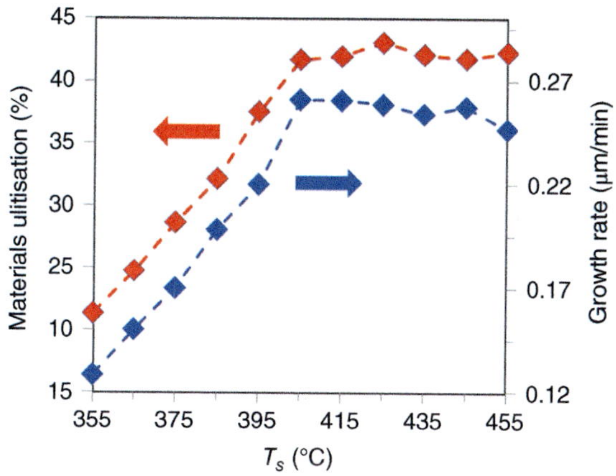

Figure 10.9 *CdTe growth rate and utilization of metalorganic precursors for MOCVD CdTe films as a function of the substrate temperature. While up to 400 °C the growth is kinetically limited, it becomes mass-transport limited at higher temperatures. Data obtained using a vertical-injector MOCVD chamber (described in Section 10.11). Source: Reprinted from Barrioz et al. 2012 [57] with permission from Elsevier.*

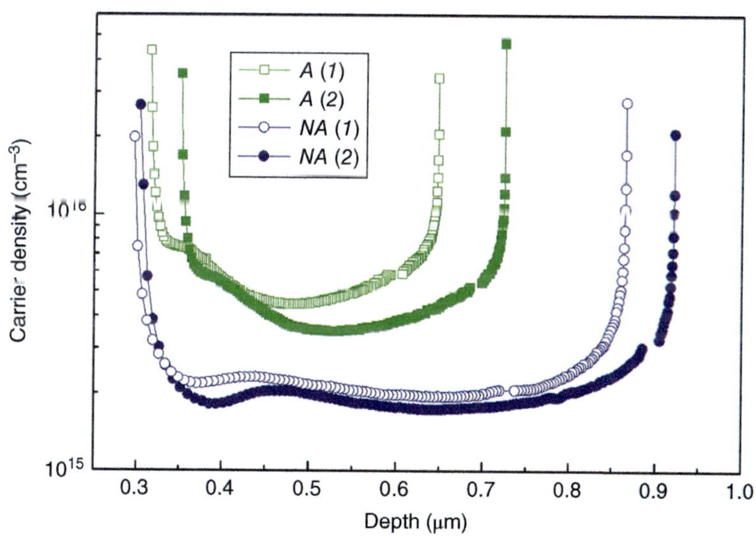

Figure 10.10 *Carrier density in MOCVD CdTe absorber films (produced in H_2 ambient) without (NA) and with (A) a postgrowth anneal in air at 170 °C. Carrier density in the bulk of the CdTe film (i.e. flatter region of the U-shaped curves) increases by two- to threefold upon the air anneal. Narrowing of the depletion observable for annealed films is in agreement with the carrier-density increase. Data for two different contacts per sample are plotted. Source: Reprinted from Kartopu et al. 2016 [59] with permission from Wiley.*

10.9.3 $CdCl_2$ Treatment Layer

It is widely known that the polycrystalline CdS/CdTe solar-cell structure needs to undergo an *activation* step, entailing $CdCl_2$ deposition on the CdTe surface and firing at 390–450 °C for 10–30 min. This is usually referred to as the *$CdCl_2$ heat treatment* (CHT) and applies to all types of polycrystalline CdTe solar cells regardless of the cell-deposition method(s) employed. Without the CHT, polycrystalline CdTe solar cells perform very poorly due to excessive carrier recombination. In this process, the Cl ions diffuse into the entire cell structure through the CdTe grain boundaries, driving various changes such as CdTe recrystallization and grain enlargement, surface-passivation provided by Cl, and S and Te interdiffusion. An excess of $CdCl_2$ treatment layer is often preferred to ensure completeness and homogeneity of the reaction, whilst unreacted $CdCl_2$ residue is rinsed off from the CdTe surface, using water as the solvent, before depositing the back-contact. When the annealing atmosphere contains air or O_2, some additional passivation and p-type doping provided with oxygen can also take place.

For a typical polycrystalline thin-film solar cell, as the grain boundaries (GBs) are the main recombination sites, a drop in their areal density, achievable by grain growth, would be beneficial for the cell performance. Thus, by providing both grain growth and surface passivation, the CHT step significantly reduces the defect-density and carrier recombination in the CdTe absorber.

Additionally, since the Cl ions can diffuse down to the transparent front contact via the grain boundaries, recrystallization of the window (CdS or CdZnS) layer as well as promotion of S and Te interdiffusion at the heterojunction interface can take place.

The exact mechanism of the CHT process and enhancement of device performance are still being debated and actively investigated in the literature.

For the activation of CdTe solar cells, MOCVD of a $CdCl_2$ treatment layer was also reported, using dimethylcadmium and tetrabutylchloride precursors [60]. The crystallinity and surface coverage of $CdCl_2$ layers were found to be the best at deposition temperatures 200–300 °C. Microstructure of exemplar films grown on a single-crystalline Si substrate is shown in Figure 10.11. The film grown at 200 °C was continuous but formed of platelets. The density of these particles reduces, with the film becoming more compact, as the temperature was raised to above 300 °C. The optimum deposition temperature for obtaining dense MOCVD $CdCl_2$ films when using a glass substrate was somewhat lower [60].

The change in surface morphology of a MOCVD-grown CdTe film is depicted in Figure 10.12, where a ~1-μm MOCVD $CdCl_2$ activation layer was first grown at 200 °C and then the device structure annealed at 420 °C for 10 min in H_2. Enlargement and recrystallization of the CdTe grains are clearly evident.

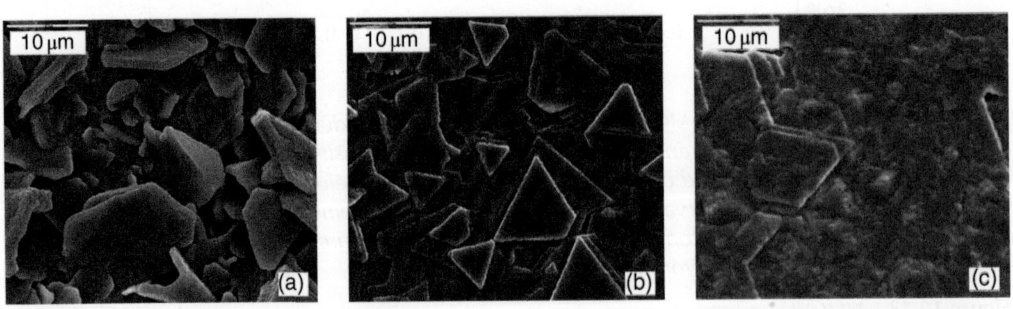

Figure 10.11 $CdCl_2$ layers deposited on a silicon (111) substrate by MOCVD at (a) 200, (b) 300, and (c) 350 °C substrate temperature. Source: Reprinted from Barrioz et al. 2007 [60] with permission from Elsevier.

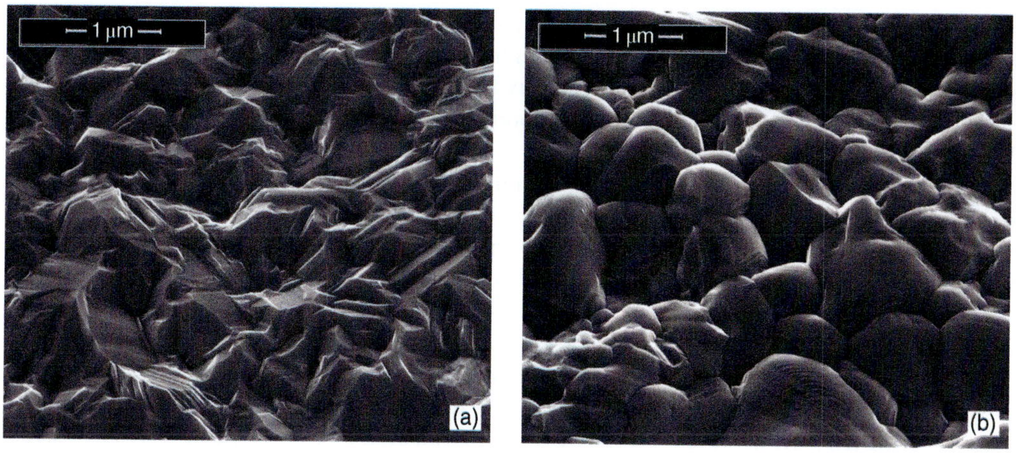

__Figure 10.12__ SEM images MOCVD CdTe film (a) before and (b) after CHT performed using a MOCVD $CdCl_2$ activation layer in H_2 ambient. Source: Reprinted with the kind permission of Dr Sachit Grover, First Solar.

Finally, other Cl-containing compounds are also being researched, as nontoxic alternatives to $CdCl_2$, with thus far only $MgCl_2$ being found to be a reasonable substitute [61]. The motivation of such efforts stems also from the high water solubility of $CdCl_2$.

10.9.4 Photovoltaic Planar Devices

MOCVD was first evaluated for polycrystalline CdTe PV devices in the early 1990s. The best solar cells developed at the Georgia Institute of Technology, which were doped by *ex situ* Cu treatment, attained 11.9% efficiency (J_{sc}=24.5 mA/cm^2, FF=64%, and V_{oc}=0.76 V), albeit on a small scale (0.077 cm^2 cell area) [62]. In their process, CHT was performed by dipping the CdS/CdTe structure in a $CdCl_2$/methanol solution and then heat treating in air at 400 °C for 30 min. The CdS thickness was varied to enhance the J_{sc} (Figure 10.13a). At the lower thickness of 100 nm, the blue response improved with the J_{sc} enhancement of ~2 mA/cm^{-2}; however, the other cell parameters (V_{oc}, FF)

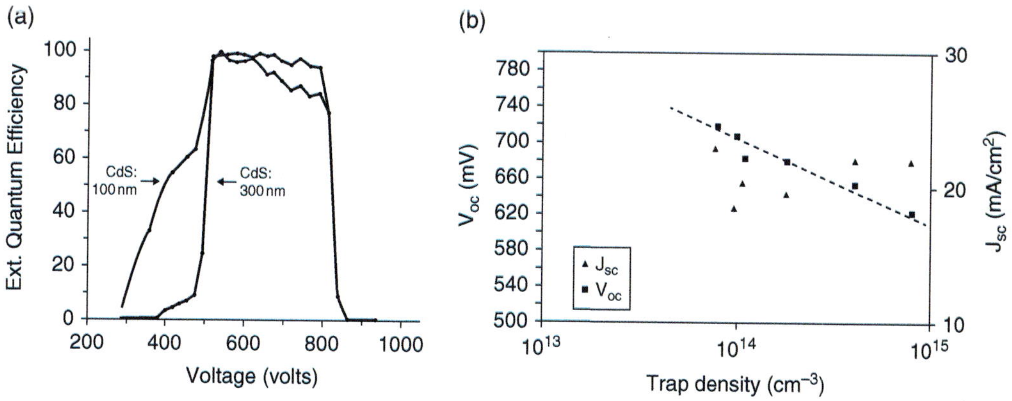

__Figure 10.13__ MOCVD CdTe solar cell made at the Georgia Institute of Technology (1992). (a) Spectral response of CdS/CdTe cells showing the effect of CdS thickness; (b) V_{oc} (▬) and J_{sc} (▲) of measured cells as a function of the V_b+0.64 eV trap density. Source: Reprinted from Rohatgi 1992 [62] with permission from Taylor and Francis.

suffered slightly, negating the gain in the J_{sc}. Therefore, the CdS thickness was adjusted to optimize overall cell performance. Additionally, deep-level transient spectroscopy was used on a number of solar-cell devices, which showed that the density of deep-level trap states at ~0.6 eV above the valence band (V_b) of CdTe was inversely proportional to the V_{oc} but not correlated to the J_{sc} (Figure 10.13b).

In subsequent years, physical vapor deposition methods (CSS, VTD) dominated over MOCVD for CdTe PV research, driven by the industrial interest in easily scalable manufacturing methods, exploiting rapid film growth and lower cost of ownership associated with these techniques. After enjoying years of research and investment, the VTD method has recently given rise to world-record PCEs of 22.1% for a ~0.5 cm^2 cell and 18.6% for ~0.7 m^2 module size [63]. This tendency slowed the research and further development of MOCVD for CdTe PV technology. Nevertheless, MOCVD CdTe solar cells having larger cells and improved performance were also reported recently.

By 2007, fully MOCVD-grown CdTe solar cells developed by the Centre for Solar Energy Research (CSER), UK, achieved 13.3% PCE (J_{sc} = 26 mA/cm^2, FF = 74%, and V_{oc} = 0.69 V) for 0.25-cm^2 cell size. This was achieved introducing As doping *in situ* to the CdTe absorber (depicted in Figure 10.14) and replacing the CdS window layer with the wider bandgap CZS [64]. In the optimum structure, As doping levels for the bulk and back-contact-surface of the CdTe absorber were $\sim 2\times10^{18}$ and $\sim 1\times10^{19}$ atom/cm^3, respectively, while the CZS window was 240 nm thick. In this process, the layers of CdS, CdTe:As and $CdCl_2$ were applied and the CHT process was carried out in a continuous manner in the same MOCVD chamber.

More recently, MOCVD-grown CZS/CdTe solar cells with a performance approaching the 15% mark for 0.25-cm^2 cell size was also demonstrated by CSER. Introduction of a low-temperature (170 °C) postgrowth air anneal, which was shown to form a thin oxide layer at the CdTe back surface [65], caused significant improvements to the FF and V_{oc}, whilst slightly reducing the J_{sc}. These improvements were attributed to the enhancement of p-type carrier density (illustrated in Figure 10.10) and to the reduction of the back-contact barrier height [59]. The current loss, on the other hand, was associated with deteriorating long-wavelength spectral response, indicative of increased bulk recombination. In order to compensate for this loss, the thickness of the CZS window layer was reduced to 150 nm (optimum), giving the hitherto best MOCVD solar-cell parameters: J_{sc} = 26 mA/cm^2, FF = 79%, and V_{oc} = 0.79 V [59].

Following the latter solar-cell improvements, minimodule MOCVD CdTe devices were also fabricated and studied for the first time. Initial devices, prepared by a mixture of laser and mechanical

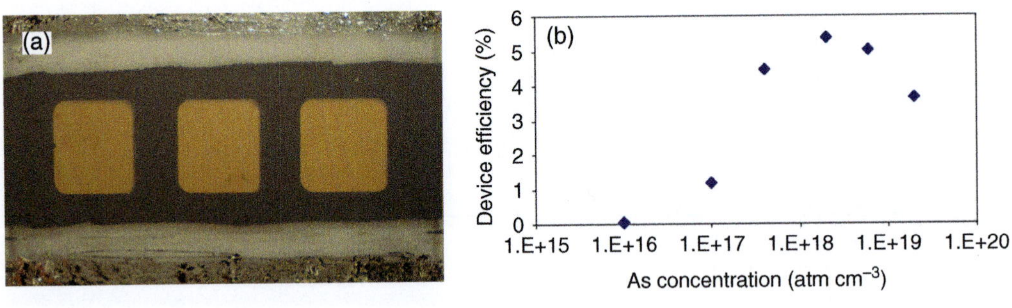

Figure 10.14 *MOCVD CdTe solar cell developed by CSER (2007): (a) test device with three 0.5 × 0.5 cm^2 gold contacts (i.e. individual cells), and (b) effect of As concentration in CdTe bulk on the cell efficiency (unoptimized). Source: Reprinted from Irvine et al. 2008 [64] with permission from Elsevier.*

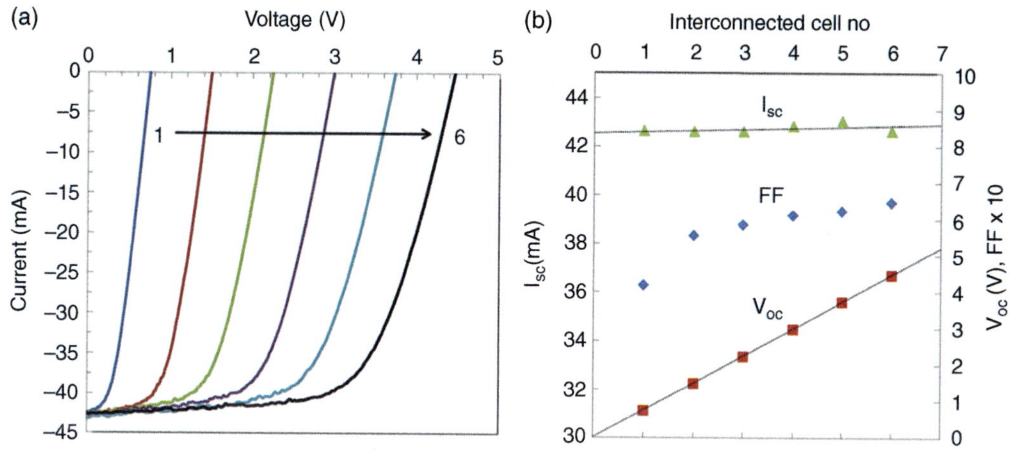

Figure 10.15 *The first MOCVD CdTe minimodule reported with up to six subcells connected in series. (a) Light I–V curves and (b)* V_{oc}, I_{sc} *and FF parameters extracted. Source: Reprinted from Kartopu et al. 2016 [59] with permission from Wiley.*

scribes for cell isolation, consisted of monolithically integrated 2 cm^2 subcells (1 cm cell width) in series connection (Figure 10.15). As can be expected from the serial connection, the device current (I_{sc}) remains constant, whilst the voltage output increases linearly with more cells being interconnected. For a six-cell minimodule, 11.1% best efficiency was reported for an active area of 11.7 cm^2. The high series resistance was observed to be limiting the FF to ~63%. This originates mainly from the lateral conductance of the transparent front contact. With greater cell widths, the series resistance due to the front contact dominates the electrical behavior. Reduction of the cell width, e.g. to ~0.5 cm, is expected to provide improvement in the efficiency and power output [66].

10.10 Core-Shell Nanowire Photovoltaic Devices

Utilization of a nanowire array architecture for PV devices could reduce material usage by increasing light absorption via light trapping. In this structure, since the carriers would be captured orthogonal to the wire axis, the absorber layer thickness can be made comparable to even the very short minority carrier diffusion lengths for a low-electronic-quality material. In other words, the requirement for having high-quality absorbers with relatively long carrier lifetimes in making efficient planar solar devices is quite relaxed when using the nanowire architecture.

MOCVD was utilized in forming core-shell nanowire arrays for CdTe PV devices. For example, deposition of a CdS layer (shell) onto CdTe nanowires showed that CdS coverage was uniform, with no damage to the CdTe nanowire scaffold observable [67]. Comparatively, alternative wet-chemical deposition methods, e.g. chemical bath deposition (a common method for growing CdS thin films in planar CdTe devices), resulted in significant loss of substrate coverage by breaking off the CdTe nanowires scaffold. On the other hand, MOCVD of CZS on CdTe nanowires results in the growth of secondary (dendritic) nanowhiskers, perpendicular to the CdTe wire axis, as well as providing lateral surface coverage to the CdTe cores (Figures 10.16a and b). This is apparently related to the presence of Zn precursor (diethylzinc) in the gas mixture, which leads to phase-separated growth with ZnS like nanowhiskers. Such structures may be explored for applications requiring larger surface areas, such as gas sensors and catalyzers for water splitting and water cleaning. For PV applications, however, core-shell nanowires with smooth surfaces are

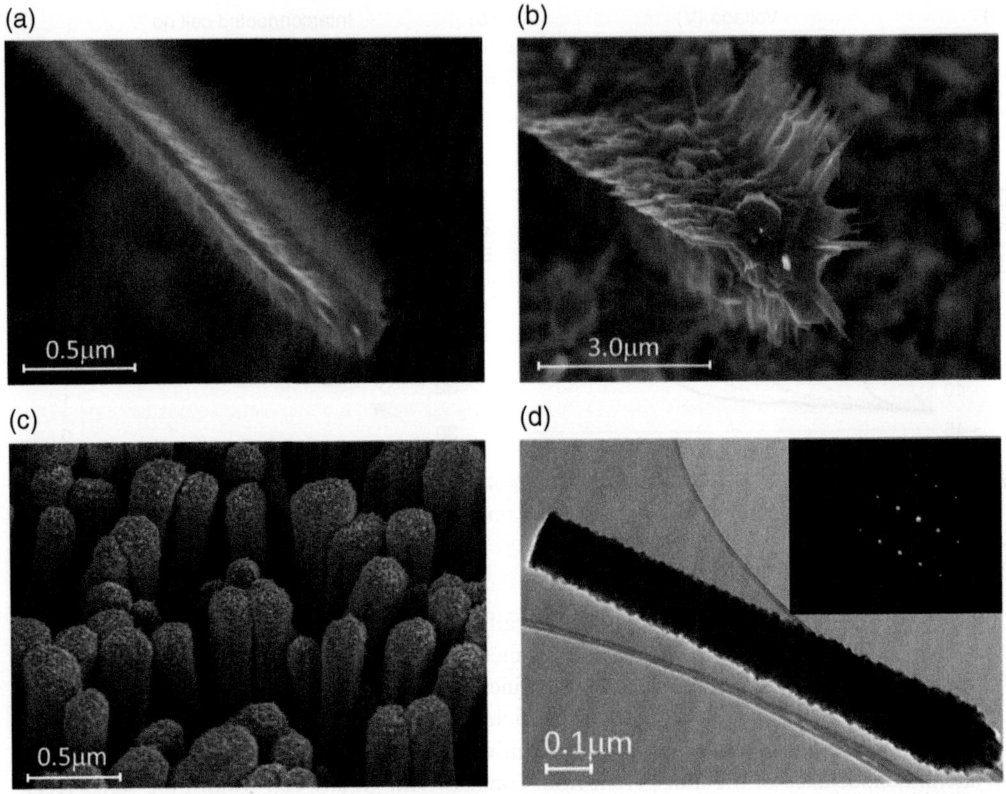

Figure 10.16 *Examples of MOCVD film growth on nanowire surfaces: CdZnS (CZS) on CdTe nanowires (a, b) and ZnO:Al (AZO) on ZnO nanowires (c, d). Additional nanowhiskers are also evident in the case of CZS/CdTe wires, the thickness of which increase with deposition time (longer for (b)). AZO provides conformal coverage and epitaxial growth on ZnO wires. Inset in (d) is the electron-diffraction pattern of the AZO shell.*

desirable to prevent electrical shunts. Conformal growth of ZnO:Al, CdS, and CdTe onto ZnO nanowires by MOCVD is relatively straightforward, leading to highly crystalline and smooth films. AZO also grows epitaxially on the ZnO surfaces (Figures 10.16c and d), which could potentially increase the surface conductivity of ZnO nanowires when utilized as the transparent front electrode in PV nanostructures.

As with planar PV devices, both substrate and superstrate type nanowire-based CdTe solar cells have been attempted in the literature. As notable examples, ~3% PCE nanowire cells were reported where indium-tin-oxide front contact/CdS window layers were deposited by reactive sputtering onto CdTe nanowire arrays [68]. In this method, the CdTe nanowires were grown by the CSS method using a planar CdTe seed film decorated with catalyst (gold) nanoparticles. The contribution of this seed CdTe layer to the PV response, however, was not determined. Additionally, metal impurities within the nanowire due to the catalyst particles is a concern for device stability, as with any other semiconductor devices. To this end, for example, ZnO nanowire arrays, which can be self-catalytically grown on a pure ZnO seed layer deposited onto glass superstrate, appear to be more attractive for improved CdTe nanowire solar cells with higher performance and lower stability issues. ZnO nanowires can be grown from a few nm thick ZnO seed using facile, inexpensive, and low-temperature chemical synthesis methods such as the hydrothermal

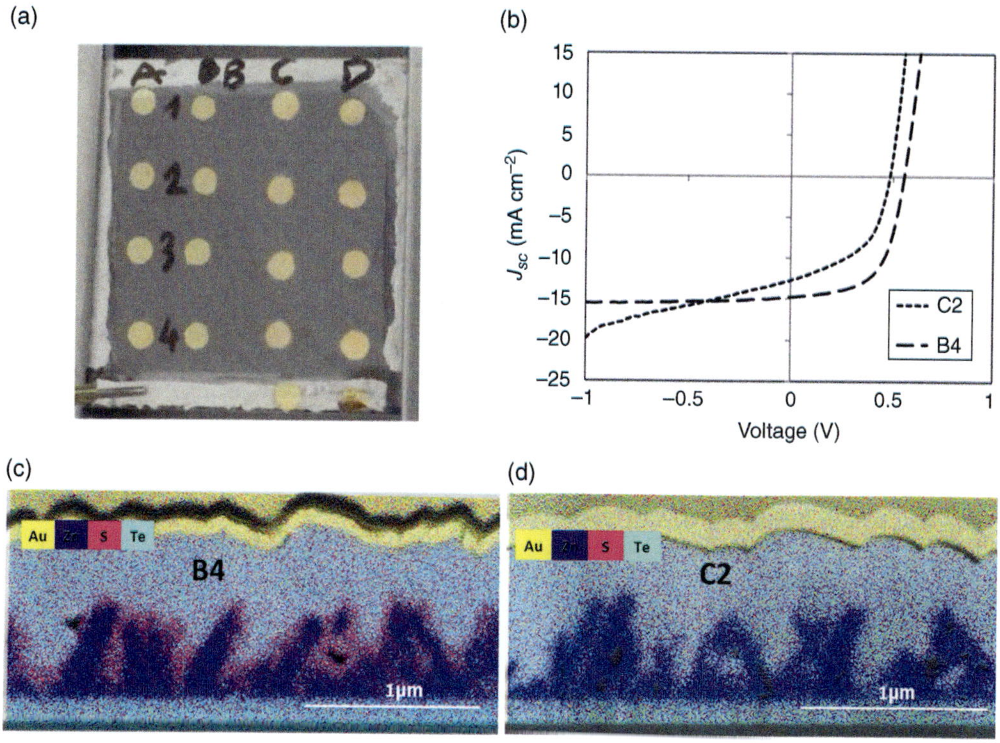

Figure 10.17 A 2.5 × 2.5 cm² CdTe/(CdS)/ZnO nanowire device with 2-mm diameter dot contacts (i.e. individual solar cells) (a). Cross-sectional energy-dispersive X-ray (EDX) elemental (Au, Zn, S, and Te) maps of (c) CdTe/CdS/ZnO and (d) CdTe/ZnO nanowire cells, with corresponding light I–V curves (b). In both cases, the ZnO nanowires are buried within the CdTe absorber, and gold is the back-contact material. Structure with a CdS interlayer exhibits clear improvements in all device parameters. Source: Reprinted from Kartopu et al. 2018 [70] with permission from Elsevier.

route [69]. Using such superstrate nanowire substrates, 4.5% PCE CdTe solar cells were obtained by conformally growing a CdS/CdTe:As shell heterojunction via MOCVD [70]. The beneficial role of the CdS window was also demonstrated (Figure 10.17) by comparison to equivalent nanowire solar cells having only the CdTe:As shell present. For these ZnO/CdS/CdTe core–double shell solar cells, the absorber thickness was in fact overgrown, with the ZnO nanowire buried inside; this was preferred to avoid shunting between the front- and back-contact materials. Using this fabrication route, it should be possible to explore the extremely thin absorber (ETA) device concept in the future, which could provide higher performance compared to equivalent planar devices using equivalent, or even less, material [71].

10.11 Inline MOCVD for Scaling of CdTe

The suitability of MOCVD for inline deposition of CdTe thin films was first assessed using a vertical coating head to deliver the metalorganics via a slit-type injector at a short distance to a heated-translation stage carrying the deposition substrate (Figure 10.18a) [57, 72]. An improved precursor utilization rate of 40% was demonstrated (shown in Figure 10.8) compared with a conventional horizontal MOCVD reactor, which typically has a utilization rate of ~5%. This

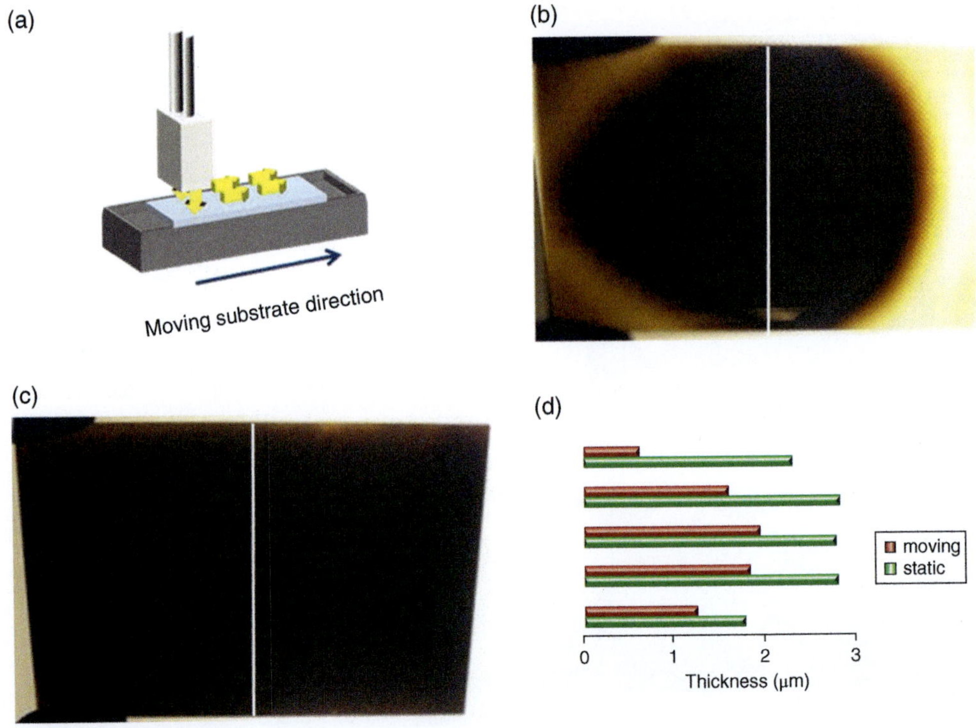

Figure 10.18 (a) Schematic of inline MOCVD process developed for growing CdTe thin films over 5-cm wide substrates; (b) plume spread in static mode; (c) film deposited on a moving substrate; (d) spatial distribution of layer thickness across the width of the coatings shown in (b) and (c). Films are thinner near the edges due to nonuniform delivery of precursors to the substrate. Source: Reprinted from Barrioz et al. 2012 [57] with permission from Elsevier.

difference is chiefly due to the ability to deliver the precursors at a much smaller substrate-nozzle distance with the inline approach, which in turn demands much lower (by a factor of 10) carrier and dilution gas flows. Use of lowered processing gas enables a higher precursor concentration (by a factor of 10) to arrive at the substrate surface and proportionally higher thin-film growth rates. This process was first carried out in an enclosed stainless-steel chamber that was controlled just below the atmospheric pressure, across 5-cm wide substrates (Figures 10.18b and c) [57]. These films were found to be thinner on the edges and thicker near the center (Figure 10.18d), highlighting the need for improvements to the homogeneity of the precursor delivery and extraction near the substrate. An activation energy of 49 kJ mol^{-1} for inline MOCVD CdTe films, derived from the Arrhenius plot of the growth rate versus temperature data, agreed well with data obtained by conventional (horizontal tube) MOCVD growth [57]. Furthermore, solar cells made using an inline-deposited CdTe absorber achieved 11% best device efficiency, which was slightly below those made using the conventional horizontal chamber MOVCD route [72].

In recent years, a "chamberless" inline MOCVD process (CIP) was also developed, with various semiconducting thin films, including CdTe, applied to over 15-cm wide glass substrates [73–75]. By this inline approach, the chemical reaction is contained within the desired volume over the substrate using a curtain gas flow made of nitrogen rather than a physical enclosure (Figure 10.19a). The curtain flow has two major functions: (i) preventing ingress of oxygen and moisture from ambient to the reaction zone, which otherwise could cause oxidation or impurity effects to the growing film; (ii) leakage of carrier and precursor gases from the reaction zone to

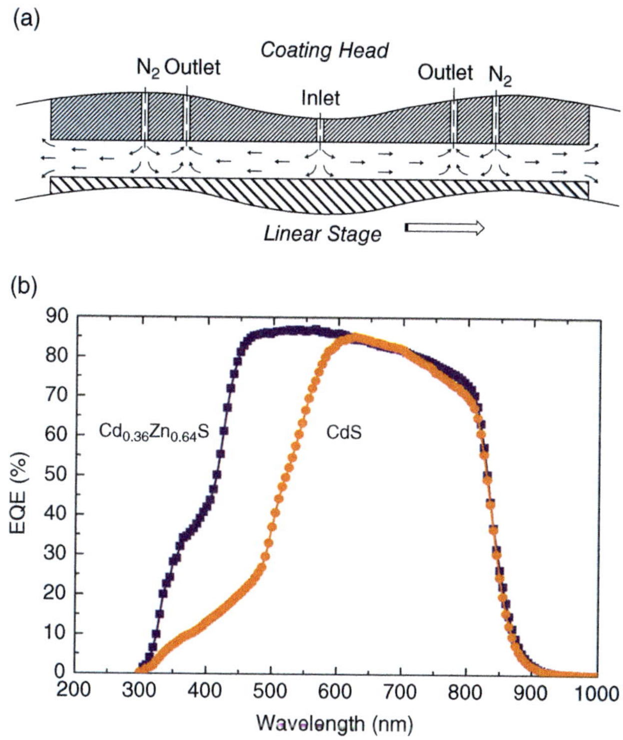

Figure 10.19 *(a) Schematic of a chamberless inline MOCVD process (CIP) developed for growing CdTe thin films over 15-cm wide substrates; (b) spectral response of CdS/CdTe and $Cd_{1-x}Zn_xS$/CdTe solar cells deposited using the CIP with 10% and 13.6% best PCE, respectively. The $Cd_{1-x}Zn_xS$ (x=0.64) layer permits significantly more blue photons to the CdTe film, increasing the J_{sc} by 5–6 mA/cm², in comparison to the CdS window. Source: Reprinted from Kartopu et al. 2015 [75] with permission from Elsevier.*

the ambient, which could compromise safe operation. These functions are supported by a controlled active exhaust (outlet), typically 10 Torr below the ambient, to assist with extraction of gaseous byproducts. Balance of the strength of the curtain flow versus gas extraction is typically achieved by monitoring the oxygen/moisture content within the chemical constituents in the extract using a mass spectrometer as well as utilizing relevant gas sensors fitted near the deposition area (outside the nitrogen curtain) for the processing gas such as hydrogen. Using the CIP method, CdTe growth rates of ~0.4 µm/min were demonstrated with ~50% precursor utilization [74]. Solar cells prepared using this technique achieved results comparable to those obtained using the conventional horizontal reactor (batch) MOCVD process, with 10% and 13.6% best PCEs for CdS/CdTe and CZS/CdTe inline-deposited solar cells, respectively [75]. The quality of the very thin (~100 nm) CZS layers was found to be satisfactory in terms of film coverage and homogeneity, as they provided high transmittance to short-wavelength photons in the blue region (Figure 10.19b).

In future, MOCVD-based CIP also needs to demonstrate much higher deposition rates at high translation rates (on the order of ~1 µm/min at ≥5 cm/min translation speed) and lower cost of fabrication to become a competitor with the readily scaled-up processes (mainly VTD) for CdTe PV manufacturing. This calls for the use of additional measures, such as growth using photo- or plasma-assistance (discussed in Sections 10.5 and 10.6) and lower-purity metalorganic precursors.

10.12 MOCVD of CdTe for Radiation Detectors

Semiconductor radiation detectors rely on absorption of the X-ray photon in the semiconductor material by the photoelectric or Compton effects, where the number of electron–hole pairs generated per X-ray photon is proportional to the photon energy. The high atomic number of CdTe makes it a good absorber for X-ray photons, but thicknesses from tens of micrometers to millimeters are needed, depending on the photon energy, to absorb the X-ray. An electric field is applied across the CdTe cell to collect the electrons and holes at their respective electrodes. If the carrier collection is efficient, then the detector will act as a spectrometer, where the charge collected in each burst is proportional to the X-ray energy [76]. For effective spectrometer function, the device needs excellent carrier collection, so the crystal quality and purity must be very high. In the past, this has limited the use of CdTe in favor of high-quality crystals of germanium and silicon. Improvements in single-crystal CdTe and the alloy $Cd_{1-x}Zn_xTe$ (at dilute Zn concentrations) have increased the commercial potential for CdTe. For soft X-rays, an absorption depth of less than 100 μm can be used, which brings it into the realms of thick-layer epitaxy. Vapor-phase epitaxy has the following advantages over bulk growth:

- Lower temperature growth, reducing the point defect concentration
- Larger-area growth for large detector arrays
- Good uniformity
- Doping control for carrier compensation

The attraction of MOVPE comes from the growing interest in p-i-n devices as opposed to the traditional Schottky contact devices. This has the advantage of reducing the dark current under high voltage bias. A schematic of a CdTe p-i-n radiation detector is shown in Figure 10.20.

A p-i-n MOVPE structure was grown by Noda et al. [77], using a bulk CdZnTe crystal and growing p-type CdZnTe onto one face and n-type CdSeTe onto the other face. This takes advantage of the thick CdZnTe bulk crystal and the doping ability of the MOVPE layers. The growth was carried out at low temperature (200 °C) using a remote RF plasma to generate H radicals. Yasuda et al. reported on MOVPE of CdTe from DMCd and DETe at growth temperatures of 415–560 °C to obtain high growth rates and achieve epitaxial thickness of up to 200 μm for detection of X-ray energy up to 100 keV [78]. The device structure used an n-type CdTe layer on an n-type GaAs substrate as the anode with an Au top contact. Good energy discrimination was obtained with the 59.5-keV line from a ^{241}Am gamma source. More recently, Niraula et al. [79] grew a heterojunction

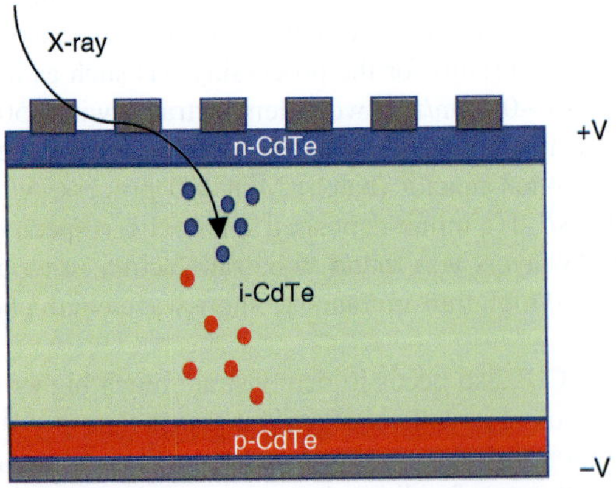

Figure 10.20 *Schematic of a p-i-n CdTe radiation detector.*

with n-CdTe/p-CdTe/p-CdZnTe where the wider-bandgap alloy, CdZnTe, was used, with the objective to reduce the dark current. These layers were grown onto Si (211) substrates at the relatively high temperature of 650 °C.

MOVPE of radiation detectors has been studied for some time, but the potential for growing very thick layers with low defect density has been demonstrated and offers some exciting potential for advanced radiation detectors when combined with II-VI heterostructures and progress with doping control. A growing demand for sensitive radiation detectors with good energy resolution for medical diagnostics and for security will drive continuing research in II-VI MOVPE.

References

1. Manasevit, H.M. and Simpson, W.I. (1971). The use of metal-organics in the preparation of semiconductor materials. *Journal of the Electrochemical Society: Solid State Science* 118: 644–647.
2. Cockayne, B. and Wright, P.J. (1984). Metalorganic vapour deposition of wide band gap II-VI compounds. *Journal of Crystal Growth* 68: 223–230.
3. Krajewski, J.J. and Kisker, D.W. (1989). Investigation of the efficiency of activated carbon for the removal of diethyltellurium from carrier gas stream. *Journal of the Electrochemical Society* 136: 745–747.
4. Kisker, D.W. (1989). Issues in the OMVPE growth of II-VI alloys for optoelectronics. *Journal of Crystal Growth* 98: 127–139.
5. Yong, K., Martinez, H., Gellman, A.J., and Sides, P.J. (1999). Homoepitaxial growth of CdTe on vicinal CdTe (100) surfaces: reaction kinetics and mechanism. *Journal of Vacuum Science and Technology A* 17: 1–8.
6. Cinader, G., Raizman, A., and Oron, M. (1990). Effect of substrate orientation on crystalline microstructure of HgTe, CdTe and HgCdTe. *Journal of Crystal Growth* 101: 167–170.
7. Sochinskii, N.V., Munoz, V., Bellani, V. et al. (1997). Substrate effect on CdTe layers grown by metalorganic vapor phase epitaxy. *Applied Physics Letters* 70: 1314–1316.
8. Farrow, R.F.C., Jones, G.R., Williams, G.M., and Young, I.M. (1981). Molecular beam epitaxial CdTe films on InSb (001). *Applied Physics Letters* 39: 12, 954.
9. Hoke, W.E., Lemonias, P.J., and Traczewski, R. (1984). Metalorganic vapor deposition of CdTe and HgCdTe epitaxial films on InSb and GaAs substrates. *Applied Physics Letters* 44 (11): 1046–1048.
10. Taskar, N.R., Bhat, I.B., Borrego, J.M., and Ghandhi, S.K. (1986). OMVPE growth of CdTe on InSb substrates. *Journal of Electronic Materials* 15: 165.
11. Wang, C.H., Cheng, K.Y., Yang, S.J., and Hwang, F.C. (1985). The growth and characterisation of CdTe epitaxial layers on CdTe and InSb by metalorganic chemical vapor deposition. *Journal of Applied Physics* 58: 757–762.
12. Zuniga-Perez, J., Tena-Zaera, R., and Munoz-Sanjose, V. (2004). Structural characterisation of CdTe layers grown on (0001) sapphire by MOCVD. *Journal of Crystal Growth* 270: 309–315.
13. Glass, H.L., Appleby-Woods, M.R., Buehnerkemper, D.L. et al. (1993). Control of microtwins in MOCVD of CdTe on sapphire. *Journal of Crystal Growth* 128: 617–622.
14. Giess, J., Gough, J.S., Irvine, S.J.C. et al. (1987). Orientation effects on the heteroepitaxial growth of $Cd_xHg_{1-x}Te$ onto CdTe and GaAs. *Materials Research Society Symposium Proceedings* 90: 389–395.
15. Ekawa, M., Yasuda, K., Sone, S. et al. (1990). X-ray photoelectron spectroscopy and Auger electron spectroscopy analyses of the initial growth mechanism of CdTe layers on (100) GaAs by metalorganic vapor phase epitaxy. *Journal of Applied Physics* 67: 6865–6870.
16. Petruzzello, J., Olego, D., Ghandhi, S. K. et al. (1987). Transmission electron microscopy of (001) CdTe on (001) GaAs grown by metalorganic chemical vapour deposition. *Applied Physics Letters* 50: 1423.
17. Hamilton, W.J., Vigil, J.A., Konkel, W.H. et al. (1993). The effect of substrate tilt on MOCVD growth of (100) CdTe on (100) GaAs. *Journal of Electronic Materials* 22: 879–885.
18. Leo, G., Longo, M., and Lovergine, N. (1996). Influence of a ZnTe buffer layer on the structural quality of CdTe epilayers grown on (100)GaAs by metalorganic vapor phase epitaxy. *Journal of Vacuum Science and Technology B* 14: 1–6.
19. Su, P-Y., Lee, C., Wang, G-C. et al. (2014). CdTe/ZnTe/GaAs heterostructures for single-crystal CdTe solar cells. *Journal of Electronic Materials* 43: 2895–2900.
20. Szeles, C., Soldber, S.A., Vydrin, S. et al. (2008). CdZnTe semiconductor detectors for spectroscopic X-ray imaging. *IEEE Transactions on Nuclear Science* 55 (1): 572–582.
21. Rao, S.R., Shintri, S.S., Markunas, J.K. et al. (2011). High-quality (211)B CdTe on (211)Si substrates using metalorganic vapor-phase epitaxy. *Journal of Electronic Materials* 40 (8): 1790–1794.

22. Carmody, M., Mallick, S., Margetis, J. et al. (2010). Single-crystal II-VI on Si single-junction and tandem solar cells. *Applied Physics Letters* 96: 153502.
23. Irvine, S.J.C. and Mullin, J.B. (1981). The growth by MOVPE and characterisation of $Cd_xHg_{1-x}Te$. *Journal of Crystal Growth* 55: 107–115.
24. Hoke, W.E., Lemonias, P.J., and Korenstein, R. (1988). An examination of organometallic thermal stability and its relevance to low-temperature MOCVD growth of HgCdTe. *Journal of Materials Research* 3: 329–334.
25. Hoke, W.E. and Lemonias, P.J. (1985). Metalorganic growth of CdTe and HgCdTe epitaxial films at a reduced substrate temperature using diisopropyltelluride. *Applied Physics Letters* 46: 398–400.
26. Berrigan, R.A., Maung, N., and Irvine, S.J.C. et al. (1988). Thin film CdTe/CdS grown by MOCVD for photovoltaics. *Journal of Crystal Growth* 195: 718–724.
27. Hails, J.E., Cole-Hamilton, D.J., Stevenson, J., and Bell, W. (2000). Allyl-iso-propyltelluride, a new MOVPE precursor for CdTe, HgTe and (Hg, Cd)Te. *Journal of Crystal Growth* 214–215: 45–49.
28. Irvine, S.J.C., Mullin, J.B., Robbins, D.J., and Glasper, J.L. (1985). A study of the UV absorption spectra and photolysis of some Group II and Group VI alkyls. *Journal of the Electrochemical Society* 132: 968–972.
29. Liu, B., Hicks, R.F., and Zinck, J.J. (1992). Chemistry of photo-assisted organometallic vapor-phase epitaxy of cadmium telluride. *Journal of Crystal Growth* 123: 500–518.
30. Fujita, Y., Terada, T., Fujii, S., and Iuchi, T. (1991). Reaction measurements and growth of HgTe/CdTe by photo-assisted MOCVD using an excimer laser. *Journal of Crystal Growth* 107: 621–625.
31. Ahlgren, W.L., Smith, E.J., James, J.B. et al. (1988). Photo-MOCVD growth of HgTe-CdTe superlattices. *Journal of Crystal Growth* 86: 198–209.
32. Irvine, S.J.C., Mullin, J.B., and Tunnicliffe, J. (1984). Photosensitisation: a stimulant for low temperature growth of epitaxial HgTe. *Journal of Crystal Growth* 68: 188–193.
33. Fujita, S., Tanabe, A., Sakamoto, T. et al. (1988). Investigations of photo-association mechanism for growth rate enhancement in photo-assisted OMVPE of ZnSe and ZnS. *Journal of Crystal Growth* 93: 259–264.
34. Irvine, S.J.C., Hill, H., and Brown, G.T. et al. (1989). Selected area epitaxy in II-VI compounds by laser-induced photo-metalorganic vapor phase epitaxy. *Journal of Vacuum Science and Technology B* 7: 1191–1199.
35. Niraula, M., Nakamura, A., Aoki, T., and Hatanaka, Y. (2002). Low-temperature growth and doping of CdTe epilayers on CdTe substrates in a remote-plasma-assisted MOCVD system for nuclear radiation detector applications. *Physica Status Solidi* 229: 83–87.
36. Lu, Y-F., Ye, Z-Z., Zenf, Y-J. et al. (2007). Low-temperature growth of p-type ZnO thin films via plasma-assisted MOCVD. *Chemical Vapor Deposition* 13: 295–297.
37. Niraula, M., Mochizuki, D., Aoki, T. et al. (2000). Low-temperature growth and n-type doping of CdTe by the remote-plasma-assisted metalorganic chemical vapor deposition method. *Vacuum* 59: 678–685.
38. Zilko, J.L. (2012). Metal organic chemical vapor deposition: technology and equipment. In: Handbook of Thin Film Deposition 3e (ed. K. Seshan). Elsevier.
39. Zoppi, G. (2005). Studies of CdTe thin films and solar cells grown by MOCVD. Durham theses, Durham University. http://etheses.dur.ac.uk/2616/.
40. Irvine, S.J.C. and Bajaj, J. (1993). In situ characterization techniques for monitoring and control of VPE growth of $Hg_{1-x}Cd_xTe$. *Semiconductor Science and Technology* 8: 860–871.
41. Kisker, D.W., Fuoss, P.H., Brennan, S. et al. (1990). In situ characterization of organometallic growth of ZnSe using grazing incidence X-ray scattering. *Journal of Crystal Growth* 101: 42–47.
42. Richter, W. (1993). Optical in situ surface control during MOVPE and MBE growth. *Philosophical Transactions A* 344. DOI: 10.1098/rsta.1993.0100.
43. Johs, B., Herzinger, C., and Dinan, J.H. (1998). Real-time monitoring and control of epitaxial semiconductor growth in a production environment by in situ spectroscopic ellipsometry. *Thin Solid Films* 313–314: 490–495.
44. Sankur, H., Southwell, W., and Hall, R. (1991). In situ optical monitoring of OMVPE deposition of AlGaAs by laser reflectance. *Journal of Electronic Materials* 20: 1099–1104.
45. Zinck, J.J., Brewer, P.D., Jensen, J.E. et al. (1988). Excimer laser-assisted metalorganic vapor phase epitaxy of CdTe on GaAs. *Applied Physics Letters* 52: 1434.
46. Bajaj, J., Irvine, S.J.C., Sankur, H.O., and Svoronos, S.A. (1993). Modeling of in situ monitored laser reflectance during MOCVD growth of HgCdTe. *Journal of Electronic Materials* 22: 899–906.
47. Irvine, S.J.C. and Baja, J. (1994). A study of the growth kinetics of II–VI metalorganic vapour phase epitaxy using in situ laser reflectometry. *Journal of Crystal Growth* 145: 74–81.
48. Macleod, H.A. (1986). *Thin Film Optical Filters* 2e. New York: Macmillan.
49. Irvine, S.J.C., Hartley, A., and Stafford, A. (2000). In situ monitoring of the MOCVD growth of CdS/CdTe. *Journal of Crystal Growth* 221: 117–123.
50. Wu, X. (2004). High-efficiency polycrystalline CdTe thin-film solar cells. *Solar Energy* 77: 803–814.
51. Xia, W., Welt, J.A., Lin, H. et al. (2010). Fabrication of $Cd_{1-x}Zn_xS$ films with controllable zinc doping using a vapor zinc chloride treatment. *Solar Energy Materials in Solar Cells* 94: 2113–2118.

52. Zhou, Z., Zhao, K., and Huang, F. (2010). Optical properties of $Cd_{1-x}Zn_xS$ thin films for $CuInGaSe_2$ solar cell application. *Materials Research Bulletin* 45: 1537–1540.
53. Kartopu, G., Clayton, A.J., Brooks, W.S.M. et al. (2014). Effect of window layer composition in $Cd_{1-x}Zn_xS$/CdTe solar cells. *Progress in Photovoltaics: Research Applications* 22: 18–23.
54. McCandless, B.E. and Sites, J.R. (2010). Cadmium telluride solar cells. In: *Handbook of Photovoltaic Science and Engineering* 2e (eds. A. Luque and S. Hegedus). John Wiley & Sons, Ltd.
55. Hanket, G.M., McCandless, B.E., Buchanan, W.A. et al. (2006). Design of a vapor transport deposition process for thin film materials. *Journal of Vacuum Science and Technology A* 24: 1695–1701.
56. Perrenoud, J., Kranz, L., Gretener, C. et al. (2013). A comprehensive picture of Cu doping in CdTe solar cells. *Journal of Applied Physics* 114: 174505 (1–10).
57. Barrioz, V., Kartopu, G., Irvine, S.J.C. et al. (2012). Material utilisation when depositing CdTe layers by inline AP-MOCVD. *Journal of Crystal Growth* 354: 81–85.
58. Rowlands, R.L., Irvine, S.J.C., Barrioz, V. et al. (2008). SIMS analysis of intentional in situ arsenic doping in CdS/CdTe solar cells. *Semiconductor Science and Technology* 23: 015017.
59. Kartopu, G., Phillips, L.J., Barrioz, V. et al. (2016). Progression of metalorganic chemical vapour-deposited CdTe thin-film PV devices towards modules. *Progress in Photovoltaics: Research Applications* 24: 283–291.
60. Barrioz, V., Irvine, S.J.C., Jones, E.W. et al. (2007). In situ deposition of cadmium chloride films using MOCVD for CdTe solar cells. *Thin Solid Films* 515: 5808–5813.
61. Major, J.D., Treharne, R.E., Phillips, L.J., and Durose, K. (2014). A low-cost non-toxic post-growth activation step for CdTe solar cells. *Nature* 511: 334–337.
62. Rohatgi, A. (1992). A study of efficiency limiting defects in polycrystalline CdTe/CdS solar cells. *International Journal of Solar Energy* 12: 37–49.
63. Green, M.A., Hishikawa, Y., Warta, W. et al. (2016). Solar cell efficiency tables (version 50). *Progress in Photovoltaics: Research Applications* 25: 668–676.
64. Irvine, S.J.C., Barrioz, V., Lamb, D. et al. (2008). MOCVD of thin film photovoltaic solar cells – Next generation production technology? *Journal of Crystal Growth* 310: 5198–5203.
65. Rugen-Hankey, S.L., Clayton, A.J., Barrioz, V. et al. (2015). Improvement to thin film CdTe solar cells with controlled back surface oxidation. *Solar Energy Materials in Solar Cells* 136. 213–217.
66. Rowell, M.W. and McGehee, M.D. (2011). Transparent electrode requirements for thin film solar cell modules. *Energy and Environmental Science* 4: 131–134.
67. Halliday, D.P., Williams, B.L., Durose, K. et al. (2011). CdTe/CdS core-shell nanowire structures on glass substrates for photovoltaic applications – Growth and characterization. In: *Proceedings of the 37th IEEE Photovoltaic Specialists Conference*, Seattle, WA, 19–24 June, 234–238. IEEE.
68. Williams, B.L., Taylor, A.A., Mendis, B.G. et al. (2014). Core-shell ITO/ZnO/CdS/CdTe nanowire solar cells. *Applied Physics Letters* 104: 053907 (1–5).
69. Akgun, M.C., Kalay, Y.E., and Unalan, H.E. (2012). Hydrothermal zinc oxide nanowire growth using zinc acetate dihydrate salt. *Journal of Materials Research* 27: 1445–1451.
70. Kartopu, G., Turkay, D., Ozcan, C. et al. (2018). Photovoltaic performance of CdS/CdTe junctions on ZnO nanorod arrays. *Solar Energy Materials in Solar Cells* 176: 100–108.
71. Özcan, C., Kartopu, G., Hadibrata, W. et al. (2017). Performance of CdS/CdTe junctions on ZnO nanorod arrays by metalorganic chemical vapour deposition. *E-MRS Spring Meeting*, Strasbourg.
72. Kartopu, G., Barrioz, V., Irvine, S.J.C. et al. (2014). Inline atmospheric pressure metal-organic chemical vapour deposition for thin film CdTe solar cells. *Thin Solid Films* 558: 374–377.
73. Barrioz, V., Lamb, D.A., Monir, S. et al. (2014). UK Patent Application (GB1302306.4); PCT patent application (PCT/GB2014/050386).
74. Barrioz, V., Monir, S., Kartopu, G. et al. (2015). MOCVD for solar cells, a transition towards a chamberless inline process. *Journal of Crystal Growth* 414: 223–231.
75. Kartopu, G., Barrioz, V., Monir, S. et al. (2015). CdTe thin film solar cells produced using a chamberless inline process via metalorganic chemical vapour deposition. *Thin Solid Films* 578: 93–97.
76. Del Sordo, S., Abbene, L., Caroli, E. et al. (2009). Progress in the development of CdTe and CdZnTe semiconductor radiation detectors for astrophysical and medical applications. *Sensors* 9: 3491–3526.
77. Noda, D., Aoki, T., Nakanishi, Y., and Hatanaka, Y. (2000). Growth of CdZnTe and CdSeTe crystals for p–i–n radiation detectors. *Journal of Crystal Growth* 214–215: 1121–1124.
78. Yasuda, K., Kasama, H., Yamamoto, Y. et al. (2005). Development of nuclear radiation detectors with energy discrimination capabilities based on thick CdTe layers grown by metalorganic vapor phase epitaxy. *IEEE Transactions on Nuclear Science* 52: 1951–1955.
79. Niraula, M., Yasuda, K., Namba, S. et al. (2013). MOVPE growth of thick single crystal CdZnTe epitaxial layers on Si substrates for nuclear radiation detector development. *IEEE Transactions on Nuclear Science* 60: 2859–2863.

11

ZnO and Related Materials

V. Muñoz-Sanjosé[1] *and S.J.C. Irvine*[2]
[1] *Dpto. Física Aplicada y Electromagnetismo, University of Valencia, Spain*
[2] *Centre for Solar Energy Research, College of Engineering, Swansea University, OpTIC Centre, St. Asaph, UK*

11.1 Introduction

The early work on MOCVD (MOVPE) of ZnO (1980–1990) concentrated on the selection of precursors and prevention of prereaction, the zinc organometallics being very reactive with oxygen [1]. Thus, the choice of precursors and how prereaction was minimized is covered in the next section on precursors. It is worth noting that the choice of precursors also depends on growth temperature, and ZnO can be grown over a very wide range of temperatures (300–700 °C or even more), depending on the required properties. High-temperature growth is preferred for the growth of ZnO nanorods, while planar epitaxial layers are typically grown at much lower temperatures.

The extensive interest in MOCVD (MOVPE) of ZnO comes from its remarkable properties as both a photonic and piezoelectric material. It has a direct bandgap at room temperature of 3.3 eV and normally exists in the hexagonal wurtzite structure. It has a high exciton binding energy of 60 meV that enables bright room-temperature photoluminescence. The crystal structure, along with its ionic nature, enables a high piezoelectric coefficient suitable for energy harvesting. Other applications include transparent conductors (with n-type dopants) and gas sensors; and, when alloyed with magnetic ions, ZnO can be used for spintronics. Although it has been seen as a challenger to GaN for blue and UV LEDs, the difficulties in p-type doping due to self-compensation have resulted in very few reports of electroluminescent p-n devices. The application to photonic devices can, to some extent, circumvent this limitation by forming heterojunctions with other p-type materials and even combining III-V with II-VI materials. The bandgap can be widened by alloying with Mg or narrowed by alloying with Cd. CdO has a low density of states

Metalorganic Vapor Phase Epitaxy (MOVPE): Growth, Materials Properties, and Applications,
First Edition. Edited by Stuart Irvine and Peter Capper.

in the conduction band that results in a very strong Burstein–Moss effect where the optical bandgap ranges from 2.3–3.4 eV and the electronic bandgap is estimated to be 2.23 eV [2]. A consequence of this band structure is the relatively high electron mobility at high n-type carrier concentrations and high transmission in the infrared for transparent conducting oxide (TCO) applications [3].

This chapter highlights some of the key research in MOCVD (MOVPE) of ZnO and related II-VI materials, mainly focusing on the crystal growth experimental work for obtaining layers and nanostructures of this family of materials. This covers precursors (see also Chapter 13 in this book), which is still an active field of research when compared with III-V MOCVD; basic setups; substrates and energy-assisted MOCVD; as well as the analysis of some effects of growth conditions on the process and consequently on the results. The problems associated with prereaction of the metal and oxygen sources has stimulated a lot of research on the precursor chemistry, reactor design, and determining optimized growth conditions. This has resulted in higher-quality ZnO and improved doping control that will be covered in some depth in this chapter. For simplicity of notation, we will use the term MOCVD as including MOVPE processes.

11.2 Sources for the MOCVD Growth of ZnO and Related Materials

In order to promote successful use of the MOCVD technique, the availability of pure, volatile, and thermally stable precursors is fundamental. In addition, it will be necessary that these precursors accomplish the requirements emphasised in Section 13.2 of this book, in particular [4]:

i. Have sufficient high vapor pressures.
ii. Provide clean delivery of species.
iii. Decompose and react in order to achieve a crystalline pure phase on the substrate.

Finding a metalorganic precursor for zinc, cadmium, or magnesium should not be a difficult task. Many precursors have been developed, which could be potentially useful if their reactivity with oxygen's precursors allows a suitable growth process under defined conditions. Hence, this section will deal with some of the metalorganic and oxygen precursors that have been proposed for the growth of ZnO and its related materials.

11.2.1 Metalorganic Zinc Precursors

More than 300 metalorganic volatile zinc compounds can be found [5]. However, it is worth noting that the most common zinc precursors for MOCVD processes have been the simple metal alkyl, pyrophoric compounds: Diethylzinc ($Zn(C_2H_5)_2$) and dimethylzinc ($Zn(CH_3)_2$). Thus, this subsection will start by analyzing some of the properties of these two precursors in a complementary form to the ones discussed in Section 13.6.

11.2.1.1 Diethylzinc ($Zn(C_2H_5)_2$)

Diethylzinc (DEZn or DEZ) is a colorless liquid, highly pyrophoric organo-zinc compound that reacts violently with water. The average bond strength to break the first carbon–metal bond, which is typically the activation energy for pyrolysis when radical mechanisms dominate, is 35 kcal mol^{-1} [6]. DEZn is an electron-deficient compound with two vacant orbitals available for bonding. This explains its high chemical reactivity, which allows it to form complexes with compounds containing free electron pairs. The DEZn molar mass is 123.50 g mol^{-1}, its density is 1.205 g cm^{-3}, and its melting and boiling points are −28 °C and 118 °C, respectively.

Fulem et al. [7] published a review on the vapor pressure data of DEZn, and a $P(T)$ in the form of Antoine's law was determined as:

$$\mathrm{Log}_{10}\, P(\mathrm{Torr}) = 7.331 - \frac{1.512 \times 10^3}{T(\mathrm{K}) - 50.797}$$

Another approach to the temperature dependence of the saturated vapor pressure can be found in Table 13.7.

Regarding decomposition, hydrogen gas reacts with DEZn as soon as it begins to decompose and promotes the formation of methane with increasing temperature through hydrogenation and hydrocracking of the various released hydrocarbons [8]. It can be concluded that diethylzinc decomposes, releasing ethane and n-butane as major species with small amounts of ethylene, trans-2-butene, and cis-2-butene [8].

As pointed out by Thiandoume et al. [9], the decomposition temperature is dependent on the gas environment, including carrier gas and other precursors there present, as well as on the reactor wall cleaning conditions. Maejima et al. [10], working on the study of the DEZn dissociation, suggested that the dissociation reaction is due to the β-hydride elimination,

$$\mathrm{Zn}(C_2H_5)_2 \rightarrow \mathrm{ZnH_2} + 2C_2H_4$$

nevertheless, the chemical reactions between the pyrolyzed precursors are crucial to determine the decomposition paths [10].

Note that, in spite of its relatively low saturated vapor pressure, diethylzinc has been extensively present as zinc precursor. In fact, it is the most frequently used for the growth of ZnO-based materials, due to, among other reasons, the lower carbon contamination that ethyl radicals have as compared to other metalorganics (the significance of carbon presence will be discussed in the frame of doping in Section 11.5.4.2) and the easy availability and high purity with which this precursor can be found.

11.2.1.2 Dimethylzinc ($Zn(CH_3)_2$)

Despite DEZn being the most used zinc precursor, dimethylzinc (DMZn or DMZ) has also been extensively proposed (e.g. [11–13]) due to its relatively lower reactivity with oxygen sources: reactivity that, if it is too high, can be deleterious for the growth process in forming undesired prereactions in the gas phase.

Dimethylzinc is a colorless liquid at room temperature and atmospheric pressure, with a disagreeable odor; its melting and boiling points are −42 °C and 46 °C, respectively. The average bond strength to break the first carbon–metal bond is 42 kcal mol^{-1} [6]. The vapor pressure temperature dependence can be found in Table 13.7.

Regarding DMZn thermal decomposition, the study of Kuniya et al. [8] indicates that dimethylzinc decomposes in the vapor phase, releasing methane as a major species with small amounts of ethane, ethylene, propane, and propylene. Nevertheless, as already noted, the pyrolysis of the precursors can be modified by the gas in which the process takes place [9, 14].

Fujita [15] studied the thermal decomposition of DMZn by in situ ultraviolet absorption spectra measurements. A mixture of nitrogen and hydrogen was used as carrier gas. DMZn decomposition enhancement was proportional to the H_2 concentration, concluding that one hydrogen molecule contributed to the reaction for the decomposition of DMZn.

Some authors have conducted comparative studies using DEZn and DMZn as zinc precursors, showing that they exhibit different growth mechanisms and properties (see [16] for details). In addition, surface reactions and adsorption at the substrates can be different, as pointed out in

[17, 18], who studied the surface reactions and adsorption of DEZn and DMZn on silicon and GaAs. Surface reactions, due to their significance and correlated effects, should be taken into account depending on the particular application.

11.2.1.3 Some Metalorganic Zinc Precursors as Alternatives to the Alkylmetals

Due to the reactivity with common oxygen sources of the pyrophoric alkylmetals, safety reasons, and the need for relatively high temperatures to promote their decomposition, alternative zinc precursors have been proposed. In the long list of these alternative zinc precursors, zinc acetate ($Zn(C_2H_3O_2)_2$) [19, 20] and zinc acetylacetonate, ($Zn(acac)_2$, $Zn(C_5H_7O_2)_2$) [21–24] are considered low-cost, easy-to-handle zinc precursors. In fact, they have been the preferred options as solid precursors at room temperature for the growth of ZnO and its related materials. In some cases, zinc acetate dihydrate ($Zn(C_2H_3O_2)_2 \cdot 2H_2O$) has been used as a source for zinc acetate [19]. The decomposition paths of zinc acetate dihydrate were studied by McAdie [25].

Different sublimation temperatures in the range of 115–160 °C for zinc acetylacetonate [26, 27], and in the range of 100–200 °C for zinc acetate [19, 20], have been used to generate the vapors carried out to the reaction chamber with or without oxygen, depending on the experiments.

Other precursors have also had some presence for different reasons in the MOCVD growth of the ZnO family. Some of them are discussed in the given references, as e.g. diisopropylzinc (Di-PrZn) [28–30], which was suggested to enable low-temperature growth of ZnO. Another approach was published by Kumar et al. [31], where zinc 2-ethyl hexanoate was used as the zinc source. Zinc malonates of the type $[Zn\{C(H)(R_1C{=}O)(R_2C{=}O)\}_2]$ (R_1, R_2: OMe, OEt, O^iPr, and O^tBu) were synthesized and evaluated as possible precursors for MOCVD of ZnO in [32] as well as two bis(ketoiminato)zinc(II) complexes $[Zn\{[(CH_2)_xOCH_3]NC(CH_3){=}C(H)C(CH_3){=}O\}_2]$($x$ =2 or x =3) [33], just to name some of these less-used precursors.

11.2.1.4 Zinc Precursor Adducts

Different reasons have been invoked in the literature for using adducts in MOCVD processes, e.g. because adducts can avoid or reduce premature reactions in the gas phase [34], because they can allow easy purification and can be easier to handle [35], because they offer the availability of more convenient vapor pressures for the growth experiment [36], etc. On this basis, several adduct-based precursors for growing ZnO were synthesized and tested. In Table 11.1, some of them are listed together with the reference(s) in which details of the synthesis and/or growth experiments can be found.

11.2.1.5 Zinc Single Precursors for the Growth of ZnO and Related Materials

Defining a zinc single precursor for the growth of ZnO as one that does not need any additional feeding of oxygen sources to react and produce ZnO, it is worth noting that some zinc precursors can be used with or without an additional oxidizing agent. Thus, Khan et al. [52] used zinc acetate dihydrate dried under reduced pressure before its use in the growing experiment. Other examples of using this single precursor can be found in [19].

Basic zinc acetate ($Zn_4O(CH_3CO_2)_6$) was proposed as a single precursor for the deposition of ZnO by low-pressure metalorganic chemical vapor deposition [53], and zinc acetylacetonate ($Zn(C_5H_7O_2)_2$) without additional oxygen was the choice in [54]. Another zinc acetate compound ($Zn_4(OH)_2(O_2CCH_3)_6 \cdot 2H_2O$) can also be used as single source, as demonstrated by Chen et al. [55].

Alkylzinc alkoxides like methylzinc isoprotoxide ($MeZn(OPr^i)$) [36, 52] and methyl-zinc tert-butoxide ($MeZn(OBu^t)$) [36, 56] were proposed as single-source precursors for the growth of ZnO films and the mechanism of pyrolysis analyzed [56]. In the search for new options, Matthews et al.

Table 11.1 *Adduct-based precursors for growing ZnO.*

Zn precursor	Reference(s)	Remarks
DMZn:triethylamine (DMZn-TEN)	[35, 37]	$Zn(CH_3)_2$:$(C_2H_5)_3N$
DMZn:1,3,5-trimethyl	[35]	$Zn(CH_3)_2$: $(CH_3)_3$
DMZn:1,3,5 triethylhexahydro 1,3,5 triazine	[35]	$Zn(CH_3)_2$:$C_9H_{21}N_3$
DMZn:1,4-dioxane	[38, 39]	$Zn(CH_3)_2$:$C_4H_8O_2$
DMZn:1,4-thioxane	[38, 39]	$Zn(CH_3)_2$:C_4H_8OS
DMZn:2THF	[40]	THF = tetrahydrofuran (C_4H_8O)
DMZn:tmeda	[41]	tmeda = N,N,N′,N′-tetramethylethylendiamine; $C_6H_{16}N_2$
$Zn(tta)_2$:tmeda	[42, 43]	Zinc Bis(2-thenoyl-trifluoroacetonate); tta = deprotonated of 1-thenoyl - 4, 4, 4 –trifluoroacetone
DMZn:THF	[34, 39, 44]	THF = tetrahydrofuran (C_4H_8O)
DMZn:THP	[39, 44]	THP = tetrahydropyran ($C_5H_{10}O$)
DMZn:Furan	[39, 44]	Furan (C_4H_4O)
$Zn(hfa)_2$ $2H_2O$ diglyme	[45, 46]	diglyme = bis(2-methoxyethyl)ether($C_6H_{14}O_3$) hfa = 1,1,1,5,5,5-hexafluoro-2,4- pentanedionate ($C_5F_6HO_2$)
$Zn(hfa)_2$ $2H_2O$ triglyme	[45, 46]	triglyme – 2,5,8,11 tetraoxadodecane
$Zn(hfa)_2$ $2H_2O$ tetraglyme	[45, 46]	tetraglyme =2,5,8,- 11,14-pentaoxapentadecane ($C_8H_{18}O_4$)
$Zn(tmp)_2$	[47]	Bis(2,2,6,6-tetramethylpiperidinyl)zinc ($C_{18}H_{36}N_2$)
$Zn(dpm)_2$(NN′DEA)	[48]	dpm = 2,2,6,6- tetramethyl-3,5-heptanedionato
DEZn-THF	[34]	THF = tetrahydrofuran (C_4H_8O)
DEZn-n-hexane (83%)	[19]	
Ethyl zinc diethylamide	[50]	$(C_2H_5)ZnN(C_2H_5)_2$
$Zn(tmhd)_2$	[51]	Bis (2,2,6,6,tetramethyl-heptanedionate)

[57] synthesized and structurally characterized two zinc precursors: $Zn(CH_3C(NCH_2CH_2CH_2CH_3)CHCOCH_3)_2$ and $Zn(CH_3C(NCH(CH_3)_2)CHC(O)OCH_2CH_3)_2$. Likewise, Holmes et al. [58] studied the Zn β-ketoiminate system, based on acetylacetimine with n-propyl, isopropyl, and butyl groups, for preparing ZnO thin films. Surface carbon was a primary impurity with these precursors. Similarly, Cosham et al. [59] studied the synthesis and characterization of a family of zinc bis(β-ketoiminate) complexes; one of them $[Zn\{MeC(O)CHC(NCH_2CH_2OMe)CF_3\}_2]$ was applied for the MOCVD growth of ZnO:F at atmospheric pressure in the absence of additional oxidant. The use of 1:1 adducts between DMZn and the oxygen-donor heterocyclic ligands tetrahydrofuran, tetrahydropyran, and furan was considered an option to avoid the DMZn/O_2 prereaction [44]. A further argued advantage associated with the use of dimethylzinc adducts is that they are nonpyrophoric, in marked contrast to DMZn [44]. Analogously, Ashraf et al. [38] proposed the use of the dimethylzinc adduct DMZn:1,4-dioxane as a single source. In this paper, the use of DMZn: 1,2-dimethoxyethane and DMZn:1,4-thioxane in solution with n-octane as single source for the growth of nanowires was also studied. More recently, bis[(pentylnitrilomethylidine)(pentylnitrilomethylidine-μ-phenalato)]dizinc(II) [60] has also been synthesized, characterized, and proposed as a single source for the growth of ZnO by using plasma-assisted MOCVD.

11.2.2 Metalorganic Cadmium Precursors

As in the case of zinc precursors, a large variety of organometallic cadmium precursors can be found. The metal alkyl, pyrophoric compounds: dimethylcadmium ($Cd(CH_3)_2$ or DMCd) and diethylcadmium ($Cd(C_2H_5)_2$ or DECd) appear as an initial option, but the presence of DECd as precursor for the growth by MOCVD of cadmium-related materials is not frequent, in particular for the growth of oxides. The low vapor pressure of DECd as regards DMCd and the reactivity with oxygen sources [61] can be some of the reasons for its unusual presence in the growth of the cadmium oxide family by MOCVD. In addition, when concerned with ternary compounds, $Zn_{1-x}Cd_xO$ in our case, the choice of precursors must take into account the possibility of ligand exchange reactions between the Group II precursors [62]. In this context, DMCd has been the main option. An interesting analysis of the organometallic chemistry regarding cadmium can be found in [61].

Dimethylcadmium is a colorless, highly toxic liquid that fumes in air; its melting and boiling points are −4.5 °C and 106 °C, respectively. The average bond strength to break the first carbon–metal bond is 33 kcal mol^{-1} [6].

The temperature dependence of evaporation for DMCd can be found in Table 13.7. On the other hand, the decomposition of DMCd appears as an increase of the pressure in the temperature range 180–290 °C [8]. Likewise, the study of the pyrolysis of DMCd in the presence of hydrogen showed that DMCd decomposes in the range of 350–380 °C [14].

11.2.3 Metalorganic Magnesium Precursors

Several magnesium organometallic precursors can be found as tentative candidates for the growth of magnesium oxides. Some of them are briefly mentioned here.

Bis(cyclopentadienyl)magnesium ($(C_5H_5)_2Mg$ or Cp_2Mg), with a melting point of 177 °C and a boiling point of 222 °C, has been used by different authors: [63, 64], just to name a few. Cyclopentadienyl metal compounds have been suggested for the vapor deposition of metals, since many vaporize easily and the metal–ring bonds are usually weaker than those within the ring, suggesting the possibility of a clean thermal decomposition [65]. This clean decomposition was also reported in [66] by using Cp_2Mg as a Mg doping agent, with no carbon incorporation into the crystal. The vapor pressure of bis(cyclopentadienyl)magnesium was studied by Hull et al. [67]. The vapor pressure can be found in Table 13.9 or in [68–70].

Bis(methylcyclopentadienyl)magnesium ($C_{12}H_{14}Mg$ or $MeCp_2Mg$) was the precursor chosen by Gruber et al. [71] for the growth of $Zn_{1-x}Mg_xO$ on GaN/sapphire templates. The vapor-pressure dependence with temperature can be described [72] as

$$\mathrm{Log}_{10}\, P(\mathrm{Torr}) = 7.30 - \frac{2.358\times10^{3}}{T(\mathrm{K})}$$

Bis(ethylcyclopentadienyl)magnesium ($C_{14}H_{18}Mg$ or $EtCp_2Mg$) has been another option as proposed by Chiba et al. [73] and Nakamura et al. [74]. The vapor pressure of $EtCp_2Mg$ is one order of magnitude lower than that of $MeCp_2Mg$. Therefore, low values of the growth rate can be attained if that is the goal.

Other magnesium precursors have been studied for the growth of MgO, but their presence in MOCVD experiments in growing $Zn_{1-x}Mg_xO$ is rather low. Thus, in spite of the number of Mg precursors studied in the MOCVD literature, it must be noted that bis(cyclopentadienyl)magnesium (Cp_2Mg), bis(methylcyclopentadienyl)magnesium ($MeCp_2Mg$), and bis(ethylcyclopentadienyl) magnesium ($EtCp_2Mg$) have been largely the option in MOCVD experiments for the growth of $Zn_xMg_{1-x}O$ alloys.

11.2.4 Precursors for Oxygen

11.2.4.1 Oxygen (O_2)

The most obvious source for oxygen is oxygen itself. In spite of its high reactivity with some metalorganic precursors it has been one of the most frequently used oxidizers in conjunction with zinc, cadmium, and magnesium precursors. Parasitic reactions have been avoided or reduced by adequate combination of reactor design, methodology, and growth parameters. The direct control of the O_2 flow and the high purity of molecular oxygen have been exploited advantages that have justified its use as oxidizer source.

Maejima et al. [10], making reference to the works of Sosnovsky et al. [75] and Coates et al. [76], comment that an oxygen molecule is inserted between Zn and an ethyl ligand on the reaction of DEZn and O_2, and it is supposed that this is the first reaction to take place:

$$Zn(C_2H_5)_2 + O_2 \rightarrow C_2H_5ZnOOC_2H_5$$

This first product will follow a decomposition path to finally deposit ZnO on the substrate by the elimination of hydrocarbon molecules in a temperature-dependent process.

11.2.4.2 Water (H_2O)

Water has been another of the choices as oxidant agent for different zinc precursors. Just as a reference, the vapor pressure for water can be calculated [77] from:

$$\mathrm{Log}_{10} P(\mathrm{Torr}) = 8.07131 - \frac{1.730 \times 10^3}{T(\mathrm{K}) - 39.574}; \text{in the range } 1 \rightarrow 100°C$$

DEZn with H_2O forms complexes from temperatures near ambient [78]. Thus, as discussed in the paper of Lau et al. [79], in the gas stream, the reaction at room temperature is:

$$(C_2H_5)_2 Zn + 2H_2O \rightarrow Zn(OH)_2 + 2C_2H_6$$

and on the surface substrate at high temperature (400 °C),

$$Zn(OH)_2 \rightarrow ZnO + H_2O$$

Similar to the case of using O_2, due to the formation of $Zn(OH)_2$ in the DEZ/H_2O system at low temperature, the reactants must be kept separate until they are a short distance from the substrate. In [80], a computational study of the reactions between the products of the thermal decomposition of diethylzinc and water was investigated. Three possible reaction pathways led to ethyl-hydridozinc (ZnC_2H_5OH), which further proceeded to dihydridozinc ($Zn(OH)_2$) with additional water. Thus, $Zn(OH)_2$ would play a key role in the deposition of ZnO.

11.2.4.3 Alcohols

The use of an alcohol as oxygen source instead of, e.g. O_2 or H_2O, has been found to prevent or significantly reduce prereactions, which can lead to deposition of particles upstream from the substrate, both at low pressure [81] and at atmospheric pressure [82]. Alcohols have lower reactivity and decomposition temperatures [13, 83] than other oxygen sources. Indeed, it has been suggested that by using an alcohol and DEZn, an intermediate product (an alkylzinc alkoxide) can be formed in the gas phase prior to the ZnO deposition on the substrate surface [36, 76]. In this frame, several

Table 11.2 Coefficients for the vapor pressure Antoine equation.

Alcohol	*A*	*B*	*C*	Temperature range	Melting point (°C)	Boiling point (°C)
Tert-butanol	7.15711	1080.55	103.00	−24 -> 235 °C	25.81	83
Iso-propanol	8.00308	1505.52	61.55	10 -> 90 °C	−89.5	82.3
n-butanol	7.92484	1617.52	69.85	<118 °C	−89.8	117.7
Methanol	8.08097	1582.27	33.45	15 ->100 °C	−97.8	64.7
Ethanol	8.20417	1642.89	42.85	−57-> 80 °C	−114.1	78.2

Data from [77].

alcohols like ethanol, methanol, tert-butanol, iso-propanol, and n-butanol have been proposed in combination with several zinc precursors, usually DEZn or DMZn.

Table 11.2 includes the coefficients for the vapor pressure Antoine equation

$$\mathrm{Log}_{10} P(\mathrm{Torr}) = A - \frac{B}{T(\mathrm{K}) - C}$$

corresponding to some of the more commonly used alcohols as oxidizing agent in MOCVD experiments.

A study of the thermal decomposition of tert-butanol in the presence of DEZn can be found in [9].

11.2.4.4 Other Oxygen Sources

Other less-reactive oxygen sources such as nitrous oxide (N_2O), nitrogen dioxide (NO_2), carbon dioxide (CO_2), furan (C_4H_4O), tetrahydrofuran (THF, C_4H_8O), tetrahydrophyran ($C_5H_{10}O$), and heavy water (D_2O) can be alternatives as oxidizing agents. Their use can be justified because, under defined conditions, they could allow a better control of the reactivity and growth process. It has been claimed that these less-reactive agents suppress (or reduce) homogeneous reactions, especially when working at lower pressure [79]. If a low-temperature growth is necessary when using these less-reactive agents, assisted techniques like plasma or ultraviolet light, simultaneously with thermal excitation, can be an option. Maejima et al. [84] conducted an interesting study on the dissociation of DEZn and N_2O by in situ monitoring Fourier transform infrared absorption. From their results, the following reaction model was proposed: DEZn is essentially dissociated by two-step β-hydride elimination processes at 300–550 °C and 650–900 °C, where first and second ethyl groups are eliminated, respectively. N_2O dissociates into N_2 and an oxygen atom gradually, from nearly room temperature, with the increase of temperature. Given the supply of both DEZn and N_2O to grow ZnO, DEZn first reacts with an oxygen atom generated by the dissociation of N_2O to form DEOZn. This is followed by the elimination of C_2H_4. As a result, C_2H_5OZnOH is transferred to the substrate in the lower-growth-temperature regime (around 500 °C), while $Zn(OH)_2$ is the transferred compound in the higher-temperature regime (700–900 °C).

On the other hand, the use of heterocyclic compounds like furan, tetrahydrofuran, and tetrahydrophyran was proposed by Wright [12]. The vapor pressure of furan can be approached [85] by

$$\mathrm{Log}_{10} P(\mathrm{Torr}) = 6.9752 - \frac{1.061 \times 10^3}{T(\mathrm{K}) - 45.42}$$

in the range 2.7–61.6 °C; its melting point is −85.6 °C, and its boiling point is 31.4 °C.

With respect to tetrahydrofuran, the vapor pressure can be obtained [77] from

$$\mathrm{Log}_{10}P(\mathrm{Torr}) = 6.99515 - \frac{1.2023 \times 10^3}{T(\mathrm{K}) - 43.746}$$

in the range 23–100 °C. Its melting point is –180 °C, and the boiling point 67 °C.

For its part, tetrahydrophyran has a vapor pressure of 71.5 mm Hg at 25 °C, and its melting and boiling points are –45 °C and 88 °C, respectively [86].

Again, the possible reactions with the atmosphere, in which the decomposition and reactions take place, must be considered, particularly when the oxidizing agents have paths of decomposition in which reactions with ambient atmosphere can be present.

Wenas et al. [87] introduced the use of heavy water (D_2O) mixed with water (H_2O) as oxidizing agent, but only a few results (see, for example, [88]) using this oxidizing agent in MOCVD experiments can be found in the current literature.

11.2.5 Precursors for Doping

Several doping agents for ZnO and related materials have been claimed for MOCVD processes. Some of the most frequently used are listed in Tables 11.3 and 11.4 with some reference(s) in which the precursor was used.

Other less frequently used precursors for n-type doping are introduced in Section 11.5.4.1.

Similarly as for donors, other less frequently used precursors to act as acceptor agent are introduced in Section 11.5.4.2.

Table 11.3 *n-Type doping agents used for ZnO and related materials.*

Precursors potentially acting as donors	Reference(s)	Remarks
Diborane (B_2H_6)	[88, 89]	Diluted in H_2, Ar, or He.
Tetramethyltin (TMT)	[90]	Used as doping agent for CdO $P(T)$: $\mathrm{Log}_{10}P(\mathrm{Torr}) = 7.495 - \frac{1.620 \times 10^3}{T(\mathrm{K})}$
Trimethylaluminum (TMAl)	[91]	$P(T)$: $\mathrm{Log}_{10}P(\mathrm{Torr}) = 8.224 - \frac{2.2349 \times 10^3}{T(\mathrm{K})}$
Triethylaluminum (TEAl)	[92]	$P(T)$: $\mathrm{Log}_{10}P(\mathrm{Torr}) = 10.784 - \frac{3.625 \times 10^3}{T(\mathrm{K})}$
Triethylgallium (TEGa)	[93]	$P(T)$: $\mathrm{Log}_{10}P(\mathrm{Torr}) = 9.165 - \frac{2.530 \times 10^3}{T(\mathrm{K})}$
Trimethylgallium (TMGa)	[94]	$P(T)$: $\mathrm{Log}_{10}P(\mathrm{Torr}) = 8.07 - \frac{1.730 \times 10^3}{T(\mathrm{K})}$ In [95], a comparative study between TEGa and TMGa can be found.
Triisopropylgallium (TIPGa)	[96]	$P(T)$: 1 Torr @ 30°C [6]
Gallium acetylacetonate	[97]	
Trimethylindium (TMIn)	[98]	$P(T)$: $\mathrm{Log}_{10}P(\mathrm{Torr}) = 10.52 - \frac{3.014 \times 10^3}{T(\mathrm{K})}$

The vapor pressure versus temperature equations in Table 11.3 are from [35] and [68].

Table 11.4 *p-Type doping agents used for ZnO and related materials.*

Precursors acting potentially as acceptors	Reference(s)	Remarks
Cyclopentadienyl (tributylphosphine)-copper	[34]	
Trimethylantimony (TMSb)	[99, 100]	*P*(*T*): $\text{Log}_{10}P(\text{Torr}) = 7.730 - \frac{1.709 \times 10^3}{T(\text{K})}$
Diallylamine ($C_6H_{11}N$)	[37]	Vapor pressure of 18 Torr @20 °C
Monomethylhydrazine (MMHy) CH_6N_2	[101]	Boiling point 87.5 °C and vapor pressure of 50 Torr @ 25 °C
Ammonia (NH_3)	[102–104]	
Unsymmetrical-dimethyl-hydrazine (UDMHy)	[105]	Boiling point of 64 °C and vapor pressure of 167 Torr @ 25 °C
Nitric oxide (NO)	[106]	Diluted 2% in argon
Nitrous oxide (N_2O)	[107, 108]	
Nitrogen dioxide (NO_2)	[109]	
Arsine (As_2H_3)	[105]	*P*(*T*): $\text{Log}_{10}P(\text{Torr}) = 6.569 - \frac{0.762 \times 10^3}{T(\text{K})}$
Diisopropyltelluride (DIPTe)	[110]	For co-doping experiments with NH_3 *P*(*T*): $\text{Log}_{10}P(\text{Torr}) = 8.290 - \frac{2.309 \times 10^3}{T(\text{K})}$

The vapor pressure versus temperature equations in Table 11.4 are from [35] and [68].

11.3 Substrates for the MOCVD Growth of ZnO and Related Materials

Substrates play a crucial role in inducing high-quality deposition. If we concentrate on epitaxy, the ideal substrate would be one that is perfectly lattice-matched to the epitaxial layer, as in homoepitaxy [111]. However, homoepitaxy is not always possible or suitable, and other substrates must be considered.

In order to reduce strains, cracks, and dislocation density in the growth, substrates with similar crystallographic symmetry, the smallest lattice mismatch, and similar inplane linear expansion coefficients should be chosen [5]. In addition, no chemical reactions and/or diffusion processes on the interface should occur. Nevertheless, the availably of substrates, cost, and uses can drive the choice, and not only academic criteria. Different substrates, for different reasons and with different properties, drawbacks, and advantages, have been chosen for the growth of the ZnO-based materials.

Zinc oxide crystallizes in the wurtzite lattice (space group: $P6_3mc - C_{6v}^{\ 4}$). The a, r, m, and c planes of the wurtzite structure are pictured in Figure 11.1.

In Table 11.5, some data of interest for the growth process corresponding to ZnO and some of the more commonly used substrates are presented. In the cited references, more details can be found.

In the following subsections, we will discuss some of these most-used substrates for the MOCVD growth of the ZnO family, emphasizing their interest and properties, as well as some cleaning, thermal, and preparation procedures. Other substrates also present in MOCVD experiments will be introduced in Section 11.5.

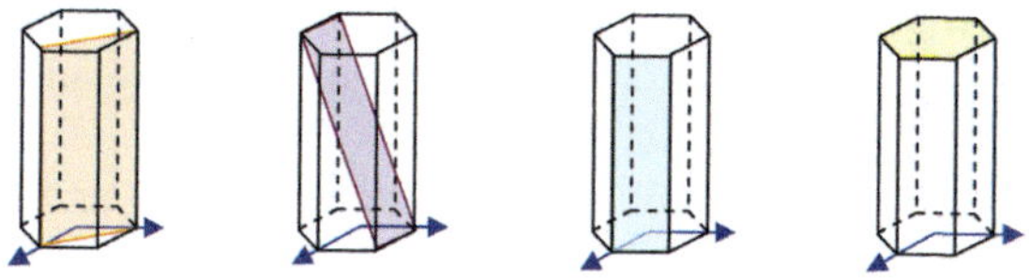

Figure 11.1 *The a, r, m and c planes of the wurtzite structure.*

Table 11.5 *Data of interest on potential substrate materials for ZnO growth.*

Material	Crystal structure	Lattice constant(s) (Å)	Thermal expansion coefficient × 10^6 (K^{-1})	Thermal conductivity x10^{-2} W m^{-1} K^{-1}	Melting point (K)
ZnO	Wurtzite	a=b=3.249(6) [112]	$\alpha_{\perp c}$ =4.75 @ 300 K [112] (6.51 [113])	0.54 @ 300 K [112] (0.6 [113])	2242 [112]
		c=5.2042(20) [112]	$\alpha_{//c}$ = 2.92 @ 399 K [112] (3.02 [113])		
Sapphire (Al_2O_3)	Corundum/ Rhombohedral	a=b=4.758 [113]	$\alpha_{\perp c}$ 7.7 @ 310–610 K [114] (7.50 [113]; 7.1 @ 773 K [115]).	⊥c 0.30 @ 298 K [135] (0.5 [113])	2323 [114]
		c=12.991 [113]	$\alpha_{//c}$ 7.0 @ 310-610 K [114] (8.50 [113]; 8.0 @ 773 K [115])	//c 0.33 @ 298 K [114] (0.252 @ 300 K [116])	
GaN	Wurtzite	a=b=3.190(1) [112] (a=b=3.186 [113])	3.17 [112] (5.59 [113])	1.3 [112] 1.35 [113]	2791 [112]
		c=5.189(1) [112] (c= 5.178 [113])	5.59 [112] (7.75 [113])		
Silicon	Cubic	a=5.431 @ 295.7 K [112]	2.92(6) [112]	1.56 [112]	1687 [112]

11.3.1 ZnO Single Crystals and ZnO Templates as Substrates

11.3.1.1 Interest

The use of ZnO as substrate, either as a single crystal or as a template on some foreign substrate, offers significant advantages for the growth of ZnO and its alloys in the Zn-rich region, namely [117]:

i) The absence of substantial mismatches in lattice parameters and in coefficients of thermal expansion between the wafer and the grown material, and, consequently, the absence or reduction of threading dislocations; ii) the substantial reduction in the diffusion of unwanted dopants and other impurities from the wafer into the grown material, which will allow a marked reduction in both the density of threading dislocations and the residual stresses relative to those observed in heteroepitaxial grown.

Given these considerations, different cut planes (primarily (0001) and $(11\bar{2}0)$) of ZnO single crystals have been used for the homoepitaxial growth of layers and nanostructures. For c-oriented substrates, O- and Zn-terminated surfaces exhibit a different behavior, as discussed by Gu et al. [118] and references therein; the electron affinities and chemical properties depend on the polarity of the material. Thus, different polar surfaces will result in different behavior of electrons and different stabilized configuration and chemical activity of surfaces, inducing different kinetics and material properties.

Commercial ZnO single-crystal substrates grown by various techniques are available. On the other hand, ZnO templates, primarily on silicon and sapphire as an alternative to ZnO bulk substrates, have also been widely exploited in the literature. We would like to distinguish between templates and buffer layers, just to emphasize that templates can be grown in a separate run by using other growth techniques, while a buffer layer will be referred to here as a layer grown in the same run as the main structure.

11.3.1.2 Cleaning and Thermal Treatments

In some cases, ZnO bulk substrates present damage induced by sawing, and some residues from mechanical polishing can remain on them. Thus, they are often submitted to a degreasing bath in organic solvents. A usual procedure is a bath in acetone followed by a bath in methanol, 5 min each, after which they are blown dry using high-purity nitrogen [117]. Other variants include, for example, using only methanol and blowing dry with N_2, as in [119], or cleaning with trichloroethylene, acetone, and methanol in sequence before the growth [120].

Adequate surface preparation of ZnO substrates is fundamental to take advantage of the homoepitaxy. One difficulty in the ZnO surface preparation is that ZnO can react with both acids and alkalis, which could make a surface-preparation process based on wet chemical etching unsuitable [121]. Differences in behavior of ZnO by chemical etching with HCl have been reported on both semipolar [122] and polar faces [118].

Gu et al. [121] reported a cleaning and thermal treatment that can substantially improve the surface of both O and Zn faces of the c-plane and make them appropriate for homoepitaxy. ZnO substrates underwent a degreasing procedure in acetone by using ultrasonic agitation for 3 min, followed by methanol, ultrasonic agitation again for 3 min, followed by a 3-min rinse in deionized water. The cleaned ZnO substrate was then placed in a furnace for annealing in ambient atmosphere. After annealing, the ZnO substrate was cleaned by using the same procedure. A 950 °C annealing temperature was able to remove scratches, when present, at the ZnO surface. Furthermore, after 30 min of annealing at 1050 °C, a terrace-like surface began to form on the ZnO substrate. The terraces coalesced, and their boundaries became straighter with increasing annealing time. Increasing the annealing time to 3 h resulted in the formation of straight and parallel terraces with a low roughness, < 1 nm, which is indicative of a surface free of damage, which is appropriate for epitaxy and facilitates smooth two-dimensional growth. In addition, atomically Zn-terminated flat surfaces of ZnO (0001) single-crystal substrates without scratches were obtained [123] by high-temperature annealing between 800 and 1300 °C in an oxygen atmosphere. After annealing, the ZnO surfaces were cleaned by using Ar^+ ion etching at room temperature.

Ogata et al. [124, 125], working on ZnO/sapphire and ZnO/silicon templates, studied different thermal treatments in N_2 atmosphere. Thermal treatments were adopted to obtain a flat surface. The surface roughness of the c-ZnO/a-sapphire template did not change significantly when annealed at 800 °C in N_2 atmosphere, while an atomically flat surface was observed when annealed at 1000 °C. Considering that a similar flat surface was obtained after annealing MBE-grown ZnO layers at 1000 °C in O_2 atmosphere [126], it was concluded that morphological changes could be attributed to giving sufficient thermal energy for the atoms on the surface to migrate and form a flat surface. For the ZnO growth on Si, in contrast to the previous results, surface roughness of ZnO/Si(0001) did not improve with thermal treatment. Annealing at 1000 °C degraded the optical qualities, probably due to heavier diffusion at the ZnO/Si interface, the generation of oxygen vacancies, and/or the reaction at the interface forming any compound. Thus, in the growth on ZnO/Si templates, the ZnO/Si should be treated, if necessary, at lower temperatures than, at least, 1000 °C. Liang et al. [127] highlighted that spinel $ZnAl_2O_4$ forms at the interface of sapphire and ZnO during high-temperature annealing (>850 °C). Similarly, Zn_2SiO_4 with a trigonal structure is formed at temperatures higher than 770 °C on the ZnO/Si interface. These facts can limit the annealing temperature for ZnO templates. The temperature must be maintained at sufficiently low values to avoid reactions and forming any compound at the interface, if, depending on the application, this could have some deleterious effect.

11.3.2 Sapphire (Al_2O_3)

In addition to the properties depicted in Table 11.5, an interesting compendium of sapphire properties can be found in [114].

11.3.2.1 Interest and Applications

Significant efforts for growing ZnO and its alloys have concentrated on the use of sapphire as substrate. Indeed, sapphire has probably been the most commonly used substrate for ZnO-based materials, not only for the growth of layers but also in obtaining nanostructures.

Several reasons have been invoked, as pointed out by Muthukumar et al. [115] and Gorla et al. [128], amongst others:

- Al_2O_3 has oxygen sublattices with hexagonal symmetry.
- High-quality, relatively low cost Al_2O_3 substrates are commercially available in differently oriented cut planes.
- Al_2O_3 is available in large sizes.
- Al_2O_3 has a high acoustic velocity and low attenuation, and it is therefore the preferred choice as substrate for high-frequency, due to its high acoustic velocity, and low-loss surface acoustic wave (SAW) devices.

The use in SAW devices may be the origin of the choice of this substrate in some of the seminal papers [79, 129], as well as for acousto-optic devices [115].

Sapphire provides several useful epitaxial relationships with ZnO, as we will see later in Section 11.5. While ZnO (space group P6^{3}mc– C $_{6v}$^{4} with a = 3.249 Å and c = 5.204 Å) and Al_2O_3 ($R\bar{3}c$ with a = 4.758 Å and c = 12.991 Å) have large lattice mismatch in some orientations, other orientations provide the necessary conditions for directional growth. The inplane alignments between ZnO films and sapphire substrates can be analyzed (see Moriyama et al. [130]) in terms of the lattice mismatch along certain directions. On m-plane sapphire, the lattice constant along the c-axis (0001) sapphire is almost four times that along the $[\bar{2}110]$ ZnO, with a mismatch of only 0.077%. This seems to contribute to fixing the inplane alignment ZnO$[\bar{2}110]$//sapphire[0001]. On r-plane sapphire, the lattice constant along the $[01\bar{1}\bar{1}]$ axis is 15.380 Å, which is almost three times that along the c-axis [0001] ZnO with a mismatch of 1.68%, while the mismatch along the direction perpendicular to the c-axis is 18.3% [128, 131]. This may allow the inplane alignment of ZnO[0001]//sapphire$[01\bar{1}\bar{1}]$. Thus, inplane alignment is considered to be determined by the smallness of lattice mismatch to a certain direction [130]. Nevertheless, the growth conditions are of primary significance to determine the epitaxial relationships, as reported in the literature (see, for instance, [132]). The interface energy is composed mainly of strain energy caused by the mismatch between the film and the substrate. Hence, competition between the force of keeping smaller strain energy, or smaller lattice mismatch, and that of self-texture will occur [132], thus determining the epitaxy.

ZnO films grown on r-plane sapphire substrates can have inplane anisotropic optical, electrical, and piezoelectric properties, which provides multiple device applications, making this option of particular technological interest. It is worth noting that by using r-plane sapphire substrates, the growth of the $(11\bar{2}0)$ ZnO orientation can be promoted. In this plane, the cations (Zn) and anions (O) are in stoichiometric ratio, resulting in zero net charge on the surface. In contrast, the ZnO (0001) orientation has alternating layers of cations and anions, resulting in a nonzero dipole moment perpendicular to the surface [131].

With respect to the application of sapphire as substrate, it is worth noting that sapphire is insoluble in water, HNO_3, H_2SO_4, HCl, HF (at 300 °C), alkalis (at 800 °C), and melts of metals like Mg, Al, Cr, Co, Ni, Na, K, Bi, Zn, and Cs (at 800–1000 °C) [116].

11.3.2.2 Cleaning/Etching and Thermal Treatments

Regarding cleaning and etching processes, sapphire has been extensively used as provided by commercial suppliers. Nevertheless, some authors have proposed chemical polishing and/or chemical etching processes to give a better surface finish by using phosphoric acid; for example, Quon et al. [133] performed this etching at temperatures between 290 and 360 °C, giving a good surface finish for $(01\bar{1}2)$-oriented substrates but enlarging the scratch marks that were initially present on the (0001) orientation. Other authors, like Lau et al. [79], used a standard procedure consisting of polishing for 10 min followed by cleaning with trichloroethylene, acetone, and methanol. Deionized water was used for the final wash before a dry nitrogen blow-off. The choice of organic-based cleaning has been widely used as a degreasing agent, as in e.g. [120, 134, 135]. Based on the action of phosphoric acid, different solutions have been reported as H_2SO_4:H_3PO_4 (1:1), which was studied by Lu et al. [134] performing this etching at 200 °C for 1 h prior to organic solvent cleaning; or heating the solution (3:1) at 160 °C, as in [136, 137]. The effects of mechanical polishing, chemical mechanical polishing (CMP), as well as CMP and subsequent chemical etching on the properties of sapphire substrate surfaces were studied by Wang et al. [138].

In addition to cleaning with organic solvents or even chemical etching, thermal processes are proposed as an option to modify the surface of the sapphire substrate. For example, Ogata et al. [139] submitted the sapphire substrates to a thermal etch at 900 °C to obtain atomically flat surfaces. Another reason for thermal treatment is to overcome the fluctuation of surface quality found in some commercially available substrates. Thus, Wang et al. [140] undertook a systematic study on the annealing effects of sapphire substrates. Atomic steps were formed on the (0001) sapphire substrate surface by annealing at high temperatures, 1000–1400 °C. The formation of steps was probably the origin of the initial growing film, because well-defined steps at the substrate can be nucleation sites for film growth, due to the reduction of surface energy. Thus, the nucleation of islands at the edge of steps formed on the annealed sapphire substrate is favored. We will return to this in Section 11.5.

11.3.2.3 Templates on Sapphire

Sapphire substrates have also been used for the growth of templates. Here, we will comment on some of those most frequently used in the growth of ZnO and its alloys with cadmium and magnesium.

As discussed by Dadgar et al. [141], GaN is a suitable heteroepitaxial substrate for ZnO because it shares the same wurtzite-type symmetry, a close lattice match ($a_{ZnO} = 3.249$ Å and $a_{GaN} = 3.190(1)$ Å), and a small difference between the thermal expansion coefficients ($\alpha_{ZnO} = 6.51\times10^{-6}$ K^{-1} and $\alpha_{GaN} = 5.59\times10^{-6}$ K^{-1}); see Table 11.5.

Due to the lack of low-cost GaN bulk crystals and the availability of growing high-quality GaN films on Al_2O_3 substrates, GaN epilayers have been used in the growth of ZnO and its alloys. The (0001) sapphire plane was the choice amongst others in [103, 141, 142].

GaN/Al_2O_3(0001) substrates have also been of interest in the growth of nanostructures [143]. As reported in [120], the alignment of ZnO nanorods greatly depends on the substrates, and in order to obtain a narrow inplane and out-of-plane mosaic distribution, the growth on lattice-matched substrates such as GaN/Al_2O_3 (0001) and ZnO (0001) seems to be crucial. In addition, the use of lattice-matched substrates favors the growth of nanostructures directly on the substrate without the presence of a continuous interfacial ZnO layer that appears, when using less-matched substrates, before the growth of nanostructures.

ZnO buffer layers are used in order to induce a better growth process of the main layer. This could be seen as a kind of template on foreign substrates. Muthukumar et al. [115] reported that a ZnO buffer layer was critical for the growth of single-crystal hexagonal wurtzite-type $Mg_{1-x}Zn_xO$ films on r-sapphire substrates. Similar critical effects were found by Zhang [137] on the growth of ZnO layers and Zuñiga-Perez et al. [144] in the case of $Zn_{1-x}Cd_xO$.

It is worth noting that high-temperature pretreatments of sapphire lead to a significant modification of the surface, which can result in enhanced growth nucleation and consequent improvement of the surface morphology and quality of the ZnO layers. As reported by Huang et al. [145], the evolution of the surface morphology – investigated by atomic force microscopy (AFM) – indicates a growth-mode transition from three-dimensional to quasi-two-dimensional as the sapphire substrate pretreatment temperature increases. In their study, a minimum surface roughness was obtained at a pretreatment temperature of 1150 °C. This pretreatment temperature or higher may lead to a conversion of the surface polarity from O-face to Zn-face in the overlying ZnO.

Other templates such as GaAs/Al_2O_3 [146], Ti/Al_2O_3 [147], and Ti/r-Al_2O_3, SiO_2/r-Al_2O_3, and silicon-on-sapphire (SOS) substrates [148] have also been studied.

11.3.3 Silicon

11.3.3.1 Interest and Applications

Silicon (Si) substrates are an alternative option for the growth of ZnO-based materials because of its relatively low cost, excellent thermal conductivity, high crystallinity, and availability in large size and in all types of conductivity [149, 150]. No less important is the opportunity for integration of ZnO-based devices in silicon-based technologies [151].

As discussed by Haga et al. [24], in the preparation of epitaxial ZnO films with high crystallinity, sapphire substrates have been widely employed due to their stability at typical growth temperatures. However, sapphire substrates are unsatisfactory when used as electrodes, because of their high resistivity. On the contrary, the conductivity of silicon wafers makes them interesting as substrates on which electrical contacts can be fabricated. However, the lattice mismatch between Si(111) and ZnO(0001) is very large because of the difference in the lattice constant and crystalline structure. In order to relieve this mismatch, a ZnO buffer layer on the Si(111) substrate can be useful [24]. As Shimizu et al. [152] point out, silicon substrates have high potential for many technological applications, such as optical waveguides, surface acoustic wave devices, and heterojunction devices, as well as for integration in current technologies. It is well known [31] that *c*-axis oriented ZnO thin films show piezoelectric properties, and they are useful in surface acoustic wave, bulk acoustic wave, and acousto-optic devices. These *c*-axis oriented layers can be obtained by depositing ZnO on Si (100) under the experimental conditions reported by, among others, Kumar et al. [31]. In addition, Muthukumar et al. [153, 154] report on the ZnO/SiO_2/Si system. High-quality piezoelectric ZnO films grown on Si substrates also pave the way for integration of SAW devices with silicon technologies. For some technological applications, however, the anisotropy of the thermal expansion coefficient for wurtzite ZnO can be an important element to be taken into account.

The growth of high-quality ZnO on Si substrates is not an obvious task due to the easy oxidation of the Si surface during growth. The oxidation of the Si surface can result in the formation of an amorphous SiO_x layer [149] or a very thin interfacial SiO_2 layer between the deposited ZnO film and Si substrate, probably due to the larger value of the formation enthalpy of SiO_2 (–910.7 ±1.8 kJ mol^{-1}) compared to that of ZnO (–350.46 ±0.27 kJ mol^{-1}), which makes it easier to form silicon dioxide than ZnO [155]. The formation of this interfacial SiO_2 layer is usually responsible for obtaining polycrystalline ZnO [91].

Silicon substrates have been used not only for depositing ZnO but also for MgO. Consequently, silicon appears as a potential substrate for the growth of cubic $Mg_{1-x}Zn_xO$.

Several nanostructures have also been grown on SiO_2/Si, e.g. self-organized nanometer-sized ZnO islands [156], ZnO nanopillars [157], nanorods [120], and nanotubes [158], as we will discuss in detail in Section 11.5. Nevertheless, significant strains are likely to develop in the initial stages of ZnO nanorod growth on Si (111) and Si (001) substrates due to the large lattice mismatches, which may prevent direct growth of ZnO nanorods on those substrates. Instead, a continuous interfacial layer of a certain critical thickness is formed before the ZnO nanorods start to grow [120]; in contrast, when using lattice-matched substrates, this interfacial layer does not appear, and nanorods grow directly on the substrate, as already mentioned in Section 11.3.2.3.

11.3.3.2 Cleaning and Thermal Procedures

As noted earlier, the purity of wafer surfaces is an essential requisite for their successful use as substrates. Different cleaning/etching procedures for silicon substrates can be found in the literature. Some of them will be presented here.

Boo et al. [135] submitted their (100)-oriented silicon substrates to a cleaning process with trichloroethylene, acetone, deionized water, and methanol. Then, the substrate was treated with 20% HF and 36% HCl solutions in order to remove the surface oxide layer. Other authors, such as Zhaochun et al. [91], preferred just a thermal treatment in a H_2 atmosphere to remove adsorbed water. The Radio Corporation of America (RCA) cleaning procedures [159, 160] are another option to remove foreign matter from the surface of the silicon wafers (dirt, scum, silicon dust, etc.) prior to processing (see e.g. [158]). The procedure implies the use of two solutions that contain hydrogen peroxide (H_2O_2) to remove residual organic, ionic, and metallic contamination. The decontamination is based (RCA-1 [159]) on a sequential oxidative desorption and complexing with H_2O_2-NH_4OH-H_2O (1:1:6) [161]. RCA-1 clean is used to remove organic residues; in the process, it oxidizes the silicon and leaves a thin oxide on the surface of the wafer, which should be removed if a pure silicon surface is desired. A second process (RCA-2 [160]) is often used with H_2O_2-HCl-H_2O (1:1:6) [161] to further clean the surface. An interesting review and analysis of the silicon cleaning processes can be found in [162].

Dadgar et al. [163] etched their Si(111) substrates in a solution (3:1:1) of H_2SO_4:H_2O_2:H_2O followed by etching in diluted HF (2%) in order to produce an oxide-free hydrogen-terminated Si surface, and no high-temperature annealing step had to be applied. On the same basis, Chen et al. [164] etched the silicon substrates with a 3:1 solution of H_2SO_4:H_2O_2 followed by a solution of diluted HF, washed it with deionized water, and finally dried it with nitrogen. Lin et al. [165] cleaned their substrates using HCl:H_2O mixed solutions (1:150) and HF dip to remove the Si native oxide layers. Further cleaning of the Si surface was performed by heating at 800 °C for 30 min in a vacuum. In the frame of cleaning Si substrates, Fu et al. [166] submitted their p-type Si(111) to an ultrasonic bath with carbon tetrachloride (tetrachloromethane, CCl_4), toluene, acetone, and ethanol in sequence; after that, the substrates were etched in aqua regia (HNO_3+3HCl) for approximately 3 min. Finally, the substrates were treated with a 5% HF solution, then washed with deionized water and blown off with nitrogen.

Ar^+ plasma bombardment before the growth has also been proposed, and the critical influence of time and plasma power has been discussed [167] as a way to eliminate remnants of amorphous silica from the substrate surface. Nevertheless, the SiO_2 layer does not have a detrimental effect on the growth in all cases, as discussed by Kim et al. for the growth of ZnO quantum dots [168] and self-organized ZnO nanosize islands [156] on SiO_2/Si (111) substrates.

11.3.3.3 Templates on Silicon

Thin layers of different materials have been introduced on silicon substrates to prevent the Si substrate surface from being oxidized at the initial growth stage of the ZnO thin films. We will comment on some of the more commonly used templates in the growth by MOCVD. More details can be found in the references.

Chen et al. [164] proposed the use of an Al layer with a thickness of 10 Å in order to protect the Si surface from oxidation. It was claimed that the Al buffer layer produces a superior template for the growth of a ZnO epilayer. Li et al. [149] deposited a thin titanium layer to protect the Si surface from oxidation and reduce the lattice mismatch between the Si substrates and the growing ZnO thin films. Ti has high chemical resistance, a high melting point (1668 °C), and the same hexagonal-close-packed structure as ZnO. In addition, the lattice mismatch between Ti and ZnO is 10%, which is smaller than the lattice mismatch between Si(111) and ZnO (15%).

The advantages of using GaN templates on sapphire have already been briefly commented upon. Motived by similar arguments, the growth on GaN templates on silicon has been studied (see e.g. [13]). For the growth of high quality GaN templates on silicon, a previous buffer layer has been shown to be very useful, as reported by Takeuchi et al. [169], who proposed the use of an intermediate layer of 3C-SiC for the growth of GaN on Si; or, as reported by Watanabe et al. [170], using an AlN buffer layer (30 nm thick) because the lattice mismatch between GaN and AlN is as small as 2.5%, much smaller than that between GaN and Si (17%). Likewise, GaN templates directly grown on Si(111) were studied by Krost et al. [171].

AlN has the wurtzite structure and a relatively small lattice mismatch (5%) to ZnO. The thermal expansion coefficient of AlN is in the middle, between ZnO and Si, and AlN single crystals can be grown directly on silicon [170]. The lattice parameters for AlN are a= 3.111 Å and c= 4.979 Å, and the linear thermal expansion coefficient perpendicular and parallel to the c-axis are $\alpha_\perp = 4.35\times10^{-6}$ K^{-1} and $\alpha_\parallel = 3.48\times10^{-6}$ K^{-1}, respectively [112]. These properties suggest that AlN can be an appropriate template for ZnO films grown on Si(111) substrates, as demonstrated by Wang et al. [150] and Jiang et al. [172], who successfully proposed the use of a thin AlN on Si(111) to accommodate the growth of high-quality ZnO.

Y_2O_3 is a cubic material, thermodynamically more stable than SiO_2. Its density is 5.03 g cm^{-3}, and its melting point is 2410 °C. The thermal conductivity at 20 °C is 8–12 $Wm^{-1}K^{-1}$, and the thermal expansion coefficient in the range 20–1000 °C is 8.1×10^{-6} K^{-1}. The lattice constant of Y_2O_3 is 10.604 Å, which compared to 5.431 Å for silicon gives a lattice mismatch of Y_2O_3 (222)/Si(111) and Y_2O_3 (222)/Si(100) of about 2.4% and 43.6%, respectively. Y_2O_3 not only is thermodynamically more stable than SiO_2, which can avoid formation of silicon oxide on Si, but also can easily be epitaxially grown on Si. Details of using Y_2O_3/Si(111) and Y_2O_3/Si(100) as substrates for the growth of highly oriented ZnO can be found in [165, 173].

On the other hand, the lattice and the thermal expansion mismatch between ZnO and SiC are smaller than those between ZnO and Si. Thus, 3C-SiC templates on silicon were the substrates proposed by Zhu et al. [174] for the growth of ZnO, which is another available option.

Finally, with less presence in the literature, other templates such as Si_3N_4/Si(100) [175] and Lu_2O_3/Si(111) [176] have been also proposed for the use in MOCVD experiments in order to reduce the lattice mismatch and avoid the effects of SiO_2 on the growing ZnO.

11.3.4 Glass Substrates

11.3.4.1 Interest and Applications

The growth of highly textured polycrystalline thin films on amorphous substrates is of significant scientific interest and practical application. Glass has been widely used in the growth of the members of the ZnO family. One of the more frequently invoked reasons is the use in solar cells that need low-cost substrates able to induce the growth of polycrystalline films. These films can be used when the reflection should be reduced in order to improve the efficiency of solar cells. Additionally, glass has been used as a substrate for templates as in [177].

Some of the most frequently used glass substrates are listed here, together with some seminal references in which their application was demonstrated:

- Pyrex glass [21]
- Corning 7059 glass [152, 178]
- Soda lime glass [31]
- Standard float glass [82]
- Corning 1737 glass [106, 177, 179]
- Alumino-silicate glass [180]
- Common glass: CaO:NaO:6SiO_2 [181]
- Fluorinated tin oxide coated glass (FTO) film (TEC15, Hartford Glass Co.) [182]
- Indium tin oxide coated glass (ITO) [183]
- AF45 Schott glass [184]

11.3.4.2 Glass Properties

The characteristics of some of the more commonly used glasses for the growth of ZnO and its alloys are described in the following and in Table 11.6 (technical data from the Corning catalog and [185], in which more details can be found). It is worth noting that some of these glasses are no longer commercially available, but the data could help to understand the literature and to find an equivalent if necessary.

Soda lime glass and Pyrex 7740 are general-purpose glasses. In particular, soda lime can be used as microscope slides, with a chemical composition (0215 Corning Glass Slides) of SiO_2 73%,

Table 11.6 *Some properties of some commercial glasses.*

	S.L.0215	Pyrex (7740)	7059F	1737F	Eagle 2000	Eagle XG
TEC *, α x10^7 (°C^{-1})	89	32.5	46	37.6	31.8	31.7
Strain point (°C)	511	510	539	666	666	669
Anneal point (°C)	545	560	639	721	722	722
Softening point (°C)	724	821	844	975	985	971
Density (g cm^{-3})	2.40	2.23 @ 20 °C	2.76	2.54	2.37 @ 25 °C	2.38 @ 20 °C

*TEC is the thermal expansion coefficient.

Na_2O 14%, CaO 7%, MgO 4%, and Al_2O_3 2%. Other slightly different compositions can be found, depending on the supplier. Pyrex 7740 is a Corning trade-name borosilicate glass with good heat-shock resistance.

Corning 7059 and 7059 F are barium-borosilicate flat glasses, almost alkali-free, that also feature a good thermal shock resistance. They have been a standard of the electronics industry for thin-film circuits. 7059 F glass is fusion drawn into sheets to final thickness, obtaining a flat, smooth surface finish directly from the molten glass. 7059, however, is slot drawn and is not as flat as 7059 F. Both 7059 glass materials have a low level of alkali, which is less than 0.3%. This can be important for many electronic applications, because alkali ions are known to lower the performance, reliability, and longevity of thin-film devices.

On the other hand, Corning 1737 F is an alkaline-earth boro-aluminosilicate glass with high temperature and shock resistance. Low alkali content allows special manufacturing processes required for various electronic components. A coefficient of thermal expansion closely matching that of silicon opens up the option for chip-on-glass applications. In general, the natural surfaces of 1737 F glass have lower microroughness. An alternative to Corning 1737 is Corning's Eagle 2000.

Eagle 2000 glass is a lightweight alkaline earth boro-aluminosilicate glass with low density, a low coefficient of thermal expansion, and a total alkali content of approximately 0.1 wt-% (typically <0.05 wt-%). The fusion drawing provides excellent flatness and a high surface quality that is indispensable for high-resolution applications. The low coefficient of thermal expansion allows the use of Eagle 2000 in chip-on-glass processes and reduces the effect of thermal shock and glass breakage.

Another option as substrate is Corning Eagle XG, which is a borosilicate glass. It is considered environmentally friendly as it contains no heavy metals (arsenic, antimony, barium, or halides). It also features high surface quality, excellent thermal properties, low density, and high resistance to chemicals.

11.4 Some Techniques for the MOCVD Growth of ZnO and Related Materials

From the works of Manasevit regarding the "use of metalorganics in the preparation of semiconductors" and in particular the paper on the II-VI compounds [186], several conventional horizontal and vertical reactor designs, ranging from commercial equipment to homemade ones, have been considered in the literature. Figure 11.2 shows the scheme of the reactors adopted by Lau et al. [79], which are representative of the conventional MOCVD configuration.

An interesting analysis of different growth parameters and their fundamentals in the conventional MOCVD setup has been developed in detail by Betsch [187]. In this paper, the gas flows in standard designed equipment in atmospheric as well as low-pressure regimes are discussed. Particular emphasis is given to the control of composition with respect to the gas manifold operation. Significant parts of this analysis can be applied to the different techniques discussed in Section 11.4.

Premature reaction between Zn metalorganic precursors and oxidants can appear in the gas phase. These reactions can lead to unwanted deposits upstream from the susceptor, which is one of the problems that need to be solved in order to achieve successful MOCVD growth of ZnO and its alloys. Acting on the chemical reaction itself by using less-reactive precursors can be an option, but the growth rate and/or growth conditions can be seriously compromised. Another way to solve the problem of premature reactions can be the use of growth techniques, growth parameters, and reactor designs that allow better control of the process.

Roth et al. [188] addressed the problem of prereactions by avoiding the direct contact between the Zn precursor and oxygen before attaining the substrate. Thus, separate inlets for the metalorganic

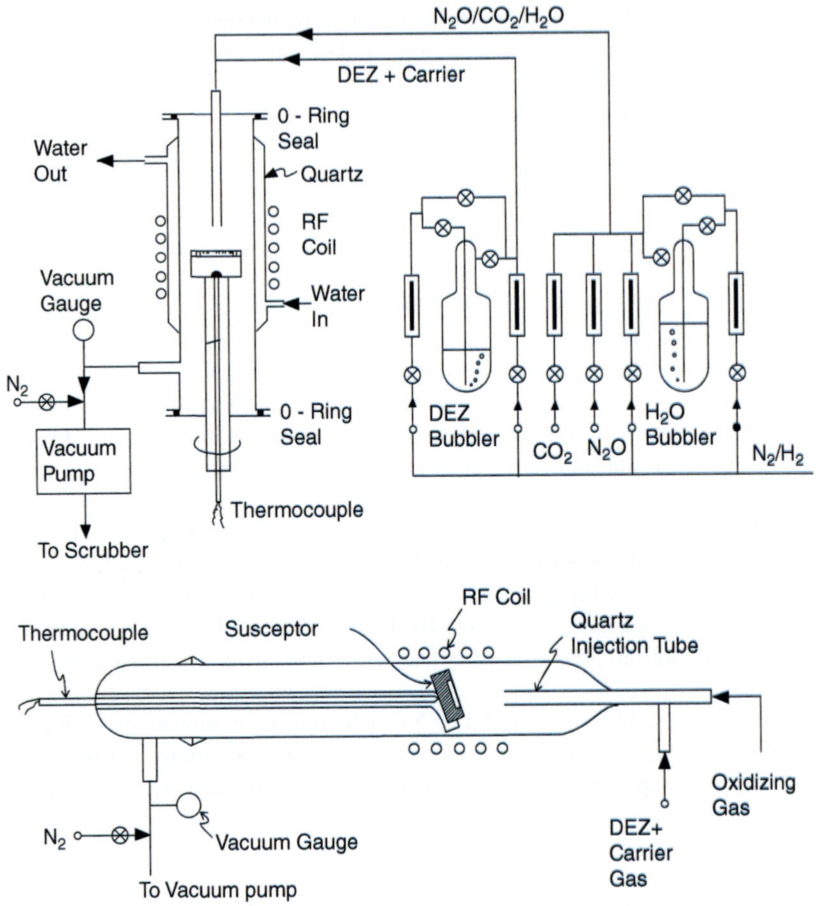

Figure 11.2 *Schematic of the growth system used for ZnO MOCVD growth (top) vertical geometry and (bottom) horizontal geometry. Source: Reproduced from Lau et al. 1980 [79] with the permission of The Electrochemical Society.*

compound and the oxidant were adopted. Since then, this option has been generalized for the growth of the ZnO family of compounds in horizontal or vertical reactors working at low or atmospheric pressure. Several variants can be found in the literature with different systems for distributing the flow, e.g. with the gas flow parallel to the substrate [189], forming a certain angle between the flow and the substrate [128, 190], or completely vertical, e.g. [191, 192], as some of these variants. In the references, schematic drawings and/or descriptions of different setups can be found.

11.4.1 Atmospheric and Low-Pressure Conditions in Conventional MOCVD Systems

The work at low pressure (LP) has been invoked to improve control of the chemical reactions involved in the deposition process, as well as the fact that the vapors are mixed very near to the substrate; thus, it is expected to improve the thickness homogeneity [193]. As discussed by Fay and Shah (Chapter 6 in [193]), usually, but depending on the growth temperature, the atmospheric-pressure (AP) MOCVD processes work in the diffusion-limited regime, and the transport of the reactants to the growing surface is the limiting factor. Conversely, the low-pressure regime is usually kinetics-limited rather than diffusion-limited. This means the growth of ZnO films is mainly limited by the kinetics of the reactions that occur at the growing surface. Low pressures can help to minimize or reduce premature reaction by decreasing the number of collisions between molecules in the time between gas mixing and contact with the substrate. In other words, by creating conditions for high velocity of the reactant molecules, the value of resident time and the number of collisions between molecules in this period will be reduced [111]. The LP option has been largely used in the literature for the growth of the ZnO family (e.g. [118, 129, 174, 178, 194]), and in particular for the growth of nanostructures [195–198]. However, good results can also be obtained at atmospheric pressure. Consequently, the AP regime has been the object of extended work (see e.g. [82, 191, 199, 200]). Atmospheric-pressure methodology is easier to use, and by using a two-inlet configuration and under well-defined growth conditions, the premature reactions can be suppressed or reduced until no significant effects are detected.

Rotation of the substrate can help to obtain better uniformity acting on the boundary layer on the substrate and reactants distribution, and giving an additional control parameter impinging directly on the hydrodynamics of the process. The rotation of the substrate has been usually applied with the main flow perpendicular to the substrate but also in other configurations [197]. Regarding the substrate heating, radio frequency (rf) induction, lamp heaters, and more recently laser heating [201] have been the sources for MOCVD systems.

On the basis of the conventional MOCVD processes at low or atmospheric pressure, it is possible to adopt some changes in the growth protocol such as atomic layer deposition (ALD) (see e.g. [202]) or pulsed MOCVD (see e.g. [203]), protocols in which several steps in the injection of precursors act on the control of the process.

Directly related to conventional MOCVD is the *hybrid-physical-chemical-vapor deposition* (HP-CVD) method (see e.g. [204]). HP-CVD involves three processes: vaporization of a solid source, transport of this vapor to the reaction chamber by a carrier gas, and chemical decomposition of the precursor. This methodology has been applied with solid precursors, e.g. zinc acetate, zinc acetate dihydrate, and zinc acetylacetonate (see Section 11.2.1.3). The solid precursor can be dissolved in a solvent to favor vaporization. Likewise, in some cases, the process has been helped by using ultrasonic spray-assisted methodology [205] in which the precursors are ultrasonically atomized and formed into mists by an ultrasonic transducer and transported to the reaction chamber.

Other methods based on the conventional MOCVD scheme have been also introduced. This is the case of the direct liquid injection MOCVD (DLI-MOCVD), which has proved to be very useful when precursors with low vapor pressures are adopted. Regarding oxides, Roeder et al. [206] worked with this DLI-MOCVD technique based on a "flash vaporization" of metalorganic precursors. As discussed in [206], in the liquid delivery approach, the solid sources can be dissolved in a solvent and maintained at a fixed temperature prior to the injection into a vaporizer. The vaporizer transforms the liquid mixture into a gas in a very short period of time (<10 ms). This process, referred to as *flash vaporization*, preserves the integrity of the chemical precursor species. Moreover, a high degree of process precision results, because liquids can be precisely metered volumetrically. The principle of a liquid-injection system (see Figure 11.3) is based on a sequential computer-controlled injection of microamounts of a solution of precursors into a vaporization chamber, where the flash vaporization occurs. The solution is pushed by a carrier gas toward the injector, which is a high-speed micro-electro-valve driven by a computer. The flow rate of the precursor, which defines the vapor pressure of the active species, is controlled [207] by:

- The volume of each injection, determined by the duration of the aperture of the valve (on the order of ms and directly related to the volume of each injection)
- The frequency of injection (which can vary between 1 and 10 Hz)
- The concentration of the solution
- The differential pressure between the liquid inside the injector and the evaporator
- The viscosity of the solution

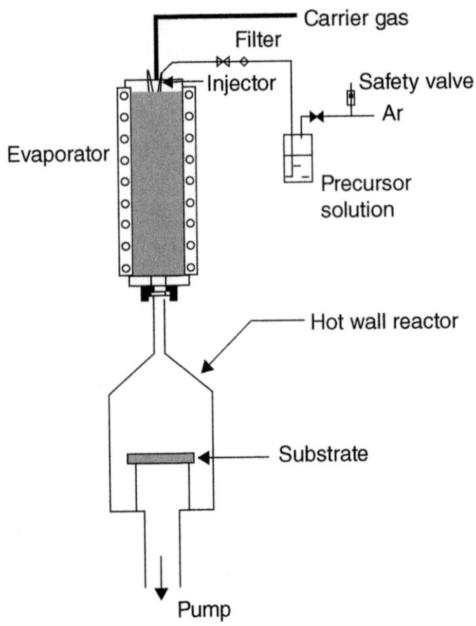

Figure 11.3 *Schematic representation of the liquid-injection CVD system. Source: Reproduced from Manin et al. 2005 [207] with the permission of Elsevier.*

Several contributions in the literature confirm the applicability of the DLI-MOCVD methodology in growing ZnO and related compounds: see e.g. [51] for the growth of $Co_xZn_{1-x}O$ or [38, 208] for the growth of ZnO nanostructures. For Al-doped $Co_xZn_{1-x}O$, Tuan et al. [92] successfully used this technique in conjunction with plasma-enhanced metalorganic chemical vapor deposition (see Section 11.4.2.2 for a description of this technique).

In addition to the conventional MOCVD and its variants, the need for optimized control of the process and the convenience, in some cases, of low-temperature processes have been the origin of the use of other methodologies. Some of them will be briefly presented in the following subsections.

11.4.2 MOCVD-Assisted Processes

As analyzed by Alexandrov and Hitchman in [209], most chemical reactions in CVD are thermodynamically endothermic and/or have an associated apparent kinetic energy of activation. This can be seen as an advantage, since the reactions can be controlled by regulating the energy input. Thus, the energy has to be supplied to the gaseous precursor system to initiate the chemical reactions leading to deposition, and consequently the CVD processes can be controlled by using heat as a source of energy. The need for using some substrates that are thermally sensitive, and the generation of defects induced by temperature have motivated the search for processes (methods) that work at low temperatures and are able to maintain good-quality deposition and acceptable growth rates. In particular, the energy of photons and plasmas has been widely used for activation of CVD processes. However, the use of such sources of energy does not prevent, in all cases, additional heating of the substrate.

11.4.2.1 Photoassisted Processes

As pointed out by Irvine and Lamb in [209], the use of photons to stimulate reactions in MOCVD has been very significant in the development of crystal growth because it provides a route to

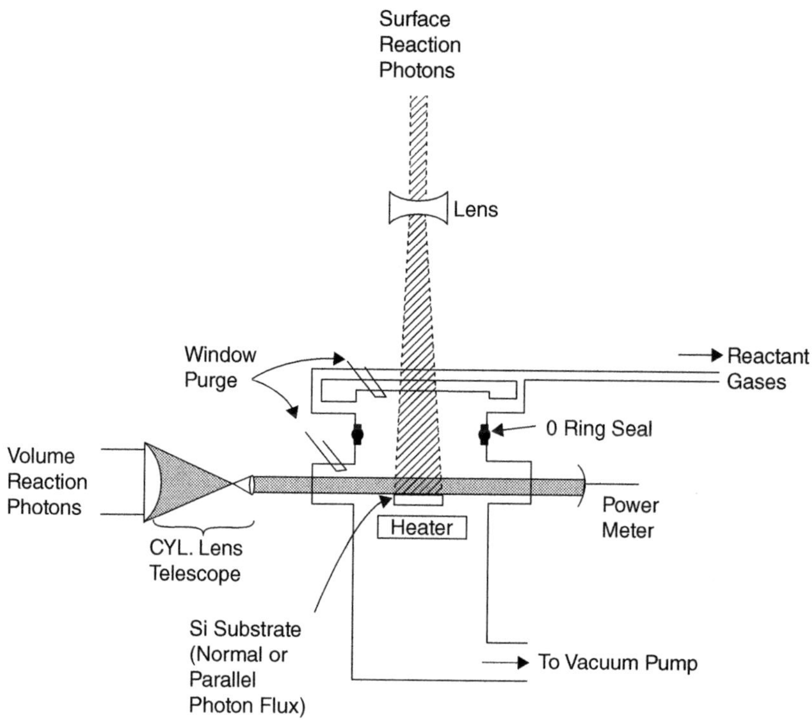

Figure 11.4 *Experimental approaches for providing either a parallel or a perpendicular photon beam for the laser photodeposition of ZnO films. Source: Reproduced from Solanki and Collins 1983 [11] with the permission of AIP Publishing.*

low-temperature processes. Photoassisted methodologies can be seen as a way of modifying the deposition kinctics. It is for this reason that the properties of the grown material can also be modified. There are several reviews available on photoepitaxy in which the fundamentals are clearly exposed and discussed ([35, 209, 210]). Here, the subject will be briefly dealt with.

A number of potential advantages can be invoked for photoassisted processes, including [210]:

- Low-temperature growth
- Selected-area growth
- Modifications to the properties of grown material
- Independent control of the substrate temperature and dissociation of the precursor
- Compatibility with existing laser deposition methods used for *in situ* fabrication

All the previous availabilities are highly suitable not only to obtain good crystals but also to put them into the technological-application field.

Figure 11.4 shows the scheme proposed by Solanskii et al. [11] for photoassisted processes in which the laser beam can be perpendicular or parallel to the substrate.

In the photoassisted methodology, three kind of processes can be distinguished: photothermal, photolysis, and photocatalysis.

The principle of photothermal processes is to create local heating on the substrate by using a focused light source. Thus, photothermal processes can generate local controlled deposition while keeping the background temperature lower. The light beam can be scanned across the surface in a suitable pattern or focused on a singular region. The choice of wavelength will be determined by the absorption characteristics of the substrate and the growing material.

Regarding photolysis, low-temperature deposition below the pyrolysis threshold temperature can be attained by photolytic reaction of the precursors. The photolysis reactions are employed as a means to yield metal atoms that will diffuse to the surface. The process can be achieved at low temperature because the breakage of ligands can occur without the input of heat energy. Photolysis efficiency depends on the absorption cross section of the precursor to create an electronically excited state. For many of the simple alkyl organometallics, this absorption occurs in the UV with a wavelength between 200 and 300 nm. These absorption bands are broad, so that any wavelength within the band can be effective in photolysis. The energy required to produce photolysis is usually obtained from a laser source or wavelength-equivalent discharge lamp. Lamps such as mercury, deuterium, and xenon provide a broad spectral range with intensities in the UV spectral range. Amongst these, mercury lamps are more commonly used due to their intense peak at 254 nm.

Regarding photocatalysis, semiconductor substrates can absorb the incident radiation by using a photon energy that is greater than their bandgap. The absorption of light will create electron–hole pairs that will eventually recombine, either by emission of light or by release of energy through heat. However, if the minority carriers can diffuse to the surface before recombining, then they will change the charge state of the surface. This can induce a catalytic action of the surface by enhancing the desorption rate of organic products or by promoting reaction between the organic ligands to create a stable organic group.

The growth of ZnO and its related materials has taken advantage of the photoassisted processes, as highlighted in the literature. From seminal papers like [11, 89, 211–216] to more recent contributions, e.g. [217–219], the methodology has been successfully applied by using different precursors and light conditions, and decomposition processes and effects of the UV irradiation have been analyzed in depth.

11.4.2.2 Plasma-Assisted Processes

Promotion of chemical reactions by plasmas is founded on two basic functions, as discussed in Chapter 12 of [209]: (i) the formation of chemically active species used for the growth as a result of inelastic collisions of precursor molecules with energetic particles (mainly electrons) formed in the plasma, and (ii) delivery of energy to the substrate surface for enhancement of surface processes such as nucleation, particle migration, and heterogeneous kinetics. The plasma parameters will be crucial to the composition, structure, and properties of the deposited films.

By using plasmas in MOCVD processes, it is possible to overcome high deposition temperatures. Most plasma-enhanced chemical vapor deposition (PE-CVD) systems are based on the use of a radio-frequency glow discharge sustained in an appropriate gas mixture, and the substrates are usually placed on one of the electrodes in the region of discharge generation. This conventional PE-CVD (Figure 11.5a) could present some drawbacks, as analyzed by Alexandrov et al. (Chapter 12, [209]):

- Electrically active defects can be generated near the surface of the substrate because of bombardment by energetic particles.
- Plasma particle bombardment of the growing film may cause generation of pinholes and charges trapped in the film.
- Complex reactions can result from the simultaneous production of many different reacting species, and this can cause unpredictable compositions of the deposited films.
- The main process parameters, such as the total pressure, rf power, gas composition, and gas flow rate are all interdependent, and their individual influences can be difficult to isolate.

One of the alternatives has been using remote PE-CVD (RPE-CVD) (Figure 11.5b). In RPE-CVD, the plasma-generation region and the substrate are spatially separated, and only some of the

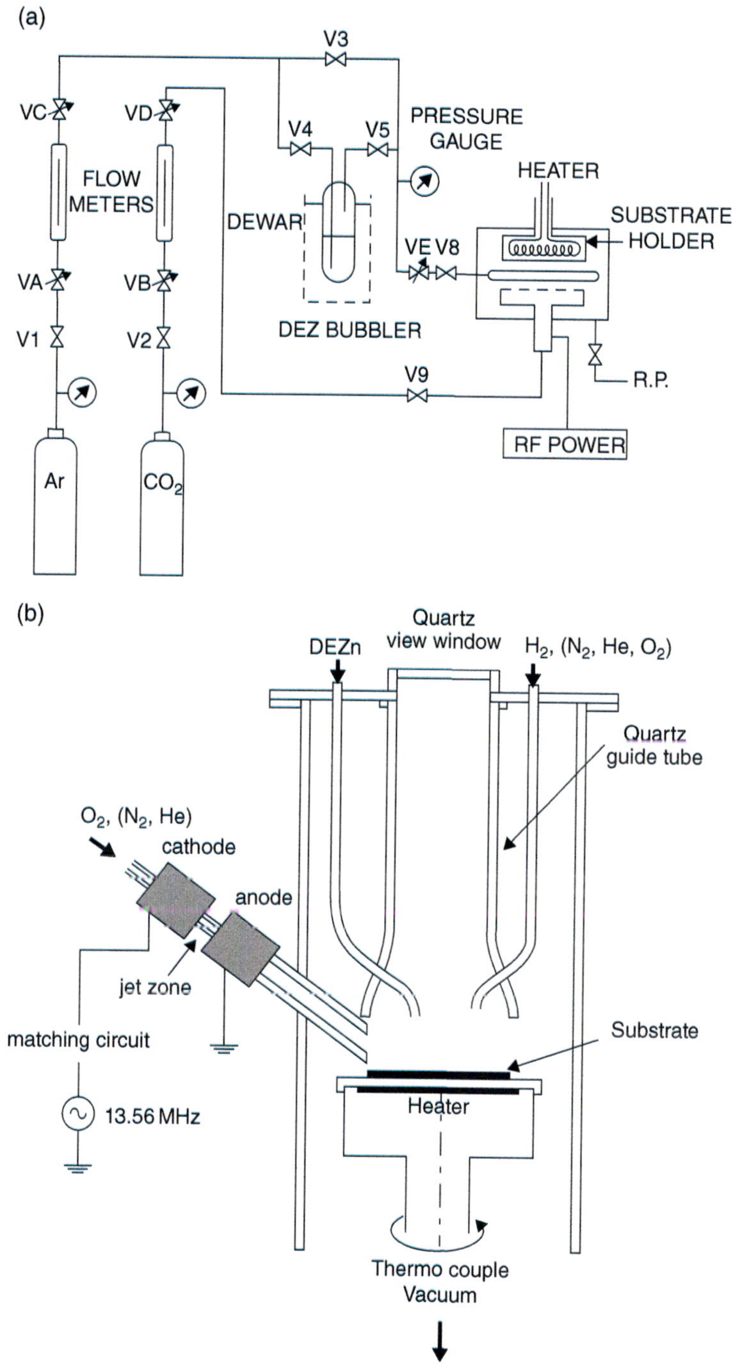

Figure 11.5 *(a) Schematic diagram of the PE MOCVD system. Source: Reproduced from Shimizu et al. 1982 [221] with the permission of Elsevier. (b) Schematic diagram of a remote plasma MOCVD system. Source: After Nakamura et al. 2004 [249], copyright The Japan Society of Applied Physics.*

reactants are excited by the glow discharge; then, the formed active species are transported to the deposition zone. Other reactants are introduced separately from the discharge region to the deposition zone. The method of rf power coupling into the reaction chamber gives rise to two usual approaches for the RPE-CVD: inductively or capacitively coupled rf plasma.

RPE-CVD offers a potential method for minimizing the density of electrically active defects generated by bombarding energetic particles during deposition, because the surface undergoes much less degradation as compared with conventional PE-CVD ([209], Chapter 12).

If we focus now on the application of this technique to the growth of our materials, a seminal paper from Shiosaki et al. [220] reported the effectiveness of plasma discharge to decrease the growth substrate temperature. This method has been subsequently applied to the growth of ZnO, e.g. in [42, 221–223], and also to the growth of $Mg_xZn_{1-x}O$ [64].

In addition, some modifications such as *dual-plasma-enhanced metal organic chemical vapor deposition* (DPE-MOCVD) have been proposed for growing oxides of the family. The DPE-MOCVD method [224] makes use of direct voltage (DC)-driven capacitor-coupled electrodes and a rf plasma system. The DC plasma system enhances the dissociation of oxygen gas (O_2), while the rf plasma system maintains the dissociated free radicals in the chamber. The substrate can be oriented with its normal parallel to the electrodes of the rf plasma system, allowing a reduction in the ion-bombardment damage enhancing the dissociation of oxygen gas. Similarly, on the basis of plasma-assisted processes, electron cyclotron resonance MOCVD has been used in the growth of ZnO on polyethylene terephthalate substrates [225].

11.5 Crystal Growth of ZnO and Related Materials

In Section 11.5, we will analyze the growth of ZnO layers and nanostructures, focusing on the effects of different pre-, during, and postgrowth actions and conditions (Sections 11.5.1 and 11.5.2). Section 11.5.3 is devoted on the growth of related alloys, mainly with cadmium and magnesium, followed in the last subsection by n- and p-type doping experimental work.

An interesting review of the state of the art in the early 2000s can be found in [226].

11.5.1 Crystal Growth by MOCVD of ZnO Layers

11.5.1.1 Epitaxial Growth: Layers with Inplane and Out-of-Plane Defined Orientation

In order to promote epitaxy of ZnO, conventional MOCVD methodology at low and atmospheric pressure has been currently applied. Regarding precursors, DEZn has consolidated as the preferred zinc source in conjunction with water or tert-butanol as oxidizing agents for atmospheric work, and O_2 or N_2O when working at low pressure. When using the HP-CVD methodology, zinc acetylacetonate and O_2 have been the most usual precursors. On the other hand, the etching effect of H_2 on ZnO [227], and the reactions that can appear with some products of the precursors' pyrolysis, have induced the use of alternative carrier gases like nitrogen, helium, and argon.

We will analyze some of the ways to promote ZnO epitaxy as regards the more usual substrates on which it can be attained. Then, the effect of some annealing processes and growth parameters and conditions will be discussed.

11.5.1.1.1 Homoepitaxy of ZnO

The most obvious route for epitaxy is homoepitaxy, which has been carried out on ZnO bulk substrates as well as on ZnO templates or by using ZnO buffer layers. c-Oriented ZnO substrates (e.g. [118, 125, 205, 228–230]) and a-oriented ZnO substrates (e.g. [231–233]) have been the preferred orientations in order to obtain polar and nonpolar ZnO layers.

When growing on the c-oriented plane of ZnO substrates, differences appear in using the Zn and the O face of the ZnO substrate, as discussed by Gu et al. [118] (see Section 11.3). Generally, the Zn-polar samples have better crystalline quality and smoother surfaces than the O-polar ones, and carbon impurities tend to be suppressed on Zn-polar samples. The significant difference of the chemical configuration on the different polarized surfaces is the origin of the distinctive properties for Zn-polar and O-polar samples [118, 234]. In addition, the effects of miscut on the (0001) orientation [205] and on the $(11\bar{2}0)$ orientation [232], and thermal treatments on the ZnO substrates [118], must be considered in analyzing the results.

As a representative example of experimental work in promoting epitaxy, in [231], ZnO layers were grown at atmospheric pressure by the HP-CVD method in the vertical configuration on single-crystal ZnO $(11\bar{2}0)$substrates by using zinc acetylacetonate ($Zn(C_5H_7O_2)_2$) and O_2 gas. Oxygen was introduced into the chamber separately to avoid direct reaction with $Zn(C_5H_7O_2)_2$. The structural study from X-ray diffraction (XRD) measurements showed that ZnO $(11\bar{2}0)$ films without rotation domains were grown. The full width at half-maximum (FWHM) of the diffraction peak corresponding to the $(1\bar{1}00)$ plane was dependent on the thickness, and for a thickness of 0.95 μm, the FWHM was lower than that of the substrate. Another example can be found in [228], where (0001) ZnO layers were grown at 200 Torr in a horizontal MOVPE reactor. DEZn and N_2O were used as source materials, and high-purity N_2 gas was applied as carrier gas. ZnO substrates with +c orientation (Zn face) were used. By AFM observation, the formation of an atomically flat surface with atomic steps was confirmed on the ZnO substrate after a thermal treatment at 1000°C, and a significant decrease of the root mean square (RMS) roughness with increasing temperature and/or VI/II ratio was reported.

The influence of the oxygen/zinc precursor's ratio, growth temperature, and reactor pressure on the structural properties and surface morphology, by using a conventional MOCVD technique, has been emphasized by, among others, Heinze et al. [229], who compared N_2O and O_2 as oxygen precursors by growing on the O-face of ZnO substrates. With a special focus on the effects of the growth temperature, the work of Ive et al. [230], by using conventional MOCVD, and Nishinaka et al. [205], by using HP-CVD, can give a picture of these effects.

Homoepitaxy can be also attained on ZnO templates. Sato et al. [235] studied the growth by HP-CVD at AP of epitaxial ZnO films on ZnO/sapphire (0001) (5–40 nm thick) templates deposited by rf sputtering. $Zn(C_5H_7O_2)_2$ (99.999% purity) and O_2 (99.99% purity) were used as precursors. Similarly, as in [231], O_2 was introduced into the reaction chamber separately in order to avoid direct reaction with the zinc source precursor. From the XRD analysis of the ϕ-scan patterns, it was demonstrated that rotation domains were completely eliminated. The epitaxial relationship between ZnO (0001) film and sapphire (0001) substrate with the ZnO template, as regards the inplane orientation, was ZnO $[10\bar{1}0]$//sapphire $[11\bar{2}0]$ (the cited crystallographic directions refer to the crystallographic axes used in the corresponding reference, as we will do hereinafter). Thicknesses larger than 10 nm were sufficient to improve and stabilize the crystallinity inplane and out-of-plane of the layer.

As a second example of the growth on ZnO templates, zinc oxide layers were grown by MOVPE [124, 125] on c-ZnO/a-sapphire substrates pretreated at 1000 °C. The growth conditions were similar to those used by the same group in [228] already described. Thermal treatment for the ZnO template was adopted to obtain a flat surface, as commented in Section 11.3, and it was essential for improving the crystalline quality of the growing layers. Choi et al. [236] reported on the influence of the VI/II precursor's ratio on the growth of ZnO layers by using conventional LP-MOCVD on ZnO/sapphire templates, and it was demonstrated that the surface of the ZnO film becomes very smooth by adequate selection of the VI/II ratio.

These examples are representative of the good results for epitaxy that can be attained by using templates as a feasible alternative for ZnO single crystals as substrates.

With respect to the use of buffer layers, as Briot discusses in Chapter 3 of [237] regarding GaN, a high density of nucleation can be obtained when a main layer is grown onto a buffer of the same material, due to the reduced interfacial energy between the main layer and its buffer. Then, the nuclei grow as small units, but at an early stage of the units' growth, the lateral growth mode can be enhanced and coalescence can occur, as also commented by Amano et al. [238].

Following the way opened by the studies on the growth of GaN, it has been well established that the use of an adequate buffer layer can improve the crystalline and physical properties of the ZnO overlying films. Under this overall principle, the growth of buffer layers or promoting multistep growth processes [239] have been some of the protocols followed to induce epitaxy.

ZnO buffer layers can be grown at low temperatures on c-sapphire [240], on GaN/sapphire [103, 141], and on an AlN/Si (111) template [150], to name just a few options. High-temperature buffer layers have been also proposed on c-sapphire [191] in order to reduce the nucleation density, but a smoothing step by annealing at 850 °C was necessary. A comparison between ZnO layers grown at low or high temperatures by using PE-MOCVD, with respect to the buffer-layer growth temperature, was reported by Khranovskyy et al. [241], showing that a low-temperature ZnO buffer layer significantly decreases the inplane biaxial stress.

The growth of the buffer layer can imply differences not only in the growth temperature, regarding the main layer, but also in the flow ratio and even in the precursors as shown by, amongst others, Dadgar et al. [103] and Dai et al. [242].

As in the case of GaN (Briot in [237]), the effects of annealing the buffer layer at high temperature prior to the growth of the main layer have also been reported [191, 242], and their benefits for the morphology and crystallinity have been demonstrated. By studying the significance of the low-temperature buffer thickness, Tang et al. [243] demonstrated that the quality of the epitaxial layer is mainly determined by the surface morphology of the buffer, rather than by its structural, optical, or electrical properties.

Ogata et al. [228] showed that the ZnO growth mode tends to be two-dimensional (2D) at higher temperatures (provided sufficient nucleation points) and three-dimensional (3D) at lower temperatures and/or a lower VI/II ratio. They proposed a two-step growth process in growing on sapphire substrates: a first growth was done at low temperature (500 °C in their case), resulting in columnar structures, whereas the second growth process was done at high temperature (800 °C) so that high-quality layers could be formed by the lateral growth from the columns, which, in a sufficiently high density, provided the needed nucleation points to have a low average distance between nuclei favoring the coalescence. On the other hand, Park et al. [244, 245] demonstrated that when using a reduced growth temperature (300 °C) and relatively high reactor pressure, lateral growth was induced, which facilitated the formation of an epitaxial 2D layer, while the use of a higher growth temperature (390 °C) and lower reactor pressure enhanced the vertical growth and resulted in the growth of 1D nanostructures, which could be used as buffer.

11.5.1.1.2 Heteroepitaxy of ZnO

As previously commented in Section 11.3, the use of sapphire as substrate is widely extended. Regarding the epitaxial relationships of ZnO on sapphire, Kasuga et al. [246] reviewed six different modes on the orientation relationships between ZnO and sapphire, as can be seen in Figure 11.6.

The epitaxial relationship depends not only on the substrate but also on the growth parameters. Focusing on the MOCVD technique, the epitaxy of ZnO on r-sapphire has been attained mainly by using LP-MOCVD and DEZn and O_2 as precursors [128, 129, 131, 247] or by using HP-CVD and acetylacetonate and O_2 at AP [22, 23, 233]. As Figure 11.6 shows, the nonpolar a-plane of ZnO can be epitaxially grown on r-sapphire, as demonstrated by, amongst others, Chen et al. [55]. Likewise Quon et al. [133] explain that the ease growth of $(11\bar{2}0)$ ZnO films on $(01\bar{1}2)$ sapphire

Sapphire substrate		Epi-type	Planes ZnO	Directions ZnO//sapphire
R $(1\bar{1}02)$		R_1	$(11\bar{2}0)$	$[\bar{1}100]//[11\bar{2}0]$
		R_2	$(\bar{1}102)$	$[1\bar{1}00]//[\bar{1}2\bar{1}0]$
		R_3	(0001)	$[\bar{1}2\bar{1}0]//[11\bar{2}0]$
		R_4	$(1\bar{2}16)$	$[\bar{1}010]//[\bar{3}302]$
A $(2\bar{1}\bar{1}0)$		A_1	(0001)	$[\bar{1}100]//[01\bar{1}0]$
		A_2 $(=R_1)$	$(01\bar{1}\bar{1})$	$[\bar{1}100]//[11\bar{2}0]$
C (0001)		C_1	(0001)	$[10\bar{1}0]//[11\bar{2}0]$
		C_2 $(=R_2)$	$(11\bar{2}4)$	$[1\bar{1}00]//[\bar{1}2\bar{1}0]$

Figure 11.6 *Previously known ZnO-sapphire orientation relationships. Source: Reproduced from Kasuga and Mochizuki 1981 [246] with the permission of Elsevier.*

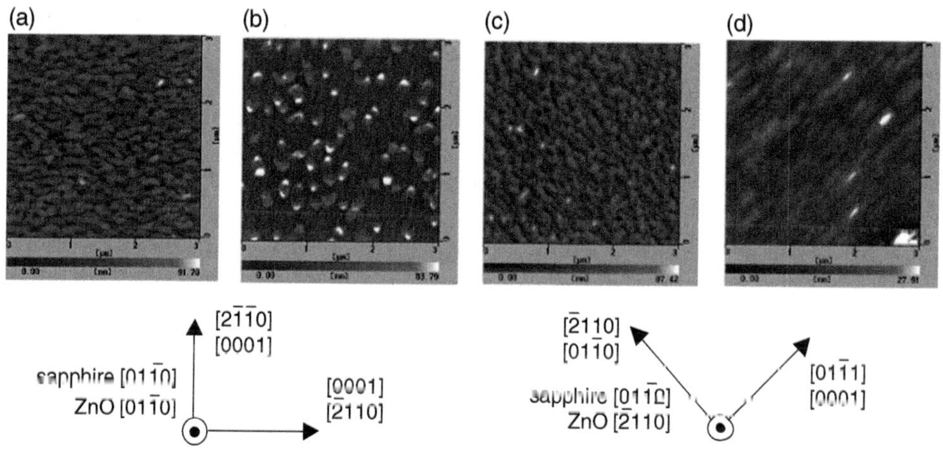

Figure 11.7 *AFM images of ZnO surfaces on m-plane sapphire (a) at high growth temperature (800 °C) and (b) at low growth temperature (500 °C) and on r-plane sapphire (c) at high growth temperature (800 °C) and (d) at low growth temperature (500 °C). Source: Reproduced from Moriyama and Fujita 2006 [130] with the permission of John Wiley and Sons.*

substrates is due to the close lattice fit between both planes (see also [248]). Furthermore, on the growth of nonpolar m and a planes of ZnO on sapphire, Moriyama et al. [130] studied the epitaxial growth characteristics of ZnO on m- and r-plane sapphire substrates. The growth of ZnO along the *c*-axis direction, which is preferable for smooth lateral growth and aligns parallel to the [2$\bar{1}\bar{1}$0] and [01$\bar{1}\bar{1}$] directions of m and r-plane sapphire, was achieved under lower growth temperature and/or lower VI/II ratio by using LP-MOCVD and DEZn and N_2O as precursors. On m-plane sapphire where m-plane ZnO was grown, the inplane alignment was found to be ZnO[0001]//sapphire[2$\bar{1}\bar{1}$0] and ZnO[$\bar{2}$110]//sapphire [0001]. On the other hand, on r-plane sapphire, where a-plane ZnO was grown, the inplane alignment was ZnO[0001]//sapphire[01$\bar{1}\bar{1}$] and ZnO[01$\bar{1}$0]//sapphire[$\bar{2}$110].

In Figure 11.7, after Moriyama et al. [130], the significant effect of the growth temperature can be seen.

Table 11.7 *Other substrates used in the growth of epitaxial ZnO. Source: Adapted from Table 2 of Kaul et al. 2005 [251].*

Substrate	Epitaxial relationship out-of-plane	Epitaxial relationship inplane
(111)$SrTiO_3$	(0001)ZnO//(111)$SrTiO_3$	$[11\bar{2}0]$ZnO//$[1\bar{1}0]$$SrTiO_3$
(111)$Gd_3Ga_5O_{12}$	(0001)ZnO//(111)$Gd_3Ga_5O_{12}$	$[11\bar{2}0]$ZnO//$[1\bar{1}0]$$Gd_3Ga_5O_{12}$
(111)MgO	(0001)ZnO//(111)MgO	$[10\bar{1}0]$ZnO//$[1\bar{1}0]$MgO
(111)$ZrO_2(Y_2O_3)$	(0001)ZnO//(111)$ZrO_2(Y_2O_3)$	$[10\bar{1}0]$ZnO//$[1\bar{1}0]$$ZrO_2(Y_2O_3)$
(111)$MgAl_2O_4$	(0001)ZnO//(111)$MgAl_2O_4$	$[11\bar{2}0]$ZnO//$[1\bar{1}0]$$MgAl_2O_4$, $[10\bar{1}0]$ZnO//$[1\bar{1}0]$$MgAl_2O_4$

The growth of epitaxial c-oriented ZnO can be attained by using DEZn and O_2 as precursors in LP-MOCVD processes on c-sapphire [128]. Likewise, inducing ZnO epitaxy on c-sapphire is also possible by using a plasma-assisted methodology with DEZn and CO_2 as precursors [221]. The epitaxial relationships are (0001)ZnO//(0001)Al_2O_3 and $[10\bar{1}0]$ZnO//$[2\bar{1}\bar{1}0]$$Al_2O_3$. This rotation inplane results in a continuation of the oxygen sublattices of the two materials. Based upon these considerations, the lattice misfit along the $[2\bar{1}\bar{1}0]$ direction of ZnO is on the order of 18.3% [128]. Nakamura et al. [249] used plasma-assisted MOCVD and reported on the ZnO epitaxy on c, r, and a-sapphire substrates and, in comparison with Fons et al. [250], who used MBE, inferred the epitaxial relationship on a-sapphire as (0001) ZnO//$(11\bar{2}0)$$Al_2O_3$.

In addition to sapphire, other substrates allow the growth of epitaxial ZnO, as shown in Table 11.7 [251].

Yttria-stabilized zirconia $ZrO_2(Y_2O_3)$ as substrate was studied by Chao et al. [252], and the epitaxial relationships were reported as (0001)ZnO//(111)$ZrO_2(Y_2O_3)$ and $[11\bar{2}0]$ZnO//$[1\bar{1}0]$ $ZrO_2(Y_2O_3)$. The growth of ZnO on single-crystal $SrTiO_3$ was analyzed in [253], where it was concluded, from X-ray φ-scan measurements, that there is no epitaxial relationship between ZnO and (011)$SrTiO_3$ substrates; nevertheless, on (111)$SrTiO_3$, there is a single domain with an epitaxial relationship of $[\bar{1}100]$ ZnO//$[\bar{1}10]$ $SrTiO_3$. In addition to these substrates, γ-$LiAlO_2$(100) has also been proposed as substrate in order to promote the MOCVD growth of nonpolar m-oriented ZnO $(10\bar{1}0)$ [254].

On the other hand, some templates have been proposed as an alternative to single crystals in order to promote heteroepitaxy [255]. As discussed in Section 11.3, using GaN as substrate has several advantages for the growth of ZnO. Thus, the epitaxy of (0001)ZnO on (0001)GaN templates has been demonstrated by using LP-MOCVD and DEZn and iso-propanol (i-PrOH) as precursors [142, 256] or DEZn and N_2O [257]. This same epitaxy was also reported by Ataev et al. [258] using a ZnO buffer layer grown at low temperature. With GaN/Si(111) templates, Oleynik et al. [259] obtained tensile-stressed (0001)-oriented ZnO by using LP-MOCVD and DMZn and tert-butanol as precursors. A strain reduction was observed on increasing the growth temperature. Krost et al. [171], under similar experimental conditions, showed that a 2D flat morphology of ZnO could be accomplished by including in situ annealing at 900 °C. As a function of the annealing time, there was a transition from 3D to 2D morphology that was attributed to an Ostwald-ripening process with a subsequent flattening of the crystal surface.

With less presence in the literature, GaN/γ-$LiAlO_2$(100) has proved [260] of interest in obtaining m-ZnO layers, and Y_2O_3/Si(111) was able to induce ZnO epitaxy, as shown by Lin et al. [173] by using HP-CVD at AP with zinc acetylacetonate and O_2 as precursors.

Just to reiterate, the ability to obtain epitaxial ZnO on a defined substrate is directly related to the growth conditions, a question which will be addressed in the following sections, where some effects related to annealing and growth parameters are discussed.

11.5.1.1.3 Induced Effects on Epitaxy by Thermal Annealing and Chemical Etching Processes on Substrates

As noted earlier, the single-crystal quality of an epitaxial layer is directly related to the substrate surface. Consequently, annealing processes, preparation, and chemical etching have been shown to be essential in improving the properties and characteristics of the growing ZnO. In addition to the questions noted in Section 11.3, it is worth highlighting some contributions that show the huge effects thermal annealing on the substrate has on the epitaxy.

In the late 1980s, Souletie et al. [261] emphasized that, under their experimental conditions, the growth of single-crystalline ZnO layers was only possible after well-defined c-sapphire chemical etching and heating at 750 °C for 15 min in a flowing hydrogen ambient. We will present some examples of thermal treatments of the substrate in obtaining ZnO epitaxy on sapphire (a-r-c oriented) and on ZnO (0001) single-crystal substrates.

Achieving atomically flat substrate surfaces has been reported as significant in obtaining high-quality epitaxial ZnO. Ogata et al. [228] performed thermal annealing on (0001) ZnO substrates in a nitrogen atmosphere at 1000 °C, and the formation of atomically flat surfaces with atomic steps was confirmed, by AFM measurements, on the ZnO substrate. The RMS roughness was only 0.5 nm, and at high growth temperature and VI/II high ratio, 2D growth was enhanced. Wang et al. [140] studied the effects of annealing on (0001) sapphire substrates at temperatures in the range 1000–1400 °C. The substrates were cleaned by a conventional procedure for removing contaminants (see Section 11.3). The effects of annealing on the substrate properties and how they affect the as-deposited ZnO thin films was explained as follows: the substrates annealed at high temperature presented a high density of atomic steps on the substrate surface, and the formation of the steps was probably the origin of the growth, as well-defined steps on the sapphire substrate can be nucleation sites for film growth due to the reduction of surface energy. It was highly favorable for the islands to nucleate at the edge of steps formed on the annealed sapphire substrate, and the nucleated grains on the atomic steps had better interconnections between neighbouring grains, showing a single dominant orientation.

As another example, in Nakamura et al. [249], the a plane $(11\bar{2}0)$, c-plane (0001), and r-plane $(01\bar{1}2)$ sapphire substrates were cleaned in organic solvents, chemically etched in a solution of 3:1 H_2SO_4:H_2PO_4 at 160 °C, rinsed in deionized water, and dried, before undergoing a heating process at 600 °C in hydrogen at a pressure of 0.1 Torr for 30 min. To improve the crystal quality of the (0001)-oriented ZnO, an initial monolayer of Zn was deposited by self-limiting growth on the sapphire substrate. The XRD pattern of a ZnO film without Zn termination on a sapphire substrate exhibited a (0002) line regardless of substrate direction. Regarding the ZnO film grown on the Zn-termination substrate, the XRD pattern exhibited a (0002) line when the growth was on the a- and c-plane sapphire substrates, and a $(11\bar{2}0)$ line when the process was on the r-plane substrate. From the XRD pattern, the ZnO film grown on sapphire with Zn termination was shown to be single crystalline.

11.5.1.1.4 Some Effects of Growth Temperature, Layer Thickness, and Postgrowth Annealing on ZnO Epitaxy

Growth temperature is one of the key parameters in the MOCVD growth process. The effect on the growth rate, morphology, and crystalline quality will be briefly discussed in this subsection.

Provided that the precursors have been completely pyrolyzed, the effects of substrate temperature on the properties are caused by adsorption, evaporation, surface migration of atoms, etc. For fixed other growth parameters and experimental conditions, depending on the growth temperature, the growth rate in the MOCVD methodology is determined [6, 237]:

- By surface reactions, in which case the growth rate will be thermally activated and the limiting step in the reactions occurring during the growth will be reflected by the activation energy. In this regime, the growth rate will be very sensitive to the substrate surface (orientation, roughness, chemical composition, etc.).
- As the growth temperature increases, provided the crystal does not decompose, the chemical reactions will be fast enough, and the growth will be limited by the supply of reactants to the interface. The growth rate is then almost independent of the growth temperature and of the substrate orientation. Since the reactant mass transport occurs in MOCVD by diffusion of the species through the carrier gas, this regime is called *diffusion-limited.*
- At higher temperatures, a decrease in the growth rate appears due to thermodynamic factors, although other explanations, such as depletion of reactants [6] or homogeneous nucleation, as well as re-evaporation of the layer [251], must be considered.

The overall behavior of the growth rate as a function of temperature previously described is mainly followed by the growth of epitaxial ZnO layers, as can be seen, for example, in the work of Kaul et al. [251], and in [247], in which the high-temperature regime was particularly studied. On the other hand, it is interesting to note that the optimum value of the growth rate is determined by the bond strength in the crystal, its thermal stability, the precursor chemistry, etc.

In addition to the incidence of the growth temperature on the growth rate, other effects, more or less directly related to it, such as effects on the morphology and structural properties, have been demonstrated. The detailed studies of Liu et al. [262] and Zang et al. [263] showed that the epitaxial relationships between the orientation of ZnO and substrate unit cells varied from no-twist for $T_g \cong 200\,°C$ to 30° twist for $T_g \geq 300\,°C$, where T_g is the growth temperature; whereas for growths at 250 °C, the ZnO film consisted of these two types of inplane orientations. Thus, it was revealed that in MOCVD-grown ZnO on sapphire (0001) substrates, the inplane orientation can be controlled by selecting the growth temperature. It was argued that different growth modes at the initial stage of growth were responsible for different epitaxial relationships. Then, under their experimental conditions [263], the position of the first layer of zinc atoms would determine the epitaxial relationship.

Adequate selection of the growth-temperature window can significantly reduce the FWHM and increase the intensity of the out-of-plane diffraction peak, as evidence of a better alignment along the preferred orientation [264].

In heteroepitaxial processes, it can be expected that the effect of the thickness of the main layer could be significant on the crystalline and structural quality of the epitaxy. Indeed, relaxation of the lattice parameters can be expected, depending on the thickness. Kashiwaba et al. [22, 23] studied the evolution of the a-lattice constant, the FWHM, and the peak intensity of X-ray diffraction from the ZnO ($11\bar{2}0$) plane grown on sapphire as a function of the film thickness. A smaller a-lattice parameter value, as regards the lattice constant for ZnO powder, at smaller thicknesses, was due to tensile stress along the *c*-axis. This stress is reported to be relaxed with increasing thickness, and consequently the stress probably comes from the lattice mismatch between ZnO and sapphire. Moreover, the a-lattice constant is slightly larger than the standard value for a film with a thickness of around 0.5 μm. This can be explained by stress due to the difference between the thermal expansion coefficients of ZnO and sapphire. The thermal expansion coefficient along the *c*-axis for ZnO is smaller than that of sapphire (as seen in Section 11.3). Therefore, a compressive stress along the *c*-axis is induced, which results in an enlarged value of the a-lattice parameter. The effects of layer thickness on the lattice parameters, strain, and strain anisotropy have also been studied by Saraf et al. [265] on a-ZnO layers grown on r-sapphire by using the LP-MOCVD methodology.

Postgrowth annealing processes have been another way to induce better crystalline quality on the layers, acting on dislocations or point defects, or even inducing recrystallization. Nevertheless, the thermal conditions for postgrowth annealing must avoid, as far as possible, interdiffusion processes or crystallization of undesired compounds at the interface layer. Positive effects on the properties of the epilayers have been largely reported and could be considered as a way to increase the layer quality (see e.g. [266]), also reducing the density of dislocations, as reported by Wang et al. [267], who showed that, in their ZnO epilayers, edge dislocations were more sensitive to annealing processes (decreasing significantly) than to screw dislocations.

11.5.1.2 Nonepitaxial Growth: Textured and Randomly Out-of-Plane Oriented ZnO Layers

Motived by the interest in applications for which epitaxy is not fundamental, the research on ZnO nonepitaxial layers, both textured and randomly oriented, has been widely developed and reported. Crystalline substrates were the first option in obtaining textured ZnO, while amorphous substrates were the main choice for nontextured layers. Due to the structural and chemical mismatch of ZnO with the common heterosubstrates, randomly oriented layers are easily obtained. To achieve highly oriented layers, a correct selection of the growth parameters, as well as the use of buffer layers and/or annealing processes, has become the way to follow.

Taking advantage of the easier ZnO growth in the basal plane direction, highly oriented ZnO layers have been grown with c- or a-planes parallel to the substrate plane. With respect to precursors, alkyls (DMZn and DEZn) together with alcohols or H_2O in experiments at AP and O_2, N_2O, or NO_2 when using low pressure are the main option in the literature. The HP-CVD technique using solid zinc sources (at room temperature) usually takes zinc acetylacetonate and O_2 as precursors. Relatively low growth temperatures (<450 °C) work well at AP with the previously cited precursors, and higher temperatures are more usual at LP by using O_2, N_2O, and NO_2 as oxidizing agents. Nevertheless, the lowest temperatures can be attained in plasma- or photoassisted processes, as discussed in Section 11.4.

Using some seminal papers in the literature, we will analyze the effects of some growth parameters and experimental conditions that can promote the growth of nonepitaxial ZnO layers.

11.5.1.2.1 Some Comments on Experimental Setup Conditions: Effect of Reactor Pressure and Substrate Rotation Rate on the Growth of Nonepitaxial ZnO Layers

In conventional MOCVD at low pressure, but also in other methodologies, the working pressure is the first parameter to be selected. The effects on the hydrodynamics and precursor reactivity must be considered, as they have a direct influence on the growth rate. The significance of reactor pressure has been emphasized by, amongst others, Waag et al. [142] and Gruber et al. [256], who demonstrated that the reactor pressure was a crucial parameter for growing ZnO layers at reasonable growth rates. In their experimental work, the maximum growth rate was obtained at a pressure of 400 mbar, and the growth rate decreased for lower pressures. It was assumed that the mean free path of the adducts of DEZn and i-PrOH was too long for low pressures (equivalent to low concentrations of the adducts in the reactor), resulting in a small formation of the intermediate alkylzinc alkoxide necessary for the ZnO deposition. A further pressure increase led to a decrease in the growth rate. A depletion of the gas phase at higher pressure was inferred from the observation of a white powdery fall-out in the reactor upstream from the substrate.

As discussed by Zhu et al. [268], low reactor pressures should usually benefit high-quality ZnO growth by depressing the gas-phase prereaction between precursors. However, for ZnO growth, the case could be different (depending on the other growth parameters), because by increasing the reactor pressure, a longer residence time of the reaction precursors is induced; this could result in enhanced dissociation of zinc species, and the growth rate would be higher, as analyzed in [268].

The results of Zhu et al. [268] show that the surface morphology of ZnO layers changes significantly from 3D to quasi-2D growth mode with increasing reactor pressure. When the reactor pressure was increased, enhanced dissociation of DMZn led to improved initial nucleation, with density and size increased. The coalescence of nucleation islands was finished to achieve quasi-2D growth mode.

Also impinging on the hydrodynamics are the effects related to the rotation speed of substrates. Pan et al. [227] studied the growth window in which the gas flow was stable in the laminar-flow regime. This was necessary to form an efficient and stable boundary layer and maintain an efficient heat exchange between the heated substrate and the boundary layer. The rotation speed was in the range of 600–1200 rpm in their work. Rotation speed affected the crystal quality, decreasing it at rotation speeds below 750 rpm and above 1050 rpm. This suggested that the boundary layer was too thick below 750 rpm, leading to less adsorption of reactants onto the surface and nonuniform heat distribution, thus reducing the crystal quality. Reduction in the crystal quality at rotation speeds above 1050 rpm was explained because the boundary layer, at this rotation speed, began to break down, causing nonuniform gas flow and heat transfer.

Zhang et al. [269], working at slightly higher pressure and temperature than Pan et al. in [227], studied the effect of the rotation speed, between 80 and 750 rpm, on the growth rate. They found an increase in the growth rate as the rotation increases at the mass-transport growth regime. In vertical reactors, the gas flows downward, impinging directly on the susceptor, which is oriented perpendicular to the direction of the gas flow, leading to the formation of a boundary layer. An analysis of the boundary layer thickness can be found in Stringfellow [6]: the boundary layer thickness (δ_0) under ideal conditions, in which the disk radius is much greater than δ_0, is proportional to $\omega^{-1/2}$, where ω is the angular rotation speed. From this proportionality, the following analysis was proposed: at very low rotation rate speed, the δ_0 thickness is larger, and molecules have to diffuse through a thick boundary layer, so the growth rate is low. As the rotation rate increases, the boundary layer thickness decreases, and then the reactants diffuse through the boundary layer more easily and the growth rate improves quickly. When the rotation rate is higher than a certain value (530 rpm in the case of Zhang et al. [269]), the gas flow can begin to be greatly affected by the hot susceptor, since the boundary layer is very thin. Nevertheless, Nishimoto et al. [29] reported that high speed (400 rpm) disk rotation induces laminar flow by suppressing the thermally induced buoyancy flow, and it was possible to inhibit the oxygen vacancies because the source gases were attracted on the substrate by the centripetal force. Defect formation was suppressed, and a high growth rate was obtained because the gases flowed toward the substrate surface. Once again, it is worth noting that the effects of a particular parameter are dependent on the overall experimental conditions, and care must be taken to analyze in a comparative form the results in the literature.

11.5.1.2.2 Some Comments on the Experimental Setup Work Conditions in Assisted Processes and Their Effects on the Growth of Nonepitaxial ZnO Layers

When using assisted techniques to obtain nonepitaxial layers, parameters like the rf power in plasma-assisted or the light intensity in photoassisted processes can be of significance. The effects of rf input power in PE-MOCVD were analyzed by Shimizu et al. [152, 221], where highly *c*-axis-oriented ZnO thin films were grown on glass substrates. The effects of the rf input power level on the film properties are presumably caused by the change in the generation rate of active species and in the kind of species generated with changing rf input power. Thus, from standard deviation measurements of the X-ray rocking curve and from *reflection high-energy electron diffraction* (RHEED) measurements, it was found [221] that the films obtained at the higher substrate temperature and at the higher rf input power had the better *c*-axis orientation. Films grown at a substrate temperature lower than 200 °C and an rf input power less than 220 W had a mixed orientation. Other studies of this effect of the plasma power input are also reported by Seeomon et al. [270].

In photoassisted processes, care must be taken regarding the influence not only of the wavelength of the assisting light but also of its intensity, as analyzed by Yoshino et al. [214], who showed that high values of intensity limited obtaining good-quality films. Additional results on the effect of the light intensity on the morphology and structural and optical properties of the ZnO layers can be found in [271], where the increase of the grain size on increasing the light irradiation intensity was analyzed. It was suggested that the photoinduced nonequilibrium carriers enhance the surface migration of adsorbed atoms, and the grains mix together much more easily, so that much larger grains are obtained. Shimizu et al. [211] studied the influence of UV light irradiation on the growth rate, crystallinity, and resistivity, due to its role in the gas-phase reactions and surface reactions. The effects on the morphology and its evolution as a function of the light intensity have also been reported by Shi et al. [217], with special attention on the nucleation step.

11.5.1.2.3 Some Comments on Using Buffer Layers and Their Effects on the Growth of Nonepitaxial ZnO Layers

The effect of a buffer layer on epitaxy has been already discussed. In order to improve the crystalline quality on ZnO nonepitaxial layers when using mismatched substrates, the presence of a buffer layer has shown its potential. Buffer layers grown in a two-step process have usually been obtained at lower temperatures than the main layer. Likewise, the use of buffer layers that are thermally annealed at higher temperature than the main growth has been suggested in order to improve the surface smoothness and crystalline quality [124, 149, 155]. In addition, it has been reported that the buffer layer allows control of the orientation and morphology of the ZnO films [153, 154].

Multistep processes have been also studied, e.g. in [272], where Moriyama et al. proposed a three-step process to improve the quality of ZnO grown on silicon substrates. First, a nucleation layer was grown at 300 °C (during 15 min); second, a recovery layer was grown at 500 °C (15 min); and third, the main layer was grown at 800 °C (60 min). The VI/II (N_2O/DEZn) ratio was kept constant during the overall process.

In the extensive literature on the growth of buffer layers, the effectiveness of these buffer layers on different substrates (sapphire, silicon, templates, diamond, GaAs, glass, etc.) has been demonstrated in a wide range of experimental protocols: by using the same or different temperatures with respect to the main growth (lower than [273]; higher than [153, 154]; as well as by using the same growth temperature as in [272]), or different precursor ratios [230, 255], or even different precursors, as in [274].

The significance of the thickness of the buffer layer has been studied by Zhang et al. [137] on *c*-axis oriented ZnO films, and it was found that the surface morphology and structural and optical properties of the films depended on that thickness. The ZnO layer grown on a 15-nm thick buffer showed a flat and dense surface with good structural and optical properties. This buffer layer presented the roughest surface as regards other thicknesses, favoring the growth of a high density of ZnO nuclei. High-density nucleation causes coalescence among different grains, which facilitates the formation of a flat, dense surface.

11.5.1.2.4 Some Comments on the Effect of the Growth Temperature on the Growth of Nonepitaxial ZnO Layers

The kinetic behavior in the growth of nonepitaxial layers by using different techniques, substrates, and experimental conditions (see e.g. [19, 188, 227, 270, 275]) is not too different from the kinetic behavior in the growth of epitaxial layers. In addition, other, more-complex temperature dependence in the growth rate of ZnO can be found (see e.g. [276]). Likewise, it is worth noting that the growth kinetics are dependent on the selected precursors, as demonstrated, for example, by Kauffman et al. [34] or in [12, 81, 82].

Growth temperature has a direct effect on the process itself and on the properties of the growing layer (see e.g. [211–213]). In the temperature range in which no modifications of the substrate and no reactions at the interface level are present, we can expect better crystallinity as temperature increases, until a critical value is attained (see e.g. [21, 31, 179, 227]). Thus, a significant reduction of the FWHM as a function of temperature is not a surprise, as reported, for example, by Park et al. [264], corresponding to c-oriented samples grown on sapphire. Similar overall behavior can be seen also in the work of Roro et al. [277] growing ZnO on glass substrates.

The growth temperature can induce a change in texture, as demonstrated by Chen et al. [181] and in [278, 279]. Chen et al. [181] detected that the microstructure of ZnO films evolves from (0002) crystal plane growth mode to a $(10\bar{1}1)$ mode, following the sequence (0002)->$(11\bar{2}0)$->$(10\bar{1}0)$->$(10\bar{1}1)$ crystal plane as a function of growth temperature. This behavior was attributed to different surface free energies, and hence the substrate temperature was able to activate the ZnO film growth from relative lower surface free energy to higher surface free energy ($1.6\,J\,m^{-2}$ for (0002), $2.0\,J\,m^{-2}$ for $(11\bar{2}0)$, and $3.4\,J\,m^{-2}$ for $(10\bar{1}0)$, respectively), opening the opportunity to change from c-oriented to randomly polycrystalline film. Nicolay et al. [184] showed that by using LP-MOCVD, the preferential orientation of ZnO on glass evolves from *c*-axis to *a*-axis as the growth temperature is increased from 110 to 220 °C, and then the preferential orientation goes back to *c*-axis at a growth temperature of 380°C. This behavior can be understood based on the temperature-dependent selection of different preferential orientations present at the initial layer. Different effects of the growth temperature on the morphology have been reported by using different substrates [81, 179, 181, 275, 277, 279–281]. Likewise, transition from films to nanostructures as a function of growth temperature has been reported [282], as well as different morphologies in low-dimensional structures [283], as we will see later.

As demonstrated by Oda et al. [81], using glass as substrate, a clear dependence of the grain size on the growth temperature and the oxidizing agent for DEZn can be found. Similar results with respect to the growth temperature were obtained by Khranovskyy et al. [284] on silicon substrates, and increased RMS values of the surface roughness as a function of the growth temperature were reported by Kim et al. [280]. Focusing on crystalline quality, we should bear in mind that differences in thermal expansion coefficient between substrate and layer can be of significance, depending on the growth temperature, as reported by Gandhi et al. [1].

It is worth emphasizing that kinetics and pyrolysis must be considered together when analyzing the thermal behavior of the growth rate. Lamb et al. [179] studied the deposition rate as a function of substrate temperature for two different VI/II ratios (1 and 1/3). For both VI/II ratios, the growth rate steadily increased with increasing temperature, reaching a maximum around 240 °C. This increase in growth rate was attributed to the decomposition of the precursors being thermally activated. Above 240 °C, the growth rate decreased, possibly as a result of thermal cracking of the precursors before reaching the reaction zone or desorption of unreacted precursors. For (VI/II)=1, the growth rates were approximately five times lower than for (VI/II)=(1/3), but they followed the same general trend. A rapid change in surface morphology was noted between macrosmooth, below 260 °C, to a rough textured film above 260 °C.

11.5.1.2.5 Some Comments on the Effect of the Carrier Gas on the Growth of Nonepitaxial ZnO Layers

In crystal-growth processes, the carrier gas can be of significance by the reactivity it can have in the gas phase, acting on the precursor decomposition and/or on the layer (etching effects), on the hydrodynamics, etc. Fujita et al. [28] reported on the effect of H_2 as carrier gas on the growth rate. Thus, a-oriented $(11\bar{2}0)$ ZnO films were grown on a-plane sapphire substrates using diisopropylzinc as a zinc source and iso-propanol or tert-butanol as oxygen sources. Nitrogen and hydrogen were used as carrier gases. These authors found significant differences in the growth kinetics due to the presence, when using hydrogen, of a reducing atmosphere or by other related reactions. Nevertheless, the growth temperature and the used zinc precursor must be taken into account, because other authors [83] reported a different behavior of the growth kinetics at lower temperatures, by using DEZn, iso-propanol, or tert-butanol as oxygen sources and hydrogen as carrier gas, showing once again the interplay of the overall process.

Differences in choosing H_2 or He as carrier gas were analyzed by Wang [285], and it was demonstrated that the films grown with He carrier gas showed improved crystallinity, smooth surface microstructure, and reduced hydrogen defects concentration. The decrease in the growth rate, if H_2 was used as carrier gas, was attributed to a reduced reaction rate. Mao et al. [286] analyzed the influence of H_2 addition to the oxygen source gas line on the properties of ZnO films grown by MOCVD with DMZn and DEZn as zinc precursors and N_2O and O_2 as oxygen sources, respectively. It was reported that H_2 addition significantly improved the structural quality of ZnO thin films grown by DMZn for both oxidizer agents with no carbon impurity incorporation. An opposite effect was found to occur on the ZnO thin films grown with DEZn and H_2 addition, where the improvement in the quality of the ZnO film with full suppression of carbon impurity was obtained only with N_2O as oxygen source; but a deterioration in the crystalline quality of the film with high carbon impurity incorporation was obtained when using O_2. It can be concluded that H_2 can play a significant role in the chemistry of the growth process.

11.5.1.2.6 Some Comments on the Effect of the Precursor Ratio on the Growth of Nonepitaxial ZnO Layers

It is expected that the molar ratio between the precursors controls the material stoichiometry and consequently has a significant effect on the electrical and optical properties, as well as on the crystalline ones; this is due to the influence on the density of vacancies and interstitials, as well as promoting, or not, the lateral growth rate (Briot in Chapter 3 of [237]). The literature offers a wide spectrum of results about the effect of VI/II precursor ratio on the growth rate. These results can be understood if we take into account the experimental conditions in which rich or deficient conditions of one of the precursors are present. Indeed, if, for example, the experimental work is under oxygen-deficient conditions, the growth rate will be controlled by the reactivity of the oxygen precursor. Experimental results on that can be found, for example, in [276]. The reactivity can follow different paths, involving different reactions. On this basis, differences in the growth rate as a function of precursor ratio corresponding to different precursors can appear, as can be seen, for example, in [82].

A saturation in the growth rate as the limiting precursor increases (the other growth parameters being fixed) is expected. Then, the increase in the flow can block the availability of reactions, and the growth rate could decrease. Lamb et al. [179], studying the deposition of thin-film ZnO from diethylzinc and n-butanol, demonstrated that increasing the II/VI ratio led to a pseudo-first-order growth rate model, while increasing the VI/II ratio resulted in a decrease in growth rate. It was proposed that, when deposition takes place at above equimolar VI/II ratios, excess n-BuOH acts as a site-blocking molecule, according to a Langmuir–Hinshelwood mechanism, lowering the growth rate. Similarly, Fujita et al. [28], by using i-PrOH with hydrogen as carrier gas, observed that the growth rate decreased with increasing VI/II ratio. It was considered that the decrease in the growth rate was caused by a prereaction between i-PrOH and Di-PrZn or hydrogen. By contrast, when tert-butanol was used with hydrogen or nitrogen as carrier gas, the growth rates increased with increasing VI/II and saturated at higher VI/II ratios. These results point to the significance of considering the overall process and reactions.

Other examples of the effect of VI/II ratio on the growth rate by using other substrates and/or precursors can be found, for example, in [139, 287]. In [287], the authors studied the growth dynamics of ZnO from a modelling of competitive adsorption and two-site occupation effects. The model, based on the Langmuir–Hinshelwood mechanism, explains the growth-rate dependence on the VI/II ratio for different growth regimes. At lower temperatures, it was found that the growth rate (normalized to the unit DEZn molar partial pressure) decreased gradually with the increase of the VI/II ratio. At higher temperatures, the normalized growth rate increases sublinearly with the increase of the VI/II ratio until a saturation for larger ratios is found. This behavior can be understood in the frame of the proposed model because with the surface reaction-controlled regime (low temperature), oxygen species prevent the adsorption of Zn-related species, resulting in the decrease in growth rate. By contrast, in the mass-transport limited regime, a two-site adsorption mechanism is expected, and the growth rate is determined by mass transport of Zn precursor.

Focusing now on the crystallinity, Hahn et al. [82] observed that when the VI/II ratio was lowered by either increasing the DMZn or reducing the tert-butanol flow, the average grain size increases. Their samples showed a textured surface that appears milky due to light scattering at the grains. Below (VI/II) = 2, the layer surface became rough and the layers appeared brownish. The films grown with DMZn at high VI/II ratios showed a significant *c*-axis orientation perpendicular to the substrate surface in most grains, while the films grown at high temperatures with low VI/II ratio exhibited various reflection peaks characteristic of unoriented polycrystalline layers. Furthermore, regarding the impact of the VI/II ratio, the work of Kim et al. [288] analyzes the effects on the morphology; on the other hand, defect generation at the crystal lattice, oxygen vacancies, and other oxygen-induced defects directly related to the VI/II ratio affect the films' tensile strain, stress, grain size, bandgap structure, and dispersion properties, as discussed by Pagni et al. [289], consequently affecting the crystalline quality. The effects on structural, morphological, and preferred orientation as a function of VI/II ratio in ZnO films is an open subject in the literature [290, 291]. Usually, ZnO shows a strong tendency toward 3D growth mode with higher flow rates of the zinc source. Obviously this has a direct influence on the morphology, as evidenced in [288] amongst others. Morphology transitions have been particularly studied, and some examples can be found in [291]. Other authors have focused more directly on the impact of the precursor ratio on physical properties, such as in [142, 277, 278, 290, 292–294], to name just a few, and the clear correlation between the precursor ratio and the physical properties has been corroborated due to the direct incidence on stoichiometry and generation of defects.

11.5.1.2.7 Some Comments on the Effect of Pre- and Postgrowth Treatments

In addition to the comments on the thermal treatments on the substrates discussed in Section 11.3, it is worth noting that similar to the case of epitaxy, the ZnO film quality is strongly affected by annealing treatment and miscut direction of the substrates. In fact, depending on the miscut and thermal treatment, a low quantity of defects and good vertical and lateral alignment can be selectively obtained [295].

On the other hand, postgrowth annealing, both conventional and rapid thermal annealing (RTA) processes, as tools to increase the sample's qualities, can also be used on as-grown textured and nontextured samples [190, 296, 297]. As pointed out by Sun et al. [298], it is generally accepted that annealing is an effective method to improve the qualities of thin films. However, the results depending on the annealing atmosphere and pressure could be thought to be controversial. The as-grown state of the samples is the variable that justifies different annealing procedures and their effects, as can be seen in the literature. Annealing temperatures for 30–60 min in the range 600–850 °C have shown their effectiveness in introducing changes in the film properties in correlation with the atmosphere in which the annealing takes place. Oxygen and nitrogen at different pressures, air, and vacuum can be used. Oxygen can be a good option in modifying not only structural but also electrical properties; however, the effects are pressure dependent [298], and too much O_2 pressure can introduce defects more than remove them. Beneficial action of other postgrowth annealing processes has been also reported, as in Ye et al. [299], who proposed a two-step annealing process at 850 °C in pure oxygen (1 atm) first for 30 min and then in pure nitrogen (1 atm) for 30 min. It was demonstrated that the effects of oxygen and nitrogen annealing on the properties of undoped ZnO films were reversible due to the creation and annihilation of extrinsic trap states of antisite oxygen O_{Zn} and oxygen vacancies V_O, which result from the chemisorption and desorption of oxygen, respectively. Regarding the etching effects of H_2, Myong et al. [215, 216] investigated the effect of low-temperature annealing under H_2 atmosphere on polycrystalline ZnO thin films prepared by the photoassisted MOCVD technique. The H_2 post-treatment increased the surface roughness; very small and weak grains growing on the surface were etched by activated hydrogen radicals. This roughness can be of interest in some applications, as invoked by the authors. Plasma-etching processes have been proposed as paths to modify the morphology and physical properties, as reported by Wang et al. [300], who demonstrated that, under an optimized hydrogen and methane mixed gas plasma post-treatment process, it was possible to obtain a surface texture constituted of pyramids with enlarged feature size and shrunken feature size, which can improve the light-trapping capability in both short- and long-wavelength regions. Likewise, H_2 plasma postgrowth treatment of LP-MOCVD ZnO films was proposed by Ding et al. [301] as a way to improve electrical properties without altering their transparency in the visible range, which can be attractive in TCO applications.

As noted earlier, the substrate surface state impinges directly on the quality and properties of the overlying ZnO, so changes in the preferred orientation by chemical etching on the substrate can be induced, as done by Cui et al. [302] working on the growth of ZnO films on GaAs(001) substrates. The GaAs surfaces were etched in a solution of H_2SO_4:H_2O_2:H_2O (5:1:1) at 70 °C prior to growth. The etching times were 30, 60, 90, and 120 s. In the XRD spectra, only strong ZnO (0002) diffraction peaks were observed when the etching times were 30 and 60 s, while ZnO $(10\bar{1}0)$ diffraction peaks were observed when the etching times were 90 and 120 s. X-ray photoelectroscopy shows that the variations in the preferred orientation of ZnO films take place because an As-rich layer is formed in the process of chemical etching. This result shows not only the opportunity of modifying the preferred orientation acting on the substrate but also the significance of the substrate interface reactivity on the properties of the overlying film. Another example was reported by Hongsingthong et al. [303], who used reactive ion etching (RIE) treatment to modify the soda-lime substrate surface morphology before the deposition of ZnO films. On increasing the glass-etching pressure, significant increases in the RMS roughness and the haze ratio of the films was induced. Different RIE etching conditions, particularly rf power and process duration, have also been used in order to obtain different surface modifications on acrylic polymer substrates [304].

11.5.2 Crystal Growth of ZnO Nanostructures

Prompted by the huge interest in nanoscience, in a general form, and in nanotechnology in particular, the MOCVD technique has been involved in obtaining nanostructures of ZnO and related materials with different dimensionalities and morphologies. As is well demonstrated, the physical properties of nanostructures are dependent on size, orientation, and morphology, opening up new opportunities in technological and academic fields [305–307]. These nanostructures can be grown with high chemical purity and low defect concentration, and they offer the possibility to be used as building blocks for electronics and photonics, as well as for life-science applications [308]. As commented by Kim et al. [309], ZnO nanostructures that are self-assembled or induced by patterning, such as nanodots, nanotips, nanowires, nanorods, nanobelts, nanowalls, and nanotubes, have attracted a great deal of attention due to their practical applications, as well as for fundamental academic research [310–313].

Interesting reviews of the state-of-the-art at the time can be found, among others, in Zhang et al. [313] about 1D ZnO nanostructures and their application in the field of solar cells, Gomez et al. [311] on the growth techniques and the applications of nanostructures, Panda et al. [312] on the synthesis of 1D nanostructures and optical properties, Willander et al. [314] on ZnO nanorod-based photonic devices, and Yi et al. [308] on ZnO nanorods/nanowires.

11.5.2.1 Significance of the Substrate on the Growth of ZnO Nanostructures

Bearing in mind that the physical properties of nanomaterials are dependent on orientation and morphology, the role played by the substrate has been given particular attention. Regarding the substrates, the well-known ones for the growth of layers have been considered in the growth of nanostructures by using the MOCVD technology. Thus, the growth of ZnO nanostructures on crystalline substrates such as sapphire [196, 310, 315–317] and silicon [43, 120, 318, 319] as well as amorphous substrates like glasses [199, 208, 320, 321] has been frequently studied. However, the large literature on nanostructures grown by MOCVD has analyzed other less-current substrates too, such as 4H-SiC [322], $SrTiO_3$ [157], Ti/Al_2O_3 [147, 323], AlN/Si [324], AlN/Al_2O_3 [325], Pt patterned SiO_2/Si [326], Ni-W alloy [327], $ZrO_2(Y_2O_3)$ and $MgAl_2O_4$ [328], graphene [329], γ-$LiAlO_2$(100) [330], F:SnO_2/glass [38], ITO/glass [331], GaN/Al_2O_3 and GaN/AlN/Si(111)[332], GaAs [333], and even polymeric substrates [334].

Well-matched substrates like ZnO, GaN/Al_2O_3(0001), and ZnO buffer layers on different substrates can favor the growth of aligned nanostructures as nanocolumns [335], nanorods [316], or nanoneedles [336] under defined experimental conditions. On the other hand, in using buffer layers, the alignment of nanostructures can be controlled by the crystallographic alignment of the predeposited ZnO buffer [195, 337]. In contrast, less-matched substrates like Al_2O_3 and silicon are prone to form a continuous interfacial layer of a certain critical thickness due to the significant strain developed in the initial stage of growth (a detailed study of the interfacial layer can be found in [338]). The nanostructures will start to grow on the top of the interfacial layer as analyzed on Si (111) and Si(001) in [120, 317], or in the case of Al_2O_3 substrates as studied by Perillat-Merceroz et al. [339].

The substrate orientation has shown its influence on the morphology of nanostructures, as reported by Kar et al. [307], work where zinc oxide nanostructures were grown on a-, c-, and r-oriented sapphire. The shape of nanostructures was greatly influenced by the underlying sapphire substrate. Vertical-aligned nanowires were observed on a- and c-sapphire planes, whereas nanopencils were grown on r-plane sapphire. The effects of the polarity of the substrate face on which growth takes place have been analyzed by Perillat-Merceroz et al. [339], who investigated the growth of ZnO nanowires grown on O-polar ZnO and on sapphire substrates. The nanostructures mainly consisted of pyramids with nanowires sitting on top. It was shown by convergent beam electron diffraction (CBED) that, whatever the substrate, nanowires and pyramids had opposite polarities: nanowires were Zn-polar and pyramids O-polar. Thus, for growth on O-polar ZnO, nanowire nucleation implies the creation of an inverted domain. For growth on sapphire substrates, the nucleation of Zn-polar domains occurred either on the O-polar pyramids or directly on sapphire.

Scalisi et al. [43] report on the growth of ZnO nanostructures on different substrates ($SrTiO_3$(100), Si(100), and Al_2O_3 (0001)). It was demonstrated that both the morphology and crystalline structure of the obtained ZnO nanostructured samples strongly depend upon the crystalline mismatch with the substrate, as similarly found by Ternon et al. [208] by using other substrates like glass (alumina borosilicate glass), (100) silicon, SnO_2, and (001) 6H-SiC.

The growth of self-arranged periodic ZnO nanostructures on sapphire significantly depends on the initial surface structure of the substrate and growth kinetics. Hence, a faceted surface of annealed sapphire can provide a natural template for the epitaxial growth of self-organized periodic nanostructures [340]. In addition, nanopatterned substrates can be very useful to manipulate the self-assembled mechanism in growing nanostructures, e.g. in the growth of nanodots [309], nanotips [148], or ultrafine ZnO nanowires [341].

The correlation between physical properties and morphology has been the basis of several works using different substrates, etching processes, thermal treatments, and growth parameters. A different option, proposed by Kim et al. [198] for a morphology-controlled selective growth, is a microheating method, which provided controlled local heating. ZnO nanostructure morphology depends on the local growth temperature, so various nanostructure morphologies were obtained selectively at specific positions on the glass substrates.

On the other hand, in order to induce vapor-liquid-solid (VLS) processes and consequently to have an additional control on the growth of nanostructures, the use of gold as catalyst has been proposed, among others in [342], where details can be found.

11.5.2.2 Significance of Growth Conditions and Parameters on Structural and Morphological Properties of ZnO Nanostructures

In the growth of layers, inducing a 2D growth is the goal. For the growth of nanostructures, the point of view should be changed in order to induce a 3D growth mode. As already discussed, the control of the density of nucleation and the ratio between vertical and lateral growth rate are crucial for the control of morphology. A key parameter will be the growth temperature. Effects of the precursor partial pressures on the nucleation density are also expected. In addition, carrier gas flow and reactor pressure can impinge on the nucleation due to reactions, if any, with precursors, etching on the nanostructures, and/or effects on the hydrodynamics.

As regards the control of the lateral growth, the process will occur by propagation of steps (classical step-flow epitaxy) if the surface diffusion dominates and it is fast enough compared to the arrival of atoms. Conversely, if the surface mobility is low and the on-axis growth rate is sufficiently fast, the atoms arriving on the top of the nuclei will stick there, while desorption will be important between them as a result of the low bonding between the substrate and the arriving species, as analyzed by Briot (Chapter 3 in [237]). In order to grow individual nanostructures, low nucleation as well as low lateral growth rate are needed (just the opposite to promote the growth of smooth layers). In order to act on nucleation density and lateral growth rate, a balance of the growth conditions will be necessary. In addition, selective growth can be induced by using buffer layers or patterned substrates or even particular growth protocols.

11.5.2.2.1 Some Effects Regarding the Reactor Pressure

Zhang et al. have shown the effects of reactor pressure to activate different growth modes in several papers [263, 343, 344] working on the growth of 1D nanostructures. A nanostructure can be regarded as a small-volume crystal. The crystal shape depends on the surface energy. If an equilibrium process is assumed at each pressure, the condition for equilibrium is given by the minimum of the Helmholtz free energy of the system, which results in [343, 345]:

$$P_c - P_v = \sum_i \delta(\sigma_i S_i)/\delta V_c$$

where P_c is the inner pressure of the crystal phase, P_v the reactor pressure, σ_i and S_i are the density of surface energy and the area of the surface i, respectively, V_c is the volume of the crystal, and *i* indicates different surfaces for a small crystal. As can be seen, when the reactor pressure is decreased, the total surface energy should be increased (assuming V_c is constant). Thus, the equation indicates that lower chamber pressure facilitates a higher surface-to-volume ratio, as Zhang et al. observed. It was demonstrated that it is possible to obtain ZnO nanorods, nanotubes, or nanowalls on sapphire (0001) substrates by adjusting the growth pressure. The appearance of different nanostructures was due to different growth modes, and the variation of nanostructure shape with change in pressure was explained in terms of surface-to-volume ratio. In order to study the initial nucleation, growths under different pressures were performed [343] with changes in the growth time. The initial nucleation depended on both the growth condition (substrate temperature, reactor pressure, etc.) and the film–substrate lattice mismatch, and it was similar for different nanostructures. Not only the morphology, but also the density, is pressure dependent, as studied by Zhang et al. [263], who found a significant decrease of the nanotubes density as the pressure increased in the range 0.3–3 Torr. Nevertheless, the diameters of the nanotubes were almost independent of the growth pressure.

Effects on the morphology and/or structural properties with respect to the reactor pressure have been reported too, in [33, 192, 315] among others.

11.5.2.2.2 Some Effects of Using a ZnO Buffer Layer in the Growth of ZnO Nanostructures

The use of a buffer layer has been invoked not only to avoid direct growth on the substrate (see Section 11.5.2.1) and to act on the alignment of the nanostructures, but also as a way to control the density of nanorods on r-sapphire [346]. Kim et al. [347] reported on the analysis of the effect of the buffer-layer thickness on the microstructural and optical properties of ZnO nanorods grown on Si substrates. Under their experimental growth conditions, the increase of the thickness of the buffer layer significantly enhanced the vertical alignment and density of the nanorods. Nevertheless, thick buffer layers induced a rugged morphology with a pyramidal shape that promoted the growth of inclined nanorods. The work of Mohanta et al. [320] impinges on the effect of the thickness of the buffer layer on the aspect ratio and alignment of the ZnO nanorods. By using sapphire as substrate, Wu et al. [348] demonstrated that the density of vertically aligned nanorods is governed by the morphology and thickness of the ZnO buffer layer. That is to say, the ZnO buffer layer can be used as a nucleation template to control the growth alignment and density of the nanostructures.

11.5.2.2.3 Some Effects Related to the Growth Temperature

Growth temperature can act not only on the growth rate but also on the nucleation density. As discussed by Yang et al. [349], working on self-assembled ZnO quantum dots (QDs) grown on Si(111) substrates, the densities and sizes of ZnO QDs can be well controlled by adjusting the growth temperature. Higher temperature results in increasing the QDs sizes and decreasing the QDs density. By contrast, when the growth temperature is low, the ZnO QDs are easily deposited with a high density.

The selective process of nucleation, induced by the growth temperature, can be understood if we admit that when the growth temperature is higher, zinc atoms become more diffusive on the substrate surface. Zinc atoms can select sites having smaller lattice mismatch. Thus, high-quality growth with a lower density of misfit dislocation can be achieved [350]. Similarly, Zhou et al. [351] reported on the variation of densities and sizes of ZnO nanoislands grown on c-sapphire with respect to the growth temperature, showing that the island densities decrease markedly on increasing this growth parameter. Adatom diffusion length usually increases with the growth temperature, which led to the increase of island size [352]. Other authors, such as Kim et al. [156], by growing ZnO islands on SiO_2, have reported that the islands' width and height decrease and the density increases with increased growth temperature. The difference in the chosen substrate and other growth parameters can be determinant for the controversial result between these two papers ([156, 351]), but they serve to show that the growth results come from a convolution of the overall growth experiment including substrate, technique, growth conditions, etc. The effects of the growth temperature on the density and dimensions of nanotubes grown on sapphire (0001) have been also demonstrated, e.g. [344], where higher growth temperature and lower reactor pressure facilitated a high-density formation of ZnO nanotubes.

Growth temperature has also a significant effect on morphology, as studied by Park et al. [353], who reported on morphological changes in the growth of ZnO on Al_2O_3 (0001) substrates. The morphology of ZnO changes with the growth temperature (T_g): at low growth temperatures, $T_g \leq 200$ °C, no significant ZnO was formed. At temperatures in the interval 200 °C $< T_g \leq 260$ °C, columnar-grained textured ZnO films having a large number of crystalline defects were present. At higher growth temperatures, in the range 260 °C $< T_g \leq 320$ °C, arrays of vertically well-aligned ZnO nanorods were grown; the nanorods were epitaxially aligned with a 30° rotation of the ZnO basal planes with respect to Al_2O_3 ones. At further higher growth temperatures of 320 °C $< T_g \leq 380$ °C, ZnO nanoneedles were grown. However, at $T_g > 380$ °C, ZnO nanowires started to grow on top of a continuous ZnO layer. Nanoneedles and nanowires were not only well aligned both in the out-of- and inplane directions, but also showed a very low density of crystalline defects.

The same conclusion was pointed out by Park et al. [318] when asserting that, although the reactant gas flow rates, the growth time, and the growth temperature affect the diameter and the aspect ratio of the ZnO nanorods, the nanorod size can be controlled by changing the growth temperature. Other authors have worked on the same subject, and the influence of the growth temperature on the nanostructure selection has been demonstrated, such as e.g. in Kharanovskyy et al. [319] by using HP-CVD at atmospheric pressure in the temperature range 200–500 °C. In their case, a regular transformation of the material from conventional polycrystalline layers to hierarchically arranged sheaves of ZnO nanowires was observed. The growth evolution can be explained in terms of the difference in the free energy and growth kinetics between different ZnO crystal planes. Other examples of inducing different morphologies as a function of the growth temperature can be found in [33, 44, 321, 328, 342, 354].

11.5.2.2.4 Some Effects Regarding the Carrier Gas

The selection of the carrier gas as well as its flow rate can have significant effects on the growth of nanostructures, as we can see in the following examples. Thus, differences in the behavior of nitrogen and helium as carrier gases have been reported by Sallet et al. [355], who found that well-defined arrays of 1D structures were difficult to achieve in the temperature range 640–780 °C, by using DMZn-TEN and N_2O as precursors and N_2 as carrier gas. By contrast, ZnO nanorods and nanotubes were obtained under the same growth conditions when nitrogen was replaced by helium. Likewise, Kim et al. [356], in growing ZnO nanorods, found that an increase in the Ar carrier flow rate through a separate gas line promoted growth along the vertical direction, resulting in the formation of ZnO nanorods with an increased aspect ratio at relatively low temperature. The enhanced aspect ratio of nanorods was attributed to the increase in adsorption efficiency and the high directionality of the deposited atoms through the enhanced transfer velocity of the reactive gas.

11.5.2.2.5 Some Effects Regarding the Precursor Ratio

Changes in morphology can be attained by changing the precursor ratio, as demonstrated by Park et al. [335], who reported that thick nanocolumns grew at a high O/Zn precursor ratio, and thin nanocolumns grew at a lower O/Zn precursor ratio. Moreover, effects on alignment can be observed, as in [357], where at a high O/Zn precursor ratio, well-aligned thick ZnO nanorod arrays were produced. In the same context, Kim et al. [175] reported that, depending on MOCVD-growth conditions, nanowalls with extremely small wall thicknesses, below 10 nm, were formed into nanowalls with a thickness of about 20 nm, resulting in the formation of two-dimensional nanowall networks. The density of the nanowalls increased and their size decreased significantly by increasing the DEZn flow rate (higher II/VI ratio). Similar effects have been reported for nanorods [358] and nanowires [359]. Thus, the density of ZnO nanowires increased with zinc concentration at low concentration levels and decreased with zinc concentration at a high concentration level. The growth process could be divided into three stages, as controlled by growth kinetics. The first stage was characterized by lateral growth, the second was dominated by axial growth, and the third was described as proportional growth.

Induced transition from nanostructures to films due to changes in precursor ratio has been demonstrated, e.g. by Jeong et al. [360] and Ogata [228]. Jeong et al. [360] reported that as the zinc precursor (DEZn in their case) flow rate increases, the morphology evolves from nanowires to films at a growth temperature of 500 °C, because supersaturation rises, promoting 2D growth. However, at a growth temperature of 360 °C, morphology of the deposited ZnO was not affected by the precursor ratio. It was inferred that continued growth of nanowires was driven by the surface diffusion of ZnO, and a low level of supersaturation was required to prevent 2D growth.

Induced morphology transitions (nanorods–nanowalls and nanorods–nanotubes–layer) due to changes in the VI/II precursor ratio have also been studied by, among others, Thiandoume et al. [283], Song et al. [361], and Montenegro et al. [362]. Wu et al. [348] studied the effect on the size, crystal, and optical quality of nanorods by controlling the precursor zinc flow rate (DEZn in their case), whereas Behrends et al. [192] studied the effects on aspect ratio in nanopillars. Likewise, regarding the growth of nanowalls, Wu et al. [354] found that the wall width and network size of ZnO nanowall-network structures depended on DEZn flow rate.

Additionally, it is worth noting that different morphologies under the same growth parameters and substrate can be induced by using different precursors, as reported by Sallet et al. [355].

11.5.3 Crystal Growth of ZnO-Related Materials

The growth of ZnO-based alloys, in particular with magnesium and cadmium, has been undertaken as a means of bandgap engineering and to enlarge potential applications of the ZnO family. A first problem appears due to the different crystalline stable structure between the wurtzite ZnO and the rocksalt of CdO or MgO. This points to a limited range of solubility in maintaining a single phase.

Experimental work to obtain these alloys could take, as a starting point, the growth conditions of the binary parent matrix from which a third element could be incorporated. Nevertheless, the vapor pressure and the pyrolysis behavior of the usual precursors for cadmium, zinc, and magnesium, as well as the adatom and desorption phenomena, are different. Thus, the material-science community working on MOCVD methodologies has been given an attractive challenge.

11.5.3.1 Crystal Growth of $Zn_{1-x}Cd_xO$

Conventional MOCVD at low or atmospheric pressure has been used for the growth of $Zn_{1-x}Cd_xO$. The usual precursors for zinc and cadmium have been DEZn and DMCd, respectively. As oxidizer, tert-butanol (at low growth temperatures) and N_2O (at higher ones) are the main choice in conventional MOCVD experiments in which the most frequently used substrates have been sapphire with different cut planes: r-plane [194, 363, 364], a-plane [363, 364], c plane [365, 366], and ZnO [365].

Despite the similarity in the radii of Zn^{2+} and Cd^{2+} [223], the literature shows that relatively low values of Cd content in the wurtzite structure can be attained by using conventional MOCVD at low or atmospheric pressure, without the presence of mixed phases. Mixed phases appear as the Cd content increases, as reported in [367]. In fact, the thermodynamic solubility limit of Cd in ZnO is 2% [367], but different growth techniques can extend this range. The growth by AP-MOCVD with DEZn, DMCd, and tert-butanol as precursors of single phase a-oriented $Zn_{1-x}Cd_xO$ ($x \leq 0.085$) on a thick ZnO buffer layer on r-sapphire was reported by Zuñiga-Perez et al. [144], work where the structural properties were analyzed in depth. Higher Cd concentrations can be attained by using RPE-MOCVD, as reported by Shigemory et al. [368], who demonstrated values up to 69.7% of cadmium in c-oriented $Zn_{1-x}Cd_xO$. In these experiments, DEZn, DMCd, and plasma O_2 were the precursors and a-sapphire the substrate.

The review of $Zn_{1-x}Cd_xO$ by Zuñiga-Perez [369] accurately presents the state of the art on this subject regarding growth, physical properties, and related devices.

Until now, there is a reduced presence of published studies on the growth by MOCVD of the rocksalt $Cd_{1-x}Zn_xO$ alloy, on which Zhang et al. [370] report about the evolution of the preferred growth orientation in terms of the surface/strain energy balance, as a function of Zn content and film thickness.

11.5.3.2 Crystal Growth of $Zn_{1-x}Mg_xO$

Alloys of $Zn_{1-x}Mg_xO$ have been synthesized by MOCVD as layers and as nanostructures, and also forming part of heterostructures. The usual zinc precursor has been DEZn and Cp_2Mg, as well as MCp_2Mg or $EtCp_2Mg$, for magnesium, and O_2 and N_2O as oxidizers. The significant difference in the vapor pressures and pyrolysis temperature between zinc and magnesium precursors has been one of the main difficulties to be solved in order to promote the growth of the ternary alloy without losing crystalline quality and feasibility of the process.

The lattice constant could be expected not to change too much with Mg substitution on the Zn site in wurtzite ZnO crystals, because the ionic radii of Zn^{2+} and Mg^{2+} are similar [371]. However, the results [371] show that the c-lattice constant reduces linearly with the Mg composition ratio. Thus, strain-distorted lattice and limits in the solubility are expected.

11.5.3.2.1 Crystal Growth of $Zn_{1-x}Mg_xO$ Layers

Conventional LP-MOCVD and RPE-MOCVD are the most-applied techniques to obtain good layers of the alloy; sapphire with the c, a, and r orientations have been usually chosen to induce c- and a-oriented $Zn_{1-x}Mg_xO$ layers. Depending on the growth technique and conditions, different limits of solubility for the wurtzite single phase have been obtained. In the overall range of contents, three regions can be distinguished with a wurtzite and rocksalt crystalline structure in the zinc- and magnesium-rich content regions and, between them, a mixed zone in which both phases coexist, as reported among others by Zhang et al. [63].

The role of a ZnO buffer layer (both thin or thick) has proved to be crucial in obtaining good crystalline layers on sapphire in order to avoid interdiffusion and also to maintain the wurtzite structure, since the presence of Al_2O_3 could favor the formation of cubic rocksalt-type ($Mg_{1-x}Zn_xO$), as Mg atoms prefer the octahedral sites. A detailed study of the epitaxy of this alloy on r-sapphire has been carried out by Mutukumar et al. [115], who report that the epitaxial relationships are identical to those for ZnO films grown on r-plane sapphire for magnesium contents up to 33%. The direct growth on sapphire without a buffer layer generally produces a polycrystalline structure.

Most studies have focused on the zinc-rich part or the alloy (wurtzite structure), and the Mg-rich region (cubic structure) grown by MOCVD has been given less attention. Some results on this cubic alloy can be found in [372] on MgO substrates.

As already noted, different techniques such as conventional MOCVD [373, 374], pulse-injection MOCVD [375], and plasma-assisted-MOCVD [64] can be useful in obtaining the alloy. On the other hand, the overall criteria and effects of different parameters, already analyzed in relation to the growth of ZnO, can be of application here, as we can see in detail in, among others, the following references: effects of the substrate [373] and substrate orientation [374] on the growth of $Zn_{1-x}Mg_xO$; effects of a ZnO buffer layer [115] or a MgO buffer layer [376]; effects of growth temperature [377]; effects of partial pressure of precursors [378]; effects of using different precursors [73]; effects of magnesium composition on physical properties [379]; and effects of postgrowth annealing [380].

As regards substrates, ZnO templates [381], β-$LiGaO_2$ [382], fused silica [383], sapphire [374], and MgO [372] have been used for the growth of layers of the alloy.

11.5.3.2.2 Crystal Growth of $Zn_{1-x}Mg_xO$ Nanostructures

Most efforts relating to $Zn_{1-x}Mg_xO$ nanostructures have focused on obtaining heterostructures rather than on nanostructures arrays or individual nanosize elements. Silicon (001) and a-sapphire have shown to be capable of inducing c-oriented well-aligned nanorods/nanopillars. HP-CVD by using acetylacetonates of Zn, Mg, and O_2 has demonstrated its possibilities to obtain $Zn_{1-x}Mg_xO$ nanorods on Si (001) at relatively low temperature (500 °C) [27], while higher temperatures (>700 °C) were used in obtaining nanopillars on sapphire by LP-MOCVD with N_2O, DEZn, and $MeCp_2$ as precursors [157]. Different morphologies such as nanowires [384, 385], nanorods [386], nanowalls [384, 385], QDs [387], and hexagonal-nanotowers/films [388] have been reported for this alloy. Morphology transitions from nanowires to nanowalls as a function of the growth temperature and Zn/Mg flux ratio were demonstrated by Kim et al. [384]. Likewise, Shi et al. [374] studied the effects of different orientations of sapphire substrates, and Lee et al. [385] analyzed the effects of the growth temperature on the morphologies of $Zn_{1-x}Mg_xO$ grown on GaAs, to name some results in the literature.

11.5.3.3 Crystal Growth of Transition-Metal ZnO-Based Materials

In the frame of enlarging ZnO potential applications, several ZnO alloys with different materials have been obtained and analyzed. Only focusing on the MOCVD technology, some examples of these alloys (in some cases with only a few percent of foreign element) are: Er,Yb:ZnO thin films [389]; Er:ZnO [390]; Cr:ZnO [391]; and V:ZnO [392]. Likewise, $Zn_{1-x}Mn_xO$ [393], $Zn_{1-x}Co_xO$ [51, 92], $Zn_{1-x}Cu_xO$ [394], and $Zn_{1-x}Ni_xO$ [218] have been the subject of interesting studies.

11.5.4 Doping of ZnO and Related Materials

Doping of ZnO and related materials can be a hard task due to the significant role that complexes play in this, also complex chemistry. The study in depth of physical properties is beyond the scope of this chapter, and thus we will only focus on the potential doping induced in the MOCVD growth process, by analyzing separately some several potential donors or acceptors.

Undoped ZnO is an n-type material due to the effects of oxygen vacancies and zinc interstitials. On the other hand, the role of hydrogen as a shallow donor has been largely recognized [226]. Thus, the presence of hydrogen and carbon as products of pyrolysis, or, in the case of H_2, as carrier gas or as atmosphere for annealing processes when used, must be considered in analyzing the results of MOCVD experiments. Some studies of intentional incorporation of hydrogen by postdeposition processes exposing the samples to hydrogen from Hg-sensitized photodecomposition of H_2 gas [216] or from H_2 plasma [395] demonstrate the significance of H_2 on the electrical properties of ZnO. The effects of carbon, far from being unimportant, can be fundamental due to the, in some cases, easy formation of complexes, as reported by Nickel et al. [396] and Li et al. [397] with respect to nitrogen doping, as we will discuss in Section 11.5.4.2.

11.5.4.1 Some Elements Acting as Potential Donors

Elements from Groups III and VII of the periodic table are potential agents as donors in ZnO. In comparison with Group VII elements (Cl, Br), Group III elements (Al, Ga, and In) can be more suitable for n-type doping of ZnO materials due to their lower vapor pressure [398], which allows better control for small incorporation without perturbing the growth of the main matrix. Among the dopants, Ga doping seems to be the most successful and promising due to a similar ionic and covalent radius, as well as covalent bond lengths with respect to Zn. These similarities should result in small ZnO lattice deformations even in the case of high Ga concentrations [96]. Also, Ga is less reactive and more resistive to oxidation, as compared to Al [399].

Ga doping of ZnO can be approached by using the same growth techniques already discussed. In addition, induced epitaxial stabilization by the substrate can help to expand the range of solid solutions [251]. As precursors, triisopropylgallium (TIGa) [96], TEGa [93], TMGa [400], and gallium acetylacetonate [97, 401] were shown to be useful.

Thi et al. [95] determined that total carbon incorporation into Ga:ZnO films prepared by TEGa was slightly lower than that of using TMGa source, indicating that the formation of stable C_2H_4 species during reaction when using TEGa can reduce the carbon incorporation into the Ga:ZnO films compared to the less-stable CH_3 species presents when using TMGa as gallium source.

In spite of the easy oxidation of aluminum, its ability to act as a doping agent has been demonstrated, for example, by using trimethylaluminum (TMAl) and AP-MOCVD at moderate temperatures (350 °C) and DEZn and THF as zinc and oxygen precursors, respectively [91], in order to obtain Ga:ZnO layers. Similarly, other authors have also reported on the doping effects of TMAl in films and nanostructures by using conventional MOCVD at atmospheric or low pressures [402, 403] or plasma-assisted MOCVD [404].

Tuan et al. [92] chose triethylaluminum over trimethylaluminum because TMAl would be prone to introduce carbon contamination due to the strong aluminum–carbon bond. Other precursors such as trimethyllamina-alanne (TMAA) (P(T): 1.5 Torr at 25 °C) [405], $Al(thd)_3$ [406], and $Al(acac)_3$ [407] have been also studied. Based on the use of Zn(tta)2·tmeda and $Al(acac)_3$ precursors, Fragala et al. [408] implemented a single multimetal (Al/Zn) liquid source from $Al(acac)_3$ solved in $Zn(tta)_2$ ·tmeda and thus forming a liquid source that evaporates with the required stoichiometric ratio with a constant rate of mass transport. The experimental setup consisted of a reduced-pressure, horizontal hot-wall MOCVD reactor equipped with a single sublimation zone maintained at a temperature of 170°C for the multimetal source. Ar was used as carrier gas, while O_2 was the oxidizer gas.

Indium, as a member of Group III, has been also studied as a doping agent for ZnO. TMIn has been the most selected precursor [98], together with DEZn and O_2, or by using DMZn and N_2O [409]. With a lesser presence, $In(thd)_3$ [401] can be another alternative as indium precursor.

Boron doping, by using diluted B_2H_6, has been applied to control the conductivity of ZnO mainly in the area of solar cells, as studied amongst others in [88, 89, 410, 411].

11.5.4.2 Some Elements Acting as Potential Acceptors

p-Doping of ZnO could be achieved, in principle, by the incorporation of foreign atoms either from the Group I elements, such as Li, Na, and K, for the Zn sites or from the Group V elements, like N, P, As, and Sb, for the O sites. In fact, Group I elements as p-type dopants in ZnO should be better than Group V elements. However, the doping efficiency of Group I elements is generally limited by the formation of compensating interstitials [412].

As pointed out by Kun et al. [413], several exciting results encouraged work on the p-doping of ZnO, especially the improved performance of light-emitting devices. Some findings like light-emitting diodes (LEDs) based on nitrogen-doped $Zn_{1-x}Mg_xO$/ZnO heterojunctions reported by

Nakahara et al. [414] and the electrically pumped waveguide lasing observed in a p-nanowire-n-film ZnO homojunction [415], to name but a few, seem to point to evidence of true p-type ZnO. Remarkable progress has been achieved, both theoretically and experimentally, and the chemical configuration and formation mechanism of the acceptors in ZnO have been better understood in recent years [413]. Nevertheless, the configuration of the real acceptor remains an object of study, and the experimental challenge is open. In Fan et al. [412], a theory of p-type doping in ZnO is developed, and growth techniques for p-type ZnO are reported.

Here, an overview of some experimental work to achieve the incorporation of nominal acceptor levels will be presented.

A major obstacle in achieving p-type material is generally supposed to be the self-compensation induced by hydrogen and native defects such as Zn_i or/and V_O. [416–418]. Some of the problems associated with the p-doping of ZnO have been reviewed by Maksimov [419].

Focusing once again only on MOCVD experiments, efforts to make p-ZnO thin films have involved different dopants (e.g. nitrogen, phosphorus, arsenic, antimony, etc.). For nitrogen, codoping with aluminum, gallium, indium, or tellurium has been reported with varied degrees of success. In fact, the window of the deposition parameters in achieving p-type material seems to be narrow [106, 161], and postdeposition annealing has been proposed, in some cases, to induce p-type conduction.

In spite of theoretical works like that of Zhang et al. [417] that claim the impossibility to obtain p-doping via native defects (O_i, V_{Zn}), some experimental work has been focused in obtaining intrinsic p-doping via the control of the oxygen partial pressure during growth [136, 222]. Intrinsic p-doping has been also reported by Tan et al. [107] and ascribed to intrinsic acceptor-like defects, V_{Zn}.

Moving to doping by incorporation of foreign atoms in the MOCVD growth, copper acceptors have been invoked as a way to compensate the electron concentration, as reported by Kaufmann et al. [34] by using cyclopentadienyl (tributy1phosphine)-copper (I) at AP-MOCVD. Other authors [420] have proposed the use of $Cu(tmhd)_2$ as a precursor, and the effects on the structural and optical properties have been analyzed [420].

Phosphorus has also been considered as a dopant agent [421]. The presence of Zn–P and O–P complexes was detected, as well as the amphoteric nature of phosphorus in ZnO. Limpijumnong et al. [422], argued that dopants such as P, As, and Sb prefer to occupy Zn sites rather than O sites, and the resulting complex is an acceptor with both low formation energy and low ionization energy. In order to induce phosphorus doping, phosphorous pentoxide (P_2O_5) powder can be evaporated to act as P agent [423]. A phosphorus-related acceptor, probably a P_{Zn}–$2V_{Zn}$ complex, with an energy level of 163 meV was identified from free-to-neutral-acceptor transitions [423].

Attention should be paid to the fact that the MOCVD technique involves a variety of sources containing carbon and hydrogen, which may introduce a high concentration of nitrogen-related defects such as NO-H, (CN)O, and HCN. Nevertheless, nitrogen has been regarded as the most promising acceptor dopant because of its low ionization energy, suitable ionic radius, ease in handling, low toxicity, and abundance. Consequently, much effort has been devoted to nitrogen-related doping and codoping.

Although a variety of nitrogen sources and growth techniques have resulted in reports about the p-type nitrogen-doped ZnO, the creation of p-type N:ZnO is not routine. Several nitrogen agents can be envisaged as possible candidates, such as N_2, N_2O, NO_2, NO, and even organic compounds. We will start with the latter.

Organic compounds like monomethylhydrazine (MMHy) [101], diallylamine [37, 396], tert-butylamine ($C_4H_{11}N$) [30], and unsymmetrical-dimethyl-hydrazine (UDMHy) [103] have been proposed as candidates to induce nitrogen doping in ZnO. By using diallylamine, Nickel et al. [396] studied from Raman backscattering and Fourier transform infrared measurements the vibrational spectroscopy of undoped and nitrogen-doped ZnO samples, and the presence of carbon sp^2 clusters was detected. In nitrogen-doped ZnO, local vibrational modes caused by NO, NNO, Zn–H, CO_2, and C≡N complexes were observed. It was concluded that the formation of these complexes reduces the doping efficiency of nitrogen.

Ammonia (NH_3) has also been proposed as an option to nitrogen doping [247]. In Wang et al. [102], p-type conductivity was reported, but as the NH_3 flux increased, some hydrogen binding to nitrogen was introduced into the ZnO films, which promoted changes from p- to n-type conductivity in the ZnO sample at high NH_3 flux. As Dadgar et al. [103] point out, high ammonia flows are likely to introduce not only nitrogen but also lattice defects. On the other hand, at elevated growth temperatures, hydrogen has a strong etching effect on the ZnO surface. Consequently, hydrogen from ammonia decomposition might degrade the lattice, causing the brownish color of the samples reported by Dadgar et al. [103].

Das et al. [424] determined that NH_3 concentration creates a few donor levels (due to structural defects) as well as acceptor levels (due to nitrogen doping). With the increase of NH_3 concentration during deposition, the oxygen substitution by nitrogen increased, and more N_O defects appeared in the PL spectrum. The effect of oxygen vacancies was compensated by the effect of generated defect levels due to nitrogen doping, which caused the p-type nature of the samples. Photoassisted MOCVD has been also used with NH_3, as done by Li et al. [104], who reported on the conductivity transition from n-type to p-type via the introduction of in situ light irradiation during the growth process.

By using different nitrogen-doping agents, Pan et al. [425] concluded that under their experimental conditions, although p-type ZnO epilayers can be obtained from both NH_3 and N_2O, N_2O is more suitable for MOCVD growth of p-type ZnO.

The p-type conductivity induced by N_2O has been justified as due to impurities (N_O acceptors) formed during the doping process [107]. Sun et al. [108] studied nitrogen-doped ZnO films by using plasma-assisted MOCVD on insulating Si (111) substrates; N_2O plasma was used as nitrogen precursor of N:ZnO films, which was activated by a radio-frequency generator. The X-ray photoelectron spectroscopy (XPS) study revealed a competition between the acceptor N_O and the donor $(N_2)_O$ during doping, which influences the properties of ZnO films in correlation with the rf power. The comparison between NO_2 and N_2O as sources for nitrogen doping carried out by Egerton et al. [109] shows that, under their experimental conditions, using N_2O as precursor resulted in a lower atomic concentration of nitrogen compared to that using NO_2 as nitrogen source. However, this last precursor does not have much presence in the MOCVD literature.

Molecular nitrogen (N_2) has been reported to act as an acceptor dopant by using plasma-assisted MOCVD [222]. N_2 is characterized by neutral and nonreactive molecules, and the rf source should be fundamental in order to produce chemical species with high reactivity. Nevertheless, an in situ annealing process was necessary to activate the dopant [222].

Yan et al. [426] analyzed theoretically the problem of NO as nitrogen source for doping and predicted that NO gas could be a better nitrogen dopant source than N_2O or N_2. Their model indicates that the defect-formation energy of nitrogen on an oxygen site (N_O) from NO should be lower than that from N_2O. Calculation indicates that the formation energy of NO is negative under Zn-rich condition and increases linearly to positive values under O-rich conditions. Therefore, if the Zn-rich growth conditions are maintained during growth, forming the N_O defect from NO gas could not require adding energy (e.g. high temperature or plasma) [427]. Sustained by this prediction, several authors have worked on the NO doping effects, but the compensation/passivation role of hydrogen and carbon seems to limit the effectiveness of the doping [106, 397, 428].

As for the effects of hydrogen, Keyes et al. [429] provided evidence of hydrogen incorporation in nitrogen-doped ZnO through the observation of C–H, O–H, and N–H bonds. The observed O–H band is influenced by the presence of nitrogen, and it is predicted to contribute as a donor state to the material [430]. The observed N–H bond represents the passivation of a nitrogen acceptor [428]. In addition to the possibility of diatomic molecular donor complexes, these configurations are detrimental to the creation of p-type ZnO and help to explain the relatively low hole-carrier concentration that can occur with the incorporation of nitrogen. Carbon concentrations have been reported to increase greatly in nitrogen-doped samples, which suggests that the unintentionally doped carbon impurities prefer to form defect complexes with nitrogen [431]. Due to the strong bonding between hydrogen and carbon with nitrogen, the $(N_O–H)^0$ and $(NC)_O^{1+}$ defect complexes have quite low formation energies and can exist in large concentrations. The low activity of NO has also been reported by Dadgar et al. [103].

Lu et al. [432], by using PE-MOCVD, studied the effect of rf power in N:ZnO samples with NO as doping agent. It was demonstrated that as the rf power increases, the morphologies of the films became more compact, and the crystallinity of the films downgraded due to incorporation of more nitrogen atoms as the rf power increases.

By also using PE-MOCVD in the nitrogen-ZnO doping with NO plasma, Zeng et al. [433] demonstrated that the growth temperature was fundamental in optimizing the p-type conductivity of the nitrogen-doped ZnO thin films. Their results corroborate previous findings in the literature, which point to a decrease in the nitrogen concentration in ZnO with increasing growth temperature [434]. Defects such as $(N_2)_O$ or $(NC)_O$ form readily and can compensate the NO acceptor.

The activity of carbon has been additionally discussed in unintentional carbon-doped ZnO thin films grown by MOCVD through postgrowth annealing treatments [435]. MOCVD, as already noted, involves a variety of sources containing carbon and hydrogen, which may introduce a high concentration of related defects depending on the environment in which the process is carried out, a situation that should be considered in analyzing experimental results, in particular if doping effects must be explained.

Doping with arsenic can be another option to promote p-doping in ZnO. Dadgar et al. [105] report on the use of arsine as arsenic doping. The physical mechanisms for As-doped ZnO thin films can be explained as arsenic substitution for oxygen (As_O) or As substitution for Zn and As combined with two Zn vacancies ($As_{Zn}–2V_{Zn}$) [436]. These authors reported on the in situ annealing at different temperatures into two environments: Zn-rich, using diethylzinc (DEZn) as ambient gas, and O-rich, using water vapor (H_2O) as ambient gas, respectively. This should help to create As_O and $As_{Zn}–2V_{Zn}$, as sustained by the published results.

On the other hand, organic precursors for arsine like tert-butylarsine ($C_4H_{11}As$) can be also a source for As doping, as Ye et al. [437] studied in correlation with the arsenic effects on the growth kinetics. We should note also the arsenic thermal-induced diffusion from GaAs substrates or templates [200, 438] as a potential source for p-doping.

Zhao et al. [99] report on the growth and characterization of p-type Sb-doped ZnO thin films on c-plane sapphire substrates by using trimethylantimony as doping precursor together with DEZn and O_2; see also [100]. Another option as antimony source is triethylantimony (TESb), as proposed in [439]. Likewise, growth of Na-doped ZnO was achieved by using cyclopentadienylsodium (CpNa) (C_5H_5Na) as the sodium source in films [440] as well as in ZnO nanorods [441].

We would like to briefly discuss the codoping technique, which has been invoked in order to enhance acceptor incorporation by simultaneously doping with acceptors (e.g. nitrogen) and donors (e.g. Al, Ga, or In), as proposed by Yamamoto et al. [442]. Subsequently, much effort has been devoted to pursuing this approach (see e.g. [412] and references therein).

Examples of Ga-N codoped ZnO experimental MOCVD work can be found in [443], on the study of ZnO QDs by using DEZn, O_2, TEGa, and NO as precursors, or in ZnO films [94] by using dimethylhydrazine (DMHy), DEZn, N_2O, and TMGa and LP-MOCVD. Other options such as In-P codoped ZnO films on quartz substrates have been reported [444] by using LP-MOCVD and phosphine (PH_3) and TMIn together with DEZn and O_2 as precursors. Su et al. [445] studied Al-N codoping and found that the rf plasma was a key factor for the improvement of crystal quality. Finally, we will note that the influence of defect physics and the Te-N codoping mechanism for p-type conduction have been reviewed by Kun et al. [413], and examples of this Te-N codoping can be found in ref [110].

11.6 Conclusions

MOCVD (MOVPE) of ZnO and related materials has progressed considerably over the past three decades with high-quality epitaxy and polycrystalline films deposited onto a variety of different substrates, depending on the application. Doped ZnO is widely used as a TCO for photovoltaic solar cells and is increasingly produced by MOCVD where high-quality films are required. This chapter has described in some detail the range of precursors used and combinations that will reduce the prereaction of oxygen with the Zn organometallics. Despite considerable progress with the precursor chemistry, the process is very sensitive to the design of the reaction chamber and controlling the mixing of precursors. Energy-assisted growth, through either photoassistance or plasma-assistance, provides additional control over the reaction processes and can enable lower-temperature deposition of ZnO films. Kumar et al. [405] demonstrated extremely sharp photoluminescence from doped ZnO nanowires on sapphire substrates that were more indicative of very-high-quality bulk single-crystal ZnO. This illustrates how nanowires of ZnO can combine high-quality single-crystal properties with the flexibility of different substrates that would otherwise introduce significant strain in 2D layers.

Some of the exciting opportunities for MOCVD of ZnO in the future are likely to be through the formation of ZnO nanostructures that combine the advantages of the bulk properties of ZnO with functional nanostructured devices. The formation of ZnO nanowires for photocatalysis has also been demonstrated successfully by MOVPE and has the advantage over more commonly used TiO_2 in being a direct-bandgap semiconductor with a higher excitonic energy. This has proved to be effective in neutralizing combinations of harmful xenobiotics commonly found in surface water [446].

The potential for ZnO nanowires to overcome the limitations experienced with 2D films is demonstrated by the work of Shi et al. [447], where electroluminescence was achieved in core-shell nanowires with the structure ZnO/MgO/p-NiO. An advantage of ZnO nanowires is the ability to grow without the use of a catalyst that might otherwise contaminate the material and reduce luminescent yield. The small dimensions enable better accommodation of strain and lattice mismatch that would introduce large concentrations of defects on 2D layers. ZnO nanowires provide an excellent template for other materials to form functional nanowire devices. The combination of p-GaN layers with n-ZnO nanowires has successfully demonstrated blue and UV electroluminescence [448]. It is observed that nanowire arrays can be more easily formed in ZnO rather than in GaN, and these devices take advantage of the stable p-type doping of GaN with the high exciton energy of ZnO. The ZnO nanowires can be grown by a variety of different techniques: electrodeposition, hydrothermal, and pulsed laser deposition, as well as MOVPE. Increasing interest in MOVPE of the ZnO nanowires for light-emitting applications is coming from the ability to control the doping of the ZnO nanowires.

These emerging applications depend on the past three decades of research on MOCVD of ZnO that has developed a detailed knowledge of precursor chemistry and kinetic processes, as detailed in this chapter. This will encourage a wider exploration of the application of binary oxides and alloys of ZnO, CdO, and MgO for sensors, emitters, light management structures, photovoltaic solar cells, and piezoelectric energy harvesting.

Acknowledgments

One of us, VMS, would like to acknowledge the Ministry of Economy and Competitiveness (MINECO) (Spain) and the Generalitat Valenciana for financial support under projects TEC2014-60173 and TEC2017-85912-C2-2-R, and Prometeo II 2015/004, respectively.

References

1. Ghandhi, S.K., Field, R.J., and Shealy, J.R. (1980). Highly oriented zinc oxide films grown by the oxidation of diethylzinc. *Applied Physics Letters* 37 (5): 449–451.
2. Segura, A., Sánchez-Royo, J.F., García-Domene, B., and Almonacid, G. (2011). Current underestimation of the optical gap and Burstein-Moss shift in CdO thin films: A consequence of extended misuse of α^2-versus-hν plots. *Applied Physics Letters* 99 (15): 2011–2014.
3. Lamb, D. and Irvine, S.J.C. (2009). Near infrared transparent conducting cadmium oxide deposited by MOCVD. *Thin Solid Films*, 518 (4): 1222–1224.
4. Wang, L., Yang, Y., Ni, J. et al. (2005). Synthesis and characterization of low-melting, highly volatile magnesium MOCVD precursors and their implementation in MgO thin film growth. *Chemistry of Materials* 17 (23): 5697–5704.
5. Oleynik, N. (2007). MOVPE growth and characterisation of ZnO properties for optoelectronic. *Thesis dissertation*, Otto-von-Guericke-Universität Magdebur.
6. Stringfellow, G.B. (1999). *Organometallic Vapor-Phase Epitaxy: Theory and Practice*. Academic Press.
7. Fulem, M., Ruziicka, K., Ruziicka, V. et al. (2003). Vapor pressure of metal organic precursors. *Journal of Crystal Growth* 248, 99–107.
8. Kuniya, Y., Deguchi, Y., and Ichida, M. (1991). Physicochemical properties of dimethylzinc, dimethylcadmium and diethylzinc. *Applied Organometallic Chemistry* 5(4): 337–348.
9. Thiandoume, C., Sallet, V., Triboulet, R., and Gorochov, O. (2009). Decomposition kinetics of tertiarybutanol and diethylzinc used as precursor sources for the growth of ZnO. *Journal of Crystal Growth* 311 (5): 1411–1415.
10. Maejima, K. and Fujita, S. (2006). Chemical vapor reactions of ZnO growth by metal-organic vapor phase epitaxy. *Journal of Crystal Growth* 293 (2): 305–310.
11. Solanki, R. and Collins, G.J. (1983). Laser induced deposition of zinc oxide. *Applied Physics Letters* 42 (8): 662–663.
12. Wright, P.J., Griffiths, R.J.M., and Cockayne, B. (1984). The use of heterocyclic compounds in the organometallic chemical vapour deposition of epitaxial ZnS, ZnSe and ZnO. *Journal of Crystal Growth* 66: 26–34.
13. Oleynik, N., Adam, M., Krtschil, A. et al. (2003). Metalorganic chemical vapor phase deposition of ZnO with different O-precursors. *Journal of Crystal Growth* 248: 14–19.
14. Mullin, J.B., Irvine, S.J.C., and Ashen, D.J. (1981). Organometallic growth of II-VI compounds. *Journal of Crystal Growth* 55: 92–106.
15. Fujita, Y. (2000). Reaction mechanisms for precursors in photo-assisted metalorganic-vapor-phase epitaxy growth of ZnSe. *Journal of Crystal Growth* 221 (1–4): 382–387.
16. Ye, J., Gu, S., Zhu, S. et al. (2005). Comparative study of diethylzinc and dimethylzinc for the growth of ZnO. *Journal of Crystal Growth* 274(3–4): 489–494.
17. Rueter, M.A. and Vohs, J.M. (1992). The surface reactions of ethyl groups on Si(100) formed via dissociation of adsorbed diethylzinc. *Surface Science* 262 (1–2): 42–50.
18. Lam, H.-T., Venkateswaran, N., and Vohs, J.M. (1997). Reaction of dimethylzinc and diethylzinc on the As-rich GaAs(100)-c(4×4) surface. *Journal of Vacuum Science and Technology A* 15 (3): 1159–1162.
19. Maruyama, T. and Shionoya, J. (1992). Zinc oxide thin films prepared by chemical vapour deposition from zinc acetate. *Thin Solid Films* 11: 170–172.
20. Kobayashi, K., Matsubara, T., Matsushima, S., and Okada, G. (1993). Preparation of ZnO films for low pressure organometallic chemical vapor deposition. *Chemistry Letters*: 2133–2136.
21. Ogawa, M.F., Natsume, Y., Hirayama, T., and Sakata, H. (1990). Preparation and electrical properties of undoped zinc oxide films by CVD. *Journal of Materials Science Letters* 9 (11): 1351–1353.
22. Kashiwaba, Y., Katahira, F., Haga, K. et al. (2000). Hetero-epitaxial growth of ZnO thin films by atmospheric pressure CVD method. *Journal of Crystal Growth* 221 (1–4): 431–434.
23. Kashiwaba, Y., Haga, K., Watanabe, H. et al. (2002). Structures and photoluminescence properties of ZnO films epitaxially grown by atmospheric pressure MOCVD. *Physica status solidi (b)* 229 (2): 921–924.

24. Haga, K., Suzuki, T., Kashiwaba, Y. et al. (2003). High-quality ZnO films prepared on Si wafers by low-pressure MO-CVD. *Thin Solid Films* 433: 131–134.
25. McAdie, H.G. (1966). The pyrosynthesis of strontium zincate. *Journal of Inorganic and Nuclear Chemistry* 28: 2801–2809.
26. Yasuda, T. and Segawa, Y. (2004). Zinc oxide thin films synthesized by metal organic chemical reactions. *Physica status solidi (b)* 241 (3): 676–679.
27. Ku, C.H., Chiang, H.H., and Wu, J.J. (2005). Bandgap engineering of well-aligned $Zn_{1-x}Mg_xO$ nanorods grown by metalorganic chemical vapor deposition. *Chemical Physics Letters* 404 (1–3): 132–135.
28. Fujita, Y. and Nakai, R. (2004). Growth of ZnO films by MOVPE using diisopropylzinc and alcohols. *Journal of Crystal Growth* 272 (1–4): 795–799.
29. Nishimoto, N., Senthilkumar, O., Yamamae, T. et al. (2008). Growth of ZnO thin films by using MOCVD with a high-speed rotating disk reactor. *Journal of the Korean Physical Society* 53 (5): 2951–2954.
30. Nishimoto, N., Matsuo, Y., and Fujita, Y. (2010). Effect of annealing on nitrogen-doped ZnO grown by MOVPE with a high-speed rotating disk reactor. *Physica status solidi (c)* 7 (6): 1542–1544.
31. Kumar, N.D., Kamalasanan, M.N., and Chandra, S. (1994). Metalorganic chemical vapor deposition technique for growing c-axis oriented ZnO thin films in atmospheric pressure air. *Applied Physics Letters* 65 (11): 1373–1375.
32. Bekermann D., Pilard, D., Fischer, R., and Devi, A. (2009). Zinc malonate based precursors for MOCVD of ZnO. *ECS Transactions* 25 (8): 601–608.
33. Bekermann, D., Gasparotto, A., Barreca, D. et al. (2010). Highly oriented ZnO nanorod arrays by a novel plasma chemical vapor deposition process. *Crystal Growth and Design* 10: 2011–2018.
34. Kaufmann, T., Fuchs, G., Webert, M. et al. (1989). MOCVD layer growth of ZnO using adducts of dimethylzinc and diethylzinc. *Crystal Research and Technology* 24 (3): 269–274.
35. Jones, A.C. and O'Brien, P. (1997). *CVD of Compound Semiconductors : Precursor Synthesis, Development and Applications*. Weinheim (Federal Republic of Germany): VCH.
36. Jones, A.C., Rushworth, S.A., and Auld, J. (1995). Recent developments in metalorganic precursors for metalorganic chemical vapour deposition. *Journal of Crystal Growth* 146 (1–4): 503–510.
37. Rommeluère, J.F., Svob, L., Jomard, F. et al. (2003). Electrical activity of nitrogen acceptors in ZnO films grown by metalorganic vapor phase epitaxy. *Applied Physics Letters* 83 (2): 287–289.
38. Ashraf, S., Jones, A.C., Bacsa, J. et al. (2011). MOCVD of vertically aligned ZnO nanowires using bidentate ether adducts of dimethylzinc. *Chemical Vapor Deposition* 17 (1–3): 45–53.
39. Hindley, S., Jones, A.C., Ashraf, S. et al. (2011). Metal organic chemical vapour deposition of vertically aligned ZnO nanowires using oxygen donor adducts. *Journal of Nanoscience and Nanotechnology* 11 (9): 8294–8301.
40. Kaufmann, Th., Webert, M., Fuchs, G. and Arnold, H. (1991). Copper Incorporation in ZnO during LP-MOCVD. *Crystal Research and Technology* 26 (3): 263–272.
41. Kim, H.J., Sung, K., An, K.S. et al. (2004). ZnO nanowhiskers on ZnO nanoparticle-deposited Si(111) by MOCVD. *Journal of Materials Chemistry* 14 (23): 3396–3397.
42. Losurdo, M., Giangregorio, M.M., Capezzuto, P. et al. (2005). Reactivity of ZnO: Impact of polarity and nanostructure. *Superlattices and Microstructures* 38 (4–6): 291–299.
43. Scalisi, A.A., Toro, R.G., Malandrino, G. et al. (2008). Growth of ZnO nanostructures produced by MOCVD: A study of the effect of the substrate. *Chemical Vapor Deposition* 14 (5–6): 115–122.
44. Black, K., Chalker, P.R., Jones, A.C. et al. (2010). A new method for the growth of zinc oxide nanowires by MOCVD using oxygen-donor adducts of dimethylzinc. *Chemical Vapor Deposition* 16 (1–3): 106–111.
45. Gulino, A., Castelli, F., Dapporto, P. et al. (2000). Synthesis and characterization of novel self-generating liquid MOCVD precursors for thin films of zinc oxide. *Chemistry of Materials* 12 (2): 548–554.
46. Gulino, A., Lupo, F., and Fragalá, M.E. (2008). Substrate-free, self-standing ZnO thin films. *Journal of Physical Chemistry C* 112 (36): 13869–13872.
47. Suh, S., Miinea, L.A., Hoffman, D.M. et al. (2001). Atmospheric pressure chemical vapor deposition of undoped zinc oxide films from a zinc amide precursor. *Journal of Materials Science Letters* 20: 115–118.
48. Babcock, J.R., Wang, A., Edleman, N.L. et al. (2000). Development and implementation of new volatile Cd and Zn precursors for the growth of transparent conducting oxide thin films via MOCVD. *Materials Research Society Symposium Proceedings* 623: 317–328.
49. Velasco, A., Mitsuji, H., Minami, W. et al. (2008). Determination of the molecular size of growth species in the AP-MOCVD of ZnO from DEZ and H_2O. *Journal of Crystal Growth* 310 (16): 3837–3842.
50. Suh, S., Hoffman, D., Atagi, L., and Smith, D. (1999). A new metal-organic precursor for the low-temperature atmospheric pressure chemical vapor deposition of zinc oxide films. *Journal of Materials Science Letters* 18: 789–791.
51. Zukova, A., Teiserskis, A., Van Dijken, S. et al. (2006). Giant moment and magnetic anisotropy in Co-doped ZnO films grown by pulse-injection metal organic chemical vapor deposition. *Applied Physics Letters* 89 (23): 87–90.

52. Khan, O.F.Z. and O'Brien, P. (1989). On the use of zinc acetate as a novel precursor for the deposition of ZnO by low-pressure metal-organic chemical vapour deposition. *Thin Solid Films* 173 (1): 95–97.
53. Gyani, A.K., Khan, O.F.Z., O'Brien, P., and Urch, D.S. (1989). On the use of basic zinc acetate $Zn_4O(CH_3CO_2)_6$ as a novel precursor for the deposition of ZnO by low-pressure metallo organic chemical vapour deposition: their characterization by low energy electron induced X-ray emission spectroscopy. *Thin Solid Films* 182 (1–2): 3–6.
54. Purica, M., Budianu, E., Rusu, E. et al. (2002). Optical and structural investigation of ZnO thin films prepared by chemical vapor deposition (CVD). *Thin Solid Films* 403–404: 485–488.
55. Chen, J., Deng, H., Li, N. et al. (2011). Realization of nonpolar a-plane ZnO films on r-plane sapphire substrates using a simple single-source chemical vapor deposition. *Materials Letters* 65 (4): 716–718.
56. Auld, J., Houlton, D.J., Jones, A.C. et al. (1994). Growth of ZnO by MOCVD using alkylzinc alkoxides as single-source precursors. *Journal of Materials Chemistry* 4 (8): 1249–1253
57. Matthews, J.S., Onakoya, O.O., Ouattara, T.S., and Butcher, R.J. (2006). Synthesis and characterization of zinc AP-MOCVD precursors and their utility in the growth of ZnO. *Dalton Transactions* (31): 3806–3811.
58. Holmes, J., Hohnson, K., Zhang, B. et al. (2012). Metal organic chemical vapor deposition of ZnO from β-ketoiminates. *Applied Organometallic Chemistry* 26: 267–270.
59. Cosham, S.D., Kociok-Köhn, G., Johnson, A.L. et al. (2015). Synthesis and characterization of fluorinated β-ketoiminate zinc precursors and their utility in the AP-MOCVD growth of ZnO:F. *European Journal of Inorganic Chemistry* 2015 (26): 4362–4372.
60. Chandrakala, C., Sravanthi, P., Raj Bharath, S. et al. (2017). Synthesis, structure, vapour pressure and deposition of ZnO thin film by plasma assisted MOCVD technique using a novel precursor bis[(pentylnitrilomethylidine)(pentylnitrilomethylidine-μ-phenalato)]dizinc(II). *Journal of Molecular Structure* 1130: 1–9.
61. Gupta, N.N. and Atwood, D.A. (2011). *Cadmium: organometallic chemistry*. In: *Encyclopedia of Inorganic and Bioinorganic Chemistry*. John Wiley & Sons, Ltd.
62. Kappers, M.J., Wilkerson, K.J., and Hicks, R.F. (1997). Effects of ligand exchange reactions on the composition of $Cd_{1-y}Zn_yTe$ grown by metalorganic vapor-phase epitaxy. *Journal of Physical Chemistry B* 101: 4882–4888.
63. Zhang, Y., Du, G., Liu, D. et al. (2004). Structural and optical properties of $Mg_xZn_{1-x}O$ thin films grown by metal-organic chemical vapor deposition. *Journal of Crystal Growth* 268 (1–2): 140–143.
64. Asahara, H., Takamizu, D., Inokuchi, A. et al. (2010). Characterization of MgZnO films grown by plasma enhanced metal-organic chemical vapor deposition. *Thin Solid Films* 518 (11): 2953–2956.
65. Turnbull, A.G. (1967). Thermochemistry of biscyclopentadienyl metal compounds. *Australian Journal of Chemistry* 20 (10): 2059–2067.
66. Lewis, C.R., Dietze, W.T., and Ludowise, M.J. (1983). The growth of magnesium-doped GaAs by the OM-VPE process. *Journal of Electronic Materials* 12 (3): 507–524.
67. Hull, H.S., Reid, A.F., and Turnbull, A.G. (1965). Vapour pressures of cyclopentadiene and bis(cyclopentadienyl) magnesium. *Australian Journal of Chemistry* 18 (2): 249–252.
68. SAFC. (2008). High performance chemicals SAFC for advanced semincoductor applications.
69. ELMOS. (2003). Organometallics and chemicals: Bis (cyclopentadienyl) magnesium. http://www.metalorganics.ru/files/PR-ME-E/PR-MG1-E.htm.
70. Liu, A. (2015). Part III. Metal-Organic Chemical Vapor Deposition. https://www.slideshare.net/auod590/epitaxy3?from_action=save. S12.
71. Gruber, T., Kirchner, C., Kling, R. et al. (2004). ZnMgO epilayers and ZnO-ZnMgO quantum wells for optoelectronic applications in the blue and UV spectral region. *Applied Physics Letters* 84 (26): 5359–5361.
72. ELMOS. (2003). Organometallics and chemicals: Bis (methylcyclopentadienyl) magnesium. http://www.metalorganics.ru/files/PR-ME-E/PR-MG2-E.htm.
73. Chiba, Y., Meng, F., Yamada, A., and Konagai, M. (2007). Growth of polycrystalline $Zn_{1-x}Mg_xO$ thin films using $EtCp_2Mg$ and $MeCp_2Mg$ by metal organic chemical vapor deposition. *Japanese Journal of Applied Physics Part 1* 46 (8 A): 5040–5043.
74. Nakamura, A., Ishihara, J., Shigemori, S. et al. (2005). Growth of $Mg_xZn_{1-x}O$ films using remote plasma MOCVD. *Applied Surface Science* 244 (1–4): 385–388.
75. Sosnovsky, G. and Brown, J.H. (1966). The chemistry of organometallic and organometalloid peroxides. *Chemical Reviews* 66 (5): 529–566.
76. Coates, G.E. and Ridley, D. (1965). Alkoxy-, thio-, and amino-derivatives of methylzinc. *Journal of the Chemical Society* 0: 1870–1877.
77. Dortmund Data Bank (DDBST GmbH) (2018). Saturated vapor pressure: calculation by Antoine Equation. http://ddbonline.ddbst.de/AntoineCalculation/AntoineCalculationCGI.exe.
78. J. G. Noltes. (1973). Organozinc compounds. In: *Zinc Chemicals* (ed. C.H. Farnsworth). London: MK Zinc Institute.
79. Lau, C.K., Tiku, S.K., and Lakin, K.M. (1980). Growth of epitaxial ZnO thin films by OMCVD. *Journal of the Electrochemical Society* 127 (8): 1843–1847.

80. Kim, Y.S. and Won, Y.S. (2009). Investigation on reaction pathways for ZnO formation from diethylzinc and water during chemical vapor deposition. *Bulletin of the Korean Chemical Society* 30 (7): 1573–1578.
81. Oda, S., Tokunaga, H., Kitajima, N. et al. (1985). Highly oriented ZnO films prepared by MOCVD from diethylzinc and alcohols. *Japanese Journal of Applied Physics* 24 (12): 1607–1610.
82. Hahn, B., Heindel, G., Pschorr-Schoberer, E., and Gebhardt, W. (1998). MOCVD layer growth of ZnO using DMZn and tertiary butanol. *Semiconductor Science and Technology* 13 (7): 788–791.
83. Kirchner, C., Gruber, T., Reuß, F. et al. (2003). MOVPE growth of ZnO using various oxygen precursors. *Journal of Crystal Growth* 248: 20–24.
84. Maejima, K., Kawabata, H., and Fujita, S. (2007). Quantum chemical study on interactions of diethylzinc with nitrous oxide and water for ZnO growth by metal-organic vapor phase epitaxy. *Japanese Journal of Applied Physics Part 1* 46 (12): 7885–7887.
85. Guthrie, G.B.J., Scott, D.W., Hubbard, W.N. et al. (1952). Thermodynamic properties of furan. *Journal of the American Chemical Society* 74 (18): 4662–4669.
86. NIH U.S. National Library of Medicine. (2019). Tetrahydropyran. https://pubchem.ncbi.nlm.nih.gov/compound/8894.
87. Wenas, W.W., Yamada, A., Konagai, M., and Takahashi, K. (1994). Metalorganic chemical vapor deposition of ZnO using D_2O as oxidant. *Japanese Journal of Applied Physics* 33 (2 (3A)): L-283–L-285.
88. Hongsingthong, A., Afdi Yunaz, I., Miyajima, S., and Konagai, M. (2011). Preparation of ZnO thin films using MOCVD technique with D_2O/H_2O gas mixture for use as TCO in silicon-based thin film solar cells. *Solar Energy Materials and Solar Cells* 95 (1): 171–174.
89. Sang, B., Yamada, A., and Konagai, M. (1998). Textured ZnO thin films for solar cells grown by a two-step process with the atomic layer deposition technique. *Japanese Journal of Applied Physics Part 2 Letters* 37 (2B): L206–L208.
90. Zhao, Z., Morel, D.L., and Ferekides, C.S. (2002). Electrical and optical properties of tin-doped CdO films deposited by atmospheric metalorganic chemical vapor deposition. *Thin Solid Films* 413 (1–2): 203–211.
91. Zhaochun, Z., Baibiao, H., Yongqin, Y., and Deliang, C. (2001). Electrical properties and Raman spectra of undoped and Al-doped ZnO thin films by metalorganic vapor phase epitaxy. *Materials Science and Engineering: B* 86 (2): 109–112.
92. Tuan, A.C., Bryan, J.D., Pakhomov, A.B. et al. (2004). Epitaxial growth and properties of cobalt-doped ZnO on α-Al_2O_3 single-crystal substrates. *Physical Review B* 70: 054424.
93. Zhong, J., Muthukumar, S., Chen, Y. et al. (2003). Ga-doped ZnO single-crystal nanotips grown on fused silica by metalorganic chemical vapor deposition. *Applied Physics Letters* 83 (16): 3401–3403.
94. Wang, H., Ho, H.P., Lo, K.C., and Cheah, K.W. (2007). Growth of p-type ZnO thin films by (N, Ga) co-doping using DMHy dopant. *Journal of Physics D: Applied Physics* 40 (15): 4682–4685.
95. Thi Thanh Ho, V., Giang Bach, L., Thanh, T. et al. (2015). Effect of gallium source material on the transparent conducting properties of Ga:ZnO thin films through metalorganic chemical vapor deposition. *Molecular Crystals and Liquid Crystals* 623 (1): 433–443.
96. Ye, J.D., Gu, S.L., Zhu, S.M. et al. (2005). Gallium doping dependence of single-crystal n-type ZnO grown by metal organic chemical vapor deposition. *Journal of Crystal Growth* 283 (3–4): 279–285.
97. Khranovskyy, V., Grossner, U., Lazorenko, V. et al. (2007). Conductivity increase of ZnO:Ga films by rapid thermal annealing. *Superlattices and Microstructures* 42 (1–6): 379–386.
98. Ben-Yaacov, T., Ive, T., Van De Walle, C.G. et al. (2010). Properties of In-doped ZnO films grown by metalorganic chemical vapor deposition on GaN(0001) templates. *Journal of Electronic Materials* 39 (5): 608–611.
99. Zhao, J.Z., Liang, H.W., Sun, J.C. et al. (2008). p-Type Sb-doped ZnO thin films prepared by metallorganic chemical vapor deposition using metallorganic dopant. *Electrochemical and Solid-State Letters* 11 (12): H323–H326.
100. Liang, H., Chen, Y., Xia, X. et al. (2015). Influence of Sb valency on the conductivity type of Sb-doped ZnO. *Thin Solid Films* 589: 199–202.
101. Saito, K., Hosokai, Y., Nagayama, K. et al. (2004). MOCVD growth of monomethylhydrazine-doped ZnO layers. *Journal of Crystal Growth* 272 (1–4): 805–809.
102. Wang, J., Du, G., Zhao, B. et al. (2003). Epitaxial growth of NH_3-doped ZnO thin films on <0 2-2 4> oriented sapphire substrates. *Journal of Crystal Growth* 255 (3–4): 293–297.
103. Dadgar, A., Oleynik, N., Bläsing, J. et al. (2004). Heteroepitaxy and nitrogen doping of high-quality ZnO. *Journal of Crystal Growth* 272 (1–4): 800–804.
104. Li, X., Shen, R., Zhang, B. et al. (2011). Nitrogen doped ZnO thin films prepared by photo-assisted metal-organic chemical vapor deposition. *Journal of Nanoscience and Nanotechnology* 11 (11): 9741–9744.
105. Dadgar, A., Krtschil, A., Bertram, F. et al. (2005). ZnO MOVPE growth: From local impurity incorporation towards p-type doping. *Superlattices and Microstructures* 38 (4–6): 245–255.
106. Li, X., Yan, Y., Gessert, T.A. et al. (2003). p-Type ZnO thin films formed by CVD reaction of diethylzinc and NO gas. *Electrochemical and Solid-State Letters* 6 (4): C56–C58.
107. Tan, S.T., Chen, B.J., Sun, X.W. et al. (2005) Realization of intrinsic p-Type ZnO thin films by metal organic chemical vapor deposition. *Journal of Electronic Materials* 34 (8): 1172–1176.

108. Sun, J.C., Bian, J.M., Wang, Y. et al. (2012). Influence of radical power on the electrical and optical properties of ZnO:N films grown by metal-organic chemical vapor deposition with N_2O plasma doping source. *Thin Solid Films* 521: 253–256.
109. Egerton, E.J., Sood, A.K., Singh, R. et al. (2005). p-Type doping utilizing nitrogen and Mn doping of ZnO using MOCVD for ultraviolet lasers and spintronic applications. *Journal of Electronic Materials* 34 (6): 949–952.
110. Kun, T., Gu, S., Wu, K. et al. (2010). Tellurium assisted realization of p-type N-doped ZnO. *Applied Physics Letters* 96 (24): 2010–2013.
111. Mullin, J.B., Irvine, S.J.C., Giess, J., and Royle, A. (1985). Recent developments in the MOVPE of II-VI compounds. *Journal of Crystal Growth* 72: 1–12.
112. Madelung, O. (2004). *Semiconductors: Data Handbook*. Berlin Heidelberg: Springer.
113. Vispute, R.D., Talyansky, V., Choopun, S. et al. (1998). Heteroepitaxy of ZnO on GaN and its implications for fabrication of hybrid optoelectronic devices. *Applied Physics Letters* 73 (3): 348–350.
114. Dobrovinskaya, E.R. and Lytvynov, L.A. (2009). *Sapphire Material, Manufacturing, Applications*. New York: Springer.
115. Muthukumar, S., Chen, Y., Zhong, J. et al. (2004). Metalorganic chemical vapor deposition and characterizations of epitaxial $Mg_xZn_{1-x}O$ films on r-sapphire substrates. *Journal of Crystal Growth* 261 (2–3): 316–323.
116. MoltTech GmbH. (2016). Sapphire (Al_2O_3). http://www.mt-berlin.com/frames_cryst/descriptions/sapphire.htm.
117. Pierce, J.M., Adekore, B.T., Davis, R.F., and Stevie, F.A. (2005). Homoepitaxial growth of dense ZnO(0001) and ZnO (11-20)films via MOVPE on selected ZnO substrates. *Journal of Crystal Growth* 283 (1–2): 147–155.
118. Gu, R., Tang, K., Gu, S. et al. (2016). Substrate polarity and surface pretreatment temperature dependence of ZnO homoepitaxy. *Applied Surface Science* 361: 33–40.
119. Smith, T.P., McLean, H., Smith, D.J., and Davis, R.F. (2004). Homoepitaxial growth of (0001)- and (000-1)-oriented ZnO thin films via metalorganic vapor-phase epitaxy and their characterization. *Journal of Crystal Growth* 265 (3): 390–398.
120. Lee, D.J., Park, J.Y., Yun, Y.S. et al. (2005). Comparative studies on the growth behavior of ZnO nanorods by metalorganic chemical vapor deposition depending on the type of substrates. *Journal of Crystal Growth* 276 (3–4): 458–464.
121. Gu, X., Sabuktagin, S., Teke, A. et al. (2004). Effect of thermal treatment on ZnO substrate for epitaxial growth. *Journal of Materials Science: Materials in Electronics* 15 (6): 373–378.
122. Palacios-Lidón, E., Pérez-García, B., Vennéguès, P. et al. (2009). Anisotropic chemical etching of semipolar {10-1-1}/{10-1+1} ZnO crystallographic planes: polarity versus dangling bonds. *Nanotechnology* 20 (6): 065701.
123. Haga, K., Shishido, T., and Nakajima, K. (2007). Homo-epitaxial growth on ZnO substrate by MO-CVD using $(C_5H_7O_2)_2$. *Surface Review and Letters* 14 (4): 783–787.
124. Ogata, K., Kawanishi, T., Maejima, K. et al. (2002). ZnO growth using homoepitaxial technique on sapphire and Si substrates by metalorganic vapor phase epitaxy. *Journal of Crystal Growth*: 237–239, 553–557.
125. Ogata, K., Kawanishi, T., Sakurai, K. et al. (2002). Homoepitaxial growth of ZnO by metalorganic vapor phase epitaxy. *Physica status solidi (b)* 229 (2): 915–919.
126. Ohtomo, A., Kimura, H., Saito, K. et al. (2000). Lateral grain size and electron mobility in ZnO epitaxial films grown on sapphire substrates. *Journal of Crystal Growth* 214: 284–288.
127. Liang, S., Gorla, C.R., Emanetoglu, N. et al. (1998). Epitaxial growth of (11-20) ZnO on (01-12). Al_2O_3 by metalorganic chemical vapor deposition. *Journal of Electronic Materials* 27 (11): 72–76.
128. Gorla, C. and Emanetoglu, N. (1999). Structural, optical, and surface acoustic wave properties of epitaxial ZnO films grown on (01-12). sapphire by metalorganic chemical vapor deposition. *Journal of Applied Physics* 85 (5): 2595–2602.
129. Emanetoglu, N.W., Gorla, C., Liu, Y. et al. (1999). Epitaxial ZnO piezoelectric thin films for saw filters. *Materials Science in Semiconductor Processing* 2 (3): 247–252.
130. Moriyama, T. and Fujita, S. (2006). Epitaxial growth of nonpolar ZnO by MOVPE. *Physica status solidi (c)* 3 (4): 726–729.
131. Sheng, H., Muthukumar, S., Emanetoglu, N.W., and Lu, Y. (2002). Schottky diode with Ag on (11-2 0) epitaxial ZnO film. *Applied Physics Letters* 80 (12): 2132–2134.
132. Zhang, B.P., Waaktsuki, K., Binh, N.T. et al. (2004). Effects of growth temperature on the characteristics of ZnO epitaxial films deposited by metalorganic chemical vapor deposition. *Thin Solid Films* 449: 12–19.
133. Quon, H.H. and Malanka, D.P. (1975). Chemical vapour deposition of epitaxial ZnO. *Materials Research Bulletin* 10 (5): 349–354.
134. Lu, Z., Feigelson, R.S., Route, R.K. et al. (1993). Solid source MOCVD for the epitaxial growth of thin oxide films. *Journal of Crystal Growth* 128 (1–4): 788–792.
135. Boo, J., Lee, S., Yu, K. et al. (1999). Growth of magnesium oxide thin films using single molecular precursors by metal-organic chemical vapor deposition. *Thin Solid Films* 341: 63–67.

136. Ma, Y., Du, G.T., Yang, S.R. et al. (2004). Control of conductivity type in undoped ZnO thin films grown by metalorganic vapor phase epitaxy. *Journal of Applied Physics* 95 (11): 6268–6272.
137. Zhang, Y., Du, G., Liu, B. et al. (2004). Effects of ZnO buffer layer thickness on properties of ZnO thin films deposited by low-pressure MOCVD. *Journal of Crystal Growth* 262 (1–4): 456–460.
138. Wang, Y., Chu, B., and He, Q. (2008). Effects of sapphire substrates surface treatment on epitaxial ZnO thin films grown by MOCVD. *Vacuum* 82 (11): 1229–1232.
139. Ogata, K., Maejima, K., Fujita, S., and Fujita, S. (2001). ZnO growth toward optical devices by MOVPE using N_2O. *Journal of Electronic Materials* 30 (6): 659–661.
140. Wang, Y., Wang, S., Zhou, S. et al. (2006). Effects of sapphire substrate annealing on ZnO epitaxial films grown by MOCVD. *Applied Surface Science* 253 (4): 1745–1747.
141. Dadgar, A., Oleynik, N., Forster, D. et al. (2004). A two-step metal organic vapor phase epitaxy growth method for high-quality ZnO on GaN/Al_2O_3 (0001). *Journal of Crystal Growth* 267 (1–2): 140–144.
142. Waag, A., Gruber, T., Kirchner, C. et al. (2002). ZnO metal organic vapor phase epitaxy: present state and prospective application in optoelectronics and spin electronics. *Advances in Solid State Physics* 42: 81–91.
143. Zhong, J., Saraf, G., Chen, H. et al. (2007). Structural and optical properties of ZnO nanotips grown on GaN using metalorganic chemical vapor deposition. *Journal of Electronic Materials* 36 (6): 654–658.
144. Zúñiga-Pérez, J., Muñoz-Sanjosé, V., Lorenz, M. et al. (2006). Structural characterization of a-plane $Zn_{1-x}Cd_xO$ ($0 \leq x \leq 0.085$). Thin films grown by metal-organic vapor phase epitaxy. *Journal of Applied Physics* 99 (2): 023514.
145. Huang, S., Gu, S., Zhu, S. et al. (2013). Thermal pretreatment of sapphire substrates prior to ZnO buffer layer growth. *Journal of Vacuum Science and Technology B* 31 (5): 051203.
146. Xia, X., Zhao, W., Jian, Y. et al. (2009). The structure and optical characters of the ZnO film grown on GaAs/Al_2O_3 substrate. *Applied Surface Science* 255 (12): 6313–6317.
147. Kwak, C.-H., Kim, B.-H., Park, C.-I. et al. (2010). Structural and electrical properties of ZnO nanorods and Ti buffer layers. *Applied Physics Letters* 96 (5): 051908.
148. Chen, H., Zhong, J., Saraf, G. et al. (2005). Properties of ZnO nanotips selectively grown by MOCVD. *Proceedings of the SPIE* 5592: 164–169.
149. Li, F., Li, D., Dai, J. et al. (2006). Effect of the initial thin Ti buffer layers on the quality of ZnO thin films grown on Si(111) substrates by MOCVD. *Superlattices and Microstructures* 40 (1): 56–63.
150. Wang, L., Pu, Y., Chen, Y.F. et al. (2005). MOCVD growth of ZnO films on Si(111) substrate using a thin AlN buffer layer. *Journal of Crystal Growth* 284: 459–463.
151. Wang, X., Yang, S., Yang, X. et al. (2002). ZnO thin film grown on silicon by metal-organic chemical vapor deposition. *Journal of Crystal Growth* 243 (1): 13–18.
152. Shimizu, M., Matsueda, Y., Shiosaki, T., and Kawabata, A. (1985). Growth of ZnO films by the plasma-enhanced metalorganic chemical vapor deposition technique. *Journal of Crystal Growth* 71 (1): 209–219
153. Muthukumar, S., Emanetoglu, N.W., Patounakis, G. et al. (2001). Two-step metalorganic chemical vapor deposition growth of piezoelectric ZnO thin film on SiO_2/Si substrate. *Journal of Vacuum Science and Technology A* 19 (4): 1850–1853.
154. Muthukumar, S., Gorla, C.R., Emanetoglu, N.W. et al. (2001). Control of morphology and orientation of ZnO thin films grown on SiO_2/Si substrates. *Journal of Crystal Growth* 225: 197–201.
155. Ogata, K., Kim, S.-W., Fujita, S., and Fujita, S. (2002). ZnO growth on Si substrates by metalorganic vapor phase epitaxy. *Journal of Crystal Growth* 240 (1–2): 112–116.
156. Kim, S.W., Fujita, S., and Fujita, S. (2002). Self-assembled three-dimensional ZnO nanosize islands on Si substrates with SiO_2 intermediate layer by metalorganic chemical vapour deposition. *Japanese Journal of Applied Physics Part 2 Letters* 41 (5 B), 39–42.
157. Kling, R., Kirchner, C., Gruber, T. et al. (2004). Analysis of ZnO and ZnMgO nanopillars grown by self-organization. *Nanotechnology* 15 (8): 1043–1046.
158. Xu, W.Z., Ye, Z.Z., Ma, D.W. et al. (2005). Quasi-aligned ZnO nanotubes grown on Si substrates. *Applied Physics Letters* 87 (9): 224–227.
159. INRF application note. (1960). RCA-1 silicon wafer cleaning. Process name: RCA01.
160. INRF application note. (1962). RCA-2 silicon wafer cleaning. Process name: RCA02.
161. Yuan, G.D., Ye, Z.Z., Zhu, L.P. et al. (2005). Control of conduction type in Al- and N-codoped ZnO thin films. *Applied Physics Letters* 86 (20): 1–3.
162. Kern, W. (1990). The evolution of silicon wafer cleaning technology. *Journal of the Electrochemical Society* 137 (6): 1887–1892.
163. Dadgar, A., Bläsing, J., Diez, A. et al. (2000). Metalorganic chemical vapor phase epitaxy of crack-free GaN on Si (111) exceeding 1 μm in thickness. *Japanese Journal of Applied Physics Part 2 Letters* 39 (11B): L1183–L1185.
164. Chen, Y., Jiang, F., Wang, L. et al. (2005). Structural and luminescent properties of ZnO epitaxial film grown on Si(111) substrate by atmospheric-pressure MOCVD. *Journal of Crystal Growth* 275 (3–4): 486–491.
165. Lin, C.W., Cheng, T.Y., Chang, L., and Juang, J.Y. (2004). Chemical vapor deposition of zinc oxide thin films on Y_2O_3/Si substrates. *Physica status solidi (c)* 1 (4): 851–855.

166. Fu, Z., Lin, B., and Zu, J. (2002). Photoluminescence and structure of ZnO films deposited on Si substrates by metal-organic chemical vapor deposition. *Thin Solid Films* 402 (1–2): 302–306.
167. Zhu, J., Yao, R., Zhong, S. et al. (2007). Plasma pretreatment on Si(111) substrates for the growth of ZnO thin films. *Journal of Crystal Growth* 303 (2): 655–658.
168. Kim, S.-W., Fujita, S., and Fujita, S. (2002). Self-organized ZnO quantum dots on SiO_2/Si substrates by metalorganic chemical vapor deposition. *Applied Physics Letters* 81 (26): 5036–5038.
169. Takeuchi, T., Amano, H., Hiramatsu, K. et al. (1991). Growth of single crystalline GaN film on Si-substrate using 3C-SiC as an intermediate layer. *Journal of Crystal Growth* 115 (1–4): 634–638.
170. Watanabe, A., Takeuchi, T., Hirosawa, K. et al. (1993). The growth of single crystalline GaN on a Si substrate using AlN as an intermediate layer. *Journal of Crystal Growth* 128 (1–4): 391–396.
171. Krost, A., Christen, J., Oleynik, N. et al. (2004). Ostwald ripening and flattening of epitaxial ZnO layers during in situ annealing in metalorganic vapor phase epitaxy. *Applied Physics Letters* 85 (9): 1496–1498.
172. Jiang, F.Y., Zheng, C.D., Wang, L. et al. (2007). The growth and properties of ZnO film on Si(111) substrate with an AlN buffer by AP-MOCVD. *Journal of Luminescence* 122: 905–907.
173. Lin, C.W., Cheng, T.Y., Chang, L., and Juang, J.Y. (2005). Growth of zinc oxide thin films on Y_2O_3/Si substrates by chemical vapor deposition. *Journal of Crystal Growth* 275 (1–2): 2481–2485.
174. Zhu, J., Lin, B., Sun, X. et al. (2005). Heteroepitaxy of ZnO film on Si (111) substrate using a 3C-SiC buffer layer. *Thin Solid Films* 478 (1–2): 218–222.
175. Kim, S.-W., Fujita, S., Yi, M.-S., and Yoon, D.H. (2006). Catalyst-free synthesis of ZnO nanowall networks on Si_3N_4/Si substrates by metalorganic chemical vapor deposition. *Applied Physics Letters* 88 (25): 253114.
176. Guo, W., Allenic, A., Chen, Y.B. et al. (2008). ZnO epitaxy on (111) Si using epitaxial Lu_2O_3 buffer layers. *Applied Physics Letters* 92 (7): 3–6.
177. Wang, L., Yang, Y., Jin, S., and Marks, T.J. (2006). MgO(100) template layer for CdO thin film growth: Strategies to enhance microstructural crystallinity and charge carrier mobility. *Applied Physics Letters* 88 (16): 10–13.
178. Li, X., Gessert, T.A., and Coutts, T. (2004). The properties of cadmium tin oxide thin-film compounds prepared by linear combinatorial synthesis. *Applied Surface Science* 223 (1–3): 138–143.
179. Lamb, D.A. and Irvine, S.J.C. (2004). Growth properties of thin film ZnO deposited by MOCVD with n-butyl alcohol as the oxygen precursor. *Journal of Crystal Growth* 273 (1–2): 111–117.
180. Ellis, D.M. and Irvine, S.J.C. (2004). MOCVD of highly conductive CdO thin films. *Journal of Materials Science: Materials in Electronics* 15 (6): 369–372.
181. Chen, X.L., Geng, X.H., Xue, J.M. et al. (2006). Temperature-dependent growth of zinc oxide thin films grown by metal organic chemical vapor deposition. *Journal of Crystal Growth* 296 (1): 43–50.
182. Du Pasquier, A., Chen, H., and Lu, Y. (2006). Dye sensitized solar cells using well-aligned zinc oxide nanotip arrays. *Applied Physics Letters* 89 (25): 1–4.
183. Chen, X.L., Geng, X.H., Xue, J.M., and Li, L.N. (2007). Two-step growth of ZnO films with high conductivity and high roughness. *Journal of Crystal Growth* 299 (1): 77–81.
184. Nicolay, S., Fay, S., and Ballif, C. (2009). Growth model of MOCVD polycrystalline ZnO. *Crystal Growth and Design* 9 (11): 4957–4962.
185. SPG (Speciality Glass products), (2019). Borosilicate. http://www.sgpinc.com/materials/borosilicate/.
186. Manasevit, H.M. and Simpson, W.I. (1971). The Use of Metal-Organics in the Preparation of Semiconductor Materials II. II-VI Compounds. *Journal of the Electrochemical Society* 118 (4): 644–647.
187. Betsch, R.J. (1986). Parametric analysis of control parameters in MOCVD. *Journal of Crystal Growth* 77 (1–3): 210–218.
188. Roth, A.P. and Williams, D.F. (1981). Semiconducting zinc oxide films prepared by metal organic chemical vapor deposition from diethyl zinc. *Journal of the Electrochemical Society* 128 (12): 2684–2686.
189. Wenas, W.W., Yamada, A., Konagai, M., and Takahashi, K. (1991). Textured ZnO thin films for solar cells grown by metalorganic chemical vapor deposition. *Japanese Journal of Applied Physics Part 1*, 30 (3B): 441–443.
190. Ye, J., Gu, S., Zhu, S. et al. (2002). The growth and annealing of single crystalline ZnO films by low-pressure MOCVD. *Journal of Crystal Growth* 243 (1): 151–156.
191. Wang, L., Pu, Y., Fang, W. et al. (2005). High-quality ZnO films grown by atmospheric pressure metal- organic chemical vapor deposition. *Journal of Crystal Growth* 283 (1–2): 87–92.
192. Behrends, A., Bakin, A., and Waag, A. (2009). Investigation of ZnO nanopillars fabrication in a new Thomas Swan close coupled showerhead MOCVD reactor. *Microelectronics Journal* 40 (2): 280–282.
193. Ellmer, K., Klein, A., and Rech, B. (2008). *Transparent Conductive Zinc Oxide: Basics and Applications in* Thin Film Solar Cells. Springer.
194. Boukadhaba, M.A., Chebil, W., Fouzri, A. et al. (2015). Microstructural and optical properties of $Zn_{(1-x)}Cd_xO$ thin films grown on r-plane sapphire substrates: Effect of growth temperature and position of sample in MO-CVD reactor. *Ceramics International* 41 (1): 1677–1685.

195. Park, W.I., Kim, D.H., Jung, S.W., and Yi, G.C. (2002). Metalorganic vapor-phase epitaxial growth of vertically well-aligned ZnO nanorods. *Applied Physics Letters* 80 (22): 4232–4234.
196. Muthukumar, S., Sheng, H., Zhong, J. et al. (2003). Selective MOCVD growth of ZnO nanotips. *IEEE Transactions on Nanotechnology* 2 (1): 50–54.
197. Zhang, Y., Du, G., Wang, X. et al. (2003). X-ray photoelectron spectroscopy study of ZnO films grown by metal-organic chemical vapor deposition. *Journal of Crystal Growth* 252 (1–3): 180–183.
198. Kim, Y.-J., Jeon, J.-M., Choi, J.H. et al. (2010). ZnO nanostructures with controlled morphologies on a glass substrate. *Nanotechnology* 21 (26): 265603.
199. Yuan, H. and Zhang, Y. (2004). Preparation of well-aligned ZnO whiskers on glass substrate by atmospheric MOCVD. *Journal of Crystal Growth* 263 (1–4): 119–124.
200. Biswas, P., Kundu, S., and Banerji, P. (2013). A study on electrical transport vis-à-vis the effect of thermal annealing on the p-type conductivity in arsenic-doped MOCVD grown ZnO in the temperature range 10-300K. *Journal of Alloys and Compounds* 552: 304–309.
201. Fujimoto, E., Sumiya, M., Ohnishi, T. et al. (2012). Development of a new laser heating system for thin film growth by chemical vapor deposition. *Review of Scientific Instruments* 83 (9): 25–31.
202. Bakke, J.R., Hagglund, C., Jung, H.J. et al. (2013). Atomic layer deposition of CdO and $Cd_xZn_{1-x}O$ films. *Materials Chemistry and Physics* 140 (2–3): 465–471.
203. Alema, F., Ledyaev, O., Miller, R. et al. (2016). Growth of high Mg content wurtzite MgZnO epitaxial films via pulsed metal organic chemical vapor deposition. *Journal of Crystal Growth* 435: 6–11.
204. Lamborn, D.R. (2007). *Experimental and modeling studies of the hybrid physical-chemical vapor deposition of superconducting magnesium diboride thin films*. Thesis dissertation, Pennsylvania State University.
205. Nishinaka, H. and Fujita, S. (2008). Step-flow growth of homoepitaxial ZnO thin films by ultrasonic spray-assisted MOVPE. *Journal of Crystal Growth* 310 (23): 5007–5010.
206. Roeder, J.F., Bilodeau, S.M., Carl, R.J.J. et al. (1997). MOCVD of polycrystalline and epitaxial complex oxides by liquid delivery. *Materials Research Society Symposium Proceedings* 474: 21–30.
207. Manin, M., Thollon, S., Emieux, F. et al. (2005). Deposition of MgO thin film by liquid pulsed injection MOCVD. *Surface and Coatings Technology* 200 (5–6): 1424–1429.
208. Ternon, C., Rey, G., Labeaub, M. et al. (2009). Growth of ZnO nanowires by MOCVD: fundamental role of the substrate. *ECS Transactions* 25: 437–443.
209. Jones, A.C. and Hitchman, M.L. eds. (2008). *Chemical Vapour Deposition*. Royal Society of Chemistry Cambridge.
210. Irvine, S.J.C.C. (1994). Chapter 18, *Handbook of Crystal Growth* (ed. D.T.J. Hurle). Amsterdam: Elsevier.
211. Shimizu, M., Kamei, H., Tanizawa, M. et al. (1988). Low temperature growth of ZnO film by photo MOCVD. *Journal of Crystal Growth* 89 (4): 365–370.
212. Shimizu, M., Monma, A., Shiosaki, T. et al. (1989). Effects of UV light irradiation on the growth of ZnO films. *Journal of Crystal Growth* 94 (4): 895–900.
213. Maruyama, T. and Nakai, A. (1989). Plasma metalorganic chemical vapor deposition of indium oxide thin films. *Japanese Journal of Applied Physics* 28 (3): L346–L348.
214. Yoshino, M., Wenas, W.W., Yamada, A. et al. (1993). Large-area ZnO thin films for solar cells prepared by photo-induced metalorganic chemical vapor deposition. *Japanese Journal of Applied Physics Part 1* 32 (2): 726–730.
215. Myong, S.Y. and Lim, K.S. (2005). Improvement of electrical and optical properties of ZnO thin films prepared by MOCVD using UV light irradiation and in situ H_2 post-treatment. *Solar Energy Materials and Solar Cells* 86 (1): 105–112.
216. Myong, S.Y., Park, S. Il., and Lim, K.S. (2006). Improvement of electrical stability of polycrystalline ZnO thin films via intentional post-deposition hydrogen doping. *Thin Solid Films* 513 (1–2): 148–151.
217. Shi, Z., Wu, B., Cai, X. et al. (2012). Photofacilitated controllable growth of ZnO films using photoassisted metal organic chemical vapor deposition. *Crystal Growth and Design* 12 (9): 4417–4424.
218. Wang, J., Dong, X., Zhang, B. et al. (2013). P-Type NiZnO thin films grown by photo-assist metal-organic chemical vapor deposition. *Journal of Alloys and Compounds* 579: 160–164.
219. Zhao, Y.D., Dong, X., Ma, Z.Z. et al. (2016). High hole concentration Li-doped NiZnO thin films grown by photo-assisted metal-organic chemical vapor deposition. *Journal of Crystal Growth* 454: 30–34.
220. Shiosaki, T., Yamamoto, T., Yagi, M., and Kawabata, A. (1981). Plasma-enhanced metalorganic chemical vapor deposition of c-axis oriented and epitaxial films of ZnO at low substrate temperatures. *Applied Physics Letters* 39 (5): 399–401.
221. Shimizu, M., Horii, T., Shiosaki, T., and Kawabata, A. (1982). Preparation of ZnO thin films by plasma-enhanced organometallic chemical vapour deposition. *Thin Solid Films* 96 (2): 149–154.
222. Du, G., Ma, Y., Zhang, Y., and Yang, T. (2005). Preparation of intrinsic and N-doped p -type ZnO thin films by metalorganic vapor phase epitaxy. *Applied Physics Letters* 87 (21): 1–3.

223. Nakamura, A., Ishihara, J., Shigemori, S. et al. (2004). Characterization of wurtzite $Zn_{1-x}Cd_xO$ films using remote plasma-enhanced metalorganic chemical vapor deposition. *Japanese Journal of Applied Physics Part 2: Letters* 43 (11 A): 1452–1454.
224. Lei, P.H., Wu, H.M., and Hsu, C.M. (2012). Zinc oxide (ZnO) grown on flexible substrate using dual-plasma-enhanced metalorganic vapor deposition (DPEMOCVD). *Surface and Coatings Technology* 206 (14): 3258–3263.
225. Park, J.H., Byun, D., Jeon, B.J., and Lee, J.K. (2008). Effect of hydrogen content on the ZnO thin films on the surface of polyethylene terephthalate substrate through electron cyclotron resonance-metal organic chemical vapor deposition. *Journal of Materials Science* 43 (10): 3417–3423.
226. Triboulet, R. and Perrière, J. (2003). Epitaxial growth of ZnO films. *Progress in Crystal Growth and Characterization of Materials* 47: 65–138.
227. Pan, M., Fenwick, W.E., Strassburg, M. et al. (2006). Metal-organic chemical vapor deposition of ZnO. *Journal of Crystal Growth* 287 (2): 688–693.
228. Ogata, K., Maejima, K., Fujita, S., and Fujita, S. (2003). Growth mode control of ZnO toward nanorod structures or high-quality layered structures by metal-organic vapor phase epitaxy. *Journal of Crystal Growth* 248: 25–30.
229. Heinze, S., Krtschil, A., Bläsing, J. et al. (2007). Homoepitaxial growth of ZnO by metalorganic vapor phase epitaxy in two-dimensional growth mode. *Journal of Crystal Growth* 308 (1): 170–175.
230. Ive, T., Ben-Yaacov, T., Murai, A. et al. (2008). Metalorganic chemical vapor deposition of ZnO(0001) thin films on GaN(0001) templates and ZnO(0001) substrates. *Physica status solidi (c)* 5 (9): 3091–3094.
231. Kashiwaba, Y., Kato, H., Kikuchi, T. et al. (2005). Homoepitaxial growth of ZnO films on ZnO (11-20) substrates. *Applied Surface Science* 244 (1–4): 373–376.
232. Abe, T., Kashiwaba, Y., Onodera, S. et al. (2007). Homoepitaxial growth of non-polar ZnO (11-20) films on off-angle ZnO substrates by MOCVD. *Journal of Crystal Growth* 298: 457–460.
233. Kashiwaba, Y., Abe, T., Onodera, S. et al. (2007). Comparison of non-polar ZnO (11-20) films deposited on single crystal ZnO (11-20) and sapphire (01-12). substrates. *Journal of Crystal Growth* 298: 477–480.
234. Allen, M.W., Zemlyanov, D., Waterhouse, G.I.N. et al. (2011). Polarity effects in the X-ray photoemission of ZnO and other wurtzite semiconductors. *Applied Physics Letters* 98: 101906.
235. Sato, D., Kashiwaba, Y., Haga, K. et al. (2004). Effect of ZnO buffer layer prepared by rf sputtering on heteroepitaxial growth of high-quality ZnO films. *Vacuum* 74 (3–4): 601–605.
236. Choi, Y.-S., Hwang, D.-K., Kwon, B.-J. et al. (2011). Effect of VI/II gas ratio on the epitaxial growth of ZnO films by metalorganic chemical vapor deposition. *Japanese Journal of Applied Physics* 50 (10): 105502.
237. Gil, B. (1998). *Group III Nitride Semiconductor Compounds: Physics and Applications*. Clarendon Press.
238. Amano, H., Akasaki, I., Hiramatsu, K. et al. (1988). Effects of the buffer layer in metalorganic vapour phase epitaxy of GaN on sapphire substrate. *Thin Solid Films* 163: 415–420.
239. Pierce, J.M., Adekore, B.T., Davis, R.F., and Stevie, F.A. (2005). Growth of dense ZnO films via MOVPE on GaN (0001) epilayers using a low/high-temperature sequence. *Journal of Crystal Growth* 277 (1–4): 345–351.
240. Feng, Z., Yao, S., Hou, L., and Jin, R. (2005). Depth dependent elastic strain in ZnO epilayer: Combined Rutherford backscattering/channeling and X-ray diffraction. *Nuclear Instruments and Methods in Physics Research, Section B* 229 (2): 246–252.
241. Khranovskyy, V., Minikayev, R., Trushkin, S. et al. (2007). Improvement of ZnO thin film properties by application of ZnO buffer layers. *Journal of Crystal Growth* 308 (1): 93–98.
242. Dai, J., Su, H., Wang, L. et al. (2006). Properties of ZnO films grown on (0001) sapphire substrate using H_2O and N_2O as O precursors by atmospheric pressure MOCVD. *Journal of Crystal Growth* 290 (2): 426–430.
243. Tang, K., Huang, S., Gu, S. et al. (2016). The roles of buffer layer thickness on the properties of the ZnO epitaxial films. *Applied Surface Science* 388: 557–564.
244. Park, D.J., Kim, D.C., Lee, J.Y., and Cho, H.K. (2007). Epitaxial growth of ZnO layers using nanorods with high crystalline quality. *Nanotechnology* 18 (39): 395605.
245. Park, S.H., Seo, S.Y., Kim, S.H., and Han, S.W. (2007). Homoepitaxial ZnO film growth on vertically aligned ZnO nanorods. *Journal of Crystal Growth* 303 (2): 580–584.
246. Kasuga, M. and Mochizuki, M. (1981). Orientation relationships of zinc oxide on sapphire in heteroepitaxial chemical vapor deposition. *Journal of Crystal Growth* 54 (2): 185–194.
247. Liu, Y., Gorla, C.R., Liang, S. et al. (2000). Ultraviolet detectors based on epitaxial ZnO films grown by MOCVD. *Journal of Electronic Materials* 29: 69–74.
248. Galli, G. and Coker, J.E. (1970). Epitaxial ZnO on sapphire. *Applied Physics Letters* 16 (11): 439–441.
249. Nakamura, A., Shigemori, S., Shimizu, Y. et al. (2004). Hydroxyl-radical-assisted growth of ZnO films by remote plasma metal-organic chemical vapor deposition. *Japanese Journal of Applied Physics Part 1*, 43 (11 A): 7672–7676.
250. Fons, P., Iwata, K., Niki, S. et al. (2000). Uniaxial locked growth of high-quality epitaxial ZnO films on (1 1 -20)α-Al_2O_3. *Journal of Crystal Growth* 209 (2–3): 532–536.

251. Kaul, A.R., Gorbenko, O.Y., Botev, A.N., and Burova, L.I. (2005). MOCVD of pure and Ga-doped epitaxial ZnO. *Superlattices and Microstructures* 38 (4–6): 272–282.
252. Chao, Y.C., Lin, C.W., Ke, D.J. et al. (2007). Growth of epitaxial ZnO thin film on yttria-stabilized zirconia single-crystal substrate. *Journal of Crystal Growth* 298: 461–463.
253. Jia, C.H., Chen, Y.H., Liu, G.H. et al. (2009). Structural and optical properties of ZnO films on $SrTiO_3$ substrates by MOCVD. *Journal of Physics D: Applied Physics* 42 (1): 015415.
254. Lin, H., Zhou, S., Teng, H., and Wang, J. (2009). Growth and in-plane optical anisotropy of crystalline quality enhanced non-polar m-plane ZnO(GaN) films on trenched (100) $LiAlO_2$ substrates. *Journal of Crystal Growth* 311 (3): 456–458.
255. Ive, T., Ben-Yaacov, T., Van de Walle, C.G. et al. (2008). Step-flow growth of ZnO(0001) on GaN(0001) by metalorganic chemical vapor epitaxy. *Journal of Crystal Growth* 310 (15): 3407–3412.
256. Gruber, T., Kirchner, C., and Waag, A. (2002). MOCVD growth of ZnO on different substrate materials. *Physica status solidi (b)* 229 (2): 841–844.
257. Wang, J., Harada, Y., Sumiya, M. et al. (2015). Cathodoluminescence study of optical properties along the growth direction of ZnO films on GaN substrate. *Physica status solidi (c)* 12 (8): 1129–1131.
258. Ataev, B.M., Lundin, M.W., Mamedov, V.V. et al. (2001). Low-pressure chemical vapour deposition growth of high-quality ZnO films on epi-GaN/α -Al_2O_3. *Journal of Physics: Condensed Matter* 13: L211–L214.
259. Oleynik, N., Dadgar, A., Jurgen Blasing, M.A. et al. (2003). Metal organic vapor phase epitaxy of ZnO on GaN/Si (111) using tertiary-butanol as O- precursor. *Japanese Journal of Applied Physics* 7474 (Part 1 12): 7474–7477.
260. Lin, H., Zhou, S., Teng, H. et al. (2009). Structural and optical properties of nearly stress-free m-plane ZnO film on (100) γ-$LiAlO_2$ with a GaN buffer layer by metal-organic chemical vapor deposition. *Applied Surface Science* 255 (22): 9146–9148.
261. Souletie, P., Bethke, S., Wessels, B.W., and Pan, H. (1988). Growth and characterization of heteroepitaxial ZnO thin films by organometallic chemical vapor deposition. *Journal of Crystal Growth* 86 (1–4): 248–251.
262. Liu, C.Y., Zhang, B.P., Binh, N.T., and Segawa, Y. (2004). Second harmonic generation in ZnO thin films fabricated by metalorganic chemical vapor deposition. *Optics Communications* 237 (1–3): 65–70.
263. Zhang, B.P., Binh, N.T., Wakatsuki, K. et al. (2004). Low-temperature growth of single-crystalline ZnO tubes on sapphire(0001) substrates. *Applied Physics A: Materials Science and Processing* 79 (7): 1711–1714.
264. Park, W.I., An, S., Yi, G., and Jang, H.M. (2001). Metal-organic vapor phase epitaxial growth of high-quality ZnO films on Al_2O_3 (00.1). *Journal of Materials Research* 16 (5): 1358–1362.
265. Saraf, G., Lu, Y., and Siegrist, T. (2008). In-plane anisotropic strain in a-ZnO films grown on r -sapphire substrates. *Applied Physics Letters* 93 (4): 2006–2009.
266. Gruber, T., Kirchner, C., Thonke, K. et al. (2002). MOCVD growth of ZnO for optoelectronic applications. *Physica status solidi (a)* 192 (1): 166–170.
267. Wang, L., Pu, Y., Fang, W. et al. (2005). Effect of high-temperature annealing on the structural and optical properties of ZnO films. *Thin Solid Films* 491 (1–2): 323–327.
268. Zhu, G., Gu, S., Zhu, S. et al. (2012). Optimization study of metal-organic chemical vapor deposition of ZnO on sapphire substrate. *Journal of Crystal Growth* 349 (1): 6–11.
269. Zhang, P., Wei, H., Cong, G. et al. (2008). Effects of disk rotation rate on the growth of ZnO films by low-pressure metal-organic chemical vapor deposition. *Thin Solid Films* 516 (6): 925–928.
270. Seomoon, K., Lee, J., Jang, P. et al. (2011). Synthesis and characterization of ZnO thin films deposited via PE-MOCVD. *Current Applied Physics* 11: S26–S29.
271. Li, X., Zhang, B., Zhu, H. et al. (2008). Properties of ZnO thin films grown on Si substrates by photo-assisted MOCVD. *Applied Surface Science* 254 (7): 2081–2084.
272. Moriyama, T. and Fujita, S. (2007). Crystal growth of ZnO on Si(111) by metalorganic vapor phase epitaxy. *Journal of Crystal Growth* 298: 464–467.
273. Lan, S.M., Uen, W.Y., Chan, C.E. et al. (2008). Morphology and optical properties of zinc oxide thin films grown on Si (100) by metal-organic chemical vapor deposition. *Journal of Materials Science: Materials in Electronics* 20 (S1): 441–445.
274. Montenegro, D.N., Hortelano, V., Martínez, O. et al. (2013). Influence of metal organic chemical vapour deposition growth conditions on vibrational and luminescent properties of ZnO nanorods. *Journal of Applied Physics* 113: 143513.
275. Agouram, S., Zuñiga Perez, J., and Muñoz-Sanjosé, V. (2007). Structural and morphological characterizations of ZnO films grown on GaAs substrates by MOCVD. *Applied Physics A: Materials Science & Processing* 88 (1): 83–87.
276. Souletie, P. and Wessels, B.W. (1988). Growth kinetics of ZnO prepared by organometallic chemical vapor deposition. *Journal of Materials Research* 3 (4): 740–744.
277. Roro, K.T., Dangbegnon, J.K., Sivaraya, S. et al. (2008). Influence of metal organic chemical vapor deposition growth parameters on the luminescent properties of ZnO thin films deposited on glass substrates. *Journal of Applied Physics* 103 (5): 053516.

278. Terasako, T., Yamanaka, T., Yura, S. et al. (2010). Photoluminescence, photoacoustic and Raman spectra of zinc oxide films grown by LP-MOCVD using diethylzinc and water as precursors. *Thin Solid Films* 519 (5): 1546–1551.
279. Wu, Y., Liu, D., Yu, N. et al. (2013). Structure and electrical characteristics of zinc oxide thin films grown on Si (111) by metal-organic chemical vapor deposition. *Journal of Materials Science and Technology* 29 (9): 830–834.
280. Kim, H.W., Kim, N.H., and Lee, C. (2004). Very low temperature growth of ZnO thin films on Si substrates using the metalorganic chemical vapor deposition technique. *Journal of the Korean Physical Society* 44 (1): 14–17.
281. Tian, J.-S., Liang, M.-H., Ho, Y.-T. et al. (2008). Growth of a-plane ZnO thin films on $LaAlO_3$ (100) substrate by metal-organic chemical vapor deposition. *Journal of Crystal Growth* 310: 777–782.
282. Kim, D.C., Kong, B.H., Cho, H.K. et al. (2007). Low temperature growth of ZnO thin film by metalorganic chemical vapor deposition. *Physica status solidi (b)* 244 (5): 1512–1516.
283. Thiandoume, C., Barjon, J., Ka, O. et al. (2009). Morphology transition of one-dimensional ZnO grown by metal organic vapour phase epitaxy on (0001)-ZnO substrate. *Journal of Crystal Growth* 311 (18): 4311–4316.
284. Khranovskyy, V., Ulyashin, A., Lashkarev, G. et al. (2008). Morphology, electrical and optical properties of undoped ZnO layers deposited on silicon substrates by PEMOCVD. *Thin Solid Films* 516 (7): 1396–1400.
285. Wang, J., Elamurugu, E., Sallet, V. et al. (2008). Influence of different carrier gases on the properties of ZnO films grown by MOCVD. *Boletin de la Sociedad Española de Ceramica y Vidrio* 47 (4): 242–244.
286. Mao, H., Gu, S., Ye, J. et al. (2015). Comparative study of the effect of H_2 addition on ZnO films grown by different zinc and oxygen precursors. *Journal of Materials Research* 30 (7): 935–945.
287. Ye, J.D., Gu, S.L., Liu, W. et al. (2007). Competitive adsorption and two-site occupation effects in metal-organic chemical vapor deposition of ZnO. *Applied Physics Letters* 90 (17): 2005–2008.
288. Kim, H. and Jeong, S. (2007). Route from ZnO thin films to nanostructures on Si substrates by metal organic chemical vapor deposition. *Journal of the Korean Physical Society* 51: 207–211.
289. Pagni, O. and Leitch, A.W.R. (2004). Influence of VI:II ratio on the properties of MOCVD-grown ZnO thin films. *Physica status solidi (c)* 1 (9): 2213–2218.
290. Kong, B.H., Kim, D.C., Mohanta, S.K., and Cho, H.K. (2010). Influence of VI/II ratios on the growth of ZnO thin films on sapphire substrates by low temperature MOCVD. *Thin Solid Films* 518 (11): 2975–2979.
291. Yang, J., Pei, Y., Hu, R. et al. (2012). Morphology controlled synthesis of crystalline ZnO film by MOCVD: from hexagon to rhombus. *Crystal Engineering Communications* 14 (24): 8345–8348.
292. Roro, K.T., Botha, J.R., and Leitch, A.W.R. (2008). Effect of deposition conditions on the growth rate and electrical properties of ZnO thin films grown by MOCVD. *Physica status solidi (c)* 5 (2): 563–565.
293. Venkatachalapathy, V., Galeckas, A., Zubiaga, A. et al. (2010). Changing vacancy balance in ZnO by tuning synthesis between zinc/oxygen lean conditions. *Journal of Applied Physics* 108 (4): 1–4.
294. Marzouki, A., Sayari, A., Jomard, F. et al. (2013). Carrier gas and VI/II ratio effects on carbon clusters incorporation into ZnO films grown by MOCVD. *Materials Science in Semiconductor Processing* 16 (3): 1022–1028.
295. Boukadhaba, M.A., Fouzri, A., Sallet, V. et al. (2015). Characterization of ZnO thin film grown on c-plane substrates by MO-CVD: Effect of substrate annealing temperature, vicinal-cut angle and miscut direction. *Superlattices and Microstructures* 85: 820–834.
296. Zhang, P.F., Liu, X.L., Wei, H.Y. et al. (2007). Rapid thermal annealing properties of ZnO films grown using methanol as oxidant. *Journal of Physics D: Applied Physics* 40 (19): 6010–6013.
297. Sayari, A., Marzouki, A., Lusson, A. et al. (2010). Annealing and partial pressure ratio effects on ZnO films grown by metal-organic chemical vapor deposition using tert-butanol as oxidant. *Thin Solid Films* 518 (23): 6870–6875.
298. Sun, J., Yang, T., Du, G. et al. (2006). Influence of annealing atmosphere on ZnO thin films grown by MOCVD. *Applied Surface Science* 253 (4): 2066–2070.
299. Ye, J.D., Gu, S.L., Zhu, S.M. et al. (2004). Production of high-quality ZnO films by the two-step annealing method. *Journal of Applied Physics* 96 (9): 5308–5310.
300. Wang, L., Zhang, X., Yang, X. et al. (2013). H_2-CH_4 mixed gas plasma treatment on LP-MOCVD ZnO:B for amorphous silicon thin film solar cells. *Solar Energy Materials and Solar Cells* 116: 231–237.
301. Ding, L., Nicolay, S., Steinhauser, J. et al. (2013). Relaxing the conductivity/transparency trade-off in MOCVD ZnO thin films by hydrogen plasma. *Advanced Functional Materials* 23 (41): 5177–5182.
302. Cui, Y., Du, G., Zhang, Y. et al. (2005). Growth of ZnO(002) and ZnO(100) films on GaAs substrates by MOCVD. *Journal of Crystal Growth* 282 (3): 389–393.
303. Hongsingthong, A., Krajangsang, T., Limmanee, A. et al. (2013). Development of textured ZnO-coated low-cost glass substrate with very high haze ratio for silicon-based thin film solar cells. *Thin Solid Films* 537: 291–295.
304. Addonizio, M.L. and Fusco, L. (2015). Preparation method of double-textured ZnO:B films deposited by MOCVD on plasma etched polymer buffer. *Journal of Alloys and Compounds* 622: 851–858.
305. Chiu, H.-M. and Wu, J.-M. (2013). Opto-electrical properties and chemisorption reactivity of Ga-doped ZnO nanopagodas. *Journal of Materials Chemistry A* 1 (18): 5524–5534.
306. Park, W.I., Jun, Y.H., Jung, S.W., and Yi, G.C. (2003). Excitonic emissions observed in ZnO single crystal nanorods. *Applied Physics Letters* 82 (6): 964–966.

307. Kar, J.P., Das, S.N., Choi, J.H. et al. (2010). Study of the morphological evolution of ZnO nanostructures on various sapphire substrates. *Applied Surface Science* 256 (16): 4995–4999.
308. Yi, G.-C., Wang, C., and Park, W. Il. (2005). ZnO nanorods: synthesis, characterization and applications. *Semiconductor Science and Technology* 20 (4): S22–S34.
309. Kim, S.W., Ueda, M., Funato, M. et al. (2005). Artificial control of ZnO nanodots by ion-beam nanopatterning. *Journal of Applied Physics* 97: 104316.
310. Biswas, P., Kundu, S., Banerji, P., and Bhunia, S. (2013). Super rapid response of humidity sensor based on MOCVD grown ZnO nanotips array. *Sensors and Actuators, B* 178: 331–338.
311. Gomez, J.L. and Tigli, O. (2013). Zinc oxide nanostructures: From growth to application. *Journal of Materials Science* 48 (2): 612–624.
312. Panda, D. and Tseng, T.Y. (2013). One-dimensional ZnO nanostructures: Fabrication, optoelectronic properties, and device applications. *Journal of Materials Science* 48 (20): 6849–6877.
313. Zhang, Y., Heng, L., and Jiang, L. (2014). Chemically controllable fabrication of one-dimensional ZnO nanostructures and their applications in solar cells. *Journal of Nanoscience and Nanotechnology* 14 (8): 5597–5613.
314. Willander, M., Nur, O., Zhao, Q.X., and Yang, L.L. (2009). Zinc oxide nanorod based photonic devices: recent progress in growth light emitting diodes and lasers. *Nanotechnology* 20 (33): 332001.
315. Park, D.J., Lee, J.Y., Kim, D.C. et al. (2007). Defects in interfacial layers and their role in the growth of ZnO nanorods by metallorganic chemical vapor deposition. *Applied Physics Letters* 91 (14): 2005–2008.
316. Kwak, C.H., Kim, B.H., Park, S.H. et al. (2009). In-situ and ex-situ ZnO nanorod growth on ZnO homo-buffer layers. *Journal of Crystal Growth* 311 (20): 4491–4494.
317. Jung, I.O., Park, J.Y., and Kim, S.S. (2012). Substrate dependent growth modes of ZnO nanorods grown by metalorganic chemical vapor deposition. *Journal of Crystal Growth* 355 (1): 78–83.
318. Park, W.I., Yoo, J., and Yi, G.C. (2005). Catalyst-free metalorganic chemical-vapor deposition of ultrafine ZnO nanorods. *Journal of the Korean Physical Society* 46 (5): L1067–L1070.
319. Khranovskyy, V. and Yakimova, R. (2012). Morphology engineering of ZnO nanostructures. *Physica B: Condensed Matter* 407 (10): 1533–1537.
320. Mohanta, S.K., Kim, D.C., Cho, H.K. et al. (2008). Influence of buffer layer on structural and optical properties of ZnO nanorods on glass substrates. *Electrochemical and Solid-State Letters* 11 (6): H143–H146.
321. Kim, D.-S., Lee, D., Lee, J.-H., and Byun, D. (2014). Growth and characterization of a multi-dimensional ZnO hybrid structure on a glass substrate by using metal organic chemical vapor deposition. *Journal of the Korean Physical Society* 64 (10): 1524–1528.
322. Tsiaoussis, I., Khranovskyy, V., Dimitrakopulos, G.P. et al. (2011). Structural characterization of ZnO nanopillars grown by atmospheric-pressure metalorganic chemical vapor deposition on vicinal 4H-SiC and SiO_2/Si substrates. *Journal of Applied Physics* 109 (4): 043507.
323. Kwak, C.H., Kim, B.H., Park, C.I. et al. (2011). Synthesis and microstructural properties of ZnO nanorods on Ti-buffer layers. *Journal of Crystal Growth* 314 (1): 264–267.
324. Kar, J.P., Das, S.N., Choi, J.H. et al. (2011). Growth modulation and photoresponse characteristics of vertically aligned ZnO nanowires. *Applied Surface Science* 257 (11): 4973–4977.
325. Wei, H.Y., Cong, G.W., Zhang, P.F. et al. (2007). Combined structure of ZnO vertical well-aligned nanorods and net-like structures on AlN/sapphire. *Journal of Crystal Growth* 306 (1): 12–15.
326. Fujisawa, H., Imi, Y., Nakashima, S. et al. (2012). Selective growth of ZnO nanorods and their applications to ferroelectric nanorods. *Journal of Applied Physics* 112 (3): 034111.
327. Amiri, G., Halaoui, M., Montenegro, D.N. et al. (2016). MOCVD growth of ZnO nanowires on Ni–W metallic substrates. *Physica status solidi (c)* 13 (7–9): 430–434.
328. Burova, L.I., Perov, N.S., Semisalova, A.S. et al. (2012). Effect of the nanostructure on room temperature ferromagnetism and resistivity of undoped ZnO thin films grown by chemical vapor deposition. *Thin Solid Films* 520 (14): 4580–4585.
329. Kim, Y.-J., Lee, J.-H., and Yi, G.-C. (2009). Vertically aligned ZnO nanostructures grown on graphene layers. *Applied Physics Letters* 95 (21): 213101.
330. Zhang, G., Adachi, M., Ganjil, S. et al. (2007). Vertically aligned single-crystal ZnO nanotubes grown on γ-$LiAlO_2$ (100) substrate by metalorganic chemical vapor deposition. *Japanese Journal of Applied Physics Part 2: Letters* 46 (29): L730–L732.
331. Jung, B.O., Kim, D.C., Kong, B.H. et al. (2011). Fully transparent vertically aligned ZnO nanostructure-based ultraviolet photodetectors with high responsivity. *Sensors and Actuators, B: Chemical* 160 (1): 740–746.
332. Yoo, J., Picraux, S.T., and Yi, G.-C. (2011). Growth of ZnO-based nanorod heterostructures and their photonic device applications. *Proceedings of the SPIE* 8106: 81060N1–81060N8.
333. Lee, W.N., Jeong, M.C., and Myoung, J.M. (2004). Fabrication and application potential of ZnO nanowires grown on GaAs(002) substrates by metal-organic chemical vapour deposition. *Nanotechnology* 15 (3): 254–259.
334. Grassi, M., Soares, D.A.W., de Queiroz, A.A.A. et al. (2004). Organometallic chemical vapor deposition of compound semiconductors. *Materials Science and Engineering B* 112 (2–3): 179–181.

335. Park, J.Y., Yun, Y.S., Hong, Y.S. et al. (2006). Synthesis and electrical properties of aligned ZnO nanocolumns. *Composites Part B: Engineering*, 37 (6): 408–412.
336. Park, D., Tak, Y., Kim, J., and Yong, K. (2007). Low-temperature synthesized ZnO nanoneedles: XPS and PL analysis. *Surface Review and Letters* 14 (6): 1061–1065.
337. Park, D.J., Kim, D.C., Lee, J.Y., and Cho, H.K. (2006). Synthesis and microstructural characterization of growth direction controlled ZnO nanorods using a buffer layer. *Nanotechnology* 17 (20): 5238–5243.
338. Park, J.Y., Chang, C.H., and Kim, S.S. (2005). Early stage growth behavior of laser-deposited heteroepitaxial ZnO films on Al_2O_3 (0001). *Journal of Crystal Growth* 281 (2–4): 440–445.
339. Perillat-Merceroz, G., Jouneau, P.H., Feuillet, G. et al. (2010). MOCVD growth mechanisms of ZnO nanorods. *Journal of Physics: Conference Series* 209 (1): 012034.
340. Chen, H.G., Jian, S.R., Li, Z.W. et al. (2010). Epitaxial growth of self-arranged periodic ZnO nanostructures on sapphire substrates grown by MOCVD. *Journal of Alloys and Compounds* 504S: S368–S371.
341. Lee, D.H., Son, K., and Park, W.I. (2010). Patterned synthesis of laterally oriented ultrafine ZnO nanowires by controlled metalorganic chemical vapour deposition. *Journal of Physics D: Applied Physics* 43 (24): 245402.
342. Falyouni, F., Benmamas, L., Thiandoume, C. et al. (2009). Metal organic chemical vapor deposition growth and luminescence of ZnO micro- and nanowires. *Journal of Vacuum Science and Technology B* 27 (3): 1662–1666.
343. Zhang, B.P., Binh, N.T., Wakatsuki, K. et al. (2004). Pressure-dependent ZnO nanocrsytal growth in a chemical vapor deposition process. *J Phys Chem B* 108: 10899–10902.
344. Zhang, B.P., Binh, N.T., Wakatsuki, K. et al. (2004). Formation of highly aligned ZnO tubes on sapphire (0001) substrates. *Applied Physics Letters* 84 (20): 4098–4100.
345. Markov, I.V. (2003). Chapter 1 In: *Crystal Growth for Beginners. Fundamentals of Nucleation, Crystal Growth and Epitaxy*. World Scientific.
346. Fujisawa, H., Kobayashi, C., Nakashima, S., and Shimizu, M. (2013). Two-step growth of ZnO nanorods by using MOCVD and control of their diameters and surface densities. *Journal of the Korean Physical Society* 62 (8): 1164–1168.
347. Kim, D.C., Kong, B.H., Cho, H.K. et al. (2006). Effects of buffer layer thickness on growth and properties of ZnO nanorods grown by metalorganic chemical vapour deposition. *Nanotechnology* 18 (1): 015603.
348. Wu, C.C., Wuu, D.S., Lin, P.R. et al. (2010). Realization and manipulation of ZnO nanorod arrays on sapphire substrates using a catalyst-free metalorganic chemical vapor deposition technique. *Journal of Nanoscience and Nanotechnology* 10 (5): 3001–3011.
349. Yang, L.M., Ye, Z.Z., Zeng, Y.J. et al. (2006). Density controllable growth of ZnO quantum dots by MOCVD. *Solid State Communications* 138 (12): 577–579.
350. Chin, S.-C., Chi, C.-Y., Lu, Y.-C. et al. (2006). Nano-structure study of ZnO thin films on sapphire grown with different temperature conditions. *Journal of Crystal Growth* 293 (2): 344–350.
351. Zhou, X., Gu, S., Qin, F. et al. (2004). MOCVD growth of self-arranged ZnO nanosize islands. *Journal of Crystal Growth* 269 (2–4): 362–366.
352. Oshinowo, J., Nishioka, M., Ishida, S., and Arakawa, Y. (1994). Area density control of quantum-size InGaAs/Ga(Al)As dots by metalorganic chemical vapor deposition. *Japanese Journal of Applied Physics Part 2* 33 (11B): L1634–L1637.
353. Park, J.Y., Lee, D.J., Yun, Y.S. et al. (2005). Temperature-induced morphological changes of ZnO grown by metalorganic chemical vapor deposition. *Journal of Crystal Growth* 276 (1–2): 158–164.
354. Wu, C.-C., Wuu, D.-S., Chen, T.-N. et al. (2008). Growth and characterization of epitaxial ZnO nanowall networks using metal organic chemical vapor deposition. *Japanese Journal of Applied Physics* 47 (1): 746–750.
355. Sallet, V., Falyouni, F., Zeuner, A. et al. (2008). Some aspects of the MOCVD growth of ZnO nanorods by using N_2O. *Journal of the Korean Physical Society* 53 (925): 3051–3054.
356. Kim, D.C., Kong, B.H., and Cho, H.K. (2009). Synthesis and growth mechanism of catalyst free ZnO nanorods with enhanced aspect ratio by high flow additional carrier gas at low temperature. *Journal of Physics D: Applied Physics* 42 (6): 65406.
357. Park, J.Y., Lee, D.J., and Kim, S.S. (2005). Size control of ZnO nanorod arrays grown by metalorganic chemical vapour deposition. *Nanotechnology* 16 (10): 2044–2047.
358. Kim, H.W., Kim, N.H., Shim, J.-H. et al. (2005). Catalyst-free MOCVD growth of ZnO nanorods and their structural characterization. *Journal of Materials Science: Materials in Electronics* 16 (1): 13–15.
359. Xu, S., Lao, C., Weintraub, B., and Wang, Z.L. (2008). Density-controlled growth of aligned ZnO nanowire arrays by seedless chemical approach on smooth surfaces. *Journal of Materials Research* 23 (8): 2072–2077.
360. Jeong, M.-C., Oh, B.-Y., Lee, W., and Myoung, J.-M. (2004). Comparative study on the growth characteristics of ZnO nanowires and thin films by metalorganic chemical vapor deposition (MOCVD). *Journal of Crystal Growth* 268 (1–2): 149–154.

361. Song, T., Choung, J.W., Park, J.-G. et al. (2008). Surface polarity and shape-controlled synthesis of ZnO nanostructures on GaN thin films based on catalyst-free metalorganic vapor phase epitaxy. *Advanced Materials* 20: 4464–4469.
362. Montenegro, D.N., Souissi, A., Martínez-Tomás, C. et al. (2012). Morphology transitions in ZnO nanorods grown by MOCVD. *Journal of Crystal Growth* 359 (1): 122–128.
363. Fouzri, A., Sallet, V., and Oumezzine, M. (2011). A comparative structure and morphology study of $Zn_{(1-x)}Cd_xO$ solid solution grown on ZnO and different sapphire orientations. *Journal of Crystal Growth* 331 (1): 18–24.
364. Fouzri, A., Boukadhaba, M.A., Touré, A. et al. (2014). Structural, morphological and optical properties of Cd doped ZnO film grown on a- and r-plane sapphire substrate by MOCVD. *Applied Surface Science* 311: 648–658.
365. Fouzri, A., Boukadhaba, M.A., Oumezzine, M., and Sallet, V. (2012). Structural properties and morphology of $Zn_{(1-x)}Cd_xO$ solid solution grown on ZnO and c-plane sapphire substrate. *Thin Solid Films* 520 (7): 2582–2588.
366. Debbichi, M., Souissi, M., Fouzri, A. et al. (2014). Room temperature ferromagnetism in Cd-doped ZnO thin films through defect engineering. *Journal of Alloys and Compounds* 598: 120–125.
367. Bertram, F., Giemsch, S., Forster, D. et al. (2006). Direct imaging of phase separation in ZnCdO layers. *Applied Physics Letters* 88 (6): 129–132.
368. Shigemori, S., Nakamura, A., Ishihara, J. et al. (2004). $Zn_{1-x}Cd_xO$ film growth using remote plasma-enhanced metalorganic chemical vapor deposition. *Japanese Journal of Applied Physics* 43 (8B): L1088–L1090.
369. Zúñiga-Pérez, J. (2017). ZnCdO : Status after 20 years of research. *Materials Science in Semiconductor Processing* 69: 36–43.
370. Zhang, T.C. and Kuznetsov, A.Y. (2011). Surface/strain energy balance controlling preferred orientation in CdZnO films. *Journal of Applied Physics* 110 (5): 1–6.
371. Park, W.I., Yi, G.C., and Jang, H.M. (2001). Metalorganic vapor-phase epitaxial growth and photoluminescent properties of $Zn_{1-x}Mg_xO(0 \leq x \leq 0.49)$. thin films. *Applied Physics Letters* 79 (13): 2022–2024.
372. Han, S., Zhang, J., Zhang, Z. et al. (2010). $Mg_{0.58}Zn_{0.42}O$ thin films on MgO substrates with MgO buffer layer. *ACS Applied Materials and Interfaces* 2 (7): 1918–1921.
373. Torres-Huerta, A.M., Domínguez-Crespo, M.A., Brachetti-Sibaja, S.B. et al. (2011). Effect of the substrate on the properties of ZnO–MgO thin films grown by atmospheric pressure metal-organic chemical vapor deposition. *Thin Solid Films* 519 (18): 6044–6052.
374. Shi, K., Yang, A.L., Wang, J. et al. (2011). The effect of different oriented sapphire substrates on the growth of polar and non-polar ZnMgO by MOCVD. *Journal of Crystal Growth* 314 (1): 39–42.
375. Alema, F., Hertog, B., Ledyaev, O. et al. (2016). Temperature and pulse duration effects on the growth of MgZnO via pulsed metal organic chemical vapor deposition. *Japanese Journal of Applied Physics* 55: 035501.
376. Kim, D.C., Kong, B.H., Ahn, C.H., and Cho, H.K. (2009). Characteristics improvement of metalorganic chemical vapor deposition grown MgZnO films by MgO buffer layers. *Thin Solid Films* 518 (4): 1185–1189
377. Thiandoume, C., Lusson, A., Galtier, P., and Sallet, V. (2010). Temperature dependence of $Zn_{1-x}Mg_xO$ films grown on c-plane sapphire by metal organic vapor phase epitaxy. *Journal of Crystal Growth* 312 (9): 1529–1533.
378. Shi, Z., Liu, D., Yan, X. et al. (2008). Effect of oxygen partial pressure on conductivity type of MgZnO nanocrystalline thin films prepared by metal-organic chemical vapor deposition. *Microelectronics Journal*, 39 (12): 1583–1586.
379. Wu, C.C., Wuu, D.S., Lin, P.R. et al. (2010). Characterization of $Mg_xZn_{1-x}O$ thin films grown on sapphire substrates by metalorganic chemical vapor deposition. *Thin Solid Films* 519: 1966–1970.
380. Yang, A.L., Song, H.P., Liang, D.C. et al. (2010). Photoluminescence spectroscopy and positron annihilation spectroscopy probe of alloying and annealing effects in nonpolar m -plane ZnMgO thin films. *Applied Physics Letters* 96 (15): 4–7.
381. Ito, S., Sumiya, M., Mieno, M., and Koinuma, H. (2010). Growth of $Mg_xZn_{1-x}O$ film by MOCVD equipped laser heating system. *Materials Science and Engineering B* 173 (1–3): 11–13.
382. Lin, W.-H., Chou, M.M.C., and Wu, J.-J. (2011). Growth characterization, and polarity dentification of (0001) $Zn_{1-x}Mg_xO$ epitaxial films on lattice-matched β-$LiGaO_2$ (001) substrates. *Journal of the Electrochemical Society* 158 (1): D28–D32.
383. Meng, F.Y., Chiba, Y., Yamada, A., and Konagai, M. (2007). Growth of $Zn_{1-x}Mg_xO$ films with single wurtzite structure by MOCVD process and their application to $Cu(InGa)(SSe)_2$ solar cells. *Solar Energy Materials and Solar Cells*, 91 (20): 1887–1891.
384. Kim, D.C., Lee, J.H., Mohanta, S.K. et al. (2010). Structural transition from MgZnO nanowires to ultrathin nanowalls by surface separation: growth evolution and gas sensing properties. *Nanotechnology* 21 (42): 425503.
385. Lee, J.H., Kim, D.C., Hwang, S. et al. (2011). A facile method for morphological control of MgZnO nanostructures on GaAs substrates and their optical properties. *Journal of Nanoscience and Nanotechnology* 11 (8): 7327–7330.
386. Yang, A.L., Wei, H.Y., Liu, X.L. et al. (2009). Synthesis and characterization of well-aligned $Zn_{1-x}Mg_xO$ nanorods and film by metal organic chemical vapor deposition. *Journal of Crystal Growth* 311 (2): 278–281.

387. Zeng, Y.J., Ye, Z.Z., Lu, Y.F. et al. (2007). ZnMgO quantum dots grown by low-pressure metal organic chemical vapor deposition. *Applied Physics Letters* 90 (1): 2005–2008.
388. Yang, A.L., Wei, H. Y., Liu, X.L. et al. (2008). Characterization of ZnMgO hexagonal-nanotowers/films on m-plane sapphire synthesized by metal organic chemical vapour deposition. *Journal of Physics D: Applied Physics* 41 (20): 205416.
389. Elleuch, R., Salhi, R., Deschanvres, J.-L., and Maalej, R. (2015). Antireflective downconversion ZnO:Er^{3+},Yb^{3+} thin film for Si solar cell applications. *Journal of Applied Physics* 117 (5): 055301.
390. Terai, Y., Yamaoka, K., Yamaguchi, T., and Fujiwara, Y. (2009). Structural and luminescent properties of Er-doped ZnO films grown by metalorganic chemical vapor deposition. *Journal of Vacuum Science and Technology B* 27 (5): 2248–2251.
391. Saraf, L. V., Zhu, Z.H., Wang, C.M., and Engelhard, M.H. (2009). Microstructure and secondary phase segregation correlation in epitaxial/oriented ZnO films with unfavorable Cr dopant. *Journal of Materials Research* 24 (2): 506–515.
392. Schlenker, E., Bakin, A., Postels, B. et al. (2007). Magnetic characterization of ZnO doped with vanadium. *Superlattices and Microstructures* 42 (1–6): 236–241.
393. Yan, W., Sun, Z., Liu, Q. et al. (2007). Zn vacancy induced room-temperature ferromagnetism in Mn-doped ZnO. *Applied Physics Letters* 91 (6): 1–4.
394. Hu, B., Adachi, M., Yamamoto, K. et al. (2010). Growth of $Zn_{1-x}Cu_xO$ films by remote plasma-enhanced metalorganic chemical vapor deposition with $Cu(dibm)_2$. *Physica status solidi (c)*7 (6): 1571–1573.
395. Theys, B., Sallet, V., Jomard, F. et al. (2002). Effects of intentionally introduced hydrogen on the electrical properties of ZnO layers grown by metalorganic chemical vapor deposition. *Journal of Applied Physics* 91 (6): 3922–3924.
396. Nickel, N.H., Friedrich, F., Rommeluère, J.F., and Galtier, P. (2005). Vibrational spectroscopy of undoped and nitrogen-doped ZnO grown by metalorganic chemical vapor deposition. *Applied Physics Letters* 87 (21): 211905.
397. Li, X., Asher, S.E., Limpijumnong, S. et al. (2006). Unintentional doping and compensation effects of carbon in metal-organic chemical-vapor deposition fabricated ZnO thin films. *Journal of Vacuum Science and Technology A* 24 (4): 1213–1217.
398. Kato, H., Sano, M., Miyamoto, K., and Yao, T. (2002). Growth and characterization of Ga-doped ZnO layers on a-plane sapphire substrates grown by molecular beam epitaxy. *Journal of Crystal Growth* 237–239 (1–4): 538–543.
399. Ko, H.J., Chen, Y.F., Hong, S.K. et al. (2000). Ga-doped ZnO films grown on GaN templates by plasma-assisted molecular-beam epitaxy. *Applied Physics Letters* 77 (23): 3761–3763.
400. Nguyen, N.G., Ho, V.T.T., and Hong, L.S. (2013). Low-resistivity, high-transmittance Ga:ZnO films prepared through metalorganic chemical vapor deposition using an inexpensive solution of diethylzinc in n-hexane as the Zn precursor. *Applied Physics Letters* 102 (18): 10–14.
401. Kuprenaite, S., Murauskas, T., Abrutis, A. et al. (2015). Properties of In-, Ga-, and Al-doped ZnO films grown by aerosol-assisted MOCVD: Influence of deposition temperature, doping level and annealing. *Surface and Coatings Technology* 271: 156–164.
402. Illiberi, A., Simons, P.J.P.M., Kniknie, B. et al. (2012). Growth of ZnO_x:Al by high-throughput CVD at atmospheric pressure. *Journal of Crystal Growth* 347 (1): 56–61.
403. Brochen, S., Lafossas, M., Robin, I.C. et al. (2014). Residual and intentional n-type doping of ZnO thin films grown by metal-organic vapor phase epitaxy on sapphire and ZnO substrates. *Journal of Applied Physics* 115 (11): 113508.
404. Lei, P.H., Hsu, C.M., and Fan, Y.S. (2013). Flexible organic light-emitting diodes on a polyestersulfone (PES) substrate using Al-doped ZnO anode grown by dual-plasma-enhanced metalorganic deposition system. *Organic Electronics* 14 (1): 236–249.
405. Kumar, E.S., Anderson, I.P., Deng, Z. et al. (2013). Effect of Group-III donors on high-resolution photoluminescence and morphology of ZnO nanowires grown by metalorganic vapour phase epitaxy. *Semiconductor Science and Technology* 28 (4): 045014.
406. Kuprenaite, S., Abrutis, A., Kubilius, V. et al. (2016). Effects of annealing conditions and film thickness on electrical and optical properties of epitaxial Al-doped ZnO films. *Thin Solid Films* 599: 19–26.
407. Manzi, J.A., Knapp, C.E., Parkin, I.P., and Carmalt, C.J. (2016). Aerosol assisted chemical vapour deposition of transparent conductive aluminum-doped zinc oxide thin films from a zinc triflate precursor. *Thin Solid Films* 616: 477–481.
408. Fragala, M.E., Malandrino, G., Giangregorio, M.M. et al. (2009). Structural, optical, and electrical characterization of ZnO and Al-doped ZnO thin films deposited by MOCVD. *Chemical Vapor Deposition* 15 (10–12): 327–333.
409. Watkins, S.P., Deng, Z.W., Li, D.C., and Huang, H. (2011). High resolution photoluminescence spectroscopy of donors in undoped and In-doped ZnO grown by metalorganic vapor phase epitaxy. *Journal of Applied Physics* 110 (8): 083506.
410. Tabuchi, K., Wenas, W.W., Yamada, A. et al. (1993). Optimization of ZnO films for amorphous silicon solar cells. *Japanese Journal of Applied Physics Part 1*, 32 (9A): 3764–3769.

411. Maejima, K., Koida, T., Sai, H. et al. (2014). Influences of deposition temperature on characteristics of B-doped ZnO films deposited by metal-organic chemical vapor deposition. *Thin Solid Films* 559: 83–87.
412. Fan, J.C., Sreekanth, K.M., Xie, Z. et al. (2013). P-type ZnO materials: theory, growth properties and devices. *Progress in Materials Science* 58 (6): 874–985.
413. Kun, T., Shulin, G., Jiandong, Y., and Shunming, Z. (2016). High-quality ZnO growth, doping, and polarization effect. *Journal of Semiconductors* 37 (3): 1–13.
414. Nakahara, K., Akasaka, S., Yuji, H. et al. (2010). Nitrogen doped $Mg_xZn_{1-x}O$/ZnO single heterostructure ultraviolet light-emitting diodes on ZnO substrates. *Applied Physics Letters* 97 (1): 013501.
415. Chu, S., Wang, G., Zhou, W. et al. (2011). Electrically pumped waveguide lasing from ZnO nanowires. *Nature Nanotechnology* 6 (8): 506–510.
416. Look, D.C., Hemsky, J.W., and Sizelove, J.R. (1999). Residual native shallow donor in ZnO. *Physical Review Letters* 82 (12): 2552–2555.
417. Zhang, S.B., Wei, S.-H., and Zunger, A. (2001). Intrinsic n -type versus p -type doping asymmetry and the defect physics of ZnO. *Physical Review B* 63 (7): 075205.
418. Hofmann, D.M., Hofstaetter, A., Leiter, F. et al. (2002). Hydrogen: A relevant shallow donor in zinc oxide. *Physical Review Letters* 88 (4): 045504.
419. Maksimov, O. (2010). Recent advances and novel approaches of p-type doping of zinc oxide. *Reviews on Advanced Materials Science* 24: 26–34.
420. Liu, Y., Liang, H., Xu, L. et al. (2010). Cu related doublets green band emission in ZnO:Cu thin films. *Journal of Applied Physics* 108 (11): 1–5.
421. Ye, J.D., Gu, S.L., Zhu, S.M. et al. (2006). Raman study of lattice dynamic behaviors in phosphorus-doped ZnO films. *Applied Physics Letters* 88 (10): 20–23.
422. Limpijumnong, S., Zhang, S.B., Wei, S.H., and Park, C.H. (2004). Doping by large-size-mismatched impurities: The microscopic origin of arsenic or antimony-doped p-type zinc oxide. *Physical Review Letters* 92 (15): 155504.
423. Pan, X.H., Jiang, J., Zeng, Y.J. et al. (2008). Electrical and optical properties of phosphorus-doped p -type ZnO films grown by metalorganic chemical vapor deposition. *Journal of Applied Physics* 103 (2): 1–5.
424. Das, S., Patra, S., Prakash Kar, J. et al. (2015). Origin of p-type conductivity for N-doped ZnO nanostructure synthesized by MOCVD method. *Materials Letters* 161: 701–704.
425. Pan, M., Nause, J., Rengarajan, V. et al. (2007). Epitaxial growth and characterization of p-type ZnO. *Journal of Electronic Materials* 36 (4): 457–461.
426. Yan, Y., Zhang, S.B., and Pantelides, S.T. (2001). Control of doping by impurity chemical potentials: Predictions for p-type ZnO. *Physical Review Letters* 86 (25): 5723–5726.
427. Li, X., Yan, Y., Gessert, T.A. et al. (2003). Chemical vapor deposition-formed p-type ZnO thin films. *Journal of Vacuum Science and Technology A* 21 (4): 1342–1346.
428. Li, X., Keyes, B., Asher, S. et al. (2005). Hydrogen passivation effect in nitrogen-doped ZnO thin films. *Applied Physics Letters* 86 (12): 1–3.
429. Keyes, B.M., Gedvilas, L.M., Li, X., and Coutts, T.J. (2005). Infrared spectroscopy of polycrystalline ZnO and ZnO:N thin films. *Journal of Crystal Growth* 281 (2–4): 297–302.
430. Lavrov, E. V., Weber, J., Börrnert, F. et al. (2002). Hydrogen-related defects in ZnO studied by infrared absorption spectroscopy. *Physical Review B* 66 (16): 165205.
431. Li, X., Asher, S.E., Limpijumnong, S. et al. (2006). Impurity effects in ZnO and nitrogen-doped ZnO thin films fabricated by MOCVD. *Journal of Crystal Growth* 287 (1): 94–100.
432. Lu, Y., Ye, Z., Zeng, Y. et al. (2007). Effects of RF power variation on properties of N-doped p-type ZnO thin films grown by plasma-assisted MOCVD. *Optical Materials* 29 (12): 1612–1615.
433. Zeng, Y.J., Ye, Z.Z., Xu, W.Z. et al. (2007). Study on the Hall-effect and photoluminescence of N-doped p-type ZnO thin films. *Materials Letters* 61 (1): 41–44.
434. Tsukazaki, A., Saito, H., Tamura, K. et al. (2002). Systematic examination of carrier polarity in composition spread ZnO thin films codoped with Ga and N. *Applied Physics Letters* 81 (2): 235–237.
435. Tan, S.T., Sun, X.W., Yu, Z.G. et al. (2007). p-type conduction in unintentional carbon-doped ZnO thin films. *Applied Physics Letters* 91 (7): 91–94.
436. Hung, S.C., Wang, K.J., Lan, S.M. et al. (2012). Investigation of arsenic-doped ZnO thin films grown on Si substrate by atmospheric-pressure metal-organic chemical vapor deposition. *Physica status solidi (a)* 209 (6): 1053–1058.
437. Ye, J.D., Tan, S.T., Pannirselvam, S. et al. (2009). Surfactant effect of arsenic doping on modification of ZnO (0001) growth kinetics. *Applied Physics Letters* 95 (10): 1–4.
438. Feng, T.-H. and Xia, X.-C. (2016). Growth of arsenic doped ZnO films using a finite surface doping source by metal organic chemical vapor deposition. *Optical Materials Express* 6 (12): 3733–3740.
439. Wang, L., Sallet, V., Sartel, C., and Brémond, G. (2016). Cross-section imaging and p-type doping assessment of ZnO/ZnO:Sb core-shell nanowires by scanning capacitance microscopy and scanning spreading resistance microscopy. *Applied Physics Letters* 109 (9): 092101.

440. Lu, Y.F., Wu, K.W., Zeng, Y.J. et al. (2013). Local p-type conduction of Na-doped ZnO thin films grown by MOCVD. *Chemical Physics Letters* 582: 82–85.
441. Lei, M., He, H., Yu, Q. et al. (2015). Optical properties of Na-doped ZnO nanorods grown by metalorganic chemical vapor deposition. *Materials Letters* 160: 547–549.
442. Yamamoto, T.Y. and Oshida, H.K.A. (1999). Solution using a codoping method to unipolarity for the fabrication of p -type ZnO. *Japanese Journal of Applied Physics* 38 (2): L166.
443. Ye, Z.Z., Zeng, Y.J., Lu, Y.F. et al. (2007). Donor/acceptor doping and electrical tailoring in ZnO quantum dots. *Applied Physics Letters* 91 (11): 2005–2008.
444. Ye, J.D., Gu, S.L., Li, F. et al. (2007). Correlation between carrier recombination and p-type doping in P monodoped and In-P codoped ZnO epilayers. *Applied Physics Letters* 90 (15): 88–91.
445. Su, J., Zang, C., Cheng, C. et al. (2010). Structural, optical and electrical properties of Al-N codoped ZnO films by RF-assisted MOCVD method. *Applied Surface Science* 257 (1): 160–164.
446. Roge, V., Guignard, C., Lamblin, G., et al. (2018). Photocatalytic degradation behaviour of multiple xenobiotics using MOCVD synthesized ZnO nanowires. *Catalysis Today* 306: 215–222.
447. Shi, Z.-F., Zhang, Y.-T., Cui, X.-J. et al. (2015). Epitaxial growth of vertically aligned ZnO nanowires for bidirectional direct-current driven light-emitting diodes applications. *Crystal Engineering Communications* 17 (1): 40–49.
448. Lupan, O., Pauporté, T., and Viana, B. (2010). Low-voltage UV-electroluminescence from ZnO-Nanowire array/p-GaN light-emitting diodes. *Advanced Materials* 22 (30): 3298–3302.

12

Epitaxial Systems for III-V and III-Nitride MOVPE

W. Lundin[1] *and R. Talalaev*[2]
[1]*Ioffe Institute, Russia*
[2]*STR Group Ltd, Russia*

12.1 Introduction

In contrast to its main competitor – molecular beam epitaxy (MBE) – metalorganic vapor phase epitaxy (MOVPE) is extremely hardware dependent. Moreover, MOVPE is one of the most (if not the most) hardware-dependent techniques in semiconductor technology.

In MBE, molecules or atoms evaporated from the sources directly move in high vacuum toward the substrate without any interference. In contrast, in MOVPE, a carrier gas mixed with precursors passes through the reactor and is heated. As a result, complex chains of chemical reactions (some of which even start at room temperature) develop in the gas phase before the gas reaches the grown surface. The simplest is MO precursor pyrolysis, while a more complex example is formation of an adduct and its consequent transformation. In some cases, there are even multiple competing parallel chains of such reactions. As a result, species reaching the surface of epitaxial layer are often not the same molecules that were injected into the reactor. Indeed, development of these processes strongly depends on reactor design. Even in reactors that are similar but different in size, process development may be different. It should be noted that in spite of the great progress of MOVPE as an industrial technology, many details of reactor chemistry are still under discussion.

The second reason for such a strong dependence of practical results of the MOVPE process on equipment design is the real complexity of the whole MOVPE machine in engineering terms. A block diagram of an MOVPE tool is shown in Figure 12.1. Besides the reactor, an MOVPE tool is composed of a gas-blending system where precursors are mixed with carrier gases for injection into the reactor, and an exhaust system that not only conducts gases from the reactor but also is responsible for maintaining pressure in the reactor. Typical engineering solutions used in these

Metalorganic Vapor Phase Epitaxy (MOVPE): Growth, Materials Properties, and Applications,
First Edition. Edited by Stuart Irvine and Peter Capper.

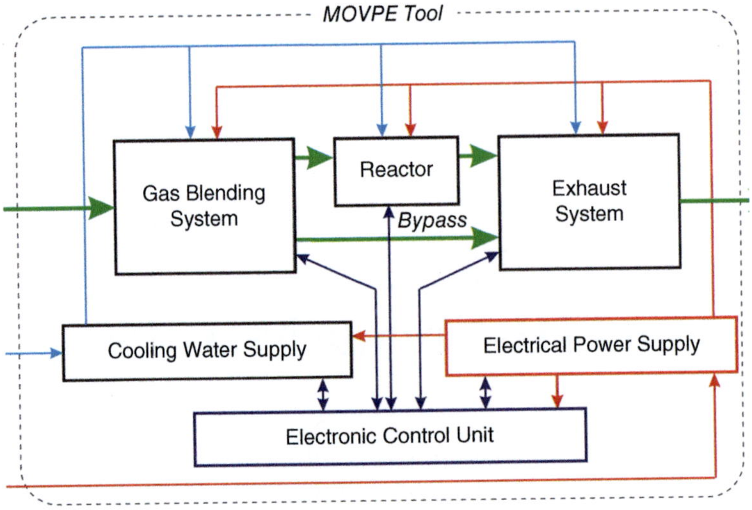

Figure 12.1 *Main components and systems of an MOVPE tool.*

systems will be described in the first part of the chapter. The reactor, gas-blending, and exhaust systems are fed with electrical power and cooling water and controlled by an electronic control unit. These systems are not so much MOVPE specific and thus are beyond the scope of this chapter.

In contrast to the reactor, there is nothing unknown in the operation of other parts of an MOVPE tool. However, we have never seen any MOVPE tool where all of these systems are perfect. Sometimes imperfection comes from compromises, sometimes from mistakes during design or manufacturing; but despite their origin, these issues often influence the MOVPE tool operation as strongly as reactor design. Moreover, usually they determine the MOVPE systems users' preferences. At the same time, only those who have developed MOVPE tools themselves may criticize other tools.[1]

During the first years of MOVPE development, everybody used homemade systems. As a result, it was nearly impossible to compare results, and the process transfer between different tools was extremely complex. Scientists spent huge amounts of time in MOVPE tool construction, and sometimes the results were imperfect. In the mid-1980s, some companies started to offer MOVPE systems. In the beginning, the effect of the appearance of standard tools with much better engineering than that of typical homemade systems was clearly positive. As the result of hard competition, currently most of the market is controlled by two companies: AIXTRON and Veeco. TNSC is a third power, popular in Japan. In this chapter, we will describe some of the reactors offered by these three main players. Unfortunately, the small number of players has led to a small variety of MOVPE tools on the market. We will touch on this issue at the end of the chapter.

12.2 Typical Engineering Solutions Inside MOVPE Tools

12.2.1 Gas-Blending System

The function of the gas-blending system (GBS) is to prepare mixtures of precursors with carrier gas for injection into the reactor. The compositions of these mixtures should be well controlled, and for the growth of complex epitaxial structures, the GBS should allow gradual and abrupt

[1] We may.

change of the injected gas compositions. General requirements for the design of GBS are to use only stainless-steel pipes with internal electropolishing. Any dead ends in the pipeline system should be avoided; inevitable dead volumes inside active parts (valves, flow controllers, and pressure controllers) should be minimized. Metal gasket face seal fittings are used for part connections. The worldwide standard for these fittings, supported by most high-purity gas-handling equipment manufacturers, is VCR (so-called VCR-compatible fittings should be used with caution). Argon-arc welding is used for permanent joints. Orbital welding is usually used, but manual welding done by highly skilled welders provides comparable quality.

In the description of some typical parts and schemes used in GBSs given in this chapter, we assume that all gases (hydrogen, nitrogen, hydrides, etc.) are supplied to the MOVPE tool already purified and at pressures of a few bar. Thus, systems providing this (cylinder cabinets and purifiers) will not be described, in spite of the fact that they are sometimes integrated into GBS.

Any vapor-phase epitaxial process requires precise control over all gas flows injected into the reactor. Gas (or liquid) flow measurement may be performed in various ways: using Doppler or Coriolis effects, measuring the pressure drop across an orifice, etc. However, in the semiconductor technology, only thermal flow-meters and flow-controllers became popular.

The function of the device called a *mass-flow controller* (MFC; the sense of this name will become clear for the reader a little later) is to receive a value of the desired gas flow to be maintained (*setpoint*) from the electronic control unit, measure an actual flow, send this value back the electronic control unit (*MFC reading*), and correct the actual flow to be equal to the setpoint.

A simplified design of a thermal MFC is shown in Figure 12.2. The sensor of the device is a very fine (~0.3 mm) thin-wall stainless-steel pipe with a heating wire coil around its central part. In the absence of a gas-flow temperature distribution along the pipe is symmetrical. With the increase of gas-flow temperature upstream the heater is decreased due to cooling by the gas and increased downstream due to transfer of heat by the gas. This difference is detected with two temperature sensors TS1 and TS2. (Sometimes a simplified design with only two elements acting at the same time as heaters and temperature sensors is used.) At relatively low gas flows, the temperature difference is proportional to the product of the gas flow and its heat capacity. The measured electrical temperature signal in the electronic board is compared with the setpoint, and a command to the valve is generated to correct the actual gas flow.

The signal measured by the sensor is proportional to the gas flow only at relatively low flows. When the gas flow increases above some critical value, the power of the heater is not enough to heat the gas, and thus the temperature difference is reduced with gas flow. This negative slope of

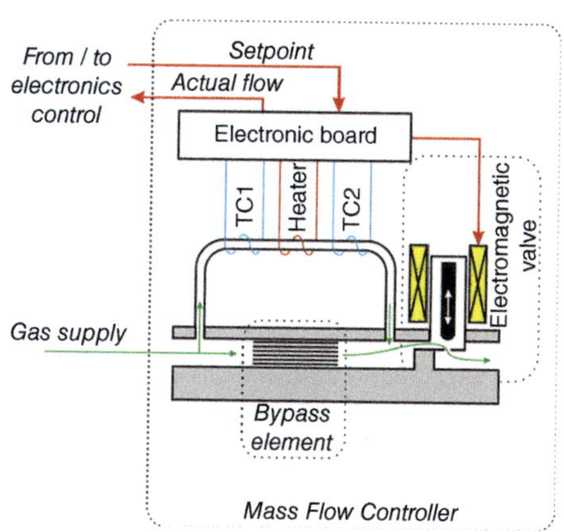

Figure 12.2 *Simplified design of a thermal mass flow controller.*

the dependence may be observed in practice in the case of either a valve or an electronic board malfunctioning. A manifestation of this problem is a reduction of the MFC reading with increase of pressure difference across the device. Indeed, the sensor should operate only in the linear range.

Straightforward methods of extending the range by increasing the heater power result in too high a power, overheating problems, and an increase in size, which also lead to a slower response, etc. A smarter solution is used. In parallel with the sensor, an element with much higher gas conductivity is installed. Various manufactures call it *bypass*, *laminar flow device*, *restrictor*, etc. In all MFCs we have disassembled, this element is designed as a stack of identical standard parts that allows easy assembly of the element with any necessary gas conductivity, or changing it if necessary (by well-qualified personnel). Hence, most of the gas passes via this bypass element, and only a small amount goes via the sensor. The principle is absolutely the same as that of measuring of an ampere-range current with a microammeter using a shunt resistor.

It is clear that physically, a MFC sensor provides output in some electrical units. It is proportional to the product of mole or mass gas flow and its heat capacity, and thus, to determine a real flow, a MFC should be calibrated in some units of flow of some type of gas. This may be mass per time or mole per time. However, other units have become standard for electronics technology: *standard liters per minute* (SLM) and *normal liters per minute* (LN/M). These values correspond to the quantity of gas occupying one liter at standard or normal conditions for temperature and pressure. Submultiple units are correspondingly standard cubic centimeters per minute (sccm) and normal milliliters per minute (nml/m). Thus, physically, mass-flow controllers should be named mole-flow controllers.

Now, after defining gas-flow units, we can note that the linear range for sensor is up to 5–60 sccm, and the maximal flow for MFCs used for MOVPE may be in the range of 10 sccm to 100 SLM or even wider.

It should be kept in mind that different definitions of standard (or normal) conditions are currently used by various organizations. Standard temperature varies in the range 0–25 °C, and standard pressure in the range 100–101.33 kPa. Hence, the difference may reach 10%. Misunderstanding may result in multiplicative errors, but the number is not as high (a few percent). To determine precisely what is measured by a purchased MFC, one should learn what definition of normal or standard conditions is used by the MFC manufacture.

A more important issue is the influence of the gas type on the MFC readings, because the sensor signal is proportional to the product of gas mole flow and its molar heat capacity (J/°C mol), which depends on the type of gas. Usually, MFCs are calibrated with nitrogen or clean, dry air. Replacing it with another gas results in a change of molar heat capacity. If, during purchasing, an MFC customer denotes the working gas, the delivered MFC will be calibrated in SLM (or other similar units) of the requested gas; but if the type of working gas is changed, rescaling should be done by using appropriate gas-conversation factors. For air, nitrogen, hydrogen, and some other two-molecule gases, these factors are close to 1; for other gases, the user should refer to the tables of gas-conversation factors that are accessible, for example, on the websites of MFC manufactures.

Commonly, the valves of MFCs are electromagnetic, but other types (piezoactuator valves, thermal actuator valves, motorized bellow needle valves, etc.) are also used for special applications. A substantial difference of the valve's operation is its *normal state*. The valves may be *normally open* (NO) or *normally closed* (NC); the same definitions are used for MFCs with corresponding names. During normal operation, there is no difference between NO and NC MFCs, but a power failure results, respectively, in much more than the full-scale gas flow or no gas flow through the MFC. This difference should be taken into consideration for safety.

The other element for conditions in the gas-blending system adjustment is the pressure controller (PC; see Figure 12.3). Its design and functionality are similar to those for a MFC. Moreover, most manufacturers produce MFCs and PCs that are so alike that they may be distinguished only by

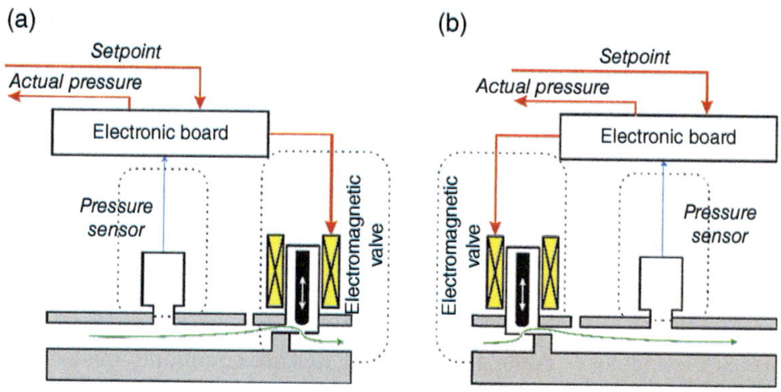

Figure 12.3 *Simplified design of pressure controllers stabilizing pressure upstream (a) and downstream (b).*

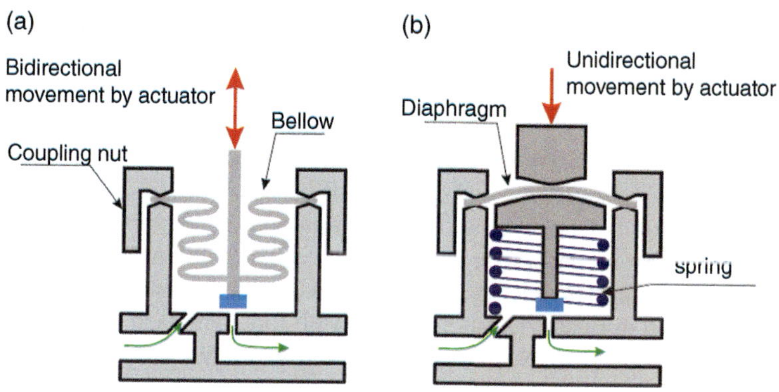

Figure 12.4 *Sketches of bellow (a) and diaphragm (b) valves. Screw-on bodies and coupling nuts are not shown.*

reading the label. A PC should control the pressure, so some kind of pressure sensor is used, and clearly there is no necessity for a bypass line. An important issue is that two different types of PCs are manufactured. The first type controls the pressure upstream (Figure 12.3a), the second – downstream (Figure 12.3b). Correspondingly, they are mounted in the gas pipeline with the sensor before the valve and the sensor after the valve.

For decades, electronic boards of MFCs and PCs were analog, and data exchange with electronic control units of MOVPE tools was performed via a 0–5 V or 4–20 mA interface. Nowadays, this class of devices is used for low-cost solutions only; high-end MFCs and PCs are digital. Some of them allow analog interface emulation, but normally data exchange is performed via some form of digital interfaces. Using digital electronic boards and interfaces results not only in noise reduction but also in much higher long-time zero and calibration stability, better linearity, ability to issue additional commands from the electronic control unit, etc.

Both MFCs and PCs are elements for tuning the regimes. They are relatively slow (a few seconds is taken for stabilization after abrupt setpoint changes) and do not guarantee the absence of leaks across the element even in the shutoff state. Valves are used in gas-blending systems for fast gas-flow switching. Clearly, any type of valve with a stuffing box stem sealing is unsuitable for MOVPE applications due to poor tightness. Bellow- and diaphragm-sealed valves are used (see Figure 12.4). The only place air sealing is used is at a joint of a bellow or a diaphragm with a valve body – and it may be metal-to-metal, which results in very high levels of tightness.

Almost all valves used in MOVPE systems are pneumatically actuated. Valves with direct electromagnetic actuation are used occasionally, but they may cause heating of the whole gas-blending system. Moreover, rather high current or voltage supply to the elements is not desirable in MOVPE gas-blending systems controlling hydrogen, spontaneously flammable MOs, and toxic hydrides.

Valves used in MOVPE tools may be discrete or united in blocks (two, three, or more valves with a common body, with interconnections between valves drilled inside this body). Some examples of such valves will be described later.

Components of the first generations of MFCs, PCs, and valves were coupled with elastomer seals. During the last two decades, these devices for MOVPE have become all-metal sealed to promote higher purity. Elastomers are used for internal seals only (e.g. as the material of stem tips).

All metal parts of these devices contacted by pure gases are commonly made of stainless steel, usually AISI 316 group. These parts should be internally electropolished. Magnetic plungers of electromagnetically actuated valves of MFCs/PCs, and sometimes elastically deformable parts (springs, diaphragms) are made of special alloys with appropriate combinations of properties.

The configuration of the gas-blending system strongly depends not only on the peculiarities of the materials to be grown and the reactor design, but also on the basic concept of safety management. One of two concepts is commonly used. At alarm, either the whole system is to be purged with high flows of nitrogen, or all elements are closed for isolation of different parts of the system. Indeed, these solutions inside one tool should be matched. Detailed discussion of this issue is beyond the scope of this chapter. We hope that the reader will not design an MOVPE tool using only the information below on some typical parts of GBS.

The simplest configuration of sources used to supply gaseous precursors requires just one MFC and NC valve. However, the second valve, which may be NO or NC depending on the basic safety concept, is always used for purging the line with nitrogen or hydrogen in the idle state. A sketch of such a source is presented in Figure 12.5a. Some MOVPE tool manufacturers prefer to use a single-body block of these two valves. Such blocks are offered by a few manufacturers, and to avoid favoritism, we present only a sketch of such a valve instead of a photo (see Figure 12.5b). Gas flow from the source before being injected in the reactor is mixed with a carrier gas. Very low gas flows (a few tens of sccm and below) in the long pipe between a MFC and the point of mixing with a large flow of carrier gas may be strongly influenced by pressure fluctuations in the system. To make the precursor supply more stable, the flow in the pipe from the source may be accelerated using an extra flow (*pushing flow*) from an additional MFC supplied with hydrogen or nitrogen (depending on the carrier gas used in the specific epitaxial process). Sometimes, in addition, one or two check valves are added to prevent backflow under pressure misbalance.

The guaranteed dynamic range (ratio of maximum to well-controlled minimum flows) of MFC is about 50. It is better to avoid operation of MFCs with flows below 3–5% of the maximal range,

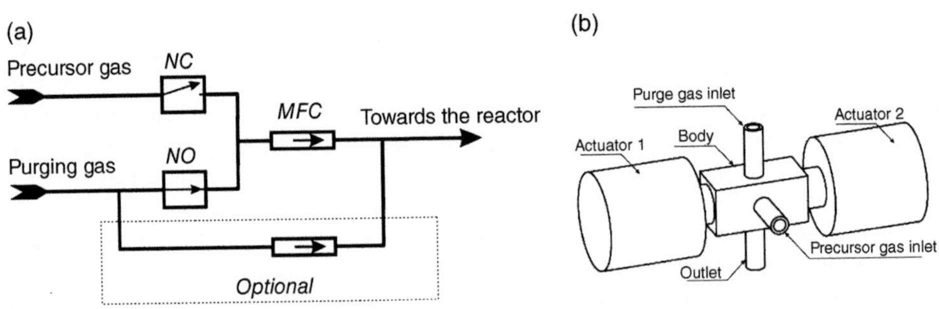

Figure 12.5 *(a) A chart of a simple gaseous precursor source; (b) a sketch of the single-body block with two valves.*

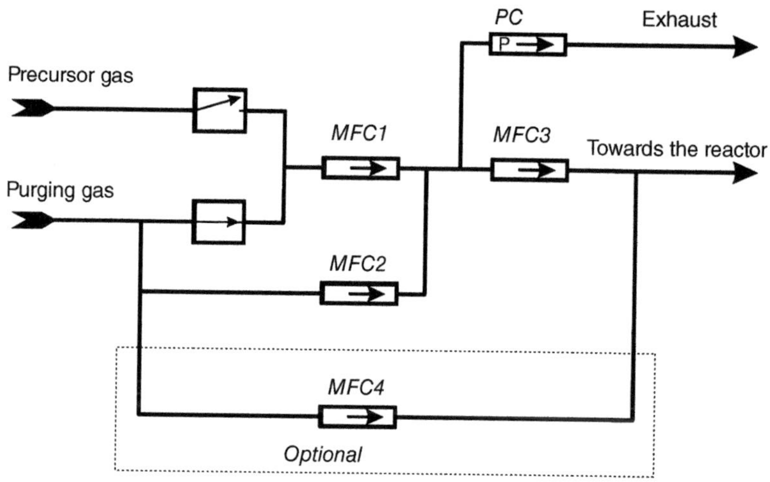

Figure 12.6 *A chart of a gaseous precursor source with double dilution.*

which means a dynamic range of 20–30. Usually this range is enough for supplying Group V precursors (AsH_3, PH_3, etc.). A wider range is necessary for very specific applications only, and two lines for the same precursor may be installed in this case.

A much more complex configuration of gas source, known as a *double-dilution system* is used for doping from gaseous precursors, e.g. from SiH_4 (see Figure 12.6). Doping usually requires a very wide range of precursor supply. Even more specifically, light doping requires very low precursor concentrations in the reactor, much lower than may be reached using a combination of simple gas source and diluted precursor in the bottle. The precursor gas passing via MFC 1 with the flow f_1 is diluted by the carrier gas passing via MFC 2 with the flow f_2, and then a portion of this gas (f_3) is taken into the reactor via MFC 3; all the rest drains through a pressure controller toward the system exhaust.

Thus, the precursor flow from the source is:

$$f_{\text{dopant}} = c_{\text{cylinder}} \times \frac{f_1}{f_1 + f_2} \times f_3,$$

where:

f_{dopant} – Mole flow of the doping gas from the source.

f_1, f_2, f_3 – Mole flows via MFC 1, MFC 2, MFC 3, respectively.

The parameter c_{cylinder} is the concentration of dopant gas in the cylinder (usually already diluted gas is used).

Usually, the ranges of MFC 1 and MFC 3 are low (10–100 sccm), and the range of MFC 2 is high (~1 SLM). Simple analysis of this formula shows that the dynamic range of the source may easily exceed 10^5, which is more than enough even for doping purposes. If the SiH_4 concentration in the cylinder is 100 ppm, the mole flow from the source may be in the range of 0.1 nmol/min without running MFCs at flows below 5 sccm.

It should be pointed out that the precursor flow to the reactor is independent of the pressure maintained by the pressure controller before it. This device is necessary to maintain the pressure drop for the proper operation of all MFCs. In a simplified configuration, the PC may be replaced by an orifice, but in this case stable operation of the scheme is achieved in a constricted range of flows.

The functions of the valves upstream MFC 1, and MFC 4, are the same as in a simple gas source.

It should be noted that the gas draining through the pressure controller is wasted. Typically, for this scheme, $f_2 >> f_1, f_3$, which means most of the precursor is lost. It is not a problem for cheap diluted silane; but for more-expensive precursors, if very low flows are not required, it is better to extend the range of flows by using two simple gas lines.

The design and operation of an MO source is a bit more complex. The MOs used in MOVPE are volatile liquids or solids. Vapors of MO compounds coming from the surface are transferred to the reactor and mixed with a carrier gas (this gas flow usually is called *bubbling gas* to avoid confusion with the carrier gas flow in the reactor). Most of the MOs used in MOVPE are spontaneously flammable, so handling them is not easy. Standardized cylindrical containers with MO (bubblers) delivered from manufacturers are attached to GBS lines via two connecting tubes with manual valves at the ends, and empty bubblers are disconnected and sent to the MO manufacturer for reloading. A typical bubbler for R&D applications contains from 100 g to 2 kg of MO compound (Figure 12.7). Larger bubblers are sometimes used in tools for mass production.

The simplest schematic of a liquid MO source is shown in Figure 12.8a. The carrier gas flow, controlled by a MFC, is fed into the bubbler via an inlet tube that ends close to the bottom of the bubbler. This tube is usually called the *dip leg*. This gas bubbles through the MO, saturates with MO vapors, and then leaves the bubbler. With the assumption of equilibrium conditions in the bubbler (the partial pressure of the MO above the liquid is equal to the saturated pressure, and the temperature is uniform), the MO flow from the source is:

$$f_{MO} = f_{bubbling} \times \frac{P_{MO}}{P_{total} - P_{MO}},$$

where:

f_{MO} – Mole flow of MO from the bubbler
$f_{bubbling}$ – Mole flow of bubbling gas
P_{MO} – MO partial pressure in the bubbler
P_{total} – Total pressure in the bubbler

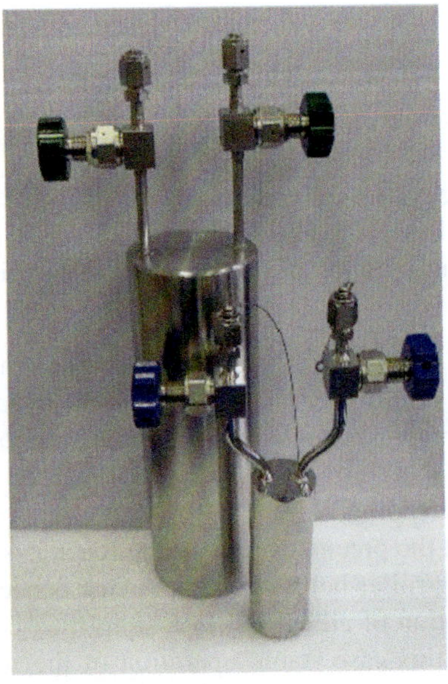

Figure 12.7 *Standard bubblers for MO. Large one (100 mm in diameter) may contain 1.5–2 kg of TMGa. Small one (50 mm) – 150–200 g.*

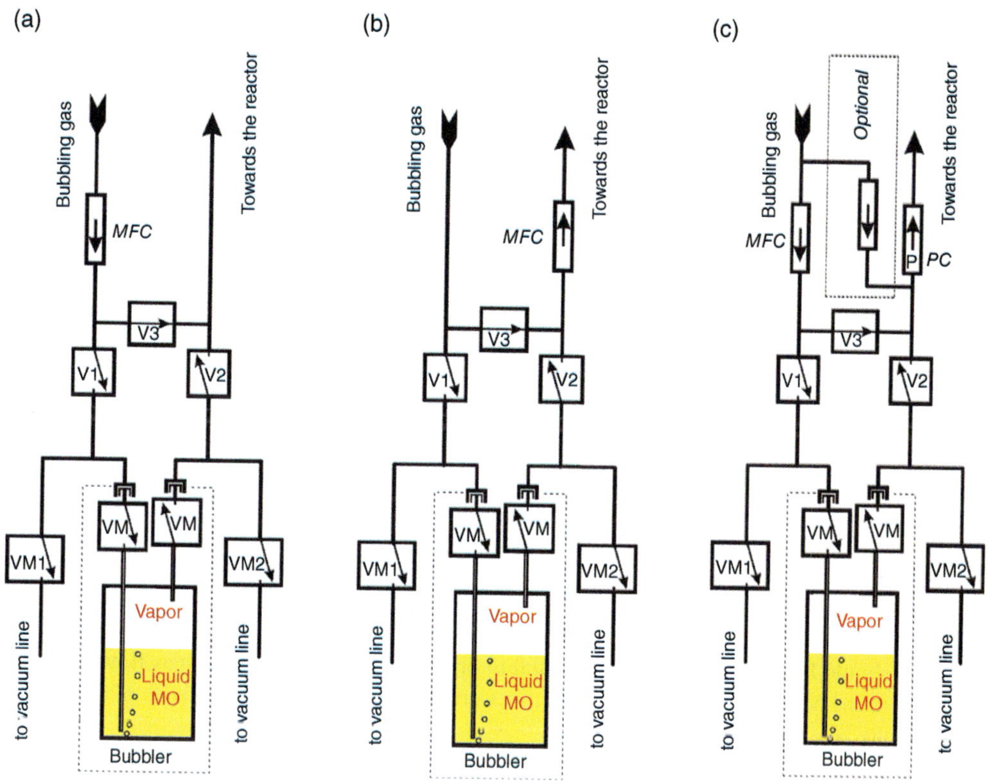

Figure 12.8 *Schematics of typical MO sources.*

The MO partial pressure depends on the temperature, so the bubbler should be placed in a thermostatic bath. A linear dependence of MO flow on bubbling gas and MO partial pressure is quite natural. The denominator requires more detailed discussion. If the total pressure is high and/or the partial pressure of MO is low, the denominator is ~ P_{total}. This means the higher the total pressure, the lower the MO flow. This dependence provides an additional parameter to control the MO flow supplied to the reactor. As the total pressure gets close to the MO partial pressure, the MO flow from the source increases significantly. Conditions when $P_{total} = P_{MO}$ result in the physical process known as *boiling*. Under this condition, the MO vapors leave the bubbler without a carrier gas flow. Feeding any bubbling gas in this case results in violation of the assumption of equilibrium conditions in the bubbler, making the formula incorrect.

If the exhaust of the bubbler is connected to the reactor (directly or via valves), as shown in Figure 12.8a, the pressure in the bubbler is equal to the reactor pressure. This solution may be used in simple atmospheric-pressure MOVPE tools, where the reactor is connected to the atmosphere and thus the reactor pressure is stable, at least during one epitaxial process. (Indeed, the reactor is never directly connected to the atmosphere; but in simple atmospheric-pressure tools, a check valve, a hydraulic syphon, a scrubber [if used], and exhaust ventilation give some constant small [positive or negative] additive to the pressure of ambient air.) Day-to-day fluctuations of atmospheric pressure are not as great. If necessary, the drift of MO flow from the bubbler with atmospheric pressure may be compensated for by tuning the bubbling gas flow.

In low-pressure MOVPE (also called reduced-pressure MOVPE), as well as in super-atmospheric pressure MOVPE, these simple MO sources are inapplicable. First, this is due to the strong dependence of the MO flow on the total pressure in the bubbler. Then, the existence of the relatively large (and increasing with MO usage) gas volume above the MO level leads to strong, fast fluctuations of MO flow with even small reactor pressure fluctuations, which are unavoidable. Other schemes of MO sources, presented in Figures 12.8b and c are used, instead. The simplest solution is to place the MO after the bubbler and maintain the pressure in the bubbler using any appropriate device connected to the dip leg (e.g. by the same simple mechanical pressure regulator feeding the whole gas-blending system). This scheme really works, but it has a few disadvantages. First, the heat capacity of the mixture of bubbling gas and MO vapors differs from the heat capacity of the pure bubbling gas, which depends on the MO concentration. This complicates the evaluation of a real MO flow. Then, inside a MFC, gas passes via long capillary channels. Various contaminants in the bubbling gas or MO compounds, very small leaks, etc., result in the formation of fine deposits on the walls of the pipes. For pipes a few millimeter in diameter, this may be negligible. For capillaries, sooner or later, well before the blockage, flow resistance is changed, which means the gas flow can no longer be precisely controlled.

In an alternative scheme (Figure 12.8c), the bubbler pressure is independent of the reactor pressure, being stabilized by a PC installed downstream of the bubbler, while pure bubbling gas is fed via a MFC. Parametric analysis (which is beyond the scope of this chapter) shows that this scheme is a little more sensitive to reactor pressure fluctuations than shown in Figure 12.8b, but in contrast to a MFC, a PC is much less affected by fine deposits. This scheme became popular, and it is used in the majority of MOVPE tools. An additional MFC in Figure 12.8 acts as a source of a pushing gas like an optional MFC in Figure 12.5. An additional gas flow from this MFC also leads to more stable operation of the PC. Finally, and most important for MO precursors with low vapor pressure, which thus should be kept at temperatures equal to or above room temperature, gas flow from this MFC reduces the MO concentration. It suppresses condensation of MO on surfaces of valves and pipes that may be colder than the bubbler.

Valves V1–V3 are used for switching the source from operating conditions (V1,V2 – open, V3 – closed) to the idle state (an inverse state of valves). V1 and V2 are always NC; V3 may be NC or NO depending on the safety concept of the system. As for the source of gaseous precursors, these valves may be discreet or combined in a three-valve single-body block.

In the description of gas sources, we assumed that cylinders with compressed or liquefied precursor gas are installed in the cylinder cabinet that contains all systems necessary for maintenance and cylinder exchange. Bubblers with MO precursors are installed inside the MOVPE GBS. Manual valves VM1 and VM2 connecting the bubbler legs with vacuum lines are used for He leak testing and evacuation before and after bubbler exchange. This procedure is usually executed using many repetitions of evacuation and refilling with nitrogen using the V1 and V3 valves.

For doping from a MO source (e.g. from DEZn), sometimes a combination of the schemes shown in Figures 12.6 and 12.8c is used.

Some of the popular MO precursors are solid under conditions suitable for use. In theory, everything we have said so far about liquid precursors is correct for solid ones, but technically the situation is more complex. Solid precursors are used in the form of small granules. If the conventional bubbler is used, bubbling gas penetrates into the bubbler via the dip leg, infiltrates through the layer of these granules, saturates with vapors, and leaves the bubbler. For dopant precursors (Cp_2Mg, Cp_2Fe), it works well due to the very low flow of precursor leaving the bubbler. Most of the problems occur for TMIn. In the beginning, a bubbler filled with TMIn provides a high flow, as expected. However, after some time, the granules stick together. Most of the gas passes through the layer of granules in the vicinity of the dip leg, and TMIn mostly evaporates there. Agglomerated granules do not fill the formed cavity, and the bubbling gas passes above the TMIn instead of infiltrating through it. This leads to undersaturation of gas, and the TMIn flow from the bubbler may significantly decrease.

Various solutions for this problem are used. Some groups disconnect the bubbler from time to time and shake it to restore initial conditions. Others reverse bubble (the dip leg is the output). Installation of two or more bubblers in a cascade strongly reduces the problem. Most of the material is consumed from the first bubbler, and final saturation occurs in the last bubbler without significant material consumption from it. Special bubblers with an internal design that prevents the problems described and a liquid source that is a suspension in a saturated solution of TMIn in an appropriate solvent are also available.

These methods were proposed many years ago, but neither of them is free of disadvantages. It seems that the most reliable way to be sure of the TMIn concentration downstream of the bubbler is to use a MO vapor concentration-monitoring tool. It may be used for other MO compounds, as well.

Various methods have been proposed, and systems measuring MO concentration via its influence on sound velocity are commercially available. These systems not only allow monitoring of the MO vapor concentration but also may be integrated into the electronic controls of MOVPE tools to promote repeatable MO flow from the source.

The subsystems described earlier allow for control of the desired precursors flows. If a single thick epitaxial layer is to be grown, these precursor flows may be simply mixed with the desired flow of carrier gas and injected into the reactor. However, almost all MOVPE assemblies are designed for growth of multilayer structures. Flows from the sources cannot be changed immediately. Formation of abrupt interfaces is possible only if the desired precursor flows are stabilized, being directed to the bypass to the reactor and then, for an appropriate time, switched into the reactor (Figure 12.9). This switching of gas flows between the reactor and the bypass line is done in a special net of valves. Various equipment manufactures call it an *injection manifold*, a *run-vent stack*, etc. Depending on the design, this system may be assembled from discreet valves, joined by VCR fittings or welding, a stack of single-body coupled valves, or even a single-body block of many valves. A very important requirement is to minimize dead volumes in this system and promote good purging. For proper operation, it is important to promote pressure balance in the reactor and bypass lines. Active balancing is sometimes used, which means control of the bypass MFC flow or conductivity of a motorized needle valve in the bypass line on the basis of the differential pressure transducer reading. These elements are framed in Figure 12.9 by dotted lines.

Gas mixtures formed in the described manifold are then injected into the reactor directly, in the case of single-point injection, or distributed between a number of injectors by orifices, needle valves, or MFCs.

12.2.2 Exhaust System

Even in the case of atmospheric-pressure MOVPE, gases from the reactor together with those bypassing the reactor usually are not directly bled into the exhaust. As a minimum, a simple device preventing backflow (a check valve or a hydraulic syphon) is necessary. A high V/III ratio typical

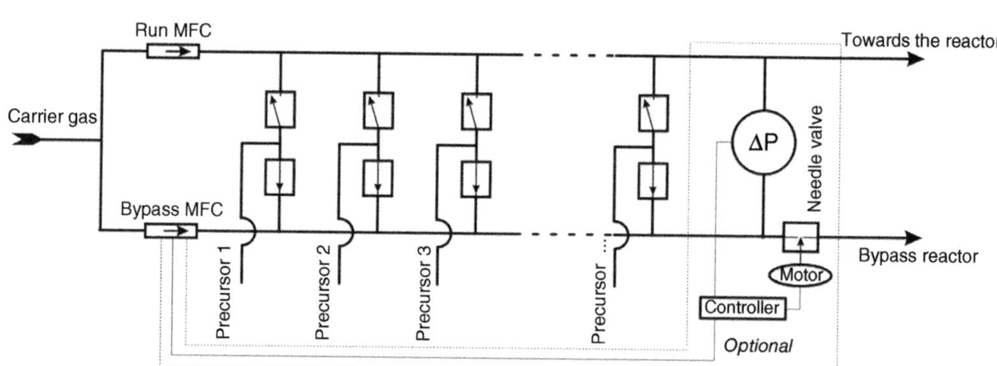

Figure 12.9 *A schematic of a valve network upstream of the reactor (run-vent stack).*

for MOVPE means high Group V precursor flows. In contrast to nitrides, these precursors are mostly decomposed inside the reactor, forming vapors of the corresponding elements. After leaving the hot reactor, these vapors condense on the cold walls of the pipelines. These condensed solids are toxic, are flammable, and may cause total blockage of the exhaust pipes. For safe handling of these vapors, special water-cooled traps with isolation valves are installed. After being filled, these traps may be disconnected and sent for refurbishment. Indeed, the exhaust pipes upstream of the trap should be heated to prevent condensation.

If the reactor pressure differs from ambient, some active system of pressure stabilization is used. Its main components are a pressure transducer, an electronic controller, and a throttle valve (see Figure 12.10). The latter, typical for MOVPE tools, is a turning circular shutter in a cylindrical or conical conduit driven by a stepper motor. The majority of MOVPE systems with variable reactor pressure are low-pressure MOVPE systems. Even most of the rare super-atmospheric systems allow operation with reactor pressure below atmospheric. Thus, a pump installed downstream of the throttle valve is an essential part of the exhaust system. Two types of pumps are used in MOVPE for reactor pressure reduction: rotary vane pumps and multistage Roots pumps. The first are filled with oil. They are extremely reliable and long lasting. However, the presence of oil means a risk of oil-vapor penetration upstream and reactor contamination, as well as oil leaks outside the pump. These pumps are considered to be outdated for MOVPE applications. Multistage Roots pumps are regarded as *dry pumps*. Oil is used in these pumps to grease the bearings, but the design of the pump ensures the impossibility of oil penetration inside the evacuated volume.

Another essential part of an MOVPE exhaust system is the filter installed upstream of the throttle valve to protect both the throttle valve and pump from dust generated inside the reactor. If an arsenic/phosphorus trap is installed in the system, the filter may be integrated into this trap; in III-N MOVPE tools, it is a separate part.

In addition to the elements shown in Figure 12.10, exhaust systems of real MOVPE tools contain a number of bypass pipes, pneumatically actuated valves, and check valves. These valves guarantee safety during reactor reloading and maintenance, and under various alarm conditions. The general task is to prevent excessive reactor pressure and back streams from the exhaust into the reactor. However, the exact design of the exhaust system network depends on many details and thus is not presented here.

As in GBS, most elements of an exhaust system are made of stainless steel. Much lower requirements for purity allow dead volumes and use of elastomeric seals. The standard type of seal is ISO-KF, also known as ISO-NW or ISO-QF. However, the proper elastomeric material choice sometimes is challenging (this is also true for elastomers used in gas-blending systems and

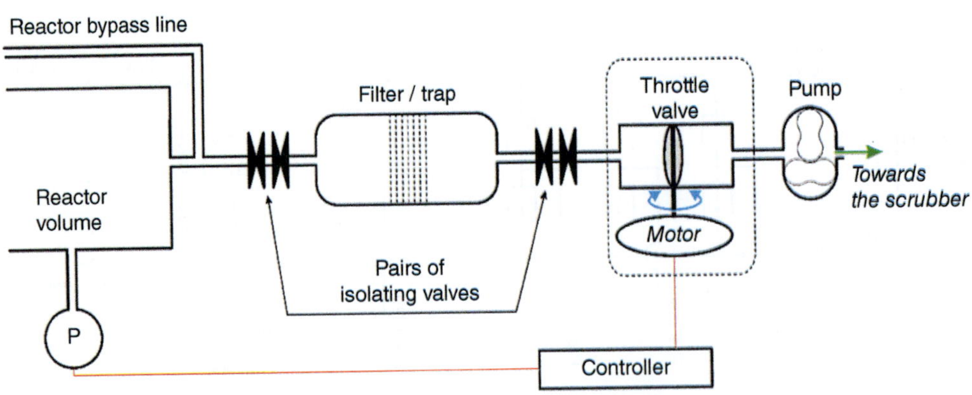

Figure 12.10 *Greatly simplified schematic of an exhaust system.*

reactors). Many grades of rubber are unstable in hydrogen. Viton is the most popular elastomer for MOVPE, but in the presence of ammonia it loses elasticity. Neoprene is stable in an ammonia ambient, but its maximum operating temperature sometimes is insufficient.

12.2.3 Reactors

In the previous description of MOVPE systems, reactors were skipped. Various reactor designs will be described in the following subsections. Here, we will provide only some general information about materials used for building reactors and methods of heating and rotating the substrates.

Quartz glass (fused silica) is one of the most convenient materials for MOVPE reactors. It tolerates strong, non-uniform heating and thermal stress, does not react with any of the precursors or carrier gases used in MOVPE, is transparent over a wide optical range, and is an insulator. Most of the early MOVPE reactors were fabricated from this material. However, quartz is fragile. This means not only that it is sensitive to mechanical impacts but also that crack extension is very fast and tensile stress is critical for it. Nowadays, quartz is used in some MOVPE reactors as a material for internal elements. For many years, the outside cases of MOVPE reactors were made of metal. Usually, this is stainless steel from the 316 group. Other metals are used only in some parts of reactors: e.g. internal parts heated above 600–800 °C are made of molybdenum or tungsten.

At first glance, some aluminum alloys are very attractive due to much easier machining. However, corrosion takes place in the water channels inside aluminum parts. This problem is especially troublesome if the water contains ions of copper, which is typical for inductive heaters (discussed later). For reactors containing water-cooled aluminum and copper, separate water circuits for these two materials are obligatory. Moreover, gallium (a final product of TMGa or TEGa pyrolysis) that may condense on cold walls dissolves in aluminum, which make it unstable in air and can even lead to total loss of mechanical strength. However, sometimes use of aluminum alloys is unavoidable. The low thermal conductivity of stainless steel may lead to warping of water-cooled stainless-steel walls. Penetration of electromagnetic fields (if induction heating is used) into the wall leads to an additional temperature gradient that increases warp. Usually this is not a problem, but for planetary reactors, described in the next subsection, requirements for flatness of the upper lid are very high; so, usually they are completely or partially made from an aluminum alloy.

MOVPE reactors can be *cold-wall* or *warm-wall*, so a furnace-like design is not acceptable. The three basic types of heating systems are *induction heating*, *resistive heating*, and *lamp heating*.

In the first type of heating system, an electromagnetic field from an inductor induces current in the susceptor. These induced currents generate Joule heat. Typical frequencies used in these systems are 40–60 kHz. The inductor is always a specially shaped water-cooled copper tube, sometimes gold or silver plated. In quartz-made reactors, the inductor is usually coiled around the reactor. In contrast, the inductor should be placed inside metal-walled reactors in a position that minimizes induction heating of the reactor walls.

The susceptor should be electrically conductive with a not-very-low specific resistance. Good thermal conductivity is very desirable for temperature-uniformity improvement. Graphite is one of the best materials for this. For reduction of contamination of the reactor ambient from impurities contained in the graphite, it should be of a special grade and strongly annealed. Coating of graphite susceptors with silicon carbide is widely used. For III-N MOVPE, coating with SiC or materials with similar chemical stability under reactor conditions (AlN, TaC) is obligatory, because at high temperatures graphite is attacked by ammonia.

Induction systems are very reliable and do not contain parts with a limited lifetime. However, dynamic temperature distribution control (multizone heating) in these systems is so complex that it is never used in practice.

In the other popular technical solution, heat is generated in the resistive heater. A multizone resistive heater is easy realized by installation of a number of smaller heaters with individual power controls instead of a single large one, and this is the main advantage of these systems. For relatively low-temperature epitaxy of arsenides, phosphides, and antimonides, a number of materials may be used for heaters. Graphite, molybdenum, tungsten, and even inexpensive and easily machined alloys similar to Nichrome give good results with a long heater lifetime. Usually, the heater is placed below the susceptor and shaped as a flat spiral, either cut from sheet material or bent from spiral wire.

In the majority of MOVPE reactors, the susceptor is rotated while the heater is fixed. As a result, some gap between the heater and the susceptor exists, and heat transfer to the susceptor is mostly radiative. The susceptor should have high optical absorption in the IR range, and graphite again is one of the best materials.

In III-N MOVPE, operating conditions for heaters are much harsher. First, the required susceptor temperature is 1000–1200 °C, and the heater should be much hotter, which reduces its lifetime. Then, at high temperatures, ammonia attacks graphite, molybdenum, and tungsten. Perhaps the best material for resistive heaters for III-N MOVPE is rhenium. However, it is very expensive, its production is strictly limited, and it has typically been in short supply in recent decades. For some time, rhenium heaters were popular in MOVPE reactors for blue LEDs; but with the increase of volume of these reactors that has coincided with the increase in production volume of these MOVPE systems, rhenium was replaced with tungsten. Now, rhenium is mostly used for heaters of small high-temperature research reactors. Graphite and tungsten are still the most widely used materials for III-N reactors' resistive heaters, and their lifetime is limited. This is a significant improvement compared to the first generations, but heater replacement that is a complex procedure is still a part of routine maintenance.

The third type of heating systems is *lamp heating*. Its operation is similar to resistive heating, but incandescent lamps are used for IR-radiation generation instead of simple resistive heaters. Usually, strip lamps are used. Again, there is no problem with multizone heating realization. If a quartz reactor is used, lamps may be placed outside the reactor. This simple solution was popular due to simple replacement and low cost of standard lamps. In a reactor with a metal case, lamps are installed inside the reactor, and this configuration is not as beneficial as resistive heating. Also, it is very difficult to maintain the high temperature necessary for III-N epitaxy for a few hours using lamp heating. Nowadays, lamp heating is almost never used in MOVPE, in contrast to CVD of silicon, where it is the most popular type of heating system due to the much shorter growth process.

As was already mentioned, the susceptor of most MOVPE reactors is rotated. Two methods of realization are popular. Mechanical rotation feedthrough is typically realized using a magnetic fluid seal. This component is relatively reliable; however, it requires regular technical inspection during maintenance. In rundown conditions, a breakthrough may occur, resulting in magnetic fluid and air penetrating into the reactor with high flow.

An alternative method, used by AIXTRON, is gas-foil rotation. To the best of our knowledge, it was described for the first time in [1]. In this technique, the substrate is placed on a small, thin wafer carrier, which levitates above a main susceptor on a thin layer of gas injected via a few (typically three) fine holes in the main susceptor. This effect is similar to that used in the well-known "air-hockey" game. Rotation is generated by spiral groves in the main susceptor below the rotating one. Due to the absence of friction between the movable parts (excluding a pin centering the upper disc), no dust is generated, and no mechanical feedthrough is necessary, which simplifies the reactor design. As will be described later, gas-foil rotation is usually combined with mechanical rotation in the same reactor to provide different kinds of motion.

Precise temperature control of the substrate holder is very important for epitaxial process repeatability. In our opinion, the most reliable method for temperature measurements is using a thermocouple embedded in the hole in the controlled object. This configuration is standard for MOVPE without substrate rotation. However, in the majority of MOVPE reactors, the substrate carrier is rotated, which means it is impossible to achieve good thermal contact with the thermocouple. Two solutions are used. The first is to place the thermocouple near the susceptor. The measured temperature differs from the real one, sometimes by a few hundred degrees; but if the gap and surface conditions of the thermocouple and the susceptor are constant, the thermocouple temperature will be a one-to-one function of the real temperature of the susceptor, which is enough for process repeatability.

The second technique is use of a pyrometer, which may be placed outside the reactor and focused on the susceptor surface via a small window. However, the emissivity of the surface strongly influences the measured value.

The pyrometer may be focused on three different zones of the susceptor: on the wafer surface, on the wafer side of the susceptor between wafers, and on the susceptor surface opposite to the wafers (backside). The last one is always clean, the emissivity of this surface is rather constant, and this surface seems to be the best for temperature measurements. However, for many reactor designs, other factors influence the correctness of measurements. In reactors with resistive or lamp heating, these red-hot objects screen the susceptor. Even if a pyrometer is focused at the susceptor backside surface between lamps or lamellas, light reflected or scattered by the susceptor surface penetrates into the pyrometer and blinds it.

In planetary reactors with gas-foil rotation (and induction heating), the temperature of the bottom part of the main susceptor may be measured precisely by a pyrometer, and this value is typically used for process control and even given in papers. However, an actual temperature of the upper rotating disc is 50–150 °C lower due to the thin gas layer below, and this difference depends on many factors.

Hence, focusing on the backside of the susceptor provides correct results only for systems with induction heating without gas-foil rotation. Unfortunately, there are very few such systems on the market.

To focus a pyrometer on the wafer side of the susceptor, an appropriate window should be available in the reactor wall, but not every type of reactor geometry allows this optical access without disturbing the epitaxial process. Then, the emissivity of the wafer side of the susceptor, between wafers, strongly varies with polycrystalline material deposition during the process. The temperature measurements in this region give incorrect and irreproducible results. The third zone where the pyrometer may be focused is the wafer surface. The emissivity of the wafer depends on the substrate material, epitaxial layer composition, and even its thickness, but at least these parameters are known and now may even be measured during an epitaxial process, which allows the correction of the signal from the pyrometer. This technique – *emissivity-corrected pyrometry* – seems to be the most universal method providing high accuracy of measurements.

Although it was invented as an accessorial method of epitaxial process characterization, in some present-day MOVPE systems emissivity-corrected pyrometry is primary, providing data for the temperature-control system. However, depending on the substrate material, either the wafer temperature or the temperature of the substrate holder below the wafer, or even their combination, is measured and it should be taken into consideration. New methods allowing separation of these temperatures, e.g. UV pyrometry for GaN-based structures, are developing now, but they are beyond the scope of this chapter.

To conclude, temperature measurement during an MOVPE process is challenging. In many MOVPE systems, the real substrate holder temperature strongly differs from that used for process control and unfortunately often reported as a process temperature. Thus, when comparing data from various papers, the reader should take into consideration the method of temperature measurement in the MOVPE system used.

12.3 Reactors for MOVPE of III-V Materials

12.3.1 General Features of III-V MOVPE

For the first time, MOVPE was applied for epitaxy of GaAs [2] for electronics applications, and in one or two decades was adapted for growth of device heterostructures based on various arsenides, phosphides, and antimonides like AlGaAs on GaAs substrates, InGaAsP on InP or GaAs, and InGaAlP on GaAs. These materials are widely used in optoelectronics for spectral ranges from the mid-infrared to orange, electronics, and photovoltaics. Details may be found in numerous books and review papers; optoelectronic applications as well as properties of heterostructures composed of these materials are well described in [3, 4].

The common features of epitaxial conditions for classical III-Vs are as follows. They are grown in the intermediate temperature range of 500–800 °C, which allows the use of relatively simple heating and cooling systems. In spite of the significant difference of typical temperatures used for growth of various materials, most conventional heterostructures may be grown at about a constant temperature through the whole epitaxial process. For some of these materials, the variation of reactor pressure over a wide range does not influence the material properties, while for the others some specific pressure is believed to promote better material properties; but for all of them, all layers composing the whole heterostructure may be grown while maintaining a constant reactor pressure. The standard carrier gas for all classical III-Vs is hydrogen, and usually there is no reason to change the value of the carrier gas flow. Finally, the hydride precursor concentration in the carrier gas does not exceed a few percent; usually it is less than 1% (the MO or liquid Group V precursor [if used] concentrations are in the range of 10^{-3}–10^{-4}). All of this means reactors used for MOVPE of any of the classical III-V based heterostructures should be optimized for one set of reactor conditions: temperature, pressure, and gas flow, and this gas (in terms of fluid dynamics) may be considered to be hydrogen. It is very important to understand that all types of MOVPE reactors dominating at present were designed in the 1980–1990s in order to fulfill those simple requirements.

One additional element of almost all MOVPE reactor designs is a separate injection of Group III and V precursors. This solution became standard at the very beginning of MOVPE development for several reasons. Chemists preferred it due to the tradition of keeping precursors unmixed until introduced into the reactor chamber (at least it was so in the very first Russian MOVPE reactors). If MO and hydrides are mixed in the presence of traces of oxygen or water (insufficient purity of precursors and/or carrier gas was typical for the first years of MOVPE development), they form low-volatile compounds deposited on the walls of pipes if mixing occurs in the gas-blending unit prior to injection. Sometimes, this may result in full blockage of the pipe. Some exotic precursors (e.g. *dimethylethylamine alane*) form low-volatility adducts with hydrides, again deposited on the walls or forming dust in the gas phase. Finally, there was some uncertainty concerning parasitic prereactions between indium precursors and phosphine. In [5] and references therein, the reader can find that pure TMIn does not react with PH_3 at room temperature; however, cooperative pyrolysis of TMIn and PH_3 strongly depends on precursor concentration, indicating the existence of prereactions. Moreover, in [6] and references therein, it is directly indicated that room-temperature adducts of TMIn and PH_3 may form polymeric coatings on the cold walls.

As a consequence of these observations, and to ensure the possibility of using the same MOVPE reactor for different applications, it was commonly decided to design reactors with separate injection.

Now, these reasons are outdated. If of sufficient purity, mixing of standard MO precursors (TMGa, TMAl, TMIn) with hydrides used for classical III-V growth does not result in any deposits; other precursors are rarely used. Homemade MOVPE reactors with premixed precursor injection are successfully used for classical III-V materials epitaxy in a number of laboratories. However, all commercially available MOVPE systems are equipped with reactors with separate injection. Due

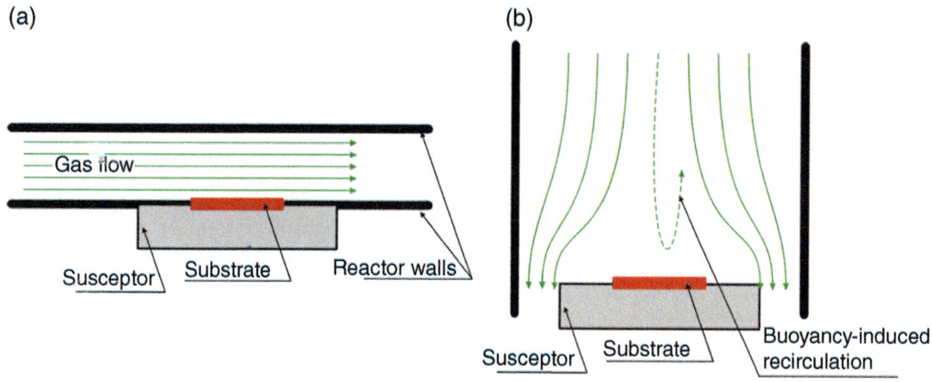

Figure 12.11 *Gas flow in a conventional simple horizontal flow (a) and vertical flow (b) reactors.*

to the low concentration of precursors in the carrier gas, the physical properties of the gas mixtures injected via different injectors are close to the physical properties of the carrier gas (hydrogen). This allows simple coupled adjustment of gas flows and injectors cross sections to ensure equal gas velocities and momentum, promoting smooth injection without generation of vortices. Moreover, for some reactors, the presence of at least two separate injectors allows adjusting the growth-rate uniformity by admixing nitrogen in the carrier gas (H_2) [7].

Two basic vapor-phase epitaxy reactor concepts are horizontal and vertical. They were well known before MOVPE was invented, and they were used in MOVPE from the very beginning.

In a conventional simple horizontal flow reactor (Figure 12.11a), the carrier gas with precursors passes in a horizontal direction above the horizontal or slightly inclined top surface of a heated substrate holder (susceptor). In this design, the susceptor disturbs the gas flow only by its heating, resulting in potential buoyancy effects, which may result in gas-flow recirculation in the case of improper design and/or reactor conditions; but this is atypical for horizontal reactors.

In a conventional simple vertical flow reactor (Figure 12.11b), the carrier gas with precursors passes vertically down toward a horizontally oriented susceptor. Flow around the wide susceptor strongly disturbs the gas flow. Due to the much larger (compared to the horizontal reactor) gap between the hot susceptor and the cold ceiling/injector, buoyancy effects are much stronger.

The combined action of these two effects makes uniform epitaxy almost impossible unless special measures are taken. These two basic reactor concepts underlie all types of MOVPE reactors described in this chapter.

12.3.2 From Simple Horizontal Flow to Planetary Reactors

A simplified sketch of a real horizontal flow reactor is shown in Figure 12.12. The carrier gas with precursors is passed above the susceptor inside an internal reactor, usually made of quartz. This reactor part typically has a rectangular cross section providing a constant value of the gap between the susceptor and ceiling in the transverse direction. If the external reactor is made of quartz, it should have an axisymmetric shape (e.g. a simple tube) to ensure durability under low reactor pressure. This is necessary not only for low-pressure operation but also for helium leak checking. If an external reactor is made of stainless steel, it may be almost any shape convenient for manufacturing and maintenance. Due to the relatively low epitaxial temperatures, lamp heating may be used as well as resistive or induction heating.

Access into the reactor for reloading and maintenance is usually performed through the lid located in the downstream edge of the reactor. In most systems, the lid is opened into a volume of

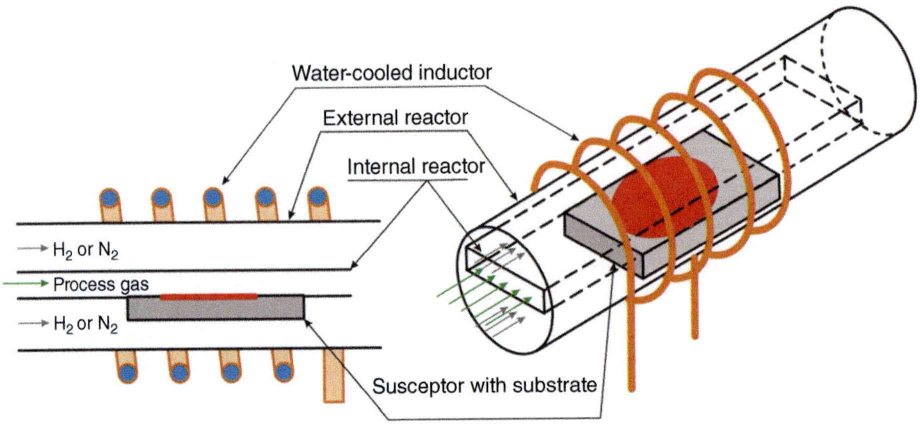

Figure 12.12 *A simplified sketch of a real horizontal flow reactor.*

a glovebox filled with pure nitrogen to prevent the reactor volume from contamination with air and protect personnel from toxic contaminants. In rare cases, an evacuated loadlock is used instead of the simple lid, which allows eliminating the glovebox and reloading without the reactor cooling down to room temperature and purging the toxic gases out.

A couple of examples of real horizontal flow reactors are presented in Figure 12.13.

According to flow-dynamics considerations, the gas mixture velocity reaches a maximum between the susceptor and the ceiling and falls to zero near the walls (including the susceptor). The gas mixture is heated by the susceptor, and the closer to the susceptor, the higher the gas temperature. Finally, the precursor concentration is also nonuniform. MOVPE is carried out under an excess of Group V precursors, and the growth temperature is high enough to promote fast surface chemical reactions, so the growth rate depends on the Group III precursor concentration.

In the vicinity of the susceptor (as well as any other hot surface) Group III precursors are consumed by the grown layer, this region is depleted and precursors are transported along the reactor midstream by the carrier gas flow (convective transport) and toward the grown layer by the diffusion through this depleted layer.

Hence, we can speak about a velocity boundary layer, a temperature boundary layer, and a concentration boundary layer.

In the simple horizontal flow reactor, the growth rate is reduced with distance passed by the gas mixture above the susceptor due to consumption of precursor by the growing surface and gradual reduction of its gas-phase concentration in the downstream direction (curve 1 in Figure 12.14). This curve may be called a *depletion type of growth-rate profile*. If Group III and V precursors are injected separately via horizontal levels (this solution is typical for commercially available reactors), and Group III precursors are not supplied via the bottom level, the lowest gas layer is virtually free from this precursor from the beginning of the reactor. Downstream in the reactor, Group III precursors penetrate into this layer by diffusion. As a result, if the overall velocity is high enough, an inverted growth-rate profile may take place (curve 2 in Figure 12.14). Indeed, an intermediate case is also possible, when the Group III precursor concentration in the bottom gas layer in the upstream region is low due to separate injection, in the downstream region it is low due to conventional depletion, and the growth rate reaches a maximum in the central region (curve 3 in Figure 12.14). If the gas-flow velocity is low enough, the growth rate is similar to the depletion type, even in the case of separate injection. It should also be noted that the depleted type of growth-rate profile indicates that most of the Group III precursors are consumed inside the reactor, and this growth regime is economically preferable. In the case of the inverted growth-rate profile, most of the Group III precursors are purged out of the reactor and wasted.

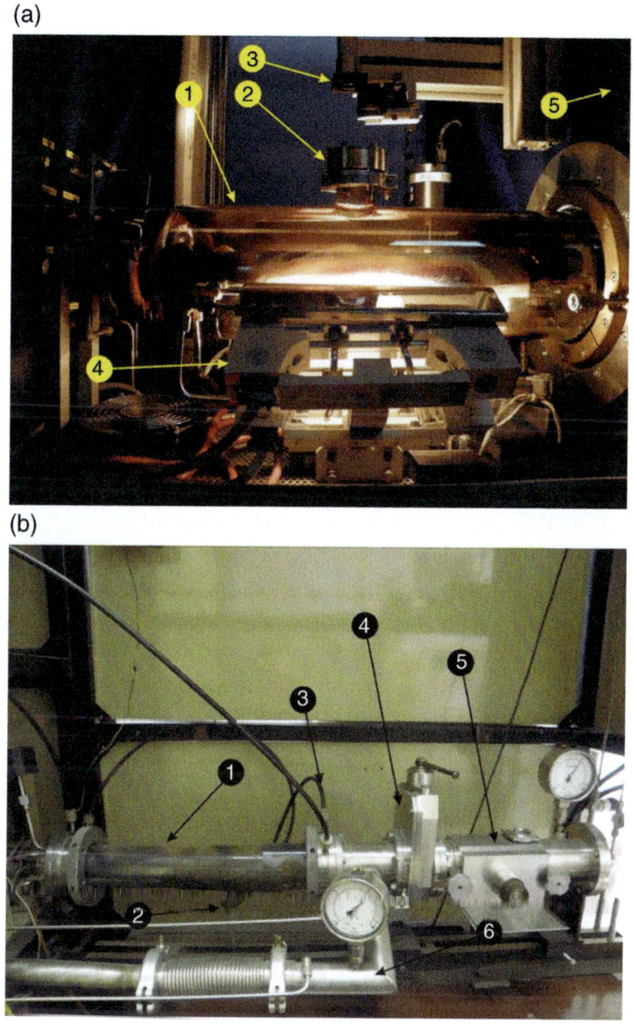

Figure 12.13 MOVPE reactors for classical III-V. (a) AIX 200/4 system (AIXTRON): 1 – reactor; 2 – viewport; 3 – in situ optical monitoring system (EpiRAS TT by LayTec); 4 – lamp heating system; 5 – wall of the glovebox. Reactor is used for R&D of multijunction solar cells. (b) Homemade low-pressure reactor with vacuum loadlock and resistive heater (direct thermal contact of heater and substrate holder): 1 – reactor; 2 – thermocouple; 3 – water supply; 4 – gate valve; 5 – vacuum loadlock; 6 – exhaust. Reactor was built in 1987–88. InGaAlAs LDs with J_{th} ~100 A/cm^2 were routinely grown in this reactor in the early 1990s. The reactor is still in operation for narrow-bandgap III-V materials. Source: Ioffe Institute; Courtesy of Dr. Nikolay Kalyuzhnyy.

If the growth rate is of the depletion type or inverted type, the uniformity of epitaxial layer thickness may be significantly improved by substrate rotation. Two types of rotation promotion are used: mechanical rotation (e.g. in reactors manufactured by TNSC) and gas-foil rotation (in reactors manufactured by AIXTRON).

The processes described take place not only above the susceptor with wafers but also near any hot surface of the reactor. Quartz surfaces of the internal reactor are not as hot as the susceptor but in the typical horizontal flow reactor are hot enough to promote III-V material deposition. This process above the susceptor, on the ceiling, does not directly influence growth. However, it results

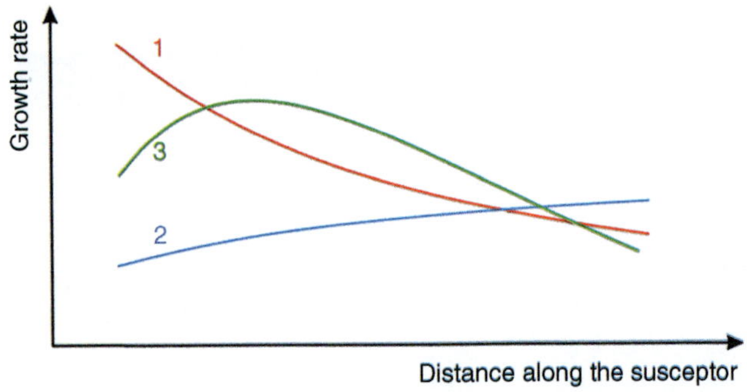

Figure 12.14 *Possible growth rate profiles in a horizontal reactor (see comments in the text).*

in gas-phase depletion; but until the thickness of the gap between the susceptor and the ceiling is larger than the sum of the depleted layers' thicknesses, it does not affect the growth rate, only the material consumption. (The other problems – peeling of the deposits from the ceiling and falling on the wafers as well as changes in the optical properties of the reactor walls – will be discussed later in detail.) However, similar processes on the reactor side walls strongly affect growth-rate uniformity. Due to deposition on the side walls, the growth rate in the transverse direction is also nonuniform: it is higher near the central line of the reactor and falls toward the sides. This nonuniformity cannot be totally suppressed by simple susceptor rotation.

The other reason for the growth-rate nonuniformity in horizontal flow reactors is a problem of uniform injection. In the simple designs of reactors of the 1980s, gases were supplied to the reactor through one or two (for separate injection) gas pipelines, and then gas flows were expanded to the whole reactor width by diffusers. Such a solution may promote more or less uniform gas velocity at the midstream of the reactor but not near the side walls. At a given pressure, temperature, and type of carrier gas, variation of the total gas flow through the reactor in combination with wafer-carrier rotation allows some reasonable level of uniformity to be reached that was enough at that time for scientific purposes but not for industrial application. Nowadays, uniformity typical for simple horizontal flow reactors sometimes is not enough even for R&D. Much better results were obtained by using relatively wide reactors with only one small substrate placed at the center of the wafer carrier or by using double rotation (see Figure 12.15a), when the entire wafer-holder rotation is combined with individual rotation of every substrate. The first solution is very ineffective in terms of material utilization, while the second is rather complex and cannot be used for growth on one large substrate. (It should be noted that this double rotation of wafers, also called *planetary rotation*, was known well before the invention of MOVPE and widely used for uniformity improvement e.g. in metal evaporation vacuum systems.)

The other smart solution of the side-wall problem became popular instead. This is the *planetary reactor*. In this reactor, a similar type of double rotation is used, but in combination with the other reactor geometry: wafers are placed individually or in groups on relatively small wafer carriers (also called *planetary discs*) that themselves are placed in the pockets of the main susceptor around the center of the reactor (Figures 12.15b and 12.16). The main susceptor is rotated around the reactor center, and every planetary disc is rotated as well. Gases are passed from the reactor center toward its periphery or (a much rarer solution but also used, e.g. in one MOVPE system family produced in USSR) in the reverse direction from the periphery toward the center.

In both cases, the gas streamlines are parallel to the susceptor, so this reactor may be described as a translation of a simple horizontal flow reactor around the vertical axis. This solution allows elimination of the side walls as well as promotion (in theory) of axisymmetric gas flow. The rotation of the main susceptor (main rotation) is necessary mostly to suppress wafer-to-wafer nonuniformities originating from nonsymmetrical design of the rector heater and possible nonideality of the shape of the gas injector.

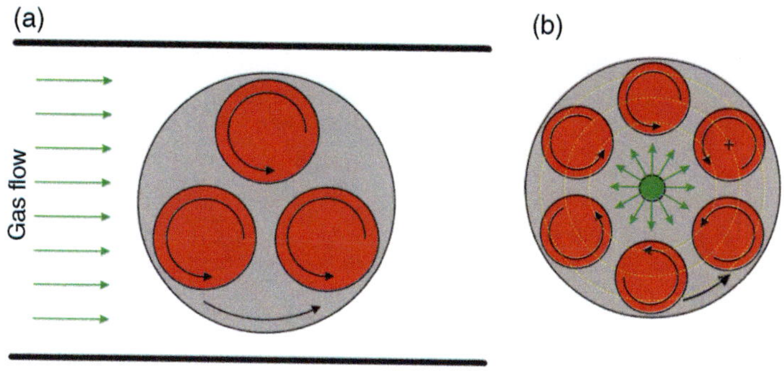

Figure 12.15 Gas flow and substrates movement. (a) Horizontal reactor with double rotation (planetary rotation of substrates). (b) Planetary reactor (or horizontal radial flow reactor with planetary substrates motion). Yellow dotted circles represent equal distance of gas passed above the susceptor (see Section 12.4.3).

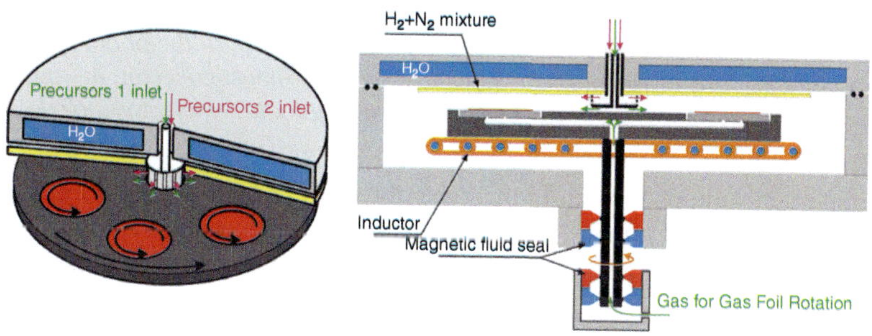

Figure 12.16 Sketch of a planetary reactor.

Two types of rotation are used in planetary reactors. The first is mechanical rotation (used e.g. in reactors produced by TNSC), which is rather complex in this case and may produce dust inside the reactor volume in the case of improper design or maintenance. The second solution is gas-foil rotation, used by AIXTRON for rotation of planetary discs in combination with mechanical rotation of the main susceptor. (Both rotations were gas foil in the first planetary reactors by AIXTRON, but this solution was too complex.)

An external reactor is always metal with water-cooled walls. Access into the reactor for reloading and maintenance is reached by opening the upper lid. To prevent the reactor volume from contamination with air and to protect personnel from toxic contaminants, the reactor is placed inside a glovebox filled with pure nitrogen (our reactor for III-N materials is shown in Figure 12.17). In some planetary reactors for industrial applications, an additional wide gate in the side wall allows automatic reloading without opening the reactor. This system also allows reloading at elevated temperatures (up to ~400 °C), which reduces the growth-cycle duration.

In the popular case of central gas injection, gases may be supplied to the injector via the center of the upper lid (solution used by AIXTRON) or from the bottom of the reactor through the hole in the main susceptor (used in reactors by TNSC). Originally, injectors were usually made of quartz; now stainless-steel injectors are used. Typically, at least two-zone injection with Group III and V precursor separation is used.

As in horizontal flow reactors, in planetary reactors for classical III-V materials, all three common types of heaters may be used. The first generations of planetary reactors produced by AIXTRON used lamp heating; this was replaced in later generations with induction heating. TNSC uses resistive heaters.

Figure 12.17 AIX 2000HT planetary reactor (AIXTRON) installed at the Ioffe Institute. Note that it is a reactor for III-N materials, but the main components are the same, so the photo is placed here for the sake of clarity. 1 – main susceptor; 2 – satellite disc; 3 – exhaust collector; 4 – quartz ceiling; 5 – ceiling deposits; 6 – two holes for in situ optical monitoring tool; 7 – injector (nozzle).

One extra important solution typical for this type of reactors allows precise temperature control of the ceiling. A flat ceiling (for many decades made of quartz, nowadays from graphite in some reactors) is placed close to the water-cooled metal lid of the reactor, which results in effective reduction of the ceiling temperature. Moreover, control of the gas composition is this gap using hydrogen–nitrogen mixtures allows the thermal conductivity to be controlled across this gap, which in turns allows the ceiling temperature to be controlled. It was shown that the temperature range 300–400 °C promotes the lowest deposition rate on the ceiling. Use of quartz ceilings with a matted (lapped, frosted) surface instead of polished prevents deposits peeling off.

The most efficient way to suppress deposits peeling off and influence the epiwafer quality is to use a *face-down geometry* for the reactor; this may be used both in horizontal flow and planetary reactors (and in vertical reactors as well). However, technical decisions for realization of this geometry are always very complex – one should assure good contact of the wafer with the wafer carrier without using elements that interfere with the gas flows. Reactors of this type, both horizontal flow and planetary, are manufactured by TNSC (e.g. model HR-6000, BRC, and BMC series) for classical III-V materials only. Much higher temperatures and aggressive ambients in III-N reactors makes such solutions too complex for real use. Indeed, only mechanical rotation may be used in face-down reactors.

A characteristic feature of both horizontal flow and planetary reactors is the long time necessary for gases to pass above the heated susceptor (residence time). It may have no significant influence on the epitaxial process. However, if undesirable reactions occur in the heated gas phase, this long residence time may cause negative effects, so it should be taken into consideration.

It should be pointed out that a planetary rotation system takes extra space in the reactor because the substrates cannot be placed side by side, so surface-utilization efficiency of these reactors is lower than in other types, However, individual rotation of satellite discs allows the reactor to be run in regimes resulting in very deep depletion of the gas phase, when the growth rate at the periphery of the reactor is many times lower than that close to the injector. This means almost all Group III precursors are consumed inside the reactor and not wasted.

Temperature control of the ceiling allows significant reduction of the ceiling deposition rate, which not only lessens the problem of deposits peeling off but also reduces material consumption

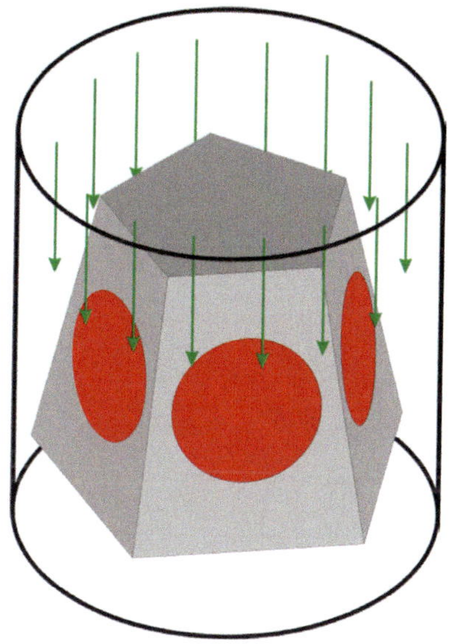

Figure 12.18 A sketch of a barrel reactor. Substrates are located in the pockets on the side walls of a truncated pyramid. Gases pass down in the gap between the pyramid and the walls of a cylindrical reactor.

for these deposit formations. Finally, the absence of side walls (in contrast to horizontal flow reactors) means no material is lost because of deposit formation on them. These factors make planetary reactors the most efficient across all types in terms of material utilization efficiency. They are dominating mass production for such applications as epiwafers for LEDs, where material utilization efficiency is one of the most important parameters. A nice, detailed description of a planetary reactor with gas-foil rotation was given in [1].

It may be added that an alternative solution for elimination of side walls, while maintaining the benefits of gas streamlines being parallel to the susceptor, is the *barrel reactor* (see Figure 12.18). These reactors were widely used in CVD of silicon; however, their use for MOVPE has never been popular.

12.3.3 Close-Coupled Showerhead (CCS) Reactors

As was already pointed out, gas flow around the wide susceptor and strong buoyancy effects make uniform epitaxy in simple vertical reactors almost impossible. Two significantly different solutions for these problems are used.

Buoyancy in CCS (see Figure 12.19) is suppressed by significant (down to 5–20 mm) reduction of the gap between the hot susceptor and the cold injector. In such a geometry, gas injection should be very uniform across the reactor because, due to the very small distance from the injectors to the top to the susceptor, gases have no chance to intermix if injected in a nonuniform manner. Uniform injection is reached by supplying gases through the dense array of tubes (up to 100/sq. inch).

They are combined into two plenums for separate injection of Group III and V precursors, both diluted in carrier gas to ensure equal gas velocities. Tubes of these two plenums are located one-by-one. If no prereactions between precursors take place, the premixed precursors may be supplied via all tubes, as was done in the first reactors of this type [8, 9]. The showerhead is water cooled. There are tens of thousands of joints in the tubes and walls in the showerhead, and even a small leak in any joint means water penetration into the reactor. Hence, manufacturing of this unit is challenging.

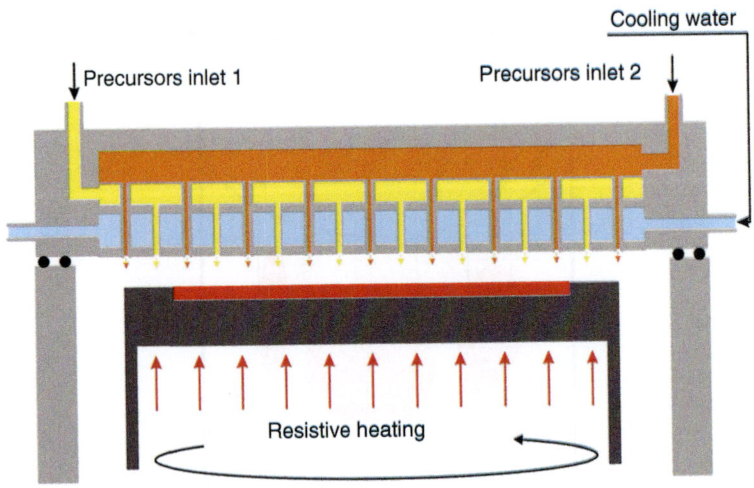

Figure 12.19 *A sketch of a CCS reactor with two injection plenums.*

The momentum of the vertically injected gas jets relaxes a few millimeters from the top, and then the gas passes in the radial direction toward the periphery of the reactor. Hence, at first glance, in terms of gas streamlines, this reactor is similar to a planetary one. However, what really influences the epitaxial process – precursor distribution – is controlled by another mechanism. This type of precursor injection results in a uniform concentration distribution just below the ceiling. Then, the precursors diffuse across the passing gas toward the depleted layer close to the hot susceptor. This results in a very uniform growth-rate distribution across the whole wafer carrier following distribution of precursor diffusion flux. (By the way, a CCS reactor may be described as horizontal flow with uniform feeding of precursors from the ceiling.) The narrow gap between the hot susceptor and the cooled showerhead results also in a temperature distribution that is uniform across the reactor and linear in the vertical direction.

Thus, operation of a CCS reactor is controlled by the diffusion from the top, where the precursor concentration is uniform, down to the susceptor, and is virtually insensitive to such reactor conditions as total gas flow, pressure, etc. On the one hand, it is beneficial, especially for newcomers. On the other hand, there is not much room for process parameter optimization.

CCS MOVPE reactors were initially invented by Thomas Swan Ltd. (later, this business was acquired by AIXTRON) for classical III-V growth, but did not become popular for this application. One reason is that there was no large market – other types of reactors were already well developed, and there was no need for new types. The other reason is formation of showerhead deposits. In addition to the already-described diffusion of injected precursors and products of their pyrolysis toward the susceptor, there is an additional diffusion flux in the counter direction – diffusion of pyrolysis products toward the showerhead. Those that are not sufficiently volatile or tend to polymerization form a layer of deposits on the surface of the water-cooled showerhead. This effect will be discussed later, in the section devoted to III-N material, for which CCS reactors are used much more widely. Here, we would like to point out that in the case of classical III-V MOVPE in CCS reactors, there is a high probability of arsenic and/or phosphorus condensation on the showerhead if the concentration of arsine or phosphine in the reactor is too high and the distance between the susceptor and the showerhead is small. With the latter's increase, the problem disappears, but the advantages of close spacing also vanish. Flakes of arsenic, phosphorus, or their mixture peeling off and dropping on the hot wafer surface leads to flash-like evaporation, resulting in brief local changes in Group V concentration. It is unfavorable in any case and is detrimental to the epitaxy of alloys containing more than one Group V element, because it results in local variation of alloy composition. This problem was described in one of the first publications on CCS reactors [8]. In the same publication, other problems with CCS reactors were noted: significant narrowing of the growth-parameter window, the necessity of using precursors with low pyrolysis temperature (e.g. TEG instead of TMGa) for growth of some alloys, and lack of reactor flexibility. However, these were only the first steps of CCS reactor development. Geometry and growth-condition optimization, including use of heated water for showerhead cooling, allows these problems to be minimized, and a wider growth-parameter window was realized. These reactors were commercialized and are available on the market. A few research teams have shown excellent results in classical III-V epitaxy using CCS reactors [10–12].

The majority of CCS MOVPE reactors used worldwide are produced by Thomas Swan Ltd. or AIXTRON. These reactors have stainless-steel, water-cooled external housings with a graphite susceptor placed on the top of a quartz support – a wide vertical tube surrounding the multizone resistive heater. Access inside the reactor for reloading and maintenance is achieved by opening the top lid, which contains a showerhead. This solution is similar to that used in planetary reactors, and for the same reasons, the reactor is enclosed into glovebox with pure nitrogen ambient. An example of a CCS reactor (for III-N materials) is presented in Figure 12.20.

It should be noted also that in many early papers, CCS reactors were called *rotating-disk reactors*; however, it is more correct to use this name for another type of MOVPE reactors, described next.

12.3.4 Rotating-Disk Reactors

Another solution to the problems of the simple vertical reactor is used in rotating-disk reactors (RDRs). These reactors generally differ from a simple vertical reactor only by very fast rotation of the susceptor (up to 1000 rpm and beyond). A rotated disk acts as a pump (indeed, such pumps exist), evacuating the gas from the reactor volume. A thin boundary layer above the rotating disk

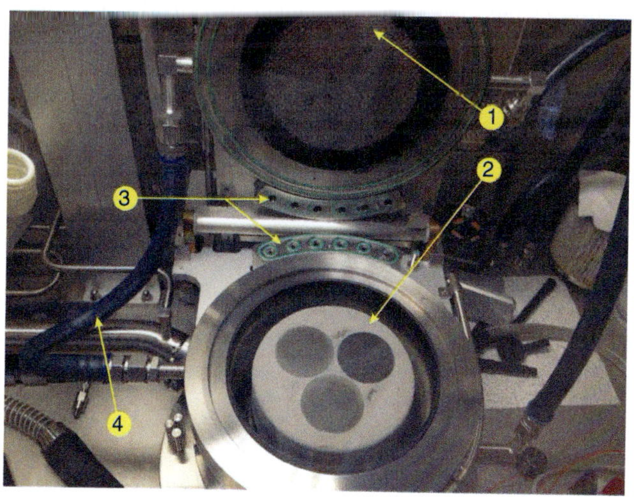

Figure 12.20 *CCS reactor for III-N materials with 3 × 2-inch capacity produced by AIXTRON. Note that this is a reactor for III-N materials, but the main components are the same, so the photo is placed here for the sake of clarity. 1 – showerhead; 2 –susceptor; 3 – separable lines for gases supply to the showerhead; 4 – water supply. Source: Courtesy of Dr. Tomas Grinys, Faculty of Physics, Vilnius University, Lithuania.*

(susceptor) is formed. Gas above this boundary layer passes downward undisturbed as if there were no susceptor. Inside the boundary layer, gases rotate quickly (note the difference from, for example, the horizontal flow reactor) and pass toward the outer region along flat spiral lines. As in a horizontal reactor, the boundary layer is heated and depleted.

Precursors are transported by the carrier gas toward the boundary layer and then diffuse through it toward the susceptor/wafers. The faster the rotation, the thinner the boundary layer, and the more effective the diffusion transport. The boundary layer is rather uniform, and if the precursor concentration in the carrier gas above the boundary layer is uniform, the growth rate will be uniform as well, in spite of the absence of secondary rotation. A general description of this reactor operation as well as very nice photographs of smoke streamers visualizing the flow are given in [13]. However, everything is fine only in theory. The central region of a RDR is a special point, and the epitaxial process is not necessarily uniform in the vicinity of this point. Side walls also influence the fluid dynamics of this reactor. Finally, separate injection of Group III and V precursors means nonuniform precursor concentration above the boundary layer. Indeed, inside this boundary layer, precursors are intermixed due to fast rotation, but not necessarily uniformly enough. Thus, during the development of RDRs, injection schemes become more and more complex. Significant uniformity improvement was reached using multipoint injection. In this case, precursors are supplied in a mixture with carrier gas through a number of inlets placed at various distances from the reactor axis. The precursor concentrations in these injection lines are adjusted individually to compensate for nonuniformity of epitaxial growth. This complicates the reactor and gas-blending system but works well. Only the latest version of the RDR for large-volume classical III-Vs produced by Veeco (TurboDisc K475i system) uses the company's multi-slit injection geometry, which was invented initially for III-N reactors (discussed shortly); it significantly simplifies uniformity adjustment.

The main disadvantage of the RDR for industrial applications is the much higher gas consumption than in planetary reactors. First, the same value of growth rate at the periphery of the wafer carrier as in the central region means at least some precursors are purged away from the reactor and not consumed by the grown layer. A more important reason for the high material consumption of RDRs is that operation of the reactor in the manner described (called the *plug-flow regime*) occurs only if the total gas flow, the rotation rate, and the reactor pressure are matched. This coupling is described by a *pressure-rotating rate* stability diagram (Figure 12.21) [14]. If (at a given

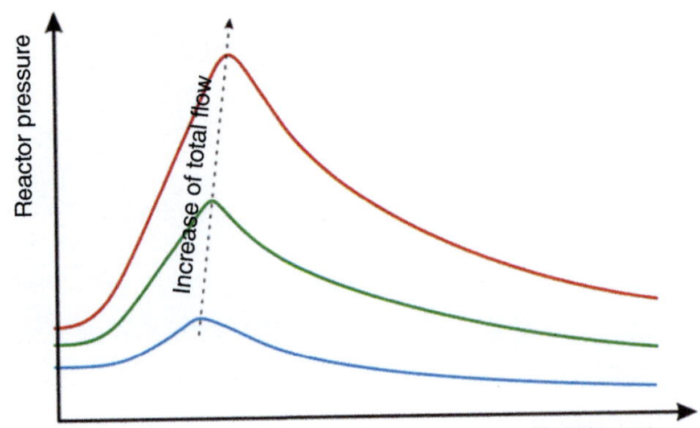

Figure 12.21 *Pressure–rotation rate stability diagram for three total flows through the reactor. Qualitative hand-drawing representation. Plug-flow regime for a given total flow occurs only below the corresponding curve.*

reactor pressure and susceptor temperature) the rotation rate is too low, rotation-induced effects are too weak to suppress buoyancy, and an upstream flow is formed above the central part of the susceptor. If (at given pressure and rotation rate) the total gas flow is below some critical value, the productivity of the "pump" exceeds the gas supply, and recirculation of gas inside the reactor is started, resulting in an upstream flow above the susceptor edges. Both cases result in reactor conditions that are unacceptable for epitaxial processes. First, any upstream transport precursor pyrolysis products to the cooled walls and injection flange leads to deposit formation. Also, these upward flows are not stable, which leads to unstable growth conditions.

The higher the reactor pressure, the higher the total gas flow necessary to maintain the drag-flow regime. For classical III-V materials, the whole epitaxial process may be carried out at low pressure; but even under these conditions, material consumption of RDRs is typically much higher than in planetary reactors. High material consumption means not only higher usage of expensive precursors, but also more exhaust gases should be utilized. Depending on the location (which means logistics, local safety, ecological regulations, etc.), this emission purification and exhaust material utilization may cost even more than the high-purity precursors. Thus, for mass production of classical III-V materials, RDRs are used mostly in those cases where the cost of the epitaxial process is not as important: laser diodes, solar cells for space applications, etc. Small RD reactors are also widely used for research purposes in universities and research centers.

Horizontal flow, planetary, and CCS reactors may be called *intuitively clear reactors*. CFD simulations are very useful, but gas flows are very stable in these types of reactors. Even significant modifications of reactor conditions typically do not lead to recirculation or other problems. This is not the case for RDRs. Nowadays, when CFD simulations are accessible, it is strongly recommended to use them during any process optimization that requires significant modification of reactor conditions. A stability diagram may be roughly estimated using simple equations; however, precise boundaries of the stability region in parameter space may be defined using 3D reactor modeling only. An experimental determination of the stability regimes is wrong for RDR users. Even one epitaxial run under conditions resulting in gas recirculation may require a total reactor disassembly for cleaning. Moreover, in the majority of multipoint injection schemes (e.g. in popular schemes when Group III precursor injectors are placed along reactor radii or diameter), significant redistribution of gas flows between injectors may result in very strong and nonobvious gas-flow nonuniformities.

The largest worldwide supplier of RDRs is Veeco, which acquired this business from EMCORE. Veeco's reactors are being sold under the name TurboDisc. However, a few other companies offer reactors of this type as well, which is why we use the more general term *rotating-disk reactor* in this chapter. General engineering solutions used in all of these reactors (Figures 12.22 and 12.23) are similar and strongly differ from those used in horizontal flow, planetary, and CCS reactors. These reactors are always a cylinder shape with a mechanical rotation feed-through at the bottom and an injection flange on the top. A graphite wafer carrier is placed on top of a vertical axis. The majority of RDRs are equipped with multizone resistive heaters placed below the wafer carrier. A gate valve between the reactor side wall and the evacuated loadlock allows for reloading without opening the reactor. Thus, there is no necessity for a glovebox with pure nitrogen surrounding the reactor. While only substrates (or planetary discs, in planetary reactors with reloading automation) are loaded and unloaded in horizontal flow, planetary, and CCS reactors, in RDRs, a whole wafer carrier is moved. In the current reactors, this is performed automatically. These solutions allow for exchanging wafer carriers at relatively high temperatures (~400 °C) and performing wafer-carrier cleaning and filling with substrates while the next epitaxial run is being executed. These solutions lead to a significant rise in productivity. Rotating-disk reactors including loadlock and manipulators for reloading are made of stainless steel and are compliant with high-vacuum standards. In appearance, these reactors are similar to MBE systems.

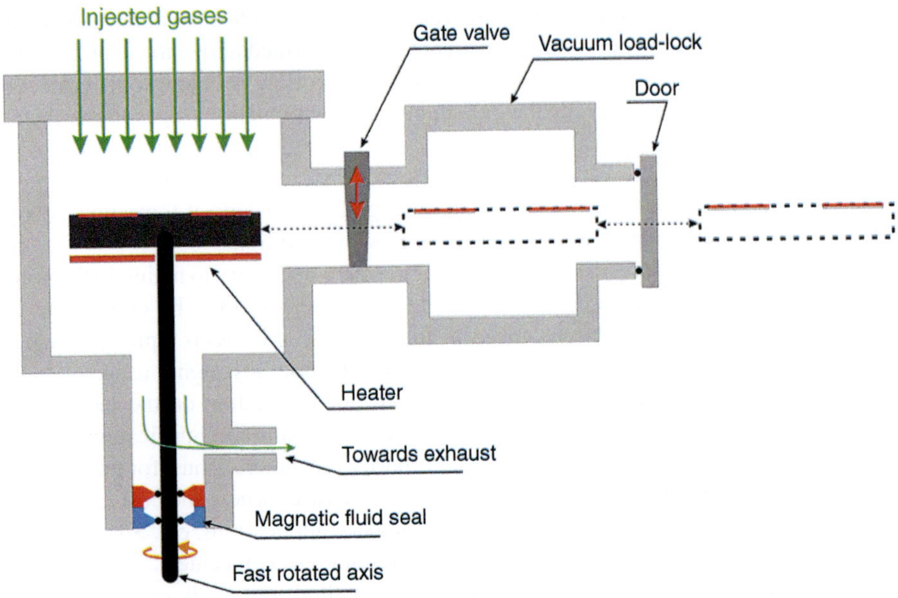

Figure 12.22 *A sketch of a typical rotating-disk reactor.*

Figure 12.23 *Reactor of E300 (GaNZilla) III-N MOVPE system produced by Veeco. 1 – reactor vessel; 2 – injection flange; 3 – viewport; 4 – water supply; 5 – MFCs of multizone precursor distribution system; 7 – gate valve; 8 – vacuum loadlock. Source: Courtesy of Dr. Dmitry A. Zakheim, Ioffe Institute; and Svetlana-Optoelectronics, St. Petersburg, Russia.*

12.4 Reactors for MOVPE of III-N Materials

12.4.1 Principal Differences between MOVPE of Classical III-Vs and III-Ns

At first glance, MOVPE of GaN, AlN, and alloys based on these two binary compounds looks very similar to MOVPE of classical III-V compounds like GaAs and AlAs. It seems that one should just replace arsine with ammonia. Moreover, the majority of reactors used for III-N MOVPE are of the same types as used for MOVPE of classical III-Vs. However, some significant peculiarities of III-N MOVPE result in a number of very specific requirements for the reactor design.

The first two issues – high growth temperature and high chemical activity of ammonia against most materials – were already discussed. They are important but rather obvious. They were hot topics during the first years of III-N MOVPE reactor development, but now they are more or less solved. The other differences were realized much later.

As we already stressed, MOVPE of any of the classical III-V–based heterostructures may be performed under reactor conditions (temperature, pressure, gas flow, type of carrier gas) that are practically constant during the entire epitaxial process. Moreover, in terms of fluid dynamics, gas that passes through a reactor may be considered like hydrogen. All of the three (or four, if one accounts for horizontal flow and planetary reactors separately) types of reactors dominating the market for all III-Vs, including III-N, were invented and developed for growth of classical III-V materials.

The requirements for III-N MOVPE reactors are much more complex. The first important issue is the very high ammonia flow rate necessary for growth of most of III-N materials. The required ammonia fraction in the reactor ambient also strongly varies. While for GaN MOVPE, the typical value is in the range of 20–30%, for AlGaN, it is reduced from that normal for GaN to a few percent with an increase of the AlN mole fraction toward the binary AlN compound. In contrast, for InGaN growth, the ammonia flow typically exceeds the carrier gas flow, so the ammonia concentration is above 50%. The physical properties of ammonia strongly differ from those of hydrogen and nitrogen. All of this means ammonia is not only a precursor but also a carrier gas, strongly influencing gas-flow patterns and temperature distribution in the reactor.

Separate injection of Group III and V precursors in the case of III-N MOVPE results in some additional problems.

Generally, separate injection means there are at least two injection slits or holes in the reactor. Low concentrations of both Group III and V precursors in the case of III-V MOVPE means that both of them may be well diluted in hydrogen before injection; thus, in terms of flow dynamics, hydrogen is supplied though both injection slits. The velocity of these two hydrogen flows may be well matched to avoid any vortices. Alternatively, the hydride precursor may be supplied undiluted via a very narrow injection slit, and it also will not disturb the overall gas-flow pattern.

In the case of III-N MOVPE, this matching is complicated. The ammonia flow is high and varies significantly for different materials grown. This means if the injection slit cross section is optimized for a given ratio of ammonia-to-carrier gas to ensure matching of injected gases' velocity, changing the conditions to grow the other materials composing the device structure will mismatch these velocities. Moreover, even if the velocities are matched, momenta will be mismatched due to the difference in molecular weights of ammonia and hydrogen or nitrogen. These mismatches promote the appearance of vortices in the reactor, which is very undesirable. Both velocity and momentum may be matched if a hydrogen/nitrogen mixture is used as a carrier gas. However, as will be described shortly, carrier-gas composition is not a free parameter for III-N MOVPE. This solution is inapplicable for InGaN and has an impact on GaN and AlGaN growth. An alternative to hydrogen-nitrogen mixtures are mixtures with helium and/or argon, but these expensive gases are rarely used in MOVPE. Finally, the addition of nitrogen to hydrogen for GaN or Al(Ga)N growth commonly results in a reduction in the growth rate due to reduction of diffusion coefficients.

The second issue is the active role of hydrogen in the growth chemistry. In the case of classical III-Vs, hydrogen is simply a carrier gas. Using nitrogen instead of hydrogen may require some modification of other growth conditions or even reactor geometry but does not have a significant effect on the material growth process. This is not so for nitrides. Generally, it is well known that GaN and Al(Ga)N should be grown using hydrogen as a carrier gas, while for InGaN, a hydrogen-free ambient is necessary. However, the situation is even more complex.

At high temperatures, GaN is very unstable under hydrogen. GaN epitaxial layers of a few micrometers thickness at 1000 °C under hydrogen ambient totally decompose in a few minutes. Under hydrogen-ammonia mixtures, the process is less dramatic and may be characterized as etching [15, 16]. It takes place during epitaxial growth, and it is approximately true to consider the growth process a result of two independent processes: direct process of GaN synthesis and the reverse process of GaN etching. Under ammonia-to-hydrogen concentrations and temperatures typical for GaN epitaxy, the GaN etching rate is about 0.2–0.5 μm/h, which is up to 10% of the typical growth rate, so the overall growth process is in quasiequilibrium. The lower the ammonia-to-hydrogen ratio, the stronger the etching, and thus the more the growth process is in equilibrium. If hydrogen is replaced with nitrogen, the process of GaN growth becomes much more nonequilibrium. Thus, by varying ammonia, hydrogen, and nitrogen compositions in the reactor, one can control the etching rate, the level of the growth process reversibility, and related effects like growth-rate anisotropy, doping efficiency, background impurity incorporation, point-defect concentration, interface abruptness, etc. This means a universal MOVPE reactor should accept using hydrogen, nitrogen, and their mixtures as carrier gases for GaN growth.

When AlGaN alloys are grown, the effect of hydrogen may be even stronger. The AlN binary component is stable under hydrogen, at least at typical epitaxial temperatures, while the GaN component is decomposed by hydrogen [17]. Hence, the AlGaN composition is strongly affected by the carrier gas used, and if the hydrogen-to-nitrogen composition is used for gas-flow pattern stabilization as described earlier, it affects the epitaxial process.

The effect of hydrogen on the InN binary compound of InGaN or InAlN alloys is very strong. Most of the indium is etched off and evaporated if hydrogen is used as a carrier gas. This is why these alloys usually are grown under a hydrogen-free ambient. However, some controllable addition of hydrogen during bulk InGaN growth or during some steps of InGaN/GaN multiquantum-well (MQW) structures formation may be useful for fine control over the material properties.

The third issue is the variety of strong parasitic reactions in the gas phase resulting in nanoparticle formation. Such phenomena were also observed in some other MOVPE and CVD processes, but in III-N MOVPE, these reactions are very pronounced and may take place during growth of all III-N materials but via different mechanisms [18]. It is well established that during AlN growth, AlN nanoparticles are formed; GaN nanoparticles are formed during GaN growth, while metal indium nanoparticles are formed during InGaN growth. A universal attribute of well-developed parasitic reactions is a sublinear, eventually saturating dependence of growth rate on tthe Group III-precursor concentration.

The strongest parasitic reactions take place between TMAl and ammonia. It is generally believed that the long chain of reactions starts with formation of TMAl:NH_3 adduct and results in nucleation of AlN nanoparticles. If generated, these nanoparticles grow by a mechanism similar to epitaxial growth, depleting the gas mixture and consuming precursors. If AlGaN is grown, the initial formation of nanoparticles is identical to the case of AlN growth; but then, during enlargement, they consume not only TMAl but also TMGa [17, 19]. Universal methods of suppressing these reactions are reduction of TMAl and ammonia concentration, and reduction of pressure. This is why the typical reactor pressure for Al(Ga)N growth is about 100–200 mbar or lower. Other methods are reactor-specific and will be described later. For TMAl concentrations below some limit, the AlN growth rate is proportional to TMAl concentration; but with further increase of the TMAl concentration, the parasitic reactions become stronger and stronger, most of the TMAl is consumed by nanoparticles, and the dependence of the growth rate on the TMAl concentration deviates from the linear dependence predicted by the mass-transport process and saturates. At higher TMAl concentrations, the dependence may even demonstrate a negative slope. All of this means that unless the reactor pressure or the ammonia concentration is optimized, the growth rate of AlN or AlGaN of the desired composition is limited to a rather low value.

The mechanism of indium nanoparticle formation is different from that just described. The typical InGaN growth temperature is in the range of 700–800 °C and is a compromise between what is optimal for InN and GaN. This temperature is too high for the InN binary component of InGaN, and significant indium re-evaporation takes place. To suppress this re-evaporation and obtain InGaN of the desired concentration, a significant excess of TMIn is supplied into the reactor. At this relatively high temperature, TMIn is decomposed to indium; and if the indium partial pressure in the gas phase exceeds the equilibrium vapor pressure, nucleation of indium particles takes place [20].

The mechanism of GaN nanoparticle formation is not as well defined. Either it is started with the formation of gallium nanoparticles similar to indium ones and then they react with ammonia, or direct reaction between gallium-containing MO radicals and ammonia takes place. Perhaps both processes occur. As in the AlN case, this effect limits the growth rate; the higher the pressure, the lower the achievable GaN growth rate, other conditions (including reactor geometry) being constant [21, 22].

In addition to the effect on the rate of parasitic reactions, the reactor pressure itself influences the growth process of III-N materials. To the best of our knowledge, there no general theory of these effects. It is well known that the rector pressure strongly affects the GaN growth rate anisotropy, so it significantly influences the initial stages of GaN growth on foreign substrates when 3D growth and coalescence occur. It is generally recognized that higher pressure at this stage results in higher-material quality [23, 24]. Moreover, the reactor pressure also affects InGaN QWs and even bulk GaN properties.

Finally, various layers of III-N heterostructures are grown at significantly different temperatures, from ~500 °C for GaN nucleation layers through 700–800 °C for InGaN to 1050–1100 °C for GaN and sometimes even higher for AlN and high Al-containing AlGaN (wafer carrier temperatures just below the wafer are given).

Hence, we have to conclude that a universal III-N MOVPE reactor should maintain stable, vortex-free flow for a wide range of pressure, temperature, and ammonia flows, with hydrogen or nitrogen as a carrier gas, and the epitaxial process should be affected as little as possible by parasitic reactions.

Two additional issues also should be mentioned. In situ optical-characterization tools are useful in epitaxy of any materials, but for nitrides they are especially important. The majority of III-N structures are grown on foreign substrates. On the one hand, this strongly complicates the growth process. On the other hand, it allows use of in situ optical reflectance monitoring for growth-rate measurements and surface roughness estimation due to the optical interference in the III-N structure sandwiched between the substrate and gas ambient. Control of the initial nucleation is also strongly simplified by using this technique. Strain engineering is an important part of almost any III-N heterostructure optimization, especially if the structures are grown on Si or SiC substrates. In situ optical wafer curvature measurement provides data about stress evolution during growth and is now a must-have tool for process development. Hence, good optical access to the substrates via optical ports is an absolute requirement for any modern III-N reactor, including MOVPE reactors. This does not mean in situ optical characterization is useless in MOVPE of classical III-V. Specific methods, e.g. reflectance anisotropy spectroscopy (RAS) as well as those described earlier and similar to that used for III-N epitaxy, are very effective. However, if for III-V MOVPE in situ techniques are helpful, then for III-N MOVPE they are essential. Hence, the feasibility of optical access to the wafers in various reactors, including those used for classical materials, is discussed here.

Last but not least is the issue of reactor-wall deposits. In III-N MOVPE, due to the very high temperatures, more than 90% of energy is transferred by radiation. Deposits reduce reflectance of stainless steel and transparency of quartz walls and thus significantly affect the temperature distribution in the reactor.

In the next section, we describe how the problems and challenges listed here were solved during the adaptation of reactors developed for classical III-V epitaxy to make them usable for III-N MOVPE.

12.4.2 Rotating-Disk Reactors

Rotating-disk reactors (especially the TurboDisc reactors produced by Veeco) are extremely popular in the field of III-N epitaxy. Generally, reactors of this type suffer from most of the problems described earlier more than others, and a long road of optimization was travelled before at least some of these problems were minimized.

Nanoparticles formed in the reactor volume by parasitic reactions as described previously are impacted by thermophoresis, which pushes them in the direction opposite to the temperature gradient. The hottest surface in an MOVPE reactor is the substrate carrier, so the thermophoretic force is directed toward the normal to the wafer carrier surface. It should be stressed that this effect is very advantageous. Without thermophoresis (e.g. in the hot-wall reactor with zero temperature gradient), nanoparticles may drop on the wafers, deteriorating material crystallinity.

The impact of gravity on nanoparticles is negligible due to their very small size, and the only extra force impacting them comes from the gas that pushes them in the direction of the flow. All of these conclusions are common for any reactor type. The specific problem with RDRs is that thermophoresis pushes the particles from the rotating wafer carrier at distances larger than the thickness of boundary layer. In this region, the gas flow is directed toward the wafer carrier. This means the main two forces impacting nanoparticles act in opposite directions: they balance each other, and the cloud of nanoparticles is formed a few millimeters above the wafer carrier [25]. The temperature in this region is high enough to promote growth of nanoparticles, consuming almost all MO precursors. As a result, if reactor conditions lead to nanoparticle formation, their impact on the epitaxial growth is extremely strong. This problem looks to be a distinctive characteristic of RDRs: high growth rate in these reactors may be reached only at low pressure where nanoparticles are not generated. It should be mentioned that the first detailed study of all three types of nanoparticles typical for III-N epitaxy (AlN, GaN, In) was carried out just in a RDR, where the stable cloud of nanoparticles is formed that allows them to be studied by optical methods [18].

The second specific problem of RDRs comes from the coupling of reactor conditions given by the pressure–rotating rate stability diagram. This coupling results in rather high gas consumption of RDRs for any materials. However, as mentioned earlier, for growth of some layers of III-N structures, the reactor pressure should be increased to values close to atmospheric pressure. Under these conditions, the gas consumption of RDRs becomes very high. However, this problem is not as difficult as it may seem.

On the one hand, reactor optimization produces a reduction in gas consumption. The left-upper border of stable operation in the pressure–rotation rate diagram is pushed by reduction of the reactor height, which reduces the buoyancy force (the same idea as the basis of close-coupled showerhead reactors, but for them the reactor height is much smaller). The right-upper border is pushed by fine optimization of the shape of the bottom part of the side walls, resulting in deflection of gases from the undesirable upward direction toward the exhaust. On the other hand, nowadays the cost of carrier gases and ammonia is not as high as 10–20 years ago, and high consumption is not a great disadvantage.

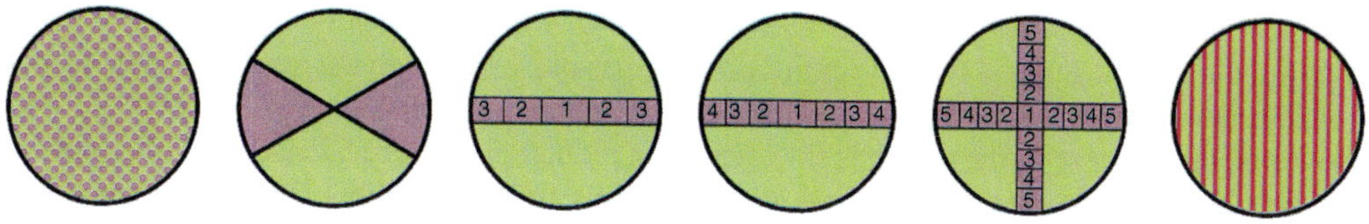

Figure 12.24 Various kinds of MO injector arrangements on the top flange of III-N rotating-disk reactors. Purple – MO; green – hydride zones. Equal digits in a few zones indicate equality of flows split by tees. The leftmost figure represents uniform injection of premixed precursors (some very first reactors), while the rightmost is a multislit injector.

The third problem for many years was imperfect operation of the reactor resulting from the velocity and/or momentum mismatch of injected ammonia and MO carrier gases. Gas injection into these reactors should be performed through a wide top flange that, in combination with the wide reactor volume below this flange, leads to easy formation of vortices if gas flows are significantly mismatched. Through approximately 15 years of optimization of III-N reactor design, gas-injection systems became more and more complex (see Figures 12.23 and 12.24), using multizone injection of nitrogen-hydrogen-ammonia mixtures of adjustable composition in order to match both the velocity and the momentum for GaN and AlGaN growth and reach some compromise for InGaN growth [26]. However, with the development and implementation of the multislit injector concept (see e.g. Figure 1 in [27]), this problem mostly disappeared.

It should be pointed out that the basic design of RD reactors allows for excellent optical access for various in situ characterization tools. The large surface of the top injector provides enough space for optical windows, and their existence does not disturb the gas flows significantly. Usually, they are purged with ammonia and carrier-gas mixtures. Besides protecting windows from deposits, purging of windows located on the vertical reactor top lid affects epitaxial process uniformity [9], which allows fine tuning of the epitaxial process.

In spite of the first two problems typical for any RDR (high gas consumption and low growth rate), they are very popular for III-N epitaxy, and these reactors manufactured by Veeco dominate the market. Apart from commercial factors, the explanation is that these reactors were primarily designed for reduction of the human factor. Undoubtedly, maintenance of these reactors is necessary from time to time and is rather complex (maybe more complex than for most other types of reactors), but routine daily operation is extremely easy and reliable and makes low demands on the qualifications of the operators. This was true even for older generations of these reactors, but for the last generation, the role of the operator is as low as it has ever been. It seems that only a few steps separate these reactors from fully automatic operation, which is standard in the silicon industry.

12.4.3 Planetary Reactors

The main advantages of planetary reactors for III-N MOVPE comparing to RDR are lower material consumption and much higher accessible growth rate. The first is similar to reactors for classical III-V and comes from stable operation of these reactors in a wide range of pressures without coupling with other reactor conditions to maintain this stability. The origin of the second advantage requires more detailed discussion.

As in RDRs, in planetary reactors, nanoparticles generated by parasitic reactions are affected by the gas flow and thermophoresis. However, in planetary reactors, the resulting forces are not contradirectional but perpendicular. Thermophoretic forces push nanoparticles from the wafer carrier to the region with the highest gas velocity directed from the reactor. Thus, these particles are pushed downstream, away from the reactor.

Practical utilization of this effect for reaching high growth rates at not-very-low pressures is complicated by the fact that almost the entire time the gas passes through the reactor, it is in the hot zone. Hence, the initial stages of parasitic reactions are rather pronounced.

However, the increase of the total gas flow results in reduction of the residence time of gas in the hot zone (which suppresses nanoparticle generation) as well as in more efficient purging the formed nanoparticles away from the reactor. Indeed, this solution reduces an advantage of planetary reactors over RDR in terms of material consumption, but it is compensated for by the significant increase of the growth rate. It is very important to understand that if, for example, growth rate is doubled by doubling the gas flows, it is very profitable, because the integral material consumption per structure remains unchanged and the reactor productivity is doubled.

Growth rates of 30 μm/h at 200 mbar and 10 μm/h at 600 mbar reactor pressure were reached in this way by AIXTRON in 8 × 4-inch wafer reactors [22], while TNSC reaches 28 μm/h in 8 × 3-inch wafer reactors even at atmospheric pressure [28]. Hence, the planetary reactor user may choose from low gas consumption with moderate growth rate or higher consumption with very high growth rates.

It should be noted that the same is true not only for planetary reactors but also for horizontal flow reactors: e.g. for any reactor where gas flows are parallel to the wafer carrier. However, in planetary reactors, the increase in total gas flow is not so simple. In reactors of this type, the gas velocity cannot be constant while passing through the reactor, but should be reduced in reverse proportionality to the radial position. At high gas flows viscous drag is insufficient for the velocity reduction, and vortices are generated. This undesirable process also may be started due to nonuniform heating of the gas by the wafer carrier and maintaining some vertical momentum in the gas leaving the injector. The last is prevented using specially optimized injector (nozzle) geometry. Finally, vortex generation may be triggered by momentum or velocity mismatch in ammonia and MO injection zones (recall that these zones, or inlets, in planetary reactors are rings placed at different distances from the ceiling). However, this mismatch may also be used to suppress the effect of nonuniform gas heating noted earlier. During the last decade, significant attention was paid to the optimization of injectors in planetary reactors by both modeling and experiments. With planetary reactor development, the injectors become more and more complex, starting from two and reaching five inlets above each other in AIX G5+ reactors [29].

The main problem with planetary reactors is the formation of deposits on the wall opposite the wafer carrier. We said earlier that this wall generally may be a floor; but due to the technical complexity of face-down placing of wafers at high temperature in III-N reactors, this wall currently is always a ceiling. Two different problems are associated with ceiling deposits. The first is flaking off and dropping on the wafers, and the second is a modification of the optical properties of the ceiling by deposits, resulting in significant drift of the temperature field from run to run.

Two different engineering solutions are used in planetary reactors. AIXTRON (before the G5 generation) and TNSC (until now) use quartz ceilings. In this case, deposits quite rapidly change the optical properties of these quartz elements during the first epitaxial process after clean ones are installed and slowly change them during consequent processes. Thus, for process reproducibility, it is necessary either to perform every process using clean quartz elements, or to perform a *coating run* after installing clean quartz elements and then perform a set of epitaxial processes until the thickness of deposits increases enough to peel them off. In any case, cleaning of removed quartz parts is performed ex situ using wet or gas chemistry or sand blasting.

The first approach means frequent cleaning of quartz parts and time losses spent exchanging them after every epitaxial process. If the second approach is used, the issue is how many processes may be completed before the peeling is started. The lowest deposition rate on the ceiling takes place at temperatures of about 300–400 °C. However, deposits formed at this low temperature are very incoherent, like frost (which is quite natural for the condensation mechanism of deposit formation). If the ceiling temperature is increased up to 550–600 °C or even more, the deposition rate is increased, but now these deposits are much more coherent, with good adhesion. Thus, peeling off occurs after a large number of epitaxial processes.

In an AIX G5 reactor, the idea of increasing the ceiling temperature was used as a basic solution. The ceiling in this reactor is made of graphite. It is black, its thermal conductivity is much higher than for quartz, and the thermal expansion coefficient is much higher, close to the value typical for the deposits. This combination of properties mean deposits do not significantly change the optical properties of the ceiling, it is heated effectively and uniformly, and deposit peeling is not pronounced. However, dismounting of the graphite ceiling for cleaning is not simple. A halide-based in situ cleaning process was developed for this reactor. It is especially useful for growth on silicon, because the entire reactor is cleaned, allowing every run to be performed in a clean reactor. However, the necessity to perform the etching process after each epitaxial process means high time losses, which diminish the advantage of the high growth rate.

It should be mentioned also that while previous generations of AIX planetary reactors were famous for very low material consumption, the last G5 and G5+ reactors – due to hotter ceilings resulting in faster gas heating – require higher gas flows to reduce parasite reactions. On the other hand, nozzle optimization allows very high injected gas flows, which in turn allow high growth rates due to purging nanoparticles downstream. The same operation mode (high growth rate with high gas flows) is typical for TNSC planetary reactors.

Arrangement of optical access for in situ characterization in planetary rectors is a little more complex than in RD reactors but does not cause any significant problems. A 2–3 mm diameter hole in the ceiling, purged with a small amount of gas, does not disturb the main flow significantly. Moreover, due to the very small ratio of the surface of this hole (holes) to the total ceiling surface, the effect of this minimal disturbance is negligible. Plenty of free space above the top lid of the reactor allows for the use of relatively large in situ characterization tools, and if a few of them are necessary, they may be placed above several optical windows. Moreover, if an in situ characterization tool is placed above the wafer center, on the assumption of axial symmetry of the rotating wafers, virtually all points of the wafer(s) pass below the window. This means something like mapping (e.g. thickness mapping) may be recorded using a proper combination of hardware and software.

Besides high achievable growth rate, one of the main advantages of planetary reactors is very high thickness and compositional uniformity. These come from individual substrate rotation, but this interrelation is much more complex that it may seem. The growth rate at any point on the rotating substrate is a result of averaging the growth rate above this point during its revolution. Thus, this profile will be hereafter called the *averaged growth rate profile*. In contrast, the growth-rate profile along the gas pass line (along a horizontal flow reactor or along the radius of a planetary rector) will be hereafter called the *instant growth rate profile*.

In Section 12.3.2, it was mentioned that at first glance, if the instant growth-rate profile is linear, rotation should result in absolutely uniform growth rate across the susceptor. Generally, this statement is incorrect. For simple horizontal flow reactors, the non-uniform growth rate is due to sidewall effects. For planetary reactors, it is due to geometrical reasons. Looking at Figure 12.15b, one can realize that some noncentral region of the substrate spends a longer time far from the reactor center than in the vicinity of the center. This means if the growth rate falls linearly with the reactor radius, the growth rate in the wafer center will be higher than at the periphery. A concave instant growth-rate profile is necessary for a constant averaged growth-rate profile. Beyond that, it is evident that any averaged growth-rate profile may be reached with an infinite set of instant growth-rate profiles (we are not sure of the mathematical correctness of this statement, but it appears to be correct in practical terms). Hence, the task of achieving uniform epilayer thickness means finding the proper instant growth-rate profile. The instant growth-rate profile, with other reactor conditions being constant, depends on the distribution of gas flows (carrier as well as precursors) between the inlets of the injector and the geometry of these inlets. The more inlets there are, the higher the level of mastery over the instant growth-rate profiles, but more parameters need to be optimized. This is challenging but eventually brings excellent results.

In addition, the composition of InGaN and InAlN ternary alloys is very temperature sensitive. In Section 12.2.3, we mentioned that in systems with gas-foil rotation, satellite discs are colder than the main susceptor by 50–150 °C. This temperature difference, apart from many other reactor conditions, depends on the rotation gas flow, which influences the gas-layer thickness. In the last generation of planetary reactors by AIXTRON, individual measurement of satellite disc temperatures is combined with a supply of rotating gas individually to each disc in the control system. This allows for precise wafer-to-wafer temperature uniformity, resulting in excellent wafer-to-wafer InGaN composition uniformity.

Thus, present-day III-N planetary reactors are very strong tools allowing fast growth with extraordinary uniformity. However, the huge number of parameters to be adjusted requires deep experience and sometimes years for fine process optimization.

12.4.4 CCS Reactors

The first primary advantages of CCS reactors are the very uniform concentration boundary layer and gas-temperature distribution, resulting in good yield without the necessity of fine process adjustments.

The second advantage, specific for nitrides, is the very short residence time of precursors in the hot zone, especially if the gap between the showerhead and the wafer carrier is reduced to a few millimeters. As a result, parasitic reactions have no time to proceed. For GaN, a linear dependence of growth rate with TMGa flow is observed to up 30 μm/h at 400 mbar, up to 20 μm/h at 600 mbar, and up to 8 μm/h at 900 mbar [30]. For AlGaN, more than 3 μm/h growth rate was reached in the entire composition range [19]. On the other hand, it should be noted that the formation of a cloud of nanoparticles above the susceptor, as in RDRs, is possible in CCS reactors as well, which results in significant growth-rate reduction [31]).

The main problem with these reactors is also common for any materials grown: too small a distance between the hot zone and the cooled showerhead results in strong parasitic deposition on the showerhead. First, this means the showerhead should be cleaned after every process. Mechanical (usually manual) brushing of the opened showerhead lid is used, because the cooled metal showerhead cannot be cleaned by in situ etching like the ceiling in an AIX G5 reactor and cannot be dismounted for ex situ cleaning like the quartz ceilings of TNSC or old AIX reactors. This procedure takes time and produces dust. There were attempts at using additional quartz screens with a high density of holes adjacent to the showerhead. It was expected that the deposition rate on this quartz screen would be lower and it may be easily used for in situ and ex situ cleaning. The results of using this solution are uncertain. After several conference reports, we can conclude that some groups use it for some applications, but the problem is not entirely solved.

Besides dust generation, there is another problem that was quite unexpected. During recent years, there were a few reports from various teams on the strong (up to 50%) background gallium incorporation in InAlN if grown just after GaN as a subsequent layer of the heterostructure [32–34]. As these structures are very attractive for high-electron-mobility transistor (HEMT) applications, a number of detailed studies of this effect were performed. What was found was that there is some source of gallium inside a CCS reactor, and the productivity of this source is proportional to the TMIn supply into the reactor. Moreover, a similar effect was reported for InGaN growth [35, 36]. The conclusions about the source of gallium in these works were contradictory, but there was strong evidence that the showerhead deposit acted as the gallium source. However, the detailed chemistry of this process is unclear as yet. We can only assume that there is some polymerization of MMGa or DMGa on the ceiling during GaN growth and exchange reactions, with indium releasing volatile gallium-containing species. It should be stressed that similar effects were not observed if InAlN/GaN heterostructures were grown in horizontal flow reactors and last-generation RDR [37, 38].

The impact of the size of the gap between the showerhead and the wafer carrier on the effects just described is the opposite. The larger the gap, the weaker the showerhead deposition, but the stronger the parasitic reactions (high growth rates for GaN and AlGaN given earlier have been reached with a very small gap). It is quite natural that AIXTRON offers CCS reactors with the gap variable during the epitaxial process. This innovation seems to be the only significant innovation for these reactors (excluding continuous scale-up).

In contrast to any other epitaxial systems, the process in CCS reactors is governed by its geometry much more than by process conditions. It is a benefit for newcomers: for new companies starting production as well as for students beginning work in the III-N MOVPE field. However, this also significantly narrows the field for fine optimization of the epitaxial process by adjustment of reactor conditions.

In situ optical characterization in CCS reactors is also possible but a little more complex than in the other types of reactors. Close proximity of the showerhead to the wafers requires a very homogeneous precursor supply. Any disturbance affects the growth-rate uniformity. Thus, windows for in situ optical characterization should be very fine in order to maintain injection hole position periodicity. A strong influence of the purging of these probes on growth process was identified quite early [9] (note that if the injector-to-wafers distance is large, the same effect allows for fine process tuning). Narrow windows strongly complicate optical access due to lack of signal and the high probability of light shadowing and/or reflecting by window edges. Moreover, fast heating of purging gas entering the reactor with a very high temperature gradient in the vicinity of the window leads to lens formation due to the dependence of the refractive index on temperature. This effect complicates optical system adjustment, especially if the distance between the susceptor and the showerhead is variable.

On the other hand, the design of CCS allows a unique in situ optical technique. Injection tubes (holes) are too small to act individually as an optical port. On the other hand, they are placed with high density, and thus the whole showerhead is semitransparent. An array of photodiodes placed above the showerhead (a simple linear window in the upper wall of the showerhead is necessary) allows temperature mapping across the entire susceptor. A corresponding device (ARGUS) is commercially available as a part of an MOVPE system by AIXTRON.

12.4.5 Horizontal Flow Reactors for III-N MOVPE

As was already noted, horizontal flow reactors were very popular at the beginning of MOVPE process development in the 1970–1980s. Reactors of that generation were described earlier. In the 1990s, when a significant portion of researchers switched from classical III-Vs to III-Ns, there were still many reactors of this type in laboratories.

At first glance, a horizontal flow reactor has the same advantage as a planetary reactor – a possibility of suppressing parasitic reactions by simple scaling of the gas flow. Moreover, in contrast to planetary reactors, even very strong gas acceleration in a horizontal flow reactor does not result in vortex formation. Gas flow in the tube with a rectangular cross section remains stable in a very wide range of temperature, pressure, gas velocity, and composition – just what is necessary for III-N. Thus it might be expected that these reactors will dominate, at least for research purposes. However, III-N horizontal flow reactors of the first generation are too similar to outdated ones developed for classical III-V materials with only a few modifications (e.g. with higher operating temperature and stability under an ammonia ambient).

Many important results were achieved using simple horizontal flow reactors, including growth of heterostructures for III-N laser diodes and HEMTs, development of GaN-on-Si technology, InAlN-based structures, etc. However, insufficient growth uniformity and some other problems become more and more pronounced compared to the properties of permanently improved reactors of other types. There were some attempts to design new horizontal flow reactors more suitable for the III-N process, but finally those studies were terminated. During the first decade of this century, the majority of the studies of the main players were concentrated on alternative reactor geometry. It should be mentioned that TNSC has continued to offer III-N horizontal flow reactors (a brief description may be found in [39]), but this did not influence the MOVPE community outside Japan. It was generally accepted that reasonable uniformity cannot be reached in a horizontal flow reactor due to the side-wall effect described in Section 12.3.2. However, this was the only one and not the main issue to be solved.

The most exact but very general explanation why development of horizontal flow reactors was interrupted for many years is that CFD modeling became strong enough to replace experiments as a main tool for reactor design a few years after horizontal flow reactors were considered to be redundant. Nowadays, these simulations, coupled with gas-phase and surface chemistry, allows a new generation of horizontal flow reactors to be developed. However, some basic ideas may be explained without any computations.

Indeed, side walls impact gas-flow patterns, temperature, and composition distribution in the reactor volume in a manner similar to the impact of the ceiling and wafer carrier, and this effect cannot be compensated for with wafer rotation. However, the influence of the side walls cannot spread from the walls to distances more than, roughly, twice the height of the reactor.

For many reasons, the height (the gap between the ceiling and substrate carrier) should be at least about 15 mm. In most old reactors, it was up to two times more. The width of the reactor should not significantly exceed the substrate size, or the growth process will be wasteful. For a 2-inch wafer, this means the entire wafer is placed in the region influenced by side-wall effects. This is why, in the mid-1980s, switching from horizontal flow to planetary reactors was an obvious and correct solution. However, at present, even research reactors should allow the use of 100-mm wafers, and the reactor height may be easily reduced to ~15–20 mm. This means the influence of side walls on the growth process is no more than a weak effect, which may be compensated for by rather simple methods.

Much more important is the geometry of the reactor entrance. In the first generation of horizontal flow reactors, gas was injected via one or two pipes (one for MO and one for hydrides) into an expanding chamber. In some reactors, it was a simple inplane diffuser; in others, the shape was more complex, but the idea was the same – to expand the injected gas from a narrow to wide parallel stream using the effect of the walls of the expanding chamber. It can be done, but only for restricted gas-flow parameters, which is acceptable for MOVPE of classical III-Vs, when the whole process may be done under one constant set of reactor conditions; but is absolutely unacceptable for III-Ns. Hence, the injected gases should enter the entire reactor width in a uniform manner as a parallel vortex-free flow. The task is complicated by the necessity of separate injection.

One of the modern horizontal flow reactors for III-N materials was designed in the framework of a joint project of teams led by the authors. The tool, named Dragon-125, has a reactor for one 100-mm substrate, is installed at the facilities of the Ioffe Institute, and has successfully operated since 2012 for R&D and pilot production. We will skip the description of technical solutions used but would like to present some epitaxial results just to prove that the potential for horizontal flow reactor geometry is much greater than it was widely considered. These results have been published and reported in [40–46].

Parasitic reactions were suppressed by gas acceleration, so very fast GaN growth was demonstrated: up to 45 μm/h at 100 mbar, 12 μm/h at 800 mbar, and even 6.5 μm/h at 1600 mbar (see Figure 12.25). In any of these regimes, the GaN growth rate nonuniformity may be tuned to less than 5% (min-to-max value), resulting in uniform epilayers and device structures (Figure 12.26). Gallium autoincorporation in InAlN is below 0.05%, which is another benefit of horizontal flow reactors. Moreover, due to the very high stability of gas flow in a wide range of reactor conditions, the reactor allowed growth of (besides various III-N heterostructures on sapphire, SiC, and Si substrates):

- Si_3N_4 for in situ III-N structures passivation and as a mask material for consequent selective area growth with deposition rates from 0.2 to 3 μm/h
- Silicon (occasionally, mostly as a joke)
- Graphene on sapphire substrates
- III-N heterostructures on graphene-on-sapphire in a single VPE-MOVPE process

The combination of high growth rate and the possibility to change the reactor conditions rapidly (due to low reactor volume and thermal mass, and high stability) results in a reduction of the growth process duration by 2–4 times comparing to typical values.

As in a planetary reactor, optical access for various in situ methods is not a problem in horizontal flow reactors. The only issue is that due to the simple rotation motion of the substrate (in contrast to double-motion in the planetary configuration), multiple optical heads are necessary for multi-point measurements. Due to the small area occupied by one substrate, the space for in situ tools is very limited, inhibiting the installation of a few standard tools. However, using in situ reflectivity and a curvimetry tool with very small optical heads specially developed together with the reactor allows multipoint growth-rate and anisotropic curvature measurements in this reactor.

The presented results prove that using CFD simulations coupled with models accounting for the specific chemistry of III-N MOVPE allows the design of a III-N horizontal flow reactor, which is free from the disadvantages generally accepted to be native but preserves all native advantages of this concept. This reactor is perfect for research purposes, but not only for those, as will be explained in the last section of the chapter.

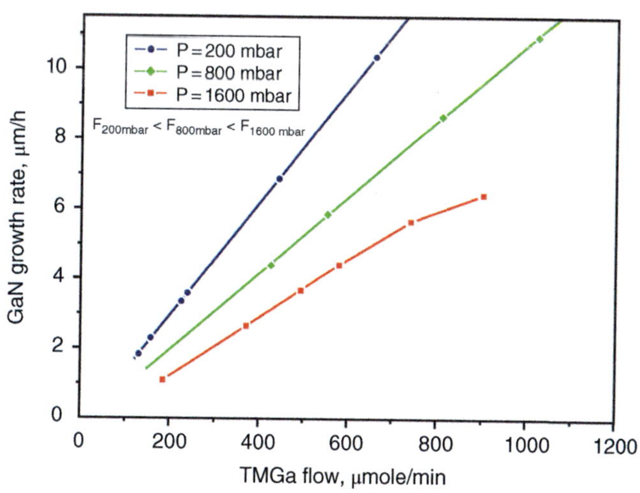

Figure 12.25 *GaN growth rate vs. TMGa flow for various reactor pressures in the Dragon-125 reactor. At 100 mbar, the growth rate reaches 45 μm/h, maintaining linearity with TMGa flow (not shown at this scale).*

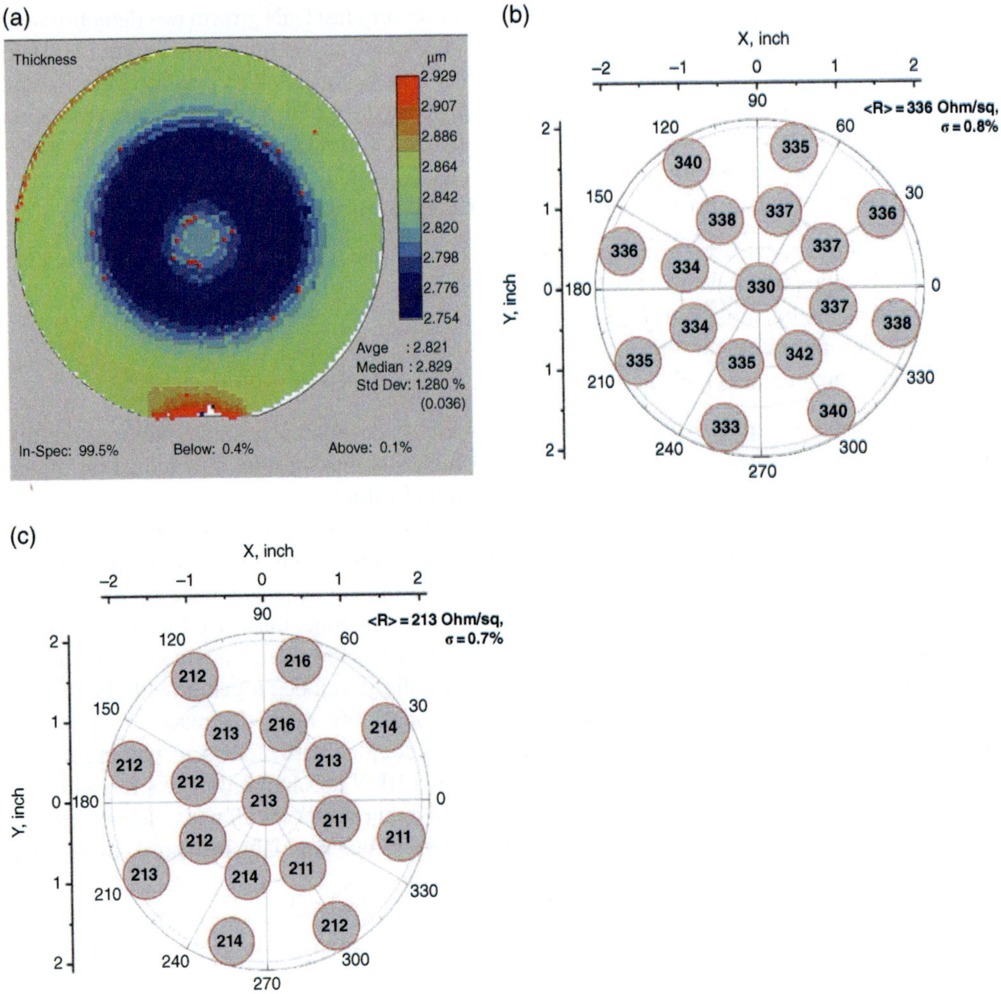

Figure 12.26 Uniformity of epitaxial structures grown in the Dragon-125. (a) Thickness map, GaN on sapphire substrate; (b, c) resistivity maps of 100 mm HEMT wafers. (b) AlGaN-based HEMT on SiC substrate; (c) InAlN HEMT on sapphire substrate.

12.5 Twenty-Five Years of Commercially Available III-N MOVPE Reactor Evolution

Summarizing, we have to conclude that much progress has been made during the last 20–25 years in the development of III-N MOVPE reactors. From generation to generation, they have become better and better, and many problems of the first III-N reactors have been solved. Moreover, a number of technical solutions, initially developed for III-N reactors, were then transferred to the reactors for classical III-Vs. However, this development was focused on large reactors for mass production of blue LEDs. In Figure 12.27, one can see that from generation to generation, not only the maximal reactor volume within a generation but also the minimal reactor volume have increased. The main reason for such a tendency is economical. The market for reactors for blue LEDs production is much larger than the market for reactors for all other III-N applications added

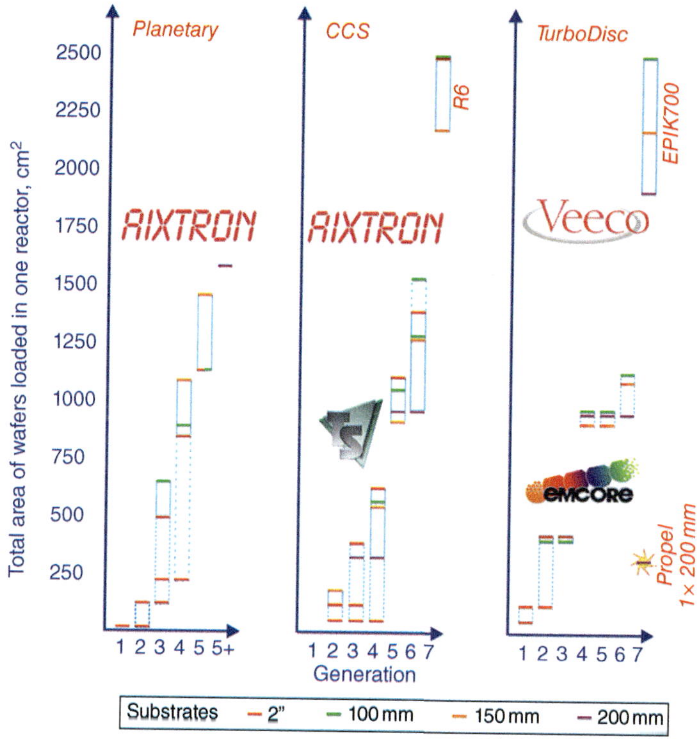

Figure 12.27 Development of III-N MOVPE reactor capacity over time. Assignment of generations is arbitrarily, based on reactor-design peculiarities (much more pronounced for planetary and TurboDisc reactors) and changes in company portfolios. Generation-to-generation transitions are time shifted for different system types and do not exactly coincide with marketing information.

together. At least a couple of extra reasons for such development also exist, but they are much less important. The larger the reactor, the less the influence of the reactor walls. This positive effect of the reactor size increase is especially pronounced for RD reactors. Also, the larger the reactor volume, the easier it is to accommodate complex reactor elements inside, like gas injectors in planetary reactors. However, the permanent increase in reactor volume results in a significant lack of state-of-the-art low-volume reactors on the market.

Indeed, this situation strongly affects R&D, but not only this area. Start-up companies require low-capacity equipment at the beginning, and not all products should be produced in large volumes. Some special types of LEDs, LDs, HEMTs, and sensors are much needed by industry but in relatively low quantities, making production of them in large reactors unprofitable.

However, we are sure that the situation is much more complex that just a vacant niche of small reactors for R&D and low-volume production. First, R&D in state-of-the-art large reactors is extremely expensive. On the other hand, R&D in old, small reactors is complicated due to the intrinsic limitations of outdated designs. Then, transfer of results of R&D from small reactors to large ones consumes time and money. In contrast, if production is performed in modern reactors with one large substrate, R&D may be easily done in the same reactors. It is already demonstrated that process time in a well-designed compact reactor may be reduced by several-fold, compared to large multiwafer reactors. Comparisons of multi- and single-wafer reactor productivity should be made, accounting for this. Thus, single-wafer reactors may be an attractive alternative for mass production of top-end devices that require frequent upgrade of technology.

Then, as was mentioned, III-N MOVPE reactors that dominate the market were developed for blue LED production: to be more exact, for production of those structures using present-day technology, which means growth on 50–150-mm sapphire substrates. This technology is well established and does not require much R&D work for further improvement. As long as it is commercially viable, the large reactors described here will be accepted by the market, and the main goal of their development will be reduction of production costs. In this paradigm, it also means an increase in reactor volume.

On the other hand, there are a number of applications that are considered the next growth drivers for MOVPE-based device production and are different from standard LEDs, such as RF and power HEMTs, UV LEDs, and laser diodes for projection and lighting. In these cases, the requirements for epitaxial productions are very different from those conventionally used for visible LEDs; existing reactors cannot provide the necessary specifications and yield, and, as a result, cost of ownership. This brings the community to the necessity to change the existing paradigm in favor of single-wafer reactors that potentially can overcome all these difficulties. The same story happened in the 1990s in silicon epitaxy, and single-wafer tools are now the industry standard for the majority of applications. It is believed that GaN will replace silicon for some of these applications, so the choice of the single-wafer concept for future development and production looks quite natural.

Only a few years ago, the situation with compact reactors development looked to be over. Most of the low-volume reactors were no longer in production. Veeco had stopped manufacturing small reactors at all, and AIXTRON offered only 3×2-inch or 6×2-inch CCS reactors. Even disregarding some disadvantages of this type of reactor, only one type of small reactor on the market looks to be insufficient. In recent years, the situation has started to change. Veeco offers a relatively small (one 200-mm wafer) Propel reactor intended for III-N HEMTs on silicon substrates, and a few companies from the United States, Japan, and Russia have demonstrated III-N MOVPE reactors developed for R&D and special applications. Thus, we forecast competition of large multiwafer reactors and compact reactors for one large substrate for mass production in the nearest future.

References

1. Frijlink, P.M. (1988). A new versatile, large size MOVPE reactor. *Journal of Crystal Growth* 93: 207–215.
2. Manasevit, H.M. (1968). Single-crystal gallium arsenide on insulating substrates. *Applied Physics Letters* 12 (4): 156–159.
3. Schubert, E.F. (2006). *Light-Emitting Diodes*, 2e. Cambridge University Press.
4. Casey, H.C., Jr, and Panish, M.B. (1978). *Heterostructure Lasers*. Academic Press.
5. Stringfellow, G.B. (1999). *Organometallic Vapor-Phase Epitaxy, Theory and Practice*, 2e. New York: Academic Press.
6. Moss, R.H. (1984). Adducts in the growth of III–V compounds. *Journal of Crystal Growth* 68 (1): 78–87.
7. Bremser, M., Dauelsberg, M., and Strauch, G.K. (2003). Method for depositing in particular crystalline layers, and device for carrying out the method. US Patent 6.972.050, filed May 15, 2003.
8. Bhat, R., Koza, M.A., Crawley, J. et al. (1995). Growth studies in a vertical rotating-disk reactor. *Booklet of the Sixth European Workshop on Metal-Organic Vapour Phase Epitaxy and Related Growth Techniques*, June 25–28, Gent, Belgium, E9.
9. Zhang, X., Moerman, I., Sys, C. et al. (1997). Highly uniform A1GaAs/GaAs and InGaAs(P)/InP structures grown in a multiwafer vertical rotating susceptor MOVPE reactor. *Journal of Crystal Growth* 170: 83–87.
10. Wang, Z., Tian, B., Pantouvaki, M. et al. (2015). Room temperature InP DFB laser array directly grown on (001) silicon. *Nature Photonics* 9: 837–842.
11. Derluyn, J., Dessein, K., Flamand, G. et al. (2003). Comparison of MOVPE grown GaAs solar cells using different substrates and group-V precursors. *Journal of Crystal Growth* 247: 237–244.
12. Zhang, X., Moerman, I., Sys, C. et al. (1997). Highly uniform A1GaAs/GaAs and InGaAs(P)/InP structures grown in a multiwafer vertical rotating susceptor MOVPE reactor. *Journal of Crystal Growth* 170: 83–87.
13. Breiland, W.G., Coltrin, M.E., Creighton, J.R. et al. (1999). Organometallic vapor phase epitaxy (OMVPE). *Materials Science and Engineering* R24: 241–274.

14. Mitrovic, B., Gurary, A., and Kadinski, L. (2006). On the flow stability in vertical rotating disc MOCVD reactors under a wide range of process parameters. *Journal of Crystal Growth* 287: 656–663.
15. Yakovlev, E.V., Talalaev, R.A., Kondratyev, A.V. et al. (2008). Hydrogen effects in III-nitride MOVPE. *Journal of Crystal Growth* 310: 4862–4866.
16. Koleske, D.D., Coltrin, M.E., Allerman, A.A. et al. (2003). In situ measurements of GaN nucleation layer decomposition. *Applied Physics Letters* 82(8): 1170–1172.
17. Lundin, W.V., Nikolaev, A.E., Rozhavskaya, M.M. et al. (2013). Fast AlGaN growth in a whole composition range in planetary reactor. *Journal of Crystal growth* 370: 7–11.
18. Creighton, J.R., Wang, G.T., Breiland, W.G., and Coltrin, M.E. (2004). Nature of the parasitic chemistry during AlGaInN OMVPE. *Journal of Crystal Growth* 261: 204–213.
19. Stellmach, J., Pristovsek, M., Savas, O. et al. (2011). High aluminium content and high growth rates of AlGaN in a close-coupled showerhead MOVPE reactor. *Journal of Crystal Growth* 315: 229–232.
20. Talalaev, R.A., Yakovlev, E.V., Karpov, S.Yu. et al. (1999). On the possible origins of low indium incorporation during MOVPE of InGaN. *Physica Status Solidi A* 176 (1): 253–256.
21. Considine, L., Thrush, E.J., Crawley, J.A. et al. (1998). Growth and in situ monitoring of GaN using IR interference effects. *Journal of Crystal Growth* 195 (1–4): 192–198.
22. Dauelsberg, M., Brien, D., Pusche, R. et al. (2011). Investigation of nitride MOVPE at high pressure and high growth rates in large production reactors by a combined modeling and experimental approach. *Journal of Crystal Growth* 315 (1): 224–228.
23. Finni, P., Wu, X., Tarsa, E.J. et al. (1998). The effect of growth environment on the morphological and extended defect evolution in GaN grown by metalorganic chemical vapor deposition. *Japanese Journal of Applied Physics* 37 (8): 4460–4466.
24. Watanabe, A., Takahashi, H., Tanaka, T. et al. (1999). Correlation between dislocation density and the macroscopic properties of GaN grown by metalorganic vapor phase epitaxy. *Japanese Journal of Applied Physics* 38 (2) (10B): L1159–L1162.
25. Yakovlev, E.V., Talalaev, R.A., and Vorob'ev, A.N. (2003). Critical issues in group-III nitride MOVPE modeling. *Electrochemical Society Proceedings* 2003–08: 258.
26. Mitrovic, B., Gurary, A., Quinn, W., and Armour, E.A. (2006). *Density-matching alkyl push flow for vertical flow rotating disk reactors*. US Patent 8980000, filed October 6, 2006.
27. Su, J., Armour, E., Lee, S.M. et al. (2016). Uniform growth of III-nitrides on 200 mm silicon substrates using a single wafer rotating disk MOCVD reactor. *Physica Status Solidi A* 213 (4): 856–860.
28. Matsumoto, K., Tokunaga, H., Ubukata, A. et al. (2008). High growth rate metal organic vapor phase epitaxy GaN. *Journal of Crystal Growth* 310 (17): 3950–3952.
29. Fahle, D., Puesche, R., Mukinovic, M. et al. (2013). Deposition control during GaN MOVPE. *The International Conference on. Compound Semiconductor Manufacturing Technology (CS MANTECH)*, May 13–16, New Orleans, Louisiana, USA, 399–402.
30. Boyd, A.Sa, Feron, O., Silva, H. et al. (2011). Development of the CRIUS® II Close Coupled Showerhead® MOCVD reactor for GaN based LED manufacturing. *Booklet of the XIV European Workshop on Metal-Organic Vapour Phase Epitaxy*, June 5–8, Wrocław, Poland, B.13.
31. Lobanova, A.V., Mazaev, K.M., Talalaev, R.A. et al. (2006). Effect of V/III ratio in AlN and AlGaN MOVPE. *Journal of Crystal Growth* 287 (2): 601–604.
32. Taylor, E., Smith, M.D., Sadler, T.C. et al. (2014). Structural and optical properties of Ga auto-incorporated InAlN epilayers. *Journal of Crystal Growth* 408: 97–101.
33. Kim, J., Ji, M.-H., Detchprohm, T. et al. (2015). Effect of Group-III precursors on unintentional gallium incorporation during epitaxial growth of InAlN layers by metalorganic chemical vapor deposition. *Journal of Applied Physics* 118: 125303.
34. Zhu, J.J., Fan, Y.M., Zhang, H. et al. (2012). Contribution of GaN template to the unexpected Ga atoms incorporated into AlInN epilayers grown under an indium-very-rich condition by metalorganic chemical vapor deposition (MOCVD). *Journal of Crystal Growth* 348 (1): 25–30.
35. Zhou, K., Ikeda, M., Liu, J. et al. (2015). Abnormal InGaN growth behavior in indium-desorption regime in metalorganic chemical vapor deposition. *Journal of Crystal Growth* 409: 51–55.
36. Zhou, K., Ren, H., Ikeda, M. et al. (2017). Unintentional gallium incorporation in InGaN layers during epitaxial growth. *Superlattices and Microstructures* 101: 323–328.
37. Ben Ammar, H., Minj, A., Gamarra, P. et al. (2017). Gallium incorporation in InAlN: role of the chamber design and history, and the effects of growth pressure, *Physica Status Solidi A* 214 (4): 1600441.
38. Lu, J., Su, J., Arif, R. et al. (2017). Epitaxial growth of InAlN/GaN heterostructures on silicon substrates in a single wafer rotating disk MOCVD reactor. *MRS Advances* 2 (5): 329–334.
39. Matsumoto, K. and Tachibana, A. (2004). Growth mechanism of atmospheric pressure MOVPE of GaN and its alloys: gas phase chemistry and its impact on reactor design. *Journal of Crystal Growth* 272 (1–4): 360–369.

40. Lundin, W.V., Davydov, D.V., Zavarin, E.E. et al. (2014). Fast MOVPE of III-N device heterostructures. *International Conference on MOVPE*, Lausanne, Switzerland, July 13–18.
41. Lundin, W.V., Zavarin, E.E., Popov, M.G. et al. (2015). A missed trend of III-N MOVPE reactor development. *Booklet of the XVI European Workshop on Metalorganic Vapour Phase Epitaxy*, Lund, Sweden.
42. Lundin, W.V., Davydov, D.V., Zavarin, E.E. et al. (2014). Optimization of MOVPE of nitride device structures for a wide range process conditions. *5th International Symposium on Growth of III-Nitrides (ISGN-5)*, Atlanta, GA, USA, May 18–22.
43. Lundin, W.V., Sakharov, A.V., Zavarin, E.E. et al. (2016). Study of GaN doping with carbon from propane in a wide range of MOVPE conditions. *Journal of Crystal Growth* 449: 108–113.
44. Lundin, W.V., Zavarin, E., Sakharov, A. et al. (2016). Growth of III-N/graphene heterostructures in single OVPE/MOVPE epitaxial process. *International Conference on MOVPE*, July 10–15, San Diego, California, USA.
45. Lundin, W.V., Rodin, S.N., Zavarin, E.E. et al. (2017). Study of silicon nitride deposition in III-N MOVPE reactors. *Booklet of the XVII European Workshop on Metalorganic Vapour Phase Epitaxy*, June 18–21, Grenoble, France, D13.
46. Lundin, W.V., Davydov, D.V., Zavarin, E.E. et al. (2015). MOVPE of III-N LED structures with short technological process. *Technical Physics Letters* 41 (3): 213–216.

13

Ultrapure Metal-Organic Precursors for MOVPE

D.V. Shenai-Khatkhate
Electronics & Imaging, DuPont de Nemours, Inc., Marlborough, Massachusetts, USA
(Formerly known as Dow Electronic Materials, Rohm and Haas Company, and Morton Metalorganics)

13.1 Introduction

As metal-organic vapor phase epitaxy (MOVPE) is becoming one of the preferred production technologies for the growth of the high-quality thin single-crystalline films of various compound semiconductors, the metal-organic (MO) precursors have been playing a significant role in the electronic industry over the past three decades. As pointed out in earlier chapters, the MOVPE technique essentially involves the decomposition of suitable volatile metal organic precursor materials to deposit the elements in their ultrapure form over a single-crystalline substrate. The MO precursors thus provide the "wings to fly" to the metals, and the elements in general, i.e. facilitate the vapor transportation of the metals and metalloid elements into the deposition chamber. As such, both the nature as well as the purity of the MO precursors become extremely crucial in determining the crystallographic and optoelectronic properties of the resulting semiconducting films. Since these epitaxial films determine the optoelectronic performance of various devices fabricated thereof, it is logical that the MO precursors, and particularly their synthesis and purification strategies, as well as their delivery techniques and associated safety aspects become critical in the MOVPE of compound semiconductors.

In this chapter, we will review some of the important developments in the chemistry of MO precursors, which are relevant to the advancement of compound semiconductors, with an emphasis on the evolution of ultrapure MO precursors since the inception of MOVPE.

The inception of the research and development of MOVPE has been widely attributed to the pioneering work by Manasevit et al. [1–4] during the late 1960s and early 1970s. Nevertheless, there have been a few patent applications, illustrating various attempts that may have laid the

Metalorganic Vapor Phase Epitaxy (MOVPE): Growth, Materials Properties, and Applications,
First Edition. Edited by Stuart Irvine and Peter Capper.

foundation for MOVPE, predating the work by Manasevit et al. by a few years [5, 6]. For example, the U.S. patent by Robert A. Ruehrwein, at Monsanto, filed in 1961, describes the production of epitaxial films using the metalorganic precursors of the Group III-B elements such as boron, aluminum, gallium, and indium [5]. Similarly, the German patent by Miederer et al., at Siemens AG (or Siemens-Schuckertwerke Aktiengesellschaft), granted in 1964, describes the deposition of GaAs using triethylgallium or trimethylgallium with organoarsenic halides [6]. One of the earlier patents on MOCVD by Manasevit is shown in Figure 13.1, which shows the schematic of a MOCVD setup during the early days. The earlier MOVPE research was, however, obstructed primarily due to the limited availability of the highly reactive and pyrophoric metal alkyl precursors, and due to the impurities in the MOVPE precursors, which in turn affected the performance of the epitaxial films. These obstacles were overcome later, during the mid-1970s, which resulted in the epitaxial GaAs films with the best low-temperature electron mobilities of the order of 100 000 $cm^2/V\,s$ [7]. The demonstration of lower minority carrier concentrations during the same time frame led to enhanced interest and an unexpected explosive growth in MOVPE during the 1980s. This greater interest in MOVPE also triggered research into the novel synthesis and purification strategies for the MOVPE precursors, leading to the successful commercial adaptation of MOVPE from the late-1980s to the present time.

The high quality offered by MOVPE is often challenged by its closest competitor, molecular beam epitaxy (MBE). Nevertheless, MOVPE is still considered more economical than MBE, as the costs associated with the maintenance of an MBE reactor are high vis-à-vis those of MOVPE.

The historic development and the applications of metalorganic precursors are covered extensively in textbooks by G.B. Stringfellow, entitled *Organometallic Vapor Phase Epitaxy* [8] and A.C. Jones and P. O'Brien, entitled *CVD of Compound Semiconductors* [9]. Hence, to avoid reiteration, our focus in this chapter is the chemistry of the synthesis and the purification of some of the most commonly used MOVPE precursors, along with relevant applications of these MOVPE precursors.

13.1.1 MOVPE Precursor Classes and Impurities

Numerous volatile MOVPE precursors are being used as precursors, such as alkylmetal compounds, alkylmetal halides, alkylmetal hydrides, metal cyclopentadienides, adducts of alkylmetal compounds with Lewis bases, and single-source precursors. The last two classes are not used extensively on a large scale for commercial applications. Most of the common MOVPE precursors are shown in Figure 13.2, while the various classes of MOVPE precursors and their applications are summarized in the Table 13.1. Also listed are the select impurities, in Table 13.2, which can conceivably be present in various MOVPE precursors; reducing these impurities, or, if possible, their complete removal for attaining the required ultrahigh purity, thus becomes the interesting challenge for all MOVPE precursor chemists [10].

Since the volatile trace metal impurities generally occupy the lattice site that normally belongs to the Group III or V element in a III-V semiconductor, or to the Group II or VI element in a II-VI semiconductor, they add an electron or a hole as the charge carriers, thereby affecting the optoelectronic properties of the semiconductors markedly. When an electron is added, the impurity acts as an n-type donor, while an addition of a hole acts as a p-type acceptor. The following examples elucidate the role of various impurities of the main Group elements on the optoelectronic properties of the III-V and II-VI semiconductors:

- The volatile trace metal impurities of Group II A, such as Be and Mg, occupy the Group III site in a III-V semiconductor and behave as a p-type shallow ionized acceptors.
- The volatile elemental impurities of Group VI A, such as S, Se, and Te, occupy the Group V A site and act as n-type shallow ionized donors, by occupying the Group V A site in a III-V semiconductor.
- Likewise, the trace volatile impurities of Group II B, such as Zn, Cd, and Hg, act as p-type shallow ionized acceptors, by occupying the Group III A site.

United States Patent [19]
Manasevit et al.

[11] **4,066,481**
[45] **Jan. 3, 1978**

[54] **METALORGANIC CHEMICAL VAPOR DEPOSITION OF IVA–VIA COMPOUNDS AND COMPOSITE**

[75] Inventors: **Harold M. Manasevit**, Anaheim; **William I. Simpson**, Orange, both of Calif.

[73] Assignee: **Rockwell International Corporation**, El Segundo, Calif.

[21] Appl. No.: **647,222**

[22] Filed: **Jan. 7, 1976**

Related U.S. Application Data

[62] Division of Ser. No. 523,599, Nov. 11, 1974, abandoned.

[51] **Int. Cl.**[2] **H01L 21/205**; H01L 21/84
[52] **U.S. Cl.** .... **148/174**; 136/89 TF; 156/610; 156/613; 156/614; 148/175; 252/62.3 S; 252/62.3 V; 357/4; 357/16; 357/30; 423/508; 423/509; 427/87; 427/90; 427/91; 427/248 B
[58] **Field of Search** .... 148/175, 174; 156/610, 156/612–614, 600, DIG. 72, 85; 252/62.35 V; 427/87, 90, 91, 248; 357/4, 16, 30; 423/508, 509; 136/89

[56] **References Cited**

U.S. PATENT DOCUMENTS

3,312,571	4/1967	Ruehrwein	148/175
3,403,133	9/1968	Frederic et al.	252/62.3 S
3,417,301	12/1968	Galli et al.	357/16 X
3,674,552	7/1972	Heywang et al.	148/175 X
3,723,177	3/1973	Toyama et al.	427/87
3,929,556	12/1975	Pandey	156/614 X

OTHER PUBLICATIONS

Rolls et al., "Preparation — Lead-Tin Telluride Photodiodes" Solid-State Electronics, vol. 13, 1970, pp 75–81.
Manasevit et al., "Use of Metal-Organics —Semiconductor Materials" J. Electrochem. Soc., vol. 116, No. 12, Dec. 1969, pp. 1725–1732.
Bylander et al., "Lead Salt Epitaxial Films With Near Bulb Properties" Proc. IEEE (correspondence), Apr. 1965, pp. 395–396.
Manasevit, H. M., "Single-Crystal Gallium Arsenide on Insulating Substrates" Applied Physics Letters, vol. 12, No. 4, Feb. 15, 1968, pp. 156–159.

Primary Examiner—L. Dewayne Rutledge
Assistant Examiner—W. G. Saba
Attorney, Agent, or Firm—H. Fredrick Hamann; G. Donald Weber, Jr.; Robert Ochis

[57] **ABSTRACT**

A composite comprising a monocrystalline substrate and one or more layers or films of monocrystalline IVA–VIA compounds and/or alloys formed thereon by a chemical vapor deposition process. The composite is formed at a preferred temperature range of approximately 450°–650° C. The IVA–VIA layer(s) are produced by the pyrolysis of a gas mixture containing metalorganic compounds. Where single crystal metallic oxide substrates of rhombohedral structure, such as sapphire, (α-Al_2O_3), or of cubic structure, such as magnesium aluminate (spinel), are used for the growth of monocrystalline lead-containing films such as $Pb_{1-x}Sn_xTe$, a nucleation layer of lead is preferably formed on the substrate prior to the pyrolysis of the mixed gaseous reactants.

Using the present process, epitaxial monocrystalline IVA–VIA compounds and/or alloys can be grown on inorganic metal oxide substrates, such as cubic and rhombohedral oxides, on alkali halides and IIA fluorides, and on II–VI and III–V compounds. The compositions of the films can be varied without removing the composites from the deposition apparatus by changing the ratio of the reactant gases and the reaction temperature. The conductivity type (n-type or p-type) of the films also can be controlled without removing the composites from the deposition apparatus by varying the reactant gas compositions and by incorporating a dopant into the reactant mixtures prior to pyrolysis.

15 Claims, 3 Drawing Figures

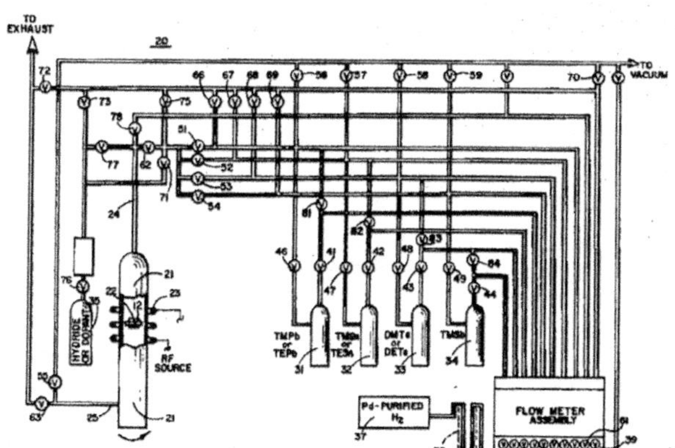

Figure 13.1 *One of the earlier US patents obtained by Rockwell International Corporation, with Harold M. Manasevit as the first inventor, disclosing the metalorganic chemical vapor deposition process. Source: Reproduced from United States Patent 4 066 481, from the USPTO patent database, with permission of USPTO (public domain).*

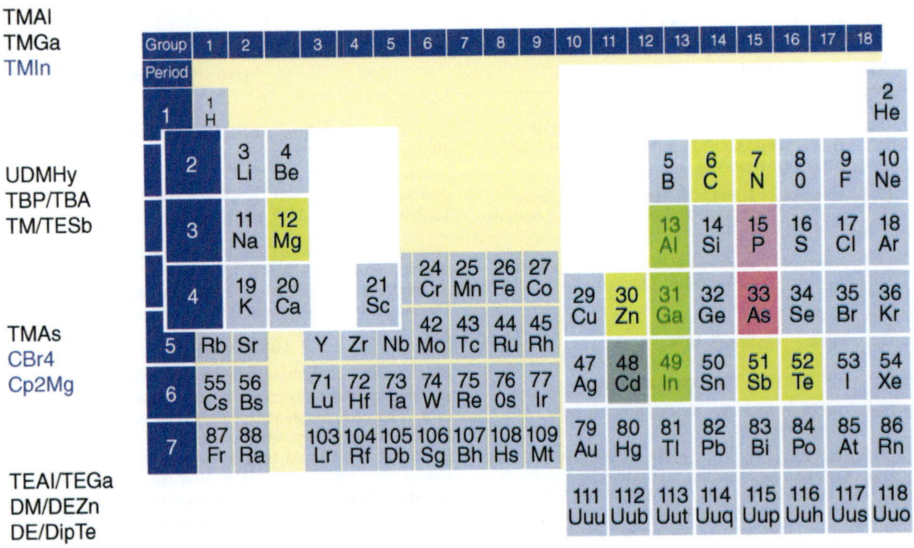

Figure 13.2 Most-commonly used MOVPE precursors for the deposition of III-V and II-VI compound semiconductors.

Table 13.1 Precursor classes and their applications.

Precursor class	Precursor examples	Applications
Metal alkyls	Trimethylaluminum, trimethylgallium, trimethylindium, dimethylcadmium dimethylzinc, trimethylantimony, trimethylarsenic, trimethylbismuth, dimethyltelluride, diethyltelluride, di-ispropyltelluride, diethylselenide, di-isopropylselenide	Epitaxial growth of single-crystalline compound semiconductors
Metal alkyl adducts	Dimethylzinc-trimethylamine, trimethylgallium-diphos	Safer and less reactive alternatives to metal alkyls; precursors for MOVPE
Metallocenes	Bis(cyclopentadienyl)magnesium, bis(methylcyclopentadienyl)magnesium	Dopant for III-V compound semiconductors
Metal alkyl hydrides	Dimethylaluminum hydride, di-isobutylaluminum hydride	Low-temperature deposition sources
Metal hydride-amine adducts	Alane-amine and gallane-amine adducts	Low-temperature metal deposition sources

- The volatile trace metal impurities of Group IV A, such as Si, Ge, and Sn, act as n-type shallow ionized donors, while occupying the Group III A site.
- Interestingly, volatile trace impurities of carbon, even though they belong to Group IV A, act as p-type shallow ionized acceptors, by occupying the Group V A site. Carbon in the epitaxial layers can also be formed because of the decomposition of the metalorganic precursor, particularly depending on the structure of the organic group attached to the metal.

Table 13.2 *Impurities and their possible concentrations at the ppb level in precursors.*

Impurity	Dopant activity	Semiconductors	Typical concentration in precursors/ppb
Si	n	InP	16
Si	n	GaAs	440
Sn	n	GaAs	0.004
Se	n	InP	2
Mg	p	InP	440
Be	p	GaAs	20

Table 13.3 *Metallic or elemental impurities and their acceptable levels in the MOVPE precursors, by ICP-OES and ICP-MS analytical techniques. Source: Shenai et al. 2003 [11], with permission from Elsevier Publications.*

Metallic or elemental impurity	Acceptable level in ppm	Metallic or elemental impurity	Acceptable level in ppm
Ag	<0.1	Mn	<0.03
Al	<0.5	Mo	<0.1
As	<0.2	Nb	<0.05
Au	<0.2	Ni	<0.3
B	<0.2	Pb	<0.6
Ba	<0.3	Pd	<0.2
Be	<0.02	Pt	<0.2
Bi	<0.3	Rh	<0.1
Ca	<0.03	Sb	<0.2
Cd	<0.1	Si	<0.1
Cr	<0.1	Sn	<0.3
Cu	<0.05	Sr	<0.03
Fe	<0.1	Tb	<0.1
Ga	<0.5	Te	<0.2
Ge	<0.2	U	<0.2
Hg	<0.5	V	<0.2
La	<0.05	W	<0.2
Li	<0.2	Y	<0.05
Mg	<0.02	Zn	<0.2

Table 13.3 shows typical metallic and elemental impurities and their acceptable levels in MOVPE precursors, as determined by inductively coupled plasma – optical emission spectroscopy (ICP-OES), and inductively coupled plasma – mass spectrometry (ICP-MS) techniques [11].

Along with trace metallic impurities, oxygen is also regarded as a deleterious impurity in MOVPE, since the presence of trace oxygen can affect optoelectronic properties such as the electron mobility and optical efficiency of III-V semiconductors. The influence of oxygen as an impurity was recognized during the early development of MOVPE, and precursor manufacturers have made painstaking purification efforts to eliminate the presence of trace oxygen in the metalorganic sources. Typically, oxygen is introduced in the precursor during the synthesis step, and it oxidizes the metalorganic precursor to form a metal alkoxide species that is volatile enough to be transported during the MOVPE process [11]: for example, the trace dimethylaluminum methoxide (Me_2AlOMe) present in trimethylaluminum (Me_3Al).

Before moving ahead with the synthesis and purification strategies for MOVPE precursors, it is worthwhile to become familiarized with the stringent requirements of MOVPE precursors for their safe application. Knowing these requirements is also important in the designing of MOVPE precursor molecules and subsequently selecting the most suitable synthesis and purification strategies for their commercial exploitation.

13.2 Stringent Requirements for Suitable MOVPE Precursors

The following are some of the stringent requirements for designing or selecting a suitable MO precursor, which apply to MOVPE as well as other types of chemical vapor depositions employed, such as atomic layer deposition (ALD), molecular layer deposition (MLD), etc. Some of these requirements have been addressed in earlier references [8–10]:

- High purity of the precursor in terms of its total metal content, organic impurities, and particularly oxygenated impurities at levels below parts per billion.
- Sufficient vapor pressure of the precursor to afford acceptable growth rates. Typically, >1 Torr vapor pressure at ambient temperature is considered acceptable for MOVPE, although vapor pressures within the range of >10–50 Torr at room temperature are preferred by the growers.
- Sufficient stability of a MO precursor is required so that no decomposition occurs during its storage and shipping.
- Sufficient stability of vapor pressure is essential to avoid batch-to-batch variation in MOVPE.
- The precursor should not decompose prematurely in the reactor, either during evaporation or in the gas phase before reaching the substrate.
- Clean decomposition of the MO precursor, either thermally in heat-assisted MOVPE or photolytically in photoassisted MOVPE, without incorporating any deleterious impurities in the deposited films.
- Chemical compatibility with the coreagents is important during the deposition process. The precursors should not form adducts with their coreagents, which can lead to an inevitable depletion of the reagents.
- Good shelf life and indefinite stability during storage and usage by the film growers.
- Liquid precursors are generally preferred due to the convenience of their handling, and their high vapor pressures, in general.
- The precursors should be safe to handle, with minimal toxicity. Nontoxic alternatives are often employed from the sustainability standpoint.
- The precursors should be available on commercial multikilogram scales and should be produced economically, reliably, and safely.
- The abatement of MOVPE precursors and their side products should be environmentally friendly.
- Last, but not least, it is most important that the precursors enable their desired purpose of MOVPE deposition by achieving the "just right" combination of the cracking temperature and the vapor pressure, as in the Goldilocks Principle [12]. As illustrated in Figure 13.3, the vapor pressure and the cracking temperature are dependent on the chemical composition and the steric bulk of the organic moieties attached to the metal center.

13.3 Synthesis and Purification Strategies for Ultrapure MOVPE Precursors

13.3.1 Synthetic Strategies for Ultrapure MOVPE Precursors

The organometallic compounds containing a metal–carbon bond, i.e. metal alkyls and metal cyclopentadienides, are some of the most important classes of MOVPE precursors. These are, in general, the high-value organometallic compounds, which are produced in high volumes to meet the

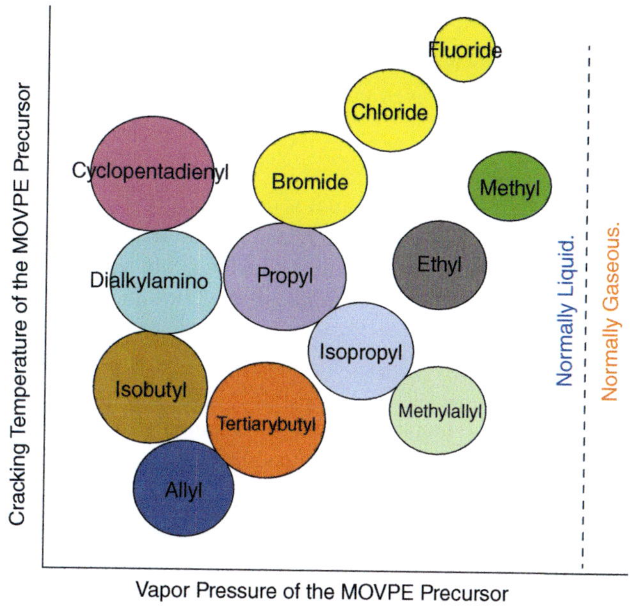

Figure 13.3 *The dependence of the cracking temperature and the vapor pressure of the precursor on the chemical composition and steric bulk of the organic moieties attached to the metal center of the MOVPE precursor.*

stringent requirements of high purity, reliable quality, and batch-to-batch consistency. Some MOVPE precursors are produced in high volumes because of their major industrial applications, such as the organoaluminum precursors, which are also employed in the production of alcohols, polymers like polyethylene and polypropylene, agricultural chemicals, olefins and polyolefins, ceramics, other specialty electronic chemicals, and recently in display materials, etc.

The formation of metal–carbon bonds can be accomplished by various synthetic strategies [13–17] that include, but are not limited to, the following:

- Exchange reactions such as transmetallation, metal exchange, metathesis, metal halogen exchange, and metallation
- Insertion reactions such as hydrometallation, carbometallation, and carbene insertion
- Oxidative addition reactions such as direct synthesis of metalorganic precursors from metal and organic reagent
- Elimination reactions such as pyrolysis of metal carboxylates

MOVPE precursors are often synthesized by using the appropriate route that offers the best advantages of ease in handling and manufacture commercially, safety of the reagents involved, consistency of product quality, productivity of the process, and overall cost of the product. The details of these synthetic strategies are as follows:

13.3.1.1 Transmetallation

Transmetallation is the reaction between a metal and an organometallic, as shown in the following equation:

$$M + RM' \rightarrow RM + M'$$

where M = Li, Cs, Be, Ba, Al, Ga, Sn, Pb, Bi, Se, Te, Zn, or Cd.
RM' is alkylmercury or alkyllithium.

13.3.1.2 Metal Exchange

Metal exchange is the reaction between two organometallic reagents where the respective metals are exchanged, as shown in the following equation:

$$RM + R'M' \rightarrow RM' + R'M$$

Metal exchange is an equilibrium reaction, and it is used where the equilibrium can be shifted to the right, to afford greater yields of the desired product. To illustrate, vinyl lithium can be produced by the reaction between aryl lithium compound and vinyl tin, where the precipitation of aryl tin byproduct shifts the equilibrium to afford high yields of vinyl lithium as the main product.

13.3.1.3 Metathesis

Metathesis is the reaction between a metal halide and an organometallic reagent where the respective metals are exchanged, as shown in the following equation:

$$RM + M'X \rightarrow RM' + MX$$

In the metathesis reaction, the equilibrium shifts to favor the product when M is more electropositive than M'. The key element of metathesis is that an electronegative halide group preferentially gets attached to a more electropositive metal, such as aluminum. Metathesis is used to produce the precursors of Ga, In, Zn, Cd, Si, Ge, and Sn, by using organoaluminum reagents.

13.3.1.4 Metal–Halogen Exchange

Metal–halogen exchange is the reaction between an organometallic reagent and an organic halide, as shown in the following equation. In this reaction, the organic moiety of the organic halide reagent exchanges with the organic group of the organometallic reagent to produce the desired organometallic product:

$$RM + R'X \rightarrow RX + R'M$$

The metal–halogen exchange reaction works effectively when the organic group of organic halide has an excellent ability to stabilize the negative charge. This reaction is used for the synthesis of various organolithiums by reacting organolithiums with organic halide reagents.

13.3.1.5 Metallation

Metallation is the reaction involving the replacement of a hydrogen (or a hydride) by a metal to produce the desired organometallic product, as shown in the following equation:

$$RM + R'H \rightarrow RH + R'M$$

The metallation strategy can be used to synthesize the organometallic product (R'M) as the resultant salt with its organic component (R') derived from the reaction between acid (R'H) and base (RM). As the acidity of R'H increases, the equilibrium shifts to the right, thereby favoring the product (R'M) formation. To illustrate, a reaction between a stronger R'H acid (benzene or C_6H_6) and organometallic reagent (n-butyllithium (C_4H_9Li)) leads to the formation of phenyl lithium (C_6H_5Li) and butane (n-C_4H_{10}), as shown in the following equation:

$$C_6H_6 + n\text{-}C_4H_9Li \rightarrow C_6H_5Li + n\text{-}C_4H_{10}$$

13.3.1.6 Hydrometallation

Hydrometallation is an insertion reaction involving the reaction between an organometallic hydride and an olefin, typically an alkyne. The hydrometallation reaction is shown in the following equation:

$$\text{M-H}+(\text{R'})(\text{R''})\text{C}=\text{C}(\text{R'})(\text{R''})\rightarrow \text{M}(\text{R'})(\text{R''})\text{C}-\text{C}(\text{R'})(\text{R''})\text{H}$$

M = B, Al, Ga, In, Si, Ge, Sn, Pb, or Zr.

R' and R' can independently be H or alkyl groups.

The reaction of hydroalumination is an example of hydrometallation, involving the reaction between diethylaluminum hydride and ethylene to produce triethylaluminum, as shown in the following equation:

$$(C_2H_5)_2\,AlH + H_2C{=}CH_2 \rightarrow (C_2H_5)_3\,Al$$

13.3.1.7 Carbometallation

Carbometallation is also an insertion reaction, like hydrometallation, where the insertion involves M–C bonds instead of M–H bonds, as shown in the following equation:

$$\text{M}-\text{R}+(\text{R'})(\text{R''})\text{C}=\text{C}(\text{R'})(\text{R''})\rightarrow \text{M}(\text{R'})(\text{R''})\text{C}-\text{C}(\text{R'})(\text{R''})\text{R}$$

M = B, Al, Ga, In, Si, Ge, Sn, Pb, or Zr.

R, R', and R' are alkyl groups.

13.3.1.8 Carbene Insertion

Carbene insertion is the reaction where carbenes are inserted between the M–H or M–X bonds to produce the desired organometallic product, as shown in the following equations:

$$C_6H_5SiH_3 + CH_2N_2 \rightarrow C_6H_5Si(CH_3)H_2 + N_2$$

$$(C_2H_5)_2\,SnCl + CH_2N_2 \rightarrow (C_2H_5)_2\,Sn(CH_2Cl)Cl + N_2$$

13.3.1.9 Direct Synthesis

Direct synthesis is the reaction between a metal and an organic halide to produce the desired organometallic compound, as shown in the following equation:

$$2M + nRX \rightarrow R_nMX_n \text{ or } R_nM + MX_n$$

The formation of organolithium reagents and Grignard reagents or organomagnesium halides are examples of direct synthesis strategy, as shown in the following equations:

$$C_4H_9Br + 2Li \rightarrow C_4H_9Li + LiBr$$

$$C_4H_9Br + Mg \rightarrow C_4H_9MgBr$$

13.3.1.10 Decarboxylation

Decarboxylation is an elimination reaction where metal carboxylates are heated, sometimes under reduced pressures, to remove carbon dioxide and yield the desired organometallic product, as shown in the following equation:

$$M(OOCR)_n \rightarrow MR_n + nCO_2$$

R is typically H or alkyl groups.

The decarboxylation of mercuric acetate to produce dimethylmercury, and the decarboxylation of tributyltin acetate to produce tributyltin hydride, are examples of decarboxylation strategy, as shown in the following equations:

$$Hg(OOCCH_3)_2 \rightarrow Hg(CH_3)_2 + 2CO_2$$

$$(C_4H_9)_3Sn(OOCH) \rightarrow (C_4H_9)_3SnH + CO_2$$

Most of these synthetic strategies are typically used for the laboratory-scale preparation of metalorganic precursors, while large-scale production is carried out using the best manufacturing practices that involve all stainless-steel equipment, high-pressure reactors, and continuous-flow reactors to ensure consistency in quality. The syntheses of various MOVPE precursors for II-VI and III-V compound semiconductors will be described in detail in the following sections.

13.3.2 Purification Strategies for MOVPE Precursors

In this section, we will review the methods of purification of metalorganic precursors that have been used industrially as preferred source materials. We will also discuss the relative merits of these purification techniques.

With the development of MOVPE, and correlating the optoelectronic properties of the epitaxial layers with the intrinsic impurities that may be conceivably present in these layers, the deleterious role of various metallic impurities present in the metalorganic sources became clearer. This understanding of the role of impurities and particularly their specific impacts has in turn encouraged synthetic organometallic chemists to evaluate various novel synthetic strategies and develop ingenious purification routes to eliminate such identified volatile trace metal impurities from metalorganic precursor sources.

The purification strategies can generally be divided into two main categories:

- Postsynthesis purification by using
  - Traditional techniques such as fractional distillation, sublimation, and zone-refining
  - Adduct purification using the thermal dissociation of the Lewis base adducts of the metalorganic precursors
- Oxygen-free synthesis using organoaluminums or the adducts of organoaluminums with appropriate Lewis bases as the alkylating agents

13.3.2.1 Postsynthesis Purification Strategies for MOVPE Precursors

The postsynthesis purification techniques often rely on the best methods to remove impurities and/or successively repeating the technique to attain the desired purity in metalorganic precursors. One of the advantages of the postsynthesis purification strategy is that regardless of the synthesis and the quality of the reagents employed, the final product can be expected to have the same degree of high purity.

Reacting with a suitable impurity scavenger, fractional distillation, crystallization, extraction, sublimation, and zone-refining [18–21] are typical postsynthesis purification strategies. Most of these methods involve homogeneous separations, i.e. the substances to be separated occur in the same phase. These techniques strongly depend on the differences in vapor pressures among the components to be separated from a mixture. As such, these techniques are found to be most suitable for purification if the vapor pressures of the impurities and the metalorganic precursor differ greatly. For example, trimethylaluminum can be separated from its low-boiling impurities containing silicon, germanium, and tin by fractional distillation [18]. Similarly, bis(cyclopentadienyl)magnesium can be purified by sublimation from relatively low-vapor-pressure impurities [19].

However, many impurities exhibit their vapor pressures in the proximity of the metalorganic precursor, thereby complicating their separation using traditional purification techniques. The second limitation of these techniques is that certain metalorganic compounds exhibit a tendency to decompose violently, and often autocatalytically, as the temperatures approach their boiling points: e.g. trimethylindium and dimethylcadmium are found to begin decomposing during their fractional distillations, rendering the technique unsafe. The third limitation, and a major drawback, is that, in general, large amounts of materials need to be discarded in these purification techniques to obtain ultrahigh-purity levels in the metalorganic precursors. This loss of product yield results in higher cost of the precursor. These physical separation techniques are also found to be ineffective to meet the stringent requirements of the semiconductor industry: i.e., the metallic impurities to be below 10 ppb, and nondetectable organic impurities.

These serious limitations of physical separation techniques by traditional distillation, sublimation, and zone-refining led British scientists (Cole-Hamilton and his research team) to explore a new chemical purification method known as *adduct purification* during the early 1980s [22].

The adduct-purification techniques for obtaining ultrapure MOVPE precursors were initially developed at the University of Liverpool by D.J. Cole-Hamilton, A.K. Holliday, J.B. Mullin, S.J.C. Irvine, A.C. Jones, and coworkers in the early 1980s. This work was then extended at Queen Mary College by D. Bradley and coworkers; and at the University of St. Andrews by D.J. Cole-Hamilton and coworkers. The successful adduct-purification strategies developed within these groups were commercialized later by a British company, Epichem Limited, based in Merseyside, UK, to produce ultrapure metalorganic precursors for MOVPE.

The adduct-purification strategy essentially differentiates the impurities from the main group metalorganic precursor based on their inability to form an adduct with a Lewis base. Typically, the deleterious metallic impurities of Group IV, such as Si, Ge, and Sn, can conceivably be removed using the strategic Lewis bases that do not form coordinate compounds, or adducts, with these metals [22–25]. The first use of adduct purification was reported by Coates et al. in 1962 to remove the ether from dimethylcadmium (DMCd) that contained an ether coordinated to DMCd, along with significant quantities of ether used as the reaction solvent [26]. An addition of equimolar quantity of 2,2'-bipyridyl to an ethereal solution of DMCd-etherate produced a yellow precipitate, which yielded DMCd upon heating in vacuo. It is worthwhile noting that ether forms a strong coordinate complex, or adduct, with DMCd. Although Coates et al. did not intend to use this process to obtain ultrapure DMCd, their work indeed illustrated the potential of adduct purification to remove ether and like oxygenated organic impurities from the metal alkyls of Group II and III.

Since the metal alkyls of Group II and III are Lewis acids, they readily form the coordinate complexes or adducts with the Lewis bases containing the donor atoms. The stability of these adducts toward thermal dissociation is determined by the nature of the metal in the metal alkyl, the nature of the donor atom, and the organic moieties present in the metal alkyl and the Lewis base [10]. The seven stringent properties necessary for the adduct purification are summarized here:

- An adduct formation between the metal alkyl and the Lewis base should occur relatively easily.
- The adduct should be crystallizable so that it can be purified: i.e. substantially isolated from all other impurities prior to the thermal dissociation.
- The Lewis base and the adduct must be of very low volatility, which is necessary to avoid contamination of the resultant metal alkyls.
- The adduct must dissociate at a practically achievable temperature, within the range of room temperature to 150 °C, which is well below the sublimation temperature for the adduct and the Lewis base, and significantly lower than the decomposition temperature of the pure metal alkyl.
- Adduct purification should yield as much of the metal alkyl as possible, to become commercially feasible.

- The Lewis base should be commercially available in a reasonably pure form, in large quantities, and at low cost.
- The Lewis base should preferably contain a Group V element for the purification of a Group III metal alkyl, and a Group VI element for the purification of a Group II metal alkyl.

The adducts, once formed between the metal alkyl and the Lewis base, can be isolated from the bulk solvent and recrystallized using standard procedures. The adducts are often nonpyrophoric and mildly air sensitive, and as such, they can be handled more easily than their parent metal alkyls. During the recrystallization step, the adducts are subjected to evacuation at reasonably high temperatures to remove all volatile organic impurities, and any noncoordinating organometallic impurities containing deleterious dopant elements. On heating, the adducts dissociate to release the pure metal alkyl at a very high yield [10].

Adduct purification has been used commercially to obtain ultrapure MOVPE precursors since its inception in the early 1980s to the present time. Adduct purification has been discussed in various references [10, 22–25].

Various Lewis bases have been used for attaining high purity in MOVPE precursors, including high-boiling ethers, various tertiary amines, and tertiary phosphines. The structures of two of the bidentate Lewis bases are shown in Figure 13.4. These Lewis bases have been used for the purification of Group III alkyls, such as trialkylindium, trialkylgallium, and trialkylaluminums [10, 23]. They are commonly known as (a) Arnold's base and (b) Diphos, with chemical nomenclatures bis(4-dimethylaminophenyl) methane and 1,2-bis(diphenylphosphino) ethane, respectively.

As can be seen from Figure 13.4, the two donor nitrogen atoms in the Arnold's base [10], or the two donor phosphorous atoms in the Diphos [23], are so far away that they cannot form a chelating-type adduct. In the purification of trimethylindium, the two nitrogen (or two phosphorous) atoms therefore coordinate with two trimethylindium molecules, and the resultant adduct is easier to dissociate thermally than its chelating analog [10]. In many cases, the chelating complexes are so volatile that they have been found to distill or sublime during adduct purification. The use of such strategically designed nonchelating adducts has enabled the highest purity in the base-free metal alkyls to be achieved. The purification of dimethylzinc using 4,4'-bipyridyl as the nonchelating Lewis base is another example of using the same strategy [27]. 4, 4' bipyridyl has also been used for the purification of various Group II MOVPE precursors, such as dimethylzinc, diethylzinc, and dimethylcadmium [27].

Figure 13.5 shows the proposed general structure between the adduct of a bidentate nitrogen containing a Lewis base and dimethylcadmium, where the two nitrogen donors coordinate with two cadmium acceptors [27], rendering a nonchelating adduct.

As a further advancement in the adduct-purification strategy, adduct purification has been used strategically by incorporating it as an integral step during the synthesis of metal alkyls. For example, strategies such as using adducts of trialkylaluminums with tertiary amines as the alkylating

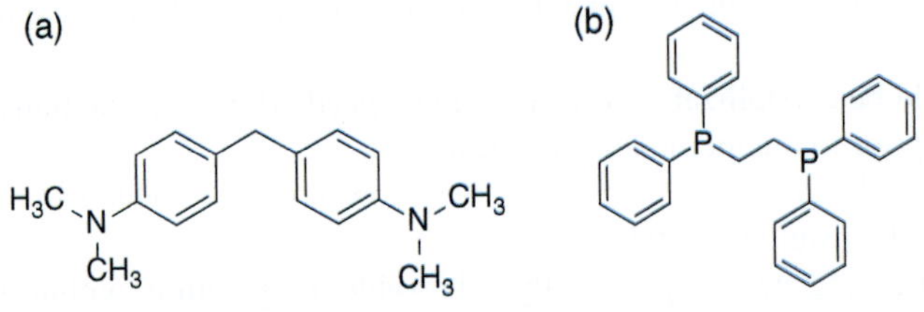

Figure 13.4 *(a) Arnold's base, or bis(4-dimethylaminophenyl) methane, the tertiary amine-based bidentate Lewis base; (b) Diphos, or 1,2-bis(diphenylphosphino) ethane, the tertiary phosphine-based bidentate Lewis base, used in the adduct purification of MOVPE precursors of Group III.*

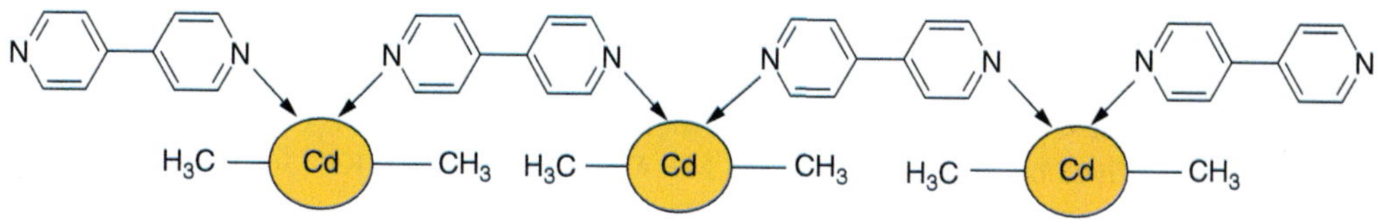

Figure 13.5 *The proposed polymeric structure of an adduct between dimethylcadmium and strategic Lewis bases, such as 4, 4′-bipyridyl. Source: Based on the discussion from Shenai-Khatkhate et al. 1986 [27], with permission from Elsevier Publications.*

agents [28–30], using the organoaluminum sesquihalides as the alkylating agents [31], or using organoaluminum sesquihalides as the alkylating agents in the presence of alkali metal halides [32, 33], have enabled oxygen removal as well as the purity enhancements in various metalorganic precursors, which will be discussed in the subsequent section.

13.3.2.2 Oxygen-Free Synthesis with in situ Purification Strategies

For the fabrication of compound semiconductors by MOVPE and their subsequent electronic device applications, the MOVPE precursors must be extremely pure and substantially free of detectable levels of both metallic impurities, such as silicon and zinc, as well as oxygenated impurities. Oxygenated impurities are typically present from the solvents, such as ethereal solvents, used to prepare such organometallic compounds, and are also present from other adventitious sources of moisture or oxygen. A challenge of using postsynthetic purification strategies is their limitation in removing trace impurities to the lowest ppb levels [34].

MOVPE precursors of Group IIB and IIIA metals are generally prepared by a variety of conventional methods, as outlined in Section 13.3.1. Such methods include reacting a Group IIB or IIIA metal halide with a Grignard reagent (alkylmagnesium halide) or an organolithium reagent in an organic solvent, or reacting an alkyl halide with a metal melt, and transalkylating a metal halide using a trialkyl aluminum, among other methods. Transalkylation reactions using organoaluminum reagents, such as those disclosed by Eisch [35] in the early 1960s, recently in the work by Shenai-Khatkhate et al. [28–30], and by Karch et al. [32, 33], have been developed to produce high-purity MOVPE precursors, as the ethereal solvents are eliminated in these reactions. The added advantage of these processes is that organoaluminums act as in situ oxygen scavengers during the synthesis. Thus, no postsynthesis purification such as adduct purification is necessary [34].

These reactions employ organoaluminums such as trialkylaluminums or organoaluminum sesquihalides as the strategic alkylating agents for the synthesis of MOVPE precursors of gallium and indium [28–33].

When transmethylation reactions are employed using a Group IIIA metal trihalide, such as indium trihalide or trimethyl aluminum, the reaction is not efficient with respect to the methyl groups transferred. For one mole of indium trihalide, three moles of trimethyl aluminum must be used, as shown in the following equation, following the strategy of Eisch [35]:

$$InCl_3 + 3Me_3Al \rightarrow Me_3In + 3Me_2AlCl$$

The resulting byproducts, namely organoaluminum halides, are generally difficult to separate from the desired reaction product, trimethyl indium, as they have very similar vapor pressures, and they often complex the trimethylindium formed. One of the innovative strategies used to separate these by-products was the use of alkali metal fluorides [20, 36] such as potassium fluoride, which form strong complexes with the byproduct dimethylaluminum chloride, as shown in the following equation:

$$Me_2AlCl + KF \rightarrow K\left[Me_2AlClF\right]$$

The complex, K[Me_2AlClF], often dissociates into potassium chloride (KCl) and dimethylaluminum fluoride (Me_2AlF), as shown next. Since Me_2AlF has a vapor pressure similar to that of trimethylindium, their separation is extremely difficult, to the extent that it is almost impossible. This limitation is solved by further complexing Me_2AlF with additional KF [36]. These reactions are shown in the following equations:

$$K\left[Me_2AlClF\right] \rightarrow KCl + Me_2AlF$$

$$Me_2AlF + KF \rightarrow K\left[Me_2AlF_2\right]$$

Figure 13.6 shows the abstract of US 5,756,786 by Power and Shenai-Khatkhate [36], disclosing an organoaluminum-based ether-free synthesis using alkali metal fluorides strategically.

Since the ether-free process strategically uses organoaluminum compounds, which are known to be potent oxygen scavengers, the products obtained by this process are claimed to be oxygen-free and significantly better in their performance than the products obtained by an ether-based process or the product obtained by postsynthesis purifications [36]. To illustrate, some of the typical Hall mobilities and carrier concentrations of the InP films grown by MOVPE were reported to provide 77 K electron mobilities as high as 287 000 $cm^2/V\,s$, and carrier concentrations as low as $8.0 \times 10^{13}\,cm^{-3}$, using trimethylindium produced by the organoaluminum reagent-based process [11].

During optimization of the KF-based process, it was discovered that when adducts of trimethylaluminum with tertiary amines (or tertiary phosphines) were used instead of neat trimethylaluminum to alkylate the Indium trihalide, not just one but two methyl groups of trimethylaluminum reagent were transferred, thus improving the process efficiency significantly [28]. According the following equation, only 1.5 moles of trimethylaluminum are required to produce the same yield of trimethylindium. Thus, the use of adducts effectively reduces the trimethylaluminum requirement by 50% per mole of indium trihalide used, or improves the yield of trimethylindium by 100% per mole of trimethylaluminum used:

$$2InCl_3 + 3Me_3Al\text{-tertiary amine} \dashrightarrow 2Me_3In + 3MeAlCl_2\text{-tertiary amine}$$

$$\text{Or}, InCl_3 + 3/2Me_3Al\text{-tertiary amine} \dashrightarrow Me_3In + 3/2MeAlCl_2\text{-tertiary amine}$$

In addition to the advantages of yield improvement, the organoaluminum–tertiary amine adduct-based synthetic strategy has been claimed to open the doors to green and sustainable chemistry for MOVPE precursor synthesis. While trimethylaluminum is highly pyrophoric, its adducts with tertiary amines and/or tertiary phosphines are far less reactive: in most cases nonpyrophoric and mildly air sensitive reagents, which is claimed to improve overall safety in metalorganics manufacturing.

Trimethylindium with substantially no oxygen-containing impurities, either as byproduct or associated solvent, is synthesized according to the reaction:

$$Me_zInX_{(3-x)} + (1-1.2)(3-z)Me_3Al + (1-1.2)(6-2z)MF \rightarrow M_e3In + (1-1.2)(3-z)M(Me_2AlF_2) + (1-1.2)(3-z)MX$$

where the Xs are the same or different and are selected from the group consisting of Cl. Br. and I; M is selected from the group consisting of Na.K. Rb, and Cs; and z is 0. 1 or 2 in a hydrocarbon solvent. such as squalance.

Figure 13.6 *The abstract for an ether-free process to synthesize trimethylindium. Source: Power. and Shenai-Khatkhate 1998 [36], with permission from USPTO (public domain).*

Table 13.4 Film performance of GaAs films grown by MOVPE. Source: Shenai-Khatkhate and Amamchyan 2007 [28], with permission from USPTO (public domain).

			300 K		77 K	
Growth reference	V/III ratio	Growth temperature (° C)	Mobility ($cm^2/V s$)	Carrier concentration (cm^{-3})	Mobility ($cm^2/V s$)	Carrier concentration (cm^{-3})
0583	150	620	6500	1.0E14	147 000	1.0E14
0585	150	620	7700	6.9E13	189 000	5.5E13
0587	150	620	7500	1.14E14	161 000	1.0E14

The quality of metalorganics, when produced by using the tertiary amine adducts of trimethylaluminum, was claimed to be consistently better in terms of organic, oxygenated, and metallic impurities [28]. The quality improvement was also claimed to improve optoelectronic performance, as shown in Table 13.4, with the Hall mobility and carrier concentration results for the epitaxial films of GaAs grown using MOVPE on a GaAs substrate heated at 620 °C using triethylgallium and arsine gas as the sources [28].

Growth references marked as 0583 and 0587 represent the results obtained by using triethylgallium that was produced by base-free triethylaluminum, which are acceptable for commercial production of GaAs by MOVPE. Growth reference 0585 represents the results obtained by using triethylgallium produced by a tertiary amine adduct of triethylaluminum. These results indicate a significant improvement in terms of the higher electron mobility and the reduced minority carrier concentration in the GaAs films grown by MOVPE, under identical conditions and process parameters.

It is worthwhile noting that the tertiary amine-trimethylaluminum adduct-based alkylation strategy is also claimed to be useful in synthesizing MOVPE precursors of Group II (i.e. Zn and Cd), Group III (Ga and In), Group IV (Si, Ge, and Sn), Group V (P, As, Sb, and Bi), and Group VI (Se and Te) [28–30]. As can be seen from Figure 13.7, a trialkylaluminum-tertiary amine-based strategy is reported to be useful for synthesizing a wide range of MOVPE precursors from Group II through Group VI.

Recently, a new strategy to synthesize trialkylmetal compounds of Group III has been developed to produce MOVPE precursors of Group III. This method is reported [32, 33] as based on the reaction between metal trichloride (MCl_3, preferably a metal of Group III such as gallium or indium) and alkyl aluminum sesquichloride ($R_3Al_2Cl_3$) in the presence of at least one alkali metal halide as the auxiliary base. The target MOVPE precursor product is separated from the reaction mixture by heating the reaction mixture to 120 °C, separating the trialkylmetal compound, recirculating the partially alkylated side-products, and finally heating the reaction mixture to a maximum temperature of 350 °C to obtain the final traces of trialkylmetal compound. This process is claimed to produce higher yields of high-purity trialkylmetal compounds of Group III, which can be used as MOVPE precursors [32, 33]. This process is also claimed to be a green process from the standpoints of safety and sustainability.

The MOVPE precursors are produced in commercial quantities using safe practices and using equipment with automated controls for all process parameters. To ensure the safety of operators, workplace, and environment, all-stainless-steel equipment is generally used. Figure 13.8 shows a typical manufacturing setup consisting of all-stainless-steel equipment [37].

A continuous process and apparatus for the same have been described [38]. The use of microreactors or compact heat exchangers has also been described for the continuous production of

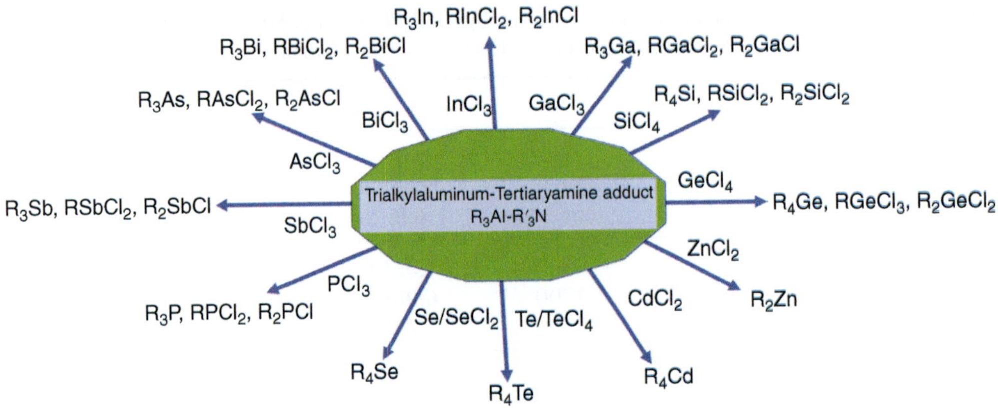

Figure 13.7 *The strategic use of trialkylaluminum-tertiary amine-based alkylation to synthesize various MOVPE precursors of Group II through Group VI. Source: Based on information from Shenai-Khatkhate and Amamchyan 2007 [28], Shenai-Khatkhate et al. 2003 [29], and Shenai-Khatkhate 2003 [30], with permission from USPTO (public domain).*

Figure 13.8 *State-of-the art equipment and automation for the safe production of MOVPE precursors on a commercial scale. Source: Shenai-Khatkhate et al. 2002 [37], with permission from IOP Publishing Ltd.*

MOVPE precursors [39–42]. The use of microchannel devices for the purification of organometallic compounds, especially MOVPE precursors, has also been described [43].

An example of such a microreactor or a compact heat exchanger, developed at Lawrence Livermore National Laboratory (LLNL), is shown in Figure 13.9. These reactors are smaller in dimensions, often comprising serpentine microchannels through which the reagents flow at a very

Figure 13.9 A microreactor design developed at LLNL for the continuous manufacturing of chemicals. Source: Reproduced with permission from United States Department of Energy, US Federal Government (public domain).

rapid rate, and react to complete the synthesis most efficiently and effectively. The general advantages of using such microfluidic devices for next-generation manufacturing are outlined here:

- Larger surface area to volume ratio available in these reactors enables efficient syntheses of highly reactive and challenging MOVPE precursors.
- Continuous process ensures cycle-time reduction, process efficiency, and no batch-to-batch variations in quality of MOVPE precursors.
- Better processing of reactive intermediates without their isolation and purification can minimize work-up delays and provide enhanced efficiency in the manufacturing environment.
- High-pressure chemistry can be handled more safely, in a continuous manner.
- Efficient heat exchange is possible with the microchannels in these reactors than in batch reactors. Typically, the heat-exchange coefficient per unit volume of a microreactor is hundreds of megawatts compared to a few kilowatts of a batch reactor, which enables microreactors to remove the heat most efficiently and improve the process safety of MOVPE precursor production.
- The minimum volume of reactive chemical reagents flowing within the reactor at any given time reduces the likelihood of safety issues, leading to green chemistry and sustainability.

In summary, the synthesis and purification strategies for MOVPE precursors are constantly evolving, with strides made towards higher purity, greater process efficiency, and safer production to ensure sustainability.

In the following sections, we will look at the most commonly used MOVPE precursors and dopants for III-V and II-VI compound semiconductors.

13.4 MOVPE Precursors for III-V Compound Semiconductors

13.4.1 Group III MOVPE Precursors

The Group III metalorganic precursors for MOVPE of compound semiconductors are, in general, highly air sensitive and pyrophoric in nature. Even though earlier MOVPE studies were focused exclusively on using trimethyl and triethyl metalorganic sources, various other sources containing different ligands have been explored as research curiosities. Some of the precursors were developed specifically for reduced carbon content, better delivery, higher vapor pressures, etc. For ensuring the focus on practical applications of MOVPE precursors, Table 13.5 lists only the select precursors that are commercially available from various manufacturers and are being used industrially, at present, for commercial-scale depositions of III-V semiconductors. Figure 13.10 shows the chemical structures of some of the Group III MOVPE precursors in their monomeric or dimeric forms, as applicable.

The vapor pressures of Group III MOVPE precursors [44] are shown in Figure 13.11. As can be seen from the vapor-pressure plots, the trimethyl sources of all Group III elements exhibit the highest vapor pressures and hence the highest throughput in MOVPE. The plots also show that the vapor pressure for trimethylgallium (TMGa) is greater than that of trimethylaluminum (TMAl). This can be attributed to the fact that TMGa exists as a monomer in the vapor phase, while TMAl exists as a dimer [13]. Trimethylindium is tetrameric in the solid state and monomeric in the vapor phase, which gives rise to its higher vapor pressure [13].

Taking into consideration thermal stability, Group III precursors exhibit decreasing stability as the atomic size of the metal increases, in the order Al, Ga, and In. The stability is also dependent on the metal–carbon bond strength, which in turn is dependent on the size of the organic group attached to the metal: e.g. trimethylindium is thermally more stable than triethylgallium. The nature of the alkyl group also plays a major role in determining the stability. The higher the alkyl group, the lower the decomposition temperature. As we traverse the alkyl groups (R) in R_3M, the threshold temperature for the thermal decomposition follows this order [8]:

$$\text{Me} > \text{Et} > \text{n-Pr} > \text{i-Pr} > \text{n-Bu} > \text{s-Bu} > \text{t-Bu} > \text{allyl}$$

This order suggests that either the increasing steric bulk of the organic group or a greater number of hydrogen atoms β to the metal leads to a lower decomposition temperature [8, 9]. Both factors

Table 13.5 *Group III MOVPE precursors.*

MOVPE precursor (Abbr.)	Formula	CAS #	Formula weight	Boiling point °C/Torr	Vapor pressure @20 °C (Torr)	Vapor pressure constants for the equation log *P* (Torr) = *A* − *B*/*T* (K): *A*	*B*
Trimethylaluminum (TMA)	$(CH_3)_3Al$	75-24-1	72.1	126/760	8.66	8.224	2135
Trimethylgallium (TMG)	$(CH_3)_3Ga$	1445-79-0	114.83	55.7/760	181.01	8.07	1703
Triethylgallium (TEG)	$(C_2H_5)_3Ga$	1115-99-7	156.91	143/760	5.06	8.083	2162
Trimethylindium (TMIn)	$(CH_3)_3In$	3385-78-2	159.93	134/760	1.11	10.98	3204
Triethylaluminum (TEA)	$(C_2H_5)_3Al$	97-93-8	114.17	194/760	8.73	8.224	2361
Tri-isobutyl gallium (TIBGa)	$(C_4H_9)_3Ga$	17150-84-4	241.07	88/10	0.08	4.769	1718
Dimethylaluminum hydride (DMAH)	$(CH_3)_2AlH$	865-37-2	58.6	154/760	1.35	8.92	2575
Triethylindium (TEIn)	$(C_2H_5)_3In$	923-34-2	202.01	144/760	0.21	8.93	2815

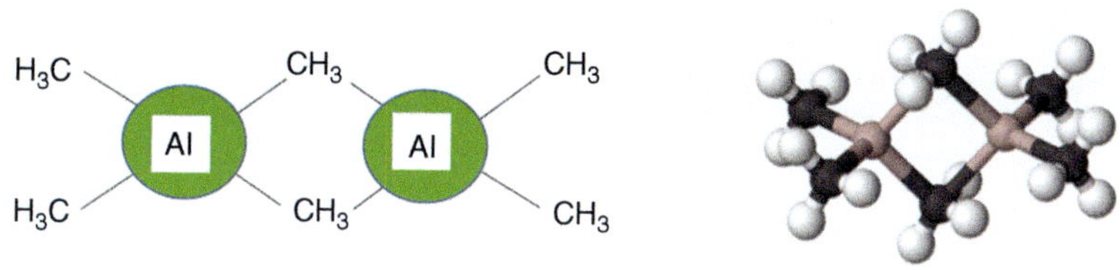

(a) Trimethylalumimum dimer (Al_2Me_6)

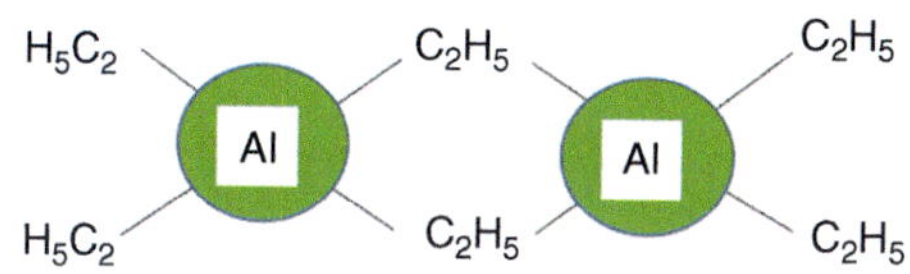

(b) Trimethylalumimum dimer (Al_2Mt_6)

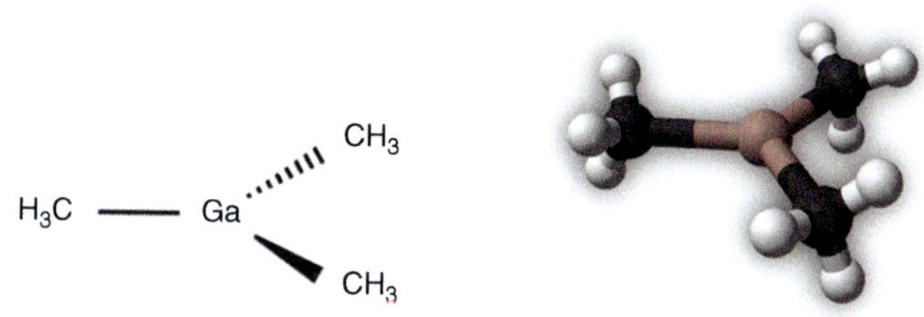

(c) Trimethylgallium monomer (Me_3Ga)

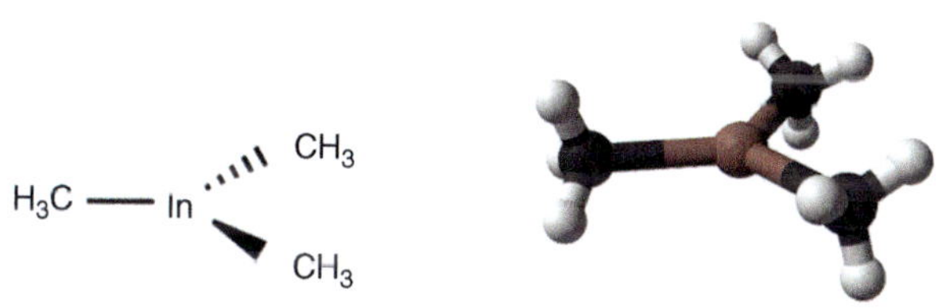

(d) Trimethylgallium monomer (Me_3In)

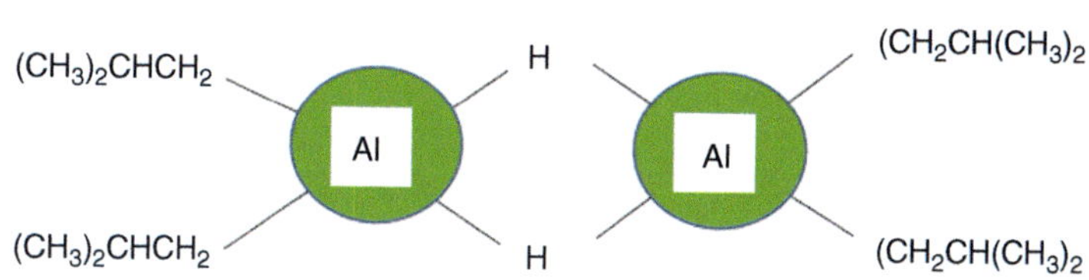

(e) Di-isobutylaluminum hydride dimer ($Al_3{}^iBu_4H_2$)

Figure 13.10 The chemical structures of Group III MOVPE precursors.

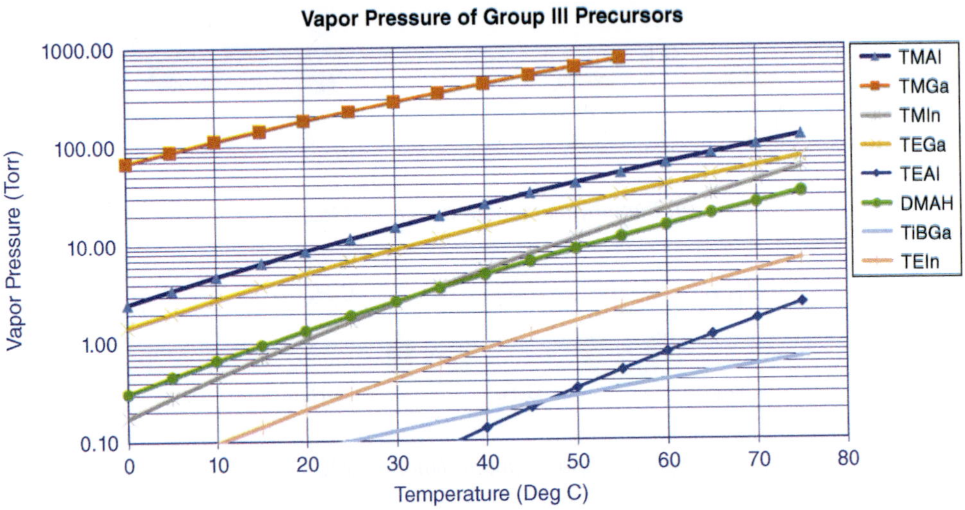

Figure 13.11 The vapor pressure–temperature plots for Group III MOVPE precursors.

are equally important, as the increased steric bulk enhances the hemolytic fission of the metal-carbon bond, while β-elimination can occur at far lower temperatures, as low as ambient temperatures. Hence, the precursor design should take into consideration these two factors while tailoring the decomposition characteristics of a MOVPE precursor.

The pyrolysis temperature, necessary to affect the epitaxial deposition, increases with the increasing metal–carbon bond strength. Metal–carbon bonds are generally weak, compared to metal–nitrogen or metal–oxygen, or metal–halogen bonds. Furthermore, the metal–carbon bond strength decreases with increasing atomic size (or atomic weight) of the metal. Also, the larger the organic moiety attached to the metal atom, the weaker becomes the bond. For example, the pyrolysis temperatures required for Group III metal alkyl precursors are found to be in the following order [8, 9]:

$$AlMe_3 > GaMe_3 > InMe_3$$

$$AlMe_3 > AlEt_3 > Al(i\text{-}Pr)_3 > Al(t\text{-}Bu)_3$$

Larger alkyl groups, such as ethyl, isopropyl, and tertiary butyl, can also undergo β-hydride elimination reactions, converting the metal alkyl into corresponding metal hydride by eliminating an alkene, as shown next:

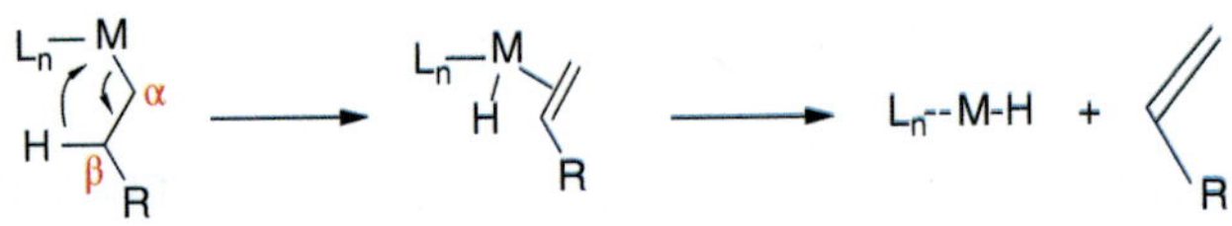

Hence, when a lower deposition temperature is desired in MOVPE, the precursors are designed by using larger alkyl groups that can facilitate β-hydride elimination, such as isopropyl, tertiarybutyl, etc. Conversely, when a higher deposition temperature is desired, the precursors are designed by using the alkyl groups that cannot undergo β-hydride elimination, such as methyl, neo-pentyl, cyclopentadienyl, cycloalkyl, etc.

The trimethyl sources have been preferred in MOVPE, mainly because of their higher vapor pressures as well as greater stability. To illustrate, triethylgallium and triethylindium have been found to decompose even at room temperatures and under inert conditions [8, 9], while their trimethyl analogs are found to be stable indefinitely, if stored under inert atmosphere, at room temperature [44]. An undesirable aspect associated with trimethyl sources is carbon incorporation in the films that they afford. The pyrolysis of methyl-containing sources leads to CH_3 radicals, which tend to decompose further to produce hydrogen and CH_2 radicals. The CH_2 radicals then further decompose to leave C in the epitaxial films. The CH_3 radicals are sometimes removed by their reaction with atomic hydrogen (H) or when they desorb, which is a slow process [8].

To avoid carbon contamination in films, precursor chemists have used the strategy of designing MOVPE precursors with larger alkyl groups, such as ethyl, isopropyl, tertiarybutyl, etc. The larger alkyl radicals bind to the surface very weakly, and they desorb before any adverse reactions that may lead to carbon incorporation in the films. More branched radicals often tend to decompose via the β-hydride elimination process, leaving behind olefins and less-reactive carbon-containing species, which reduces carbon incorporation significantly. For example, triethylgallium decomposes via the β-hydride elimination process, producing ethylene and diethylgallium hydride [8, 9]. The latter can further decompose sequentially to produce ethylgallium dihydride [8, 9] and gallium trihydride (also known as gallane). The final decomposition product is pure gallium, since gallane (or gallium trihydride) is known to be extremely unstable:

$$(C_2H_5)_3\,Ga \rightarrow (C_2H_5)_2\,GaH + C_2H_4$$

$$(C_2H_5)_2\,GaH \rightarrow (C_2H_5)GaH_2 + C_2H_4$$

$$(C_2H_5)GaH_2 \rightarrow GaH_3 + C_2H_4$$

$$GaH_3 \rightarrow Ga + 1.5H_2$$

Tri-isobutylgallium has been used for epitaxial growth of GaAs by Plass and coworkers [45]. Likewise, tritertiarybutylaluminum has been used successfully by Wang [46] for MOVPE of AlGaAs and AlGaSb. Similarly, triisopropylgallium (TIPGa) and tritertiarybutylgallium (TTBGa) were used by Jones et al. [47] for deposition of GaAs, to show an improvement in electrical and optical properties with TIPGa and unacceptably low rates of growth with TTBGa.

For indium precursors, trimethylindium (TMIn) has always been the most preferred, despite it being solid and having the potential for carbon incorporation in the films. The advantage of TMIn is in its high vapor pressure [11]. However, TMIn poses serious issues in maintaining its consistent concentration in the vapor phase. Therefore, various attempts have been made to ensure the consistent delivery of TMIn, which will be discussed later in this section.

As pointed out earlier, because of advancements in ether-free synthesis of TMIn, the quality of TMIn has improved significantly, i.e. reduced levels of metallic and oxygenated impurities, leading to more reliable MOVPE growth in commercial mode, vis-à-vis during the early MOVPE years. Liquid alternatives for TMIn, e.g. heteroleptic trialkyl indiums (or mixed alkylindium compounds) such as ethyldimethylindium ($EtMe_2In$ or EDMIn) [48] and di-isopropylmethylindium ($^iPr_2MeIn$ or DIPMeIn) [49] have been proposed, but these heteroleptic precursors were found to dissociate into the parent homoleptic trialkylindiums during MOVPE growth.

The explanation for the dissociation of EDMIn was provided by the work of Frigo et al. [50], who showed conclusively that EDMIn dissociates into DEMIn (or Et_2MeIn) and TMIn. Frigo et al. also concluded that $EtMe_2In$ and Et_2MeIn are not discrete molecules but mixtures of organoindium compounds undergoing rapid exchange of alkyl moieties [50]. Such an exchange is shown in Figure 13.12, with the reactions involving the formation and cleavage of alkyl-bridged dimers [10].

DEMIn or Et_2MeIn can be expected to dissociate further into TMIn and TEIn. As a result, when EDMIn is used as an indium precursor, the most volatile component, TMIn, is removed

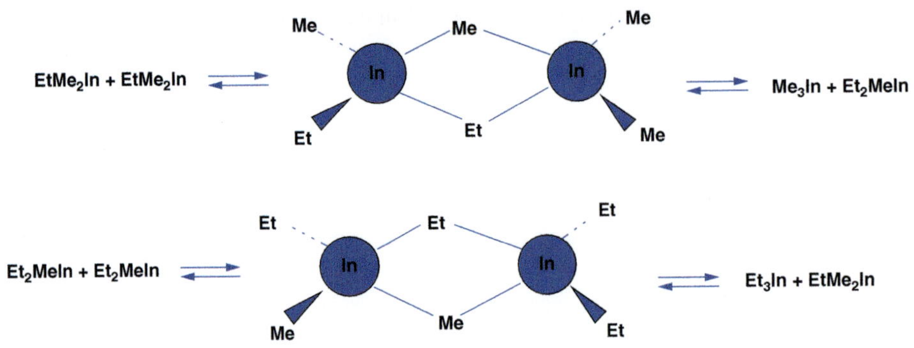

Figure 13.12 *Rapid exchange of alkyl groups via the formation and cleavage of alkyl-bridged dimers of ethyldimethylindium (EDMIn) and diethylmethylindium (DEMIn).*

preferentially from the equilibrium mixtures first. This leads to a varying rate of transportation of indium into the reactor and thus results in an uncontrollable epitaxial growth process [10, 50].

So far, solid TMI has been the most preferred indium precursor because of its high vapor pressure compared to other conventional indium precursors. However, variability in the evaporation rate for TMI has always been a major problem, especially when used in a conventional cylinder equipped with a dip tube. The inconsistency in vapor-phase concentration has been attributed to various factors, such as the reduction in total surface area during depletion, the "channels" or voids in TMI bed that limit the gas/solid contact time, and subsequent saturation of vapors, undue sublimation of TMI to regions secluded from the carrier gas flow, and the precursor surface contamination that renders TMI inaccessible for vapor transport [51].

To address this variability problem that leads to inconsistency in MOVPE operations, various improvements in TMI delivery have been tried with limited success. These improvements include depositing TMI on a porous and inert support [34], reversing the direction of carrier gas flow in a cylinder [34], dual-bubbler systems [34], and using TMI in a suspension form in either a hydrocarbon [50] or a low-volatility amine medium [50]. All these solutions were found to have inherent flaws, and the solutions have been implemented with limited success [51].

Figure 13.13 shows one of the successful dip tube-less cylinder designs [51–54], which reduces the linear velocity of the carrier gas and provides prolonged gas/solid contact time, resulting in the delivery of more than 90% of solid TMI with stable concentrations in the vapor phase, even under the high flux and high throughput conditions of growth [51–54].

To summarize, numerous advances have been made in the design, synthesis, and purification of Group III MOVPE precursors, especially the use of adduct purification to obtain ultrapure precursors, the latest use of ether-free and organoaluminum-based processes to synthesize Group III sources, and the synthesis of higher-vapor-pressure liquid alternatives to replace solid precursors such as TMIn. However, the industry has settled on conventional trialkyl Group III precursors such as TMG, TMA, and TMIn and their ethyl analogs for epitaxy that requires low carbon incorporation in the films. With increasing demands for MO precursors in compound semiconductors as much as in silicon-based semiconductors, there is still an opportunity for MOVPE precursor developers to create new processes and to implement next-generation manufacturing strategies. These would improve not only the quality and consistency, but also the cost of Group III precursors that can be transferred effectively, as well as add tremendous value for the consumers.

13.4.2 Group V MOVPE Precursors

In this section, select Group V precursors will be discussed, which are currently being used in the commercial fabrication of III-V semiconductors by MOVPE. Table 13.6 lists some of the most common Group V MOVPE precursors used for the deposition of III-V compound semiconductors.

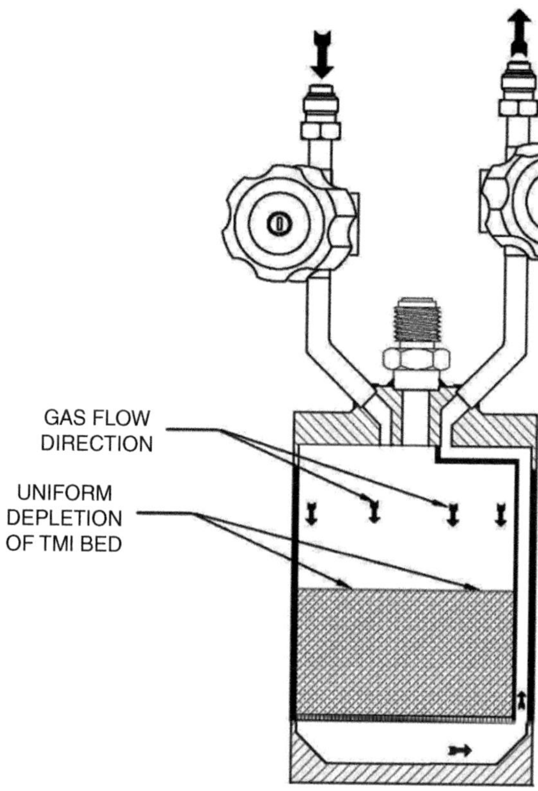

Figure 13.13 *A dip tube-less cylinder design for consistent delivery of solid MOVPE precursors in the gas phase. Sources: Shenai-Khatkhate et al. 2007 [51], Shenai-Khatkhate et al. 2010 [52], Shenai-Khatkhate 2010 [53], and Shenai-Khatkhate 2010 [54], with permission from USPTO (public domain).*

The Group V metalorganic precursors for the MOVPE of compound semiconductors are, in general, highly air sensitive, and some of them are also pyrophoric in nature. Even though earlier MOVPE studies were focused exclusively on using arsine and phosphine gases as the precursors of As and P, respectively, various other organoarsenic and organophosphorous sources containing organic ligands have been explored as research curiosities and for their potential application in MOVPE. Some of the precursors were developed specifically as safer liquid alternatives to highly toxic arsine and phosphine gases, and for safer delivery at acceptable vapor pressures.

The toxicity of Group V precursors, in general, has been a major concern, especially in large-scale production of wafers, where large quantities of arsine and phosphine gases are used routinely. The threshold limit values, or the maximum permissible limit of exposure on time weighted average for an 8-h day for arsine and phosphine are 0.05 ppm and 0.3 ppm, respectively [55]. Second, the LC_{50} value, or the total concentration for 50% rat population to die, for arsine and phosphine are reported to be 5 ppm and 11 ppm, respectively. It is worthwhile noting that antimony compounds are also considered as toxic.

The lower TLV and LC_{50} values for Group V MOVPE precursors also pose serious concerns during the storage and handling of these Group V gaseous sources. The liquid alternatives to arsine and phosphine, TBAs and TBP, are reported to be less hazardous and comparatively easy to handle, detect, and mitigate any hazards thereof [8]. In the industrial environment in fabs, toxic gas monitoring and safety alarms are used continuously to protect the workplace and environment [55].

For the MOVPE of III-V compound semiconductors, methyl-containing sources are preferred, because of their higher vapor pressures and comparatively greater stability than their higher

Table 13.6 *Group V MOVPE precursors.*

MO precursor (Abbr.)	Formula	CAS#	Formula weight	Boiling point °C	Vapor pressure @20 °C (Torr)	Vapor pressure constants for the equation log *P* (Torr) = *A* − *B*/*T* (K)	
						A	*B*
Trimethylarsenic (TMAs)	$(CH_3)_3As$	593-88-4	120.03	56	225.84	7.405	1480
Trimethylantimony (TMSb)	$(CH_3)_3Sb$	594-10-5	166.86	80.6	82.26	7.707	1697
Tert-butylphosphine (TBP)	$(C_4H_9)PH_2$	2501-43-5	90.11	54	215.5	7.586	1539
Tert-butylarsine (TBAs)	$(C_4H_9)AsH_2$	4262-43-5	134.05	69	147.56	7.243	1509
Triethylantimony (TESb)	$(C_2H_5)_3Sb$	617-85-6	208.94	156	2.84	7.904	2183
Trimethylbismuth	$(CH_3)_3Bi$	593-91-9	254.09	110	26.92	7.628	1816
Tris(dimethylamino) arsine	$((CH_3)_2N)_3As$	6596-96-9	207.15	55 (10 Torr)	1.34	8.289	2391
tris(dimethylamino) antimony	$((CH_3)_2N)_3Sb$	7289-92-1	253.98	62 (10 Torr)	2.06	6.231	1734
Triisopropyl antimony	$(C_3H_7)_3Sb$	73300-45-5	251.02	75 (10 Torr)	0.27	9.268	2881

homologs. Trimethylantimony is a preferred source for Group III antimonides. Triethylantimony [56] and tri-isopropylantimony [57] have also been used as Sb sources, when significantly lower carbon incorporation is desired. TDMASb [58] was developed for low-temperature epitaxy of Group III antimonides, particularly InSb, by chemical beam epitaxy (a subset of MOVPE), where lower vapor pressures of the order of 1 Torr at room temperature are also acceptable. TDMASb is also considered an antimony source in atomic layer deposition (ALD), which is a subset of chemical vapor deposition (CVD).

It is worthwhile noting that trimethylarsenic is primarily used as the carbon dopant source, and not as arsenic source. This is because the As–C bond in trimethylarsine is very strong, requiring higher growth temperatures for its dissociation and leading to copious incorporation of carbon during epitaxy [59]. Likewise, trimethylphosphine is also not used as a phosphorous source [59]. Arsine and phosphine have been used as preferred sources, despite challenges in managing safety aspects, for MOVPE of Group III arsenides and Group III phosphides, respectively. Tertiarybutyl arsine and tertiarybutyl phosphine have been used as the safer alternative sources, and the research so far has indicated that due to their lower pyrolysis temperatures than those required for arsine and phosphine [8], good-morphology films of Group III phosphides and Group III arsenides can be grown at V/III ratios of almost unity, vis-à-vis greater than 10 for arsine and phosphine [59]. In many cases, the carbon incorporation was found to be lower, when tertiary butyl arsine was used instead of arsine: e.g. AlGaAs layers grown using Group III methyl sources (TMA and TMGa) and TBAs instead of arsine showed remarkable reduction in carbon incorporation [8, 60]. The results obtained with TBP are similar regarding good morphology of layers, improved V/III ratios, and lower carbon incorporation during the MOVPE of Group III phosphides [59].

Hydrazine and hydrazine derivatives such as unsymmetrical dimethyl hydrazine (DMHy or UDMH) were developed and tried for GaN deposition as potential alternatives for ammonia as the nitrogen source [61]. These alternative sources were expected to lower the deposition temperatures for GaN epitaxy, eliminate the excessive consumption of ammonia (since V/III ratios as high as 1000:1 are commonly used), and enhance the nitrogen incorporation in the Group III nitride films. However, ammonia is still being used successfully for growing GaN and GaN-based alloys, despite the challenges. The alternative nitrogen precursors, such as hydrazines, are not found to be effective enough, so far, to displace ammonia as the nitrogen source. Hydrazines are also reported as highly toxic and carcinogens, with potential for injury to lungs, liver, kidneys, and the central nervous system.

Figure 13.14 shows structures of some Group V MOVPE precursors.

The vapor pressures of Group V MOVPE precursors with increasing temperatures [44] are shown in Figure 13.15.

As expected, the vapor pressures of methyl-containing sources are higher than those containing bulky alkyl groups. As discussed earlier, the larger alkyl groups offer the advantages of lower growth temperatures and less carbon incorporation, yet they suffer from lower transportation rates due to their significantly lower vapor pressures.

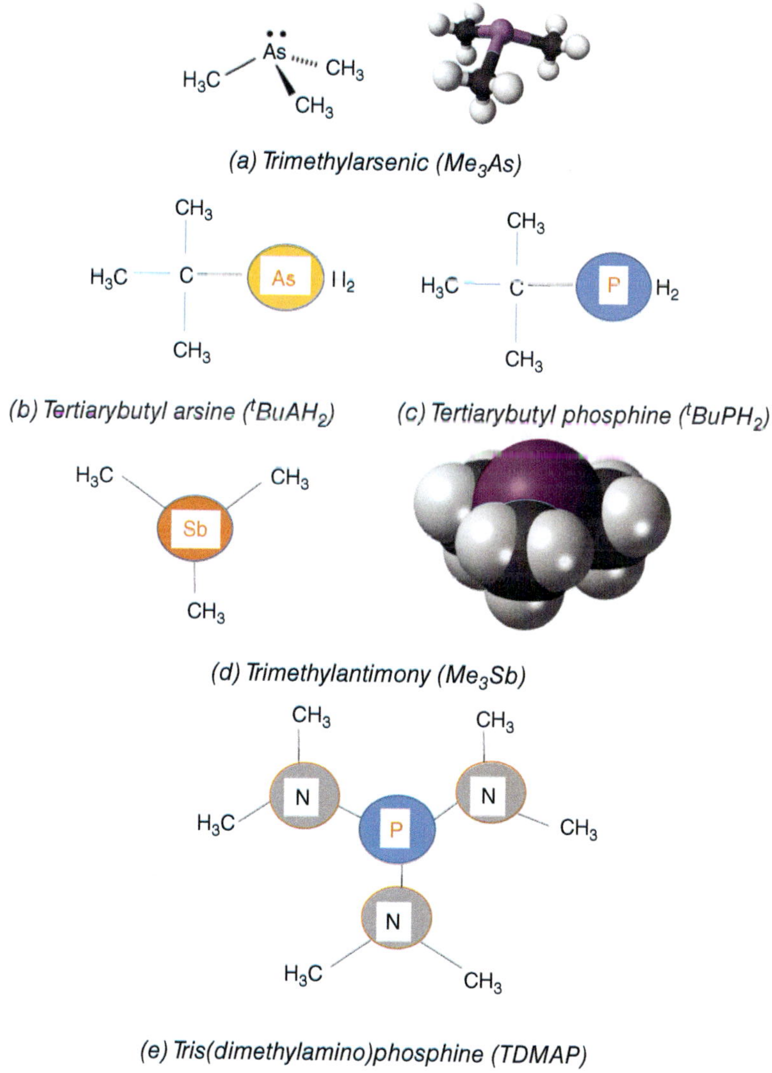

Figure 13.14 *The chemical structures of Group V MOVPE precursors.*

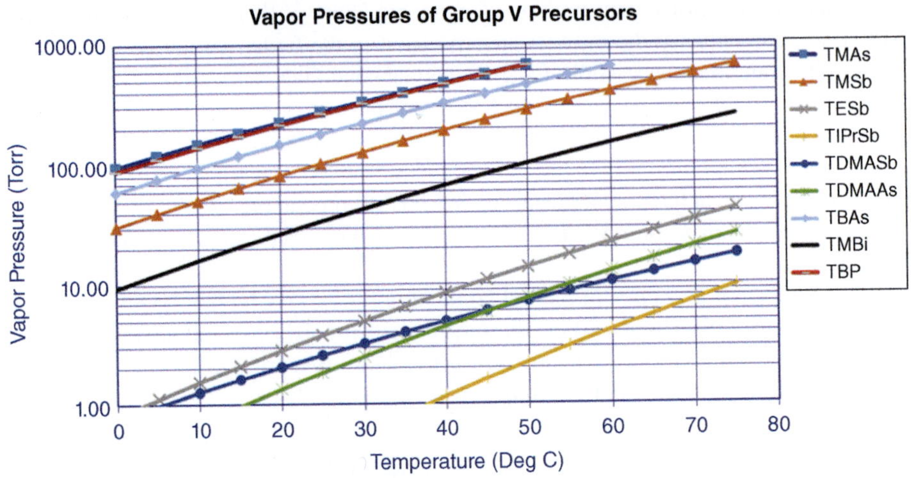

Figure 13.15 *The vapor pressure–temperature plots for Group V MOVPE precursors.*

To illustrate, for antimony sources, the vapor pressures are in this order:

$$TMSb > TESb > TIPSb$$

For the same organic moiety, the vapor pressure is inversely proportional to the atomic size of the element. For example, the vapor pressure of TBP is greater than that of TBAs, and the vapor pressure of TMAs is greater than that of TMSb. Similarly, at higher temperatures, the vapor pressure of TDMAAs is greater than that of TDMASb.

The synthesis of trialkyl Group VA compounds, such as trimethylantimony, triethylantimony, trimethylarsenic, etc., is usually carried out by using the conventional reaction of Grignard reagents or organolithium reagents [13–17] with the trichlorides of antimony and arsenic, in an ethereal solvent as shown in the following equations:

$$MCl_3 + 3MeMgBr \rightarrow Me_3M + 3MgBrCl$$

$$MCl_3 + 3MeLi \rightarrow Me_3M + 3LiCl$$

where M = As, Sb, or Bi

The resultant products are separated from the reaction mixtures by distillation to separate them from the ethereal solvents employed [13]. Further purification of the product by fractional distillation is generally required to remove trace solvents and metallic impurities.

Sawara et al. have disclosed an ether-free, trialkylaluminum-based process to produce pure trialkylarsines from aluminum alkyls and diarsenic trioxide using alkali metal halides as complexing agents in hydrocarbon solvents [62]. Shenai-Khatkhate et al. have reported a trialkylaluminum-based route [29] to synthesize the trialkyl Group V precursors, which uses the strategic organoaluminum-tertiary amine adducts as the alkylating agents. The products obtained by these ether-free and trialkylaluminum (or adducts of trimethylaluminum) based processes have been found to exhibit better quality, and better performance of the films thereof, than the precursors produced in the ethereal solvents [29]:

$$2MCl_3 + 3Me_3Al\text{-tertiary amine} \rightarrow 2Me_3M + 3MeAlCl_2\text{-tertiary amine},$$

where M = As, Sb and Bi

Tertiarybutyl phosphine and tertiarybutyl arsine are synthesized by the reaction between tertiarybutyl phosphorous dichloride ($TBPCl_2$) or tertiarybutyl arsenic dichloride ($TBAsCl_2$)

with a reducing agent, such as sodium borohydride or lithium borohydride. The reaction to form $TBPCl_2$ or $TBAsCl_2$ is by using conventional Grignard reagent or an organolithium reagent in an ethereal solvent, as shown in the following equations [13, 17]:

$$MCl_3 + {}^tBuMgBr \rightarrow {}^tBuMCl_2 + MgBrCl$$

$$MCl_3 + {}^tBuLi \rightarrow {}^tBuMCl_2 + LiCl$$

where M = As or P

$$2{}^tBuMCl_2 + LiAlH_4 \rightarrow 2{}^tBuMH_2 + LiCl + AlCl_3$$

where M = As or P

A process to produce tertiarybutyl phosphine and tertiarybutyl arsine by Shenai-Khatkhate et al. [63] discloses the reaction of MCl_3 with the Grignard reagent in the high-boiling solvents (with boiling point >175 °C) to prepare oxygen-free ${}^tBuMCl_2$ and ${}^tBuMH_2$ products, where M = As or P.

The use of organoaluminum reagents and organolithium reagents in oxygen-free solvents, optionally in the presence of a tertiary amine or a tertiary phosphine, is disclosed in a process by Shenai-Khatkhate et al. [64] for the synthesis of oxygen-free ${}^tBuMCl_2$ and ${}^tBuMH_2$ products, where M = As or P.

One of the concerns with the tertiary butyl-containing and hydride-based sources, such as TBAs and TBP, is the potential for β-hydride elimination from these sources, which may lead to the formation of highly toxic arsine and phosphine gases. Zimmermann et al. have suggested the class of nonhydride-containing Group V precursors, which are heteroleptic trialkyl Group V precursors based on ethyl-substituted arsine or phosphine, with other groups being substituted with either tertiarybutyl or isopropyl groups. It was shown that GaAs films grown with TMG and either diethyltertiarybutylarsine (DETBAs) or diethylisopropylarsine (DEIPAs) exhibited much lower carbon incorporation than those produced by using hydride-containing ethyl arsine source, such as Et_2AsH (DEAs) [65]. These heteroleptic sources have not been used in commercial applications, primarily due to their very low vapor pressures [8].

To summarize, many strides have been made in the design and synthesis of Group V MOVPE precursors, especially the use of ether-free processes to synthesize trialkyl Group V sources and the synthesis of safer liquid alternatives to replace toxic Group V hydride sources. The industry has also mitigated the risks associated with managing the toxicity hazards and the environmental impacts of Group V sources by implementing state-of-the-art sophisticated toxic-gas monitoring systems and safety alarms in facilities. Hence, arsine and phosphine have still maintained their position as the preferred sources of As and P, respectively.

13.5 MOVPE Precursors for II-VI Compound Semiconductors

13.5.1 Group II MOVPE Precursors

As the earlier MOVPE studies were focused primarily on mercury cadmium telluride (MCT), zinc selenide, or zinc sulfide selenide, the interest in the MOVPE of II-VI compound semiconductors had started to diminish in the 1990s, with the implementation of MBE for MCT-based devices. However, as pointed out in Chapter 9 by Maxey et al., that interest has been reawakened recently with the realization that MOVPE processes can be used effectively to produce high-quality, large-area IRFPA-based devices. With this encouraging renewed interest in the MOVPE of II-VI precursors, it will be interesting to revisit the Group II precursors of relevance.

The Group II metalorganic precursors for the MOVPE of compound semiconductors are, in general, highly air sensitive and pyrophoric in nature, just like Group III MOVPE precursors. It is worthwhile noting that the focus of II-VI MOVPE work has still not shifted away from dimethyl and diethyl metalorganic sources, and new synthesis and/or purification strategies have been investigated for applications that require reduced carbon contamination in the films, better delivery, lower oxygen concentrations, reduced halogen contents in the precursors, etc.

Table 13.7 lists selected precursors that are commercially available from various manufacturers, which are being used industrially, at present, for commercial-scale depositions of II-VI semiconductors. Figure 13.16 shows the vapor pressure–temperature plots for the Group II MOVPE precursors, while Figure 13.17 shows their structures [44].

Diethylzinc was the first main group organometallic compound in the world, which was synthesized by Frankland [13, 17]. On July 29, 1848, nearly 170 years ago, at Queenwood College in Hampshire, England, Frankland, then a 23-year-old researcher, had sealed finely divided zinc and ethyl iodide in a thick-walled glass tube, with an aim to produce ethyl radicals. The reaction formed ethylzinc iodide, which upon thermal dissociation produced diethylzinc, the very first organometallic compound of a main group element. This pioneering work by Frankland laid the solid foundation for synthesizing highly reactive main group organometallic compounds, the most vital components of the MOVPE process.

Table 13.7 *Group II MOVPE precursors.*

MO precursor (Abbr.)	Formula	CAS#	Formula weight	Boiling point °C	Vapor pressure @20 °C (Torr)	Vapor Pressure Constants for the equation $\log P\,(\text{Torr}) = A - B/T\,(\text{K})$	
						A	*B*
Dimethylzinc (DMZn)	$(CH_3)_2Zn$	544-97-8	98.45	46	299.1	7.80	1560
Diethylzinc (DEZn)	$(C_2H_5)_2Zn$	557-20-0	123.5	118	12.1	8.28	2109
Dimethylcadmium (DMCd)	$(CH_3)_2Cd$	506-82-1	142.88	106	28.2	7.764	1850

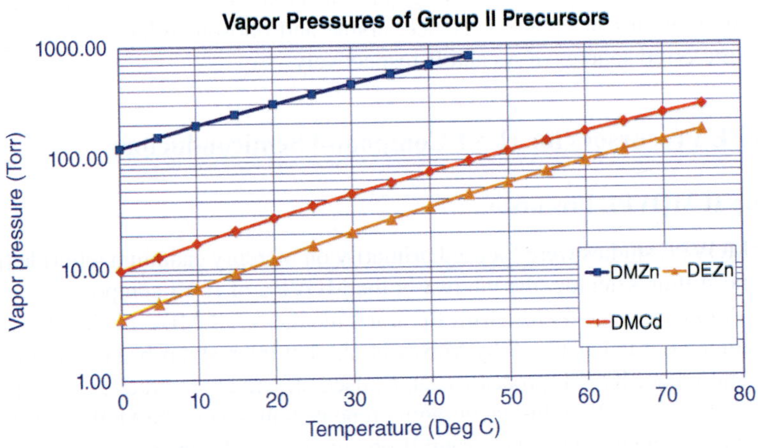

Figure 13.16 *The vapor pressure–temperature plots for Group II MOVPE precursors.*

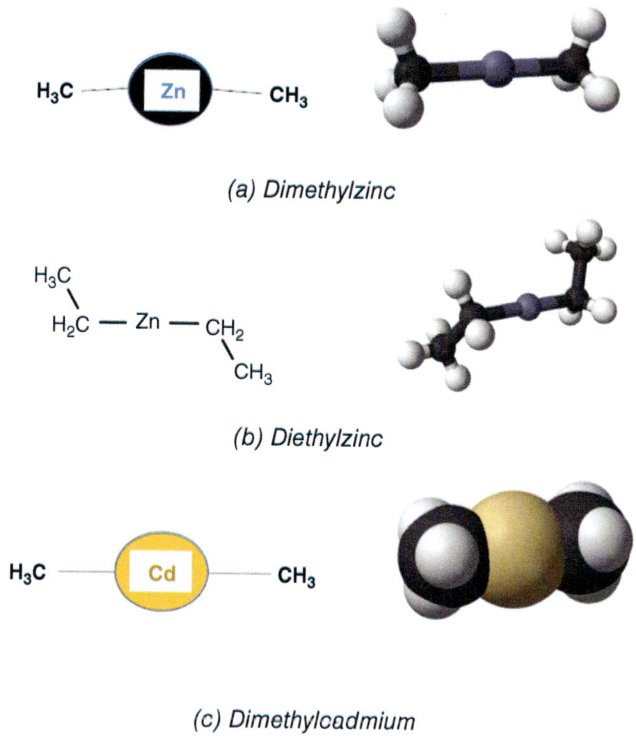

Figure 13.17 *The chemical structures of Group II MOVPE precursors.*

Dialkylzincs, such as dimethylzinc and diethylzinc, are conveniently synthesized by the reaction between anhydrous zinc chloride and a Grignard reagent, typically carried out in an ethereal solvent [13–17], as shown in the following equation:

$$ZnCl_2 + 2RMgX + (Ether) \rightarrow ZnR_2 + 2MgX_2$$

Dialkylzincs are separated from reaction mixtures by fractional distillations. The products are generally contaminated by trace quantities of ether solvent, as well as halogenated organozinc byproducts. The ether contamination can be corrected by using postsynthesis adduct-purification processes [66] or by using a high-boiling ether solvent [27].

Using ether-free processes based on using the trialkylaluminums as the strategic alkylating agents has been claimed to be effective in obtaining high-purity precursors containing lower levels of oxygenated impurities, as well as low halogen content [28]. These processes can be made safer, and more efficient, by using adducts of trialkylaluminums and tertiary amines (or tertiary phosphines) as described earlier [28–30].

Dialkylcadmiums, such as dimethylcadmium and diethylcadmium, are often synthesized by reaction between anhydrous cadmium chloride and a Grignard reagent or an organolithium reagent. These reactions are carried out in an ethereal solvent [13–17], as shown in the following equations:

$$CdCl_2 + 2RMgX + (Ether) \rightarrow CdR_2 + 2MgX_2$$

$$CdCl_2 + 2RLi + (Ether) \rightarrow CdR_2 + 2LiCl$$

Adduct purification has been used for obtaining high-quality dimethylcadmium that is ether-free, by using the adducts with tertiary amines [66]. The process uses N-donor ligands that strategically contain two nitrogen donors so far away that they form a polymer by bridging the Lewis acid (i.e. dimethylcadmium) rather than chelating complexes.

Dialkylcadmiums are less reactive, although air-sensitive, compared to their zinc analogs. They are not spontaneously flammable in air. They are typically distilled at reduced pressures, rather than at atmospheric pressures, since they tend to decompose at higher temperatures. Their decomposition can occur at 150 °C, and the materials may become explosive at such high temperatures [13, 27]. Hence, caution is necessary during the handling of dialkylcadmiums at high temperatures, particularly during their distillation or thermal dissociation of the adducts.

Using an ether-free process based on reacting the adduct of trimethylaluminum with tertiary amine with anhydrous cadmium chloride is claimed to be an effective way to obtain high-purity dimethylcadmium [28], with reduced oxygenated and halogen impurities, as described in this route.

Summarizing, the development of Group II MOVPE precursor chemistry reveals many breakthroughs that include various inventions and innovations, such as the adduct-purification process, using high-boiling ethers for the Grignard reactions, and ether-free processes based on trialkyl aluminum compounds, and adducts with tertiary amines or tertiary phosphines. With the renewed interest in MOVPE of II-VI compounds, there is an opportunity to implement the new advancements in manufacturing so that the precursors are produced safely and efficiently, with high quality and better consistency. For some of the complex device applications, there will be a compelling need for new precursor molecules that would reduce or eliminate the risks associated with thermal decomposition of the Group II precursors.

13.5.2 Group VI MOVPE Precursors

The Group VI metalorganic precursors for the MOVPE of compound semiconductors are, in general, less air sensitive and nonpyrophoric in nature, unlike the MOVPE precursors of Groups II, III, and V. During the early years of MOVPE, the major focus of II-VI MOVPE work has been on the development of best quality MCT films, especially for IR sensors and various defense-related applications.

Table 13.8 lists the Group VI MOVPE precursors that are commercially available from various manufacturers and are being used industrially, at present, for commercial scale depositions of II-VI semiconductors. Figure 13.18 shows the vapor pressure–temperature plots for the Group VI MOVPE precursors [44], while Figure 13.19 shows their structures.

Significant advancements in MOVPE of MCT were achieved when Irvine and Mullin reported the deposition of MCT in a hot-wall reactor, first by using DMCd, Hg, and DETe as the precursors [67] in 1981 and then by using a novel process called the *interdiffused multilayer process* (IMP) for the first time [68] in 1988. Their pioneering work triggered an interest and subsequent developments in optimizing MCT performance. The growth of MCT was initially limited to temperatures as high as 400–450 °C, primarily because of the higher thermal stability of the tellurium precursor, DETe. At such high temperatures, Hg/Cd interdiffusion becomes very significant, and sharp heterojunctions become difficult to achieve. Hence, efforts were focused on lowering growth temperatures by using alternative tellurium precursors during the mid-1980s. Some of the parallel efforts include precracking the tellurium precursors; using thermally unstable tellurium precursors; using UV, laser, or plasma to decompose tellurium precursors; etc. [69].

Some of the strategic elements that enabled the development of alternative precursors include:

- The development of tellurium precursors with higher alkyl groups that facilitate the homolytic fission of the Te–C bond at lower temperatures, e.g. di-n-propyltelluride, di-isopropyltelluride, ditertiarybutyltelluride, diallyltelluride, methylallyltelluride, etc.

Table 13.8 *Group VI MOVPE precursors.*

MO precursor (Abbr.)	Formula	CAS#	Formula weight	Boiling point °C	Vapor pressure @20 °C (Torr)	Vapor pressure constants for the equation log *P* (Torr) = *A* – *B*/*T* (K)	
						A	*B*
Di-isopropyl telluride (DIPTe)	$(C_3H_7)_2Te$	52492-46-3	213.77	157	2.56	8.228	2309
Diethyltelluride (DETe)	$(C_2H_5)_2Te$	627-54-3	185.72	138	7.03	7.99	2903
Di-isoprpyl selenide (DIPSe)	$(C_3H_7)_2Se$	37773-02-7	165.14	128	8.93	8.439	2194
Diethyl sulfur (DES)	$(C_2H_5)_2S$	352-93-2	90.18	92	47.37	8.184	1907
Diethyl selenide (DESe)	$(C_2H_5)_2Se$	627-53-2	137.08	108	21.8	7.905	1924
Methyl allyl telluride (MATe)	$(CH_3)Te(C_3H_5)$	114438-52-7	183.71	145	4.48	8.146	2196

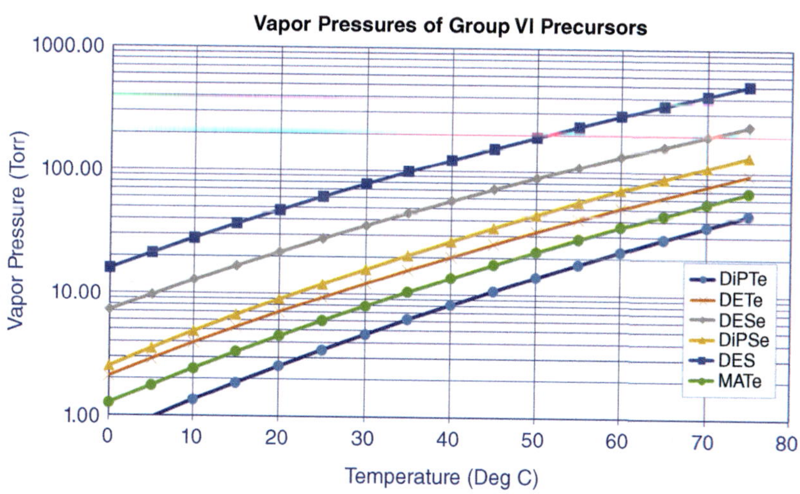

Figure 13.18 *The vapor pressure–temperature plots for Group VI MOVPE precursors.*

- The development of tellurium precursors with branched alkyl groups that facilitate β-hydride elimination at the Te–C bond at reasonably lower temperatures, e.g. di-isopropyltelluride, and ditertiarybutyltelluride.
- The development of tellurium precursors that are not thermally stable, even at lower temperatures, e.g., alkane tellurols that decompose at ambient temperatures, and heterocyclic tellurium precursors such as 2,5-dihydro-tellurophene (which required shipping at dry ice temperatures of the order of –78 °C).

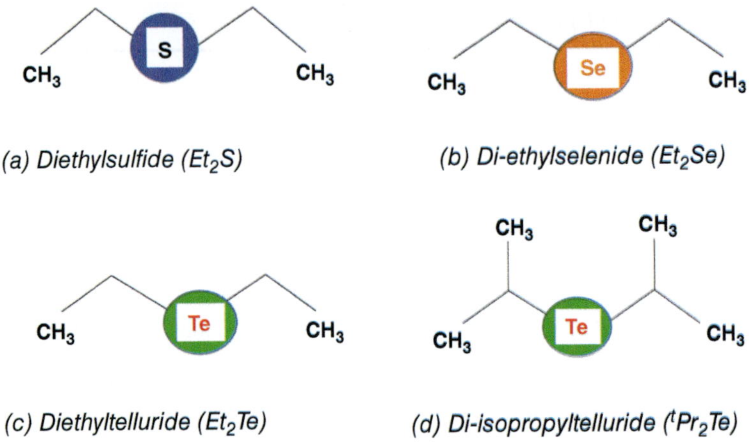

Figure 13.19 *The chemical structures of Group VI MOVPE precursors.*

- The development of tellurium precursors, either homoleptic or heteroleptic, with allyl groups, or substituted allyl groups that enable the thermal dissociation at significantly lower temperatures, e.g. diallyltelluride, methylallyltelluride, or like alkylallyltellurides.
- The development of tellurium precursors with less stable Te–Te bonds that facilitate the hemolytic fission of the Te–Te bond at lower temperatures, e.g., dimethyl ditelluride.

The work by Irvine and Mullin at the Royal Signals and Radar Establishment, England, in collaboration with GEC Hirst Research Center, England, and The University of St. Andrews, Scotland, showed conclusively that high-quality MCT layers could be deposited at 350 °C using di-isopropyltelluride (DIPTe), DMCd, and Hg as the precursors. The MCT growth temperature was thus lowered from 420 °C to 350 °C for the first time. These results were reported in the publications by Irvine et al. [70] and Thompson et al. [71] in 1988. Since then, MCT research continued to focus exclusively on DIPTe as the tellurium precursor of choice, with metallic mercury and DMCd generally being the other precursor materials [69].

Regarding the synthesis of tellurium precursors, conventional laboratory synthesis involves the alkylation of alkali metal ditelluride, such as Na_2Te [72]. Sodium telluride is synthesized by the reaction of the stoichiometric quantities of sodium and tellurium in liquid ammonia, which is further alkylated by the reaction with alkyl halide, RX, as shown here:

$$2\text{Na}+\text{Te}\left(\text{in liquid ammonia}\right)\rightarrow \text{Na}_2\text{Te}$$

$$\text{Na}_2\text{Te}+2\text{RX}\left(\text{in liquid ammonia}\right)\rightarrow \text{R}_2\text{Te}+2\text{NaX},$$

where R = alkyl group, such as methyl, ethyl, isopropyl, tertiarybutyl, etc. and X = Cl, Br, or I.

However, this route also produces ditellurides and polytellurides as side products, thereby lowering the yield of dialkyl monotellurides [72]. Furthermore, the safe handling of liquid ammonia was also a concern at industrial scale.

Hence, a modified process was developed [24, 27] to obtain Na_2Te exclusively, by the reaction between Te, sodium hydroxide, and sodium formaldehyde sulfoxylate in an aqueous medium, as shown in the following reaction sequence:

$$\text{Te}+\text{OHCH}_2\text{SO}_2\text{Na}\rightarrow \text{OHCH}_2\text{SO}_2\text{TeNa}$$

$$\text{OHCH}_2\text{SO}_2\text{TeNa}+\text{NaOH}\rightarrow \text{Na}_2\text{Te}+\text{Na}_2\text{SO}_3+\text{CH}_2\text{O}$$

$$\text{Na}_2\text{Te}+2\text{RX}\rightarrow \text{R}_2\text{Te}+2\text{NaX},$$

where R = alkyl group, such as methyl, ethyl, isopropyl, tertiarybutyl, etc. and X = Cl, Br, or I.

The product, R_2Te, is reported to be separated from the reaction mass by extraction with a suitable organic solvent (such as an aliphatic ether or an aliphatic hydrocarbon), preferably having a boiling point less than the target tellurium precursor compound. This route was reported to be suitable for diethyltelluride and di-isopropyltelluride, but unsuitable for ditertiarybutyltelluride, since tertiarybutyl chloride hydrolyzes readily to tertiary butanol in the alkaline aqueous reaction mixture, thus lowering the yield of the final product significantly.

Diethylselenide and di-isopropylselenide can also be produced using this strategy.

A novel process for ditertiarybutyltelluride [73] is reported, which is based on the reaction between Te and tertiarybutyllithium in pentane to form an intermediate complex. This intermediate can be further alkylated by its reaction with tertiarybutyl chloride, as shown in the following equations:

$$^{t}BuLi + Te\left(\text{in pentane}\right) \rightarrow {}^{t}BuTeLi$$

$$^{t}BuTeLi + 2^{t}BuCl\left(\text{in pentane}\right) \rightarrow {}^{t}Bu_2Te + 2LiCl$$

The process also discloses the synthesis of ditertiarybutylselenide.

The dialkyltelluride precursors by the aforementioned processes, particularly via the aqueous routes, were found to meet the high-quality requirements in terms of metallic and organic impurities, with the total impurity content well below 50 ppb. These high-purity dialkyltellurides were used by Irvine et al. to grow very high-quality MCT with IMP and to fabricate devices that were demonstrated to be better than those available from other sources.

Despite the high quality of the precursors, the adduct-purification strategy was also used for dialkyltellurides [74] to exceed their performance in MCT and MCT-based devices. Since dialkyltellurides are Lewis bases, they were strategically adducted with Lewis acids, such as $CdBr_2$, CdI_2, CuI, $HgCl_2$, and HgI_2. When DETe and DIPTe were used to form adducts with these Lewis acids, the adducts were found to melt at temperatures below 100 °C and dissociate at temperatures greater than 100–120 °C, to afford >60% yield of ultrapure dialkyltellurides under unoptimized conditions [74].

In summary, the development of Group VI MOVPE precursors has a two-fold focus, comprising (i) designing precursors that would lower the growth temperatures for the MOVPE of CdTe, HgTe, CdHgTe, ZnSe, ZnSeS, etc.; and (ii) developing synthesis and purification strategies to improve yield, quality, and process efficiency along with a possibility to lower the cost. The breakthroughs include various inventions and innovations, such as the aqueous process that affords the precursors free from metallic impurities, the adduct-purification process to enhance purity levels, using organolithium-based synthesis for ditertiarybutyl telluride, ether-free processes based on trialkyl aluminum compounds, and the adducts of trialkylaluminums with tertiary amines or tertiary phosphines. With the renewed interest in MOVPE of II-VI compounds, there is a strong opportunity to design and create new molecules and to advance the adduct purification to the next level.

13.6 MOVPE Dopants for Compound Semiconductors

The Group III and V precursors, described earlier, are generally used as the dopants for II-VI compound semiconductors, while the Group II and VI precursors discussed in the preceding section are used as the dopants for III-V compound semiconductors.

In addition to these source molecules, some of the common dopant sources are listed in Table 13.9. These dopant precursors are commercially available from various manufacturers and are being used industrially, at present, for the commercial scale depositions of III-V and II-VI semiconductors. Figure 13.20 shows the vapor pressure–temperature plots [44] for the three selected MOVPE dopants, while Figure 13.21 shows their structures.

Table 13.9 MOVPE dopant precursors.

MO precursor (Abbr.)	Formula	CAS#	Mol. weight	B.P. °C	Vapor pressure @20 °C (Torr)	Vapor pressure constants for the equation $\log P = A - B/T$ (K)	
						A	B
Bis(cyclopentadienyl) magnesium (Cp_2Mg)	$(C_5H_5)_2Mg$	1284-72-6	154.49	300	0.03	10.452	3522
Carbon tetrabromide (CBr)	CBr_4	558-13-4	331.63	190	0.58	8.37	2524
Isobutyl germane (IBGe)	$(C_4H_9)GeH_3$	768403-89-0	132.02	66	121.16	7.94	1715

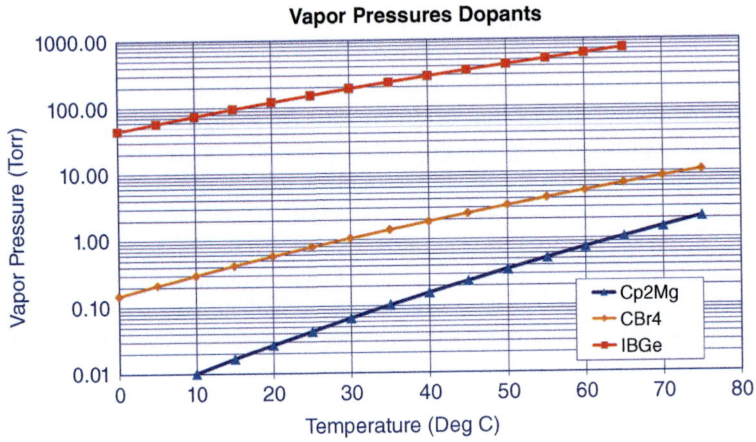

Figure 13.20 *The vapor pressure–temperature plots for MOVPE dopant precursors.*

Mg from bis(cyclopentadienyl)magnesium (Cp_2Mg) is a dopant for GaInP and Group III nitrides, such as GaN and InGaN. Carbon from CBr_4 is a shallow acceptor for III-V compound semiconductors in general [8]. Cp_2Mg is synthesized by the conventional Grignard processes [13–17], i.e. by reacting monomeric cyclopentadiene with methylmagnesium bromide. A purification process claims ultrapure Cp_2Mg, when it is purified by treating it with a strategic amine prior to its sublimation [19]. Cp_2Mg is reported to be useful in doping of GaN [8]. Isobutylgermane is recently reported as a safer alternative source to toxic germane gas, for the MOVPE of Ge and SiGe films [75, 76] and a soft dopant for III-V compounds, particularly [77].

13.7 Environment, Health, and Safety (EHS) Aspects of MOVPE Precursors

13.7.1 General Aspects and Considerations

In general terms, MOVPE precursors belong to a class of compounds comprising metals (or metalloids) and organic radicals. The environment, health, and safety (EHS) aspects of MOVPE precursors are primarily associated with their pyrophoricity and health hazards related to toxicity. Many

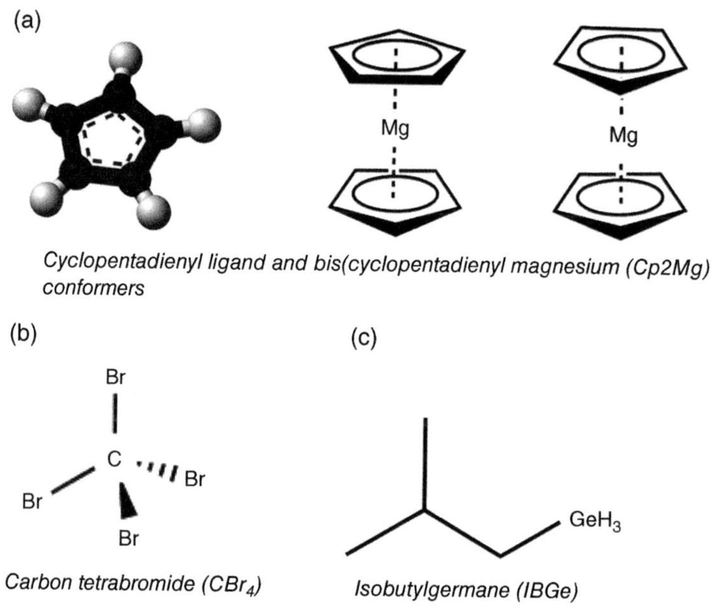

Figure 13.21 *The chemical structures of MOVPE dopant precursors.*

of the metalorganics are pyrophoric in nature, although not all of them exhibit the same extent of pyrophoricity. For example, the metalorganics of Zn, Al, Ga, and In are known to be extremely pyrophoric and are known to react violently upon contact with oxygen and moisture to afford oxides and hydroxides. On the other hand, the metalorganics of Cd, As, and Sb are pyrophoric but not as reactive as their Group III analogs. Their reactivity is found to be dependent on the quantity of material spilled or leaked, or a critical mass to be reactive and/or pyrophoric [55]. The nature of the organic moiety attached as well as the electropositive metallic character plays a major role in determining the reactivity of the MOVPE precursor under consideration. Higher metal–carbon bond energies are generally accompanied by more reactivity and less stability. Similarly, the stability of the alkyl radical formed as well as the decrease in metal–carbon bond strength with increasing steric bulk and/or branching in the alkyl radical do play a major role in the reactivity of metalorganics [69].

The spontaneous combustibility of metalorganics is well documented, yet information on their toxicity is not. Since the EHS-related information is constantly updated as new results become available, we encourage readers to refer to the list of books and references cited here [55, 78–81], wherein specific and up-to-date information can be readily accessible.

13.7.2 Employee and Environment Exposure Aspects

With the adoption of MOVPE as a mass-production process, the implementation of adequate controls to mitigate risks of possible employee and environment exposure have been standardized in routine operations: e.g. leaks or accidental release are adequately covered by advanced engineering controls during regular maintenance and repair of chambers and lines. These controls extend to incorporate continuous improvements in facility designs, employee awareness training, mandatory EHS reviews, HazOp studies with new projects, and appropriate management of change to provide greater levels of safety and total containment than those available a few years ago [55].

The likely exposure scenarios associated with nonroutine operations include inadvertent release of metalorganics into the atmosphere, release during an unforeseen hydrogen carrier gas fire, and the reaction of metalorganics upon accidental contact with water, an oxidizer, or noncompatible materials. Employee and environment exposure may involve metal oxides, metallic dust, hydrocarbons, intermediate oxidation products, reduction products as metal hydrides, and unreacted metalorganic sources. Implementing adequate engineering controls, use of personal protective equipment, and employee training have been found to be extremely effective in reducing these incidents in the metalorganics manufacturing environment as well as in compound semiconductor fabrication facilities.

Acute employee exposure may lead to possible thermal burns and irritation to skin, eyes, nose, and upper respiratory tract. Serious cases of exposure can lead to metal-fume fever, which is characterized by coughs, pains, chills, nausea, and vomiting and may last for 24–48 h. Chronic exposure to hazardous metalorganics of lead, mercury, arsenic, and cadmium is generally a function of the toxicity of the individual metallic component and the biological mechanisms associated with absorption, transportation, and excretion of that specific element in the human body [55].

13.7.3 Employee and Workplace Exposure Limits

The *workplace exposure limit* (WEL) is defined as "the airborne concentration of a chemical to which nearly all workers may be repeatedly exposed day after day, without any adverse effect." The Occupational Safety and Health Administration (OSHA) and American Conference of Governmental Industrial Hygienists (ACGIH) standards consider the maximum level of exposure to chemicals as air contaminants that may be allowed during an 8-h day and 40-h week without adverse health effects. The permissible exposure limits (PELs) as provided by OSHA and threshold limit values (TLVs) from ACGIH are updated frequently and are available at the resources provided here [78].

Summarizing, MOVPE has now become a well-established production technology for the fabrication of compound semiconductor devices. As such, concerns for the health and safety of the workers, workplace, community, and environment are constantly being addressed using advanced state-of-the-art safety systems, engineering controls, and mitigating safety controls. Extensive employee training is also provided to ensure employee awareness. Metalorganic sources are provided with adequate safety information and procedures to mitigate the risks associated with any unsafe scenarios, thus leading to high levels of safety in MOVPE facilities.

13.8 Conclusions and Future Trends

Metalorganic chemistry, since its inception, has played a crucial role in creating enabling materials, including but not limited to olefins, alcohols, polyethylene, polypropylene, agricultural chemicals, pharmaceuticals, specialty polymers, display materials, as well as more recent smart devices, power electronics, upcoming High-Speed High-Frequency 5G networks, mini- and micro-LED and electronic materials in general.

The field of MOVPE is growing rapidly, and MOVPE precursors are finding many new applications with the expansion of MOVPE into the semiconductor market. The total MOCVD market is expected to grow at a compound annual growth rate (CAGR) of about 14%, and the global market size is expected to exceed $1.2 billion by 2021, growing from $640+ million in 2016, based on reports from various market analysts.

In the lighting industry, in particular, metalorganic precursors are playing a crucial role. LED lighting has penetrated residential, retail, commercial, industrial, and roadways, after the rapid expansion in mobile devices and television backlighting. High-quality, low-cost LEDs are going to be the one of the major drivers in the market, which creates a significant opportunity for MOVPE engineers and MOVPE precursor scientists alike. Engineers are expected to improve the cost position and productivity in MOVPE, for example, with better epitaxial performance in terms of uniformity, precise control of thickness and composition, consistent optoelectronic performance of GaN epitaxial films, and devices based on GaN.

Precursor designers and manufacturers are expected to ensure consistent delivery, better efficiency, and greater utilization of enabling precursors than ever before. Good manufacturing practices will become competitive advantages, as they ensure high quality and greater consistency of quality for MOVPE precursors. Safety aspects are also going to be of paramount importance. Some of the next-generation manufacturing strategies mentioned in this chapter, such as continuous manufacturing, microfluidic reactors for manufacturing and purification, and process automation for consistency in manufacturing continuous operations will be critical success factors.

There is still a need for in situ monitoring and analyzing the compositional uniformity during the growth process. This will be important not only from the quality perspective but also from safety and reactive chemistry standpoints. It can also enlighten precursor designers regarding the mechanisms of growth and failures in MOVPE. Even though the MOVPE industry has accepted methyl-containing precursors as the preferred sources for III-Vs, carbon incorporation is still a serious concern for GaAs and GaAs-based alloys. Similarly, ammonia is still the preferred source for nitrogen in GaN, consumed at very high rates. Having an enabling nitrogen precursor as an alternative to ammonia gas is still an unmet need.

Considering the current precursor landscape and unmet needs therein, and the rapid pace at which MOVPE is expected to grow, it seems reasonable that the next decade will be an encouraging as much as a challenging phase for organometallic chemists to keep their creative juices flowing.

Acknowledgments

I would like to thank Professor David J. Cole-Hamilton, University of St. Andrews, Dr. J. Brian Mullin, and Dr. Stuart J. C. Irvine for instilling a love for MOVPE precursors in me during my post-doctoral research work at the University of Liverpool, UK, and later at the University of St. Andrews, UK. I am thankful for their encouragement and guidance, because it has enabled me to make my small contribution to the MOVPE world today. I am also thankful to Cathie Markham, Global R&D Director, and Joseph Reiser, Business Director, Electronics & Imaging, DuPont de Nemours, Inc., for their positive encouragement in my MOVPE research work. My thanks go to Ronald L. DiCarlo, Jr., who has constantly provided excellent experimental assistance and tremendous help in compiling the data whenever I needed. My personal thanks to this book's editors, Dr. Stuart J. C. Irvine and Dr. Peter Capper, for their exceptional guidance and remarkable cooperation. Last but not least, I sincerely thank my wife, Usha, and daughter, Anisha, for inspiring me and encouraging me to continue, especially during the challenging phases.

References

1. Manasevit, H.M. (1968). Single-crystal gallium arsenide on insulating substrates. *Applied Physics Letters* 12 (4): 156–159.
2. Manasevit, H.M. and Simpson, W.I. (1969). Use of metal-organics in the preparation of semiconductor materials. I. Epitaxial gallium-V compounds. *Journal of the Electrochemical Society* 116 (12): 1725–1732.
3. Manasevit, H.M. and Simpson, W.I. (1971). Use of metal-organics in the preparation of semiconductor materials. II. II-VI compounds. *Journal of the Electrochemical Society* 118 (4): 644–647.

4. Manasevit, H.M. and Simpson, W.I. (1972). Use of metallorganics in the preparation of semiconductor materials. V. Formation of indium-Group V compounds and alloys. *Journal of the Electrochemical Society* 120 (1): 135–137.
5. Ruehrwein, R. (1967). Production of epitaxial films of semiconductor compound material. US Patent 3 312 570, filed May 29, 1961.
6. Miederer, W., Ziegler, G., and Dotzer, R. (1964). Method of crucible-free production of gallium arsenide rods from alkyl galliums and arsenic compounds at low temperatures. German patent 1 176 102, filed September 25, 1962.
7. Seki, Y., Tanno, K., Iida, K., and Ichiki, E. (1975). Properties of epitaxial GaAs layers from a triethyl gallium and arsine system. *Journal of the Electrochemical Society* 122 (8): 1108–1112.
8. Stringfellow, G.B. (1999). *Organometallic Vapor Phase Epitaxy: Theory and Practice*, 2e. New York: Academic Press.
9. Jones, A.C. and O'Brien, P. (1997). *CVD of Compound Semiconductors*. New York: VCH.
10. Shenai-Khatkhate, D.V., Parker, M.B., McQueen, A.E.D. et al. (1990). Organometallic molecules for semiconductor fabrication. *Philosophical Transactions of the Royal Society of London A* 330: 173–182.
11. Shenai, D.V., Timmons, M.L., DiCarlo, R.L. et al. (2003). Correlation of vapor pressure equation and film properties with trimethylindium purity for the MOVPE grown III-V compounds. *Journal of Crystal Growth* 248: 91–98.
12. Shenai-Khatkhate, D.V., Manzik, S.J., and Wang, Q.M. (2012). Precursor compositions and methods. US Patent 8 142 847.
13. Wilkinson, G., Stone, F.G.A, and Abel, E.W. (1982). *Comprehensive Organometallic Chemistry. The Synthesis, Reactions, and Structures of Organometallic Compounds*, volumes 1–9. Oxford: Pergamon.
14. Wilkinson, G., Stone, F.G.A, and Abel, E.W. (1995). *Comprehensive Organometallic Chemistry II*. New York: Pergamon.
15. Cotton, F.A., Wilkinson, G., Murillo, C.A., and Bochmann, M. (1999). *Advanced Inorganic Chemistry*, 6e. New York: Wiley.
16. Crabtree, R.H. and Mingos, D.M.P. (2007). *Comprehensive Organometallic Chemistry III*. New York: Elsevier.
17. Elschenenbroich, C. and Salzer, A. (2006). *Organometallics, A Concise Introduction*. 3e. Weinheim, Germany: Wiley-VCH.
18. Shenai-Khatkhate, D.V. and DiCarlo, R.L. (2008). Organometallic compound purification. US Patent 7 321 048.
19. Shenai-Khatkhate, D.V. and Manzik, S.J. (2015). Organometallic compound purification. US Patent 8 962 893.
20. Shenai-Khatkhate, D.V., Power, M.B., and DiCarlo, R.L. (2004). Trialkylindium preparation. US Patent 6 770 769.
21. Shenai-Khatkhate, D.V. and Power, M.B. (2006). Metal-containing compound purification. US Patent application number 2006-0226075.
22. Jones, A.C., Holliday, A.K., Cole-Hamilton, D.J. et al. (1984). Routes to ultra-pure alkyls of indium and gallium and their adducts with ethers, phosphines and amines. *Journal of Crystal Growth* 68 (1): 1–9.
23. (a) Bradley, D., Chudzynska, H., and Faktor, M. (1990). Preparation of metal alkyls. US Patent RE 33,292. (b) Moore A.H., Scott M.D., and Davies J.J. (1988). High mobility InP epitaxial layers prepared by atmospheric pressure MOVPE using trimethylindium dissociated from an adduct with 1,2-bis(diphenylphosphino)ethane. *Journal of Crystal Growth* 77 (1–3): 19–22.
24. Shenai-Khatkhate, D.V., Webb, P., Cole-Hamilton, D.J. et al. (1988). Ultrapure organotellurium precursors for the low temperature MOVPE growth of II/VI compound semiconductor. *Journal of Crystal Growth* 93 (1–4): 744–749.
25. Foster, D.F., Rushworth, S.A., Cole-Hamilton, D.J. et al. (1988). Synthesis and thermal properties of adducts of trimethylindium with nitrogen-containing Lewis bases. *Journal of the Chemical Society Dalton Transactions*: 7.
26. Coates, G.E. and Green, S.I.E. (1962). Beryllium complexes with bipyridyl and other chelate donor groups: evidence for the bipyridyl anion as donor. *Journal of the Chemical Society*: 3340.
27. Shenai-Khatkhate, D.V., Orrell, E.D., Mullin, J.B. et al. (1986). Preparation and purification of metal alkyls for use in the MOCVD growth of II/VI compound semiconductors. *Journal of Crystal Growth* 77 (1–3): 27–31.
28. Shenai-Khatkhate, D.V. and Amamchyan, A. (2007). Method for making organometallic compounds. US Patent 7 166 734.
29. Shenai-Khatkhate, D.V., Power, M.B., and Amamchyan, A. (2003). Trialkyl Group VA metal compounds. US Patent 6 660 874.
30. Shenai-Khatkhate, D.V. (2003). Preparation of Group IVA and Group VIA compounds. US Patent 6 660 874.
31. Spencer, L.P., Amamchyan, A., Shenai-Khatkhate, D. V., and Stevens, J.C. (2014). Production of tri-alkyl compounds of Group 3A metals. PCT Int. Appl. WO A1.
32. Karch, R., Rivas-Nass, A., Frey, A. et al. (2015). Process for preparing trialkylgallium compounds. US Patent 9 108 985.
33. Karch, R., Rivas-Nass, A., Frey, A. et al. (2017). Process for preparing trialkyl compounds of metals of Group IIIA. US Patent 9 695 201.
34. Shenai-Khatkhate, D.V., Ware, R.A., DiCarlo, R.L. et al. (2006). Exceptionally stable vapor delivery of trimethylindium under intense OMVPE growth conditions. *Journal of Crystal Growth* 287 (2): 679–683.
35. Eisch, J.J. (1962). Organometallic compounds of Group III. I. The preparation of gallium and indium alkyls from organoaluminum compounds. *Journal of the American Chemical Society* 84 (19): 3605–3609.
36. Power, M.B. and Shenai-Khatkhate, D.V. (1998). High purity trimethylindium, method for synthesis. US Patent 5 756 786.
37. Shenai-Khatkhate, D.V., Timmons, M.L., and Buchler, R. (2002). Presentation at CS-Max Conference. *CS-Max 2002 Technical Digest*.

38. Lease, A.B., Williams, G., Smith, L.M. et al. (2002). Process and apparatus for the production of organometallic compounds. US Patent 6 495 707.
39. Lipiecki, F.J., Maroldo, S.G., Shenai-Khatkhate, D.V., and Ware, R.A. (2010). Method of preparing organometallic compounds. US Patent 7 659 414.
40. Lipiecki, F.J., Maroldo, S.G., Shenai-Khatkhate, D.V., and Ware, R.A. (2011). Method of preparing organometallic compounds. US Patent 7 919 638.
41. Lipiecki, F.J., Maroldo, S.G., Shenai-Khatkhate, D.V., and Ware, R.A. (2011). Method of preparing organometallic compounds. US Patent 7 939 684.
42. Lipiecki, F.J., Maroldo, S.G., Shenai-Khatkhate, D.V., and Ware, R.A. (2012). Method of preparing organometallic compounds. US Patent 8 101 787.
43. Lipiecki, F.J., Maroldo, S.G., Shenai-Khatkhate, D.V., and Ware, R.A. (2010). Purification process using microchannel devices. US Patent 8 101 048.
44. (a) Dow Metalorganic Technologies. (2015). Metalorganic technologies: precursors for epitaxial deposition. In: product catalog. (b) Shenai-Khatkhate D.V., DiCarlo R.L., and Ware R.A. (2008). Accurate vapor pressure equation for trimethylindium in OMVPE. *Journal of Crystal Growth* 310 (7–9): 2395–2398.
45. Plass, C., Heineckie, H., Kayser, O. et al. (1988). A comparative study of $Ga(CH_3)_3$, $Ga(C_2H_5)_3$ and $Ga(C_4H_9)_3$ in the low pressure MOCVD of GaAs. *Journal of Crystal Growth* 88 (4): 455–464.
46. Wang, C. (1997). Organometallic vapor phase epitaxial growth of AlSb-based alloys. *Journal of Crystal Growth* 170 (1–4): 725–731.
47. Jones, A.C., Lane, P.A., Martin, T. et al. (1992). New metalorganic gallium precursors for the growth of GaAs and AlGaAs by CBE. *Journal of Crystal Growth* 124 (1–4): 81–87.
48. Knauf, J., Schmitz, D., Strauch, G. et al. (1988). Comparison of ethyldimethylindium (EDMIn) and trimethylindium (TMIn) for GaInAs and InP growth by LP-MOVPE. *Journal of Crystal Growth* 93 (1–4): 34–40.
49. Shenai-Khatkhate, D.V. and DiCarlo, R.L. (2004). Organoindium compounds. US Patent 6 680 397.
50. Frigo, D.M., Werefridus, W., Massen, W.A.H. et al. (1992). A method for dosing solid sources for MOVPE: excellent reproducibility of dosimetry from a saturated solution of trimethylindium. *Journal of Crystal Growth* 124 (1–4): 99–105.
51. Shenai-Khatkhate, D.V., DiCarlo, R.L., Marsman, C.J. et al. (2007). Stable vapor delivery of solid sources in MOVPE of III-V compound semiconductors. *Journal of Crystal Growth* 298 (2): 176–180.
52. Shenai-Khatkhate, D.V., Timmons, M.L., Marsman, C.J. et al. (2010). Delivery device. US Patent 7 722 720.
53. Shenai-Khatkhate, D.V. (2010). Delivery device. US Patent 8 277 886.
54. Shenai-Khatkhate, D.V. (2010). Delivery device. US Patent 8 277 887.
55. Shenai-Khatkhate, D.V., Goyette, R.J., DiCarlo, R.L., and Dripps, G. (2004). Environment, health and safety issues for sources used in MOVPE of compound semiconductors. *Journal of Crystal Growth* 272 (1–4): 816–821.
56. Bhat, R. and Caneau, C.G. (2015). A study of the growth rate of GaSb using TEGa and TMSb or TESb. *Journal of Crystal Growth* 426: 82–85.
57. Biefeld, R.M. (1993). The growth of InSb using alternative organometallic Sb sources. *Journal of Crystal Growth* 128 (1–4): 511–515.
58. Shin, J., Verma, A., Stringfellow, G.B., and Gedridge, R.W. (1995). Growth of GaSb using trisdimethylaminoantimony. *Journal of Crystal Growth* 151 (1–2): 1–8.
59. Shenai, D.V., Timmons, M.L., DiCarlo, R.L., and Marsman, C.J. (2004). Correlation of film properties and reduced impurity concentrations in sources for III-V MOVPE using high purity trimethylindium and tertiarybutylphosphine. *Journal of Crystal Growth* 272: 603–608.
60. Mashita, M., Ishikawa, H., and Izumiya, T. (1995). Comparative study on carbon incorporation in MOCVD AlGaAs layers between arsine and tertiarybutylarsine. *Journal of Crystal Growth* 155 (3–4): 164–170.
61. (a) Hohnsdorf, F., Koch, J., Agert, C., and Stolz, W. (1998). Investigations of (GaIn)(NAs) bulk layers and (GaIn)(NAs)/GaAs multiple quantum well structures grown using tertiarybutylarsine (TBAs) and 1, 1-dimethylhydrazine (UDMHy). *Journal of Crystal Growth* 195 (1–4): 391–396.; (b) Onabe K. (1997). MOVPE growth and optical characterization of GaPN metastable alloy semiconductor. *Materials Research Society Symposium Proceedings* 449 (III-V Nitrides): 23–34.
62. Sawara, K., Kadokura, H., and Yako, T. (1990). Process for producing trialkylarsenic compounds. US Patent 4 906 762.
63. Shenai-Khatkhate, D.V., Amamchyan, A., Power, M.B. et al. (2005). Alkyl Group VA metal compounds. US Patent 6 939 983.
64. Shenai-Khatkhate, D.V., Power, M.B., Amamchyan, A., and DiCarlo, R.L. (2005). Alkyl Group VA metal compounds. US Patent 6 956 127.
65. Zimmermann, G., Spika, Z., Marschner, T. et al. (1994). GaAs substrate pretreatment and metalorganic vapour phase epitaxy of GaAs, AlAs and (AlGa) As using β-eliminating trialkyl-As precursors. *Journal of Crystal Growth* 145 (1–4): 512–519.
66. Mullin, J.B., Cole-Hamilton, D.J., Orrell, E.D. et al. (1989). Preparation of Group II metal alkyls. US Patent 4 812 586.
67. Irvine, S.J.C. and Mullin, J.B. (1981). The growth by MOVPE and characterization of $Cd_xHg_{1-x}Te$. *Journal of Crystal Growth* 55 (1): 107–115.

68. Irvine, S.J.C. Tunnicliffe, J., and Mullin, J.B. (1984). The growth of highly uniform cadmium mercury telluride by a new MOVPE technique. *Materials Letters* 2 (4b): 305.
69. Shenai-Khatkhate, D.V. (1990). Tellurium and selenium source review for MOVPE. Morton Technical Bulletin.
70. (a)Irvine, S.J.C., Mullin, J.B., Hill, H. et al. (1988). Photo-stimulated II–VI crystal growth: A study of low temperature epitaxy. *Journal of Crystal Growth* 86 (1–4): 188–197. (b)Irvine, S.J. C., Hill, H., Brown, S.J. et al. (1989). Selected area epitaxy in II-VI compounds by laser-induced photo-metalorganic vapor phase epitaxy, *Journal of Vacuum Science and Technology* B7 (5): 1191–1199.
71. Thompson, J., Mackett, P., Smith, L.M. et al. (1988). The growth of CdHgTe by MOCVD at reduced temperatures. *Journal of Crystal Growth* 86 (1–4): 233–239.
72. Irgolic, K.J. (1974). *The Organic Chemistry of Tellurium*. New York: Gordon Breach.
73. Mullin, J.B., Cole-Hamilton, D.J., Shenai-Khatkhate, D.V., and Webb, P. (1992). Method for preparation of dialkyltellurium and dialkylselenium. US Patent 5 091 570.
74. Mullin, J.B., Cole-Hamilton, D.J., Shenai-Khatkhate, D.V., and Webb, P. (1992). Method for purification of tellurium and selenium alkyls. US Patent 5 117 021.
75. Shenai, D.V., DiCarlo, R.L., Power, M.B. et al. (2007). Safer alternative liquid germanium precursors for relaxed graded SiGe layers and strained silicon by MOVPE. *Journal of Crystal Growth* 298: 172–175.
76. Woelk, E., Shenai-Khatkhate, D.V., DiCarlo, R.L. et al. (2006). Designing novel organogermanium OMVPE precursors for high purity germanium films. *Journal of Crystal Growth* 287: 684–687.
77. (a) Shenai-Khatkhate, D.V. (2011). Doping method. US Patent 7 989 323. (b) Shenai-Khatkhate, D.V. and Power, M.B. (2008). Method of depositing a metal containing film. US Patent 7 413 776.
78. (a) OSHA. (2017). CFR Title 29 Part 1910.1000. (b) OSHA. (2018). ACGIH TLV and BEIs.
79. Lewis, R.J. (2012) *Sax's Dangerous Properties of Industrial Materials*, 12e, volumes 1–5. New York: Wiley.
80. Rose, V.E. and Cohrssen, B. (2010). *Patty's Industrial Hygiene*, 6e, volumes 1–4. New York: Wiley.
81. Pohanish, R.P. and Greene, S.A. (2009). *Wiley Guide to Chemical Incompatibilities*, 3e. New York: Wiley.

14

Future Aspects of MOCVD Technology for Epitaxial Growth of Semiconductors

T. Detchprohm[1], J.-H. Ryou[2], X. Li[3], and R.D. Dupuis[1]
[1]Georgia Institute of Technology, School of Electrical and Computer Engineering, Atlanta, USA
[2]Department of Mechanical Engineering, The University of Houston, Texas, USA
[3]Electrical Engineering Program, Computer, Electrical, Mathematical Science and Engineering Division, King Abdullah University of Science and Technology (KAUST), Thuwal, Makkah, Saudi Arabia

14.1 Introduction – Looking Back

To imagine the future, it is often instructive to look back in time to examine history. Accordingly, we first take a look at the vector of the development of semiconductor materials technologies in the past. Naturally occurring semiconductor materials such as metal sulfides (e.g. lead sulfide, zinc sulfide, or zincblende) were widely phenomenologically known since Ferdinand Braun's pioneering work on rectifiers in 1874 [1]. These materials became widely used as "cat's whisker" rectifiers in the early twentieth century after the commercial development of radio, where such devices were used as crystal receivers. However, the first theoretical understanding of these *halbleiter* (a German word meaning "half-conductor") materials was not developed until 1931 by Alan Wilson [2, 3]. Even with this theoretical model, based upon the "new" quantum mechanics, Wilson was not certain that semiconductors really existed due to the presence of large concentrations of impurities in experimentally available semiconductors that tended to mask their intrinsic properties [4]. In that same year, 1931, Wolfgang Pauli, in a letter written to Rudolf Peierls, famously speculated that semiconductors might not even exist [5]. The application of these mysterious materials in controlled experiments and for modern semiconductor devices did not occur until around 1940, when Russell Ohl, an electrical engineer working at Bell Labs in Holmdel NJ, investigated the purification of Si and "discovered' the *p-n* junction [6] and the solar cell [7]. (See also Riordan and Hoddeson [8].)

Since these early works, many researchers have reported on the discovery of new semiconductors. Many important contributions in this field have focused on understanding and then controlling

Metalorganic Vapor Phase Epitaxy (MOVPE): Growth, Materials Properties, and Applications,
First Edition. Edited by Stuart Irvine and Peter Capper.

the properties of these materials. Much of this understanding has resulted from the constant improvement in the purity and control of the crystalline structures in which semiconductor devices are fashioned. Among the most important advances in the technology of semiconductors was the purification of the elemental sources and the development of chemical transport technologies for creating single-crystal films of the semiconductor of interest having controlled electrical, optical, and structural properties. This early work on understanding the properties of purified semiconductors set the stage for the semiconductor revolution beginning with the invention of the bipolar transistor (in purified Ge) by John Bardeen and Walter Brattain at Bell Labs in December 1947 [9, 10].

In 1960–61, the first processes were developed for the vapor-phase epitaxial (VPE) growth of Si [11], Ge [12], GaAs [13], and $GaAs_xP_{1-x}$ [14] using halogen transport. A few years later, Nelson first demonstrated liquid-phase epitaxial growth of GaAs films in 1963 [15]. In 1961, Holonyak grew the first III-V alloy semiconductor heteroepitaxial layers of any kind (in this case, alloys of $GaAs_xP_{1-x}$) by using closed-tube VPE [14, 16, 17]; and in 1964, closed-tube VPE growth of $InAs_xP_{1-x}$ alloys [18] was demonstrated, further expanding the variety of semiconductor materials that could be studied. Much of this research and development was driven by the interest in creating improved and new types of electronic and optoelectronic devices. Initially, the main device interest for III-V compound semiconductors was in the application to tunnel and laser diodes.

In the 1960s, III-V compound semiconductors became of intense interest owing to the fact that many were discovered to have direct bandgaps; and in some cases, they had very high electron mobilities relative to Si and Ge. Hall et al. reported the first demonstration of GaAs-diffused *p-n* junction infrared injection laser diodes (at 77 K) in September 1962 [19]. Subsequently, Holonyak et al. [20] demonstrated in October 1962 laser diode operation at 77 K in the visible red spectral region from $GaAs_xP_{1-x}$ *p-n* diffused junctions grown by closed-tube VPE. After these seminal results were published, the interest in improving the quality of III-V semiconductors and in the control of doping intensified. During this period, various VPE technologies for the open-tube vapor-phase growth of III-V semiconductors were developed, largely based upon the use of halogen or halide transport mechanisms [21–23] similar to that previously used for the open-tube VPE growth of Si using $SiCl_4$ [24].

In the period 1965–1970, several companies commercialized open-tube VPE processes to manufacture red LEDs, and VPE became the most common epitaxial technology used for the growth of III-V alloys of $GaAs_xP_{1-x}$, while liquid-phase epitaxy (LPE) became the most commonly used technology for $Al_xGa_{1-x}As$ alloys. $Al_xGa_{1-x}As$ films (and other Al-containing semiconductors, e.g. AlInP) were particularly difficult to grow by VPE because of the high reactivity of Al for silica and oxygen at high temperatures. Because of interest in III-V alloy growth, in 1967, Rupprecht et al. [25] modified the LPE growth of GaAs to grow $Al_xGa_{1-x}As$ alloys. LPE $Al_xGa_{1-x}As$ technology was subsequently first demonstrated for the growth of $Al_xGa_{1-x}As$-GaAs heterostructure injection lasers by Alferov et al. [26]. Thus, by the end of the 1960s, open-tube VPE and LPE were the dominant technologies being developed and improved for the growth of most III-V materials and devices with most companies worldwide devoting their resources to these technologies.

While open-tube VPE [27] and LPE [28] were quite successfully commercialized for the growth of certain elemental, binary, and ternary epitaxial films on various single-crystal semiconductor substrates, these technologies were not applicable to the heteroepitaxial growth of semiconductor films on insulating oxide substrates. While this subject was not a significant economic factor in the semiconductor devices of the 1970s, it has become an important technology of intense interest for the current era.

In the late 1960s, many III-V electronic devices, e.g. ion-implanted GaAs- and InP-based field-effect transistors (FETs), became of increasing interest and importance, and the application of these devices to high-frequency electronics required the use of high-resistivity GaAs and InP substrates. Because of the relatively high concentration of unintentional impurities in commercially available bulk GaAs and InP Czochralski-grown single crystals, these insulating substrates were created by doping with deep acceptors during growth, typically with Cr for GaAs or Fe for InP, to create deep electron traps that could render the crystals semi-insulating. However, these substrates with large, deep impurity concentrations were not stable or reliable, and the as-grown bulk crystal ingot often contained regions with markedly different resistivities. Furthermore, the high-resistivity characteristic was not stable under high-temperature processing or temperature cycling. Consequently, many companies made efforts to improve the properties of these III-V substrate materials.

Another approach, developed extensively by Manasevit et al. [29–34] used an alternative open-tube VPE process that which did not use halogen or halide transport. This process he termed *metalorganic chemical vapor deposition* (MOCVD) because it employed metalorganics for the sources of the Column III elements and hydrides for the Column V elements. Manasevit chose to use MOCVD because he was primarily interested in heteroepitaxy or, more generally, chemical vapor deposition (CVD) of semiconductor films on insulating oxide substrates. The use of single-crystal insulating oxide substrates, e.g. sapphire (Al_2O_3), spinel ($MgAl_2O_4$), beryllium oxide (BeO), etc., would solve the nonreproducible electrical conductivity and conversion problems observed for semi-insulating GaAs and InP substrates.

A few years after Manasevit began studying the heteroepitaxial growth of III-Vs on insulating oxide substrates using MOCVD, another epitaxial process for growth of III-V semiconductors, molecular-beam epitaxy (MBE), was developed by Arthur and LePore [35], Cho [36, 37], Foxon [38], and others. This process offered precise control of layer thicknesses and doping as well as the flexibility to grow a wide variety of binary and alloy semiconductor materials, including $Al_xGa_{1-x}As$, and many research laboratories adopted this process for the study of advanced semiconductor device structures, e.g. heterostructure lasers.

Owing to the emphasis on VPE, LPE, and MBE for III-V semiconductor production and research, the efforts to develop MOCVD for the growth of III-V compound heteroepitaxial structures on oxide-insulating substrates were not given broad emphasis by researchers in the period 1968–1978. Furthermore, the purity of the commercially available metalorganic precursors and their tendency to oxidize once purified (they are generally pyrophoric) caused problems in the growth of high-quality heterostructures, especially for the growth of Al-containing alloys. However, Dupuis et al.'s [39] successful demonstration of the MOCVD growth of low-threshold AlGaAs-GaAs heterostructure lasers at 300 K and solar cells in 1977 [40] caused many researchers to take another look at MOCVD for III-V epitaxial growth. The expansion of worldwide research in MOCVD during the 1980s–1990s led ultimately to the acceptance of this technology for the commercial production of a wide variety of III-V materials and devices and to today's dominance of MOCVD for the growth of virtually all III-V semiconductor devices.

The overarching themes in the development and understanding of semiconductor devices since their discovery in 1874 has been the development of improved materials, the improvement and control of material properties, and the control of device structures. At this time, MOCVD offers the most optimal characteristics to achieve the goals of improved control of the epitaxial growth of III-V compound semiconductor materials and in the commercial production of related semiconductor devices. Currently, MOCVD is widely used for the growth of virtually all III-V compound semiconductor devices, especially for large-scale commercial applications, e.g. LEDs, injection laser diodes, optical waveguide modulators, photodiodes, solar cells, heterojunction field-effect transistors (HFETs), and bipolar junction transistors (BJTs), as well as more advanced devices, e.g. quantum-cascade lasers (QCLs) and optoelectronic and photonic integrated circuits (PICs). In this chapter, we explore the characteristics of MOCVD and discuss possible future trends in the development and application of this epitaxial materials technology to future electronic and optoelectronic devices.

14.2 Future Equipment Development

MOCVD is arguably the most widely employed technology to produce high-quality compound semiconductor (CS) thin-film devices. Thus, the continuous innovation and evolution of MOCVD materials technologies are critical for further development of CS device technologies. With decades of research and development, the CS technologies have become indispensable parts of everyday life in developed societies. However, many CS technologies are still in the early stages of research and applications. While some CS semiconductors are being produced and employed, the majority are still immature and even in infancy, such as certain III-nitride technologies. Thus, the requirement for MOCVD innovation can be highly varied depending upon the specific application. This section will focus on the needs of different III-nitride materials and devices as examples to describe the state-of-the-art and emerging MOCVD technologies. In addition, MOCVD innovation for emerging related materials such as Ga_2O_3 will be discussed.

14.2.1 Production MOCVD

The current production MOCVD systems are highly complex and integrate electrical, control, mechanical, and chemical expertise. Moreover, the system has to possess superior reliability, high productivity, and run-to-run consistency. Thus, only a handful of companies were capable of designing and making successful MOCVD reactors one decade ago. The global MOCVD market was then dominated by Veeco and AXITRON, with Taiyo Nippon Sanso (TNS) being in a distant third place with single-digit market share. However, the huge investment in blue LED manufacturing in China around 2010 has altered the competition landscape dramatically. Although Veeco and AXITRON sold a record number of MOCVD systems within a short period of time, this sales bubble oversaturated the market and has led to years of MOCVD business downturn rippling through to today. More significantly, the unprecedented revenue and profit attracted dozens of new MOCVD equipment manufacturers around the world, including established semiconductor equipment giants such as Applied Materials. Many of them have subsequently ceased MOCVD system manufacture amid competition. Others have barely survived with a small number of shipments per year. However, some have grown to become major players since 2013. AMEC and TOPEC, for instance, which launched their first MOCVD systems around 2013, have grabbed considerable market share from Veeco in blue LED production in China, which is the largest market for MOCVD systems in the world. In comparison, AIXTRON's products for blue LED production have lost ground, and this partially contributed to its attempt to put itself on sale in 2016, although its reactors are popular for III-As-P and GaN-on-Si device production. With competition, the price of MOCVD systems has faced strong headwinds, falling from over $2 million USD per reactor not too long ago to around $1.5 million USD today. This forced the MOCVD equipment companies into an innovation race to develop new technologies for differentiation and customer attraction. This is beneficial for the development of the MOCVD technology in general as compared to the duopoly market controlled by Veeco and AIXTRON a few years ago. Additionally, lawsuits over intellectual property (IP) are more common to gain competitive advantage, as exemplified by the recent case filed by Veeco against SGL Carbon over SGL's alleged sale of IP-contained wafer carriers to Veeco's competitors [41].

Despite the competition in the MOCVD business, the solid-state lighting and III-V power device markets will continue to grow strongly in the long term. Also, the MOCVD systems sold around 2010 are being replaced with newer systems for better productivity and device performance. Thus, the MOCVD market is expected to slowly revive. To compete for market share, manufacturers' efforts focus on improving the productivity and flexibility of MOCVD systems for lower operational cost and thus higher profitability. This is in particular obvious in the blue LED production

Table 14.1 The wafer capacity per reactor of state-of-the-art MOCVD production systems from major MOCVD companies.

Wafer capacity	Veeco EPIK700	AIXTRON R6	AMEC A7	TOPEC ProMaxy 168	TNS UR-26 K
4 inch	31	31	34	36	NA
6 inch	12	12	14	15	10

application, as the fierce competition has transformed the LED into a commodity item, and the price has been driven down continuously. To achieve cost-performance goals, it is required that the new MOCVD systems accommodate more and larger wafers, change wafer size easily, become more efficient for utilities and precursor consumption, achieve better wafer uniformity and wafer-to-wafer consistency, be equipped with more advanced in situ monitors, have a smaller footprint and improved serviceability, and be more integrated with automation techniques, and all of these need to be accomplished in a very economical way. Table 14.1 shows the 4-inch (100-mm) and 6-inch (150-mm) wafer capacity of the current state-of-the-art MOCVD systems, most of which can accommodate 8-inch (200-mm) diameter wafers as well. It is expected that the single-wafer systems will eventually prevail in the future as the industry moves to increasingly larger wafer size as has been the case for Si epitaxial growth.

Moreover, we could see more device-oriented customization for the development of the production systems, as various devices have different growth requirements. For instance, the InGaN micro-LED technology has attracted considerable attention due to its important application in high-definition displays, as elaborated in Section 14.4. The production of the micro-LED device structure still relies on MOCVD. However, it has stricter requirements for high yield and uniformity due to much smaller device size and larger device density. Specifically, the particle contamination needs to be minimized. In addition, wavelength uniformity needs to be even better. The major MOCVD manufacturers have not launched specific products for the micro-LED market yet, since it still takes time for R&D and the micro-LED market has not yet reached prime time. Thus, most of the micro-LEDs are being produced by conventional blue LED MOCVD reactor systems. However, it is expected that the targeted MOCVD systems for this market will be available with further demonstration of the micro-LED's potentials.

14.2.2 R&D MOCVD

As the major MOCVD companies' revenue is from the sale of production systems, they have not invested heavily in the development of R&D systems. Thus, there has not been much new development, and the number of models from the major MOCVD manufacturers is limited. Among the available ones, most are minor modifications based on the legendary systems, such as AIXTRON's EpiLab and TNS's SR series. The only R&D system supplied by Veeco is the Propel reactor launched in 2014, but it is aimed at development of GaN-on-Si technologies specifically.

As universities and research laboratories still have continuing demands for R&D MOCVD systems for emerging materials and devices, the lack of products from the major players means opportunities for the smaller MOCVD reactor manufacturers, from equipment sales to modifications. These companies have employed different strategies. For instance, Structured Materials Industries (SMI) builds standard, custom, and recommissioned MOCVD systems and provides service through an application laboratory. Agnitron develops some in-house MOCVD models but mainly focuses on modifying legendary systems from Veeco and AIXTRON/Thomas Swan. EpiQuest and NuFlare Technology develop and manufacture their own high-temperature and single-wafer systems, respectively.

Because of limited resources, the smaller MOCVD companies that are active in the R&D market may not be able to keep investing heavily for their own R&D. After all, MOCVD is a highly complex process. Optimized reactor design requires in-depth research on thermodynamics, surface processes, and hydrodynamics as well as extensive testing. Even trial-and-error development would demand numerous resources, since iterative hardware modification is needed and costly. On the other hand, the major manufacturers will not focus on the R&D market as long as the production tools provide most of their revenue and their profit margin is thin. This situation may delay the development of innovative MOCVD technologies for emerging materials and cause "the valley of death" for the corresponding high-performance devices grown by MOCVD. In the long term, it would eventually impact the production MOCVD market when customers look for products for new types of materials and devices. Unfortunately, few companies and research laboratories are conducting MOCVD equipment research for emerging materials. The effect is already seen in Ga_2O_3 R&D activities: few MOCVD systems can produce high-quality Ga_2O_3 epitaxial films on non-native and economical substrates like sapphire due to the lack of R&D activities. Because the equipment technology is often the first step toward high-performance devices, government and funding agencies ought to step up to encourage more MOCVD equipment research to ensure rapid development of emerging compound semiconductor technologies.

14.2.3 MOCVD for Ultrawide-Bandgap III-Nitrides

The ultrawide-bandgap alloys such as Al(Ga)N and B(Al,Ga)N are arguably the most researched III-nitrides today for critical device applications such as UV optical sources and detectors and high-power electronics. Thus, the III-N MOCVD processes are extensively studied topics. However, growing these materials by MOCVD on commercially viable substrates is challenging primarily due to strong premature reactions between Al- and B-containing precursors and NH_3 as well as low mobility of Al and B adatoms on the epitaxial surface. This can lead to high dislocation densities and thus poor device performance. To alleviate this issue, researchers have generally utilized one, two, or all of three methods: extremely high temperature (EHT) (>1500 °C), epitaxial lateral overgrowth (ELO), and pulsed atomic layer epitaxy (PALE). These have led to relatively improved material quality. However, most discussions in academic research publications and at technical conferences has focused primarily on the method and the resulting material quality. They have seldom elaborated on whether the proposed approach with the used MOCVD reactor could eventually be employed toward practical and economical mass production of the materials and devices.

For instance, ELO requires pre-epitaxial or post-epitaxial etching regardless of the pattern size, which elongates the process time and adds cost. In addition, if the micropattern is used, a thick film has to be deposited to achieve coalescence, which further increases the process time and cost. For PALE, the growth rate would have to be reduced considerably in exchange for longer adatom movement, reducing productivity. For EHT, the MOCVD reactor generally utilizes a multizone resistance heater or an RF induction heater. The resistance heater has better temperature uniformity control. However, it is not ideal for EHT growth since it is not as stable and reliable as induction heaters at higher temperatures. Furthermore, the susceptor will partially reflect thermal radiation from the resistance heater, which reduces heating efficiency. This eventually leads to higher cost for the EHT MOCVD reactor using a resistance heater. For the induction heater, the conventional MOCVD system installs the coil outside the quartz chamber to heat up the wafer susceptor. However, because of the required gap between the coil and the susceptor, there exists an uncoupled magnetic field above the susceptor surface. Not only does the uncoupled magnetic field compromise heating efficiency, but also it can couple with the gas inlet. Specifically, the gas inlets, often made of metallic stainless steel, cannot be too close to the susceptor surface to avoid being overheated. Because of the resulting longer precursor flow path, this design can lead to considerable premature reactions, compromising the material quality and growth efficiency. Regarding premature reaction suppression amid the growth of ultrawide-bandgap III-nitrides, AIXTRON's close coupled showerhead (CCS) design may be the most superior today as evidenced by high precursor and growth efficiency [42, 43]. However, it is still not ideal, as it is difficult to clean the Al- or B-rich deposition on the wafer carrier after the growth. In addition, the optimized growth conditions are susceptible to the reactor condition. The coating on the showerhead, susceptor, and liner can alter the material quality considerably even if the same growth recipe is used [44].

Therefore, the R&D MOCVD process for the ultrawide-bandgap III-nitrides is still compromised, impeding the transfer of the research progress to practical and economical mass production. However, this is also an opportunity for MOCVD equipment scientists and engineers to come up with innovative approaches to eliminate the compromise. The ideal R&D MOCVD system for the ultrawide-bandgap III-nitrides would have low premature reactions, high growth rate, good uniformity, and stable growth conditions.

14.2.4 MOCVD for Emerging Materials

Among the many materials that can be grown by MOCVD, as discussed in the next section, Ga_2O_3 and its alloys have received the most attention recently due to their superior material properties and thus potential applications in gas sensing, power electronic devices, transparent conductive oxides, and UV detectors. MOCVD and HVPE are two of the most promising epitaxial film-depositing techniques for high-quality Ga_2O_3 with higher growth rates needed especially in power devices. The challenges of realizing either process in an optimal commercial mode, while daunting, are not insurmountable.

HVPE using gallium chlorides typically offers very high deposition rates but often results in lower-quality material and can incorporate chlorine atoms that compromise electrical properties. MOCVD can produce high-quality films and devices on Ga_2O_3 native substrates [45]. However, a high growth rate is difficult to achieve. In addition, MOCVD-grown films often require post-growth annealing to improve crystallinity or activate dopants. Common MOCVD precursors like trimethylgallium (TMG) and triethylgallium (TEG) are pyrophoric and react strongly with oxygen, meaning they have a high likelihood of prereacting and require care in reactor introduction to minimize prereactions for the growth of high-quality Ga_2O_3 films. Increasing the growth temperature may be one way to increase growth rate and surface mobility. However, this also leads to more prereactions and even combustion of the metalorganics with the oxygen precursors under some conditions.

TMG generally leaves too much carbon in the films to realize viable doping for Ga_2O_3, whereas films grown with TEG have produced high electrical quality films. Films grown with more complex large metalorganic molecular complexes are likely to also add more carbon to the grown film. Process-enhancement techniques such as using a plasma, even if upstream and only plasma activating the oxidizing species, are likely to cause prereactions of carbon-containing molecular fragments that become incorporated in the film. Oxygen seems to be the preferred oxidizer at the present time, with water not showing significant growth improvement, ozone being so reactive as to limit growth, and NO_x compounds also potentially leading to nitric acid formation.

The Ga_2O_3 material itself is an important challenge. There are five different Ga_2O_3 phases, but only the beta phase has high-temperature stability. Hence, it is generally the most sought after. However, chemical-solution techniques appear to be successful in producing device-quality alpha-phase material under some conditions. *p*-Type Ga_2O_3 material has not been achieved in any phase, and theory has not predicted a viable *p*-type dopant. Substrates for Ga_2O_3 epitaxy are themselves challenging – there is presently no clear orientation winner. Different orientations, due to the nature of how Ga_2O_3 is grown, provide different scaling limitations. While there are 5-cm and the possibility 10-cm diameter substrates available in some orientations, other orientations are limited to dimensions of a few centimeters. Growth on lattice-mismatched sapphire has only resulted in films with defect levels so high as to prevent viable doped material. Among epitaxial growth techniques, MOCVD, and perhaps ALD or MBE, are also very promising for growing nucleation and buffer layers that may allow device quality Ga_2O_3 to be grown on more commonly available large-area substrates. Nevertheless, Ga_2O_3 is still in its early stages of development, but has demonstrated significant device results by several groups. Further innovation in MOCVD science and technology would certainly facilitate Ga_2O_3 and its alloys to realize their significant potential.

14.2.5 Democratization of MOCVD

MOCVD technology was invented in the United States 50 years ago by Manasevit et al. [29] and then briefly explored by others in research community. After the successful demonstration of high-performance lasers by MOCVD by Dupuis et al. in 1977 [39], researchers from other developed countries, especially Japan, started to adopt it. When MOCVD celebrated its fortieth birthday in 2008, some 25 375 MOCVD-related journal papers had been published. As shown in Figure 14.1, researchers in the United States and Japan were leading the chart, while the paper share of most of the other countries was single digits or less.

However, the situation is changing rapidly due to growing economic applications and the need for MOCVD research and production in various other countries. As evidenced by the study of MOCVD-related publications from 2014–2017 (Figure 14.2), China has caught up with United States in terms of publication percentage thanks to the Chinese government's huge investment in MOCVD R&D and manufacturing. Surprisingly, Japan has fallen considerably to third place with a publication rate of around 10% in spite of the fact that the 2014 Nobel Prize in Physics was awarded to three Japanese scientists for their pioneering work on III-N materials and LEDs. India is rising up into the top 10. Though not in the top 10, Poland, Sweden, Switzerland, and Tunisia have improved considerably as well. Even countries like Saudi Arabia, which did not produce any MOCVD-related papers in the first four or more decades since MOCVD was demonstrated, has started to show up in the list. However, it is important to note that most of the MOCVD and related industry is still highly concentrated in the top-ranked countries, which is a result of decades of investment in technology and human capital. Therefore, the emerging countries still need to make huge efforts to catch up in this regard.

Currently, MOCVD systems still bear a high price tag for initial purchase and day-to-day operation. However, with proliferation of MOCVD systems and the resulting refurbishing business as well as emerging MOCVD equipment companies, it is expected that the cost of MOCVD systems,

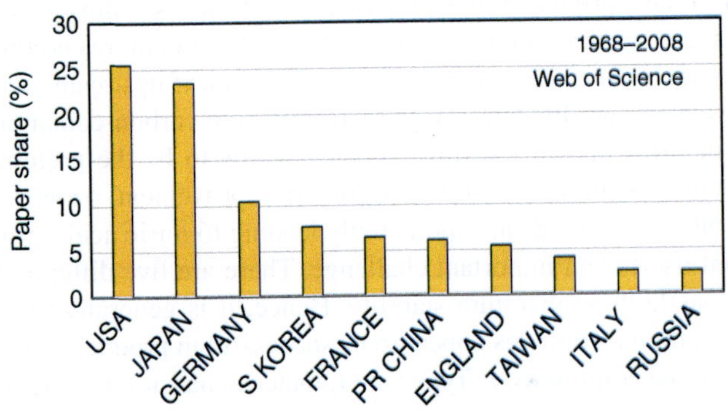

Figure 14.1 The paper share of the top 10 countries/regions in terms of MOCVD-related journal paper publication from 1968 to 2008. The total number of publications from all the countries was 25 375 in this period [46].

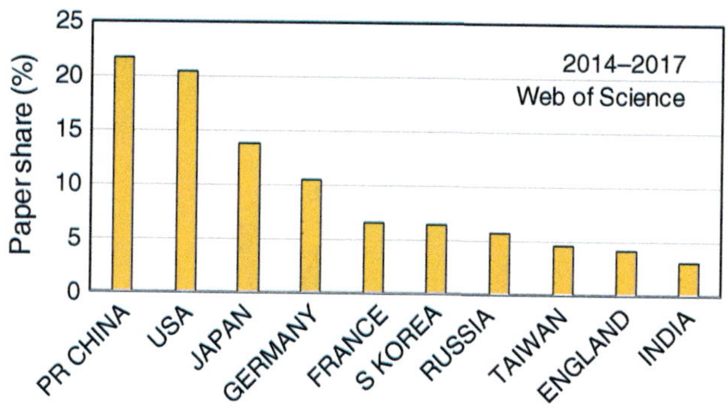

Figure 14.2 *The paper share of the top 10 countries/regions in terms of MOCVD-related journal paper publication from 2014–2017. The total number of publications from all the countries was 3978 in this period [44].*

especially the smaller R&D ones, will be more affordable in the future. In addition, more and more small-to-medium size companies are offering research epitaxial services. This situation will lead to democratization of MOCVD technologies that benefits the compound semiconductor researchers who do not have direct access to MOCVD facilities and expand the technology impact of MOCVD on a truly global scale, especially for developing countries. The democratization would in turn produce a larger high-quality labor force, leading to faster innovation of MOCVD technology the future.

14.3 Future Applications for MOCVD Research in Semiconductor Materials

14.3.1 Heteroepitaxy

14.3.1.1 III-V Semiconductor Materials Excluding III-Nitrides

For MOCVD growth of III-V semiconductor-based heterostructures, excluding III-N materials, the device performance strongly depends on the material quality. For instance, threading dislocation densities above $10^4\,cm^{-2}$ will significantly degrade the internal quantum efficiency of an optoelectronic device such as a light-emitting diode (LED). The current III-V materials commonly deployed in optoelectronic applications, including LEDs, laser diodes, photodetectors, and solar cells, have Group 13 elements Al, Ga, and In and Group 15 elements P, As, and Sb. This family of semiconductor materials form stable zincblende crystallographic structures. In order to mitigate or minimize defect formation in these materials, lattice-matching or near-lattice-matching alloys have been preferentially grown on their native bulk substrates such as GaAs, GaP, GaSb, InP, InAs, and InSb, offering low defect density. A strain-balance or graded-composition alloy structure is often employed to minimize strain in a target heterostructure of which the active part of the desired device is typically constructed with direct-bandgap alloys. The lattice mismatch (LMM) defined as LMM = [(in-plane lattice constant of substrate – in-plane lattice constant of epitaxial layer)/in-plane lattice constant of the epitaxial layer] serves as the gauge to select the substrate of choice for the heterostructure of interest. A zero value of LMM indicates the lattice-matching condition while positive, and negative values of LMM indicate tensile and compressive strain condition, respectively. In a progressive state of either tensile or compressive strain, a number of defects are generated, such as misfit dislocations. It is generally accepted that such defects are nucleated after the epitaxial layer is grown thicker than a critical layer thickness.

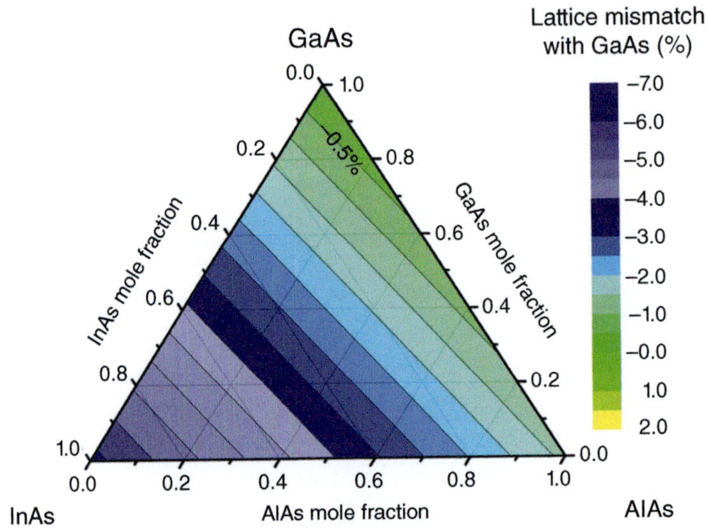

Figure 14.3 *Ternary contour map of LMM of AlInGaAs material on a GaAs substrate.*

The growth of heteroepitaxial layers using highly strained conditions often leads to the formation of threading dislocations and rough surface morphology, making it difficult to control an abrupt interface in the active part of the heterostructure, as well as achieve a highly uniform alloy composition and high doping efficiency. In III-V (V=As, P, Sb) systems, Ga-V and Al-V binary materials of the same type of compound are known to form a lattice constant relatively close to each other, while their In-V binary material counterparts typically have a much larger lattice constant. It is a significant challenge to combine a large In alloy composition in the heterostructure when a Ga-V binary substrate of the same compound is introduced as an epitaxial platform. Figure 14.3 is an exemplary mismatch situation of AlInGaAs on GaAs, in which the alloy materials fall under compressive strain for the entire composition while the LMM values are –1.3 and –6.7% for AlAs and InAs, respectively. Excluding the indirect bandgap of high Al-containing alloys, these AlInGaAs on GaAs materials form a type-I heterointerface and have been utilized mostly as GaAs-AlGaAs-based heterostructures for optoelectronic, high-speed electronic, photovoltaic applications. The corresponding optoelectronic devices can be designed to utilize both interband transitions and intraband transitions, i.e. between quantized levels in the conduction band of their heterostructure such as a quantum well (QW). For the latter, the GaAs-AlGaAs system provides the best performance for THz QCLs. However, the current target for these THz devices is to achieve room-temperature operation. Currently, the highest operating temperature is about 200 K [47]. For high-speed electronic applications, this material system was also used to initiate the research on high-electron-mobility transistors (HEMTs) or HFETs [48, 49]. Furthermore, the AlGaAs-GaAs system was used to realize the first high-performance heterojunction bipolar transistors (HBTs), allowing high doping to lower resistance in the base region by limiting minority carrier injection from the base into the emitter via a potential barrier formed at the heterointerface of the two regions [50, 51]. However, for higher-frequency applications, InGaAs/InP materials are preferred as a channel material since the free-carrier electron mobility improves significantly with the InAs mole fraction in this alloy system [52]. In the last two decades, InGaAs HEMTs grown on GaAs via strained graded-composition AlInGaAs quaternary layers, also known as metamorphic HEMTs (MHEMTs), have been developed for the sake of lower production cost and mechanical stability as compared to those grown on InP substrates [53]. Such devices demonstrate high-frequency operation with a cut-off frequency (f_T) in the range from 50 GHz up to ~700 GHz [54–56]. The gate length of 20 nm was reported for the record high-frequency MHEMT.

Due to the similarity in crystal structure of this III-V family, some substrates of one compound can, however, be utilized as a platform providing lattice-matching for a certain portion of another compound system. For instance, the quaternary AlInGaP can be grown strain-free on GaAs, as depicted in Figure 14.4. These materials in this strain-free or low-strain region are of direct bandgap and have been utilized for several applications such as LEDs, laser diodes, and triple-junction tandem solar cells in which the AlInGaP cell is designed to efficiently convert photons in the visible spectral region into electrons. There is no benefit to employing the high GaP or AlP mole fraction AlInGaP and AlGaP alloys for any active portion of an optoelectronic device due to their indirect bandgap properties. In addition, the use of MOCVD-grown InGaP as the wider-bandgap emitter layer has improved reliability and reduced health hazards for the production of III-V HBTs, as it has replaced Be-doped AlGaAs emitters that were grown by MBE [57].

In a similar manner, the larger-lattice-constant material system of AlInGaAs has been grown extensively by MOCVD heteroepitaxial growth on InP substrates, as shown in Figure 14.5. Unlike the previously mentioned heteroepitaxial system, the lattice-matching AlInGaAs alloys have smaller bandgap energies than the InP substrate, and thus optoelectronic devices with less complicated heterostructures have been invented without the issue of the substrate-induced absorption loss. With a wider selection of direct-bandgap alloys of InGaAs and InAlAs, there is vast potential to deploy this material system with several degrees of strain-balanced structures for optoelectronic devices covering the near-infrared (spectral region 800–1500 nm) via interband transitions, and midinfrared to THz (2.5 µm – 0.26 mm) spectral regions when deploying intraband transitions in QW heterostructures, i.e. between quantized levels in the conduction band. In addition, with the access to InGaAs with high InAs mole fractions, these InGaAs/InAlAs heterostructures have dominated high-speed electronic devices with the aim to achieve operation in the THz regime. The latest demonstration was reported for 1.0 THz operation of an integrated circuit consisting of a 10-stage low-noise InGaAs-channel HEMT amplifier [58]. Similar efforts [59, 60] have also been made for HBT-based circuits with the latest demonstration of the nine-stage InGaAs/InP HBT

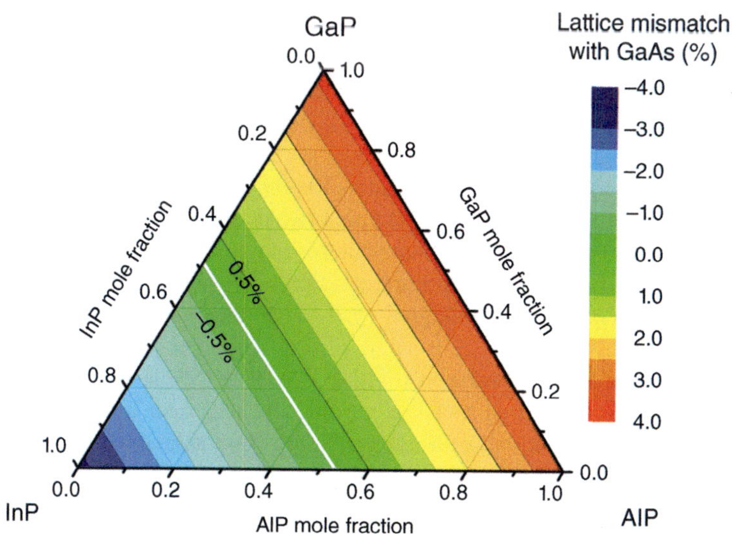

Figure 14.4 *Ternary contour map of LMM of AlInGaP material on a GaAs substrate. The white line indicates the lattice-matching alloy.*

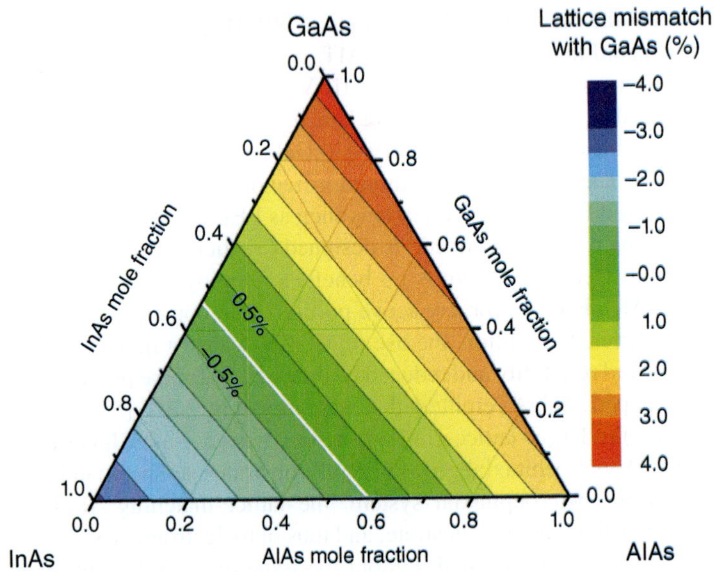

Figure 14.5 Ternary contour map of lattice mismatch of AlInGaAs material on an InP substrate. The lattice-matching alloys are highlighted by the white line.

common-base amplifier operating at 670 GHz [61]. However, the epitaxial growth technique used for these cutting-edge demonstrations is MBE. In the future, MOCVD technology will probably provide the epitaxial materials for these high-speed device applications. This is a primary challenge for a large-scale commercialization of these devices.

There is also a quaternary material system that is benefiting this lattice-matching scheme in P and As by mixing both two elements from Group 13 and two elements from Group 15, such as InGaAsP alloys on InP and GaAs substrates. As shown in Figure 14.6, low Ga composition in the alloys is preferred for lattice-matching and/or near lattice-matching heterostructures on InP, while high-Ga-composition InGaAsP alloys are required for similar purposes when growing on GaAs. The InGaAsP-on-InP-based materials are dominantly used for 1.31- and 1.55 μm laser diodes in long-distance fiber-optic telecommunication applications. However, there is not much interest in InGaAsP on GaAs for laser diode applications as the possible emission wavelengths are readily available via the AlGaAs heterostructure system.

The III-Sb materials have larger lattice parameters (0.609, 0.614, and 0.647 nm for GaSb, AlSb, and InSb, respectively) compared to those of III-P and III-As mentioned previously. There are two types of conventional commercial bulk substrates in this material system: GaSb and InSb. For the growth of III-Sb materials on a GaSb substrate, the lattice-mismatch distribution is in a similar trend as that of III-As on GaAs shown in Figure 14.3. GaSb substrates are useful for the MOCVD growth of Sb-related alloy materials with small LMM, while the InSb substrates have some potential as a platform for HgCdTe materials with LMM of 0.15% or less. There are a handful reports of optoelectronic and high-speed electronic devices using solely III-Sb material heterostructures. GaSb, AlSb, and AlGaSb are nearly lattice-matched materials when grown on InAs; however, they form either type II (staggered gap) or type III (broken gap) heterointerfaces with InAs depending on their material composition [62, 63]. These materials have been applied as a barrier layer in InAs-channel HEMTs and HBTs for high-speed electronic applications. For optoelectronic applications, the advent of the broken-gap or staggered-gap alignment at InAs/GaSb- or InAs/InGaSb-based superlattices offer effective energy gaps narrower than that of InAs,

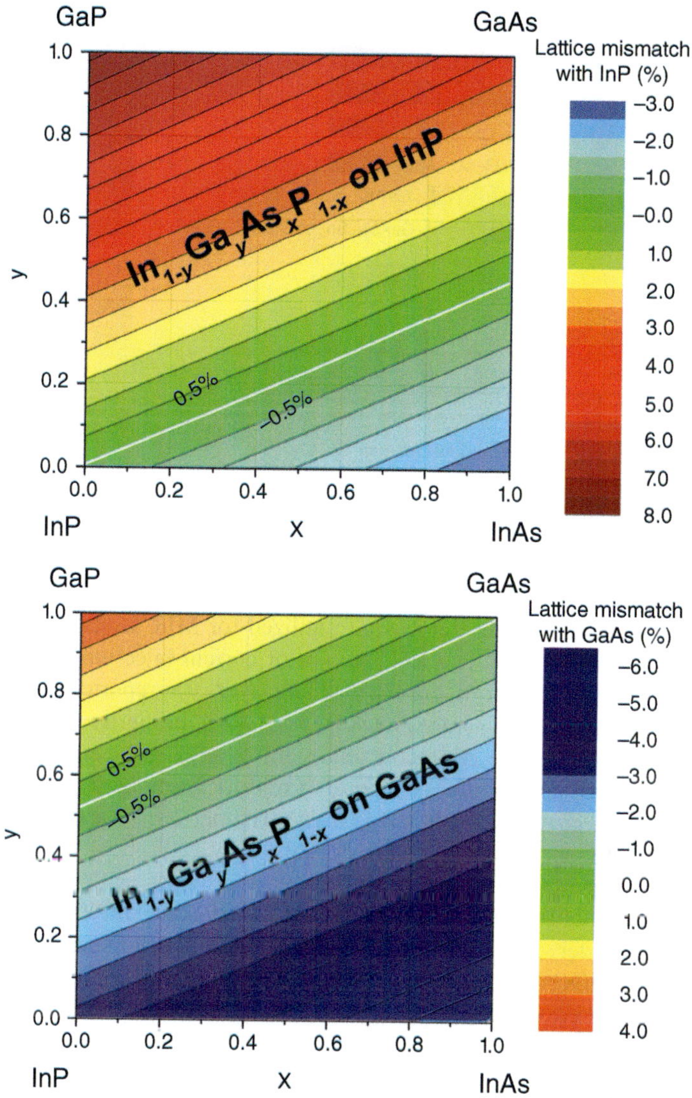

Figure 14.6 *Quaternary contour map of LMM of InGaAsP material on InP (upper) and GaAs (lower) substrates. The white lines indicate lattice-matching alloy composition to the substrates.*

and are tunable down to zero bandgap. Such narrow direct bandgaps have been sought for long-wavelength infrared applications, e.g. detectors potentially competitive to those based on HgCdTe in terms of device performance and nontoxic environment for epitaxy, device fabrication, in-field use, and post-use disposal. A number of researchers have applied InGaSb as a *p*-type base in HBTs or as a *p*-channel for HFETs as this material system exhibits excellent *p*-type characteristics with the smallest effective hole mass when doped with conventional *n*-type dopants used in III-As and III-P materials. Figure 14.7 depicts the lattice-mismatch distribution when grown on various substrates. One example is the InP/GaAsSb/InP HBT with *p*-AlGaSb lattice matching to InP, which demonstrated a f_T of 602 GHz [64]. However, the growth on GaAs has been performed by utilizing graded-alloy buffer layers such as AlInSb [65], InGaAsSb, and AlGaAsSb; however, device performance was degraded by a large density of threading dislocations, e.g. 10^7–10^8 cm^{-2}.

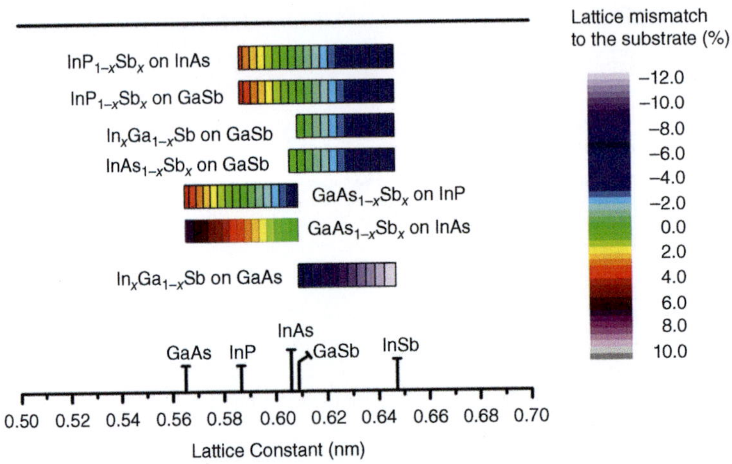

Figure 14.7 *Contour map of LMM of Sb-containing III-V ternary alloys on various III-V substrates. The substrate lattice parameters are indicated on the horizontal axis.*

In addition, many lattice-matched heterostructures utilizing InGaAsSb and AlInAsSb quaternary alloys on GaSb substrates have been successfully utilized for MIR lasers grown by LPE dating back to the late 1970s and 1980s. The LPE growth of current-injection LDs consisting of an InGaAsSb active layer with GaSb or AlGaAsSb cladding layers was reported [66–69]. GaSb-based mid-IR emitters have been predominantly grown by MBE [70], primarily because of challenging MOCVD growth of *n*- and *p*-type AlSb-containing alloys with sufficient Al content for good optical confinement. There is a handful of reports of mid-IR lasers with primitive III-Sb and/or III-AsSb heterostructures grown by MOCVD [71–75]. The performance of the diode lasers was limited by high levels of O in Al-containing layers [76].

As also shown in Figure 14.7, mixing the III-Sb with P provides an opportunity to work on a heterostructure on GaSb with lattice-matching and strain-balancing InPSb ternary and InPAsSb quaternary alloys for a wider selection of bandgap and optical characteristics. With the nonequilibrium growth nature of MOCVD, some light emitters with InPSb were demonstrated despite the immiscibility between the two binary compounds [77, 78].

In addition, theoretical studies suggested that absorber materials with bandgaps in the range of 0.3–0.4 eV are optimal for thermophotovoltaic (TPV) applications for relatively low heat-source temperatures, i.e. ~1500 K [79]. In this regard, III-AsSb alloys lattice-matched to GaSb have been reported to provide high conversion efficiency with internal quantum efficiency of >90%, J_{SC} of 1 A/cm^2, and V_{OC} of 0.3 V for GaInAsSb/AlGaAsSb TPV on GaSb substrate [80, 81]. Combining TPV with radiant-heat sources has been sought as an efficient way for waste-heat recovery applications. Moving forward, the narrower bandgap offers the capability to combine this type of TPVs with heat sources at even lower temperatures, 700–1200 K.

There have been number of studies to engineer bandgaps by incorporating N or Bi in this zincblende III-V material system, such as GaPN, GaAsN, GaAsBi, and InBiSb as alternative direct-bandgap materials for optoelectronic applications. In either case, this approach is subject to large LMM when including large concentrations of either N or Bi in the alloys. However, such incorporation results in a bandgap reduction with large bowing parameters. Most development has been reported for $InGaAs_{1-x}N_x$ material with $x<0.03$. When replacing As with N, the bandgap decreases rapidly by about 0.1 eV for every mole fraction step of 0.01 [82]. Highly efficient lasers using strained InGaAs QW structures on GaAs are very well established for emission around 1 μm. In order to achieve telecom emission wavelengths of 1.3 and 1.55 μm, a large amount of In is necessary, and therefore it induces too large a compressive strain for a high-quality heterostructure for achieving reliable lasers. However, adding N in such QWs to form diluted nitride alloys has triggered an interest in developing high-quality and reliable telecom lasers that can be developed on GaAs, which is considered for superior mechanical strength over that of InP. Such diluted InGaAsN-based telecom lasers are commercially available but have limited output power as compared to InGaAsP lasers developed on InP.

In addition, there has been some investigation of incorporating B in this III-V material system. BP and BAs can be formed as zincblende materials, however, with much smaller lattice constants. Some attempts to form BGaAs and BGaInAs had been reported with small B contents [83, 84]. There is still no device application reported for these B-containing materials.

There is also a possible option of quintary alloys such as AlInGaAsP and AlInGaAsSb; however, such alloys would be much more interesting contenders if they could offer any benefit over what is available in related ternary or quaternary alloys such as direct bandgap in high Al-containing alloys for optoelectronic applications.

14.3.1.2 III-N Semiconductor Materials

(Al, Ga, In)N materials are the most currently studied III-N semiconductor materials. They are stable in the wurtzite crystal structure, which makes them hard to form any high-quality heterostructure with the other zincblende III-V materials mentioned in the section before. Owing to their direct-bandgap characteristics, these materials can be employed in optoelectronic applications covering from mid-IR (~1.8 μm) to deep UV (~200 nm).

This material system has a long history of development on foreign substrates due to the difficulty in growing a native single-crystal boule with large diameters. For instance, commercially available bulk GaN and AlN substrates are available up to 2 in. in diameter, while freestanding GaN substrates up to 4 in. in diameter (6 in. diameter substrates are in development) are readily available for large-scale production of optoelectronic or high-speed electronic or high-power electronic devices of which relative high-power operation is demanded. For mass production of LEDs used in general lighting applications, III-N device growth is dominated by MOCVD; however, it still takes place on large-scale foreign substrates of predominantly *c*-plane sapphire and (111) Si, currently on 6 in. and 8 in. diameter substrates, respectively. There is not much recent interest in growing such LEDs on SiC wafers mainly due to the high production cost, in particular the substrate price. Unlike the other III-Vs, the In-containing III-N alloys are volatile to thermal degradation under H_2 growth ambient and typically require high NH_3 partial pressure, N_2 ambient, and much lower epitaxial temperature compared to those of GaN, AlGaN, and AlN in order to sustain the growth. Hence, it is typical to switch the type of carrier gas, e.g. H_2 and/or N_2, when growing a III-N heterostructure device that includes both In-containing and non-In-containing layers. In the last 10 years, there has been strong interest in growing light emitters on nonpolar and semipolar planes with some excellent reports on improved device efficiency in light-emitter applications [85–89]. However, the growth on these planes is highly challenging for III-N alloys due to a high degree of in-plane asymmetric crystallography and the requirement of low-dislocation-density substrates. Free-standing nonpolar and semipolar GaN substrates have been employed for this purpose with limited availability and size, while the similar version of AlN substrate is not economically viable due to the extremely high cost of the substrate. The attempt on GaN boule growth via the ammonothermal technique is expected to slash the cost of GaN bulk substrates and offer a large selection of crystallographic planes with lower dislocation density and larger substrate dimension.

Owing to the direct bandgap and type I heterointerface, III-N-based binary and alloy semiconductor materials have dominated the development of markets for light emitters and photodetectors with high quantum efficiency, covering the deep-UV to yellow-green spectral regions as compared to other existing semiconductor materials. The InGaN material quality almost reaches perfect internal quantum efficiency for emission wavelengths in the range 400–460 nm. With a down-conversion scheme, highly efficient InGaN blue LEDs and long-wavelength visible phosphor blends have been utilized as a primary light source in general lighting applications with a typical luminous efficacy (LE) of 100 lm/W nowadays. We expect even higher LE within the next 10 years. In applications requiring a high radiation flux, InGaN-based laser diodes are being employed as light engines at high driving-current density. A research frontier toward the long-wavelength visible-light regime has been focused on improving the light efficiency in green and yellow spectral regions in the past 10 years with some degree of success with semipolar materials; however, it will continue to be a challenging topic for years to come.

The frontier of AlGaN emitter applications is the task to improve the efficiency of deep-UV emitters. The external quantum efficiency (EQE) of state-of-the-art LEDs at operating current of 20 mA reached ~10, ~2, and 0.02% at emission peak wavelength of 280, 250, and 240 nm, respectively [90]. The technical challenges are (i) low hole concentrations in *p*-AlGaN due to the increasing activation energy of the acceptor, i.e. Mg, with increasing AlN mole fraction in the alloy; and (ii) shifting of optical polarization of the emitted UV light depending on the strain condition in the device, i.e. switching of transverse electric modes (TE) to transverse magnetic modes (TM) at around 245 nm. Note that no TM light is allowed to propagate through the front or back surface of the *c*-plane LEDs. To tackle the latter technological barrier, some alternative designs such as nonpolar AlGaN/AlGaN-based MQW have been attempted [91–93] and ultrathin monolayer GaN QW [94].

For electronic device applications, these III-N materials are promising for high-power and high-frequency applications owing to their high sheet-carrier density, high saturation velocity, high breakdown field, and good thermal conductivity. For instance, conventional material property indices for this application such as Keyes Figure of Merit, Baliga Figure of Merit, Johnson Figure of Merit, and Baliga High Frequency Figure of Merit are of the order of hundreds for GaN compared to tens for GaAs. With a high breakdown electric field of ~3 MV/cm and good thermal conductivity of 210 W/K m at room temperature [95], GaN-based devices can support compact circuit design for a given frequency and power level. In a planar device scheme, GaN-HEMT-based RF power amplifiers (PAs) have currently replaced GaAs-based ones as power-efficient base stations in wireless communication. Recently, GaN-based monolithic microwave integrated circuit (MMIC) PAs for the K_a-band radio have been demonstrated with the record power added efficiency (PAE) of 59% measured at a frequency of 32 GHz, bias of 3 V, and output power of 24.3 dBm [96]. The state-of-the-art E-mode GaN HEMT with 20 nm gate length has reached 350 GHz and 500 GHz for f_T and f_{max}, respectively, while the D-mode HEMT has demonstrated both f_T and f_{max} of 450 GHz [97]. To further improve high-frequency performance, epitaxial structures on low-defect native GaN bulk substrate as well as strain-managing high-quality AlInN and AlInGaN alloys will be the important factors. The dependence of the lattice constant versus alloy composition for AlInGaN alloys is shown in Figure 14.8.

In the area of vertical-conduction electronic devices, MOCVD-grown GaN-based *p-i-n* devices demonstrated excellent device performance for a high-voltage rectifier applications in the range 500–2000 V [98–100]. However, it is still quite challenging to grow by MOCVD any device that requires an alternative layer sequence of *n*- and *p*-type doping such as bipolar transistors, junction barrier Schottky rectifiers, and inverted-gate bipolar transistors. The major issue is due to the *p*-type doping process using the common Cp_2Mg precursor as the Mg atoms have a high sticking coefficient and high vapor pressure. This causes the slow turn-off of the molar concentration of the Cp_2Mg precursor and contributes to a corresponding memory effect of the Mg doping profile. There is a vast amount of residual Mg compensating other dopants in a subsequent layer, such as the Si donors in a portion near an emitter/base interface of a bipolar transistor. Many groups have

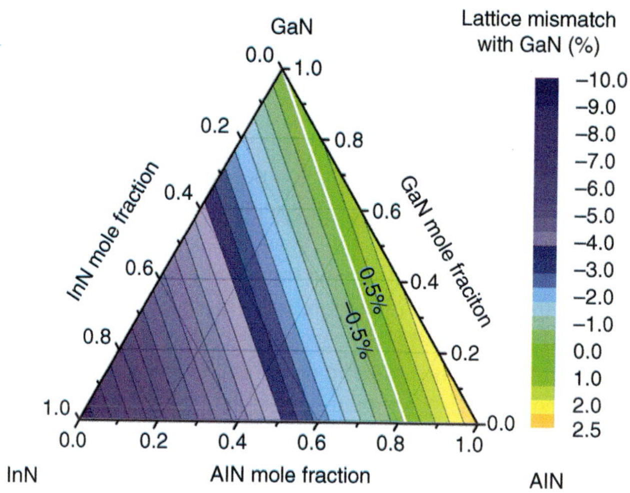

Figure 14.8 Ternary contour map of LMM of AlInGaN material on a GaN substrate. The white line indicates lattice-matching alloy.

worked on developing heterojunction bipolar transistor (HBTs); however, most of their devices were grown on foreign substrates such as sapphire and 6H-SiC with high concentrations of residual threading dislocations in the active device. Such crystallographic defects hampered the device performance. AlGaN/GaN based HBTs had been initially explored [101 103]. The marked improvement in III-N HBT performance was reported when NPN GaN/InGaN HBTs [104] employing *p*-InGaN base regions offered the higher *p*-type layer hole concentrations [105] with some success of reasonable common–emitter current gain (>40) and drive current density (J_C > 1 kA/cm^2) [106, 107]. The record performance was reported for GaN/InGaN HBTs grown by MOCVD on freestanding GaN substrates that were employed for their lower crystallographic defect densities. This is the only III-N-based HBT device structure that has achieved both large DC current gain >100 and high collector–current density >100 kA/cm^2 at room-temperature operation to date [108]. Due to the limited availability of semi insulating (SI) GaN substrates, the high frequency response was reported to reach f_T = 8 GHz for this type of HBT on sapphire substrates [109]. MOCVD of III-N HBTs on bulk SI-GaN substrates is definitely one topic to be explored in order to demonstrate the true high-frequency performance capabilities of the III-N material system.

Adding to the difficulties of creating high-performance III-N HBTs, the activation of III-N Mg-doped *p*-type layers is currently achieved by reducing Mg–H complexes inside the doped layer by electron beam irradiation or thermal annealing processes. These processes are effective when the *p*-layers locate as the topmost epitaxial layer in an epitaxial device structure. There is still no systematic study reported on how to effectively activate the *p*-type dopant when the layer is embedded deep inside the device structure. Such effort may involve optimizing the growth process, choice of dopant MO sources, growth sequence, and so on.

For III-N systems, due to limited availability of high-quality native substrates (i.e. currently only GaN and AlN are available commercially, as mentioned earlier), there are efforts by many research groups to improve the crystallographic quality of GaN and AlN including thick AlGaN and InGaN alloy templates grown on various foreign substrates. For AlN/sapphire templates, a post-growth high-temperature anneal process (up to 1750 °C) was recently reported to effectively assist recrystallization of the layer and drastically annihilated the threading dislocation density down to lower 10^8 cm^{-3} [110]. In another successful attempt, a combination of patterned sapphire substrate and an ex-situ sputtered AlN buffer deposited via physical vapor deposition (PVD) has been recently reported to reduce the threading dislocation density in the subsequently grown GaN heteroepitaxial film to as low as $5.6 \times 10^7 \text{cm}^{-2}$, which is the lowest value for such a thin GaN templates on sapphire [111]. These approaches are promising for growth of III-N on foreign substrates in order to improve performance of III-N devices.

Unlike the MOCVD growth processes of other III-V materials, III-N, and in particular GaN, AlN, and its AlGaN alloys are grown at much higher substrate temperatures. It is inevitably subject to unintentional incorporation of other elements thermally degraded from the surrounding materials such as Si, O, and C from the susceptor, quartzware, heater, as well as the MO and hydride sources. To lower these background impurities and other point defects will require novel chamber designs together with growth-process optimization, and precursor material selection and development. Another remark regarding MOCVD III-N growth is that the optimization process for growth on native III-N substrates has not been studied in detail and is often performed based on optimum growth conditions developed on a III-N template grown on a foreign substrate. A careful study of such growth is necessary to create III-N layers with low background impurities and defects in order to raise the device performance closer to its ideal characteristics.

As III-Ns, in particular GaN, have historically developed on foreign substrates, the following is a summary of past and potential substrates with mapping of their in-plane lattice parameters and thermal expansion coefficients. Some wafers may be considered as conventional substrates, such as sapphire, 4H-SiC, 6H-SiC, Si(111), $MgAl_2O_4$ (111), since there are a number of reports of good device performance. LED devices have already been reported on GaN templates developed on some nonconventional substrates such as graphene, ZrB_2 [112], and β-Ga_2O_3 (–201) [113]. An introduction of 2D materials as substrates capable of large-area and/or roll-to-roll epitaxy may hold an important breakthrough for future epitaxial growth to reduce manufacturing cost of III-N devices and new device application and design in the future.

There is very limited interaction for heterostructures composed of III-Ns and conventional III-Vs, as it is difficult to grow good-quality III-Ns on any other III-V semiconductor substrates or vice versa. There are a limited number of reports on cubic-phase GaN growth on 3C-SiC. However, low-defect-density cubic GaN can be formed through the use of carefully designed patterned substrates employing a coalescence of two GaN *c*-planes formed at a specific angle by growth on two adjacent Si {111} planes. Such work has been successful in the micrometer-scale width [114, 115]. Further development is necessary to expand it into the larger scale.

14.3.1.3 II-VI Semiconductor Materials (Excluding Oxides)

The II-VI compound semiconductors discussed in this section are formed by the Group 12 elements Zn, Cd, and Hg, and the Group 16 elements S, Se, and Te. We also include Mg from Group 2 in this family. The discussion about the related semiconducting oxides is provided in their own separate section. Manasevit and Simpson were the first to grow II-VI compound semiconductors by MOCVD in 1971, demonstrating the growth of ZnSe, ZnS, CdSe, CdS, and CdTe, on Al_2O_3, BeO, and $MgAl_2O_4$ [33]. The primary interest in these II-VI materials is the potential for highly efficient optoelectronic device applications due to their direction bandgap and type I heterointerface properties. ZnS, ZnSe, and CdSe can form an alloy semiconductor material with an engineered bandgap covering any wavelength in the visible-light spectral region. For light emitters, ZnSe has been commonly introduced to form a good epitaxial single-crystal layer of low threading dislocation density on GaAs substrate owing to its small compressive LMM of –0.3%. Addition of Cd and S into ZnSe in the active layer allows bandgap tunability to cover blue and green light emission. The first ZnSe-based LD was demonstrated under pulsed injection at 77 K for a 490-nm blue-green gain-guided type laser with thick CdZnSe QW by Haas et al. [116] using MBE. Adding Mg into the ZnSSe cladding layer improved the carrier-confinement capability and allowed a thick cladding layer to be formed with less lattice strain. However, the high resistivity of *p*-type ZnTe caused a large forward voltage drop, which was one of the major issues that limited the performance of the device. An attempt at making the *p*-type contact via the narrower bandgap materials ZnTe and ZnTeSe was carried out, but it did not sufficiently solve this issue. All the successful CdZnSe-based emitters reported to date were grown by MBE due to the fact that it is still difficult to grow any *p*-type layer for this material system by MOCVD with the same quality as those grown by MBE at lower temperatures.

Another interesting material system in this family is HgSe, which has an extremely small direct bandgap of 0.08 eV. The related HgCdSe ternary alloys formed on GaAs substrates are commonly used for photodetection in the long-wavelength infrared range (LWIR). MOCVD-grown HgCdTe heteroepitaxial films are also commonly used for LWIR photodetectors. Conventional HgCdTe IR photodetectors need to be cryogenically cooled well below ambient temperature to avoid noise and leakage currents resulting from thermal-generation processes, e.g. high dark current caused by Auger processes. Aiming for an elevated-temperature or even room-temperature operation, nonequilibrium device structures utilizing exclusion ($p+/\pi$ or $n+/\nu$) and extraction ($n+/\pi$ or $p+/\nu$) junctions in $p+/\nu/n+$ and $p+/\pi/n+$ configurations have demonstrated suppression of such Auger mechanisms by reducing the absorber carrier density below thermal equilibrium values in a reverse-bias condition [117]. The devices are often constructed with π rather than ν absorber layers for efficient Auger suppression [118] with the challenge of long minority-carrier lifetime in the low *p*-type doping sections of the absorber region. Such low *p*-type doping concentrations require metal-vacancy annihilation using two-zone annealing at temperatures above 300 °C after the growth process for both MBE [119] and MOCVD [120, 121] samples. Recently, record-performance HgCdTe uncooled photodiodes grown by MOCVD without post-growth annealing were reported with responsivities reaching 6 A/W at 300 K [122, 123]. However, a growing concern regarding the safety restriction of using the precursors and the treatment of the process byproducts will be a driving force to replace these LWIR photodetectors with ones developed by using an alternative environmentally safe small-bandgap material systems.

14.3.1.4 Novel IV Semiconductor Materials

In order to incorporate light-emitting devices in Si-based monolithic integrated circuits, several approaches have been demonstrated including hybrid integration of III-V lasers on Si, which requires complicated wafer bonding and/or selective-area growth. Another recent approach is manipulating the bandgap of strained Ge from an indirect to a direct transition by means of tensile strain and heavily *n*-type doping to compensate the energy difference between the direct and indirect conduction valleys. A demonstration of an electrically pumped laser was reported with an emission wavelength of 1590–1610 nm at room temperature [124]. This allowed the device to be selectively grown on Si with Ge, a neighboring Group 14 element; however, for large-scale production it was quite challenging to uniformly maintain such tensile strain and high doping concentration of P in Ge. To overcome such a challenge, novel Group IV single-crystal semiconductor materials of GeSn and GePb are most recently sought as direct-bandgap materials for light-emitting devices viable for integration with the Si-based CMOS process. As a pure single crystal, Sn forms a diamond structure (space group: $Fd\bar{3}m$) like C, Si, and Ge; however, Pb is stable in the face-centered cubic (fcc) structure (space group: $Fm\bar{3}m$), which may cause an additional issue if high Sn incorporation is needed. The lattice constant of Sn single crystals is 0.645 nm and thus Sn has a large compressive LMM with Si and Ge at 19.5 and 14.7%, respectively. On the other hand, the lattice constant of Pb fcc single crystals is 0.495 nm, while its virtual diamond structure would have a lattice constant of 0.680 nm [125], which points to even larger compressive LMM with Si and Ge of 20.2 and 25.3%, respectively. Theoretical work anticipated that a Sn mole fraction of $x = 0.06$ and higher in a relaxed $Ge_{1-x}Sn_x$ sufficiently lowered the bandgap at Γ valley and resulted in a direct-bandgap material due to crossing of Γ- and L- valleys [126, 127]. However, under such a compressive strain condition, higher Sn composition is needed. To couple this GeSn on Si, a Ge buffer layer has been commonly used to form a virtual Ge template that offers lower compressive strain for low Sn containing alloy, e.g. an in-plane compressive strain of 2.5% in direct-bandgap $Ge_{0.85}Sn_{0.15}$ layers on Ge. Lasing has recently been reported for optically pumped $Ga_{0.874}Sn_{0.126}$ on Ge/Si(001) with an emission wavelength of ~2.25 μm and threshold optical density of 325 kW/cm^2 at a cryogenic temperature of 20 K [128]. Additional recent progress was reported for a Ge/GeSn/Ge double heterostructure with lasing wavelength of ~2.5 μm and threshold optical pumping density of 396 kW/cm^2 at 110 K [129]. Since these GeSn alloys have been grown at temperatures below 400 °C using a low-pressure CVD system similar to MOCVD, this approach can offer several advantages such as a high-purity Sn precursor, a consistent process control with good layer thickness and composition uniformity, and impurity doping for conductivity control in any future *p-n* junction-based devices.

For $Ge_{1-x}Pb_x$ alloys, theoretical studies suggested a cross-over of the direct bandgap occurs at much lower Pb mole fraction ($x \geq 0.034$) in the relaxed alloy film. Under biaxial compressive strain conditions, the transition to direct bandgap was predicted to take place at even lower Pb mole fractions ($x \geq 0.021$) if the layer was grown on a (111) plane [130]. However, the phase diagram of the Pb-Ge system indicates no or extremely low solid solubility of Pb and Ge in each other under equilibrium conditions [131]. Well-crystalline GePb was reportedly grown by using pulsed laser-induced epitaxy, and experimental characterization suggested low Pb incorporation (Pb molar fraction was merely 0.002), which was insufficient to shift the bandgap to be a direct transition [132]. The material development of GePb is likely to be done under a nonequilibrium growth scheme in order to increase the Pb incorporation. MOCVD may be a good alternative option for this challenge.

14.3.1.5 Growth on Flexible Substrates/Templates

For flexible electronic applications, inorganic-based electronic materials have limited mechanical deformability compared to those of organic-based materials. Such constraints reduce the range of applications that the materials can contribute to. Due to the high efficiency and performance of devices already developed with the inorganic semiconductor materials mentioned in this chapter, inorganic-semiconductor-based flexible electronics can be expected to enable various new practical uses outperforming those of the organic-based flexible electronics. In straightforward approaches, the inorganic devices can be detached from their substrates via several techniques such as mechanical lapping, selective wet etching, or laser lift-off processes followed by device separation and bonding to flexible platforms such as polymer materials (polyimide, polyester, etc.) or thin metal foils (copper, etc.).

The interest in growing these semiconductors on a flexible substrate or template is driven by an economic reason to save the cost of the single-crystal substrates and post-growth processing, while maintaining a level of device performance that may be compromised but still retains a reasonable and good functionality. The substrate and template materials sought for this challenge must be thermally durable and preferably low in or free of impurities introduced during the MOCVD process. With a significant advantage as a single-crystal platform, inorganic 2D materials are being considered. Such attempts have been spearheaded by the use of graphene. The 2D material is prepared or bonded to a rigid substrate not necessarily made of the same material as that of a conventional substrate used for growing a subsequent epitaxial layer.

In addition, a uniform layer of graphene can be epitaxially formed on a single-crystalline SiC substrate via a graphitization process. The epitaxial graphene on SiC offers a wafer-scale single orientation for the entire substrate [133]. This approach overcomes the difficulty of forming a uniform monolayer-thick graphene without the necessity for bonding the 2D material on a rigid platform. Such graphene-based substrates have been applied in the growth of GaN. This approach yielded single-crystalline GaN layer with a root mean square (RMS) roughness of 0.3 nm and a low threading dislocation density (as low as 4×10^8 cm^{-2}). An InGaN-based blue LED transferred on to a flexible plastic tape was then achieved [134].

Recently, remote epitaxy has been proposed and demonstrated for GaAs, InP, and GaP materials, as only a single monolayer of graphene allows the crystallographic ordering from the semiconductors to be inherited by the epitaxial layer on top through the 2D material. Theoretical simulation suggests that such remote epitaxy is feasible if the gap between the substrate and the epitaxial layer is smaller than 0.9 nm. With a single graphene monolayer, a layer of (001) orientation was uniformly grown, while a layer of (111) orientation layer was achieved when three or more monolayers of graphene were applied. The epitaxial layers were then exfoliated from the semiconductor substrate and transferred to another flexible platform. A flexible (001) AlInGaP LED manifested device performance similar to that grown on a rigid substrate without the graphene monolayer [135]. Adoption of this concept may dominate the growth for inorganic flexible electronic devices.

Another interesting 2D material for an epitaxial growth is hexagonal BN (h-BN), also known as *white graphene* due to its wide bandgap in the deep-UV region. Layers of h-BN can be formed by MOCVD and serve as a release layer to mechanically exfoliate the III-N epitaxial device structures. Since h-BN can be grown in the standard III-N MOCVD systems, a 3-nm thick h-BN template layer was applied in growth of GaN on sapphire substrate. The GaN/h-BN/sapphire templates were utilized in growing an AlGaN/GaN heterostructure with $\mu \sim 1100$ cm^2/V s and sheet charge density of 1×10^{13} cm^{-2} at room temperature, and InGaN-based blue LEDs have been reported that were exfoliated from a sapphire substrate and bonded to another substrate [136]. A further optimization to reduce the dislocation density or by using a remote epitaxy approach could potentially make this h-BN 2D release layer the favorite choice in the future III-N flexible electronic application.

As shown in Figure 14.9, several other 2D materials can serve in MOCVD epitaxial growth for III-N materials, such as MOS_2 and WS_2 for GaN and WSe_2 for lattice-matched InGaN and AlInN alloys. The major technical challenge is how to achieve a wafer-scale template of these 2D materials with a uniform in-plane orientation.

To further reduce the manufacturing cost, there is also a strong interest in investigating high-volume growth even in larger dimensions such as roll-to-roll device epitaxy. This is a great challenge to go beyond current batch loading, which solely relies on high-quality and, in most cases, relatively expensive single-crystal wafers. Again, similar requirements of thermal stability and low impurity incorporation from the substrate material are inevitably preferred. An existing roll-to-roll CVD system deposits materials on a continuous roll of metal foils, polymer sheet, and fibers. The technology has been used in a simple coating without the need for single-crystal growth such as organic electronic materials and superconducting tape. At the current stage, however, there is no such single-crystal substrate available in a reel form that can be applied for roll-to-roll feeding MOCVD reactor configurations. As an initial step toward this mass-production dream, some research groups have already explored a possibility of MOCVD growth of reasonable quality inorganic semiconductor devices on flexible metal substrates and foils.

14.3.2 Nanostructural Materials

14.3.2.1 2D Materials

The two-dimensional (2D) materials discussed in this section are limited to materials formed by a single element or binary compound that is constructed with strong in-plane covalent bonds and weak inter-plane couplings via a van der Waals-like force. Such a property allows the 2D material to be pulled out layer by layer. Spearheaded by the discovery of graphene, the 2D allotropes of other Group IV elements, namely silicene, germanene, and stanine, have been predicted to be stable and realized

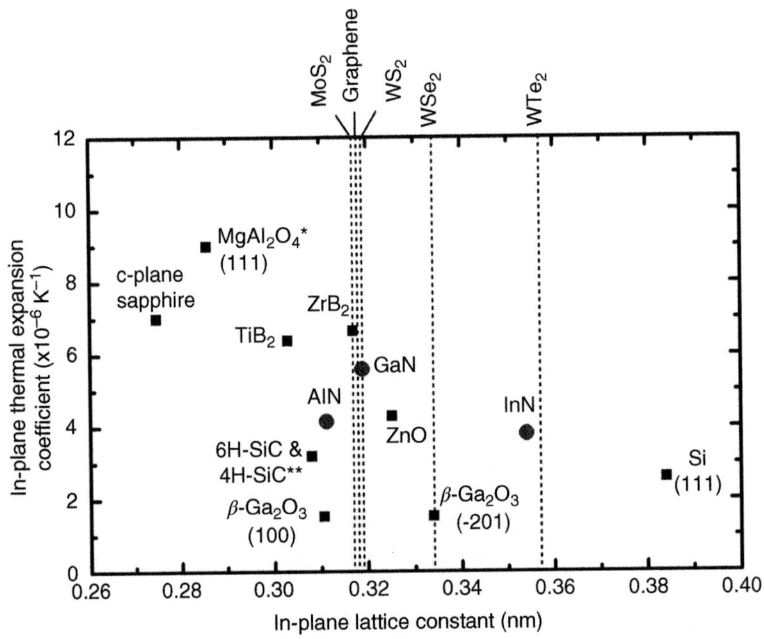

Figure 14.9 *Map of in-plane lattice constants and corresponding room-temperature thermal expansion coefficients of III-N semiconductors and various substrates. For 2D-materials-based substrates such as MoS_2 and graphene, only their in-plane lattice parameters are indicated by the vertical broken lines. Please note that one * indicates oxygen bond length, and two ** indicates materials with a high nonlinear thermal expansion coefficient.*

by ultrahigh vacuum growth schemes such as MBE growth. Other single-element allotropes are also recently reported for borophene and phosphorene. The existence of phosphorene has been demonstrated very recently through a mechanical exfoliation of black phosphorus. However, most of these single-element 2D materials, except for graphene and phosphorene, can only be formed at the nanoscale. The contribution that MOCVD can make for this endeavor seems to be limited to nanoscale devices, perhaps in a combination with other nanostructures such as nanowires and nanoribbons where the 2D materials can be selectively grown on preferred facet(s) depending on control parameters such as temperature, chamber pressure, and precursor vapor pressure. Table 14.2 summarizes the 2D materials formed by single elements as honeycomb structures, except for phosphorene, which forms a nonplanar ridge structure with properties referred to as *black phosphorous* in Table 14.3.

For binary compound-based 2D materials, the transition-metal dichalcogenides (TMDCs) are widely studied. These materials can be formed by a CVD method, and most of them are commercially available up to 10 mm by 10 mm in size. The MOCVD technique may play an important role in large-scale production of large-area TMDC films as well as impurity doping for conductivity control, in particular in an in-plane direction.

These 2D materials are promising due to their extraordinary electronic and optical properties. For instance, Dirac dispersion relations and remarkable carrier mobilities are predicted in silicene and germanene. Unlike the zero bandgap in graphene, the 2D TMDCs and phosphorene are semiconductors and predicted to exhibit excellent in-plane thermoelectric performance. Their unique 2D geometric structure and its corresponding physical properties suggest that they have an excellent potential for the next generation of optoelectronic and high-frequency electronic devices.

Table 14.3 summarizes the 2D semiconductor materials with their crystallographic properties. Most of them have indirect bandgaps; however, some exhibit a transition from indirect to direct

Table 14.2 *Single-element 2D materials with their lattice constants.*

Material	a_0 (nm)
Silicence (Si) [137]	0.3845
Graphene	0.3180
Borophene (B) [138]	0.1580
Germanene (Ge) [139]	0.3800
Stanine (Sn) [140]	0.4675
Phosphorene	a_0=0.331, b_0=1.048, c_0=0.437

Table 14.3 *2D semiconductor materials with their lattice constants, crystal structure, and electronic bandgap energy as organized by their type of crystal structure. The lattice constants and bandgap energy values are from [145–147].*

Semiconductor	a_0 (nm)	b_0 (nm)	c_0 (nm)	Crystal structure	Bandgap energy (eV)	
					Direct	Indirect
h-BN	0.2502	–	0.6617	Hexagonal	5.9	
α-GaS	0.359	–	1.544	Hexagonal	2.6	
2H-GaSe	0.374	–	1.592	Hexagonal		2.1
HfS_2	0.363	–	0.586	Hexagonal	1.2 (for MLs)	2
$HfSe_2$	0.3745	–	0.616	Hexagonal		1.1
2H-In_2Se_3	0.398	–	18.89	Hexagonal		
2H-MoS_2	0.315	–	1.229	Hexagonal	1.85 (for MLs)	1.2
2H-$MoSe_2$	0.329	–	1.289	Hexagonal	1.55 (for MLs)	1.1
2H-$MoTe_2$	0.353	–	1.396	Hexagonal		1.2
Sb_2Te_3	0.425	–	3.048	Hexagonal		
2H-SnS_2	0.365	–	0.589	Hexagonal		2.2
$SnSe_2$	0.381	–	0.614	Hexagonal		
1T-TaS_2	0.335	–	0.59	Hexagonal		
2H-WS_2	0.315	–	1.227	Hexagonal	1.99 (for MLs)	1.3
WSe_2	0.328	–	1.298	Hexagonal	1.65 (for MLs)	1.3
$ZrSe_2$	0.377	–	0.614	Hexagonal	1	
As_2Te_3	1.43	0.403	0.403	Monoclinic	0.2–0.3	
$ZrSe_3$	0.541	0.375	0.944	Monoclinic		1.1
Bi_2S_3	0.4025	1.117	1.135	Orthorhombic		
Black phosphorus	0.331	1.048	0.437	Orthorhombic	1.3 (for MLs)	~0.3
GeS	1.45	0.364	0.43	Orthorhombic		1.6
GeSe	0.383	0.44	1.078	Orthorhombic		1.1
ReS_2	0.642	0.651	0.645	Triclinic	1.55	
$ReSe_2$	0.660	0.672	0.672	Triclinic	1.32	

bandgap for a few monolayers (MLs) as their electronic bandgap energy increases inversely as the number of MLs. Most of these 2D materials can possibly be grown by MOCVD; however, some are still lacking suitable MO precursors such as rhenium (Re) and niobium (Nb). Some 2D single-crystal materials have the properties of semimetals and metals, as shown in Table 14.4 and Table 14.5, respectively. Some of them can act as superconductors below their critical temperature, as shown in Table 14.6.

In general, the hexagonal 2D materials can be grown on a single-crystal substrate or platform with a surface hexagonal atomic arrangement such as *c*-plane sapphire, GaN, AlN, and Si(111),

Table 14.4 *2D semimetal materials with their lattice constants and crystal structure, as organized by the type of crystal structure. Sources: Eichfeld et al. 2015 [143], Kang et al. 2015 [144], and Kalikhman and Ya 1973 [145].*

Semimetal	a_0 (nm)	b_0 (nm)	c_0 (nm)	Crystal structure
$HfTe_2$	0.395	–	0.666	Hexagonal
$PtSe_2$	0.375	–	0.506	Hexagonal
1T-TiS_2	0.34	–	0.57	Hexagonal
1T-$TiSe_2$	0.354	–	0.601	Hexagonal
$TiTe_2$	0.377	–	0.649	Hexagonal
WTe_2	0.348	–	0.625	Hexagonal
$ZrTe_3$	0.5864	0.3918	1.009	Monoclinic

Table 14.5 *2D metal materials with their lattice constants and crystal structure, as organized by their type of crystal structure. Sources: Eichfeld et al. 2015 [143], Kang et al. 2015 [144], and Kalikhman and Ya 1973 [145].*

Metal	a_0 (nm)	b_0 (nm)	c_0 (nm)	Crystal structure
Graphite (highly oriented pyrolytic graphite)	0.246	–	0.667	Hexagonal
Graphite (natural)	0.246	–	0.667	Hexagonal
2H-NbS_2	0.332	–	1.197	Hexagonal
3R-NbS_2	0.330	–	1.792	Hexagonal
2H-$NbSe_2$	0.344	–	1.255	Hexagonal
$PdTe_2$	0.405	–	0.511	Hexagonal
$PtTe_2$	0.400	–	0.519	Hexagonal
2H-TaS_2	0.335	–	1.207	Hexagonal
2H-$TaSe_2$	0.351	–	1.255	Hexagonal
1T-VSe_2	0.336	0.336	0.609	Hexagonal
α-AuSe	1.217	0.369	0.838	Monoclinic
β-$MoTe_2$	0.621	0.347	1.383	Monoclinic
$NbTe_2$	1.470	0.364	0.935	Monoclinic

Table 14.6 *2D superconductor materials with their lattice constants, crystal structure, and critical temperature (T_C), as organized by their type of crystal structure. Sources: Eichfeld et al. 2015 [143], Kang et al. 2015 [144], and Kalikhman and Ya 1973 [145].*

Superconductor	a_0 (nm)	b_0 (nm)	c_0 (nm)	Crystal structure	T_C (K)
2H-NbS_2	0.332	–	1.197	Hexagonal	6
3R-NbS_2	0.330	–	1.792	Hexagonal	6
2H-$NbSe_2$	0.344	–	1.255	Hexagonal	7.2
$PdTe_2$	0.404	–	0.511	Hexagonal	1.7
2H-TaS_2	0.331	–	1.207	Hexagonal	1
2H-$TaSe_2$	0.342	–	1.255	Hexagonal	0.1
$NbTe_2$	1.470	0.364	0.935	Monoclinic	0.7
$ZrTe_3$	0.586	0.392	1.009	Monoclinic	2

GaAs(111), Ge(111), and GaP (111). As shown in Figure 14.9, GaN and AlN are good candidates for MoS_2 and WS_2. AlN is superior to GaN as a growth substrate regarding thermal stability. Among these 2D materials, h-BN has been successfully grown on *c*-plane sapphire via a high-temperature MOCVD system. *p*-Type doping was also readily achieved with Mg as a dopant [141]. Recently, large-scale 50 mm-diameter free-standing h-BN foils have been demonstrated [142]. Some MOCVD activities have been recently reported for WSe_2, [143] WS_2, and MOS_2 [144]; however, there are still several issues such as limited surface mobility of transition-metal atoms, high surface desorption rate of the Group 16 chalcogen atoms, and domain coalescence with low-angle grain boundaries.

14.3.2.2 Nanopatterned Materials

The nanopatterned materials discussed in this section are mainly based on selective-area growth (SAG). In the early years, the dielectric material SiO_2 was utilized as a masking material due to its stability at the growth temperature, and it has also been widely used in SAG [148, 149]. Crystallographic facet formation is strongly dependent on MOCVD growth parameters such as substrate orientation including its off-cut angle, pressure, temperature, precursor partial pressure, and ratio of two different type of precursors involved [150–153]. As the opening of the mask decreased, it was observed that the threading dislocations from the layer underneath could be effectively blocked [149]. In III-Ns, this technique and its variation have been used to lower threading dislocation density for III-N grown on foreign substrates [154–156].

The selective growth in nanopatterned areas allows a large lattice-mismatched material to be grown with minimum defect generation or even defect-free in some case, like nanowires (NWs). Some other challenges include the limited mobility of many major precursors such as TMAl, a thermal effect on any existing layers in the growth platform of an integrated-circuit-based device. The former limitation confines material selection in SAG, while the latter limitation restrains the regrowth process window, which may compromise the regrown material quality. In addition, the consumption of precursors is subject to the regrowth surface area and the volume of material to be grown into the patterns. This will be a challenge for a discrete pattern where there are various device dimensions and spacings between them. However, the selective nanopatterned growth has been useful in fin field-effect transistor (finFET) technology [157] and NW growth [158] as it provides good controlled and uniform dimensions of the grown materials with high selectivity under optimum growth conditions, as the lithography technique has been constantly evolving since the introduction of an ArF excimer laser light source. Utilizing current extreme-UV lithography techniques, state-of-the-art finFETs in Si CMOS production technology are now reaching the 10-nm node, while 7- and 5-nm nodes are expected in the years 2019 and 2020, respectively. This paves the way for patterning capability for any other semiconductor materials that pursue the nanopatterning processes in any of their future devices. For the nanostructured devices based on nanowires and nanowalls/ribbons, the width of the unmasked area for regrowth is still of the order of several hundred nanometers.

In a recent effort toward the heterogeneous integrated circuit, InAs and InSb NWs without any dislocation defects were selectively grown inside ~100-nm SiO_2 nanotubes formed on Si [159]. Coplanar nanostructures of InAs grown in a confined SiO_2 channel were then demonstrated. The grown of InAs exhibited an electron mobility of 54 000 $cm^2/V\,s$. In addition, multiple-gate FETs with ten 55-nm wide, 23-nm thick, and 390-nm long channels were also demonstrated [160]. For III-N on Si, GaN was grown in a SiO_2 channel on Si by utilizing PVD-AlN as the nucleation layer [161]. With a patterned SiO_2 channel aligned to favor the crystal facets of these semiconductor materials, such a technique will develop into a good heterogeneous integration with the Si CMOS technology.

14.3.3 Poly, Amorphous, and Other Materials

14.3.3.1 Metals

For the MOCVD deposition of metal thin films, different types of metal chelates are used as precursors owing to their high volatility, thermal stability, and ease of reduction. Van Hemert et al. demonstrated thin films of copper (Cu), nickel (Ni), and rhodium (Rh) by hydrogen reduction of $Cu(tfacac)_2$, $Cu(hfacac)_2 \cdot H_2O$, $Ni(hfacac)_2 \cdot H_2O$, and $Rh(tfacac)_2$ (tfacac = trifluoroacetylacetonate ($C_5H_4F_3O_2$) anion and hfacac = hexafluoroacetylacetonate ($C_5HF_6O_2$) anion) from the gas phase [162]. Since then, various metal films have been deposited using metal β-diketonate precursors by CVD. The metal deposition reaction with a precursor, e.g. M(hfacac), can be expressed as:

$$2M(hfacac) \rightarrow M + M(hfacac)_2 \tag{14.1}$$

where M is metal. Two adsorbed molecules of M(hfacac) react together and produce a metal atom and a $M(hfacac)_2$ molecule. Then, a metal film is formed after the desorption of volatile products.

Cu has gained increasing attention in microelectronics, synthesis of graphene, and fuel-cell technology. In particular, Cu is a promising interconnect material to replace Al alloy in large-scale integrated circuits owing to its low RC time delay and high resistance to electromigration and stress migration [163]. Among many different techniques of depositing Cu, including CVD, sputtering, electroless plating, and electrolytic plating, CVD is the most promising [164], owing to the ability of this process to provide conformal gap filling and the possibility of selective deposition [165–167]. The deposition rate, selectivity, and gap-filling properties are dependent on deposition conditions. Awaya and Arita studied the carrier-gas effects using the most-utilized Cu MOCVD precursor, hexafluoroacetylacetonate copper(1) trimethylvinylsilane (Cu(hfacac)·(tmvs)), where tmvs is $(CH_3)_3SiCHCH_2$ [168]. The deposition rate is linearly proportional to the flow rate of precursor at a growth temperature of 220 °C, suggesting that the surface reaction rate is high enough at temperatures above 200 °C and the rate is controlled by the precursor supply: that is, the growth is mass-transfer-rate limited. At lower temperatures of 140–200 °C, the surface reactions control the deposition rate. The rate is square-root proportional to the precursor supply rate. Hydrogen as a carrier gas enhances the decomposition of the precursor and results in higher deposition rate by ~1.5 times that obtained in argon. However, the selectivity of Cu film is lower for the case of hydrogen carrier gas than the argon carrier gas. Kobayashi et al. investigated the gap-filling property and deposition rate depending on deposition conditions using hydrogen carrier gas [169, 170]. A higher deposition rate is achieved with higher temperatures. However, the gap-filling characteristics deteriorate with higher temperature at constant precursor partial pressure, and voids can be formed at the center and along the wall of the trench, possibly due to insufficient supply of the precursor. To obtain good gap filling, the precursor partial pressure should be increased, but it is restricted by the saturated vapor pressure. A satisfactory characteristic of filling a hole with an aspect ratio of 5.5 and deposition rate of 40 nm/min was demonstrated at 200 Pa of precursor partial pressure, 170 °C where the deposition rate is saturated with a Langmuir-type reaction in the surface-reaction-limited region.

Other precursors for Cu deposition have been developed. Hydrated copper formate ($Cu^{II}(OCHO)_2 \cdot xH_2O$) offers advantages of a simple and cheaper source, higher efficiency, and better safety [171]. Cu(hfcac)·(mhy), where mhy is 2-methly-1-hexen-3yne ($C_2H_5C{\equiv}CC(CH_3){=}CH_2$), offers better stability than Cu(hfacac)·(tmvs) [172].

Epitaxial growth of Cu by MOCVD has been also demonstrated. Tu et al. epitaxially grew Cu on (0001) *c*-plane of single-crystalline sapphire substrate using copper(II) acetylacetonate [$Cu(acac)_2$, $Cu(C_5H_7O_2)_2$] precursor [173]. A highly crystalline Cu film (Lotgering factor ~ 0.96) with (111) plane on the surface was grown at 350 °C. However, the surface showed a granular

Table 14.7 List of compounds used in the MOCVD deposition of pure Au. Source: Parkhomenko et al. 2015 [175].

Compound name	Chemical formula	Notes
Dimethylgold(III) dimethyldithio-phosphate	$((CH_3)_2AuSSP(OMe)_2$	Me = CH_3
Dimethylgold(III) diethyldithio-phosphate	$((CH_3)_2AuSSP(OEt)_2$	Et = C_2H_5
Dimethylgold(III) diethyldithio-phosphinate	$((CH_3)_2AuSSP^iBu_2$	$^iBu = CH_2CH(CH_3)_2)$
Dimethylgold(III)acetate	$([(CH_3)_2AuOOCCH_3]_2$, $[Me_2Au(Ac)]_2$,	Ac = $OOCCH_3$
Dimethylgold(III)pivalate	$((CH_3)_2AuOOCC(CH_3)_3$, $[Me_2Au(Piv)]_2$	Piv = $OOCC(CH_3)_3$
Dimethylgold(III)2,2,6,6-tetramethyl-3,5-heptanedionate	$((CH_3)_2AuC(CH_3)_3C(O)CHC(O)C(CH_3)_3$, $Me_2Au(tmhd$	tmhd = $C(CH_3)_3C(O)CHC(O)C(CH_3)_3)$
Dimethylgold(III) diethyldithio-carbamate	$(CH_3)_2AuSSCN(C_2H_5)_2$, $Me_2Au(dtc)$	dtc = $SSCN(C_2H_5)_2)$

morphology. Other metals can be deposited similarly using hexafluoroacetylacetone (hfacac), trifluoroacetylacetone (tfacac), dipivaloylmethane (dpm, $C_{11}H_{20}O_2$), and acetylacetone (acac) complexes of Ni, iron (Fe), hafnium (Hf), zirconium (Zr), and platinum (Pt) [174]. Pure gold (Au) thin films were recently demonstrated using various precursors including those listed in Table 14.7.

14.3.3.2 *Oxides for Transparent Conductive Electrodes (TCEs)*

Most optically transparent and electrically conducting oxides (TCOs) are impurity-doped binary or ternary compounds with one or two metallic elements. The important TCOs are impurity-doped zinc oxide (ZnO), indium oxide (In_2O_3), tin oxide (SnO_2), and cadmium oxide (CdO); ternary Zn_2SnO_4, $ZnSnO_3$, $Zn_2In_2O_5$, $Zn_3In_2O_6$, In_2SnO_4, and $CdSnO_3$; and multicomponent oxides consisting of combinations of the TCOs. Their resistivity should be as low as possible (typically of the order of 10^{-4} Ω cm or lower), and their extinction coefficient k in the optical visible spectral range should also be low (typically lower than 10^{-4}), with their bandgap energy (E_g) greater than 3 eV. This remarkable combination of conductivity and transparency is usually impossible in intrinsic stoichiometric oxides; however, it is achieved with a nonstoichiometric composition or by introducing appropriate dopants. Thin CdO films were the first materials to be identified as a TCO [176]. Later, thin films of ZnO, SnO_2, In_2O_3, and their alloys were found to function as TCOs [177]. Doping of these oxides resulted in improved electrical conductivity without degrading their optical transmission, such as aluminum-doped ZnO (ZnO:Al, AZO), gallium-doped ZnO (ZnO:Ga, GZO), tin-doped In_2O_3, (In_2O_3:Sn, ITO), antimony-doped SnO_2 (SnO_2:Sb, ATO), and fluorine-doped SnO_2 (SnO_2:SF, FTO). The actual and potential applications of TCO thin films include: (i) transparent electrodes for flat-panel displays, (ii) transparent electrodes for photovoltaic cells, (iii) low-emissivity windows, (iv) window defrosters, (v) transparent thin-film transistors, (vi) LEDs, and (vii) semiconductor lasers. As the usefulness of TCO thin films depends on both their optical and electrical properties, both parameters should be considered together with environmental stability, abrasion resistance, electron work function, and compatibility with substrate and other components of a given device, as appropriate for the application. The availability of the raw materials and the economics of the deposition method are also significant factors in choosing the most appropriate TCO material.

Tin-doped In_2O_3 (ITO) is the most widely used transparent conductive electrode (TCE) among the many utilized TCOs. ITO thin films have been mostly deposited by physical vapor deposition. The first ITO as a TCE by CVD was demonstrated by Kane et al. using dibutyl tin diacetate ($(CH_3CH_2CH_2CH_2)_2Sn(OCOCH_3)_2$) and indium chelate derived from dipivaloyl methane with oxygen as precursors [178]. The optimum temperature of deposition was 550 °C. The specific resistivity of the films was in the range of $4.8–5.8 \times 10^{-4}$ Ω cm, and the average optical transmission was higher than 80%. Akinwunmi et al. [179] demonstrated a deposition process by pyrolysis of a single-source precursor for ITO thin films. The precursor, mixed acetylacetonate, was prepared by mixing indium trichloride dissolved in methanol and tin tetrachloride and 2,4-pentanedione in methanol with the addition of sodium acetate dissolved in methanol. The film deposited at 380 °C showed an optimum value for optical transmission (>80%) and the lowest resistivity (7.2×10^{-4} Ω cm). The bandgap energy was measured to be 3.76 eV. While most ITO films are transparent in the visible spectrum, they start absorbing light in the ultraviolet region, which means they cannot be used in UV photonic devices. To overcome the UV bottleneck, Chen et al. used MOCVD to deposit textured crystalline ITO for UV-TCE [180]. Trimethylindium (TMIn, $In(CH_3)_3$), tetrakis-dimethylamino tin (TDMASn, $C_8H_{24}N_4Sn$), oxygen, and argon were used as the precursors of In, Sn, O, and carrier gas to deposit the films at a temperature of 500 °C. An optical bandgap energy of ~4.7 eV was achieved by the Burstein–Moss effect [181] from the ITO grown by MOCVD, which is significantly higher than those deposited by PVD. The average transmittance in UVA (315–400 nm) and UVB (280–315 nm) ranges is as high as 94% and 74%, respectively.

Cadmium oxide (CdO)-based TCO can be an alternative TCE. $Cd(hfa)_2(TMEDA)$ (hfa = hexafluoroacetylacetonate and TMEDA = tetramethylethylenediamine) is used as precursor for Cd [182, 183]. Ga-doped CdO (CGO) and In-doped CdO (CIO) thin films grown at 410 °C by MOCVD exhibit excellent optical transparency, with an average transmittance of >80% in the visible range. Ga and In doping widens the optical bandgap from 2.85 to 3.08 and 3.18 eV, respectively, via a Burstein–Moss shift [184].

14.3.3.3 High-Temperature Superconductors

High-temperature superconductors (HTSs) are the class of materials that show superconductivity with zero resistivity at temperatures higher than 23 K [185]. The first meaningful HTS with operating temperature higher than the boiling point of liquid nitrogen (77 K) was developed in 1987. Wu et al. reported HTS at 93 K using a $YBa_2Cu_3O_{7-\delta}$ (YBCO) system [186]. Since then, HTSs have been adopted in engineering applications such as transmission cables, transformers, motors, and generators [187]. YBCO and other HTS films can be deposited by various techniques such as pulsed laser deposition (PLD) [188], metalorganic deposition (MOD) [189], and MOCVD. Among the methods, MOCVD is the most versatile and large-scale production-friendly technique [190, 191]. The first HTS film prepared by MOCVD was reported using β-diketonate metal chelates of Y, Ba, and Cu [192–194]. Since then, the single-source liquid precursor for YBCO deposition was prepared by mixing the tetramethyl heptanedionate (tmhd: 2,2,6,6,-tetramethyl-3,5-heptanedionate) compounds of Y, Ba, and Cu in tetrahydrofuran (THF) solvent at the appropriate mole ratio [195]. Since most MOCVD deposition uses a premixed single precursor and the deposition behavior of each metallic element is dependent on growth parameters, such as temperature and oxygen flow rate [196–198], accurate control of precursor mixing for target composition of the film is important. Other precursors are described in [199]. The MOCVD method is used to fabricate second-generation HTS wire in a reel-to-reel growth chamber with I_c of 200 A/cm in kilometer lengths and 300 A/cm in 600-m lengths [200].

14.3.3.4 Silicides

Semiconducting β-$FeSi_2$ has attracted much attention over the past decade as one of the promising materials for infrared optoelectronic devices and photovoltaic solar cells on Si substrates. Akiyama et al. demonstrated MOCVD epitaxial growth of the film using iron pentacarbonyl [$Fe(CO)_5$] and monosilane (SiH_4) on various Si substrates [201, 202]. (101)-oriented epitaxial film was formed on a (111) Si substrate at 750 °C with a (202) full width at half-maximum value of 0.46°. (100)-oriented epitaxial $FeSi_2$ film with flat surfaces was formed on Si (001) substrates. The PL emission at 10 K was observed corresponding to a bandgap energy of 0.806 eV by post-annealing.

14.4 Past, Present, and Future Commercial Applications

Extrapolating from the recent past, it is relatively easy to predict the continued expansion of MOCVD materials grown for both research and commercial applications, as discussed here for many different specific materials. In this section, we address some specific commercial device applications of MOCVD-grown materials.

14.4.1 LEDs

As mentioned earlier, MOCVD was first used for the growth of high-performance AlGaAs-GaAs-based infrared laser diodes in 1977 by Dupuis et al. [39]. This was the primary focus of many of the early workers who were exploring the limits of MOCVD-grown materials. Subsequently, many workers developed MOCVD growth technologies for large-scale production of visible AlInGaP (red, amber, yellow) and InAlGaN (green, blue) LEDs. Today, virtually all visible- and infrared-emitting LEDs are grown using MOCVD including the III-N blue LEDs that are produced in the largest volumes of all LEDs. We can expect that for the foreseeable future, the expansion of MOCVD growth capacities for compound semiconductor LEDs will continue, albeit at a somewhat slower pace than has occurred in the past 10 years since much of the initial demand for standard white light bulbs for solid-state lighting applications has been met in many of the advanced markets. The future demand for MOCVD visible LEDs could dramatically increase as the standard linear fluorescent lighting fixtures common in most commercial buildings are replaced with more-efficient drop-in replacements powered by warm-white LEDs. Furthermore, new types of indoor lighting are being developed beyond the standard fixture approaches used in the past, and large buildings are using exterior lighting as an architectural element, a trend that is likely to expand greatly in the future.

However, there are potential new markets opening up as MOCVD-grown micro-LED displays [203] are being developed for watches and other small energy-efficiency personal displays, as well as more effective large-area displays. For example, Sony announced in 2012 the development of a large-area micro-LED full-color display called a *Crystal LED Display*. It was a 55-in. diagonal prototype full-color display with 6 million small LEDs – 2 million each of red, green, and blue – to create a full high-definition display with performance exceeding that of a traditional LCD [204]. Samsung is also pursuing micro-LED technology for large-area displays – called QLED technology – that could replace both LCDs and organic LEDs (OLEDs) in many applications [205]. These micro-LED displays combine III-N LEDs with quantum dots to create a large gamut of colors [206]. Apple is also reportedly developing MOCVD-grown visible micro-LED displays for its mobile products [207]. If these new displays are incorporated into both small-area portable devices and large-area displays, the demand for MOCVD-grown LED materials will expand dramatically.

14.4.2 Lasers

As noted earlier, Dupuis et al.'s (1977a) [39] report of the 300 K operation of AlGaAs-GaAs double-heterostructure lasers and subsequent demonstration by Dupuis, et al. (1978a) [208] of the first 300 K operation of quantum-well lasers established the capability of MOCVD for the growth of advanced diode lasers. In addition, in 1978, Dupuis et al. [209] grew distributed Bragg reflectors (DBRs) by MOCVD and used them in laser structures. MOCVD-grown laser diodes now dominate all applications where compact, high-efficiency sources of coherent mid-wave infrared (IR), near-IR, and visible radiation are required, including for high-speed telecommunications, BluRay players, vertical-cavity surface-emitting lasers (VCSELs), infrared sensor and monitoring systems, optical face-recognition systems, medical instruments, and laser welders. For example, both MOCVD [210] and MBE [211] have been used for the growth of VCSELs. However, at this time, MOCVD is used worldwide to grow VCSELs, and the price of these devices has dramatically decreased as a result. As another example, while the first infrared quantum-cascade lasers (QCLs) reported in 1994 by Faist et al. [212] were grown by MBE, and the first MOCVD-grown QCLs were not reported until 2003 by Green et al. [213, 214], MOCVD-grown QCLs now have equivalent performance to similar structures grown by MBE [215, 216], and it can be expected that soon, due to their high performance [217] and reduced cost of growth, MOCVD QCLs will dominate the field in the coming years.

A variety of medical applications for III-N LDs is also under development. For example, some types of medicines for the treatment for cancer can be photoactivated by using UV laser beams.

Another area of future development is the MOCVD growth of III-N lasers for LIDAR (light detection and ranging) optical sources [218]. Because of the wide field of use of MOCVD-grown lasers of all kinds and the flexibility and control of this process, it can be predicted that the dominance of MOCVD will continue and that the application of MOCVD for all types of LDs will expand.

14.4.3 OEICs

The use of integrated optical elements for information processing was first proposed by Miller [219], who coined the term *integrated optics*. In the past 10 years, optoelectronic integrated circuits (OEICs) using these early concepts – and many newer concepts – have become commercially viable, largely due to the flexibility, control, and other properties of MOCVD for the growth of complex semiconductor device structures [220]. This application of MOCVD is unique in that it is the only materials growth technology that has been demonstrated to have the full suite of capabilities required to produce the complicated structures required for high-speed PIC applications in the optical backplane of the Internet. For example, MOCVD-based PIC technology was developed in 2018, by Infinera Corporation to demonstrate system-on-chip (SOC) devices having a scalable *C*-band coherent tunable multichannel InP-based architecture for dense wavelength-division multiplexing (DWDM) applications [221]. These PICs are designed to be used in both receiver and transmitter functionality. The fully integrated SOCs consist of 14-channel transmitter and receiver PICs with independent tunable lasers operating at 33 and 44 Gbaud per channel with 16 quadrature amplitude modulation (QAM) dual-polarization modulation, enabling the operation at 1.2 Tb/s payload capacity into a single fiber. These are the most complex III-V PICs demonstrated to date. We can fully expect that MOCVD-grown PICs will continue to dominate the market for high-speed communications and that the future demand for increased capacities will call for even higher levels of integration. These challenges will be met by the development of even more highly sophisticated MOCVD growth and regrowth processes.

14.4.4 High-Speed Electronics

High-speed semiconductor electronics based upon MOCVD-grown transistors have become the backbone of many important applications, including the Internet. MOCVD was first demonstrated for the growth of AlGaAs-GaAs heterostructure bipolar transistors by Milano et al. in 1989 [222].

High-performance MOCVD-grown GaAs metal-semiconductor field-effect transistors (MESFETs) were demonstrated as early as 1981 by Kamei et al. [223]. High-performance AlGaAs-GaAs high electron mobility transistors (HEMTs) grown by MOCVD were reported by Koybashi et al. [224] in 1984. Since that time, MOCVD has become an important materials growth technology for a variety of high-speed unipolar and bipolar transistors. For example, MOCVD-grown double-heterojunction bipolar transistors (DHBTs) composed of InP emitter and collector regions and a graded GaAsSb graded-composition base layer have recently been reported by Bolognesi et al. [225] to operate at frequencies as high as $f_T = 503$ GHz and $f_{MAX} = 779$ GHz at a power density of 10 mW/μm^2 and a $BV_{CEO} = 41$ V. This work is expected to advance and has already become a commercialized technology for lower frequencies. We can project that MOCVD-grown InP-GaAsSb and InP-InGaAs-InAlAs HEMT technologies will continue to improve and provide a viable path toward multi-GHz transistors and circuits.

Another important future application is the use of MOCVD-grown III-V materials and device in integration with Si device technology. Work on the direct integration of GaAs epitaxial layers grown on Si by MOCVD has been studied at least since the work of Akiyama et al. [226]. Recently, in 2015, Kohen et al. [227] reported the fabrication of III-V virtual substrates grown by MOCVD on 200-mm Si wafers for the integration of III-Vs and Si device on the same circuit. Also, in 2016, Fitzgerald et al. [228] reported the integration of MOCVD-grown AlGaN-GaN HEMT devices on full Si CMOS wafers. This is potentially an area that could open an entirely new and expanding demand for MOCVD III-N electronic devices at the 5-nm node of Si IC technology.

14.4.5 High-Power Electronics

High-power electronics is an important, relatively new application for MOCVD-grown devices. Much of the work in this area is devoted to the exploitation of the advantageous properties of III-N materials. Khan et al. [229] were the first to report the growth of GaN-AlGaN high-electron-mobility heterostructures by MOCVD in 1991, and Khan et al. [230] also reported the first AlGaN/GaN HEMTs in 1993. Since that time, MOCVD has become the dominant materials growth technology for high-power mm-wave electronics used, for example, in advanced radar systems [231]. For example, the United States' Patriot Missile Defense Systems are being upgraded to use MOCVD-grown III-N power RF circuits [232]. Currently, high-power GaN mm-wave radar systems based upon MOCVD-grown devices are the state of the art and are being incorporated into many existing and future defense systems [233, 234]. It is obvious that future generations of these electronic systems for sensing, jamming, and other defense-related functions will be created from MOCVD-grown III-N materials.

Another important application for GaN-based high-power and high-frequency electronic devices is in LIDAR systems [235]. LIDAR is being used in autonomous vehicle systems, and this is destined to become a large and important market for MOCVD-grown GaN devices both for the light sources and the high-speed driver electronics in the LIDAR system. We can fully expect that MOCVD III-N materials will come to dominate this important application and that these systems will become ubiquitous throughout the world. As autonomous vehicles become more common, energy efficiency will be increasingly important as well, and GaN-based electronic power systems for electric vehicles will become dominant.

Another innovation is the incorporation of graphene with MOCVD-grown GaN to create a new type of hot-electron transistor for high-power electronics [236]. This work is still in its infancy, but it is clearly an area with a lot of potential for expansion in many applications in the future.

14.4.6 Solar Cells

The first MOCVD-grown solar cells were reported by Dupuis et al. in 1977 [237]. Since that time, MOCVD has come to dominate the mass production of III-V compound semiconductor solar cells, and MOCVD multijunction cells now hold the world record for efficiency. For example,

Spectrolab's NeXt Tiple-Junction (XTJ) solar cells have air-mass-zero (AM0) efficiencies as high as 29.5% with no concentration [238]. High-efficiency III-V heterostructure solar cells have become the power source of choice for many important space systems, including the Opportunity and Spirit Mars Rovers [239] launched in 2003. The next-generation Iridium satellite global communication system composed of 66 low-earth orbit (LEO) satellites and the One-Web system of 882 satellites will use MOCVD multijunction solar cells [240, 241]. We can expect that as the push to more near-earth and geosynchronous orbit space-based systems increases, the use of MOCVD solar cells will expand greatly [242]. Other MOCVD-grown materials are also being developed for photovoltaic applications. For example, novel MOCVD systems have been developed for the large-area MOCVD growth of CdTe thin films that are being developed for low-cost solar-power systems [243–245].

14.4.7 Detectors

MOCVD has been used widely to grow a broad range of both photon and particle detectors. These include high-speed PIN avalanche photodiodes (APDs) and linear-mode photodiodes. The first MOCVD grown PIN photodiodes were described by Ito et al. [246]. Since that time, a wide variety of advance III-V photodiodes have been demonstrated including avalanche photodiodes with separate absorption and gain regions [247]. Many of the existing optical communications systems employ MOCVD-grown APDs. In fact, as noted earlier, the PICs being used today for the highest-performance commercial telecommunications employ MOCVD-grown photodetectors integrated with waveguides, splitters, combiners, and modulators as well as laser diodes. As these applications increase in capacity and the demand for fiber-to-the-home (FTTH) services expands, MOCVD technology will continue to expand to meet these needs.

MOCVD-grown GaAs radiation detectors were reported as early as 1995 by Lauter et al. [248]. MOCVD has also been used to grow a variety of other types of particle detectors, including III-N HEMT devices for use in medical imaging systems [249] and GaN layers for ion-photon-emission microscopes [250]. While this area is just beginning to be developed, GaN PIN and other types of particle detectors could become increasingly important for the detection of nuclear materials in security applications.

As noted previously, an additional area of strong interest is the use of MOCVD for the growth of II-VI compound semiconductor materials for long-wavelength infrared (LWIR) photodetectors in the 3–15 μm spectral region as well as for particle detectors. As noted, Manasevit and Simpson reported the first MOCVD growth of wider-bandgap II-VI semiconductors, including ZnTe and CdTe, in 1971 [33]. Thick CdTe layers grown by MOCVD on Si substrates are being developed for nuclear imaging and for use as gamma particle detectors [251]. Mullin and Irvine first reported the MOCVD growth of the narrow-gap HgCdTe alloys on CdTe substrates in 1981 [252]. More recently, efforts have been made to develop MOCVD-growth of HgCdTe films and CdTe/sapphire substrate templates used for the LPE growth of narrow-bandgap HgCdTe materials. MOCVD growth of II-VI semiconductors has been used to supply devices for IR photodetector arrays for many important applications including surveillance and space-based earth monitoring [253, 254] as well as use in the NICMOS LWIR camera system used on the Hubble Space Telescope in studying the origins of the universe [255, 256].

14.5 Summary and Conclusions

MOCVD has been under continuous development for over 50 years since it was invented by Harold Manasevit in 1967 and now dominates most of the important research, development, and production applications for III-V compound semiconductor devices. MOCVD technology has also been applied to the creation of novel 2D materials and is expected to play an important role in the commercialization of this class of semiconductor materials. The impact of MOCVD on the development of many important systems has been immense, including in the areas of solid-state lighting, telecommunications, solar power, high-speed electronics, integrated optics, and LIDAR. It is obvious that applications for MOCVD will continue to expand in the future and, as Si-based electronics incorporates active optical functions and devices, that this technology will come to play a truly ubiquitous role in the expansion of the capabilities of the information age.

Acknowledgments

Special thanks go to Gary Tompa (Structured Materials Industries) for his valuable input on Ga_2O_3 MOCVD. In addition, the authors would like to acknowledge Shiping Guo (AMEC) and Wei Zhang (TOPEC) for their overview of the production MOCVD industry. RDD also acknowledges the support of the Steve W. Chaddick Endowed Chair in Opto-Electronics and the Georgia Research Alliance.

References

1. Braun, F. (1874). Uber die Stromleitung durch Schwefelmetalic. *Annalen der Physik and Chemie* 153 (4): 556–563. Reprinted in English as: On the current conduction in metal sulphides. In: *Semiconductor Devices: Pioneering Papers* (ed. S.M. Sze): 377–380. Singapore: World Scientific Publishing Co., 1991.
2. Wilson, A.H. (1931a). The theory of electronic semi-conductors. *Proceedings of the Royal Society of London, Series A* 133 (822): 458–491
3. Wilson, A.H. (1931b). The theory of electronic semi-conductors II. *Proceedings of the Royal Society of London, Series A* 134 (823): 277–328.
4. Wilson, A.H. (1931c). In Wilson 1931a, he concludes: "From the experimental side, therefore, the existence or non-existence of semi-conductors remains an open question… Theoretically there is no reason why semiconductors should not exist…"
5. Pauli, W. (1931). Über Halbleiter soll man nicht arbeiten, das ist eine Schweinerei; wer weiss, ob es überhaupt Halbleiter gibt. Letter to Peierls, 29 September 1931. In: Wolfgang Pauli – Wissenschaftlicher Briefwechsel mit Bohr, Einstein, Heisenberg u.a. Band II: 1930–1939. Springer, 1985. Translation: "Semiconductors are not supposed to work; this is a mess; who knows whether there are semiconductors at all."
6. Ohl, R. (1941a). Alternating current rectifier. US Patent 2 402 661.
7. Ohl, R. (1941b). Light-sensitive electric device. US Patent 2 402 662.
8. Riordan, M. and Hoddeson, L. (1997). The origins of the pn junction. *IEEE Spectrum* (June): 46–51.
9. Bardeen, J. and Brattain, W.H. (1948). The transistor, a semi-conductor triode. *Physical Review* 74: 230–231.
10. Riordan, M., Hoddeson, L., and Herring, C. (1999). The invention of the transistor. *Review of Modern Physics* 71 (2): S336–S345.
11. Wajda, E.S., Kippenhan, B.W., and White, W.H. (1960). Epitaxial growth of silicon. *IBM Journal of Research and Development* 4 (3): 288–295.
12. Marinace, J.C. (1960). Tunnel diodes by vapor growth of Ge on Ge and on GaAs. *IBM Journal of Research and Development* 4 (3): 280–282.
13. Okada, T., Kano, T., and Sasaki, Y. (1961). Epitaxial vapor growth of GaAs on Ge single crystals. *Journal of the Physical Society of Japan* 16 (12): 2591–2591.
14. Holonyak, N. Jr., Jillson, D.C., and Bevacqua, S.F. (1962). Halogen vapor transport and growth of epitaxial layers of intermetallic compounds and compound mixtures. *AIME Conf.* (August), Los Angeles. In: *Metallurgy of Semiconductor Materials* (ed. J.B. Schroeder), vol. 15. New York: Interscience Publishers.
15. Nelson, H. (1963). Epitaxial growth from the liquid state and its application to the fabrication of tunnel and laser diodes. *RCA Review* 24 (4): 603–615.
16. Closed-tube VPE was carried out in a closed sealed quartz ampoule that was evacuated after the basic compounds (e.g. small crystals of GaAs and GaP) were inserted with a small amount of a halogen – typically iodine – was added as a transport agent. The ampoule was placed in a furnace with a temperature gradient to provide for the transport of the precursor chemicals to a seed crystal or for self-nucleation and deposition on the ampoule wall. See Holonyak, N. Jr. et al. [14].

17. Holonyak, N. Jr., Bevacqua, S.F., Bielan, C.V. et al. (1963). Electrical properties of Ga($As_{1-x}P_x$) *p-n* junctions. *Proceedings of the IEEE* 51, 364.
18. Alexander, F.B., Bird, V.R., Carpenter, D.R. et al. (1964). Spontaneous and stimulated infra-red emission from indium phosphide arsenide diodes. *Physics Letters A* 4 (1): 13–15.
19. Hall, R.N., Fenner, G.E., Kingsley, J.D. et al. (1962) Coherent light mission from GaAs junctions. *Physics Review Letters* 9 (9): 366–368.
20. Holonyak, Jr., N. and Bevacqua, S.F. (1962) Coherent (visible) light emission from Ga(AsP) junctions. *Applied Physics Letters* 1 (4): 82–83.
21. Ruehrwein, R.A. (1965a). Alternating proportions in vapor deposition process to form a mixed crystal graded energy gap. US Patent 3 224 913, filed July 31, 1961.
22. Ewing, R.E, and Greene, P.E. (1964). Influence of vapor composition on the growth rate and morphology of gallium arsenide epitaxial films. *Journal of the Electrochemical Society* 111 (11): 1266–1269.
23. Ruehrwein, R.A. (1965b) Use of hydrogen halide and hydrogen in separate streams as carrier gases in vapor deposition of III-V compounds. US Patent 3 218 205, filed July 13, 1962.
24. Theuerer, H.C. (1961). Epitaxial silicon films by the hydrogen reduction of $SiCl_4$. *Journal of the Electrochemical Society* 108 (7). 649–653.
25. Rupprecht, H., Woodall, J.M., and Petit, G.D. (1967). Efficient visible electroluminescence at 300°K from $Ga_{1-x}Al_xAs$ *p-n* junctions grown by liquid-phase epitaxy. *Applied Physics Letters* 11 (3): 81–83.
26. Alferov, Zh.I., Andreev, V.M., Korol'kov, V.I. et al. (1968). Injection properties of n-$Al_xGa_{1-x}As$-p-GaAs heterojunctions. *Fizika i Tekhnika Poluprovodnikov* 2 (7): 1016–1017.
27. Craford, M.G. and Groves, W.O. (1973). Vapor phase epitaxial materials for LED applications. *Proceedings of the IEEE* 61 (7): 862–880.
28. Hayahsi, I., Panish, M.B., and Reinhart, F.K. (1971). GaAs-$Al_xGa_{1-x}As$ heterostructure injection lasers. *Journal of Applied Physics* 42 (1): 1929–1941.
29. Manasevit, H.M. (1968). Single-crystal gallium arsenide on insulating substrates. *Applied Physics Letters* 12 (4): 156–159.
30. Manasevit, H.M. and Simpson, W.I. (1969). The use of metalorganics in the preparation of semiconductor materials – I: Epitaxial gallium-V compounds. *Journal of the Electrochemical Society* 116 (12): 1725–1732.
31. Manasevit, H.M. (1971). The use of metalorganics in the preparation of semiconductor materials: III. Studies of epitaxial III-V aluminum compound formation using trimethylaluminum. *Journal of the Electrochemical Society* 118 (4): 647–650.
32. Manasevit, H.M., Erdman, F.M., and Simpson, W.I. (1971). The use of metalorganics in the preparation of semiconductor materials: IV. The nitrides of aluminum and gallium. *Journal of the Electrochemical Society* 118 (11): 1864–1868.
33. Manasevit, H.M. and Simson, W.I. (1971). The use of metalorganics in the preparation of semiconductor materials: II. II-VI Compounds. *Journal of the Electrochemical Society* 118 (4): 644–647.
34. Manasevit, H.M. (1972) The use of metalorganics in the preparation of semiconductor materials: Growth on insulating substrates. *Journal of Crystal Growth* 13/14: 306–314.
35. Arthur, J.R. and LePore, J.J. (1969). GaAs, GaP and $GaAs_xP_{1-x}$ epitaxial films grown by molecular beam deposition. *Journal of Vacuum Science and Technology* 6 (4): 545–548
36. Cho, A.Y. (1970). Morphology of epitaxial growth of GaAs by a molecular beam method: The observation of surface structures. *Journal of Applied Physics* 41 (7): 2780–2786.
37. Cho, A.Y. (1971). Growth of periodic structures by the molecular-beam method. *Applied Physics Letters* 19 (11): 467–468.
38. Foxon, C.T. (1973). Molecular beam epitaxy. *Acta Electronica* 16 (4): 323–328.
39. Dupuis, R.D. and Dapkus, P.D. (1977a) Room-temperature operation of $Ga_{(1-x)}Al_xAs$/GaAs double-heterostructure lasers grown by metalorganic chemical vapor deposition. *Applied Physics Letters* 31 (7): 466–468.
40. Dupuis, R.D., Dapkus, P.D., Yingling, R.D., and Moudy, L.A. (1977). High-efficiency GaAlAs/GaAs heterostructure solar cells grown by metalorganic chemical vapor deposition. *Applied Physics Letters* 31 (3): 201–203.
41. Veeco Instruments Inc. v. SGL Carbon, LLC et al. (2017). Case number: 1:17-cv-02217, US Civil Court Records for the Eastern District of New York.
42. Li, X., Wang, S., Wei, Y.O. et al. (2015). Growth of high-quality AlN layer on sapphire substrate at relatively low temperatures by metalorganic chemical vapor deposition. *Physica Status Solidi B* 252 (5): 1089–1905.
43. Li, X., Wang, S., Liu, H. et al. (2017). 100-nm thick single-phase wurtzite BAlN films with boron contents over 10%. *Physica Status Solidi B* 254 (8): 1600749.
44. Choi, S., Kim, H.J., Lochner, Z. et al. (2014). Origins of unintentional incorporation of gallium in AlInN layers during epitaxial growth, part I: Growth of AlInN on AlN and effects of prior coating. *Journal of Crystal Growth* 388: 137–142.

45. Green, A.J., Chabak, K.D., Heller, E.R. et al. (2016). 3.8-MV/cm breakdown strength of MOVPE-grown Sn-doped β-Ga_2O_3 MOSFETs. *IEEE Electron Device Letters* 37 (7): 902–905.
46. Web of Science. (2017). webofknowledge.com.
47. Fathololoumi, S., Dupont, E., Chan, C.W.I. et al. (2012) Terahertz quantum cascade lasers operating up to ~200 K with optimized oscillator strength and improved injection tunneling. *Optics Express* 20 (4): 3866–3876.
48. Mimura, T., Hiyamizu, S., Fujii, T., and Nanbu, K. (1980). A new field-effect transistor with selectively doped GaAs/*n*-$Al_xGa_{1-x}As$ heterojunctions. *Japanese Journal of Applied Physics* 19 (5): L225–227.
49. Mimura, T. (2002). The early history of the high electron mobility transistor (HEMT). *IEEE Transactions on Microwave Theory and Techniques* 50: 780–782.
50. Shockley, W. (1951). US Patent 2 569 374.
51. Kroemer, H. (1957). Theory of a wide-gap emitter for transistors. *Proceedings of the IRE* 45: 1535–1537.
52. Chattopadhyay, D., Sutradhar, S.K., and Nag, B.R. (1981). Electron transport in direct-gap III-V ternary alloys. *Journal of Physics C: Solid State Physics* 14: 891–908.
53. Whelan, C.S., Marsh, P.F., Hoke, W.E. et al. (2000). GaAs metamorphic HEMT (MHEMT): an attractive alternative to InP HEMTs for high performance low noise and power applications. *Conference Proceedings: International Conference on Indium Phosphide and Related Materials*.
54. Alamo, J.A.D. (2013). Nanometer-scale InGaAs field-effect transistors for THz and CMOS technologies. *Proceedings of the ESSCIRC*: 16–21.
55. Leuther, A., Tessmann, A., Doria, P. et al. (2014). 20 nm metamorphic HEMT technology for terahertz monolithic integrated circuits. *European Microwave Integrated Circuit Conference (EuMIC)*. IEEE.
56. Ajayan, J. and Nirmal, D. (2017). 20-nm enhancement-mode metamorphic GaAs HEMT with highly doped InGaAs source/drain regions for high-frequency applications. *International Journal of Electronics* 104: 504–512.
57. Takahashi, T., Sasa, S., Kawano, A. et al. (1994). High-reliability InGaP/GaAs HBTs fabricated by self-aligned process. *Proceedings of IEEE International Electron Devices Meeting*: 191–194.
58. Deal, W.R., Leong, K., Yoshida, W. et al. (2016). InP HEMT integrated circuits operating above 1,000 GHz. *IEEE International Electron Devices Meeting (IEDM)*: 29.1.1–29.1.4.
59. Lai, R., Mei, X.B., Deal, W.R. et al. (2007). Sub 50 nm InP HEMT Device with F_{max} Greater than 1 THz. *IEEE International Electron Devices Meeting*: 609–611.
60. Kim, D.H., Alamo, J.A.d., Chen, P. et al. (2010). 50-nm E-mode $In_{0.7}Ga_{0.3}As$ PHEMTs on 100-mm InP substrate with f_{max} > 1 THz. *International Electron Devices Meeting*: 30.6.1–30.6.4.
61. Urteaga, M., Carter, A., Griffith, Z. et al. (2016). THz Bandwidth InP HBT Technologies and Heterogeneous Integration with Si CMOS. *IEEE Bipolar/BiCMOS Circuits and Technology Meeting (BCTM):* 35–41.
62. Kroemer, H. (2004). The 6.1 Å family (InAs, GaSb, AlSb) and its heterostructures: a selective review. *Physica E: Low-dimensional Systems and Nanostructures* 20: 196–203.
63. Bennett, B.R., Magno, R., Boos, J.B. et al. (2005). Antimonide-based compound semiconductors for electronic devices: A review. *Solid-State Electronics* 49: 1875–1895.
64. Liu, H.G., Ostinelli, O., Zeng, Y., and Bolognesi, C.R. (2007). 600 GHz InP/GaAsSb/InP DHBTs grown by MOCVD with a Ga(As,Sb) graded-base and $f_T \times BV_{CEO}$ > 2.5 THz-V at room temperature. *IEEE International Electron Devices Meeting*: 667–670.
65. Datta, S., Ashley, T., Brask, J. et al. (2005). 85 nm gate length enhancement and depletion mode InSb quantum well transistors for ultra high speed and very low power digital logic applications. *IEEE International Electron Devices Meeting. IEDM Technical Digest*: 763–766.
66. Dolginov, L.M., Druzhinina, L.V., Eliseev, P.G. et al. (1978). Injection heterolaser based on InGaAsSb four-component solid solution. *Soviet Journal of Quantum Electronics* 8: 416.
67. Kobayashi, N., Horikoshi, Y., and Uemura, C. (1980). Room temperature operation of the InGaAsSb/AlGaAsSb DH laser at 1.8 µm Wavelength. *Japanese Journal of Applied Physics* 19: L30–L32.
68. Kano, H. and Sugiyama, K. (1980). 2.0 µm C.W. operation of GaInAsSb/GaSb d.h. lasers at 80 K. *Electronics Letters* 16: 146–147.
69. Caneau, C., Srivastava, A.K., Dentai, A.G. et al. (1985). Room-temperature GaInAsSb/AlGaAsSb DH injection lasers at 2.2 µm. *Electronics Letters* 21 (18): 815–817.
70. For example: Shterengas, L., Belenky, G., Kisin, M.V., and Donetsky, D. (2007). High power 2.4 µm heavily strained type-I quantum well GaSb-based diode lasers with more than 1W of continuous wave output power and a maximum power-conversion efficiency of 17.5%. *Applied Physics Letters* 90: 011119-1–3.
71. Wang, C.A., Jensen, K.F., Jones, A.C., and Choi, H.K. (1996). n-AlGaSb and GaSb/AlGaSb double-heterostructure lasers grown by organometallic vapor phase epitaxy. *Applied Physics Letters* 68: 400–402.
72. Allerman, A.A., Biefeld, R.M., and Kurtz, S.R. (1996). InAsSb-based mid-infrared lasers (3.8–3.9 µm) and light-emitting diodes with AlAsSb claddings and semimetal electron injection, grown by metalorganic chemical vapor deposition. *Applied Physics Letters* 69: 465–467.

73. Wang, C.A. (1997). Organometallic vapor phase epitaxial growth of AlSb-based alloys. *Journal of Crystal Growth* 170: 725–731.
74. Wang, C.A. and Choi, H.K. (1997). GaInAsSb/AlGaAsSb multiple-quantum-well diode lasers grown by organometallic vapor phase epitaxy. *Applied Physics Letters* 70: 802–804.
75. Biefeld, R.M., Allerman, A.A., Kurtz, S.R., and Baucom, K.C. (1998). Progress in the growth of mid-infrared InAsSb emitters by metal-organic chemical vapor deposition. *Journal of Crystal Growth* 195: 356–362.
76. Wang, C.A. (2004). Progress and continuing challenges in GaSb-based III–V alloys and heterostructures grown by organometallic vapor-phase epitaxy. *Journal of Crystal Growth* 272: 664–681.
77. Stein, A., Puttjer, D., Behres, A., and Heime, K. (1998). Room temperature InPSb/InAs and InPSb/InAs/InAsSb mid-infrared emitting diodes grown by MOVPE. *IEE Proceedings - Optoelectronics* 145: 257–260.
78. Biefeld, R.M., Phillips, J.D., and Kurtz, S.R. (2000). The growth of InAsSb/InAs/InPSb/InAs mid-infrared emitters by metal-organic chemical vapor deposition. *Materials Research Society Symposium Proceedings* 607: 303–308.
79. Coutts, T.J. and Ward, J.S. (1999). Thermophotovoltaic and photovoltaic conversion at high-flux densities. *IEEE Transactions on Electron Devices* 46: 2145–2153.
80. Choi, H.K., Wang, C.A., Turner, G.W. et al. (1997). High-performance GaInAsSb thermophotovoltaic devices with an AlGaAsSb window. *Applied Physics Letters* 71: 3758–3760.
81. Wang, C.A., Vineis, C.J., Choi, H.K. et al. (2003). Lattice-matched GaInAsSb/AlGaAsSb/GaSb materials for thermophotovoltaic devices. *AIP Conference Proceedings* 653: 324–334.
82. Kondow, M., Uomi, K., Niwa, A. et al. (1996). GaInNAs: a novel material for long-wavelength-range laser diodes with excellent high-temperature performance. *Japanese Journal of Applied Physics* 35: 1273–1275.
83. Geisz, J.F., Friedman, D.J., Olson, J.M. et al. (2000). BGaInAs alloys lattice matched to GaAs. *Applied Physics Letters* 76: 1443–1445.
84. Geisz, J.F., Friedman, D.J., Kurtz, S. et al. (2001). Epitaxial growth of BGaAs and BGaInAs by MOCVD. *Journal of Crystal Growth* 225: 372–376.
85. Schmidt, M.C., Poblenz, C., Chang, Y.-C. et al. (2011). High-performance blue and green laser diodes based on nonpolar/semipolar GaN substrates. *Proceedings of SPIE*: 80390D-1–8.
86. Feezell, D.F., Speck, J.S., DenBaars, S.P., and Nakamura, S. (2013). Semipolar (20-2-1). InGaN/GaN light-emitting diodes for high-efficiency solid-state lighting. *Journal of Display Technology* 9 (4): 190–198.
87. Wierer, J.J., Tsao, J.Y., and Sizov, D.S. (2013). Comparison between blue lasers and light-emitting diodes for future solid-state lighting. *Laser and Photonics Reviews* 7 (6): 963–993.
88. Masahiro, A. (2014). InGaN based green laser diodes on semipolar GaN substrate. *Japanese Journal of Applied Physics* 53 (10): 100207-1–8.
89. Yokogawa, T., Oya, M., and Fujikane, M. (2015). A high power InGaN-LED on an m-plane GaN substrate. *Electronics and Communications in Japan* 98 (5): 15–22.
90. Kneissl, M. (2016). *III-Nitride Ultraviolet Emittters: Technology and Applications* (eds. M. Kneissl and J. Rass). Springer International Publishing.
91. Taniyasu, Y., Kasu, M., and Makimoto, T. (2007). Radiation and polarization properties of free-exciton emission from AlN (0001) surface. *Applied Physics Letters* 90: 261911-1–3.
92. Taniyasu, Y. and Kasu, M. (2010). Surface 210nm light emission from an AlN p–n junction light-emitting diode enhanced by A-plane growth orientation. *Applied Physics Letters* 96: 221110-1–3.
93. Banal, RG., Taniyasu, Y., and Yamamoto, H. (2014). Deep-ultraviolet light emission properties of nonpolar M-plane AlGaN quantum wells. *Applied Physics Letters* 105: 053104-1–4.
94. Taniyasu, Y. and Kasu, M. (2011). Polarization property of deep-ultraviolet light emission from C-plane AlN/GaN short-period superlattices. *Applied Physics Letters* 99: 251112-1–4.
95. Jeżowski, A., Stachowiak, P., Suski, T. et al. (2003). Thermal conductivity of bulk GaN single crystals. *Physica B: Condensed Matter* 329–333 (2): 1531–1532.
96. Micovic, M., Brown, D., Regan, D. et al. (2016). High frequency GaN HEMTs for RF MMIC applications. *62nd IEEE International Electron Devices Meeting*, San Francisco, CA.
97. Shinohara, K., Regan, D.C., Tang, Y. et al. (2013). Scaling of GaN HEMTs and Schottky diodes for submillimeter-wave MMIC applications. *IEEE Transactions on Electron Devices* 60: 2982–2996.
98. Yoshitomo, H., Kazuki, N., Akihisa, T. et al. (2013). High-breakdown-voltage and low-specific-on-resistance GaN p–n junction diodes on free-standing GaN substrates fabricated through low-damage field-plate process. *Japanese Journal of Applied Physics* 52: 028007-1–3.
99. Kizilyalli, I.C., Edwards, A.P., Nie, H. et al. (2014). 3.7 kV vertical GaN PN diodes. *IEEE Electron Device Letters* 35: 247–249.
100. Kizilyalli, I.C., Edwards, A.P., Aktas, O. et al. (2015). Vertical power p-n diodes based on bulk GaN. *IEEE Transactions on Electron Devices* 62: 414–422.
101. Limb, J.B., McCarthy, L., Kozodoy, P. et al. (1999). AlGaN/GaN HBTs using regrown emitter. *Electronics Letters* 35: 1671–1673.

102. Han, J., Baca, A.G., Shul, R.J. et al. (1999). Growth and fabrication of GaN/AlGaN heterojunction bipolar transistor. *Applied Physics Letters* 74: 2702–2704.
103. Shelton, B.S., Huang, J.J., Lambert, D.J.H. et al. (2000). AlGaN/GaN heterojunction bipolar transistors grown by metal organic chemical vapour deposition. *Electronics Letters* 36: 80–81.
104. Makimoto, T., Kumakura, K., and Kobayashi, N. (2001). High current gains obtained by InGaN/GaN double heterojunction bipolar transistors with p-InGaN base. *Applied Physics Letters* 79: 380–381.
105. Kumakura, K., Makimoto, T., and Kobayashi, N. (2000). High hole concentrations in Mg-doped InGaN grown by MOVPE. *Journal of Crystal Growth* 221: 267–270.
106. Nishikawa, A., Kumakura, K., Kasu, M., and Makimoto, T. (2009). Low-temperature characteristics of the current gain of GaN/InGaN double-heterojunction bipolar transistors. *Journal of Crystal Growth* 311: 3000–3002.
107. Chu-Kung, B.F., Wu, C.H., Walter, G. et al. (2007). Modulation of high current gain (β>49). light-emitting InGaN/GaN heterojunction bipolar transistors. *Applied Physics Letters* 91: 232114-1–3.
108. Lee, Y.C., Zhang, Y., Kim, H.J. et al. (2010). High-current-gain direct-growth GaN/InGaN double heterojunction bipolar transistors. *IEEE Transactions on Electron Devices* 57: 2964–2969.
109. Shen, S.-C., Dupuis, R.D., Lochner, Z. et al. (2013). Working toward high-power GaN/InGaN heterojunction bipolar transistors. *Semiconductor Science and Technology* 28: 074025-1–8.
110. Miyake, H., Nishio, G., Suzuki, S. et al. (2016). Annealing of an AlN buffer layer in N_2 –CO for growth of a high-quality AlN film on sapphire. *Applied Physics Express* 9 (2): 025501-1–5.
111. Chen, S.-W., Li, H., and Lu, T.-C. (2016). Improved performance of GaN based light emitting diodes with ex-situ sputtered AlN nucleation layers. *AIP Advances* 6 (4): 045311-1–6.
112. Kamiyama, S., Takanami, S., Tomida, Y. et al. (2003). Violet and UV light-emitting diodes grown on ZrB2 substrate. *Physica Status Solidi A* 200 (1): 67–70.
113. Iizuka, K., Morishima, Y., Kuramata, A. et al. (2015). InGaN LEDs prepared on β-Ga_2O_3 (-201) substrates. *Proceedings of SPIE* 9363: 93631Z-1–6.
114. Stark, C.J.M., Detchprohm, T., Lee, S.C. et al. (2013). Green cubic GaInN/GaN light-emitting diode on microstructured silicon (100). *Applied Physics Letters* 103 (23): 232107-1–4.
115. Lee, S.C., Youngblood, N., Jiang, Y.B. et al. (2015). Incorporation of indium on cubic GaN epitaxially induced on a nanofaceted Si(001) substrate by phase transition. *Applied Physics Letters* 107 (23): 231905-1–5.
116. Haase, M.A., Qiu, J., DePuydt, J.M., and Cheng, H. (1991). Blue-green laser diodes. *Applied Physics Letters* 59 (11): 1272–1274.
117. Ashley, T. and Elliott, C.T. (1985). Nonequilibrium devices for infra-red detection. *Electronics Letters* 21: 451–452.
118. Ashley, T. and Elliott, C.T. (1991). Operation and properties of narrow-gap semiconductor devices near room temperatures using non-equilibrium techniques. *Semiconductor Science and Technology* 6: C99–105.
119. Itsuno, A.M., Emelie, P.Y., Phillips, J.D. et al. (2010). Arsenic diffusion study in HgCdTe for low p-type doping in Auger-suppressed photodiodes. *Journal of Electronic Materials* 39: 945–50.
120. Maxey C.D., Jones, C.L., Metcalfe, N.E. et al. (1997). MOVPE growth of improved non-equilibrium MCT device structures for near ambient temperature heterodyne detectors. *Proceedings of SPIE* 3122 453–464.
121. Ashby, M.K., Gordon, N.T., Elliott, C.T. et al. (2003). Novel $Hg_xCd_{1-x}Te$ device structure for higher operating temperature detectors. *Journal of Electronic Materials* 32: 667–71.
122. Madejczyk, P., Gawron, W., Piotrowski, A. et al. (2011). Improvement in performance of high-operating temperature HgCdTe photodiodes. *Infrared Physics and Technology*, 54: 310–315.
123. Madejczyk, P., Gawron, W., Martyniuk, P. et al. (2013). MOCVD grown HgCdTe device structure for ambient temperature LWIR detectors. *Semiconductor Science and Technology* 28: 105017-1–7.
124. Liu, J., Sun, X., Camacho-Aguilera, R. et al. (2010). Ge-on-Si laser operating at room temperature. *Optics Letters* 35 (5): 679–681.
125. Huang, W., Cheng, B., Xue, C., and Yang, H. (2017). The band structure and optical gain of a new IV-group alloy GePb: A first-principles calculation. *Journal of Alloys and Compounds* 701: 816–821.
126. Low, K.L., Yang, Y., Han, G. et al. (2012) Electronic band structure and effective mass parameters of $Ge_{1-x}Sn_x$ alloys. *Journal of Applied Physics* 112 (10): 103715-1–9.
127. Gupta, S., Magyari-Köpe, B., Nishi, Y., and Saraswat, K. C. (2013). Achieving direct band gap in germanium through integration of Sn alloying and external strain. *Journal of Applied Physics* 113 (7): 073707-1–7.
128. Wirths, S., Geiger, R., von den Driesch, N. et al. (2015). Lasing in direct-bandgap GeSn alloy grown on Si. *Nature Photonics* 9 (2): 88–92.
129. Al-Kabi, S., Ghetmiri, S. A., Margetis, J. et al. (2016). An optically pumped 2.5 μm GeSn laser on Si operating at 110 K. *Applied Physics Letters* 109 (17): 171105-1–4.
130. Huang, W., Cheng, B., Xue, C., and Yang, H. (2017). The band structure and optical gain of a new IV-group alloy GePb: A first-principles calculation. *Journal of Alloys and Compounds* 701: 816–821.
131. Olesinski, R.W. and Abbaschian, G.J. (1984). The Ge-Pb (germanium-lead) system. *Bulletin of Alloy Phase Diagrams* 5 (4): 374–377.

132. Zhou, Q., Ong, E.B.L., Lim, S.L. et al. (2016). Single crystalline germanium-lead formed by laser-induced epitaxy. *ECS Journal of Solid State Science and Technology* 5 (6): 353–360.
133. Emtsev, K.V., Bostwick, A., Horn, K. et al. (2009). Towards wafer-size graphene layers by atmospheric pressure graphitization of silicon carbide. *Nature Materials* 8 (3): 203–207.
134. Kim, J., Bayram, C., Park, H. et al. (2014). Principle of direct van der Waals epitaxy of single-crystalline films on epitaxial graphene. *Nature Communications* 5: 4836-1–7.
135. Kim, Y., Cruz, S.S., Lee, K. et al. (2017). Remote epitaxy through graphene enables two-dimensional material-based layer transfer. *Nature* 544 (7650): 340–343.
136. Kobayashi, Y., Kumakura, K., Akasaka, T., and Makimoto, T. (2012) Layered boron nitride as a release layer for mechanical transfer of GaN-based devices. *Nature*, 484(7393): 223–227.
137. Vogt, P., De Padova, P., Quaresima, C. et al. (2012) Silicene: compelling experimental evidence for graphenelike two-dimensional silicon. *Physical Review Letters* 108(15): 155501-1–5.
138. Piazza, Z.A., Hu, H.-S., Li, W.-L. et al. (2014). Planar hexagonal B_{36} as a potential basis for extended single-atom layer boron sheets. *Nature Communications* 5: 3113-1–6.
139. Zhang, L., Bampoulis, P., Rudenko, A.N. et al. (2016). Structural and electronic properties of germanene on MoS_2. *Physical Review Letters* 116 (25): 256804-1–6.
140. Xu, Y., Yan, B., Zhang, H.-J. et al. (2013). Large-gap quantum spin hall insulators in tin films. *Physical Review Letters* 111 (13): 136804-1–5.
141. Dahal, R., Li, J., Majety, S. et al. (2011). Epitaxially grown semiconducting hexagonal boron nitride as a deep ultraviolet photonic material. *Applied Physics Letters* 98 (21): 211110-1–3.
142. Li, X., Jordan, M.B., Ayari, T. et al. (2017). Flexible metal-semiconductor-metal device prototype on wafer-scale thick boron nitride layers grown by MOVPE. *Scientific Reports* 7 (1): 786-1–8.
143. Eichfeld, S.M., Hossain, L., Lin, Y.-C. et al. (2015). Highly scalable, atomically thin WSe_2 grown via metal–organic chemical vapor deposition. *ACS Nano* 9 (2): 2080–2087.
144. Kang, K., Xie, S., Huang, L. et al. (2015). High-mobility three-atom-thick semiconducting films with wafer-scale homogeneity. *Nature* 520 (7549): 656–660.
145. For example: Kalikhman, V.L. and Ya, S.U. (1973). Transition-metal chalcogenides with layer structures and features of the filling of their Brillouin zones. *Soviet Physics Uspekhi* 15 (6): 728–741.
146. For example: Xie, L.M. (2015). Two-dimensional transition metal dichalcogenide alloys: preparation, characterization and applications. *Nanoscale* 7 (44): 18392–18401.
147. For example: http://hqgraphene.com/index.php.
148. Tausch Jr., F.W. and Lapierre III, A.G. (1965). A novel crystal growth phenomenon: single crystal GaAs overgrowth onto silicon dioxide. *Journal of the Electrochemical Society* 112 (7): 706–709.
149. Shaw, D.W. (1966). Selective epitaxial deposition of gallium arsenide in holes. *Journal of The Electrochemical Society* 113 (9): 904–908.
150. Tatau, N., Tsuyoshi, N., and Suian, Z. (1988). Epitaxial lateral overgrowth of GaAs by LPE. *Japanese Journal of Applied Physics* 27 (6A): L964–L968.
151. Nishinaga, T. (2002) Microchannel epitaxy: an overview. *Journal of Crystal Growth* 237: 1410–1417.
152. Kato, Y., Kitamura, S., Hiramatsu, K., and Sawaki, N. (1994). Selective growth of wurtzite GaN and $Al_xGa_{1-x}N$ on GaN/sapphire substrates by metalorganic vapor phase epitaxy. *Journal of Crystal Growth* 144 (3): 133–140.
153. Dupuis, R.D., Park, J., Grudowski, P.A. et al. (1998). Selective-area and lateral epitaxial overgrowth of III-N materials by metalorganic chemical vapor deposition. *Journal of Crystal Growth* 195 (1): 340–345.
154. Akira, U., Haruo, S., Akira, S., and Yamaguchi, A. A. (1997). Thick GaN epitaxial growth with low dislocation density by hydride vapor phase epitaxy. *Japanese Journal of Applied Physics* 36 (7B): L899–L902.
155. Hiramatsu, K., Nishiyama, K., Onishi, M. et al. (2000). Fabrication and characterization of low defect density GaN using facet-controlled epitaxial lateral overgrowth (FACELO). *Journal of Crystal Growth* 221 (1): 316–326.
156. Taketomi, H., Aoki, Y., Takagi, Y. et al. (2016). Over 1 W record-peak-power operation of a 338 nm AlGaN multiple-quantum-well laser diode on a GaN substrate. *Japanese Journal of Applied Physics* 55 (5S): 05FJ05-1–3.
157. For example: https://en.wikipedia.org/wiki/10_nanometer#cite_note-20.
158. For example: Tomioka, K., Tanaka, T., Hara, S. et al. (2011). III-V nanowires on Si substrate: selective-area growth and device applications. *IEEE Journal of Selected Topics in Quantum Electronics* 17 (4): 1112–1129.
159. Pratyush Das, K., Heinz, S., Mikael, T.B. et al. (2013). Selective area growth of III–V nanowires and their heterostructures on silicon in a nanotube template: towards monolithic integration of nano-devices. *Nanotechnology* 24 (22): 225304-1–6.
160. Schmid, H., Borg, M., Moselund, K. et al. (2015). Template-assisted selective epitaxy of III–V nanoscale devices for co-planar heterogeneous integration with Si. *Applied Physics Letters* 106 (23): 233101-1–5.
161. Leung, B., Song, J., Zhang, Y., and Han, J. (2013). Evolutionary selection growth: towards template-insensitive preparation of single-crystal layers. *Advanced Materials* 25 (9): 1285–1289.

162. Van Hemert, R.L., Spendlove, L.B., and Sievers, R.E. (1965). Vapor deposition of metals by hydrogen reduction of metal chelates. *Journal of the Electrochemical Society* 112 (11): 1124–1126.
163. Li, J., Shacham-Diamand, Y., and Mayer, J.W. (1992) Copper deposition and thermal stability issues in copper-based metallization for ULSI technology. *Materials Science Reports* 9 (1): 1–51.
164. Alford, T.L., Li, J., Mayer, J.W., and Wang, S.-Q. (1995). Copper-based metallization and interconnects for ultra-large-scale integration applications. *Thin Solid Films* 262: vii–viii.
165. Blackburn, J.M., Long, D.P., Cabanas, A., and Watkins, J.J. (2001). Deposition of conformal copper and nickel films from supercritical carbon dioxide. *Science* 294: 141–145.
166. Jain, A., Gelatos, A.V., Kodas, T.T. et al. (1995). Selective chemical vapor deposition of copper using (hfac) copper(I) vinyltrimethylsilane in the absence and presence of water. *Thin Solid Films* 262: 52–59.
167. Perry, W.L., Jain, A., Kodas, T.T., and Hampden-Smith, M.J. (1995). Selective chemical vapor deposition on excimer-laser-patterned polytetrafluoroethylene from hexafluoroacetylacetonate copper (I) vinyltrimethylsilane. *Thin Solid Films* 262: 7–11.
168. Awaya, N. and Arita, Y. (1995). Carrier-gas effects on characteristics of copper chemical vapor deposition using hexafluoro-acetylacetonate-copper (1) trimethylvinylsilane. *Thin Solid Films* 262: 12–19.
169. Kobayashi, A., Sekiguchi, A., and Okada, O. (1998). Deposition rate and gap-filling characteristics in Cu chemical vapor deposition with trimethylvinylsilane hexafluoro-acetylacetonate-copper (I). *Japanese Journal of Applied Physics* 37 (12A): 6358–6363.
170. Kobayashi, A., Sekiguchi, A., Koide, T. et al. (1999). Gap-filling property of Cu film by chemical vapor deposition. *Journal of Vacuum Science and Technology B* 17 (5): 2256–2261.
171. Mouche, M.-J., Mermet, J.-K., Romand, M., and Charbonnier, M. (1995). Metal-organic chemical vapor deposition of copper using hydrated copper formate as a new precursor. *Thin Solid Films* 262: 1–6.
172. Doppelt, P. (1998). Why is coordination chemistry stretching the limits of micro-electronics technology? *Coordination Chemical Review* 178–180: 1785–1809.
173. Tu, R., Huang, J., Zhang, S., and Zhang, L. (2016). Epitaxial growth of copper film by MOCVD. *Key Engineering Materials* 680: 507–510.
174. Utriainen, M., Kroger-Laukkanen, M., Johansson, L.-S., and Niinisto, L. (2000). Studies of metallic thin film growth in an atomic layer epitaxy reactor using M(acac)2 (M = Ni, Cu, Pt) precursors. *Applied Surface Science* 157: 151–158.
175. Parkhomenko, R.G., Trubin, S.V., Turgambaeva, A.V., and Igumenov, I.K. (2015). Deposition of pure gold thin films from organometallic precursors. *Journal of Crystal Growth* 414: 143–150.
176. Baedeker, K. (1907). *Annalen der Physik (Leipzig)* 22: 749.
177. Haacke, G. (1977). Transparent conducting coatings. *Annual Review of Materials Science* 7: 73–93.
178. Kane, J., Schweizer, H.P., and Kern, W. (1975). Chemical vapor deposition of transparent electrically conducting layers of indium oxide doped with tin. *Thin Solid Films* 29: 155–163.
179. Akinwunmi, O.O., Eleruja, M.A., Olowolafe, J.O. et al. (1999). Preparation and characterization of MOCVD thin film of indium tin oxide. *Optical Materials* 13: 255–259.
180. Chen, Z., Zhuo, Y., Tu, W. et al. (2017). Highly ultraviolet transparent textured indium tin oxide thin films and the application in light emitting diodes. *Applied Physics Letters* 110 (24): 242101-1–4.
181. Hamberg, I., Granqvist, C.G., Berggren, K.-F. et al. (1984). Band-gap widening in heavily Sn-doped In_2O_3. *Physical Review B* 30 (6): 6240–3249.
182. Yang, Y., Jin, S., Medvedeva, J.E. et al. (2005). CdO as the archetypical transparent conducting oxide: Systematics of dopant ionic radius and electronic structure effects on charge transport and band structure. *Journal of the American Chemical Society* 127: 8796–8804.
183. Wang, A., Babcock, J.R., Edleman, N.L. et al. (2001). Indium-cadmium-oxide films having exceptional electrical conductivity and optical transparency: Clues for optimizing transparent conductors. *Proceedings of the National Academy of Science USA* 98 (13): 7113–7116
184. Jin, S. Yang, Y., Medvedeva, J.E. et al. (2008). Tuning the properties of transparent oxide conductors. Dopant ion size and electronic structure effects on CdO-based transparent conducting oxides. Ga- and In-doped CdO thin films grown by MOCVD. *Materials Chemistry* 20: 220–230.
185. Van Duzer, T. and Hassenzahl, W.V. (2004). Special issue on applications of superconductivity. *Proceedings of the IEEE* 92 (10): 1511–1516.
186. Wu, M.K., Ashburn, J.R., Torng, C.J. et al. (1987). Superconductivity at 93 K in a new mixed-phase Y-Ba-Cu-O compound system at ambient pressure. *Physics Review Letters* 58 (9): 908–910.
187. Sarrao, J., Kwok, W-K., Bozovic, I. et al. (2006). Basic research needs for superconductivity. *Report of the Basic Energy Sciences Workshop on Superconductivity*, May 8-11. DOI: 10.2172/899129.
188. Chang, C.C., Wu, X.D., Inam, A. et al. (1988). Smooth high T_C $Y_1Ba_2Cu_3O_x$ films by laser deposition at 650 °C. *Applied Physics Letters* 53 (6): 517–519.

189. McIntyre, P.C., Cima, M.J., and Ng, M.F. (1990). Metalorganic deposition of high-J_C $Ba_2YCu_3O_{7-x}$ thin films from trifluoroacetate precursors onto (100) $SrTiO_3$. *Journal of Applied Physics* 68 (8): 4183–4187.
190. Lu, Z., Truman, J.K., Johansson, M.E. et al. (1995). Large area double sided $YBa_2Cu_3O_{7-}\delta$ films grown by single-source metalorganic chemical vapor deposition. *Applied Physics Letters* 67 (5): 712–714.
191. Watanabe, K., Yamane, H., Kurosawa, H. et al. (1989). Critical currents at 77.3 K under magnetic fields up to 27 T for an Y-Ba-Cu-O film prepared by chemical vapor deposition. *Applied Physics Letters* 54 (6): 575–577.
192. Berry, A.D., Gaskill, D.K., Holm, R.T. et al. (1988). Formation of high Tc superconducting films by organometallic chemical vapor deposition. *Applied Physics Letters* 52 (20): 1743–1745.
193. Yamane, H., Kurosawa, H., and Hirai, T. (1988). Preparation of $YBa_2Cu_3O_{7-}x$ films by chemical vapor deposition. *Chemistry Letters* 17 (6): 939–940.
194. Yamane, H., Masumoto, H., Hirai, T. et al. (1988). Y-Ba-Cu-O superconducting films prepared on $SrTiO_3$ substrates by chemical vapor deposition. *Applied Physics Letters* 53 (16): 1548–1550.
195. Choi, J.-K., Jun, B.-H. and Kim, C.-J. (2006). Fabrication of YBCO coated conductors by MOCVD using various templates. *Physica C* 445–448: 521–524.
196. Onabe, K., Kohno, O., Nagaya, S. et al. (1996). Structure and property of YBa_2Cu_3Ox tapes formed on metallic substrate by CVD technique. *Materials Transactions* 37 (4): 893–897.
197. Choi, J.K., Kim, H.J., Jun, B.H., and Kim, C.J. (2005). Effect of the Cu/Ba ratio for the YBCO deposition onto IBAD template by the MOCVD method. *Physica C* 426–431: 920–925.
198. Heydari Gharahcheshmeh, M., Galstyan, E., Kukunuru, J. et al. (2017). MOCVD of heavily-doped 25 mol.% Zr-added (Gd, Y)$Ba_2Cu_3O_{7-\delta}$ coated conductors. *IEEE Transactions on Applied Superconductivity* 27 (4): 6602905.
199. Schulz, D.L. and Marks, T.J. (1994). MOCVD routes to thin metal oxide films for superconducting electronics. *Advanced Materials* 6 (10): 719–730.
200. Selvamanickam, V., Chen, Y., Xiong, X. et al. (2009). High performance 2G wires: From R&D to pilot-scale manufacturing. *IEEE Transactions on Applied Superconductivity* 19 (3): 3225–3229.
201. Akiyama, K., Ohya, S., Takano, H. et al. (2001). Growth of β-$FeSi_2$ thin film on Si (111) by metal-organic chemical vapor deposition. *Japanese Journal of Applied Physics* 40 (5A): L460–L462.
202. Akiyama, K., Kimura, T., Suesmasu, T. et al. (2004). Growth of β-$FeSi_2$ thin film on Si (100) by metal-organic chemical vapor deposition. *Japanese Journal of Applied Physics* 43 (4B): L551–L553.
203. Ozden, I., Diagne, M., Nurmikko, A. V. et al. (2001). A matrix addressable 1024 element blue light emitting InGaN QW diode array. *Physica Status Solidi A* 188 (1): 139–142.
204. Sony. (2010). Sony develops next-generation display, "crystal LED display," ideal for high picture quality on large screens. Press release (January 10). https://www.sony.net/SonyInfo/News/Press/201201/12-005E/.
205. Samsung. (2017). http://www.samsung.com/us/explore/qled-tv/?cid=ppc
206. Han, H.-V., Lin, H.-Y., Lin, C.-C. et al. (2015). Resonant-enhanced full-color emission of quantum-dot-based micro LED display technology. *Optics Express* 23 (25): 32504–32515.
207. Reuters. (2018). Apple developing own screens using next-generation MicroLED tech: Bloomberg. https://www.reuters.com/article/us-apple-screens/apple-developing-own-screens-using-next-generation-microled-tech-bloomberg-idUSKBN1GV04W.
208. Dupuis, R.D., Dapkus, P.D., Holonyak, Jr., N. et al. (1978). Room-temperature laser operation of quantum-well $Ga_{1-x}Al_xAs$-GaAs laser diodes grown by metalorganic chemical vapor deposition. *Applied Physics Letters* 32 (5): 295–297.
209. Dupuis, R.D. and Dapkus, P.D. (1978). Room-temperature operation of distributed Bragg-confinement $Ga_{(1-x)}Al_xAs$–GaAs lasers grown by metalorganic chemical vapor deposition. *Applied Physics Letters* 33 (1): 68–69.
210. Sakaguchi, T., Koyama, F., and Iga, K. (1988). Vertical cavity surface-emitting laser with an AlGaAs/AlAs Bragg reflector. *Electronics Letters* 24 (15): 928–929.
211. Takahasi, M., Vaccaro, P., Fujita, K. et al. (1996). An InGaAs-GaAs vertical-cavity surface-emitting laser grown on GaAs (311)A substrate having low threshold and stable polarization. *IEEE Photonics Technology Letters* 8 (6): 737–739.
212. Faist, J., Capasso, F., Sivco, D. L. et al. (1994). Quantum cascade laser. *Science* 264 (5158): 553–556.
213. Green, R.P., Krysa, A., Roberts, J.S. et al. (2003). Room-temperature operation of InGaAs/AlInAs quantum cascade lasers grown by metalorganic chemical vapor deposition. *Applied Physics Letters* 83 (10): 1921–1922.
214. Roberts, J.S., Green, R.P., Wilson, L.R. et al. (2004). Quantum cascade lasers grown by metalorganic vapor phase epitaxy. *Applied Physics Letters* 82 (24): 4221–4223.
215. Choa, F.-S., Cheng, L, Ji, X. et al. (2007). MOCVD growth and regrowth of quantum cascade lasers. *Proceedings of SPIE* 6485: 64850N.
216. Faist, J. (2013). *Quantum Cascade Lasers*. Oxford: Oxford University Press.
217. Wang, C.A., Schwarz, B., Siriani, D.F. et al. (2017). MOVPE growth of LWIR AlInAs/GaInAs/InP quantum cascade lasers: impact of growth and material quality on laser performance. *IEEE Journal on Selected Topics in Quantum Electronics* 23 (6): 1200413.

218. Najda, S.P., Perlin, P., Suski, T. et al. (2016). Advances in AlGaInN laser diode technology for defence and sensing applications. *Proceedings of SPIE* 9843: 98340K1–7.
219. Miller, S.E. (1969). Integrated optics: The introduction. *Bell System Technical Journal* 48 (7): 2059–2069
220. Nagarajan, R., Kato, M., Joyner, C.H. et al. (2005). Wide temperature (25-85C)/coolerless operation of 100 Gbit/s DWDM photonic integrated circuit, *Electronics Letters* 41 (10): 612–613.
221. Lal, V., Summers, J., Kim, N. et al. (2017). Extended C-band tunable multi-channel InP-based coherent transmitter PICs. *IEEE Journal of Lightwave Technology* 35 (7): 1320–1327.
222. Milano, R.A., Windhorn, T.H., Anderson, E.R. et al. (1979). $Al_{0.5}Ga_{0.5}As$-GaAs heterojunction phototransistors grown by metalorganic chemical vapor deposition. *Applied Physics Letters* 34 (9): 562–564.
223. Kamei, K., Kawasaki, H., Chigira, T. et al. (1981). Extremely low-noise MESFETs fabricated by metal-organic chemical vapor deposition. *Electronics Letters* 17 (13): 450–451.
224. Kobayashi, N. and Fukui, T. (1984). Improved 2DEG mobility is selectively doped GaAs/N-AlGaAs grown by MOCVD using triethyl organometallic compounds. *Electronics Letters* 20 (21): 887–888.
225. Bolognesi, C.R., Fluckiger, R., Alexandrova, M. et al. (2017). InP/GaAsSb DHBTs for THz applications and improved extraction of their cutoff frequencies. *Technical Digest - 2016 IEEE IEDM*, 29.5.1–29.5.4.
226. Akiyama, M., Kawarada, Y, and Kaminishi, K. (1984). Growth of single domain GaAs layer on (100)-oriented Si substrate by MOCVD, *Japanese Journal of Applied Physics* 23 (11): L843–L845.
227. Kohen, D., Made, R., Shuyu, B. et al. (2015). Fabrication of III-V virtual substrate on 200 mm silicon for III-V and Si devices integration. *30th CS MANTECH*, 75–79.
228. Fitzgerald, E.A., Lee, K.E., Zhang, L. et al. (2016). Novel integrated circuit platforms employing monolithic silicon CMOS + GaN devices. *ECS Transactions* 75 (12): 31–37.
229. Khan, M.A., Van Hove, J.M., Kuznia, J.N., and Olson, D.T. (1991). High electron mobility GaN/$Al_xGa_{1-x}N$ heterostructures grown by low-pressure metalorganic chemical vapor deposition. *Applied Physics Letters* 58 (21): 2408–2410.
230. Khan, M.A., Van Hove, J.M., Kuznia, J.N., Olson, D.T. et al. (1993). Microwave performance of a 0.25 μm gate AlGaN/GaN heterostructure field effect transistor. *Applied Physics Letters* 65 (9): 1121–1123.
231. Whelan, C.S., Kolias, N.J., Brierly, S. et al. (2012) GaN technology for radars. *CS MANTECH*, Boston.
232. Hambling, D. (2016). Raytheon's new radar tech could realize the Pentagon's pain ray. *Popular Mechanics*. http://www.popularmechanics.com/military/research/a20264/raytheons-gallium-nitride.
233. Military & Aerospace. (2016). Raytheon to enhance gallium nitride (GaN) semiconductor technologies for radar, RF and microwave. http://www.militaryaerospace.com/articles/2016/10/galliium-nitride-gan-radar-rf-and-microwave.html.
234. Kikuchi, K., Nishihara, M., Yamamoto, H. et al. (2015). An X-band 300-watt class high power GaN HEMT amplifier for radar applications. *SEI Technical Review* 81: 40–44.
235. Glaser, J. (2017). How GaN power transistors drive high-performance Lidar. Generating ultrafast pulsed power with GaN FETs. *IEEE Power Electronics Magazine* 4 (1): 25–35.
236. Giannazzo, F., Fishchella, G., Greco, G. et al. (2017). Graphene integration with nitride semiconductors for high power and high frequency electronics. *Physica Status Solidi A* 214 (4): 1000460.
237. Dupuis, R.D., Dapkus, P.D., Yingling, R.D., and Moudy, L.A. (1977b). High-efficiency GaAlAs/GaAs heterostructure solar cells grown by metalorganic chemical vapor deposition. *Applied Physics Letters* 31 (3): 201–203.
238. Spectrolab. (2012). 29.5% NeXt triple junction (XTJ) solar cells. http://www.spectrolab.com/DataSheets/cells/PV%20XTJ%20Cell%205-20-10.pdf.
239. Stella, P.M., Ewell, R.C., Hoskin, J.J. et al. (2005). Design and performance of the MER (Mars Exploration Rovers) solar arrays. *Conference Record of the 31st IEEE Photovoltaic Specialist Conf.*, 626–630.
240. Ovienmhada, D. (2014). Iridium NEXT enhanced solar panel design passes with flying colors. *Iridium*. http://blog.iridium.com/2014/11/11/iridium-next-enhanced-solar-panel-design-passes-with-flying-colors/.
241. http://onewebsatellites.com/; https://solaerotech.com/products/space-solar-cells-coverglass-interconnected-cells-cic.
242. Flood, D.J. and Weinberg, I. (1994). Advanced solar cells for satellite power systems. NASA Technical Memorandum 106777.
243. Irvine, S.J.C., Barrioz, V., Lamb, D. et al. (2008). MOCVD of thin film photovoltaic solar cells – Next-generation production technology. *Journal of Crystal Growth* 310: 5198–5303.
244. Lamb, D.A., Irvine, S.J.C., Clayton, A.G. et al. (2016). Characterization of MOCVD Thin-Film CdTe Photovoltaics on Space-Qualified Cover Glass. *IEEE Journal of Photovoltaics* 6 (2): 557–561.
245. Barrioz, V., Monir, S., Kartopu, G. et al. (2015). MOCVD for solar cells, a transition towards a chamberless inline process. *Journal of Crystal Growth* 414: 223–231.
246. Ito, M., Wada, O., Miura, S. et al. (1983). Planar structure AlGaAs-GaAs PIN photodiode grown by MOCVD. *Electronics Letters* 19 (14): 522–523.
247. Dupuis, R.D., Velebir, J.R., Campbell, J.C., and Qua, G.J. (1986). InP/InGaAsP/InGaAs avalanche photodiodes with separate absorption, grading, and multiplication regions grown by metalorganic chemical vapor deposition. *Electronics Letters* 22 (5): 235–236.

248. Lauter, J., Bauser, E., Forster, A. et al. (1995). Epitaxial GaAs for nuclear radiation detector applications. *Nuclear Physics B. Proceedings Supplement* 44: 381–385.
249. Schmid, M., Howgate, J., Ruehm, W., and Thalhammer, S. (2016). High mobility AlGaN/GaN devices for β-dosimetry. *Nuclear Instruments and Methods in Physics Research A* 819: 14–19.
250. Rossi, P., Doyle, B.L., Vizkelethy, G. et al. (2008). Ion-luminescence properties of GaN films being developed for IPEM. *Nuclear Instruments and Methods in Physics Research B* 206: 1264–1299.
251. Yasuda, K., Niralua, M., and Agata, Y. (2011). Development of radiation imaging devices with energy discrimination capability using thick CdTe layers grown on Si substrates by metalorganic vapor phase epitaxy. *Proceedings of the SPIE* 7995: 766952T-1–6.
252. Mullin, J.B. and Irvine, S.J.C. (1981). Vapour phase epitaxy of $Cd_xHg_{1-x}Te$ using organometallics. *Journal of Physics D: Applied Physics* 14: L149–151.
253. Mitra, P, and Reine, M. (1999). MOVPE growth of HgCdTe for bandgap engineered IR detector arrays. *Proceedings of the SPIE* 3629: 64–73.
254. Baker, I.M. (2017). Infrared engineering for the advancement of science: A UK perspective. *Proceedings of the SPIE* 10177: 1017726-1–8.
255. Space Telescope Science Institute. (2018). Hubble's NICMOS system employs three low-noise, high-quantum-efficiency, 256×256 pixel HgCdTe arrays grown by LPE on CdTe/sapphire template made by MOCVD. Information on the NICMOS can be viewed at http://www.stsci.edu/hst/nicmos/design/history and http://www.stsci.edu/hst/nicmos/design/detectors.
256. Poksheva, J.G., Tocci, L.R., Farris, M.C. et al. (1993). NICMOS flight hardware 256 x 256 MCT detector array configuration. *Proceedings of the SPIE* 1946: 161–169.

Index

Metalorganic Vapor Phase Epitaxy (MOVPE): Growth, Materials Properties, and Applications,
First Edition. Edited by Stuart Irvine and Peter Capper.

Printed and bound by CPI Group (UK) Ltd, Croydon, CR0 4YY
17/09/2026
14964303-0007